An Introduction to Sonar Systems Engineering

Important topics that are fundamental to the understanding of modern-day sonar systems engineering are featured. Linear, planar, and volume array theory, including near-field and far-field beam patterns, beam steering, and array focusing, are covered. Real-world arrays such as the twin-line planar array and a linear array of triplets, which are solutions to the port/starboard (left/right) ambiguity problem associated with linear towed arrays, are examined in detail.

Detailed explanations of the fundamentals of side-looking (side-scan) and synthetic-aperture sonars are presented. Bistatic scattering with moving platforms is explored with derivations of exact solutions for the time delay, time-compression/time-expansion factor, and Doppler shift at a receiver for both the scattered and direct acoustic paths. Time-domain and frequency-domain descriptions, and the design of CW, LFM, and Doppler-invariant HFM pulses, are explained. Target detection in the presence of reverberation and noise is examined. Time-domain and frequency-domain descriptions of MFSK, MQAM, and OFDM underwater acoustic communication signals are also discussed.

Although the book is mathematically rigorous, it is written in a tutorial style. Many useful, practical design and analysis equations for both passive and active sonar systems are derived from first principles. No major steps in the derivation of important results are skipped – all assumptions and approximations are clearly stated. Particular attention is paid to the correct units for functions and parameters. Many figures, tables, examples, and practical homework problems at the end of each chapter are included to aid in the understanding of the material covered.

New to the Second Edition

- Chapter 15 Synthetic-Aperture Sonar
- Chapter 13, Section 13.3, The Rectangular-Envelope HFM Pulse
- Chapter 10, Section 10.7, Moving Platforms, was rewritten, which allowed for the elimination of Appendix 10C from the first edition
- New explanations/discussions were added to Subsections 1.2.1 and 1.3.1 in Chapter 1
- Appendix 1A was rewritten and the new Table 1A-1 was added to Chapter 1

An Introduction to Sonar Systems Engineering

Second Edition

Lawrence J. Ziomek
Professor Emeritus
Department of Electrical and Computer Engineering
Naval Postgraduate School
Monterey, California

CRC Press
Taylor & Francis Group
Boca Raton London New York

CRC Press is an imprint of the
Taylor & Francis Group, an **informa** business

Second edition published 2023
by CRC Press
6000 Broken Sound Parkway NW, Suite 300, Boca Raton, FL 33487-2742

and by CRC Press
4 Park Square, Milton Park, Abingdon, Oxon, OX14 4RN

CRC Press is an imprint of Taylor & Francis Group, LLC

First edition published by CRC Press 2017

Library of Congress Cataloging-in-Publication Data

Names: Ziomek, Lawrence J., author.
Title: An introduction to sonar systems engineering / Lawrence J. Ziomek.
Description: Second edition. | Boca Raton : CRC Press, 2022. | Includes bibliographical references and index. | Summary: "This book discusses the fundamental topics of modern-day sonar systems engineering. It provides students and practicing engineers and scientists the tools to analyze and design both passive and active sonar systems, do basic signal design for active sonar systems, and understand the basics of underwater acoustic communication signals"-- Provided by publisher.
Identifiers: LCCN 2022012229 (print) | LCCN 2022012230 (ebook) | ISBN 9781032190037 (hardback) | ISBN 9781032195315 (paperback) | ISBN 9781003259640 (ebook)
Subjects: LCSH: Sonar.
Classification: LCC VM480 .Z49 2022 (print) | LCC VM480 (ebook) | DDC 621.389/5--dc23/eng/20220318
LC record available at https://lccn.loc.gov/2022012229
LC ebook record available at https://lccn.loc.gov/2022012230

ISBN: 978-1-032-19003-7 (hbk)
ISBN: 978-1-032-19531-5 (pbk)
ISBN: 978-1-003-25964-0 (ebk)

DOI: 10.1201/9781003259640

Publisher's note: This book has been prepared from camera-ready copy provided by the author.

To my wife *Virginia* for her many years of support and patience, *Nicole*, *Craig*, *Kirsten*, *Lachlan*, and *Poppy*

Contents

Preface to the Second Edition

In writing the second edition, I adopted the same philosophy that I used to write the first edition: In order to obtain a good understanding of the fundamentals of both passive and active sonar systems, I believe that mathematical rigor is important, especially at the graduate level, so that students can clearly see the development of important results and where assumptions and approximations are made. The second edition, like the first edition, is a graduate-level *textbook*, not a handbook. It was written for serious students – those that want to *understand* the fundamentals and how the important results were obtained.

Because sonar systems engineering is an interdisciplinary field that requires knowledge from different academic disciplines such as electrical engineering (e.g., aperture theory, array theory, digital signal processing, communication theory, random processes, and detection theory), oceanography (e.g., the ocean environment), and physics (e.g., transducers and underwater acoustic wave propagation), the second edition, like the first edition, is written in a *tutorial* style; *no* major steps in the derivation of important results are skipped. Lengthy derivations are done in appendices so as not to disrupt the flow of discussion. Although the book is mathematically rigorous, it is not theory for theory's sake. Many useful, practical design and analysis equations for both passive and active sonar systems are derived from first principles that can be evaluated by using a hand-held calculator. The book includes many figures, tables, examples, and practical homework problems at the end of each chapter to aid in the understanding of the material covered.

What's New in the Second Edition

The original 14 chapters from the first edition have been retained in the second edition, albeit with some minor changes in wording and additional short discussions added to some chapters. A summary of the topics discussed in each of the original 14 chapters can be found in the Preface from the First Edition, which is included after this Preface. There are two *new* major additions in the second edition: 1) *Section 13.3*, *The Rectangular-Envelope HFM Pulse*, in Chapter 13, and 2) *Chapter 15 Synthetic-Aperture Sonar.*

The new Section 13.3 in Chapter 13 discusses the fundamentals of hyperbolic-frequency-modulation (HFM). Chapter 13 now contains discussions of the CW (continuous-wave), LFM (linear-frequency-modulated), and HFM (hyperbolic-frequency-modulated) pulses in Sections 13.1, 13.2, and the new 13.3, respectively. These three waveforms are important, real-world, transmit signals used in active sonar systems. Hyperbolic-frequency-modulation, also known as linear-period-modulation (LPM), is an example of nonlinear frequency modulation. Two different time-domain equations that are used to describe a HFM pulse are discussed. It is shown how both of these equations can be transformed into the form of an amplitude-and-angle-modulated carrier, which makes it easy to identify the angle-modulating function and to compute the complex envelope. The relationship between LFM and HFM pulses is also covered. A HFM pulse is a very important signal because it is shown to be Doppler-invariant. As a result, it is shown that a rectangular-envelope, HFM pulse is superior to a rectangular-envelope, LFM pulse in its ability to detect a target and estimate time delay in the presence of time-compression/time-expansion effects by evaluating cross-correlation functions. The time-compression/time-expansion factor is discussed in Section 10.7 in Chapter 10. It is also shown how to design a HFM pulse to minimize time-delay estimation error.

The new Chapter 15 is devoted to the fundamentals of synthetic-aperture sonar (SAS), which relies heavily on the material covered in Chapter 5 Side-Looking Sonar. A SAS is capable of producing very high resolution underwater images of large areas on the ocean bottom, much better than a side-looking sonar (SLS). Chapter 15 begins with describing how a synthetic aperture (SA) is created using a single-element SAS (SESAS). Three-dB, Rayleigh, and null-to-null beamwidths and their approximations are discussed. Calculating the length of a SA is covered along with the Doppler history, Doppler bandwidth, and along-track (azimuthal) resolution of a SAS based on the stop-and-hop assumption. Also covered is the time-on-target, also known as the dwell time, coherent integration time, or aperture time of a SAS. The *two-way*, far-field array factor and beam pattern of a linear synthetic array (SA) are derived. Also derived are approximate formulas for computing the horizontal, Rayleigh beamwidth of the *two-way*, far-field array factor and beam pattern of a linear SA when rectangular amplitude weights are used. Slant-range and azimuthal ambiguity for a SESAS are discussed. Background for a multi-element SAS (MESAS) is provided along with a discussion of the advantages of using a MESAS instead of a SESAS. The operation of a stripmap, SAS is described.

The following additional changes were made: new, minor discussions were added to Subsections 1.2.1 and 1.3.1 in Chapter 1; Appendix 1A was rewritten and the new Table 1A-1 was added; and Section 10.7, Moving Platforms, in Chapter 10 was rewritten, which allowed for the elimination of Appendix 10C from the first edition.

Lawrence J. Ziomek
Carmel, California

Preface from the First Edition

One of the major challenges in the teaching and study of Sonar Systems Engineering is that it is an interdisciplinary field. It requires a working knowledge of topics from electrical engineering (e.g., aperture theory, array theory, digital signal processing, communication theory, random processes, and detection theory), oceanography (e.g., the ocean environment), and physics (e.g., transducers and underwater acoustic wave propagation). Another challenge is that students and/or practicing engineers and scientists who take a course in Sonar Systems Engineering, or who want to do self-study, typically have diverse undergraduate degrees.

The goal of my textbook is to discuss those important topics that are fundamental to the understanding of modern day Sonar Systems Engineering so that students and practicing engineers and scientists with diverse academic backgrounds can analyze and design both passive and active sonar systems, do basic signal design for active sonar systems, and understand the basics of underwater acoustic communication signals. My belief is that if a student has a good understanding of the fundamentals, then he can analyze and design most sonar systems. In order to obtain a good understanding of the fundamentals, I believe that mathematical rigor is important, so that students can clearly see the development of important results and where assumptions and approximations are made. Although the book is mathematically rigorous, it is not theory for theory's sake. Many useful, practical design and analysis equations for both passive and active sonar systems are derived from first principles that can be evaluated by using a hand-held (pocket) calculator.

Because Sonar Systems Engineering requires knowledge from different academic disciplines, the book is written in a *tutorial* style, *no* major steps in the derivation of important results are skipped. Lengthy derivations are done in appendices so as not to disrupt the flow of discussion. I believe that this approach will also help the instructor. I have found through experience, that by being mathematically rigorous, writing in a tutorial style, and not skipping major steps in a derivation, you can bring along all the serious students who want to learn. A consistent, mostly standard notation is used throughout the book with particular attention paid to the correct *units* for functions and parameters. The book includes many figures, tables, examples, and practical homework problems at the end of each chapter to aid in the understanding of the material covered.

The intended audiences are graduate students at universities, for example, majors in electrical engineering, ocean engineering, and physics; and practicing engineers and scientists at private defense (aerospace) companies, Navy laboratories, and university laboratories. The material in the book is *classroom tested* and *distance learning tested.* Students at the Naval Postgraduate School who take my course in Sonar Systems Engineering have diverse undergraduate degrees, as well as the distance learning students that I have taught. Ideal student prerequisites would include introductory courses in Signals and Systems (Fourier series and transforms, principles of linear, time-invariant systems, discrete Fourier transform), communication theory, random processes, and acoustics.

What to Expect Inside

Complex aperture theory is discussed before array theory, analogous to the general practice of discussing the theory of continuous-time signals before discussing discrete-time signals. Once the general principles of aperture theory are developed, array theory results can be derived quickly and easily.

Starting with an exact solution of a linear wave equation, and using transmit and receive coupling equations, general definitions of the near-field and far-field beam patterns (directivity functions) of transmit and receive volume apertures are derived in Chapter 1. The near-field and far-field beam patterns of a volume aperture are shown to be three-dimensional spatial Fresnel and Fourier transforms, respectively. Spatial frequencies and direction cosines in the X, Y, and Z directions are introduced. Equations that define both the Fresnel and Fraunhofer regions of a volume aperture are derived, along with the range to the near-field/far-field boundary. These equations are also applicable for linear and planar apertures; and linear, planar, and volume arrays. The reason Chapter 1 is devoted to volume apertures is because linear and planar apertures; and linear, planar, and volume arrays are just special cases of a volume aperture. As a consequence, the general results obtained in Chapter 1 can be used throughout the entire book.

Chapters 2 and 3 discuss the fundamental principles of complex aperture theory for linear and planar apertures. It is shown that the near-field and far-field beam patterns of linear and planar apertures are one-dimensional and two-dimensional spatial Fresnel and Fourier transforms, respectively. The classic continuous line source (or receiver) is discussed in Chapter 2. The far-field beam patterns of a continuous line source for different amplitude windows demonstrate the relationship between sidelobe levels and the 3-dB beamwidth of the mainlobe. Beam steering of the far-field beam pattern of a linear aperture is discussed, and the equation for the 3-dB beamwidth as a function of frequency, length of the aperture, amplitude window, and beam-steer angle is derived. The far-field beam patterns of rectangular and circular pistons are derived, along with equations to compute their 3-dB beamwidths in Chapter 3. Beam steering of the far-field beam pattern of a planar aperture is also discussed. The near-field beam patterns of linear and planar apertures are derived and are used to discuss beam steering and aperture focusing in the Fresnel region of linear and planar apertures. Aperture focusing is creating a far-field beam pattern at a near-field range from an aperture. The material covered in Chapters 2 and 3 also provides the theoretical background necessary for the study of linear and planar arrays.

Chapter 4 starts with the derivation of the equation for the directivity of a transmit volume aperture, where the aperture is characterized by its far-field beam pattern. The equation for the directivity and, hence, directivity index (the decibel equivalent of directivity) of a volume aperture is also applicable to linear and planar apertures, single electroacoustic transducers (e.g., a continuous line source, rectangular piston, or circular piston), and linear, planar, and volume arrays of electroacoustic transducers, since they are just special cases of a volume aperture. All that is needed is the far-field beam pattern of the aperture. Chapter 4 finishes with the derivation of the equation for the source level of a transmit volume aperture as a function of the time-average acoustic power radiated by the aperture and the directivity index of the aperture.

Using the fundamentals of linear and planar aperture theory discussed in Chapters 2 and 3, respectively, Chapter 5 discusses the design and analysis of a side-looking sonar, also known as a side-scan sonar. Side-looking sonars are active sonar systems capable of producing high resolution underwater images of large areas on the ocean bottom. Both deep water and shallow water problems are considered.

Chapters 6 and 7 are devoted to linear arrays. Equations for the far-field and near-field beam patterns of linear arrays are derived in Chapter 6. With the advent of very long towed arrays, near-field considerations are very important. Common amplitude weights, including Dolph-Chebyshev amplitude weights, are discussed in detail. Implementing complex weights (amplitude and phase weights) using digital beamforming (FFT beamforming) is covered in both Chapters 6 and 7. The theory of phased arrays, that is, using phase weights for beam steering and/or array focusing is discussed. The relationship between the far-field beam pattern of a linear array and the one-dimensional spatial discrete Fourier transform is derived. Grating lobes are discussed. Chapter 7 provides a detailed discussion of array gain and FFT beamforming for linear arrays.

Chapter 8 is devoted to planar arrays. Equations for the far-field and near-field beam patterns of planar arrays are derived. Implementing complex weights (amplitude and phase weights) using digital beamforming (FFT beamforming) is discussed. The theory of phased arrays, that is, using phase weights for beam steering and/or array focusing is covered. The relationship between the far-field beam pattern of a planar array and the two-dimensional spatial discrete Fourier transform is derived. Two very important, real-world planar arrays that are in use today – a single triplet and a twin-line planar array – are discussed in detail. A single triplet is a circular array composed of three elements whose far-field beam pattern is in the shape of a cardioid. A twin-line planar array provides a solution to the port/starboard (left/right) ambiguity problem associated with linear towed arrays. A detailed discussion of FFT beamforming for planar arrays is provided.

Chapter 9 is devoted to volume arrays. Equations for the far-field beam patterns of a cylindrical and spherical array of omnidirectional point-elements are derived. The theory of phased arrays, that is, using phase weights for beam steering are discussed. Using the results for a single triplet discussed in Chapter 8, a linear array of triplets is discussed in detail. A linear array of triplets is a very important, real-world volume array that is in use today because it also provides a solution to the port/starboard (left/right) ambiguity problem associated with linear towed arrays. A linear array of triplets can be thought of as a cylindrical array that is lying on its side.

Chapter 10 covers the important topic of bistatic scattering with fixed and moving platforms. The scattering function and target strength of a target (scatterer) are discussed, along with different scattering cross-sections such as differential, total, and bistatic. Both scattered and direct acoustic paths are analyzed. Sonar equations and broadband solutions for both the scattered and direct acoustic paths are derived. A statistical model of the scattering function is developed. For the case of moving platforms, exact equations for the time delay, time-compression/time-expansion factor, and Doppler shift at a receiver for both the scattered and direct acoustic paths are derived.

Chapter 11 provides the necessary background for analyzing real bandpass transmit signals used in active sonar systems. Chapter 11 is devoted to the theory of complex envelopes. It discusses how to represent real bandpass signals (amplitude-and-angle-modulated carriers) in terms of their lowpass (baseband) complex envelopes, and how the frequency spectra, energies, time-average powers, and power spectra of the two signals are related. It is much easier to work with real bandpass signals using their complex envelope representations. The material covered in Chapter 11 will be used in Chapters 12 through 14.

Chapter 12 is devoted to target detection in the presence of reverberation and noise. The binary hypothesis testing problem discussed in Chapter 12 is expressed in terms of the complex envelopes of the target return, reverberation return, ambient noise, and receiver noise. The decision threshold used is based on satisfying the Neyman-Pearson decision criterion, which is to maximize the probability of detection for a fixed probability of false alarm. The signal-to-interference ratio (SIR) is defined and the unnormalized and normalized auto-ambiguity functions of a transmit signal are derived. Equations for the probabilities of false alarm and detection, and the decision threshold are derived.

Receiver operating characteristic (ROC) curves are discussed.

Chapter 13 is devoted to signal design – the design of transmit signals to achieve desired range and Doppler resolutions in an active sonar system. The normalized, auto-ambiguity functions of rectangular-envelope, CW (continuous wave) and LFM (linear-frequency-modulated) pulses are derived, along with the range and Doppler profiles of both ambiguity functions. It is shown how to use the information contained in the range and Doppler profiles of both ambiguity functions to design rectangular-envelope, CW and LFM pulses to achieve desired range and Doppler resolutions. Equations to compute the bandwidth and time-average power of both signals are also derived.

Chapter 14 discusses the basic principles of three important digital modulation techniques that are not only used to transmit digital information in traditional communication systems, but are also used for underwater acoustic communication. The three digital modulation techniques are 1) M-ary Frequency-Shift Keying (MFSK), 2) M-ary Quadrature Amplitude Modulation (MQAM), and 3) Orthogonal Frequency-Division Multiplexing (OFDM). For a waveform transmitting digital information using each of the three digital modulation techniques, we shall provide a time-domain description, derive its frequency spectrum, derive bandwidth and time-average power formulas, and discuss how to demodulate the waveform. The advantages and disadvantages of each digital modulation technique are compared.

Lawrence J. Ziomek
Carmel, California

Chapter 1

Complex Aperture Theory – Volume Apertures – General Results

Let us begin by explaining the meaning of the word "aperture" as it shall be used in this and subsequent chapters. In the field of optics, a rectangular or circular hole in an opaque screen is referred to as a rectangular or circular *aperture*. The *complex aperture function* is equal to the amplitude and phase of the light distribution within the aperture. It determines the *Fresnel* and *Fraunhofer diffraction patterns* of the aperture. The complex aperture function can describe the light distribution within, for example, a single rectangular hole, or an array of rectangular holes in an opaque screen. In electromagnetics, the meaning of the word "aperture" has been extended to connote either a single electromagnetic antenna or an array of electromagnetic antennas.

Similarly, in acoustics, the word "aperture" is used to refer to either a *single electroacoustic transducer* or an *array of electroacoustic transducers*. When used in the *active mode* as a *transmitter*, an electroacoustic transducer converts electrical signals (voltages and currents) into acoustic signals (sound waves). When used in the *passive mode* as a *receiver*, an electroacoustic transducer converts acoustic signals into electrical signals.

General definitions of the near-field and far-field beam patterns of transmit and receive volume apertures, along with other important general equations, shall be derived in this chapter. Starting with an exact solution of a linear wave equation, it will be shown that the definitions of the near-field and far-field beam patterns are analogous to Fresnel and Fraunhofer diffraction integrals in optics. The reason to start with volume apertures is because linear and planar apertures, and linear, planar, and volume arrays are just special cases of a volume aperture. As a consequence, we will be able to use these general definitions and the other general results obtained in this chapter throughout the entire book.

1.1 Coupling Transmitted and Received Electrical Signals to the Fluid Medium

1.1.1 Transmit Coupling Equation

Consider a closed-surface, transmit aperture enclosing a volume V as shown in Fig. 1.1-1, where the subscripts T and M refer to the transmit aperture and fluid medium, respectively. Imagine that the closed surface is very thin

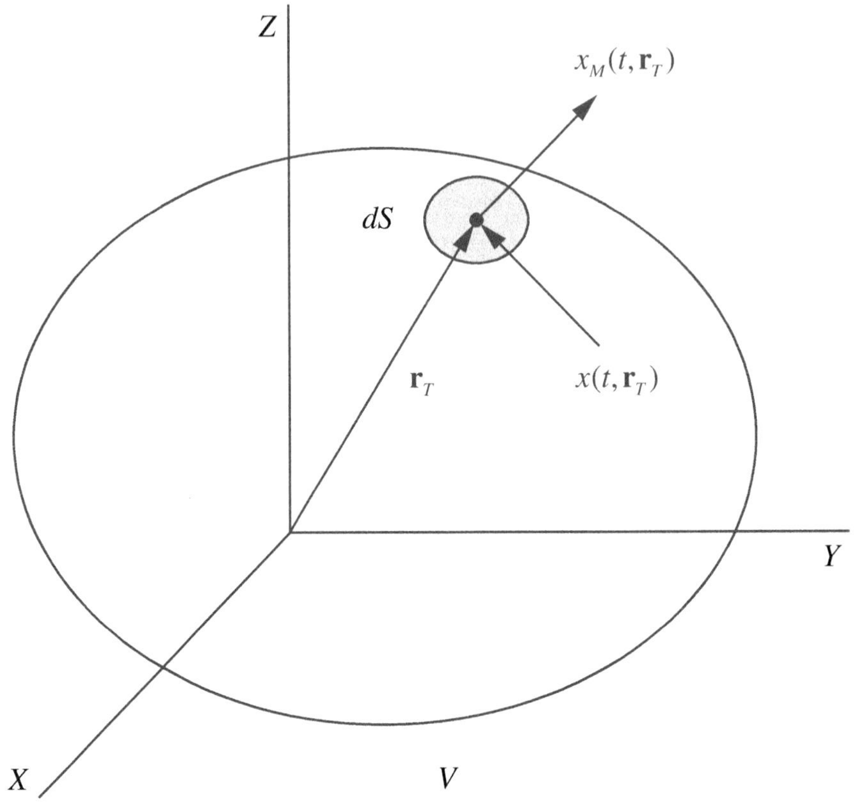

Figure 1.1-1 Closed-surface, transmit aperture enclosing a volume V .

electroacoustic transducer material and that the enclosed volume V is *empty*. This physical situation is known as a *volume aperture*. A real-world example of a volume aperture is a spherical array of electroacoustic transducers, that is, an array of electroacoustic transducers lying on the surface of a hollow sphere. Imagine applying an input electrical signal $x(t, \mathbf{r}_T)$ to the transmit aperture at time t and location $\mathbf{r}_T = (x_T, y_T, z_T)$, where $\mathbf{r}_T$ is the position vector to a point on the surface of the transmit aperture. Therefore, $x(t, \mathbf{r}_T)$ is shorthand notation for $x(t, x_T, y_T, z_T)$. Assume that $x(t, \mathbf{r}_T)$ is a voltage signal with units of volts (V). If we treat the infinitesimal surface area element dS of the transmit aperture located at $\mathbf{r}_T$ as a *linear, time-invariant filter* with *impulse response* $\alpha_T(t, \mathbf{r}_T)$ with units of $\left(\text{sec}^{-1}/\text{V}\right)/\text{sec}$, then the output acoustic signal from the transmit aperture $x_M(t, \mathbf{r}_T)$ with units of sec^{-1}, which is also the input acoustic signal to the fluid medium, can be expressed as a *time-domain convolution integral* as follows:

$$x_M(t,\mathbf{r}_T) = \int_{-\infty}^{\infty} x(\tau,\mathbf{r}_T)\alpha_T(t-\tau,\mathbf{r}_T)d\tau\,, \tag{1.1-1}$$

or

$$x_M(t,\mathbf{r}_T) = x(t,\mathbf{r}_T) \underset{t}{*} \alpha_T(t,\mathbf{r}_T), \tag{1.1-2}$$

where the asterisk denotes convolution with respect to time t. The input acoustic signal to the fluid medium $x_M(t,\mathbf{r}_T)$ is also known as the *source distribution.* It represents the volume flow rate (source strength in cubic meters per second) per unit volume of fluid at time t and location $\mathbf{r}_T$.

Taking the Fourier transform of (1.1-2) with respect to t yields

$$X_M(f,\mathbf{r}_T) = X(f,\mathbf{r}_T)A_T(f,\mathbf{r}_T), \tag{1.1-3}$$

where

$$X_M(f,\mathbf{r}_T) = F_t\{x_M(t,\mathbf{r}_T)\} = \int_{-\infty}^{\infty} x_M(t,\mathbf{r}_T)\exp(-j2\pi ft)dt \tag{1.1-4}$$

is the complex frequency spectrum of the input acoustic signal to the fluid medium (source distribution) at location $\mathbf{r}_T$ of the transmit aperture with units of $\mathrm{sec}^{-1}/\mathrm{Hz}$,

$$X(f,\mathbf{r}_T) = F_t\{x(t,\mathbf{r}_T)\} = \int_{-\infty}^{\infty} x(t,\mathbf{r}_T)\exp(-j2\pi ft)dt \tag{1.1-5}$$

is the complex frequency spectrum of the input electrical signal at location $\mathbf{r}_T$ of the transmit aperture with units of V/Hz, and

$$A_T(f,\mathbf{r}_T) = F_t\{\alpha_T(t,\mathbf{r}_T)\} = \int_{-\infty}^{\infty} \alpha_T(t,\mathbf{r}_T)\exp(-j2\pi ft)dt \tag{1.1-6}$$

is the *complex frequency response of the transmit aperture* at $\mathbf{r}_T$ (a.k.a. the *complex transmit aperture function*) with units of $\mathrm{sec}^{-1}/\mathrm{V}$, where f represents *input (transmitted)* frequencies in hertz. Since

$$x_M(t,\mathbf{r}_T) = F_f^{-1}\{X_M(f,\mathbf{r}_T)\} = \int_{-\infty}^{\infty} X_M(f,\mathbf{r}_T)\exp(+j2\pi ft)df\,, \tag{1.1-7}$$

substituting (1.1-3) into (1.1-7) yields

$$\boxed{x_M(t,\mathbf{r}_T) = \int_{-\infty}^{\infty} X(f,\mathbf{r}_T)A_T(f,\mathbf{r}_T)\exp(+j2\pi ft)df} \tag{1.1-8}$$

Compared to (1.1-1), (1.1-8) is a more useful representation of the input acoustic signal (source distribution) for our purposes. Equation (1.1-8) is the *transmit coupling equation.* It models the *coupling* of the input electrical signal to the fluid medium, that is, the production of the input acoustic signal (source distribution) due to the input electrical signal via the transmit aperture whose performance is described, in part, by the complex transmit aperture function.

1.1.2 Receive Coupling Equation

Now consider a closed-surface, receive aperture enclosing a volume V as shown in Fig. 1.1-2, where the subscripts R and M refer to the receive aperture and fluid medium, respectively. As with the transmit aperture, imagine that the closed surface is very thin electroacoustic transducer material and that the enclosed volume V is *empty.* By following the same reasoning used in the development of (1.1-1) through (1.1-8), the output electrical signal from the receive aperture $y(t, \mathbf{r}_R)$ at time t and location $\mathbf{r}_R = (x_R, y_R, z_R)$, with units of V/m^3, can be expressed as a time-domain convolution integral as follows:

$$y(t, \mathbf{r}_R) = \int_{-\infty}^{\infty} y_M(\tau, \mathbf{r}_R)\alpha_R(t-\tau, \mathbf{r}_R)d\tau, \tag{1.1-9}$$

or

$$y(t, \mathbf{r}_R) = y_M(t, \mathbf{r}_R) \underset{t}{*} \alpha_R(t, \mathbf{r}_R), \tag{1.1-10}$$

where $\mathbf{r}_R$ is the position vector to a point on the surface of the receive aperture; $y_M(t, \mathbf{r}_R)$, with units of m^2/sec, is the output acoustic signal (velocity potential) from the fluid medium, which is also the input acoustic signal to the receive aperture (i.e., it is the acoustic field incident upon the receive aperture) at time t and location $\mathbf{r}_R$; and $\alpha_R(t, \mathbf{r}_R)$, with units of $\left(\left(\text{V}/\left(\text{m}^2/\text{sec}\right)\right)/\text{m}^3\right)/\text{sec}$, is the *impulse response* of the infinitesimal surface area element dS of the receive aperture located at $\mathbf{r}_R$, which is treated as a *linear, time-invariant filter.* Note that $y(t, \mathbf{r}_R)$ is shorthand notation for $y(t, x_R, y_R, z_R)$.

Taking the Fourier transform of (1.1-10) with respect to t yields

$$Y(\eta, \mathbf{r}_R) = Y_M(\eta, \mathbf{r}_R)A_R(\eta, \mathbf{r}_R), \tag{1.1-11}$$

where

$$Y(\eta, \mathbf{r}_R) = F_t\{y(t, \mathbf{r}_R)\} = \int_{-\infty}^{\infty} y(t, \mathbf{r}_R)\exp(-j2\pi\eta t)dt \tag{1.1-12}$$

is the complex frequency spectrum of the output electrical signal at location $\mathbf{r}_R$ of

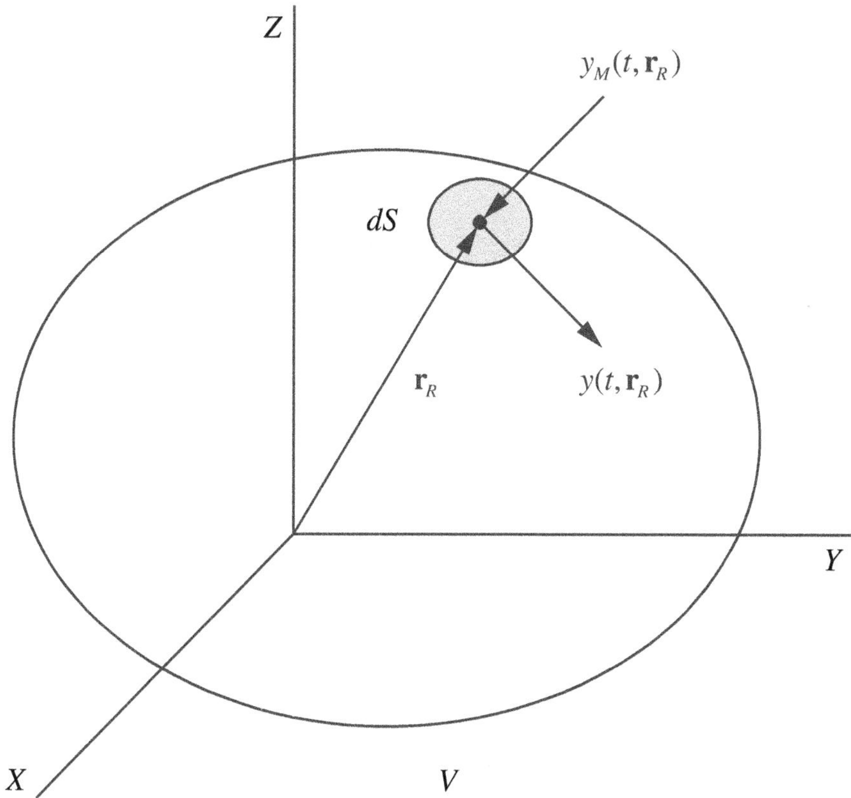

Figure 1.1-2 Closed-surface, receive aperture enclosing a volume V.

the receive aperture with units of $(\text{V/Hz})/\text{m}^3$,

$$Y_M(\eta, \mathbf{r}_R) = F_t\{y_M(t, \mathbf{r}_R)\} = \int_{-\infty}^{\infty} y_M(t, \mathbf{r}_R)\exp(-j2\pi\eta t)\,dt \tag{1.1-13}$$

is the complex frequency spectrum of the acoustic field incident upon the receive aperture at location $\mathbf{r}_R$ with units of $(\text{m}^2/\text{sec})/\text{Hz}$, and

$$A_R(\eta, \mathbf{r}_R) = F_t\{\alpha_R(t, \mathbf{r}_R)\} = \int_{-\infty}^{\infty} \alpha_R(t, \mathbf{r}_R)\exp(-j2\pi\eta t)\,dt \tag{1.1-14}$$

is the *complex frequency response of the receive aperture* at $\mathbf{r}_R$ (a.k.a. the *complex receive aperture function*) with units of $\left(\text{V}/\left(\text{m}^2/\text{sec}\right)\right)/\text{m}^3$, where η represents *output (received)* frequencies in hertz and, in general, $\eta \neq f$. For example, if there is any relative motion between transmit and receive apertures, the received frequencies will not be equal to the transmitted frequencies because of Doppler shift and bandwidth compression or expansion. Taking the inverse Fourier transform of (1.1-11) with respect to η yields

$$\boxed{y(t,\mathbf{r}_R)=\int_{-\infty}^{\infty} Y_M(\eta,\mathbf{r}_R)A_R(\eta,\mathbf{r}_R)\exp(+j2\pi\eta t)\,d\eta} \tag{1.1-15}$$

Note that (1.1-15) is analogous to (1.1-8). Equation (1.1-15) is the *receive coupling equation.* It models the *coupling* of the fluid medium to the output electrical signal, that is, the production of the output electrical signal due to the input acoustic signal (incident acoustic field) via the receive aperture whose performance is described, in part, by the complex receive aperture function.

1.2 The Near-Field Beam Pattern of a Volume Aperture

1.2.1 Transmit Aperture

The *exact* solution of the following linear, three-dimensional, lossless, inhomogeneous, wave equation

$$\nabla^2\varphi(t,\mathbf{r})-\frac{1}{c^2}\frac{\partial^2}{\partial t^2}\varphi(t,\mathbf{r})=x_M(t,\mathbf{r}) \tag{1.2-1}$$

for free-space propagation in an ideal (nonviscous), homogeneous, fluid medium is given by

$$\varphi(t,\mathbf{r})=-\frac{1}{4\pi}\int_{V_0}\frac{x_M\left(t-\left[\left|\mathbf{r}-\mathbf{r}_0\right|/c\right],\mathbf{r}_0\right)}{\left|\mathbf{r}-\mathbf{r}_0\right|}dV_0\,, \tag{1.2-2}$$

where $\varphi(t,\mathbf{r})$ is the scalar velocity potential in squared meters per second, c is the constant speed of sound in the fluid medium in meters per second, and $x_M(t,\mathbf{r})$ is the source distribution in inverse seconds. The wave equation given by (1.2-1) is referred to as being inhomogeneous because the right-hand side is nonzero. If the transmit coupling equation for the source distribution given by (1.1-8) is substituted into (1.2-2), then

$$\boxed{\varphi(t,\mathbf{r})=\int_{-\infty}^{\infty}\int_{V_0} X(f,\mathbf{r}_0)A_T(f,\mathbf{r}_0)g_f\left(\mathbf{r}\left|\mathbf{r}_0\right.\right)dV_0\exp(+j2\pi ft)\,df} \tag{1.2-3}$$

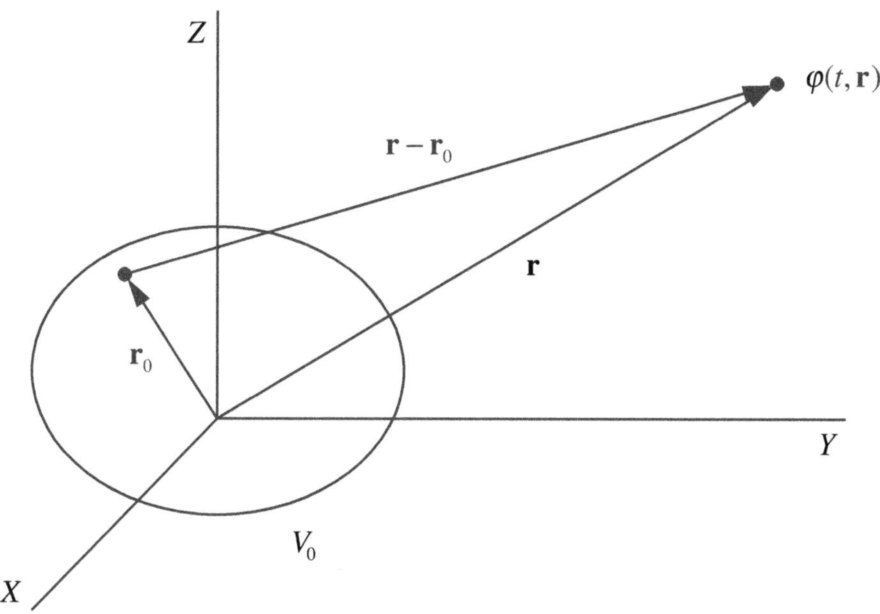

Figure 1.2-1 Closed-surface source distribution enclosing a volume V_0.

where

$$g_f\left(\mathbf{r}\,|\,\mathbf{r}_0\right) \triangleq -\frac{\exp\left(-jk\left|\mathbf{r}-\mathbf{r}_0\right|\right)}{4\pi\left|\mathbf{r}-\mathbf{r}_0\right|} \tag{1.2-4}$$

is the time-independent, free-space, Green's function (a.k.a. the spatial impulse response at frequency f hertz) of an unbounded, ideal (nonviscous), homogeneous, fluid medium with units of inverse meters,

$$k = 2\pi f/c = 2\pi/\lambda \tag{1.2-5}$$

is the wavenumber in radians per meter where $c = f\lambda$, λ is the wavelength in meters,

$$\mathbf{r} = x\hat{x} + y\hat{y} + z\hat{z} \tag{1.2-6}$$

is the position vector to a field point, and

$$\mathbf{r}_0 = x_0\hat{x} + y_0\hat{y} + z_0\hat{z} \tag{1.2-7}$$

is the position vector to a source point (see Fig. 1.2-1). The Green's function given by (1.2-4) is the *complex frequency response* of an unbounded, ideal (nonviscous), homogeneous, fluid medium at frequency f hertz and location $\mathbf{r}=(x,y,z)$ due to the application of a unit-amplitude impulse (i.e., a unit-amplitude, omnidirectional point-source) at location $\mathbf{r}_0=(x_0,y_0,z_0)$. The physical significance of (1.2-3) can be stated as follows: for each frequency component contained in the complex frequency spectrum of the input electrical signal, "sum" the contributions due to each source point contained in the source distribution in order to compute the total acoustic field $\varphi(t,\mathbf{r})$.

The acoustic pressure $p(t,\mathbf{r})$ in pascals (Pa) and the acoustic fluid-velocity-vector (a.k.a. the acoustic particle-velocity-vector) $\mathbf{u}(t,\mathbf{r})$ in meters per second can be obtained from the scalar velocity potential $\varphi(t,\mathbf{r})$ in squared meters per second as follows:

$$p(t,\mathbf{r})=-\rho_0(\mathbf{r})\frac{\partial}{\partial t}\varphi(t,\mathbf{r})=-\rho_0\frac{\partial}{\partial t}\varphi(t,\mathbf{r}), \tag{1.2-8}$$

where ρ_0 is the constant ambient (equilibrium) density of the fluid medium in kilograms per cubic meter and

$$\mathbf{u}(t,\mathbf{r})=\nabla\varphi(t,\mathbf{r}), \tag{1.2-9}$$

where

$$\nabla=\frac{\partial}{\partial x}\hat{x}+\frac{\partial}{\partial y}\hat{y}+\frac{\partial}{\partial z}\hat{z} \tag{1.2-10}$$

is the gradient expressed in the rectangular coordinates (x,y,z). Note that $1\text{ Pa}=1\text{ N}/\text{m}^2$, where $1\text{ N}=1\text{ kg-m}/\text{sec}^2$. Therefore, $1\text{ Pa}=1\text{ kg}/(\text{m-sec}^2)$. Also note that the ambient density ρ_0 and the speed of sound c are constants because we are dealing with a homogeneous fluid medium.

The solution of the wave equation given by (1.2-3) is an *exact* expression. However, by *approximating* the time-independent, free-space, Green's function using *Fresnel* and *Fraunhofer expansions*, both the near-field and far-field beam patterns (a.k.a. directivity functions) of a transmit volume aperture can be derived, respectively, along with corresponding near-field and far-field expressions for the velocity potential and acoustic pressure. We begin the analysis by noting that the range between source and field points, $|\mathbf{r}-\mathbf{r}_0|$, appears both in the amplitude and phase of the Green's function given by (1.2-4). Since

$$|\mathbf{r}-\mathbf{r}_0|=\sqrt{(\mathbf{r}-\mathbf{r}_0)\bullet(\mathbf{r}-\mathbf{r}_0)}, \tag{1.2-11}$$

it can be shown that

$$|\mathbf{r}-\mathbf{r}_0|=r\sqrt{1+b}, \tag{1.2-12}$$

where the parameter

$$b = \left(\frac{r_0}{r}\right)^2 - 2\frac{\hat{r} \bullet \mathbf{r}_0}{r} \tag{1.2-13}$$

is dimensionless,

$$r_0 = |\mathbf{r}_0| = \sqrt{x_0^2 + y_0^2 + z_0^2} \tag{1.2-14}$$

is the range to a source point in meters,

$$r = |\mathbf{r}| = \sqrt{x^2 + y^2 + z^2} \tag{1.2-15}$$

is the range to a field point in meters, and

$$\hat{r} = u\hat{x} + v\hat{y} + w\hat{z} \tag{1.2-16}$$

is the dimensionless *unit vector* in the direction of $\mathbf{r}$, that is,

$$\mathbf{r} = r\hat{r}, \tag{1.2-17}$$

where

$$u = \cos\alpha = \sin\theta\cos\psi, \tag{1.2-18}$$

$$v = \cos\beta = \sin\theta\sin\psi, \tag{1.2-19}$$

and

$$w = \cos\gamma = \cos\theta \tag{1.2-20}$$

are dimensionless *direction cosines* with respect to the X, Y, and Z axes, respectively, expressed in terms of the spherical angles θ and ψ (see Fig. 1.2-2). Note that u, v, and w take on values between -1 and 1 since $0° \leq \theta \leq 180°$ and $0° \leq \psi \leq 360°$, and the sum of their squares is equal to one:

$$u^2 + v^2 + w^2 = 1. \tag{1.2-21}$$

The Green's function given by (1.2-4) shall be approximated by approximating the range $|\mathbf{r} - \mathbf{r}_0|$.

In order to approximate $|\mathbf{r} - \mathbf{r}_0|$, we shall use a *binomial expansion* of $\sqrt{1+b}$ in (1.2-12) which yields

$$\boxed{|\mathbf{r} - \mathbf{r}_0| = r\sqrt{1+b} = r\left(1 + \frac{b}{2} - \frac{b^2}{8} + \cdots\right), \qquad |b| < 1} \tag{1.2-22}$$

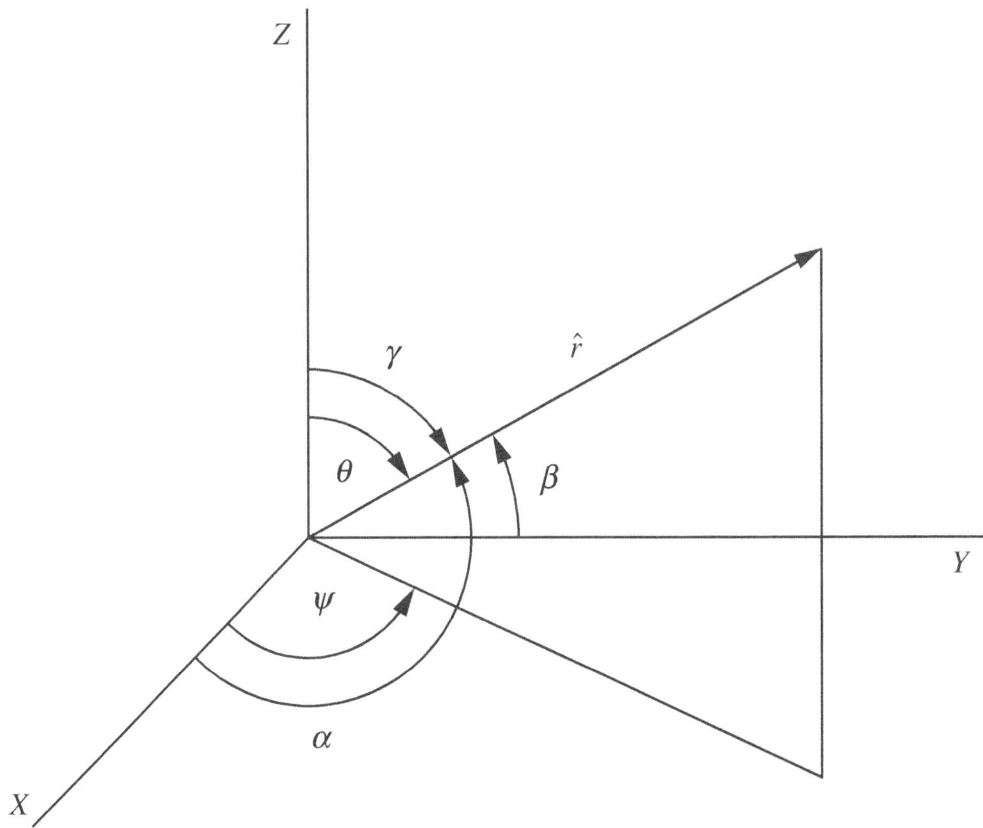

Figure 1.2-2 Unit vector $\hat{r}$ and spherical angles θ and ψ. Note that angle ψ is measured in a counter-clockwise direction.

Note that (1.2-22) is valid only if $|b|<1$, where b is given by (1.2-13). In Appendix 1A it is shown that $|b|<1$ whenever

$$r > \begin{cases} r_0\left(\sqrt{1+\cos^2(180°-\phi)}+\cos(180°-\phi)\right), & 0° \le \phi \le 90° \\ r_0\left(\sqrt{1+\cos^2\phi}-\cos\phi\right), & 90° \le \phi \le 180° \end{cases} \tag{1.2-23}$$

where ϕ is the angle between $\hat{r}$ and $\mathbf{r}_0$.

Let us first approximate the spherical spreading amplitude factor $1/|\mathbf{r}-\mathbf{r}_0|$. Assuming that $|b|<1$ and using only the *first term* of the binomial expansion in (1.2-22), $|\mathbf{r}-\mathbf{r}_0| \approx r$. Therefore,

$$\frac{1}{|\mathbf{r}-\mathbf{r}_0|} \approx \frac{1}{r}, \qquad |b|<1. \tag{1.2-24}$$

Although using only the first term of the binomial expansion is satisfactory for approximating the amplitude factor, it is *not* satisfactory for approximating the phase factor $\exp(-jk|\mathbf{r}-\mathbf{r}_0|)$ because small changes in the range $|\mathbf{r}-\mathbf{r}_0|$ can lead to large changes in phase.

In order to approximate $\exp(-jk|\mathbf{r}-\mathbf{r}_0|)$, we shall use the *first three terms* of the binomial expansion as shown in (1.2-22). If all terms involving r_0 raised to powers greater than two are neglected, then (1.2-22) reduces to

$$|\mathbf{r}-\mathbf{r}_0| \approx r - \hat{r}\bullet\mathbf{r}_0 + \frac{r_0^2 - (\hat{r}\bullet\mathbf{r}_0)^2}{2r}, \qquad |b|<1. \tag{1.2-25}$$

Therefore, by substituting (1.2-24) and (1.2-25) into (1.2-4), we obtain

$$g_f(\mathbf{r}\,|\,\mathbf{r}_0) \approx -\frac{\exp(-jkr)}{4\pi r}\exp(+jk\hat{r}\bullet\mathbf{r}_0)\exp\left[-jk\frac{r_0^2-(\hat{r}\bullet\mathbf{r}_0)^2}{2r}\right], \qquad |b|<1, \tag{1.2-26}$$

which is a *2nd-order, near-field approximation (expansion)* of the time-independent, free-space, Green's function involving *all* terms up to 2nd power in r_0.

The Fresnel approximation of $g_f(\mathbf{r}\,|\,\mathbf{r}_0)$ can also be obtained from (1.2-22) by using only the *first two terms* of the binomial expansion instead of the first three, which is equivalent to neglecting the dot product term $(\hat{r}\bullet\mathbf{r}_0)^2$ in (1.2-25) and (1.2-26). Therefore, the *Fresnel approximation (Fresnel expansion)* of the time-independent, free-space, Green's function is given by

$$\boxed{g_f(\mathbf{r}\,|\,\mathbf{r}_0) \approx -\frac{\exp(-jkr)}{4\pi r}\exp(+jk\hat{r}\bullet\mathbf{r}_0)\exp\left(-jk\frac{r_0^2}{2r}\right), \qquad |b|<1} \tag{1.2-27}$$

The Fresnel approximation of the Green's function given by (1.2-27) is *not* a true 2nd-order approximation like (1.2-26) is because it is based on neglecting the additional 2nd-order term $(\hat{r}\bullet\mathbf{r}_0)^2$.

Neglecting $(\hat{r}\bullet\mathbf{r}_0)^2$ in (1.2-26) in order to obtain (1.2-27) requires that

$$r_0^2 >> (\hat{r} \bullet \mathbf{r}_0)^2, \tag{1.2-28}$$

or, equivalently, that

$$r_0^2 \geq K(\hat{r} \bullet \mathbf{r}_0)^2, \tag{1.2-29}$$

where the constant $K > 1$ controls the accuracy of the Fresnel approximation. And by taking the positive square root of (1.2-29), we obtain

$$r_0 \geq \sqrt{K} \left| \hat{r} \bullet \mathbf{r}_0 \right|, \tag{1.2-30}$$

where the magnitude of the dot product is used because the range r_0 must be positive. Since we already designated ϕ as the angle between $\hat{r}$ and $\mathbf{r}_0$, where $\hat{r}$ is the unit vector in the direction of the position vector $\mathbf{r}$ to a field point, and $\mathbf{r}_0$ is the position vector to a source point, (1.2-30) reduces to

$$1 \geq \sqrt{K} \left| \cos\phi \right|, \qquad 0° < \phi < 180°. \tag{1.2-31}$$

From (1.2-31) it can be seen that the angle ϕ cannot be equal to $0°$ and $180°$ because $K > 1$. When $0° < \phi < 90°$, $\cos\phi > 0$ and $\left|\cos\phi\right| = \cos\phi$. When $90° < \phi < 180°$, $\cos\phi < 0$ and $\left|\cos\phi\right| = -\cos\phi = \cos(180° - \phi)$. Therefore, (1.2-31) can be rewritten as (see Fig. 1.2-3)

$$\phi_{\text{min}} \leq \phi \leq \phi_{\text{max}}, \tag{1.2-32}$$

where

$$\phi_{\text{min}} = \cos^{-1}\left(1/\sqrt{K}\right), \qquad 0° < \phi_{\text{min}} < 90°, \tag{1.2-33}$$

and

$$\phi_{\text{max}} = 180° - \phi_{\text{min}}, \qquad 90° < \phi_{\text{max}} < 180°. \tag{1.2-34}$$

Since a factor of 10 is a typical choice to represent "much greater than", if we let $K = 10$, then (1.2-32) becomes

$$\boxed{72° \leq \phi \leq 108°} \tag{1.2-35}$$

Equation (1.2-35) is the *Fresnel angle criterion*. For example, for the problem illustrated in Fig. 1.2-3, (1.2-35) indicates that the Fresnel approximation of the Green's function is very accurate inside an angular region no greater than $18°$ from the Y axis. This is analogous to the *paraxial assumption* that is made in the derivation of the Fresnel diffraction integral in optics. However, outside this angular region (which is equivalent to $K < 10$), the accuracy of the Fresnel

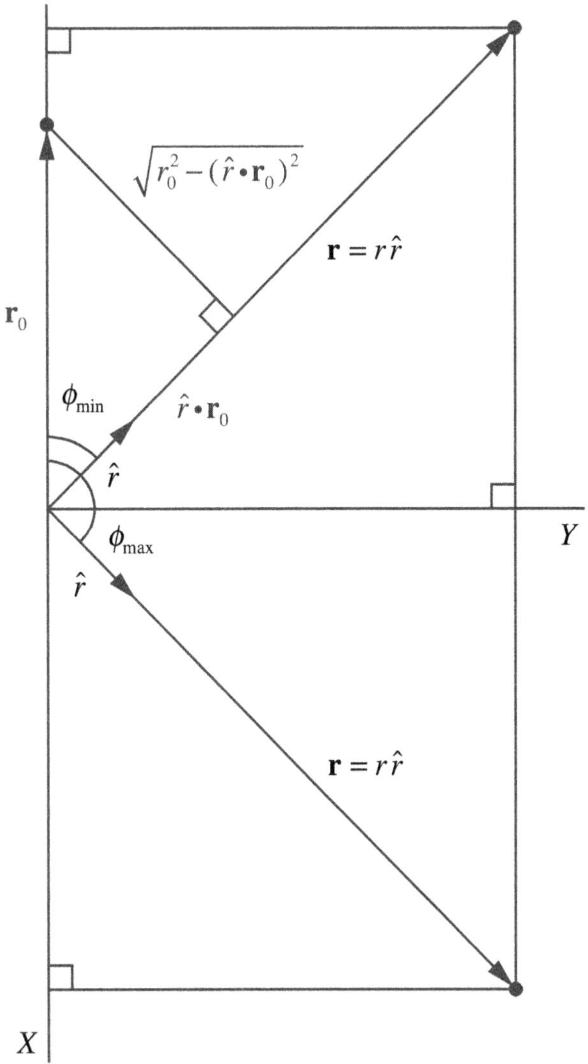

Figure 1.2-3 Illustration of the angles ϕ_{min} and ϕ_{max} for a source point on the negative X axis and two field points in the XY plane.

approximation decreases. Even if a source distribution is an arbitrarily shaped closed-surface, the inequality constraint given by (1.2-35) can easily be tested since $\phi = \cos^{-1}(\hat{r} \cdot \hat{r}_0)$, where the unit vector $\hat{r}_0 = \mathbf{r}_0 / r_0$.

Since we have just shown that in order for the Fresnel approximation of the Green's function to be accurate (1.2-35) must be satisfied, setting the range to a source point r_0 equal to its *maximum* value (call it R_0) and substituting $\phi = 72°$ and $\phi = 108°$ into (1.2-23) yields $r > 0.738 R_0$ and $r > 1.356 R_0$, respectively; thus

$$\boxed{r > 1.356 R_0} \tag{1.2-36}$$

Equation (1.2-36), which is based on $K = 10$, stipulates the *minimum* range that a field point has to be from an aperture in order to guarantee that $|b| < 1$ for $72° \le \phi \le 108°$ in order for the binomial expansion of $\sqrt{1+b}$ and, hence, the Fresnel approximation of the Green's function to be valid.

Next, by using the Fresnel approximation of the Green's function given by (1.2-27), we shall derive the Fraunhofer range criterion, the Fresnel range criterion, and the range to the near-field/far-field boundary. We begin by examining the quadratic phase factor

$$\exp\left(-jk\frac{r_0^2}{2r}\right)$$

that appears on the right-hand side of (1.2-27). This quadratic phase factor accurately models *wavefront curvature* in the Fresnel region of the aperture, and since $k = 2\pi/\lambda$, it can be rewritten as

$$\exp\left(-j\pi\frac{r_0^2}{\lambda r}\right).$$

We shall consider this quadratic phase factor to be *insignificant* if

$$\pi\frac{r_0^2}{\lambda r} < 1, \tag{1.2-37}$$

or

$$r > \pi r_0^2/\lambda. \tag{1.2-38}$$

In order to obtain a conservative criterion based on (1.2-38), consider the worst case, which is when the numerator is equal to its maximum value. This corresponds to setting r_0 equal to its maximum value R_0. Therefore, if

$$\boxed{r > \pi R_0^2/\lambda > 2.414 R_0} \tag{1.2-39}$$

then we shall consider the quadratic phase factor to be *insignificant* for all values of ϕ, that is, $0° \le \phi \le 180°$, since it will be approximately equal to $\exp(j0) = 1$, a constant, which indicates no phase variation as a function of r_0 and r. If (1.2-39) is satisfied, then $|b| < 1$ for $0° \le \phi \le 180°$. The factor $2.414 R_0$ was obtained by setting $r_0 = R_0$, $\phi = 0°$, and $\phi = 180°$ in (1.2-23) and choosing the largest factor. Equation (1.2-39) is the *Fraunhofer (far-field) range criterion.* The Fraunhofer region of an aperture is commonly referred to as the far-field region. However, if

$$\boxed{1.356R_0 < r < \pi R_0^2/\lambda} \tag{1.2-40}$$

then we shall consider the quadratic phase factor to be *significant*, that is, a nonnegligible function of r_0 and r. Equation (1.2-40) is the *Fresnel range criterion*, where the minimum range requirement given by (1.2-36) has been incorporated into (1.2-40) (see Fig. 1.2-4). The ratio $\pi R_0^2/\lambda$, which appears in both (1.2-39) and (1.2-40), is designated as the *range to the near-field/far-field boundary*, that is, $r_{\text{NF/FF}} = \pi R_0^2/\lambda$, and depends on the ratio between an aperture's size and wavelength. The numerator πR_0^2 is the *cross-sectional area* (in squared meters) of a transmit aperture with radius R_0 meters. The Fresnel region of an aperture, although commonly referred to as the near-field region, is actually only a *subset* of the near-field. For example, if $r < r_{\text{NF/FF}}$ but the Fresnel angle criterion given by (1.2-35) is *not* satisfied, then a field point is considered to be in the near-field region of a transmit aperture, *not* the Fresnel region.

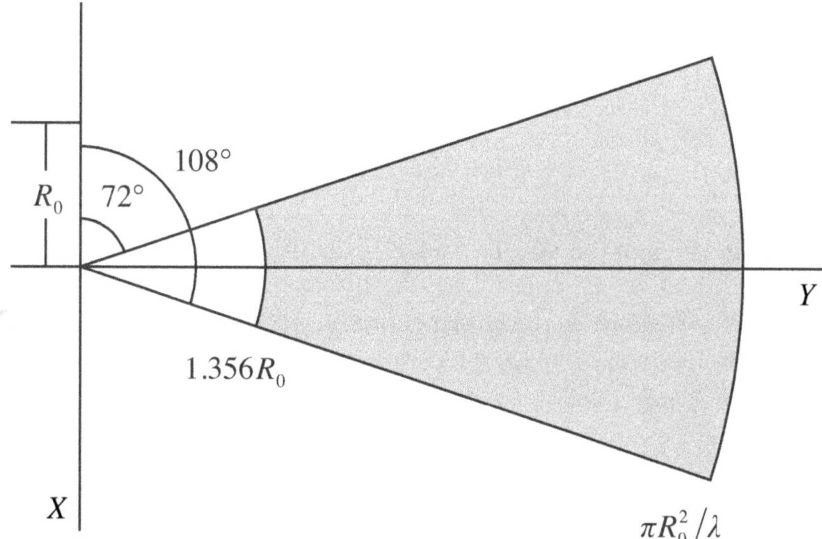

Figure 1.2-4 Illustration of the Fresnel criteria. The shaded region is the Fresnel region in the XY plane for source points lying along the X axis where $\max r_0 = R_0$ meters. Recall that the angles $72°$ and $108°$, and the factor 1.356, are based on $K = 10$.

Since we have a Fresnel approximation of the Green's function given by (1.2-27), we can proceed with the derivation of the near-field beam pattern. Substituting (1.2-27) into the exact solution of the wave equation given by (1.2-3) yields the following approximate solution:

$$\varphi(t,\mathbf{r}) \approx -\frac{1}{4\pi r}\int_{-\infty}^{\infty}\int_{V_0} X(f,\mathbf{r}_0)A_T(f,\mathbf{r}_0)\exp\left(-j\frac{k}{2r}r_0^2\right)\exp(+jk\hat{r}\bullet\mathbf{r}_0)dV_0 \times \exp\left[+j2\pi f\left(t-\frac{r}{c}\right)\right]df \tag{1.2-41}$$

for $t \geq r/c$, where the expression $t-(r/c)$ is known as the *retarded time*. Retarded time is a measure of the amount of time that has *elapsed* since the acoustic field transmitted by a source *first appears* at a receiver located r meters from the source at time $t = r/c$ seconds. Expand the exponent of $\exp(+jk\hat{r}\bullet\mathbf{r}_0)$ next.

With the use of (1.2-5), (1.2-7), and (1.2-16), we can write that

$$\exp(+jk\hat{r}\bullet\mathbf{r}_0) = \exp\left[+j2\pi(f_X x_0 + f_Y y_0 + f_Z z_0)\right], \tag{1.2-42}$$

where

$$f_X = u/\lambda = \sin\theta\cos\psi/\lambda, \tag{1.2-43}$$

$$f_Y = v/\lambda = \sin\theta\sin\psi/\lambda, \tag{1.2-44}$$

and

$$f_Z = w/\lambda = \cos\theta/\lambda \tag{1.2-45}$$

are *spatial frequencies* in the X, Y, and Z directions, respectively, with units of cycles per meter, and u, v, and w are the dimensionless direction cosines with respect to the X, Y, and Z axes, respectively, given by (1.2-18) through (1.2-20). Note that the *propagation-vector components* k_X, k_Y, and k_Z, with units of radians per meter, are related to the spatial frequencies f_X, f_Y, and f_Z, as follows: $k_X = 2\pi f_X = ku$, $k_Y = 2\pi f_Y = kv$, and $k_Z = 2\pi f_Z = kw$, where $k = 2\pi/\lambda$ is the wavenumber. If we substitute (1.2-42) into (1.2-41) and designate the volume integral by I, then

$$I = \int_{-\infty}^{\infty}\int_{-\infty}^{\infty}\int_{-\infty}^{\infty} X(f,\mathbf{r}_0)A_T(f,\mathbf{r}_0)\exp\left(-j\frac{k}{2r}r_0^2\right)\times \exp\left[+j2\pi(f_X x_0 + f_Y y_0 + f_Z z_0)\right]dx_0\,dy_0\,dz_0. \tag{1.2-46}$$

Equation (1.2-46) can be interpreted as being a *three-dimensional spatial Fourier transform*, that is,

$$I = F_{\mathbf{r}_0}\left\{X_M(f,\mathbf{r}_0)\exp\left(-j\frac{k}{2r}r_0^2\right)\right\} = F_{\mathbf{r}_0}\left\{X(f,\mathbf{r}_0)\left[A_T(f,\mathbf{r}_0)\exp\left(-j\frac{k}{2r}r_0^2\right)\right]\right\}, \tag{1.2-47}$$

where $F_{\mathbf{r}_0}\{\bullet\}$ is shorthand notation for a three-dimensional spatial Fourier transform with respect to x_0, y_0, and z_0; that is, $F_{\mathbf{r}_0}\{\bullet\} = F_{x_0}F_{y_0}F_{z_0}\{\bullet\}$, and $X_M(f,\mathbf{r}_0) = X(f,\mathbf{r}_0)A_T(f,\mathbf{r}_0)$ [see (1.1-3)]. By inspecting the right-hand side of (1.2-47), it can be seen that the complex transmit aperture function and quadratic phase factor have intentionally been grouped together, separate from the complex frequency spectrum of the transmitted electrical signal.

It is well known from Fourier transform theory that the time-domain Fourier transform of the product of two functions of time is equal to a convolution integral in the frequency domain. Similarly, the three-dimensional spatial Fourier transform of the product of two functions of three spatial variables is equal to a three-dimensional convolution integral in the spatial-frequency domain. Therefore, (1.2-47) can be expressed as follows:

$$I = \mathcal{X}_M(f,r,\boldsymbol{\alpha}) = \mathrm{X}(f,\boldsymbol{\alpha}) \underset{\boldsymbol{\alpha}}{*} \mathcal{D}_T(f,r,\boldsymbol{\alpha}), \tag{1.2-48}$$

or

$$\mathcal{X}_M(f,r,\boldsymbol{\alpha}) = \int_{-\infty}^{\infty} \mathrm{X}(f,\boldsymbol{\nu})\mathcal{D}_T(f,r,\boldsymbol{\alpha}-\boldsymbol{\nu})\,d\boldsymbol{\nu}, \tag{1.2-49}$$

where

$$\begin{aligned}\mathcal{X}_M(f,r,\boldsymbol{\alpha}) &= F_{\mathbf{r}_0}\left\{X_M(f,\mathbf{r}_0)\exp\left(-j\frac{k}{2r}r_0^2\right)\right\} \\ &= \int_{-\infty}^{\infty} X_M(f,\mathbf{r}_0)\exp\left(-j\frac{k}{2r}r_0^2\right)\exp(+j2\pi\boldsymbol{\alpha}\bullet\mathbf{r}_0)\,d\mathbf{r}_0\end{aligned} \tag{1.2-50}$$

is the near-field, complex frequency-and-angular spectrum of the input acoustic signal to the fluid medium (source distribution),

$$\mathrm{X}(f,\boldsymbol{\alpha}) = F_{\mathbf{r}_0}\{X(f,\mathbf{r}_0)\} = \int_{-\infty}^{\infty} X(f,\mathbf{r}_0)\exp(+j2\pi\boldsymbol{\alpha}\bullet\mathbf{r}_0)\,d\mathbf{r}_0 \tag{1.2-51}$$

is the complex frequency-and-angular spectrum of the transmitted electrical signal, and

$$\boxed{\begin{aligned}\mathcal{D}_T(f,r,\boldsymbol{\alpha}) &= F_{\mathbf{r}_0}\left\{A_T(f,\mathbf{r}_0)\exp\left(-j\frac{k}{2r}r_0^2\right)\right\} \\ &= \int_{-\infty}^{\infty} A_T(f,\mathbf{r}_0)\exp\left(-j\frac{k}{2r}r_0^2\right)\exp(+j2\pi\boldsymbol{\alpha}\cdot\mathbf{r}_0)\,d\mathbf{r}_0\end{aligned}} \tag{1.2-52}$$

is the *near-field beam pattern* (a.k.a. the *near-field directivity function*) of the transmit aperture, where

$$\boldsymbol{\alpha} = (f_X, f_Y, f_Z) \tag{1.2-53}$$

and

$$\mathbf{r}_0 = (x_0, y_0, z_0). \tag{1.2-54}$$

As was discussed in Subsection 1.1.1, the complex transmit aperture function $A_T(f,\mathbf{r}_0)$ is the complex frequency response of the transmit aperture at $\mathbf{r}_0$. Since the units of $A_T(f,\mathbf{r}_0)$ are $\sec^{-1}/\mathrm{V}$, the units of $\mathcal{D}_T(f,r,\boldsymbol{\alpha})$ are $\left(\mathrm{m}^3/\sec\right)/\mathrm{V}$. Equations (1.2-50) and (1.2-52) are most accurate in the Fresnel region of a transmit aperture where both the Fresnel angle criterion given by (1.2-35) and the Fresnel range criterion given by (1.2-40) are satisfied. They are less accurate in the near-field region outside the Fresnel region where $r < r_{\mathrm{NF/FF}}$, but the Fresnel angle criterion given by (1.2-35) is *not* satisfied. The right-hand side of (1.2-49) is shorthand notation for a three-dimensional convolution integral because $\boldsymbol{\nu} = (\nu_X, \nu_Y, \nu_Z)$ is a three-dimensional vector whose components are spatial frequencies in the X, Y, and Z directions, respectively, and $d\boldsymbol{\nu} = d\nu_X\, d\nu_Y\, d\nu_Z$. The right-hand sides of (1.2-50) through (1.2-52) are also shorthand notation for triple integrals because $d\mathbf{r}_0 = dx_0\, dy_0\, dz_0$. Since the spatial frequencies f_X, f_Y, and f_Z are related to the direction cosines u, v, and w, respectively, which are related to the spherical angles θ and ψ [see (1.2-43) through (1.2-45)], the near-field beam pattern given by (1.2-52) can ultimately be expressed as a function of frequency f, the range r to a field point, and the spherical angles θ and ψ; that is, $\mathcal{D}_T(f,r,\boldsymbol{\alpha}) \rightarrow \mathcal{D}_T(f,r,\theta,\psi)$. The phrase "angular spectrum" is used because spatial frequencies can be expressed in terms of spherical angles. Just as the frequency spectrum of a signal contains information as to which frequency components contain most of the signal's energy, the angular spectrum contains information as to the directions in which most of the signal's energy is propagating. The form of (1.2-52) is also known as a *Fresnel diffraction integral* in optics or as a *three-dimensional spatial Fresnel transform*. The important functions and their units at a transmit volume aperture are summarized in Table 1B-1 in Appendix 1B.

If $\mathcal{X}_M(f,r,\boldsymbol{\alpha})$ is substituted into (1.2-41) in place of the volume integral, then the approximate solution of the wave equation can be expressed as

$$\varphi(t, r, \theta, \psi) \approx -\frac{1}{4\pi r} \int_{-\infty}^{\infty} \mathcal{X}_M(f, r, \boldsymbol{\alpha}) \exp\left[+j2\pi f\left(t - \frac{r}{c}\right)\right] df \quad (1.2\text{-}55)$$

for $t \geq r/c$, and if (1.2-55) is then substituted into (1.2-8), we obtain the following equation for the acoustic pressure radiated by a transmit aperture:

$$p(t, r, \theta, \psi) \approx j\frac{\rho_0}{2r} \int_{-\infty}^{\infty} f \mathcal{X}_M(f, r, \boldsymbol{\alpha}) \exp\left[+j2\pi f\left(t - \frac{r}{c}\right)\right] df \quad (1.2\text{-}56)$$

for $t \geq r/c$, where $\mathcal{X}_M(f, r, \boldsymbol{\alpha})$ is given by (1.2-49) and $\boldsymbol{\alpha} = (f_X, f_Y, f_Z)$. The function $\mathcal{X}_M(f, r, \boldsymbol{\alpha})$ can ultimately be expressed as a function of frequency f, range r, and the spherical angles θ and ψ; that is, $\mathcal{X}_M(f, r, \boldsymbol{\alpha}) \to \mathcal{X}_M(f, r, \theta, \psi)$. Although (1.2-55) and (1.2-56) are most accurate in the Fresnel region of a transmit aperture and less accurate in the near-field region outside the Fresnel region, they are very accurate everywhere in the Fraunhofer (far-field) region as well. Recall that the quadratic phase factor is insignificant in the far-field. Near-field equations can be used to evaluate acoustic fields in the far-field. However, far-field equations can only be used to evaluate acoustic fields in the far-field – they are *not* valid in the near-field.

Example 1.2-1

An *identical* input electrical signal is applied at all locations $\mathbf{r}_0$ of a transmit aperture, that is,

$$x(t, \mathbf{r}_0) = x(t), \quad (1.2\text{-}57)$$

so that

$$X(f, \mathbf{r}_0) = X(f), \quad (1.2\text{-}58)$$

where $X(f)$ is the complex frequency spectrum (in volts per hertz) of $x(t)$. If we decide that any electronics and digital signal processing used for purposes of amplitude shading, beam steering, etc., are to be considered as part of the aperture, then stating that an identical input electrical signal is applied at all locations of a transmit aperture is *realistic* and *true* in most practical cases.

Substituting (1.2-58) into (1.2-51) yields

$$\mathrm{X}(f, \boldsymbol{\alpha}) = X(f) \int_{-\infty}^{\infty} \exp(+j2\pi \boldsymbol{\alpha} \cdot \mathbf{r}_0) d\mathbf{r}_0 = X(f) F_{\mathbf{r}_0}\{1\}, \quad (1.2\text{-}59)$$

and since

$$F_{\mathbf{r}_0}\{1\} = \delta(\boldsymbol{\alpha}), \tag{1.2-60}$$

the complex frequency-and-angular spectrum of the transmitted electrical signal given by (1.2-59) reduces to

$$\boxed{\mathsf{X}(f, \boldsymbol{\alpha}) = X(f)\delta(\boldsymbol{\alpha})} \tag{1.2-61}$$

where the impulse function $\delta(\boldsymbol{\alpha})$ is nonzero only when $\boldsymbol{\alpha} = \mathbf{0}$. Substituting (1.2-61) into (1.2-49) yields

$$\mathcal{X}_M(f, r, \boldsymbol{\alpha}) = X(f)\int_{-\infty}^{\infty} \delta(\mathbf{v})\mathcal{D}_T(f, r, \boldsymbol{\alpha} - \mathbf{v})d\mathbf{v}, \tag{1.2-62}$$

and by making use of the sifting property of impulse functions, the near-field, complex frequency-and-angular spectrum of the input acoustic signal to the fluid medium given by (1.2-62) reduces to

$$\mathcal{X}_M(f, r, \boldsymbol{\alpha}) = X(f)\mathcal{D}_T(f, r, \boldsymbol{\alpha}). \tag{1.2-63}$$

And by substituting (1.2-63) into (1.2-55) and (1.2-56), we obtain the following equations for the near-field velocity potential and acoustic pressure:

$$\varphi(t, r, \theta, \psi) \approx -\frac{1}{4\pi r}\int_{-\infty}^{\infty} X(f)\mathcal{D}_T(f, r, \boldsymbol{\alpha})\exp\left[+j2\pi f\left(t - \frac{r}{c}\right)\right]df \tag{1.2-64}$$

and

$$p(t, r, \theta, \psi) \approx j\frac{\rho_0}{2r}\int_{-\infty}^{\infty} fX(f)\mathcal{D}_T(f, r, \boldsymbol{\alpha})\exp\left[+j2\pi f\left(t - \frac{r}{c}\right)\right]df \tag{1.2-65}$$

for $t \geq r/c$, where the near-field beam pattern $\mathcal{D}_T(f, r, \boldsymbol{\alpha})$ can ultimately be expressed as a function of frequency f, the range r to a field point, and the spherical angles θ and ψ; that is, $\mathcal{D}_T(f, r, \boldsymbol{\alpha}) \rightarrow \mathcal{D}_T(f, r, \theta, \psi)$. Because an "*identical* input electrical signal" is used, the three-dimensional convolution integral equation for $\mathcal{X}_M(f, r, \boldsymbol{\alpha})$ given by (1.2-49) reduced to the simple product given by (1.2-63).

Equations (1.2-64) and (1.2-65) can be simplified further if the identical input electrical signal is *time-harmonic*, that is, if

$$x(t, \mathbf{r}_0) = x(t) = A_x \exp(+j2\pi f_0 t), \tag{1.2-66}$$

where A_x is a complex amplitude with units of volts (V), then

$$X(f, \mathbf{r}_0) = X(f) = A_x \delta(f - f_0), \tag{1.2-67}$$

where the impulse function $\delta(f - f_0)$ has units of inverse hertz. Substituting (1.2-67) into (1.2-64) and (1.2-65) yields

$$\varphi(t, r, \theta, \psi) \approx -\frac{A_x}{4\pi r} \mathcal{D}_T(f_0, r, \boldsymbol{\alpha}_0) \exp\left[+j2\pi f_0 \left(t - \frac{r}{c}\right)\right] \tag{1.2-68}$$

and

$$p(t, r, \theta, \psi) \approx jA_x \frac{\rho_0 f_0}{2r} \mathcal{D}_T(f_0, r, \boldsymbol{\alpha}_0) \exp\left[+j2\pi f_0 \left(t - \frac{r}{c}\right)\right], \tag{1.2-69}$$

respectively, for $t \geq r/c$, where $\boldsymbol{\alpha}_0 = (f_{X_0}, f_{Y_0}, f_{Z_0})$; $f_{X_0} = u/\lambda_0$, $f_{Y_0} = v/\lambda_0$, and $f_{Z_0} = w/\lambda_0$; $c = f_0 \lambda_0$; and u, v, and w are the direction cosines given by (1.2-18) through (1.2-20), respectively. The near-field beam pattern $\mathcal{D}_T(f_0, r, \boldsymbol{\alpha}_0)$ can ultimately be expressed as a function of frequency f_0, the range r to a field point, and the spherical angles θ and ψ; that is, $\mathcal{D}_T(f_0, r, \boldsymbol{\alpha}_0) \rightarrow \mathcal{D}_T(f_0, r, \theta, \psi)$.

We conclude this example by checking the units on the right-hand sides of both (1.2-68) and (1.2-69). Since A_x has units of V, the near-field beam pattern $\mathcal{D}_T(f_0, r, \boldsymbol{\alpha}_0)$ has units of $(\text{m}^3/\text{sec})/\text{V}$, and r has units of m; the product $A_x \mathcal{D}_T(f_0, r, \boldsymbol{\alpha}_0)$ has units of m^3/sec, which are the units of source strength, and $A_x \mathcal{D}_T(f_0, r, \boldsymbol{\alpha}_0)/r$ has units of m^2/sec, which are the correct units for the scalar velocity potential $\varphi(t, r, \theta, \psi)$ given by (1.2-68). And since ρ_0 has units of kg/m^3, f_0 has units of Hz, and $A_x \mathcal{D}_T(f_0, r, \boldsymbol{\alpha}_0)/r$ has units of m^2/sec; $A_x \rho_0 f_0 \mathcal{D}_T(f_0, r, \boldsymbol{\alpha}_0)/r$ has units of $\text{kg}/(\text{m-sec}^2) = \text{N}/\text{m}^2 = \text{Pa}$, which are the correct units for the acoustic pressure $p(t, r, \theta, \psi)$ given by (1.2-69). ■

1.2.2 Receive Aperture

In this subsection we shall derive the equation for the near-field beam pattern of a receive volume aperture by evaluating the receive coupling equation given by (1.1-15) for the special case of a sound-source (target) radiating sound in an unbounded, ideal (nonviscous), homogeneous, fluid medium. The sound-source (target) is modeled as an omnidirectional point-source located at $\mathbf{r}_S = (x_S, y_S, z_S)$ with arbitrary time dependence $s_0(t)$ (see Fig. 1.2-5). The source distribution for this case, with units of inverse seconds, is given by

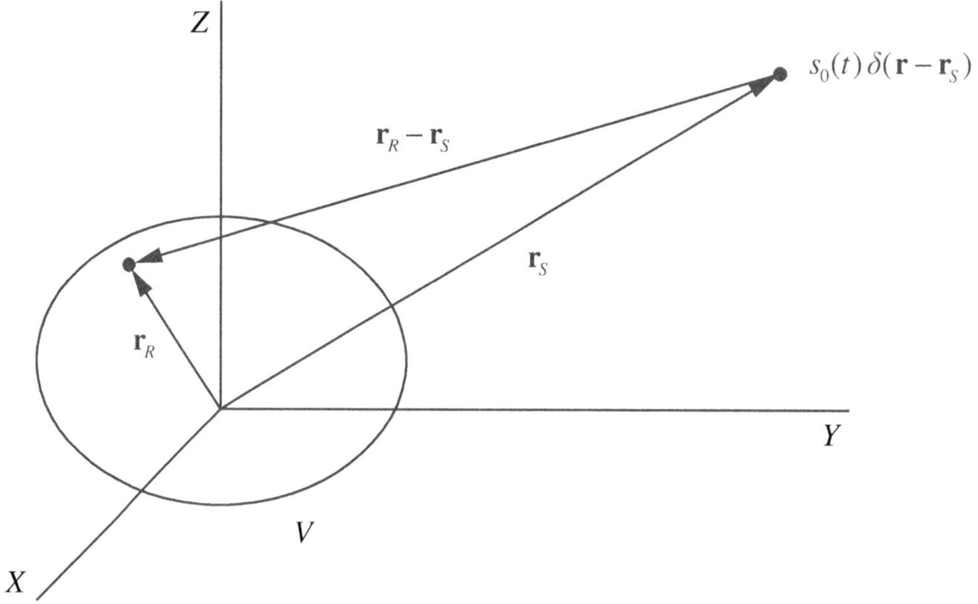

Figure 1.2-5 Omnidirectional point-source with arbitrary time dependence ensonifying a closed-surface, receive aperture enclosing a volume V.

$$x_M(t,\mathbf{r}) = s_0(t)\,\delta(\mathbf{r}-\mathbf{r}_S), \tag{1.2-70}$$

where $s_0(t)$, $t \geq 0$, is the target's *source strength* in cubic meters per second and the impulse function $\delta(\mathbf{r}-\mathbf{r}_S)$, with units of inverse cubic meters, represents a unit-amplitude, omnidirectional point-source at $\mathbf{r}_S$. Substituting (1.2-70) into the exact solution of the linear wave equation given by (1.2-2) yields

$$\varphi(t,\mathbf{r}) = -\frac{1}{4\pi}\int_{V_0} \frac{s_0\left(t-\left[\left|\mathbf{r}-\mathbf{r}_0\right|/c\right]\right)\delta(\mathbf{r}_0-\mathbf{r}_S)}{\left|\mathbf{r}-\mathbf{r}_0\right|}\,dV_0\,, \tag{1.2-71}$$

and by using the sifting property of impulse functions and letting $y_M(t,\mathbf{r}) = \varphi(t,\mathbf{r})$, we obtain

$$y_M(t,\mathbf{r}) = -\frac{1}{4\pi R}\,s_0(t-\tau), \tag{1.2-72}$$

for $t \geq \tau$, where

$$R = \left|\mathbf{r}-\mathbf{r}_S\right| \tag{1.2-73}$$

is the range in meters between the sound-source (target) and field point,

$$\tau = R/c \tag{1.2-74}$$

is the travel time or time delay in seconds, and c is the speed of sound in meters per second. Equation (1.2-72) is the velocity potential in squared meters per second of a *spherical wave with arbitrary time dependence.*

In order to evaluate the receive coupling equation given by (1.1-15), which is rewritten below for convenience,

$$y(t, \mathbf{r}_R) = \int_{-\infty}^{\infty} Y_M(\eta, \mathbf{r}_R) A_R(\eta, \mathbf{r}_R) \exp(+j2\pi\eta t)\, d\eta\,, \tag{1.1-15}$$

we need to evaluate (1.2-72) at $\mathbf{r} = \mathbf{r}_R$ and then take the Fourier transform with respect to t according to (1.1-13). Doing so yields

$$Y_M(\eta, \mathbf{r}_R) = S_0(\eta)\, g_\eta\left(\mathbf{r}_R \middle| \mathbf{r}_S\right), \tag{1.2-75}$$

where $Y_M(\eta, \mathbf{r}_R)$ is the complex frequency spectrum of the acoustic field incident upon the receive aperture at $\mathbf{r}_R = (x_R, y_R, z_R)$ with units of $\left(\text{m}^2/\text{sec}\right)/\text{Hz}$, $S_0(\eta)$ is the complex frequency spectrum of the target's source strength in $\left(\text{m}^3/\text{sec}\right)/\text{Hz}$,

$$g_\eta\left(\mathbf{r}_R \middle| \mathbf{r}_S\right) = -\frac{\exp\left(-jk\left|\mathbf{r}_R - \mathbf{r}_S\right|\right)}{4\pi\left|\mathbf{r}_R - \mathbf{r}_S\right|}, \tag{1.2-76}$$

is the time-independent, free-space, Green's function of an unbounded, ideal (nonviscous), homogeneous, fluid medium with units of m^{-1},

$$k = 2\pi\eta/c = 2\pi/\lambda \tag{1.2-77}$$

is the wavenumber in rad/m where $c = \eta\lambda$,

$$\mathbf{r}_R = x_R\hat{x} + y_R\hat{y} + z_R\hat{z} \tag{1.2-78}$$

is the position vector to an aperture point, and

$$\mathbf{r}_S = x_S\hat{x} + y_S\hat{y} + z_S\hat{z} \tag{1.2-79}$$

is the position vector to the sound-source (target) (see Fig. 1.2-5). Substituting (1.2-75) into the receive coupling equation given by (1.1-15) yields

$$\boxed{y(t, \mathbf{r}_R) = \int_{-\infty}^{\infty} S_0(\eta) A_R(\eta, \mathbf{r}_R)\, g_\eta\left(\mathbf{r}_R \middle| \mathbf{r}_S\right) \exp(+j2\pi\eta t)\, d\eta} \tag{1.2-80}$$

where $y(t,\mathbf{r}_R)$ is the output electrical signal from the receive aperture at time t and location $\mathbf{r}_R$ with units of V/m^3.

The next step is to obtain the Fresnel approximation of the Green's function given by (1.2-76) and substitute it into (1.2-80). If we follow the same procedure that was used in Subsection 1.2.1, then

$$|\mathbf{r}_R-\mathbf{r}_S|=|\mathbf{r}_S-\mathbf{r}_R|=r_S\sqrt{1+b}\,, \tag{1.2-81}$$

where now the dimensionless parameter b is given by

$$b=\left(\frac{r_R}{r_S}\right)^2-2\frac{\hat{r}_S\bullet\mathbf{r}_R}{r_S}\,, \tag{1.2-82}$$

$$r_R=|\mathbf{r}_R|=\sqrt{x_R^2+y_R^2+z_R^2} \tag{1.2-83}$$

is the range to an aperture point in meters,

$$r_S=|\mathbf{r}_S|=\sqrt{x_S^2+y_S^2+z_S^2} \tag{1.2-84}$$

is the range to the sound-source (target) in meters, and

$$\hat{r}_S=u_S\hat{x}+v_S\hat{y}+w_S\hat{z} \tag{1.2-85}$$

is the dimensionless *unit vector* in the direction of $\mathbf{r}_S$, that is,

$$\mathbf{r}_S=r_S\hat{r}_S\,, \tag{1.2-86}$$

where

$$u_S=\cos\alpha_S=\sin\theta_S\cos\psi_S\,, \tag{1.2-87}$$

$$v_S=\cos\beta_S=\sin\theta_S\sin\psi_S\,, \tag{1.2-88}$$

and

$$w_S=\cos\gamma_S=\cos\theta_S \tag{1.2-89}$$

are dimensionless *direction cosines* with respect to the X, Y, and Z axes, respectively, expressed in terms of the spherical angles θ_S and ψ_S (see Fig. 1.2-6). Therefore, the Fresnel approximation of the Green's function is given by

$$\boxed{g_\eta(\mathbf{r}_R|\mathbf{r}_S)\approx-\frac{\exp(-jkr_S)}{4\pi r_S}\exp(+jk\hat{r}_S\bullet\mathbf{r}_R)\exp\left(-jk\frac{r_R^2}{2r_S}\right),\qquad |b|<1} \tag{1.2-90}$$

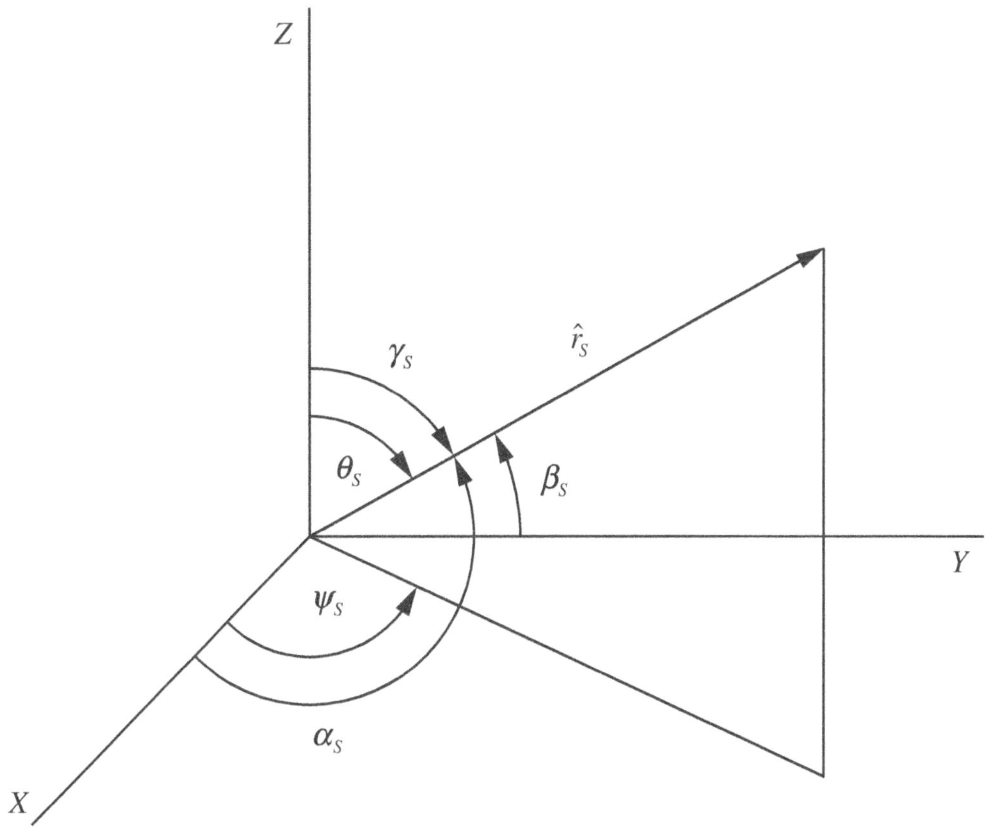

Figure 1.2-6 Unit vector $\hat{r}_S$ and spherical angles θ_S and ψ_S. Note that angle ψ_S is measured in a counter-clockwise direction.

where b is given by (1.2-82).

If we now let ϕ be the angle between $\hat{r}_S$ and $\mathbf{r}_R$, and if we follow the same procedure that was used in Subsection 1.2.1, then a sound-source (target) will be in the Fresnel region of a receive aperture if

$$\boxed{72° \leq \phi \leq 108°} \tag{1.2-91}$$

and

$$\boxed{1.356 R_R < r_S < \pi R_R^2 / \lambda} \tag{1.2-92}$$

where R_R is the maximum radial extent of the receive aperture, that is, $\max r_R = R_R$. Equation (1.2-91) is the *Fresnel angle criterion.* The Fresnel approximation of the Green's function will be very accurate in the angular region defined by (1.2-91). The Fresnel angle criterion given by (1.2-91) can easily be tested since $\phi = \cos^{-1}(\hat{r}_S \bullet \hat{r}_R)$, where the unit vector $\hat{r}_R = \mathbf{r}_R / r_R$. Equation (1.2-92) is the *Fresnel range criterion.* If $r_S > 1.356 R_R$, then $|b| < 1$ for ϕ satisfying (1.2-91), where b is given by (1.2-82). As a result, the binomial expansion of $\sqrt{1+b}$ in (1.2-81) and, hence, the Fresnel approximation of the Green's function given by (1.2-90) is valid. The range to the near-field/far-field boundary $r_{\text{NF/FF}} = \pi R_R^2 / \lambda$ depends on the ratio between an aperture's size and wavelength, where πR_R^2 is the *cross-sectional area* (in squared meters) of a receive aperture with radius R_R meters. If $r_S < r_{\text{NF/FF}}$ but the Fresnel angle criterion given by (1.2-91) is *not* satisfied, then a sound-source is considered to be in the near-field region of a receive aperture, *not* the Fresnel region.

If

$$\boxed{r_S > \pi R_R^2 / \lambda > 2.414 R_R} \tag{1.2-93}$$

then a sound-source (target) will be in the Fraunhofer (far-field) region of a receive aperture. Equation (1.2-93) is the *Fraunhofer (far-field) range criterion.* If (1.2-93) is satisfied, then $|b| < 1$ for all values of ϕ, that is, $0° \le \phi \le 180°$. The factor $2.414 R_R$ was obtained by setting $r = r_S$, $r_0 = r_R = R_R$, $\phi = 0°$, and $\phi = 180°$ in (1.2-23) and choosing the largest factor.

Now that we have the Fresnel approximation of the Green's function, substituting (1.2-90) into (1.2-80) yields

$$y(t, \mathbf{r}_R) \approx -\frac{1}{4\pi r_S} \int_{-\infty}^{\infty} S_0(\eta) A_R(\eta, \mathbf{r}_R) \exp\left(-j \frac{k}{2 r_S} r_R^2\right) \exp(+j2\pi \boldsymbol{\beta} \bullet \mathbf{r}_R) \times \exp\left[+j2\pi\eta\left(t - \frac{r_S}{c}\right)\right] d\eta \tag{1.2-94}$$

for $t \ge r_S / c$, where

$$\boldsymbol{\beta} = (\beta_X, \beta_Y, \beta_Z), \tag{1.2-95}$$

and

$$\beta_X = u_S / \lambda = \sin\theta_S \cos\psi_S / \lambda, \tag{1.2-96}$$

$$\beta_Y = v_S / \lambda = \sin\theta_S \sin\psi_S / \lambda, \tag{1.2-97}$$

and

$$\beta_Z = w_S / \lambda = \cos\theta_S / \lambda \tag{1.2-98}$$

are *spatial frequencies* in the X, Y, and Z directions, respectively, with units of cycles per meter, and u_S, v_S, and w_S are the dimensionless direction cosines with respect to the X, Y, and Z axes, respectively, given by (1.2-87) through (1.2-89). The *total* output electrical signal from the receive aperture (with units of volts) is given by

$$y(t) = \int_{-\infty}^{\infty} y(t, \mathbf{r}_R)\, d\mathbf{r}_R \tag{1.2-99}$$

and by substituting (1.2-94) into (1.2-99), we obtain

$$y(t) \approx -\frac{1}{4\pi r_S} \int_{-\infty}^{\infty} S_0(\eta)\mathcal{D}_R(\eta, r_S, \boldsymbol{\beta}) \exp\left[+j2\pi\eta\left(t - \frac{r_S}{c}\right)\right] d\eta \tag{1.2-100}$$

for $t \geq r_S/c$, where

$$\begin{aligned}\mathcal{D}_R(\eta, r_S, \boldsymbol{\beta}) &= F_{\mathbf{r}_R}\left\{A_R(\eta, \mathbf{r}_R)\exp\left(-j\frac{k}{2r_S}r_R^2\right)\right\} \\ &= \int_{-\infty}^{\infty} A_R(\eta, \mathbf{r}_R)\exp\left(-j\frac{k}{2r_S}r_R^2\right)\exp(+j2\pi\boldsymbol{\beta}\cdot\mathbf{r}_R)\, d\mathbf{r}_R\end{aligned}$$

(1.2-101)

is the *near-field beam pattern* (a.k.a. the *directivity function*) of the receive aperture, where $F_{\mathbf{r}_R}\{\bullet\}$ is shorthand notation for a three-dimensional spatial Fourier transform with respect to x_R, y_R, and z_R; that is, $F_{\mathbf{r}_R}\{\bullet\} = F_{x_R}F_{y_R}F_{z_R}\{\bullet\}$, $\mathbf{r}_R = (x_R, y_R, z_R)$, and $d\mathbf{r}_R = dx_R\, dy_R\, dz_R$. As was discussed in Subsection 1.1.2, the complex receive aperture function $A_R(\eta, \mathbf{r}_R)$ is the complex frequency response of the receive aperture at $\mathbf{r}_R$. Since the units of $A_R(\eta, \mathbf{r}_R)$ are $\left(\mathrm{V}/\left(\mathrm{m}^2/\mathrm{sec}\right)\right)/\mathrm{m}^3$, the units of $\mathcal{D}_R(\eta, r_S, \boldsymbol{\beta})$ are $\mathrm{V}/\left(\mathrm{m}^2/\mathrm{sec}\right)$. Equations (1.2-100) and (1.2-101) are most accurate when the sound-source is in the Fresnel region of a receive aperture where both the Fresnel angle criterion given by (1.2-91) and the Fresnel range criterion given by (1.2-92) are satisfied. They are less accurate when the sound-source is in the near-field region outside the Fresnel region where $r_S < r_{\mathrm{NF/FF}}$, but the Fresnel angle criterion given by (1.2-91) is *not* satisfied. Equation (1.2-100) shows that if a sound-source (target) is in the Fresnel region of a receive aperture, then the total output electrical signal depends on the near-field beam pattern of the receive aperture, *not* the far-field beam pattern. Since the spatial frequencies β_X,

β_Y, and β_Z are related to the direction cosines u_S, v_S, and w_S, respectively, which are related to the spherical angles θ_S and ψ_S [see (1.2-96) through (1.2-98)], the near-field beam pattern given by (1.2-101) can ultimately be expressed as a function of frequency η, the range r_S to a sound-source (target), and the spherical angles θ_S and ψ_S; that is, $\mathcal{D}_R(\eta, r_S, \boldsymbol{\beta}) \rightarrow \mathcal{D}_R(\eta, r_S, \theta_S, \psi_S)$. The form of (1.2-101) is also known as a *Fresnel diffraction integral* in optics or as a *three-dimensional spatial Fresnel transform.* The important functions and their units at a receive volume aperture are summarized in Table 1B-2 in Appendix 1B.

1.3 The Far-Field Beam Pattern of a Volume Aperture

1.3.1 Transmit Aperture

If a field point is in the far-field region of a transmit aperture, that is, if (1.2-39) is satisfied, then the *Fraunhofer approximation (Fraunhofer expansion)* of the time-independent, free-space, Green's function given by (1.2-4) can be obtained from its Fresnel approximation given by (1.2-27) by neglecting the quadratic phase factor. Doing so yields the following Fraunhofer (far-field) approximation:

$$\boxed{g_f(\mathbf{r}|\mathbf{r}_0) \approx -\frac{\exp(-jkr)}{4\pi r}\exp(+jk\hat{r}\bullet\mathbf{r}_0), \qquad |b|<1} \tag{1.3-1}$$

where b is given by (1.2-13). Recall that if (1.2-39) is satisfied, that is, if $r > \pi R_0^2/\lambda > 2.414R_0$, then $|b|<1$ for all values of ϕ, that is, $0° \le \phi \le 180°$, where ϕ is the angle between $\hat{r}$ and $\mathbf{r}_0$ (see Fig. 1.3-1).

Similarly, the far-field beam pattern can be obtained from the near-field beam pattern given by (1.2-52) by neglecting the quadratic phase factor. Therefore, the *far-field beam pattern* (a.k.a. the *far-field directivity function*) of the transmit aperture is given by

$$\boxed{D_T(f,\boldsymbol{\alpha}) = F_{\mathbf{r}_0}\{A_T(f,\mathbf{r}_0)\} = \int_{-\infty}^{\infty} A_T(f,\mathbf{r}_0)\exp(+j2\pi\boldsymbol{\alpha}\bullet\mathbf{r}_0)\,d\mathbf{r}_0} \tag{1.3-2}$$

where $\boldsymbol{\alpha} = (f_X, f_Y, f_Z)$, $F_{\mathbf{r}_0}\{\bullet\}$ is shorthand notation for a three-dimensional spatial Fourier transform with respect to x_0, y_0, and z_0; that is, $F_{\mathbf{r}_0}\{\bullet\} = F_{x_0}F_{y_0}F_{z_0}\{\bullet\}$, $\mathbf{r}_0 = (x_0, y_0, z_0)$, and $d\mathbf{r}_0 = dx_0\,dy_0\,dz_0$. Since the units of $A_T(f,\mathbf{r}_0)$ are $\sec^{-1}/\mathrm{V}$, the units of $D_T(f,\boldsymbol{\alpha})$ are $(\mathrm{m}^3/\sec)/\mathrm{V}$, which are the same units for the near-

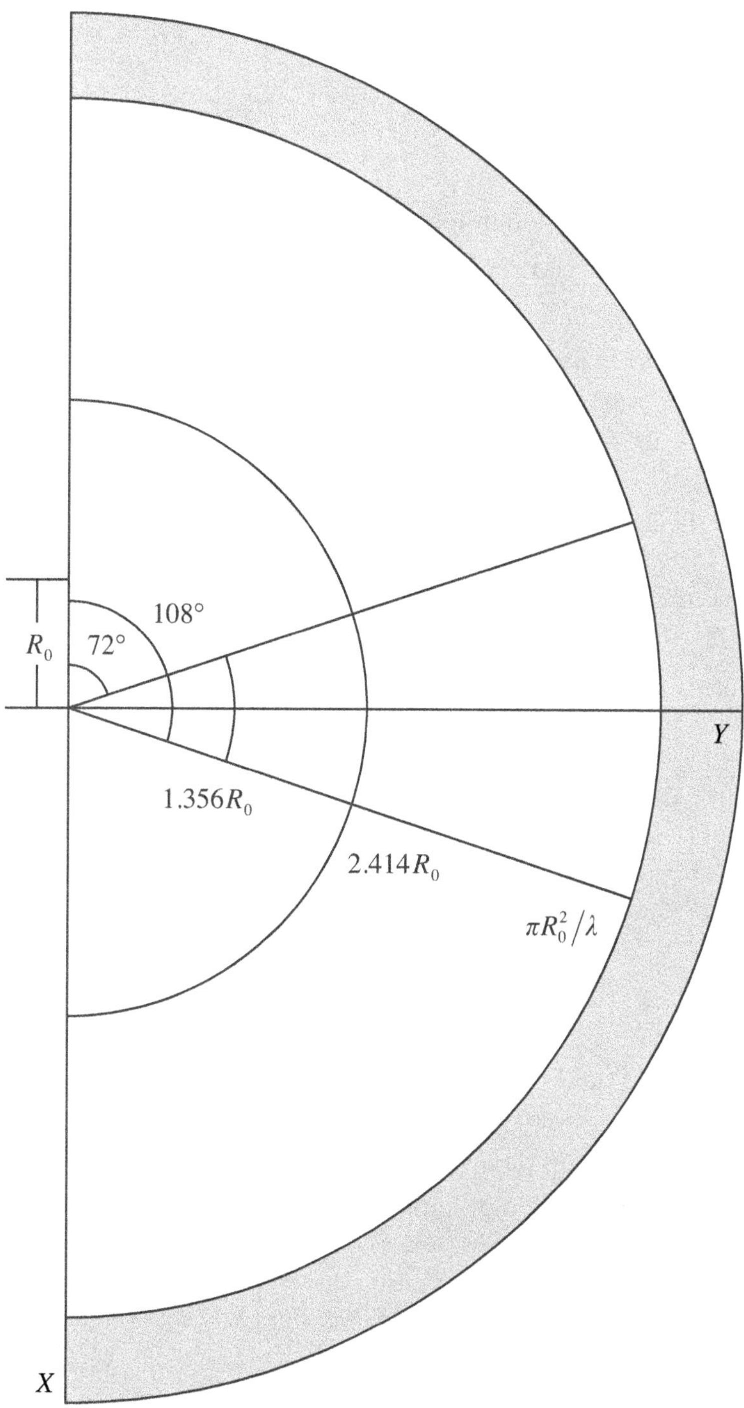

Figure 1.3-1 Illustration of the far-field criterion. The shaded region is the beginning of the Fraunhofer (far-field) region in the XY plane for source points lying along the X axis where $\max r_0 = R_0$ meters. Recall that the angles $72°$ and $108°$, and the factor 1.356, are based on $K = 10$.

field beam pattern. It also follows from (1.3-2) that

$$\boxed{A_T(f,\mathbf{r}_0) = F_{\boldsymbol{\alpha}}^{-1}\{D_T(f,\boldsymbol{\alpha})\} = \int_{-\infty}^{\infty} D_T(f,\boldsymbol{\alpha})\exp(-j2\pi\boldsymbol{\alpha}\bullet\mathbf{r}_0)\,d\boldsymbol{\alpha}} \tag{1.3-3}$$

where $F_{\boldsymbol{\alpha}}^{-1}\{\bullet\}$ is shorthand notation for a three-dimensional inverse spatial Fourier transform with respect to spatial frequencies f_X, f_Y, and f_Z; that is, $F_{\boldsymbol{\alpha}}^{-1}\{\bullet\} = F_{f_X}^{-1}F_{f_Y}^{-1}F_{f_Z}^{-1}\{\bullet\}$, and $d\boldsymbol{\alpha} = df_X\,df_Y\,df_Z$. From (1.3-2) and (1.3-3) it can be seen that *an aperture function and its far-field beam pattern form a spatial Fourier transform pair*. Note that the far-field beam pattern is *not* a function of the range r to a field point as the near-field beam pattern is. Since the spatial frequencies f_X, f_Y, and f_Z are related to the direction cosines u, v, and w, respectively, which are related to the spherical angles θ and ψ [see (1.2-43) through (1.2-45)], the far-field beam pattern given by (1.3-2) can ultimately be expressed as a function of frequency f and the spherical angles θ and ψ, that is, $D_T(f,\boldsymbol{\alpha}) \rightarrow D_T(f,\theta,\psi)$. The form of (1.3-2) is also known as a *Fraunhofer diffraction integral* in optics or as a *three-dimensional spatial Fourier transform.* The important functions and their units at a transmit volume aperture are summarized in Table 1B-1 in Appendix 1B.

The far-field, complex frequency-and-angular spectrum of the input acoustic signal to the fluid medium (source distribution) can be obtained from the near-field version given by (1.2-49) by replacing the near-field beam pattern with the far-field beam pattern. Doing so yields

$$\mathsf{X}_M(f,\boldsymbol{\alpha}) = \int_{-\infty}^{\infty} \mathsf{X}(f,\mathbf{v})D_T(f,\boldsymbol{\alpha}-\mathbf{v})\,d\mathbf{v}. \tag{1.3-4}$$

Equation (1.3-4) can also be obtained by taking the three-dimensional spatial Fourier transform of (1.1-3). Replacing $\mathcal{X}_M(f,r,\boldsymbol{\alpha})$ with $\mathsf{X}_M(f,\boldsymbol{\alpha})$ in (1.2-55) and (1.2-56) yields the following equations for the velocity potential and acoustic pressure valid in the far-field region of a transmit aperture:

$$\varphi(t,r,\theta,\psi) \approx -\frac{1}{4\pi r}\int_{-\infty}^{\infty} \mathsf{X}_M(f,\boldsymbol{\alpha})\exp\left[+j2\pi f\left(t-\frac{r}{c}\right)\right]df \tag{1.3-5}$$

and

$$p(t,r,\theta,\psi) \approx j\frac{\rho_0}{2r}\int_{-\infty}^{\infty} f\,\mathsf{X}_M(f,\boldsymbol{\alpha})\exp\left[+j2\pi f\left(t-\frac{r}{c}\right)\right]df \tag{1.3-6}$$

for $t \geq r/c$, where $\mathsf{X}_M(f,\boldsymbol{\alpha})$ is given by (1.3-4) and $\boldsymbol{\alpha} = (f_X, f_Y, f_Z)$. The function

$\mathsf{X}_M(f,\boldsymbol{\alpha})$ can ultimately be expressed as a function of frequency f and the spherical angles θ and ψ, that is, $\mathsf{X}_M(f,\boldsymbol{\alpha}) \rightarrow \mathsf{X}_M(f,\theta,\psi)$.

Example 1.3-1

This example is a continuation of Example 1.2-1. Recall from Example 1.2-1 that when an *identical* input electrical signal is applied at all locations $\mathbf{r}_0$ of a transmit aperture, the complex frequency-and-angular spectrum of the transmitted electrical signal reduces to [see (1.2-61)]

$$\mathsf{X}(f,\boldsymbol{\alpha}) = X(f)\delta(\boldsymbol{\alpha}). \tag{1.3-7}$$

Substituting (1.3-7) into (1.3-4) yields

$$\mathsf{X}_M(f,\boldsymbol{\alpha}) = X(f)\int_{-\infty}^{\infty} \delta(\boldsymbol{\nu}) D_T(f,\boldsymbol{\alpha}-\boldsymbol{\nu})\,d\boldsymbol{\nu}, \tag{1.3-8}$$

and by making use of the sifting property of impulse functions, the far-field, complex frequency-and-angular spectrum of the input acoustic signal to the fluid medium given by (1.3-8) reduces to

$$\mathsf{X}_M(f,\boldsymbol{\alpha}) = X(f) D_T(f,\boldsymbol{\alpha}). \tag{1.3-9}$$

And by substituting (1.3-9) into (1.3-5) and (1.3-6), we obtain the following equations for the velocity potential and acoustic pressure valid in the far-field region of a transmit aperture:

$$\boxed{\varphi(t,r,\theta,\psi) \approx -\frac{1}{4\pi r}\int_{-\infty}^{\infty} X(f) D_T(f,\boldsymbol{\alpha}) \exp\left[+j2\pi f\left(t-\frac{r}{c}\right)\right] df} \tag{1.3-10}$$

and

$$\boxed{p(t,r,\theta,\psi) \approx j\frac{\rho_0}{2r}\int_{-\infty}^{\infty} f X(f) D_T(f,\boldsymbol{\alpha}) \exp\left[+j2\pi f\left(t-\frac{r}{c}\right)\right] df} \tag{1.3-11}$$

for $t \geq r/c$, where the far-field beam pattern $D_T(f,\boldsymbol{\alpha})$ can ultimately be expressed as a function of frequency f and the spherical angles θ and ψ, that is,

$D_T(f,\boldsymbol{\alpha}) \to D_T(f,\theta,\psi)$. Because an "*identical* input electrical signal" is used, the three-dimensional convolution integral equation for $X_M(f,\boldsymbol{\alpha})$ given by (1.3-4) reduced to the simple product given by (1.3-9).

If the identical input electrical signal is time-harmonic, then it was shown in Example 1.2-1 that the complex frequency spectrum $X(f)$ reduces to [see (1.2-67)]

$$X(f) = A_x \delta(f - f_0), \tag{1.3-12}$$

where the amplitude factor A_x is complex with units of volts, and the impulse function $\delta(f - f_0)$ has units of inverse hertz. Substituting (1.3-12) into (1.3-10) and (1.3-11) yields

$$\boxed{\varphi(t,r,\theta,\psi) \approx -\frac{A_x}{4\pi r} D_T(f_0,\boldsymbol{\alpha}_0)\exp\left[+j2\pi f_0\left(t - \frac{r}{c}\right)\right]} \tag{1.3-13}$$

and

$$\boxed{p(t,r,\theta,\psi) \approx jA_x \frac{\rho_0 f_0}{2r} D_T(f_0,\boldsymbol{\alpha}_0)\exp\left[+j2\pi f_0\left(t - \frac{r}{c}\right)\right]} \tag{1.3-14}$$

respectively, for $t \geq r/c$, where $\boldsymbol{\alpha}_0 = (f_{X_0}, f_{Y_0}, f_{Z_0})$; $f_{X_0} = u/\lambda_0$, $f_{Y_0} = v/\lambda_0$, and $f_{Z_0} = w/\lambda_0$; $\lambda_0 = c/f_0$; and u, v, and w are the direction cosines given by (1.2-18) through (1.2-20), respectively. The far-field beam pattern $D_T(f_0,\boldsymbol{\alpha}_0)$ can ultimately be expressed as a function of frequency f_0 and the spherical angles θ and ψ, that is, $D_T(f_0,\boldsymbol{\alpha}_0) \to D_T(f_0,\theta,\psi)$.

We conclude this example by checking the units on the right-hand sides of both (1.3-13) and (1.3-14). Since A_x has units of V, the far-field beam pattern $D_T(f_0,\boldsymbol{\alpha}_0)$ has units of $(\text{m}^3/\text{sec})/\text{V}$, and r has units of m; the product $A_x D_T(f_0,\boldsymbol{\alpha}_0)$ has units of m^3/sec, which are the units of source strength, and $A_x D_T(f_0,\boldsymbol{\alpha}_0)/r$ has units of m^2/sec, which are the correct units for the scalar velocity potential $\varphi(t,r,\theta,\psi)$ given by (1.3-13). And since ρ_0 has units of kg/m^3, f_0 has units of Hz, and $A_x D_T(f_0,\boldsymbol{\alpha}_0)/r$ has units of m^2/sec; $A_x \rho_0 f_0 D_T(f_0,\boldsymbol{\alpha}_0)/r$ has units of $\text{kg}/(\text{m-sec}^2) = \text{N}/\text{m}^2 = \text{Pa}$, which are the correct units for the acoustic pressure $p(t,r,\theta,\psi)$ given by (1.3-14). Equations (1.3-13) and (1.3-14) will be used in Section 4.1 to derive the equation for the directivity of a transmit volume aperture. ■

1.3.2 Receive Aperture

If a sound-source (target) is in the far-field region of a receive aperture, that is, if (1.2-93) is satisfied, then the *Fraunhofer approximation (Fraunhofer expansion)* of the time-independent, free-space, Green's function given by (1.2-76) can be obtained from its Fresnel approximation given by (1.2-90) by neglecting the quadratic phase factor. Doing so yields the following Fraunhofer (far-field) approximation:

$$g_\eta(\mathbf{r}_R|\mathbf{r}_S) \approx -\frac{\exp(-jkr_S)}{4\pi r_S}\exp(+jk\hat{r}_S \bullet \mathbf{r}_R), \qquad |b|<1 \tag{1.3-15}$$

where b is given by (1.2-82). Recall that if (1.2-93) is satisfied, that is, if $r_S > \pi R_R^2/\lambda > 2.414 R_R$, then $|b|<1$ for all values of ϕ, that is, $0° \leq \phi \leq 180°$, where ϕ is the angle between $\hat{r}_S$ and $\mathbf{r}_R$.

Similarly, the far-field beam pattern can be obtained from the near-field beam pattern given by (1.2-101) by neglecting the quadratic phase factor. Therefore, the *far-field beam pattern* (a.k.a. the *directivity function*) of the receive aperture is given by

$$D_R(\eta,\boldsymbol{\beta}) = F_{\mathbf{r}_R}\{A_R(\eta,\mathbf{r}_R)\} = \int_{-\infty}^{\infty} A_R(\eta,\mathbf{r}_R)\exp(+j2\pi\boldsymbol{\beta}\bullet\mathbf{r}_R)\,d\mathbf{r}_R \tag{1.3-16}$$

where $\boldsymbol{\beta} = (\beta_X, \beta_Y, \beta_Z)$, $F_{\mathbf{r}_R}\{\bullet\}$ is shorthand notation for a three-dimensional spatial Fourier transform with respect to x_R, y_R, and z_R; that is, $F_{\mathbf{r}_R}\{\bullet\} = F_{x_R} F_{y_R} F_{z_R}\{\bullet\}$, $\mathbf{r}_R = (x_R, y_R, z_R)$, and $d\mathbf{r}_R = dx_R\, dy_R\, dz_R$. Since the units of $A_R(\eta,\mathbf{r}_R)$ are $\left(\mathrm{V}/\left(\mathrm{m}^2/\mathrm{sec}\right)\right)/\mathrm{m}^3$, the units of $D_R(\eta,\boldsymbol{\beta})$ are $\mathrm{V}/\left(\mathrm{m}^2/\mathrm{sec}\right)$, which are the same units for the near-field beam pattern. It also follows from (1.3-16) that

$$A_R(\eta,\mathbf{r}_R) = F_{\boldsymbol{\beta}}^{-1}\{D_R(\eta,\boldsymbol{\beta})\} = \int_{-\infty}^{\infty} D_R(\eta,\boldsymbol{\beta})\exp(-j2\pi\boldsymbol{\beta}\bullet\mathbf{r}_R)\,d\boldsymbol{\beta} \tag{1.3-17}$$

where $F_{\boldsymbol{\beta}}^{-1}\{\bullet\}$ is shorthand notation for a three-dimensional inverse spatial Fourier transform with respect to spatial frequencies β_X, β_Y, and β_Z; that is, $F_{\boldsymbol{\beta}}^{-1}\{\bullet\} = F_{\beta_X}^{-1} F_{\beta_Y}^{-1} F_{\beta_Z}^{-1}\{\bullet\}$, and $d\boldsymbol{\beta} = d\beta_X\, d\beta_Y\, d\beta_Z$. From (1.3-16) and (1.3-17) it can be seen that *an aperture function and its far-field beam pattern form a spatial Fourier transform pair*. Note that the far-field beam pattern is *not* a function of the range r_S to a sound-source (target) as the near-field beam pattern is. Since the spatial frequencies β_X, β_Y, and β_Z are related to the direction cosines u_S, v_S, and w_S, respectively, which are related to the spherical angles θ_S and ψ_S [see (1.2-96) through (1.2-98)], the far-field beam pattern given by (1.3-16) can ultimately be expressed as a function of frequency η and the spherical angles θ_S and ψ_S, that is, $D_R(\eta, \boldsymbol{\beta}) \rightarrow D_R(\eta, \theta_S, \psi_S)$. The form of (1.3-16) is also known as a *Fraunhofer diffraction integral* in optics or as a *three-dimensional spatial Fourier transform.* The important functions and their units at a receive volume aperture are summarized in Table 1B-2 in Appendix 1B.

Example 1.3-2

In this example we shall derive equations for two more functions that involve the far-field beam pattern of a receive aperture. We begin with the derivation of the equation for the complex frequency-and-angular spectrum of the output electrical signal from a receive aperture. Recall that the complex frequency spectrum of the output electrical signal from a receive aperture at $\mathbf{r}_R$ is given by (1.1-11), which is rewritten below for convenience:

$$Y(\eta, \mathbf{r}_R) = Y_M(\eta, \mathbf{r}_R) A_R(\eta, \mathbf{r}_R), \tag{1.1-11}$$

where $Y_M(\eta, \mathbf{r}_R)$ is the complex frequency spectrum of the acoustic field incident upon the receive aperture at $\mathbf{r}_R$. Taking the spatial Fourier transform of (1.1-11) with respect to $\mathbf{r}_R$ yields

$$\mathsf{Y}(\eta, \boldsymbol{\beta}) = \mathsf{Y}_M(\eta, \boldsymbol{\beta}) \underset{\boldsymbol{\beta}}{*} D_R(\eta, \boldsymbol{\beta}), \tag{1.3-18}$$

or

$$\mathsf{Y}(\eta, \boldsymbol{\beta}) = \int_{-\infty}^{\infty} \mathsf{Y}_M(\eta, \boldsymbol{\gamma}) D_R(\eta, \boldsymbol{\beta} - \boldsymbol{\gamma}) d\boldsymbol{\gamma}, \tag{1.3-19}$$

where

$$\mathsf{Y}(\eta, \boldsymbol{\beta}) = F_{\mathbf{r}_R}\{Y(\eta, \mathbf{r}_R)\} = \int_{-\infty}^{\infty} Y(\eta, \mathbf{r}_R) \exp(+j2\pi \boldsymbol{\beta} \cdot \mathbf{r}_R) d\mathbf{r}_R \tag{1.3-20}$$

is the complex frequency-and-angular spectrum of the output electrical signal from a receive aperture,

$$\mathsf{Y}_M(\eta, \boldsymbol{\beta}) = F_{\mathbf{r}_R}\{Y_M(\eta, \mathbf{r}_R)\} = \int_{-\infty}^{\infty} Y_M(\eta, \mathbf{r}_R) \exp(+j2\pi \boldsymbol{\beta} \cdot \mathbf{r}_R) d\mathbf{r}_R \tag{1.3-21}$$

is the complex frequency-and-angular spectrum of the acoustic field incident upon the receive aperture, $D_R(\eta, \boldsymbol{\beta})$ is the far-field beam pattern of the receive aperture given by (1.3-16), $\boldsymbol{\gamma} = (\gamma_X, \gamma_Y, \gamma_Z)$ is a three-dimensional vector whose components are spatial frequencies in the X, Y, and Z directions, respectively, and $d\boldsymbol{\gamma} = d\gamma_X\, d\gamma_Y\, d\gamma_Z$. The right-hand side of (1.3-19) is shorthand notation for a three-dimensional convolution integral. Equation (1.3-19) is analogous to the far-field, complex frequency-and-angular spectrum of the input acoustic signal to the fluid medium given by (1.3-4).

The second function involves the total output electrical signal from a receive aperture. If a sound-source (target) is in the far-field region of a receive aperture, then the total output electrical signal from the receive aperture can be obtained from (1.2-100) by replacing the near-field beam pattern $\mathcal{D}_R(\eta, r_S, \boldsymbol{\beta})$ with the far-field beam pattern $D_R(\eta, \boldsymbol{\beta})$. Doing so yields

$$\boxed{y(t) \approx -\frac{1}{4\pi r_S}\int_{-\infty}^{\infty} S_0(\eta)\, D_R(\eta, \boldsymbol{\beta}) \exp\left[+j2\pi\eta\left(t - \frac{r_S}{c}\right)\right] d\eta} \tag{1.3-22}$$

for $t \geq r_S/c$, where $S_0(\eta)$ is the complex frequency spectrum of the target's source strength and $D_R(\eta, \boldsymbol{\beta}) \rightarrow D_R(\eta, \theta_S, \psi_S)$. Since $S_0(\eta)$ has units of $\left(\mathrm{m}^3/\mathrm{sec}\right)/\mathrm{Hz}$, $D_R(\eta, \boldsymbol{\beta})$ has units of $\mathrm{V}/\left(\mathrm{m}^2/\mathrm{sec}\right)$, $d\eta$ has units of Hz, and r_S has units of m, $S_0(\eta) D_R(\eta, \boldsymbol{\beta})\, d\eta / r_S$ has units of V, which are the correct units for the total output electrical signal $y(t)$ given by (1.3-22). ■

Problems

Section 1.2

1-1 Verify (1.2-12) and (1.2-13).

1-2 Using Fig. P1-2, show that

$$u = \cos\alpha = \sin\theta\cos\psi ,$$

$$v = \cos\beta = \sin\theta\sin\psi ,$$

and

$$w = \cos\gamma = \cos\theta ,$$

where u, v, and w are dimensionless direction cosines with respect to the X, Y, and Z axes, respectively.

1-3 Verify (1.2-25).

1-4 Consider the derivation of the Fresnel expansion (Fresnel approximation) of the time-independent, free-space, Green's function $g_f(\mathbf{r}|\mathbf{r}_0)$. Compute the values for both $\phi_{\min}$ and $\phi_{\max}$ in degrees, and the corresponding *minimum* range that a field point has to be from a transmit aperture for

(a) $K = 2$

(b) $K = 5$

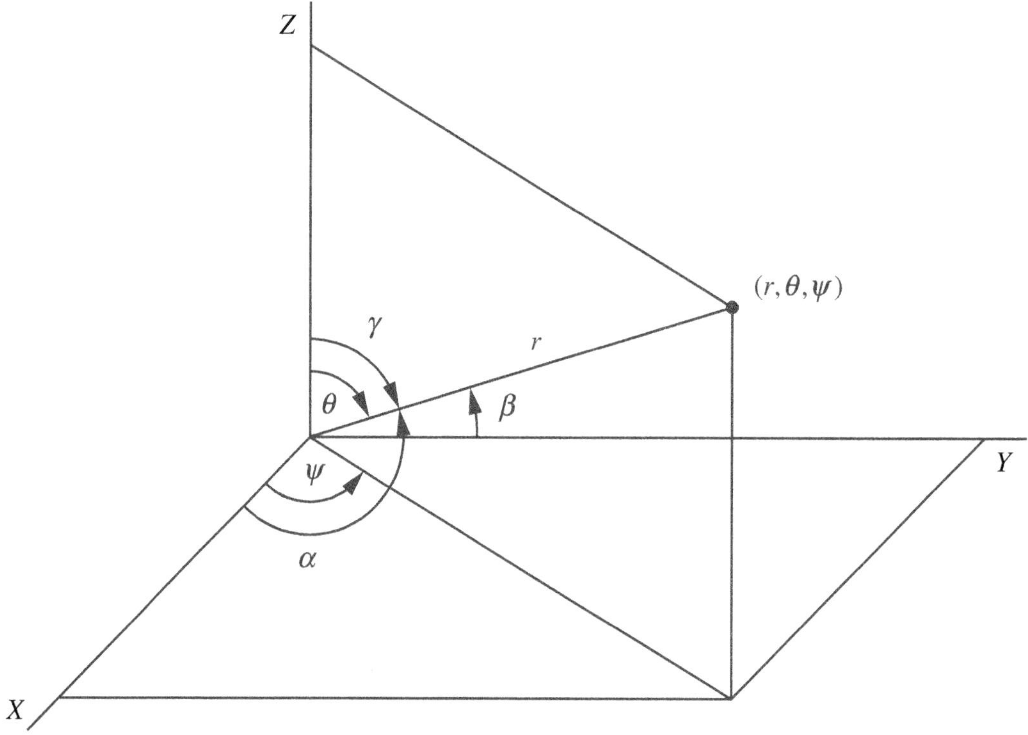

Figure P1-2

Appendix 1A

In this appendix we shall derive a criterion that will guarantee that $|b| < 1$ where [see (1.2-13)]

$$b = \left(\frac{r_0}{r}\right)^2 - 2\frac{\hat{r} \bullet \mathbf{r}_0}{r}, \tag{1A-1}$$

or, expanding the dot product in (1A-1),

$$b = \frac{r_0^2}{r^2} - 2\frac{r_0}{r}\cos\phi\,, \tag{1A-2}$$

where ϕ is the angle between $\hat{r}$ and $\mathbf{r}_0$. Therefore, with the use of (1A-2), $|b| < 1$ can be expressed as

$$\left|\frac{r_0^2}{r^2} - 2\frac{r_0}{r}\cos\phi\right| < 1. \tag{1A-3}$$

If we restrict ourselves for the moment to $90° \le \phi \le 180°$, then the absolute value sign on the left-hand side of (1A-3) can be removed because $0 \le -\cos\phi \le 1$. Therefore,

$$\frac{r_0^2}{r^2} - 2\frac{r_0}{r}\cos\phi < 1, \qquad 90° \le \phi \le 180°\,, \tag{1A-4}$$

or

$$r^2 + 2r_0\, r\cos\phi > r_0^2, \qquad 90° \le \phi \le 180°\,. \tag{1A-5}$$

Completing the square on the left-hand side of (1A-5) yields

$$r^2 + 2r_0\, r\cos\phi + r_0^2\cos^2\phi > r_0^2 + r_0^2\cos^2\phi, \qquad 90° \le \phi \le 180°\,, \tag{1A-6}$$

$$(r + r_0\cos\phi)^2 > r_0^2(1 + \cos^2\phi), \qquad 90° \le \phi \le 180°\,, \tag{1A-7}$$

and, finally,

$$r > r_0\left(\sqrt{1+\cos^2\phi} - \cos\phi\right), \qquad 90° \le \phi \le 180°\,. \tag{1A-8}$$

For $0° \le \phi \le 90°$, we substitute the trigonometric identity

$$\cos(180° - \phi) = -\cos\phi \tag{1A-9}$$

into (1A-8) which yields

$$r > r_0\left(\sqrt{1+\cos^2(180°-\phi)} + \cos(180°-\phi)\right), \qquad 0° \le \phi \le 90°\,. \tag{1A-10}$$

Combining (1A-8) and (1A-10) yields

$$r > \begin{cases} r_0\left(\sqrt{1+\cos^2(180^\circ-\phi)}+\cos(180^\circ-\phi)\right), & 0^\circ \le \phi \le 90^\circ \\ r_0\left(\sqrt{1+\cos^2\phi}-\cos\phi\right), & 90^\circ \le \phi \le 180^\circ \end{cases} \tag{1A-11}$$

If (1A-11) is satisfied, then $|b| < 1$ (see Table 1A-1). Referring to Table 1A-1, if $r/r_0 = c$ for a given value of ϕ, then $|b| < 1$ if $r > cr_0$.

Table 1A-1 Range Ratio r/r_0 using (1A-11)

ϕ (deg)	r/r_0		ϕ (deg)	r/r_0
0	0.414214		90	1.000000
5	0.415331		95	1.090947
10	0.418704		100	1.188613
15	0.424402		105	1.291770
20	0.432540		110	1.398892
25	0.443282		115	1.508255
30	0.456850		120	1.618034
35	0.473524		125	1.726395
40	0.493648		130	1.831558
45	0.517638		135	1.931852
50	0.545983		140	2.025737
55	0.579242		145	2.111828
60	0.618034		150	2.188901
65	0.663018		155	2.255898
70	0.714852		160	2.311925
75	0.774132		165	2.356254
80	0.841317		170	2.388320
85	0.916635		175	2.407720
90	1.000000		180	2.414214

Appendix 1B Important Functions and Their Units at a Transmit and Receive Volume Aperture

Table 1B-1 Important Functions and Their Units at a Transmit Volume Aperture

Function	Description	Units
$x(t,\mathbf{r}_T)$	input electrical signal to the transmit aperture at time t and location $\mathbf{r}_T=(x_T,y_T,z_T)$	V
$X(f,\mathbf{r}_T)$	complex frequency spectrum of the input electrical signal at location $\mathbf{r}_T=(x_T,y_T,z_T)$ of the transmit aperture	V/Hz
$\alpha_T(t,\mathbf{r}_T)$	impulse response of the transmit aperture at time t and location $\mathbf{r}_T=(x_T,y_T,z_T)$	$\left(\text{sec}^{-1}/\text{V}\right)/\text{sec}$
$A_T(f,\mathbf{r}_T)$	complex frequency response of the transmit aperture at $\mathbf{r}_T=(x_T,y_T,z_T)$, also known as the complex transmit aperture function	sec^{-1}/V
$\mathcal{D}_T(f,r,\boldsymbol{\alpha})$	near-field beam pattern (directivity function) of the transmit aperture	$\left(\text{m}^3/\text{sec}\right)/\text{V}$
$D_T(f,\boldsymbol{\alpha})$	far-field beam pattern (directivity function) of the transmit aperture	$\left(\text{m}^3/\text{sec}\right)/\text{V}$
$x_M(t,\mathbf{r}_T)$	output acoustic signal from the transmit aperture at time t and location $\mathbf{r}_T=(x_T,y_T,z_T)$, which is also the input acoustic signal to the fluid medium or source distribution	sec^{-1}
$X_M(f,\mathbf{r}_T)$	complex frequency spectrum of the input acoustic signal to the fluid medium (source distribution) at location $\mathbf{r}_T=(x_T,y_T,z_T)$ of the transmit aperture	$\text{sec}^{-1}/\text{Hz}$

Comments

1. The units of source strength (volume flow rate) are m^3/sec.

2. $\text{sec}^{-1}=\left(\text{m}^3/\text{sec}\right)/\text{m}^3$ source strength per unit volume

3. $\text{sec}^{-1}/\text{V}=\left(\left(\text{m}^3/\text{sec}\right)/\text{m}^3\right)/\text{V}$ source strength per unit volume, per volt

Table 1B-2 Important Functions and Their Units at a Receive Volume Aperture

Function	Description	Units
$y_M(t, \mathbf{r}_R)$	output acoustic signal (velocity potential) from the fluid medium, which is also the input acoustic signal to the receive aperture (i.e., it is the acoustic field incident upon the receive aperture) at time t and location $\mathbf{r}_R = (x_R, y_R, z_R)$	$\mathrm{m}^2/\mathrm{sec}$
$Y_M(\eta, \mathbf{r}_R)$	complex frequency spectrum of the acoustic field incident upon the receive aperture at location $\mathbf{r}_R = (x_R, y_R, z_R)$	$\left(\mathrm{m}^2/\mathrm{sec}\right)/\mathrm{Hz}$
$\alpha_R(t, \mathbf{r}_R)$	impulse response of the receive aperture at time t and location $\mathbf{r}_R = (x_R, y_R, z_R)$	$\left(\left(\mathrm{V}/\left(\mathrm{m}^2/\mathrm{sec}\right)\right)/\mathrm{m}^3\right)/\mathrm{sec}$
$A_R(\eta, \mathbf{r}_R)$	complex frequency response of the receive aperture at $\mathbf{r}_R = (x_R, y_R, z_R)$, also known as the complex receive aperture function	$\left(\mathrm{V}/\left(\mathrm{m}^2/\mathrm{sec}\right)\right)/\mathrm{m}^3$
$\mathcal{D}_R(\eta, r_S, \boldsymbol{\beta})$	near-field beam pattern (directivity function) of the receive aperture	$\mathrm{V}/\left(\mathrm{m}^2/\mathrm{sec}\right)$
$D_R(\eta, \boldsymbol{\beta})$	far-field beam pattern (directivity function) of the receive aperture	$\mathrm{V}/\left(\mathrm{m}^2/\mathrm{sec}\right)$
$y(t, \mathbf{r}_R)$	output electrical signal from the receive aperture at time t and location $\mathbf{r}_R = (x_R, y_R, z_R)$	V/m^3
$Y(\eta, \mathbf{r}_R)$	complex frequency spectrum of the output electrical signal at location $\mathbf{r}_R = (x_R, y_R, z_R)$ of the receive aperture	$(\mathrm{V}/\mathrm{Hz})/\mathrm{m}^3$
$y(t)$	total output electrical signal from the receive aperture	V

Comments

1. The units of a scalar velocity potential are $\mathrm{m}^2/\mathrm{sec}$.

Chapter 2

Complex Aperture Theory – Linear Apertures

2.1 The Far-Field Beam Pattern of a Linear Aperture

Before we begin our discussion, a few comments concerning notation are in order. We shall drop the subscript notation T and R which was used in Chapter 1 to distinguish between transmit and receive apertures, respectively, since we shall now be concerned with apertures in general. Also, the position vectors $\mathbf{r}_T$ and $\mathbf{r}_R$, which were used to identify the locations of points on the surfaces of transmit and receive apertures, respectively, shall be replaced by the position vector to an aperture point $\mathbf{r}_A$. And finally, the parameters R_0 and R_R, which were used to designate the maximum radial extent of transmit and receive apertures, respectively, shall be replaced by the maximum radial extent of an aperture R_A. Whenever we need to make a clear distinction between transmit and receive apertures, we shall reintroduce the original subscript notation established in Chapter 1.

In this section we shall demonstrate that the far-field beam pattern of a linear aperture is just a special case of the far-field beam pattern of a volume aperture. In Section 1.3 it was shown that the far-field beam pattern (directivity function) of a closed-surface, volume aperture – based on the Fraunhofer approximation of a time-independent, free-space, Green's function – is given by the following three-dimensional spatial Fourier transform:

$$D(f,\boldsymbol{\alpha}) = F_{\mathbf{r}_A}\{A(f,\mathbf{r}_A)\} = \int_{-\infty}^{\infty} A(f,\mathbf{r}_A)\exp(+j2\pi\boldsymbol{\alpha}\bullet\mathbf{r}_A)\,d\mathbf{r}_A\,, \tag{2.1-1}$$

where

$$\mathbf{r}_A = (x_A, y_A, z_A)\,, \tag{2.1-2}$$

$$\boldsymbol{\alpha} = (f_X, f_Y, f_Z)\,, \tag{2.1-3}$$

$$\boldsymbol{\alpha}\bullet\mathbf{r}_A = f_X x_A + f_Y y_A + f_Z z_A\,, \tag{2.1-4}$$

$$d\mathbf{r}_A = dx_A\,dy_A\,dz_A\,, \tag{2.1-5}$$

and $F_{\mathbf{r}_A}\{\bullet\}$ is shorthand notation for $F_{x_A}F_{y_A}F_{z_A}\{\bullet\}$. The complex aperture function $A(f,\mathbf{r}_A)$ can be expressed as

$$A(f,\mathbf{r}_A) = a(f,\mathbf{r}_A)\exp[+j\theta(f,\mathbf{r}_A)]\,, \tag{2.1-6}$$

where $a(f,\mathbf{r}_A)$ is the *amplitude response* and $\theta(f,\mathbf{r}_A)$ is the *phase response* of the aperture at $\mathbf{r}_A$. Both $a(f,\mathbf{r}_A)$ and $\theta(f,\mathbf{r}_A)$ are *real* functions. The function $a(f,\mathbf{r}_A)$ is also known as the *amplitude window*. The units of $a(f,\mathbf{r}_A)$ and, hence, $A(f,\mathbf{r}_A)$ depend on whether the aperture is a transmit aperture (see Table 1B-1 in Appendix 1B) or a receive aperture (see Table 1B-2 in Appendix 1B). The units of $\theta(f,\mathbf{r}_A)$ are radians.

Now consider the case of a *linear aperture* of length L meters lying along the X axis as shown in Fig. 2.1-1. A linear aperture can represent either a single electroacoustic transducer (e.g., a continuous line source or receiver) or a linear array of many individual electroacoustic transducers. If a linear aperture represents a single electroacoustic transducer that is being used in the active mode, that is, as a transmitter; then this physical situation corresponds to the classic problem of a continuous line source. Also shown in Fig. 2.1-1 is a field point with spherical coordinates (r,θ,ψ). The field point is in the far-field region of the aperture, where the range r satisfies the Fraunhofer (far-field) range criterion given by $r > \pi R_A^2/\lambda > 2.414R_A$ [see either (1.2-39) or (1.2-93)], and $R_A = L/2$ is the maximum radial extent of the linear aperture.

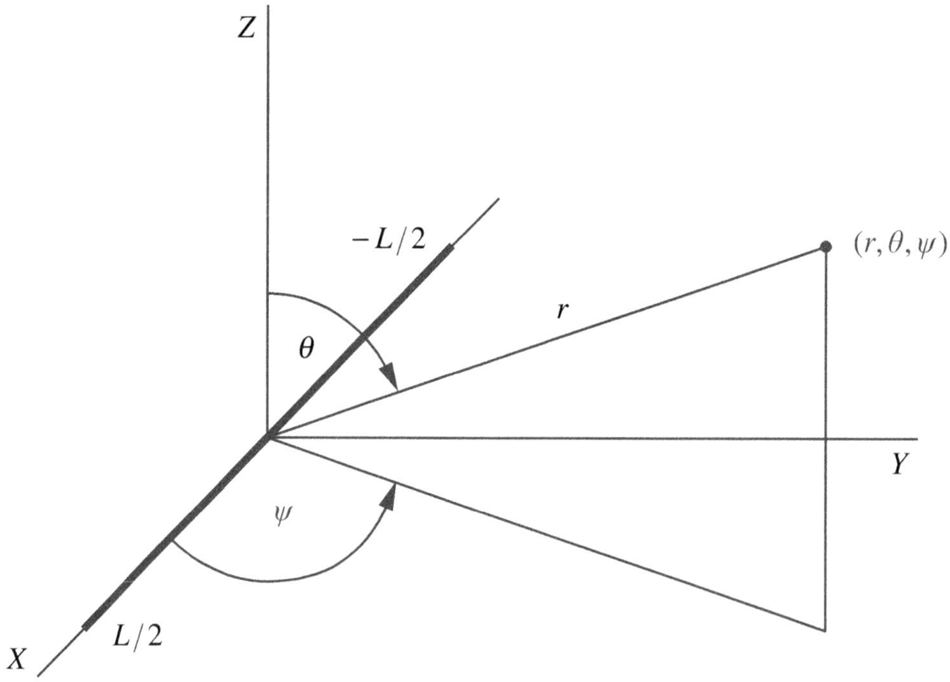

Figure 2.1-1 Linear aperture of length L meters lying along the X axis. Also shown is a field point with spherical coordinates (r,θ,ψ).

Since the linear aperture in Fig. 2.1-1 is lying along the X axis, the position vector to a point on the surface of the aperture is given by

$$\mathbf{r}_A = (x_A, 0, 0), \tag{2.1-7}$$

and as a result,

$$A(f, \mathbf{r}_A) = A(f, x_A, y_A, z_A) \rightarrow A(f, x_A)\delta(y_A)\delta(z_A). \tag{2.1-8}$$

Therefore, by substituting (2.1-8) into (2.1-1) and using the sifting property of impulse functions, we obtain the following one-dimensional spatial Fourier transform for the far-field beam pattern of a linear aperture lying along the X axis:

$$\boxed{D(f, f_X) = F_{x_A}\{A(f, x_A)\} = \int_{-L/2}^{L/2} A(f, x_A)\exp(+j2\pi f_X x_A)\, dx_A} \tag{2.1-9}$$

where

$$A(f, x_A) = a(f, x_A)\exp[+j\theta(f, x_A)] \tag{2.1-10}$$

is the complex frequency response (complex aperture function) of the linear aperture and

$$f_X = u/\lambda = \sin\theta\cos\psi/\lambda \tag{2.1-11}$$

is a spatial frequency in the X direction with units of cycles per meter [see (1.2-43)]. Since f_X can be expressed in terms of the spherical angles θ and ψ, the far-field beam pattern can ultimately be expressed as a function of frequency f and the spherical angles θ and ψ, that is, $D(f, f_X) \rightarrow D(f, \theta, \psi)$. If the linear aperture is a transmit aperture, then the aperture function $A(f, x_A)$ has units of $\left(\left(\text{m}^3/\text{sec}\right)/\text{V}\right)/\text{m}$ and the far-field beam pattern $D(f, f_X)$ has units of $\left(\text{m}^3/\text{sec}\right)/\text{V}$. However, if the linear aperture is a receive aperture, then $A(f, x_A)$ has units of $\left(\text{V}/\left(\text{m}^2/\text{sec}\right)\right)/\text{m}$ and $D(f, f_X)$ has units of $\text{V}/\left(\text{m}^2/\text{sec}\right)$. These units are summarized in Table 2.1-1. In addition to describing the performance of an electroacoustic transducer by its far-field beam pattern, values of the frequency-dependent, *transmitter* and *receiver sensitivity functions* are also used. The relationships between the complex frequency response $A(f, x_A)$ of a single electroacoustic transducer, and complex, transmitter and receiver sensitivity functions are discussed in Appendix 2A.

Finally, if a linear aperture lies along either the Y or Z axis instead of the X axis, then simply replace x_A and f_X with either y_A and f_Y, or z_A and f_Z, respectively, in (2.1-9), where spatial frequencies f_Y and f_Z are given by (1.2-44)

Table 2.1-1 Units of the Complex Aperture Function $A(f, x_A)$ and Corresponding Far-Field Beam Pattern $D(f, f_X)$ for a Linear Aperture

Linear Aperture	$A(f, x_A)$	$D(f, f_X)$
transmit	$\left(\left(\mathrm{m}^3/\mathrm{sec}\right)/\mathrm{V}\right)/\mathrm{m}$	$\left(\mathrm{m}^3/\mathrm{sec}\right)/\mathrm{V}$
receive	$\left(\mathrm{V}/\left(\mathrm{m}^2/\mathrm{sec}\right)\right)/\mathrm{m}$	$\mathrm{V}/\left(\mathrm{m}^2/\mathrm{sec}\right)$

and (1.2-45), respectively. With the use of (2.1-9), we will derive closed-form expressions for the normalized, far-field beam patterns of six different amplitude windows in Section 2.2. Although we shall be concentrating on single, continuous line sources and receivers in this chapter, as we shall discover in Chapter 6, some of the results obtained in this chapter are directly applicable to linear arrays. In other words, we are laying the foundation for linear array theory in this chapter as well.

2.2 Amplitude Windows and Corresponding Far-Field Beam Patterns

The various amplitude windows to be discussed next are used to model the frequency response of linear apertures. They are very common functions used in sonar, radar, and communication problems and will serve our purposes of demonstrating far-field beam-pattern derivations, beamwidth calculations, and the relationship between beamwidth and sidelobe levels of far-field beam patterns. Sampled versions of these continuous amplitude windows are, in fact, used to amplitude weight arrays of electroacoustic transducers, as will be discussed in Section 6.2. In addition, time-domain versions of these continuous amplitude windows can also be used to reduce the sidelobe levels of the auto-ambiguity functions of common transmitted electrical signals, as will be discussed in Chapter 13.

In the subsections that follow, it is assumed that the linear aperture represents a single electroacoustic transducer – a continuous line source or receiver – L meters in length, and that the phase response of the transducer is zero so that

$$A(f, x_A) = a(f, x_A), \qquad \theta(f, x_A) = 0\,. \tag{2.2-1}$$

Therefore, according to (2.2-1), the complex frequency response of the transducer is equal to the amplitude response or amplitude window $a(f, x_A)$. We will examine the effects of a nonzero phase response later, in Section 2.4.

2.2.1 The Rectangular Amplitude Window

In this subsection we shall derive the normalized, far-field beam pattern of the rectangular amplitude window. The rectangular amplitude window is defined as follows:

$$a(f, x_A) = A_A \operatorname{rect}\left(\frac{x_A}{L}\right) \triangleq \begin{cases} A_A, & |x_A| \le L/2 \\ 0, & |x_A| > L/2 \end{cases} \tag{2.2-2}$$

and is shown in Fig. 2.2-1 for $A_A = 1$. Therefore, if

$$A(f, x_A) = a(f, x_A) = A_A \operatorname{rect}(x_A / L), \tag{2.2-3}$$

then the complex frequency response of the transducer is constant along the entire length L of the transducer, regardless of the value of frequency f. The *real*, *positive* constant A_A carries the correct units of $A(f, x_A)$ (see Table 2.1-1). It is important to note that although A_A is shown here as a constant (and in Subsections 2.2.2 through 2.2.4), it is, in general, a *real*, *nonnegative* function of frequency, that is, $A_A \to A_A(f)$. Equation (2.2-3) is the simplest mathematical model for the complex aperture function $A(f, x_A)$.

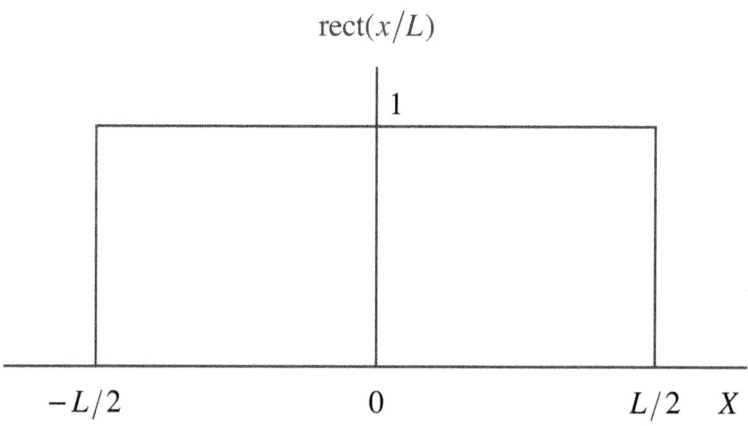

Figure 2.2-1 Rectangular amplitude window for $A_A = 1$.

Substituting (2.2-3) and (2.2-2) into (2.1-9) yields

$$D(f, f_X) = F_{x_A}\{A_A \operatorname{rect}(x_A / L)\} = A_A \int_{-L/2}^{L/2} \exp(+j2\pi f_X x_A)\, dx_A, \tag{2.2-4}$$

or

$$\boxed{D(f, f_X) = F_{x_A}\{A_A \operatorname{rect}(x_A/L)\} = A_A L \operatorname{sinc}(f_X L)} \tag{2.2-5}$$

where

$$\operatorname{sinc}(x) \triangleq \frac{\sin(\pi x)}{\pi x} . \tag{2.2-6}$$

Equation (2.2-5) is the *unnormalized*, far-field beam pattern of the rectangular amplitude window. The *dimensionless*, *normalized*, far-field beam pattern is defined as follows:

$$\boxed{D_N(f, f_X) \triangleq D(f, f_X)/D_{\max}} \tag{2.2-7}$$

where $D_{\max} = \max|D(f, f_X)|$ is the *normalization factor* – it is the *maximum* value of the *magnitude* of the unnormalized, far-field beam pattern. Therefore, by referring to (2.2-5),

$$|D(f, f_X)| = A_A L |\operatorname{sinc}(f_X L)| , \tag{2.2-8}$$

and since $\operatorname{sinc}(0) = 1$ is the maximum value of the sinc function,

$$D_{\max} = |D(f, 0)| = A_A L . \tag{2.2-9}$$

Dividing (2.2-5) by (2.2-9) yields the following expression for the *normalized*, far-field beam pattern of the rectangular amplitude window:

$$\boxed{D_N(f, f_X) = \operatorname{sinc}(f_X L)} \tag{2.2-10}$$

Note that $|D_N(f, 0)| = 1$. Since $f_X = u/\lambda$, (2.2-10) can also be expressed as

$$\boxed{D_N(f, u) = \operatorname{sinc}\left(\frac{L}{\lambda} u\right)} \tag{2.2-11}$$

where the wavelength λ was suppressed in the argument of D_N since frequency f and wavelength λ are related by $c = f\lambda$, that is, $D_N(f, f_X) = D_N(f, u/\lambda) \rightarrow D_N(f, u)$. The magnitude of the normalized, far-field beam pattern given by (2.2-11) is plotted as a function of direction cosine u in Fig. 2.2-2. The plot is shown only for positive values of u because the magnitude of the beam pattern is symmetric about $u = 0$. The level of the first sidelobe is approximately -13.3 dB.

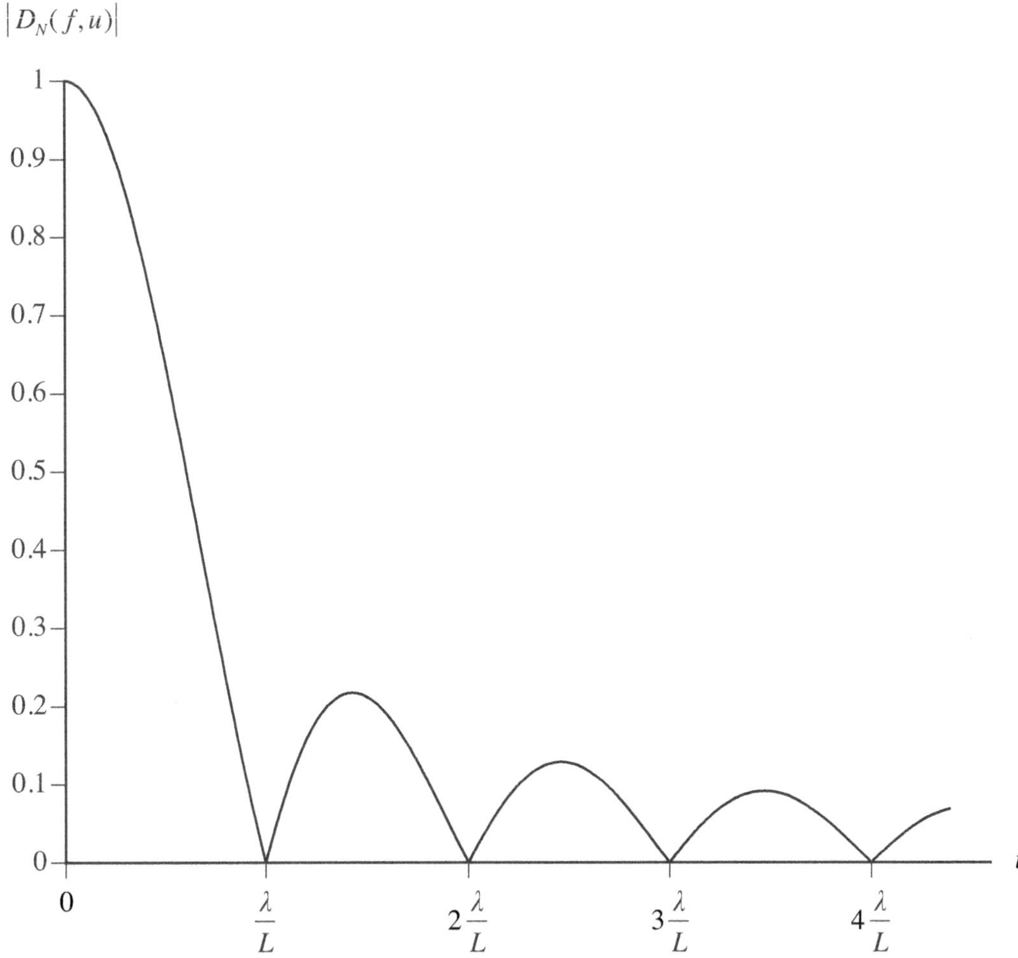

Figure 2.2-2 Magnitude of the normalized, far-field beam pattern of the rectangular amplitude window plotted as a function of direction cosine u.

From Fig. 2.2-2 it can be seen that the width of the mainlobe of the normalized (and unnormalized), far-field beam pattern depends on the ratio λ/L. Therefore, the beamwidth is directly proportional to wavelength λ (inversely proportional to frequency f) and inversely proportional to the length L of the aperture (transducer). As a result, the beamwidth can be decreased by keeping the length L of the aperture constant while increasing frequency f, or by keeping frequency f constant while increasing the length L of the aperture. Beamwidth will be discussed in detail in Sections 2.3 and 2.5.

From a theoretical point of view, the normalized, far-field beam pattern given by (2.2-11) can be evaluated at any value of u. However, it is only the interval $-1 \le u \le 1$ that pertains to the real aperture problem because u is a direction cosine given by

$$u = \sin\theta\cos\psi \,, \tag{2.2-12}$$

and as a result, $-1 \le u \le 1$ since $0° \le \theta \le 180°$ and $0° \le \psi \le 360°$. The interval $-1 \le u \le 1$ is called the *visible region* of an aperture (transducer) lying along the X axis. The value $u = 0$, which implies that $\psi = 90°$ or $\psi = 270°$, corresponds to a field point being at *broadside* relative to an aperture (transducer) lying along the X axis (see Fig. 2.1-1). And $u = \pm 1$, which implies that $\theta = 90°$ and $\psi = 0°$ for $u = 1$; and $\theta = 90°$ and $\psi = 180°$ for $u = -1$; corresponds to a field point being at *end-fire* relative to an aperture (transducer) lying along the X axis (see Fig. 2.1-1).

With the use of (2.2-12), the normalized, far-field beam pattern of the rectangular amplitude window given by (2.2-11) can finally be expressed as

$$\boxed{D_N(f, \theta, \psi) = \operatorname{sinc}\left(\frac{L}{\lambda}\sin\theta\cos\psi\right)} \tag{2.2-13}$$

By setting $\theta = 90°$ in (2.2-13), we obtain

$$D_N(f, 90°, \psi) = \operatorname{sinc}\left(\frac{L}{\lambda}\cos\psi\right), \tag{2.2-14}$$

which is the normalized, *horizontal*, far-field beam pattern of the rectangular amplitude window (see Fig. 2.2-3). As can be seen from Fig. 2.2-3, as the ratio L/λ *decreases* (λ/L *increases*), the beamwidth *increases*. In other words, as the length of the aperture becomes small compared to the wavelength, the aperture acts like an omnidirectional transducer, that is, the magnitude of the normalized, horizontal, far-field beam pattern approaches being equal to one for all values of bearing angle. Horizontal beam patterns are discussed further in Example 2.3-2.

2.2.2 The Triangular Amplitude Window

In this subsection we shall derive the normalized, far-field beam pattern of the triangular amplitude window. The triangular amplitude window (also known as the Bartlett or Fejer window) is defined as follows:

$$a(f, x_A) = A_A \operatorname{tri}\left(\frac{x_A}{L}\right) \triangleq \begin{cases} A_A\left[1 - \dfrac{|x_A|}{L/2}\right], & |x_A| \le L/2 \\ 0, & |x_A| > L/2 \end{cases} \tag{2.2-15}$$

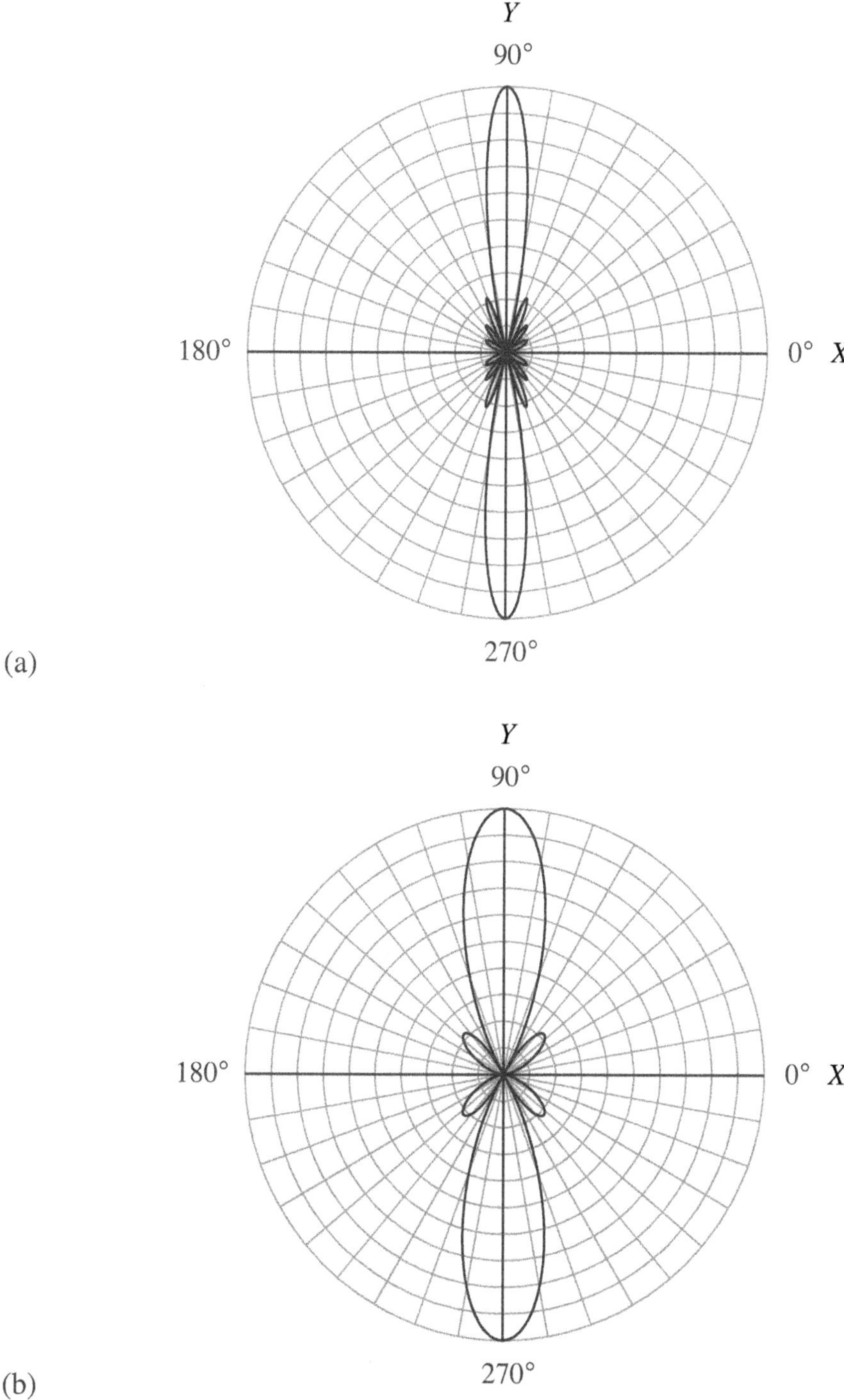

Figure 2.2-3 Polar plots as a function of the azimuthal (bearing) angle ψ of the magnitude of the normalized, horizontal, far-field beam pattern of the rectangular amplitude window given by (2.2-14) for (a) $L/\lambda = 4$, (b) $L/\lambda = 2$, (c) $L/\lambda = 1$, and (d) $L/\lambda = 0.5$.

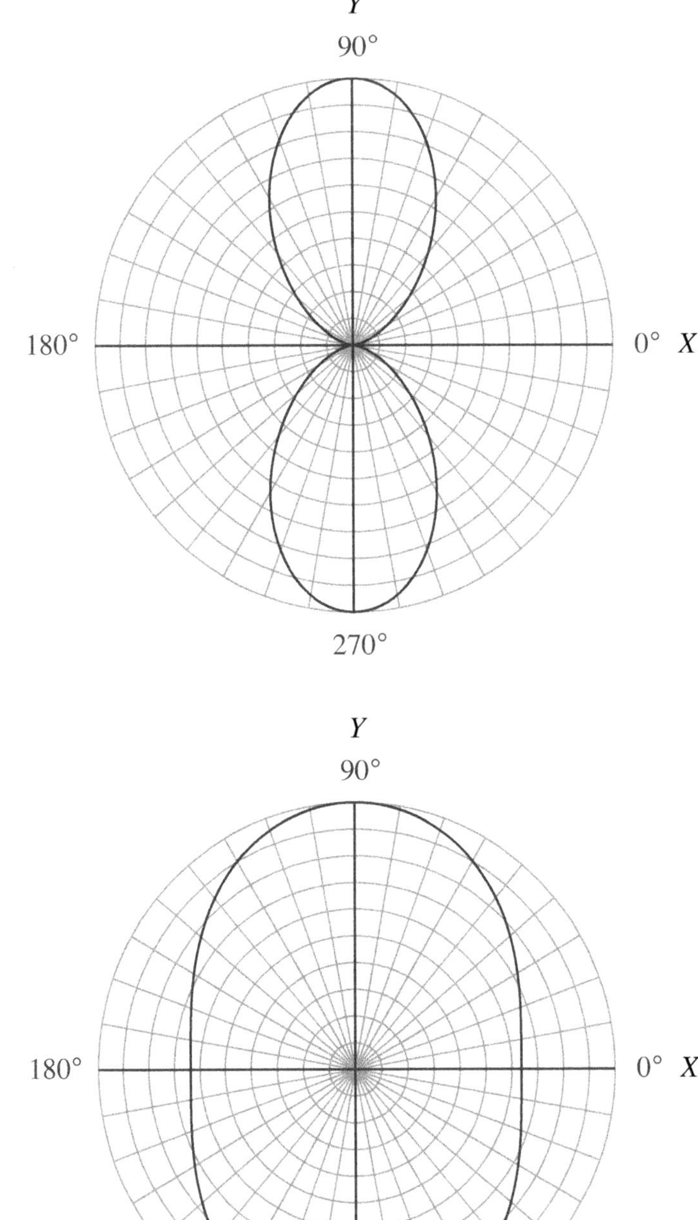

Figure 2.2-3 *continued.*

and is shown in Fig. 2.2-4 for $A_A = 1$. Therefore, if

$$A(f, x_A) = a(f, x_A) = A_A \operatorname{tri}(x_A/L), \tag{2.2-16}$$

then the complex frequency response of the transducer is triangular in shape along the length L of the transducer, regardless of the value of frequency f.

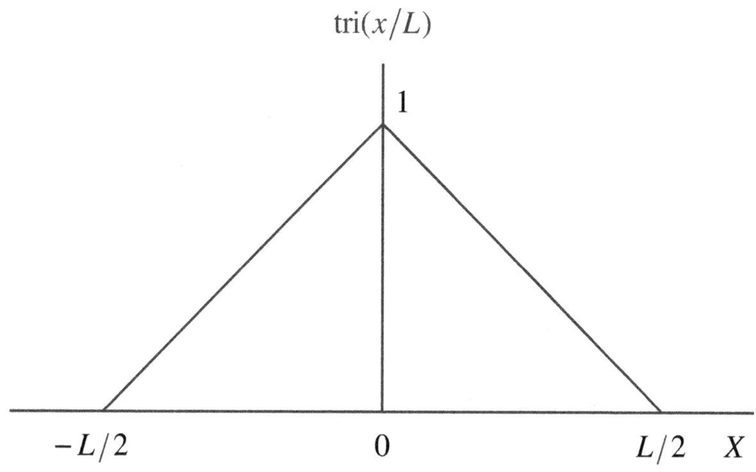

Figure 2.2-4 Triangular amplitude window for $A_A = 1$.

In order to derive the normalized, far-field beam pattern of the triangular amplitude window, we shall use the fact that a triangle is equal to the convolution of two rectangles of equal lengths, that is,

$$L_1 \operatorname{tri}\left(\frac{x_A}{2L_1}\right) = \operatorname{rect}\left(\frac{x_A}{L_1}\right) \underset{x_A}{*} \operatorname{rect}\left(\frac{x_A}{L_1}\right). \tag{2.2-17}$$

Therefore,

$$F_{x_A}\left\{L_1 \operatorname{tri}\left(\frac{x_A}{2L_1}\right)\right\} = \left[F_{x_A}\left\{\operatorname{rect}\left(\frac{x_A}{L_1}\right)\right\}\right]^2 = L_1^2 \operatorname{sinc}^2(f_X L_1), \tag{2.2-18}$$

or

$$F_{x_A}\left\{A_A \operatorname{tri}\left(\frac{x_A}{2L_1}\right)\right\} = A_A L_1 \operatorname{sinc}^2(f_X L_1). \tag{2.2-19}$$

If we let $L_1 = L/2$, then (2.2-19) reduces to

$$\boxed{D(f, f_X) = F_{x_A}\{A_A \operatorname{tri}(x_A/L)\} = (A_A L/2)\operatorname{sinc}^2(f_X L/2)} \tag{2.2-20}$$

Equation (2.2-20) is the *unnormalized*, far-field beam pattern of the triangular amplitude window. The direct evaluation of the spatial Fourier transform of (2.2-16) is left as a homework problem. By referring to (2.2-20), the normalization factor is given by

$$D_{\max} = |D(f, 0)| = A_A L/2, \tag{2.2-21}$$

since $\operatorname{sinc}(0) = 1$ is the maximum value of the sinc function. Dividing (2.2-20) by (2.2-21) yields the following expression for the *normalized*, far-field beam pattern of the triangular amplitude window (see Fig. 2.2-5):

$$\boxed{D_N(f, f_X) = \operatorname{sinc}^2(f_X L/2)} \tag{2.2-22}$$

The levels of the first sidelobes of the magnitudes of the normalized, far-field beam patterns of the rectangular and triangular amplitude windows are approximately $-13.3\,\text{dB}$ and $-26.5\,\text{dB}$, respectively. However, the mainlobe of the beam pattern of the triangular amplitude window is wider than the mainlobe of the beam pattern of the rectangular amplitude window (see Fig. 2.2-5). Note that whereas the rectangular amplitude window is *discontinuous* at the end points $x = \pm L/2$ (see Fig. 2.2-1), the triangular amplitude window approaches zero at $x = \pm L/2$ in a comparatively smooth fashion (see Fig. 2.2-4). As a general rule, the smoother an amplitude window approaches zero at the end points $x = \pm L/2$, the lower the sidelobe levels of its far-field beam pattern will be. The word "smoother" implies "with a smaller slope." However, a reduction in sidelobe levels generally results in an increase in the width of the beam pattern's mainlobe. This trend will become more apparent as we consider additional amplitude windows.

2.2.3 The Cosine Amplitude Window

In order to derive the normalized, far-field beam pattern of the cosine amplitude window, we shall first derive the far-field beam pattern of the following related aperture function:

$$A(f, x_A) = A_A \cos(2\pi x_A/d)\operatorname{rect}(x_A/L). \tag{2.2-23}$$

This function is shown in Fig. 2.2-6 for $A_A = 1$ and $L = 2.5d$, where d is the *spatial period* in meters. Equation (2.2-23) is analogous to a time-domain,

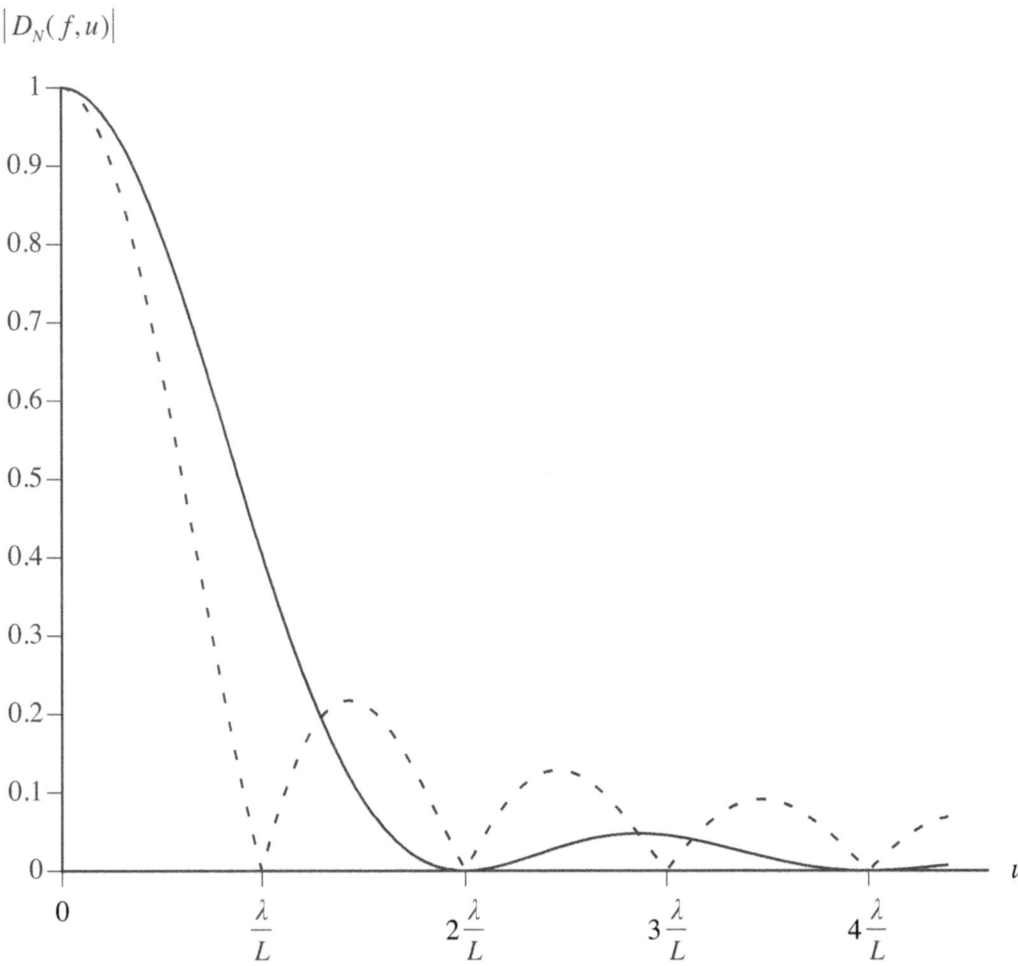

Figure 2.2-5 Magnitudes of the normalized, far-field beam patterns of the rectangular amplitude window (dashed curve) and triangular amplitude window (solid curve) plotted as a function of direction cosine u.

rectangular-envelope, continuous-wave (CW) pulse. Taking the spatial Fourier transform of (2.2-23) yields

$$\begin{aligned} D(f, f_X) &= F_{x_A}\{A_A \cos(2\pi x_A/d)\text{rect}(x_A/L)\} \\ &= A_A F_{x_A}\{\cos(2\pi x_A/d)\} \underset{f_X}{*} F_{x_A}\{\text{rect}(x_A/L)\} \\ &= A_A L F_{x_A}\{\cos(2\pi x_A/d)\} \underset{f_X}{*} \text{sinc}(f_X L), \end{aligned} \tag{2.2-24}$$

where use was made of (2.2-5).

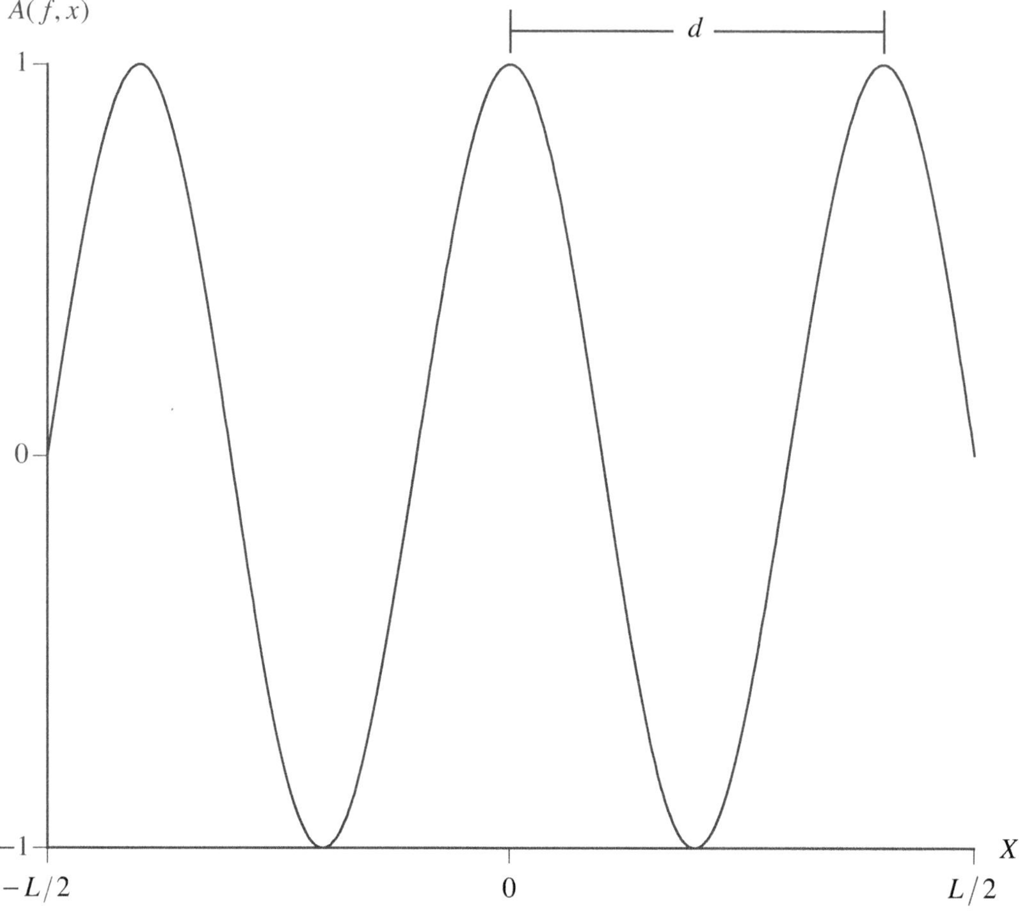

Figure 2.2-6 Plot of the aperture function given by (2.2-23) for $A_A = 1$ and $L = 2.5d$, where d is the spatial period in meters.

As can be seen from (2.2-24), what we need next is the spatial Fourier transform of the cosine function. Since

$$\cos(2\pi x_A/d) = 0.5\left[\exp(+j2\pi x_A/d) + \exp(-j2\pi x_A/d)\right] \tag{2.2-25}$$

and

$$F_{x_A}\left\{\exp(\pm j2\pi f_X' x_A)\right\} = \delta(f_X \pm f_X'), \tag{2.2-26}$$

$$F_{x_A}\left\{\cos(2\pi x_A/d)\right\} = \frac{1}{2}\left[\delta\left(f_X + \frac{1}{d}\right) + \delta\left(f_X - \frac{1}{d}\right)\right]. \tag{2.2-27}$$

Therefore, substituting (2.2-27) into (2.2-24) yields

$$
\begin{aligned}
D(f, f_X) &= F_{x_A}\{A_A \cos(2\pi x_A/d)\operatorname{rect}(x_A/L)\} \\
&= \frac{A_A L}{2}\left[\delta\left(f_X + \frac{1}{d}\right) + \delta\left(f_X - \frac{1}{d}\right)\right] \underset{f_X}{*} \operatorname{sinc}(f_X L) \qquad (2.2\text{-}28) \\
&= \frac{A_A L}{2}\left[\delta\left(f_X + \frac{1}{d}\right) \underset{f_X}{*} \operatorname{sinc}(f_X L) + \delta\left(f_X - \frac{1}{d}\right) \underset{f_X}{*} \operatorname{sinc}(f_X L)\right],
\end{aligned}
$$

or

$$
\boxed{
\begin{aligned}
D(f, f_X) &= F_{x_A}\{A_A \cos(2\pi x_A/d)\operatorname{rect}(x_A/L)\} \\
&= \frac{A_A L}{2}\left\{\operatorname{sinc}\left[\left(f_X + \frac{1}{d}\right)L\right] + \operatorname{sinc}\left[\left(f_X - \frac{1}{d}\right)L\right]\right\}
\end{aligned}
} \qquad (2.2\text{-}29)
$$

The magnitude of the unnormalized, far-field beam pattern given by (2.2-29) is shown in Fig. 2.2-7 for $A_A = 1$ and $L = 2.5d$. Note that as $d \to \infty$, the aperture function given by (2.2-23) approaches $A_A \operatorname{rect}(x_A/L)$ and the beam pattern given by (2.2-29) approaches $A_A L \operatorname{sinc}(f_X L)$, which is the unnormalized, far-field beam pattern of the rectangular amplitude window. *Equation* (2.2-29) *is a very important result because it can be used to obtain the far-field beam patterns of the cosine, Hanning, Hamming, and Blackman amplitude windows*, as we shall demonstrate next.

The cosine amplitude window is given by

$$
a(f, x_A) = A_A \cos(\pi x_A/L)\operatorname{rect}(x_A/L), \qquad (2.2\text{-}30)
$$

and is shown in Fig. 2.2-8 for $A_A = 1$. Therefore, if

$$
A(f, x_A) = a(f, x_A) = A_A \cos(\pi x_A/L)\operatorname{rect}(x_A/L), \qquad (2.2\text{-}31)
$$

then the complex frequency response of the transducer is in the shape of a half-period cosine function along the length L of the transducer, regardless of the value of frequency f. Substituting $d = 2L$ into (2.2-29) and making use of the identity

$$
\sin\left(\pi f_X L \pm \frac{\pi}{2}\right) = \pm\cos(\pi f_X L) \qquad (2.2\text{-}32)
$$

yields

$$
\boxed{
D(f, f_X) = F_{x_A}\left\{A_A \cos\left(\frac{\pi x_A}{L}\right)\operatorname{rect}\left(\frac{x_A}{L}\right)\right\} = \frac{2A_A L}{\pi}\frac{\cos(\pi f_X L)}{1-(2f_X L)^2}
} \qquad (2.2\text{-}33)
$$

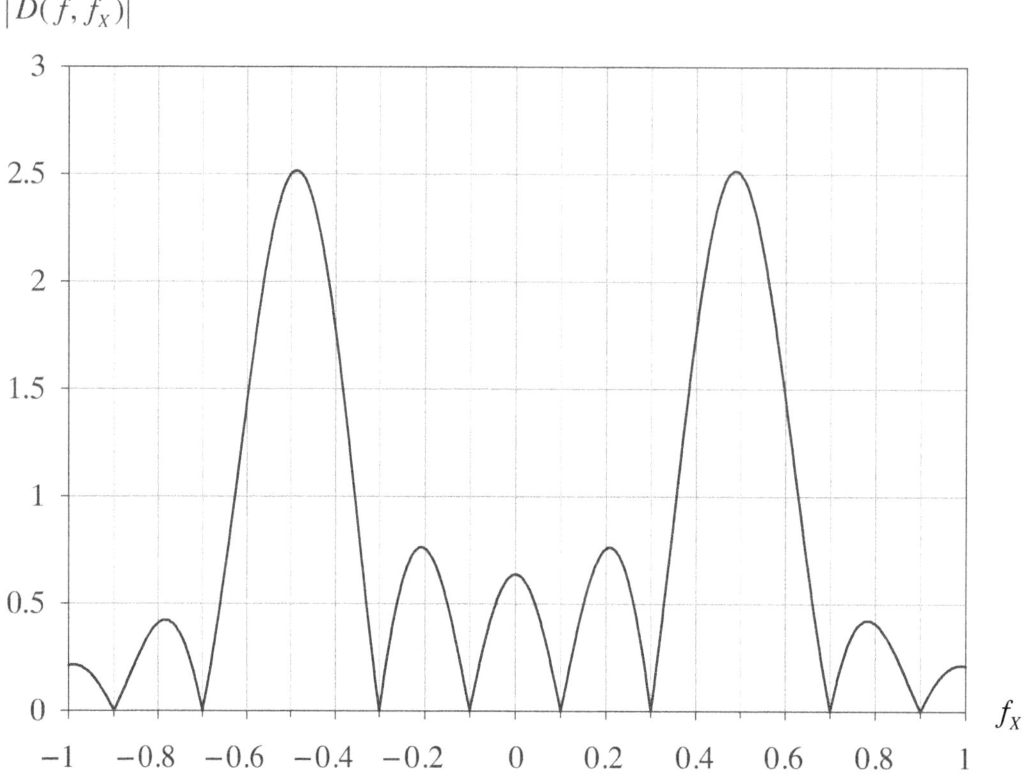

Figure 2.2-7 Magnitude of the unnormalized, far-field beam pattern given by (2.2-29) plotted as a function of spatial frequency f_X for $A_A = 1$, $L = 5$ m, and $d = 2$ m so that $L = 2.5d$. Note that in this case, the maximum value of $|D(f, f_X)|$, which is slightly greater than $A_A L/2 = 2.5$, does not occur at $f_X = \pm 1/d = \pm 0.5$ cycles/m.

which is the *unnormalized*, far-field beam pattern of the cosine amplitude window. In order to determine the normalization factor for (2.2-33), we take advantage of the following general result: If the phase response $\theta(f, x_A) = 0$ and the amplitude response (amplitude window) $a(f, x_A) \geq 0$, then the normalization factor $D_{\max} = |D(f, 0)|$ (see Appendix 2D). Therefore, by referring to (2.2-33),

$$D_{\max} = |D(f, 0)| = 2A_A L/\pi \,. \tag{2.2-34}$$

Dividing (2.2-33) by (2.2-34) yields the following expression for the *normalized*, far-field beam pattern of the cosine amplitude window (see Fig. 2.2-9):

$$\boxed{D_N(f, f_X) = \frac{\cos(\pi f_X L)}{1 - (2 f_X L)^2}} \tag{2.2-35}$$

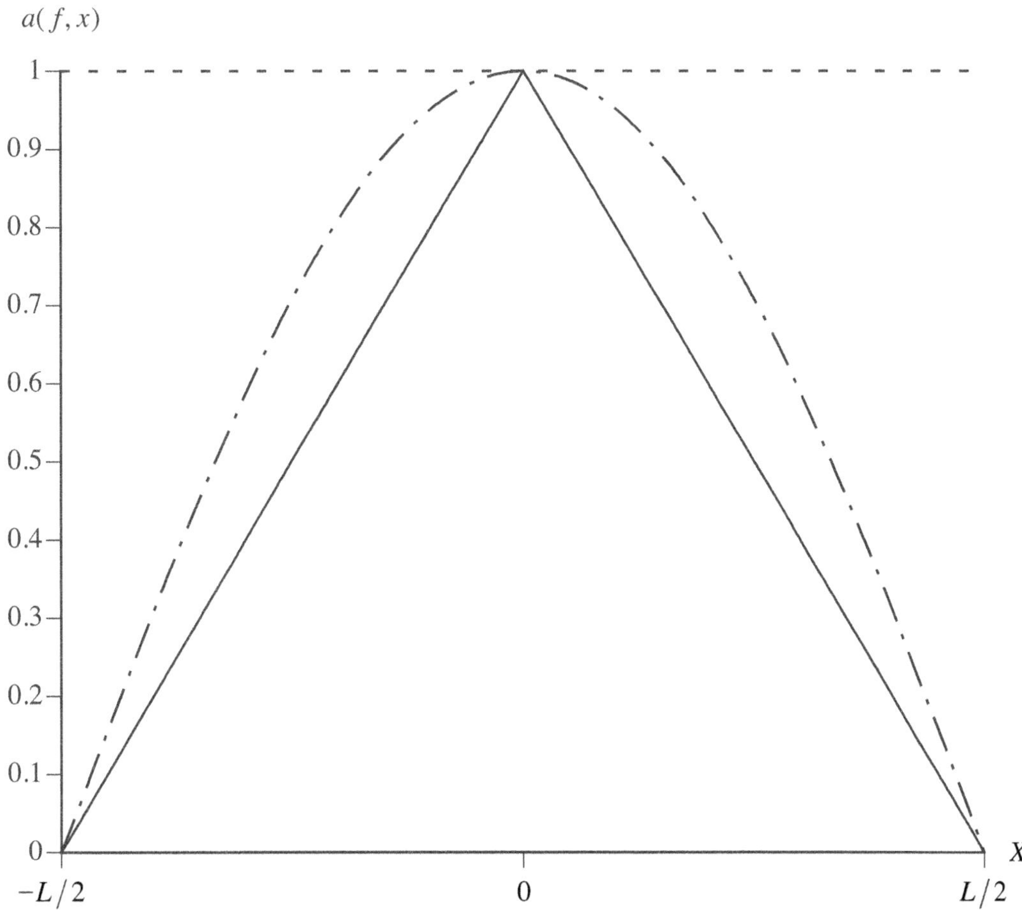

Figure 2.2-8 Rectangular amplitude window (dashed curve), triangular amplitude window (solid curve), and cosine amplitude window (dot-dash curve) for $A_A = 1$.

The level of the first sidelobe of the magnitude of the normalized, far-field beam pattern of the cosine amplitude window is approximately -23 dB, whereas the levels of the first sidelobes of the magnitudes of the normalized, far-field beam patterns of the rectangular and triangular amplitude windows are approximately -13.3 dB and -26.5 dB, respectively. This is not surprising since the rectangular amplitude window is discontinuous at the end points $x = \pm L/2$, and the cosine amplitude window approaches zero at the end points with a larger slope than the triangular amplitude window (see Fig. 2.2-8).

2.2.4 The Hanning, Hamming, and Blackman Amplitude Windows

The *Hanning* amplitude window, which is also known as the *cosine-*

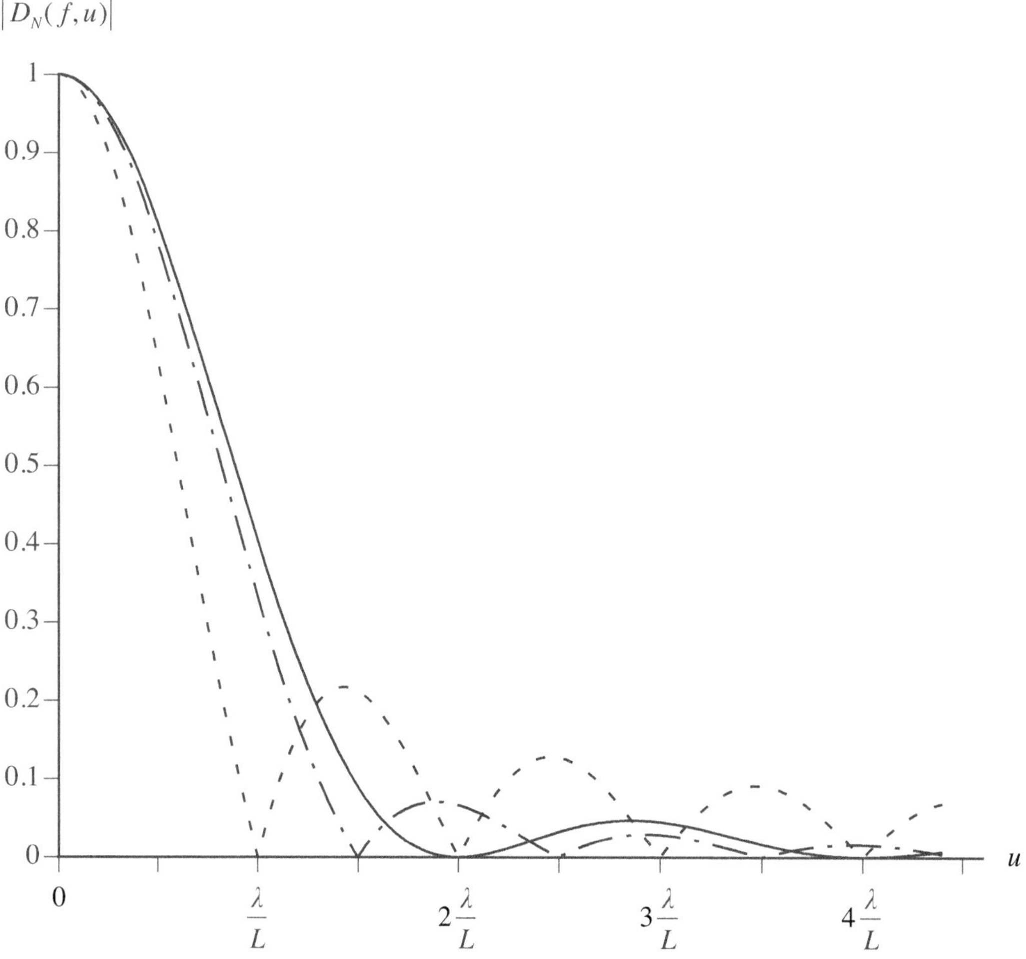

Figure 2.2-9 Magnitudes of the normalized, far-field beam patterns of the rectangular amplitude window (dashed curve), triangular amplitude window (solid curve), and cosine amplitude window (dot-dash curve) plotted as a function of direction cosine u.

squared or *raised-cosine* window, is given by

$$a(f, x_A) = A_A \cos^2(\pi x_A / L)\operatorname{rect}(x_A / L), \tag{2.2-36}$$

and is shown in Fig. 2.2-10 for $A_A = 1$. Since

$$\cos^2\alpha = 0.5\left[1 + \cos(2\alpha)\right], \tag{2.2-37}$$

(2.2-36) can be rewritten as

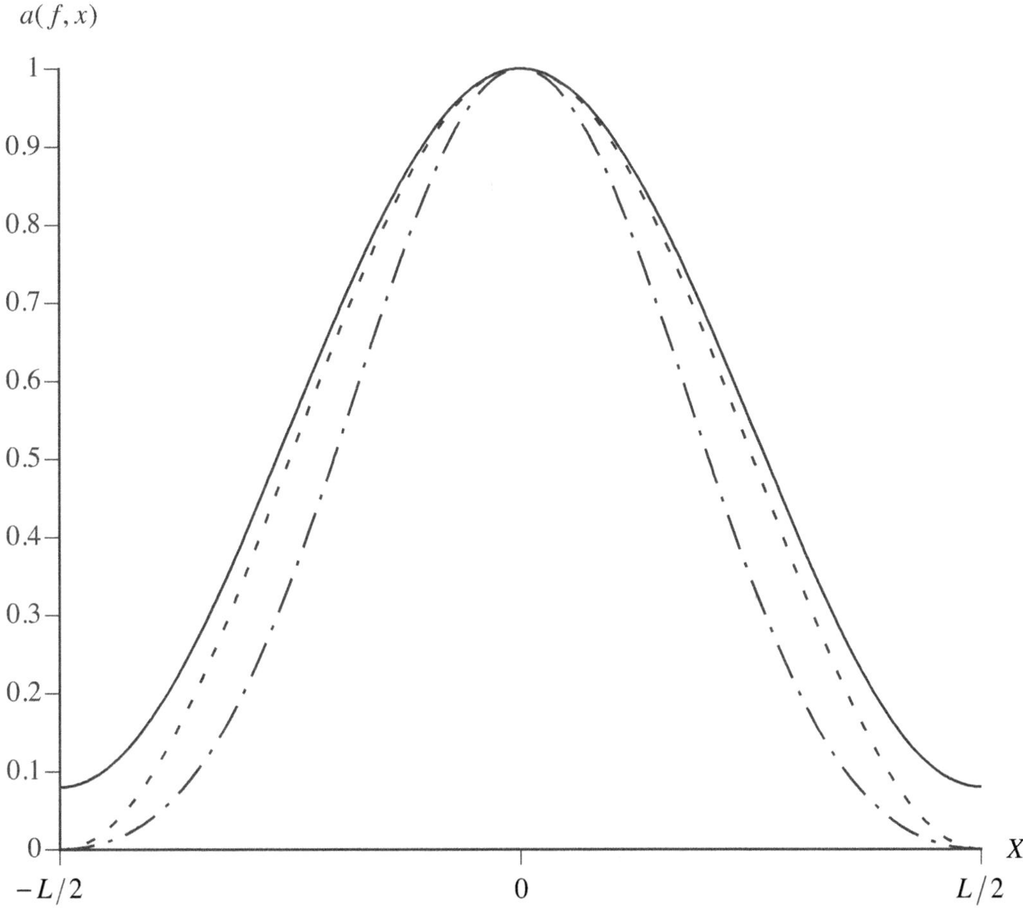

Figure 2.2-10 Hanning amplitude window (dashed curve), Hamming amplitude window (solid curve), and Blackman amplitude window (dot-dash curve) for $A_A = 1$.

$$a(f, x_A) = 0.5A_A\left[1 + \cos(2\pi x_A / L)\right]\text{rect}(x_A / L). \tag{2.2-38}$$

Therefore, if

$$A(f, x_A) = a(f, x_A) = 0.5A_A\text{rect}(x_A / L) + 0.5A_A\cos(2\pi x_A / L)\text{rect}(x_A / L), \tag{2.2-39}$$

then the complex frequency response of the transducer is in the shape of the Hanning amplitude window along the length L of the transducer, regardless of the value of frequency f.

The spatial Fourier transform of the first term on the right-hand side of (2.2-39) can be obtained by using (2.2-5), and the spatial Fourier transform of the second term can be obtained by substituting $d = L$ into (2.2-29). The unnormalized and normalized, far-field beam patterns of the Hanning amplitude window are given by

$$\boxed{D(f,f_X)=\frac{A_A L}{2}\frac{\mathrm{sinc}(f_X L)}{1-(f_X L)^2}} \tag{2.2-40}$$

and

$$\boxed{D_N(f,f_X)=\frac{\mathrm{sinc}(f_X L)}{1-(f_X L)^2}} \tag{2.2-41}$$

respectively (see Fig. 2.2-11). The level of the first sidelobe of the magnitude of the normalized, far-field beam pattern of the Hanning amplitude window is approximately -31.5 dB.

The *Hamming* amplitude window is given by

$$a(f,x_A)=A_A\left[0.54+0.46\cos(2\pi x_A/L)\right]\mathrm{rect}(x_A/L), \tag{2.2-42}$$

and is shown in Fig. 2.2-10 for $A_A=1$. Note that the Hamming amplitude window is discontinuous at the end points $x=\pm L/2$. Therefore, if

$$A(f,x_A)=a(f,x_A)=0.54A_A\,\mathrm{rect}(x_A/L)+0.46A_A\cos(2\pi x_A/L)\mathrm{rect}(x_A/L), \tag{2.2-43}$$

then the complex frequency response of the transducer is in the shape of the Hamming amplitude window along the length L of the transducer, regardless of the value of frequency f.

The spatial Fourier transform of the first term on the right-hand side of (2.2-43) can be obtained by using (2.2-5), and the spatial Fourier transform of the second term can be obtained by substituting $d=L$ into (2.2-29). The unnormalized and normalized, far-field beam patterns of the Hamming amplitude window are given by

$$\boxed{D(f,f_X)=A_A L\frac{0.54-0.08(f_X L)^2}{1-(f_X L)^2}\mathrm{sinc}(f_X L)} \tag{2.2-44}$$

and

$$\boxed{D_N(f,f_X)=\frac{1}{0.54}\frac{0.54-0.08(f_X L)^2}{1-(f_X L)^2}\mathrm{sinc}(f_X L)} \tag{2.2-45}$$

respectively (see Fig. 2.2-11). The level of the first sidelobe of the magnitude of

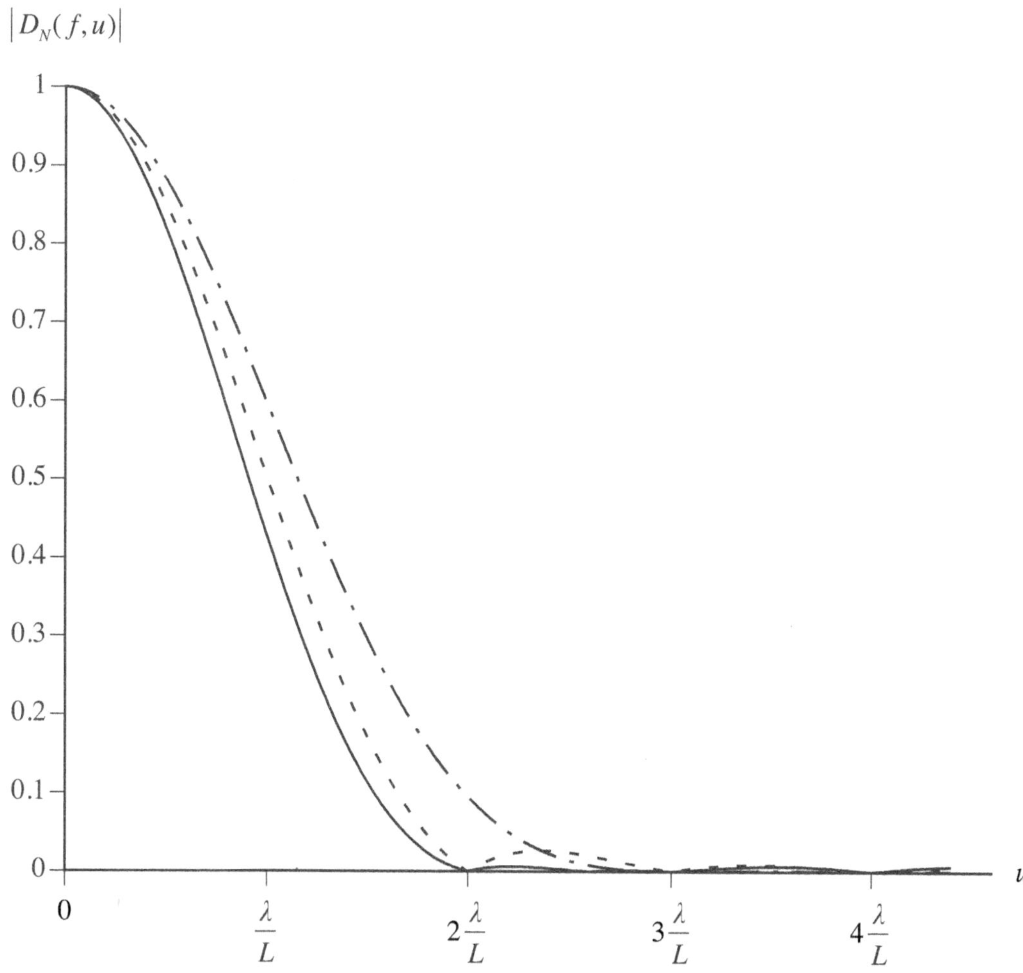

Figure 2.2-11 Magnitudes of the normalized, far-field beam patterns of the Hanning amplitude window (dashed curve), Hamming amplitude window (solid curve), and Blackman amplitude window (dot-dash curve) plotted as a function of direction cosine u.

the normalized, far-field beam pattern of the Hamming amplitude window is approximately -44 dB. From Fig. 2.2-11, it can be seen that although the level of the first sidelobe of the Hamming beam pattern is less than the level of the first sidelobe of the Hanning beam pattern, the width of the mainlobe of the Hamming beam pattern is also *less* than the width of the mainlobe of the Hanning beam pattern. This is an *exception* to the general rule that beamwidth increases as sidelobe levels decrease.

The *Blackman* amplitude window is given by

$$a(f, x_A) = A_A\left[0.42 + 0.5\cos(2\pi x_A/L) + 0.08\cos(4\pi x_A/L)\right]\text{rect}(x_A/L), \tag{2.2-46}$$

and is shown in Fig. 2.2-10 for $A_A = 1$. Therefore, if

$$\begin{aligned} A(f, x_A) &= a(f, x_A) \\ &= 0.42A_A \operatorname{rect}(x_A/L) + 0.5A_A \cos(2\pi x_A/L)\operatorname{rect}(x_A/L) + \\ &\quad 0.08A_A \cos(4\pi x_A/L)\operatorname{rect}(x_A/L), \end{aligned} \tag{2.2-47}$$

then the complex frequency response of the transducer is in the shape of the Blackman amplitude window along the length L of the transducer, regardless of the value of frequency f.

The spatial Fourier transform of the first term on the right-hand side of (2.2-47) can be obtained by using (2.2-5), and the spatial Fourier transforms of the second and third terms can be obtained by substituting $d = L$ and $d = L/2$ into (2.2-29), respectively. The unnormalized and normalized, far-field beam patterns of the Blackman amplitude window are given by

$$\boxed{D(f, f_X) = A_A L \frac{1.68 - 0.18(f_X L)^2}{\left[1-(f_X L)^2\right]\left[4-(f_X L)^2\right]} \operatorname{sinc}(f_X L)} \tag{2.2-48}$$

and

$$\boxed{D_N(f, f_X) = \frac{1}{0.42} \frac{1.68 - 0.18(f_X L)^2}{\left[1-(f_X L)^2\right]\left[4-(f_X L)^2\right]} \operatorname{sinc}(f_X L)} \tag{2.2-49}$$

respectively (see Fig. 2.2-11). The level of the first sidelobe of the magnitude of the normalized, far-field beam pattern of the Blackman amplitude window is approximately -58.1 dB.

All of the amplitude windows discussed in Section 2.2 were *even* functions of spatial variable x_A. As a result, their far-field beam patterns were *real*, *even* functions of spatial frequency f_X (direction cosine u). In general, if an amplitude window is an *even* function of a spatial variable, then the corresponding far-field beam pattern will be a *real*, *even* function of the appropriate spatial frequency (direction cosine) (see Appendix 2C). Similarly, if an amplitude window is an *odd* function of a spatial variable, then the corresponding far-field beam pattern will be an *imaginary*, *odd* function of the appropriate spatial frequency (direction cosine) (see Appendix 2C). Furthermore, if the phase response $\theta(f, x_A) = 0$ and the amplitude response (amplitude window) $a(f, x_A) \geq 0$, then the maximum value of the magnitude of the corresponding unnormalized, far-field beam pattern is at broadside. Therefore, the normalization factor in this case is $D_{\max} = |D(f, 0)|$ (see Appendix 2D). However, whether or not this is the case, D_{max} can also be found by using the standard calculus approach for finding the maximum value of a function, or by using the formulas in Appendix 2D.

A summary of the one-dimensional spatial Fourier transforms used in Section 2.2 is presented in Table 2E-1 in Appendix 2E.

2.3 Beamwidth

The term "beamwidth" refers to some measure of the width of the mainlobe of a far-field beam pattern. *The beamwidth of a far-field beam pattern is the same whether the beam pattern is normalized or unnormalized.* However, it is customary to work with normalized, far-field beam patterns. The most common measure of beamwidth is the 3-dB beamwidth. Since the maximum value of the magnitude of a *normalized*, far-field beam pattern is equal to 1, or 0 dB, the 3-dB beamwidth in two-dimensional space is defined as the width of the mainlobe between those two points that correspond to magnitude values equal to $\sqrt{2}/2$, or -3 dB. Note that

$$20\log_{10}\left(\sqrt{2}/2\right)=10\log_{10}(1/2)=-3.01\text{ dB}. \tag{2.3-1}$$

The 3-dB beamwidth is also referred to as the *half-power beamwidth* [note the argument $1/2$ in the middle expression in (2.3-1)]. Figure 2.3-1 shows the magnitude of a typical normalized, far-field beam pattern of a linear aperture lying along the X axis, plotted as a function of direction cosine u. The *dimensionless* 3-dB beamwidth Δu in direction-cosine space is given by

$$\Delta u = u_+ - u_- > 0, \tag{2.3-2}$$

or, since the magnitude of the beam pattern is symmetric about $u=0$,

$$\Delta u = 2u_+, \tag{2.3-3}$$

where $u_+ > 0$ and $u_- < 0$ as shown.

The problem now is to express the 3-dB beamwidth Δu in terms of the 3-dB beamwidths $\Delta\theta$ and $\Delta\psi$ associated with the spherical angles θ and ψ. Whereas Δu is dimensionless, $\Delta\theta$ and $\Delta\psi$ have units in *degrees*. Since $u=\sin\theta\cos\psi$, let (see Fig. 2.3-2)

$$u_+ = \sin\theta_+ \cos\psi_+ > 0, \tag{2.3-4}$$

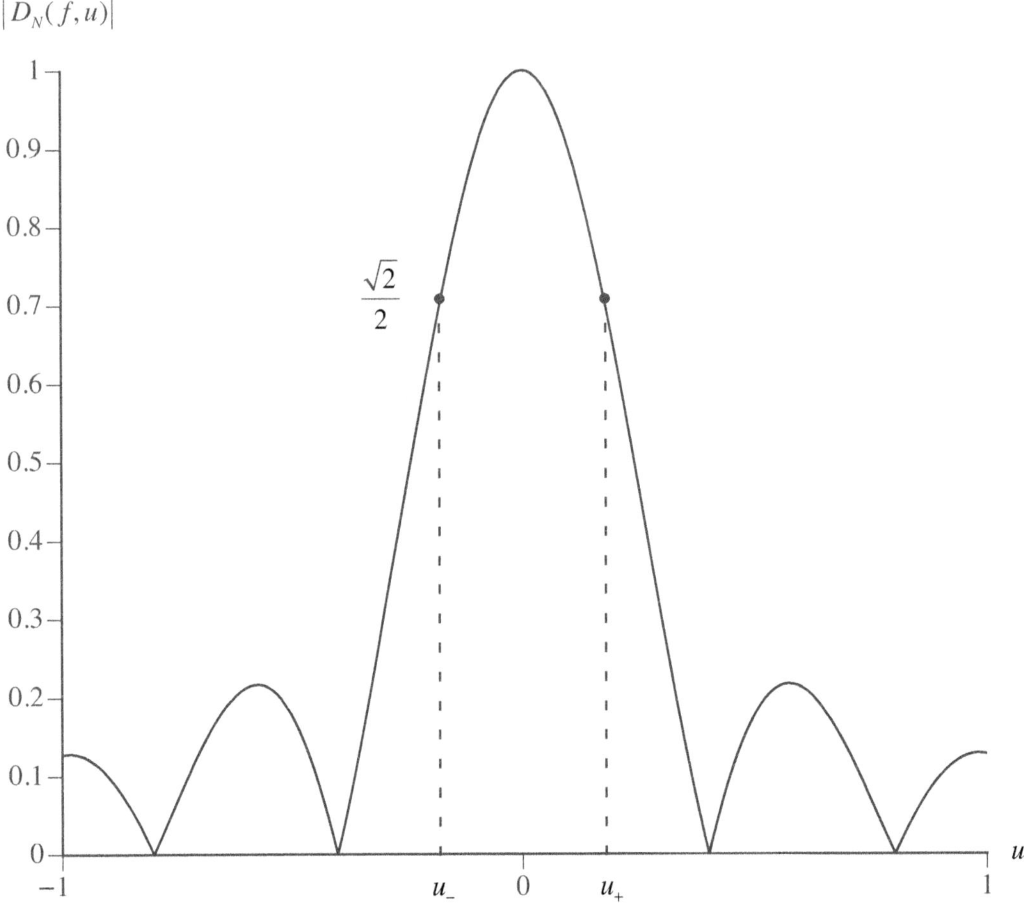

Figure 2.3-1 Magnitude of a typical normalized, far-field beam pattern of a linear aperture lying along the X axis, plotted as a function of direction cosine u.

where

$$\theta_+ = \Delta\theta/2 \tag{2.3-5}$$

and

$$\psi_+ = 90° - \frac{\Delta\psi}{2}; \tag{2.3-6}$$

and let

$$u_- = \sin\theta_- \cos\psi_- < 0, \tag{2.3-7}$$

where

$$\theta_- = \theta_+ = \Delta\theta/2 \tag{2.3-8}$$

and

$$\psi_- = 90° + \frac{\Delta\psi}{2}. \tag{2.3-9}$$

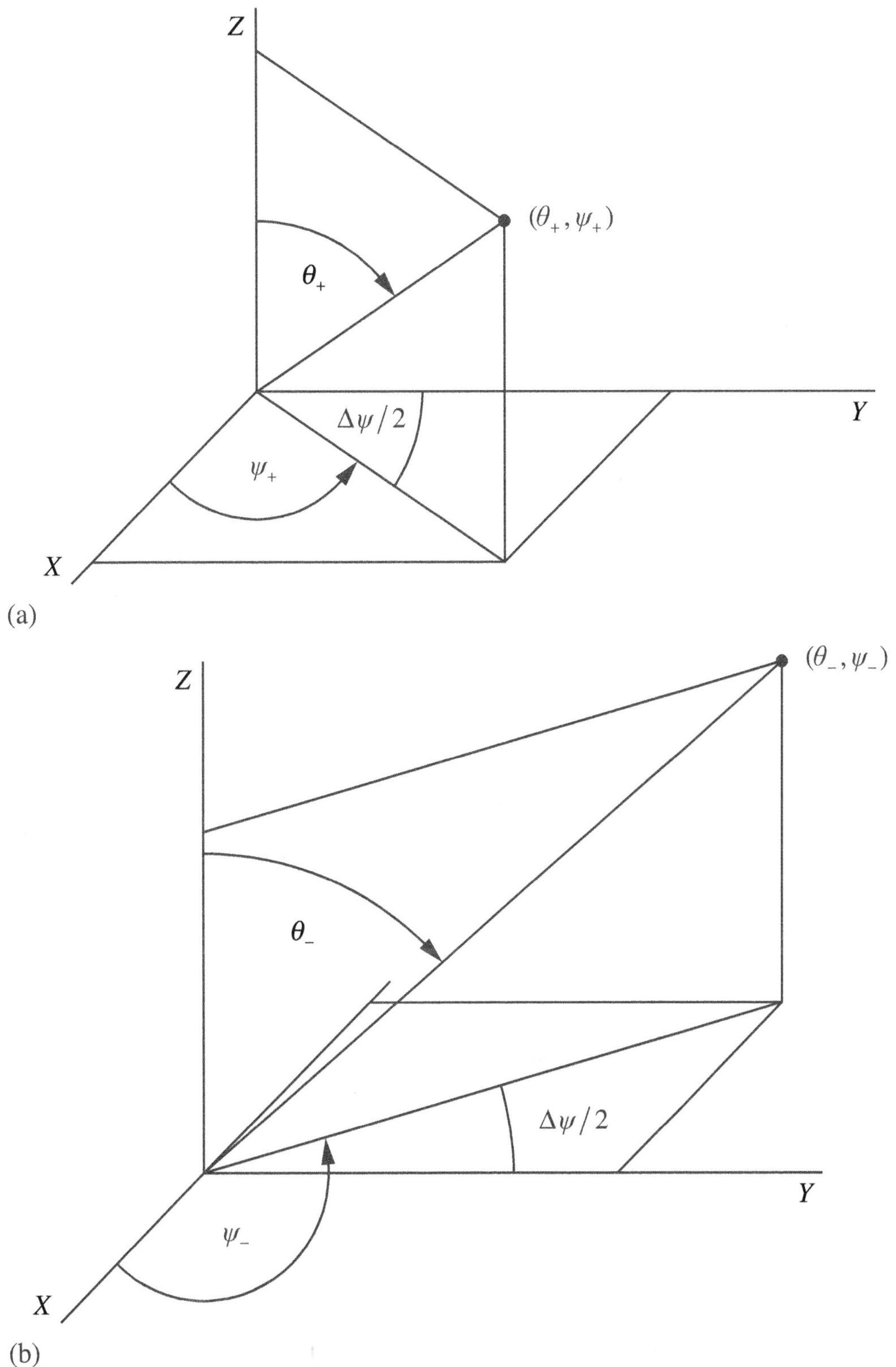

Figure 2.3-2 Representation of the spherical angles (a) (θ_+, ψ_+) and (b) (θ_-, ψ_-).

With ψ_+ and ψ_- given by (2.3-6) and (2.3-9), respectively, it is assured that $u_+ > 0$ and $u_- < 0$ since

$$\cos\psi_+ = \cos\left(90° - \frac{\Delta\psi}{2}\right) = \sin\left(\frac{\Delta\psi}{2}\right) > 0 \tag{2.3-10}$$

and

$$\cos\psi_- = \cos\left(90° + \frac{\Delta\psi}{2}\right) = -\sin\left(\frac{\Delta\psi}{2}\right) < 0 . \tag{2.3-11}$$

Substituting (2.3-5) and (2.3-10) into (2.3-4) yields

$$u_+ = \sin(\Delta\theta/2)\sin(\Delta\psi/2) , \tag{2.3-12}$$

and substituting (2.3-8) and (2.3-11) into (2.3-7) yields

$$u_- = -\sin(\Delta\theta/2)\sin(\Delta\psi/2) . \tag{2.3-13}$$

Finally, by substituting (2.3-12) and (2.3-13) into (2.3-2), we obtain

$$\Delta u = 2\sin(\Delta\theta/2)\sin(\Delta\psi/2) . \tag{2.3-14}$$

However, we now have only *one equation* but *two unknowns*, $\Delta\theta$ and $\Delta\psi$. Therefore, when dealing with a linear aperture, we must restrict ourselves to either the *vertical* or *horizontal* far-field beam pattern when making a beamwidth calculation, as is demonstrated in the next two examples.

Example 2.3-1 Vertical Beam Pattern

Consider a linear aperture lying along the X axis. Therefore, by definition, the vertical beam pattern lies in the XZ plane (see Fig. 2.3-3). In the XZ plane, $u = \sin\theta$ when $\psi = 0°$, and $u = -\sin\theta$ when $\psi = 180°$. As a result,

$$u_+ = \sin\theta_+ > 0, \qquad \psi_+ = 0° , \tag{2.3-15}$$

and

$$u_- = -\sin\theta_- < 0, \qquad \psi_- = 180° , \tag{2.3-16}$$

where

$$\theta_+ = \theta_- = \Delta\theta/2 . \tag{2.3-17}$$

Therefore,

$$\Delta u = u_+ - u_- = 2\sin(\Delta\theta/2) > 0 , \tag{2.3-18}$$

or

$$\boxed{\Delta\theta = 2\sin^{-1}(\Delta u/2), \qquad \Delta u/2 \le 1} \tag{2.3-19}$$

where $\Delta\theta$ is the 3-dB beamwidth in degrees of a vertical, far-field beam pattern in the XZ plane. ■

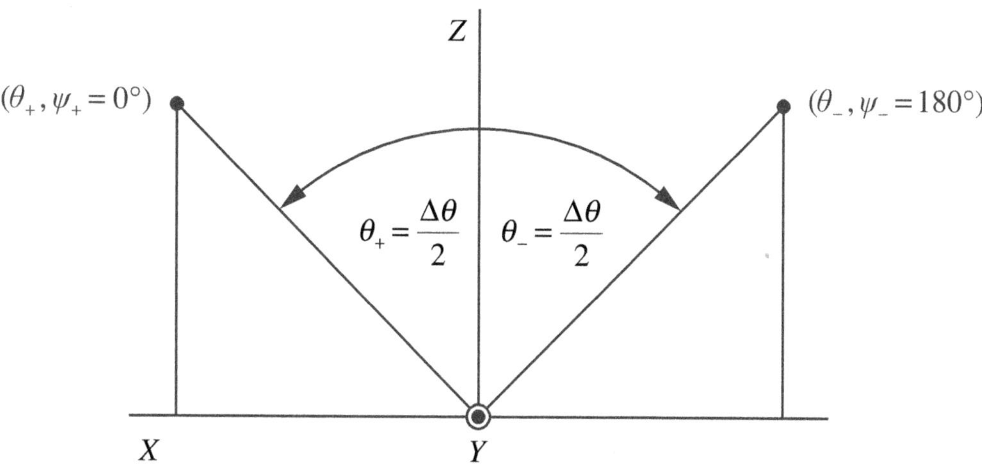

Figure 2.3-3 Representation of the spherical angles $(\theta_+, \psi_+ = 0°)$ and $(\theta_-, \psi_- = 180°)$.

Example 2.3-2 Horizontal Beam Pattern

Consider a linear aperture lying along the X axis. Therefore, by definition, the horizontal beam pattern lies in the XY plane (see Fig. 2.3-4). In the XY plane, $u = \cos\psi$ since $\theta = 90°$. As a result,

$$u_+ = \cos\psi_+ > 0, \qquad \theta_+ = 90°, \tag{2.3-20}$$

and

$$u_- = \cos\psi_- < 0, \qquad \theta_- = 90°, \tag{2.3-21}$$

where

$$\psi_+ = 90° - \frac{\Delta\psi}{2} \tag{2.3-22}$$

and

$$\psi_- = 90° + \frac{\Delta\psi}{2}. \tag{2.3-23}$$

Therefore, with the use of (2.3-10) and (2.3-11),

$$\Delta u = u_+ - u_- = 2\sin(\Delta\psi/2) > 0\,, \tag{2.3-24}$$

or

$$\boxed{\Delta\psi = 2\sin^{-1}(\Delta u/2), \qquad \Delta u/2 \le 1} \tag{2.3-25}$$

where $\Delta\psi$ is the 3-dB beamwidth in degrees of a horizontal, far-field beam pattern in the XY plane. Note that the right-hand sides of (2.3-19) and (2.3-25) are identical. ■

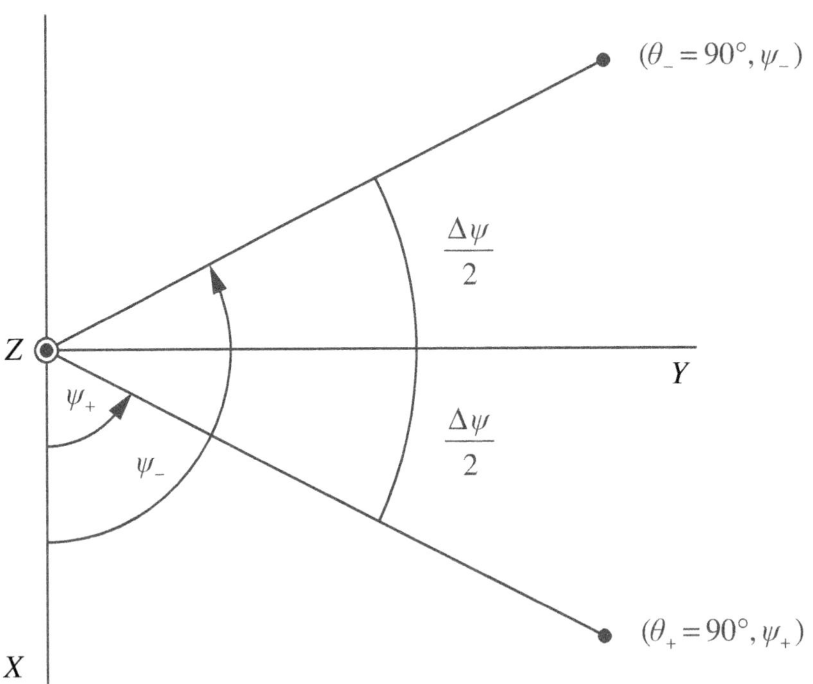

Figure 2.3-4 Representation of the spherical angles $(\theta_+ = 90°, \psi_+)$ and $(\theta_- = 90°, \psi_-)$.

By examining (2.3-19) and (2.3-25), it can be seen that in order to obtain numerical values for $\Delta\theta$ and $\Delta\psi$, we need to know Δu. Consider the next example.

Example 2.3-3

In this example we shall compute the 3-dB beamwidth Δu in direction-cosine space of the normalized, far-field beam pattern of the rectangular amplitude window given by (2.2-11). Remember that the beamwidth of a far-field beam pattern is the same whether the beam pattern is normalized or unnormalized. With the use of (2.2-6), evaluating the magnitude of (2.2-11) at $u = u_+$ yields

$$|D_N(f, u_+)| = \left|\operatorname{sinc}\left(\frac{L}{\lambda}u_+\right)\right| = \frac{\sin(\pi L u_+/\lambda)}{\pi L u_+/\lambda} = \frac{\sqrt{2}}{2} \tag{2.3-26}$$

since u_+ is the coordinate of one of the 3-dB-down points in direction-cosine space. Note that the absolute value operator was removed from $\sin(\pi L u_+/\lambda)/(\pi L u_+/\lambda)$ because this sinc function is positive when evaluated at u_+. Equation (2.3-26) can be rewritten as

$$\sin(\pi x) - \frac{\sqrt{2}}{2}\pi x = 0\,, \tag{2.3-27}$$

where

$$x = \frac{L}{\lambda}u_+\,, \tag{2.3-28}$$

and by substituting (2.3-3) into (2.3-28), we obtain

$$\Delta u = 2x\lambda/L\,. \tag{2.3-29}$$

Since the solution (root) of (2.3-27) is $x \approx 0.443$, substituting this result into (2.3-29) yields

$$\boxed{\Delta u \approx 0.886\,\lambda/L} \tag{2.3-30}$$

for the rectangular amplitude window. Therefore, by substituting (2.3-30) into (2.3-19) and (2.3-25), the 3-dB beamwidths of the vertical and horizontal, far-field beam patterns of the rectangular amplitude window can be computed as functions of frequency (wavelength) and aperture length. It is important to note that this procedure can be used to calculate Δu for *any* normalized, far-field beam pattern of a linear aperture, including the normalized, far-field beam patterns of linear arrays. ■

Table 2.3-1 summarizes the dimensionless 3-dB beamwidth Δu in direction-cosine space of the normalized, far-field beam patterns of the six different amplitude windows discussed in Section 2.2. As was previously shown in Fig. 2.2-11, and is also shown now in Table 2.3-1, although the level of the first sidelobe of the Hamming beam pattern is less than the level of the first sidelobe of the Hanning beam pattern, the width of the mainlobe of the Hamming beam pattern is also less than the width of the mainlobe of the Hanning beam pattern. This is an exception to the general rule that beamwidth increases as sidelobe levels decrease.

Table 2.3-1 The Dimensionless 3-dB Beamwidth Δu in Direction-Cosine Space of the Normalized, Far-Field Beam Patterns of Six Different Amplitude Windows

Amplitude Window	Level of 1st Sidelobe of the Magnitude of the Normalized Far-Field Beam Pattern (dB)	Δu
rectangular	−13.3	$0.886\lambda/L$
cosine	−23.0	$1.189\lambda/L$
triangular	−26.5	$1.276\lambda/L$
Hanning	−31.5	$1.441\lambda/L$
Hamming	−44.0	$1.303\lambda/L$
Blackman	−58.1	$1.644\lambda/L$

2.4 Beam Steering

In Section 2.2 we assumed that the phase response of a linear aperture was equal to zero and concentrated our efforts on deriving closed-form expressions for the far-field beam patterns of various amplitude windows. In this section we shall study what happens to a far-field beam pattern when the phase response $\theta(f, x_A)$ is nonzero, in particular, when $\theta(f, x_A)$ is a linear function of x_A.

Let us begin the analysis by expressing $\theta(f, x_A)$ in terms of the following Nth-order polynomial with frequency-dependent coefficients:

$$\theta(f, x_A) = \theta_0(f) + \theta_1(f)x_A + \theta_2(f)x_A^2 + \ldots + \theta_N(f)x_A^N, \tag{2.4-1}$$

where for a given value of frequency f, $\theta_0(f)$ represents a *constant* value of phase (constant with respect to x_A) along the length of the aperture, $\theta_1(f)x_A$ represents a *linear* phase response along the length of the aperture and is responsible for *beam steering* (*beam tilting*), and $\theta_2(f)x_A^2$ represents a *quadratic* phase response along the length of the aperture and is responsible for *focusing* in the Fresnel region of the aperture. The remaining higher-order terms in (2.4-1) are not considered. Note that (2.4-1) could be used as a mathematical model if we were trying to fit a Nth-order polynomial to actual phase response data using, for example, the method of nonlinear least-squares estimation.

Next, let $D(f, f_X)$ be the far-field beam pattern of the amplitude window $a(f, x_A)$:

$$D(f, f_X) = F_{x_A}\{a(f, x_A)\} = \int_{-\infty}^{\infty} a(f, x_A)\exp(+j2\pi f_X x_A)dx_A. \tag{2.4-2}$$

Let $D'(f, f_X)$ be the far-field beam pattern of the aperture function $A(f, x_A)$ given by (2.1-10):

$$\begin{aligned} D'(f, f_X) &= F_{x_A}\{a(f, x_A)\exp[+j\theta(f, x_A)]\} \\ &= \int_{-\infty}^{\infty} a(f, x_A)\exp[+j\theta(f, x_A)]\exp(+j2\pi f_X x_A)dx_A. \end{aligned} \tag{2.4-3}$$

Now let the phase response along the length of the aperture be a linear function of x_A, that is, let

$$\theta(f, x_A) = \theta_1(f)x_A, \tag{2.4-4}$$

where

$$\theta_1(f) = -2\pi f_X', \tag{2.4-5}$$

$$f_X' = u'/\lambda = fu'/c, \tag{2.4-6}$$

and

$$u' = \sin\theta' \cos\psi'. \tag{2.4-7}$$

The linear phase response given by (2.4-4) is an *odd* function of x_A and is a *straight line* that passes through the origin with frequency-dependent slope given by (2.4-5). If (2.4-4) and (2.4-5) are substituted into (2.4-3), then

$$\begin{aligned} D'(f, f_X) &= F_{x_A}\{a(f, x_A)\exp(-j2\pi f_X' x_A)\} \\ &= \int_{-\infty}^{\infty} a(f, x_A)\exp[+j2\pi(f_X - f_X')x_A]dx_A, \end{aligned} \tag{2.4-8}$$

and by comparing (2.4-8) with (2.4-2),

$$D'(f, f_X) = D(f, f_X - f_X'), \tag{2.4-9}$$

or

$$F_{x_A}\{a(f, x_A)\exp(-j2\pi f_X' x_A)\} = D(f, f_X - f_X'), \tag{2.4-10}$$

where $D(f, f_X)$ is given by (2.4-2). Equation (2.4-9) can also be expressed as

$$D'(f,u) = D(f,u-u') \tag{2.4-11}$$

since $D(f, f_X) = D(f, u/\lambda) \to D(f,u)$. Therefore, a linear phase response along the length of the aperture will cause the beam pattern $D(f,u)$ to be steered in the direction $u = u'$ in direction-cosine space, which is equivalent to steering (tilting) the beam pattern to $\theta = \theta'$ and $\psi = \psi'$ [see (2.4-7)]. Equation (2.4-11) indicates that the far-field beam pattern $D'(f,u)$ is simply a translated or shifted version of $D(f,u)$ along the u axis, implying that the shape and beamwidth of the mainlobe of the beam pattern remain unchanged (see Fig. 2.4-1). However, as we shall show in Section 2.5, when a beam pattern is steered, the shape and beamwidth of the mainlobe of the beam pattern *do change* when the magnitude of the beam pattern is plotted as a function of the spherical angles θ and ψ.

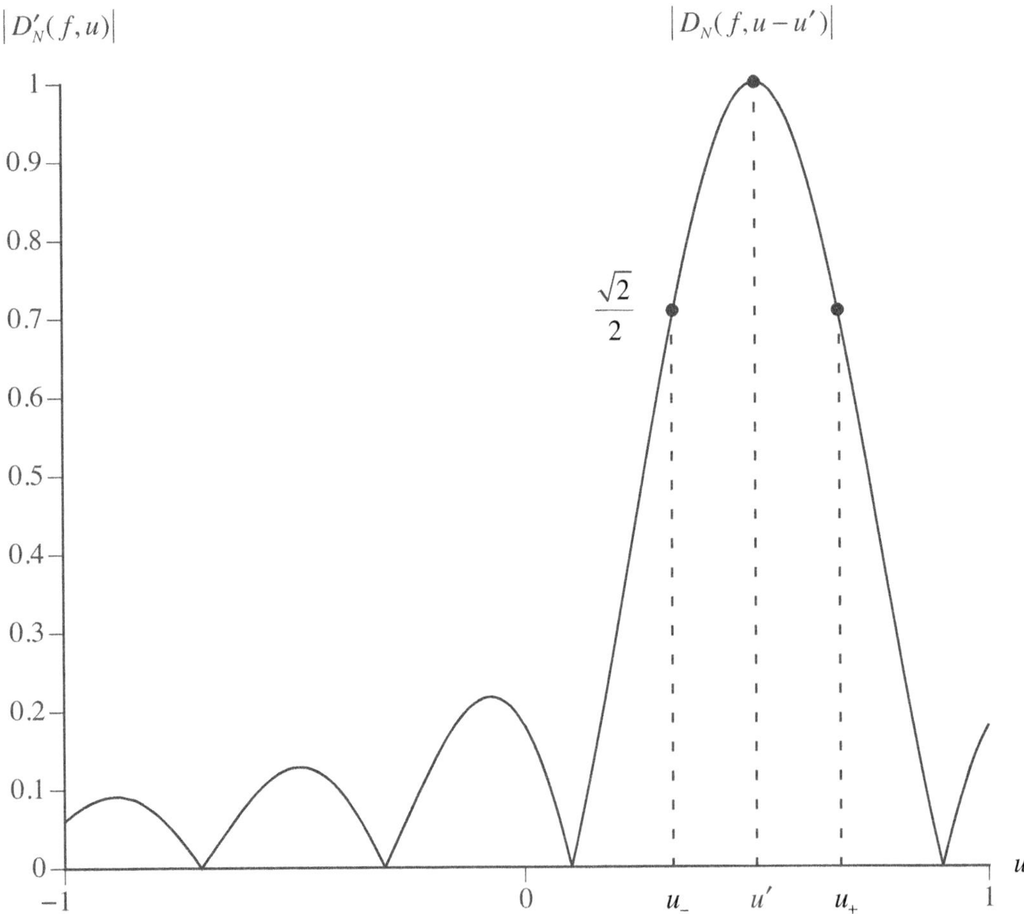

Figure 2.4-1 Magnitude of a typical normalized, far-field beam pattern of a linear aperture lying along the X axis, steered in the direction $u = u'$ in direction-cosine space.

2.5 Beamwidth at an Arbitrary Beam-Steer Angle

When a far-field beam pattern is steered (tilted) from broadside toward end-fire, its *beamwidth will increase from a minimum value at broadside to a maximum value at end-fire.*

In order to compute the beamwidth at an arbitrary beam-steer angle, let us restrict ourselves to the horizontal, far-field beam pattern of a linear aperture lying along the X axis (see Example 2.3-2), where ψ' is the beam-steer angle in the XY plane (see Fig. 2.5-1). In the XY plane, $u = \cos\psi$ since $\theta = 90°$. Therefore, by referring to Figs. 2.4-1, 2.5-1, and 2.5-2,

$$u_+ = \cos\psi_+ > 0, \qquad \theta_+ = 90°, \tag{2.5-1}$$

$$u_- = \cos\psi_- > 0, \qquad \theta_- = 90°, \tag{2.5-2}$$

and

$$u' = \cos\psi' > 0, \qquad \theta' = 90°, \tag{2.5-3}$$

where

$$u_+ > u_-, \tag{2.5-4}$$

$$u_+ = u' + (\Delta u/2), \tag{2.5-5}$$

and

$$u_- = u' - (\Delta u/2). \tag{2.5-6}$$

Substituting (2.5-1) and (2.5-3) into (2.5-5) yields

$$\cos\psi_+ = \cos\psi' + (\Delta u/2), \tag{2.5-7}$$

and by solving for ψ_+ we obtain

$$\psi_+ = \cos^{-1}\left[\cos\psi' + (\Delta u/2)\right]. \tag{2.5-8}$$

Similarly, substituting (2.5-2) and (2.5-3) into (2.5-6) yields

$$\cos\psi_- = \cos\psi' - (\Delta u/2), \tag{2.5-9}$$

and by solving for ψ_- we obtain

$$\psi_- = \cos^{-1}\left[\cos\psi' - (\Delta u/2)\right]. \tag{2.5-10}$$

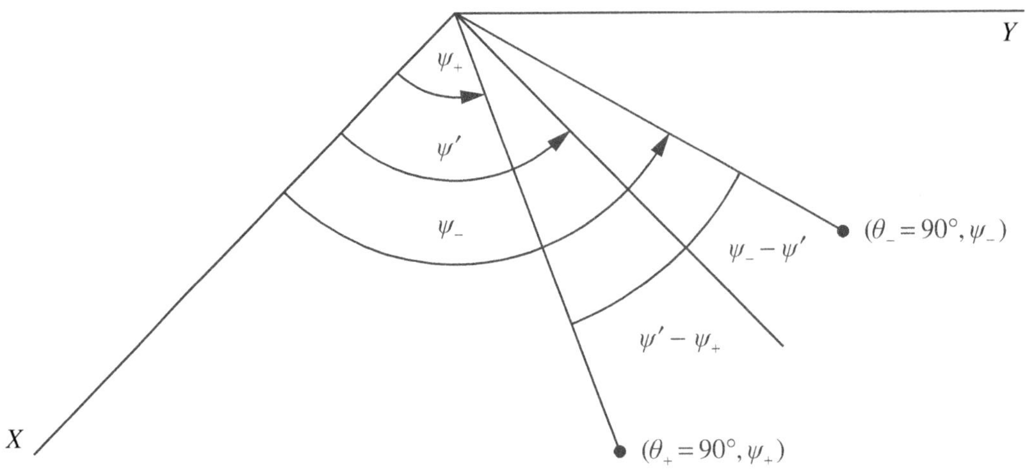

Figure 2.5-1 Representation of the spherical angles $(\theta_+ = 90°, \psi_+)$, $(\theta_- = 90°, \psi_-)$ and the beam-steer angle ψ'. Note that the *half-beamwidth angles* $\psi' - \psi_+$ and $\psi_- - \psi'$ are *not* equal in general.

From Figs. 2.5-1 and 2.5-2 it can be seen that the 3-dB beamwidth in degrees of a horizontal, far-field beam pattern in the XY plane is given by

$$\Delta\psi = \psi_- - \psi_+ . \tag{2.5-11}$$

Substituting (2.5-8) and (2.5-10) into (2.5-11) yields

$$\boxed{\Delta\psi = \cos^{-1}\left(\cos\psi' - \frac{\Delta u}{2}\right) - \cos^{-1}\left(\cos\psi' + \frac{\Delta u}{2}\right), \qquad \left|\cos\psi' \pm \frac{\Delta u}{2}\right| \le 1} \tag{2.5-12}$$

where Δu can be obtained by using the procedure outlined in Example 2.3-3 (also see Table 2.3-1). In order to use (2.5-12), the inequality on the right-hand side must hold for both the plus and minus signs.

Although the *half-beamwidth angles* $\psi' - \psi_+$ and $\psi_- - \psi'$ are not equal, in general, for arbitrary beam-steer angles ψ'; they are equal to one another and to $\Delta\psi/2$ for $\psi' = 0°$ (end-fire), $90°$ (broadside), $180°$ (end-fire), $270°$ (broadside), and $360°$ (end-fire) since $\cos\psi$ is symmetric about $\psi = \psi' = 0°$, $90°$, $180°$, $270°$, and $360°$ (see Fig. 2.5-2). Note that $\psi' = 90°$ corresponds to an *unsteered* horizontal beam pattern for a linear aperture lying along the X axis (see Fig. 2.5-1). With $\psi' = 90°$, (2.5-12) reduces to (2.3-25) since

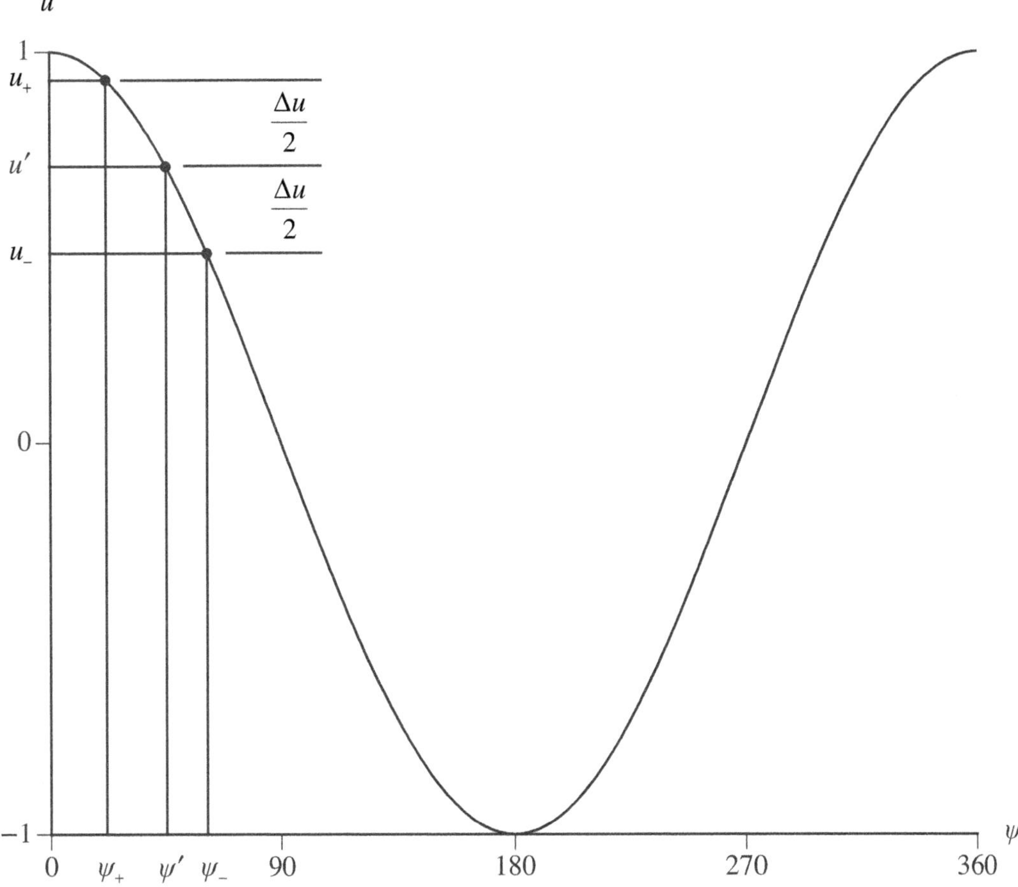

Figure 2.5-2 Direction cosine u versus bearing angle ψ (in deg) where $u = \cos\psi$ in the XY plane.

$$\cos^{-1}(x) = \frac{\pi}{2} - \sin^{-1}(x), \tag{2.5-13}$$

$$\cos^{-1}(-x) = \frac{\pi}{2} + \sin^{-1}(x), \tag{2.5-14}$$

and as a result,

$$\cos^{-1}(-x) - \cos^{-1}(x) = 2\sin^{-1}(x). \tag{2.5-15}$$

Also note that $\psi' = 0°$ and $\psi' = 180°$ corresponds to steering the horizontal beam pattern to end-fire. However, (2.5-12) is *invalid* for $\psi' = 0°$ and $\psi' = 180°$. Therefore, a different approach must be used in order to compute the beamwidth at end-fire.

Beamwidth at End-Fire

Restricting ourselves once again to the horizontal, far-field beam pattern of a linear aperture lying along the X axis, consider the case where the beam-steer angle $\psi' = 0°$ in the XY plane (see Fig. 2.5-1). With $\psi' = 0°$, $u' = 1$ [see (2.5-3)], and as a result, (2.5-6) becomes (see Fig. 2.5-3)

$$u_{-} = 1 - \frac{\Delta u}{2}. \tag{2.5-16}$$

Also, with $\psi' = 0°$, the half-beamwidth angle $\psi_{-} - \psi'$ reduces to ψ_{-} where

$$\psi_{-} = \Delta\psi/2 \tag{2.5-17}$$

since $\cos\psi$ is symmetric about $\psi = \psi' = 0°$ (see Fig. 2.5-2). Therefore, substituting (2.5-17) into (2.5-2) yields

$$u_{-} = \cos(\Delta\psi/2), \tag{2.5-18}$$

and by equating the right-hand sides of (2.5-16) and (2.5-18), we finally obtain

$$\boxed{\Delta\psi = 2\cos^{-1}\left(1 - \frac{\Delta u}{2}\right), \qquad \left|1 - \frac{\Delta u}{2}\right| \le 1} \tag{2.5-19}$$

where $\Delta\psi$ is the 3-dB beamwidth in degrees of a horizontal, far-field beam pattern in the XY plane steered to end-fire (see Table 2.3-1 for Δu).

Let us end our discussion in this section by illustrating the distortion a beam pattern undergoes due to beam steering. By referring to (2.2-11), the normalized, far-field beam pattern of the rectangular amplitude window steered to $u = u'$ is given by

$$D_N(f, u - u') = \operatorname{sinc}\left[\frac{L}{\lambda}(u - u')\right]. \tag{2.5-20}$$

The corresponding normalized, *horizontal*, far-field beam pattern is given by

$$D_N(f, 90°, \psi) = \operatorname{sinc}\left[\frac{L}{\lambda}(\cos\psi - \cos\psi')\right], \qquad \theta' = 90°. \tag{2.5-21}$$

Polar plots of the magnitude of (2.5-21) versus ψ for $L/\lambda = 4$ and for different

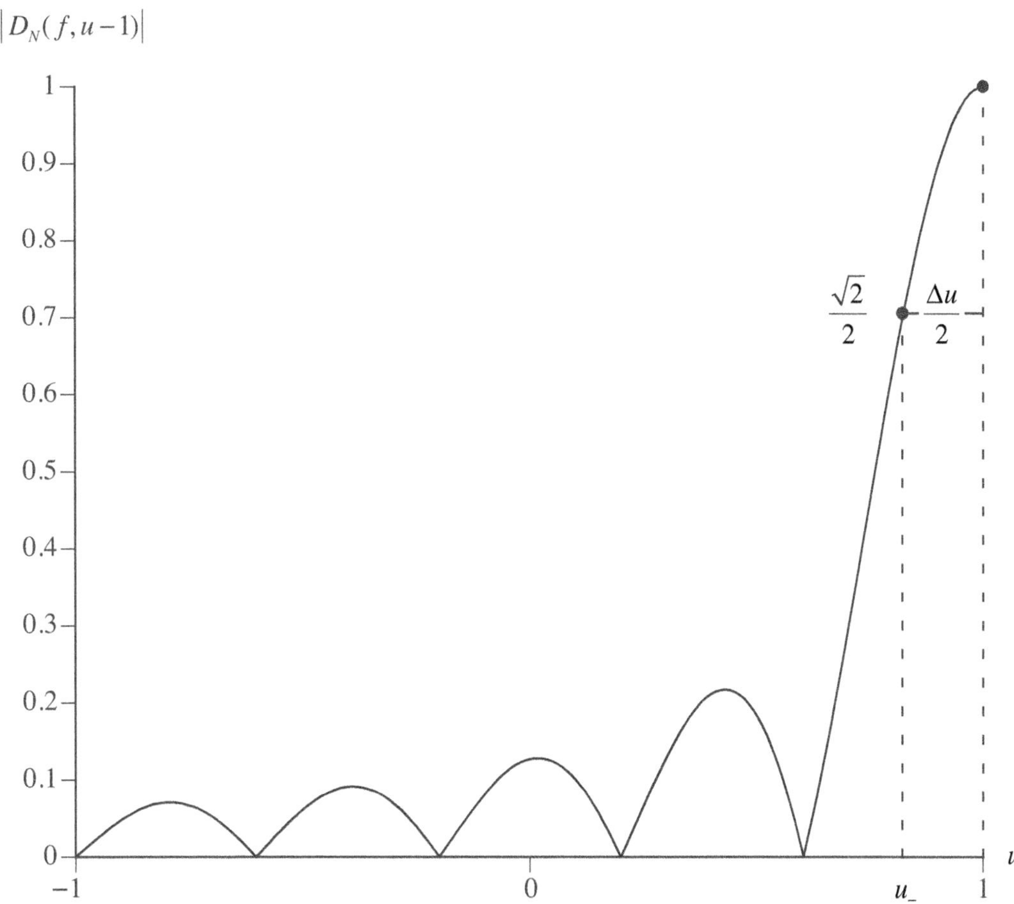

Figure 2.5-3 Magnitude of a typical normalized, far-field beam pattern of a linear aperture lying along the X axis, steered in the direction $u' = 1$ (end-fire).

values of ψ' are shown in Fig. 2.5-4. Note how *the beamwidth increases* and *the mainlobe becomes more asymmetrical about the beam-steer axis* as the beam pattern is steered from broadside ($\psi' = 90°$) to end-fire ($\psi' = 0°$).

Figure 2.5-4 also illustrates the *port/starboard (left/right) ambiguity problem* associated with linear apertures (linear arrays). For example, if the linear aperture is being used in the passive mode as a receiver trying to estimate the bearing angle to a sound-source (target), by referring to Fig. 2.5-4 (d), a sound-source at bearing angle $\psi = 45°$ or $\psi = 315°$ will produce the same output electrical signal from the linear aperture. Solutions to the port/starboard (left/right) ambiguity problem are discussed in Chapters 8 and 9, Examples 8.2-1 and 9.1-2, respectively.

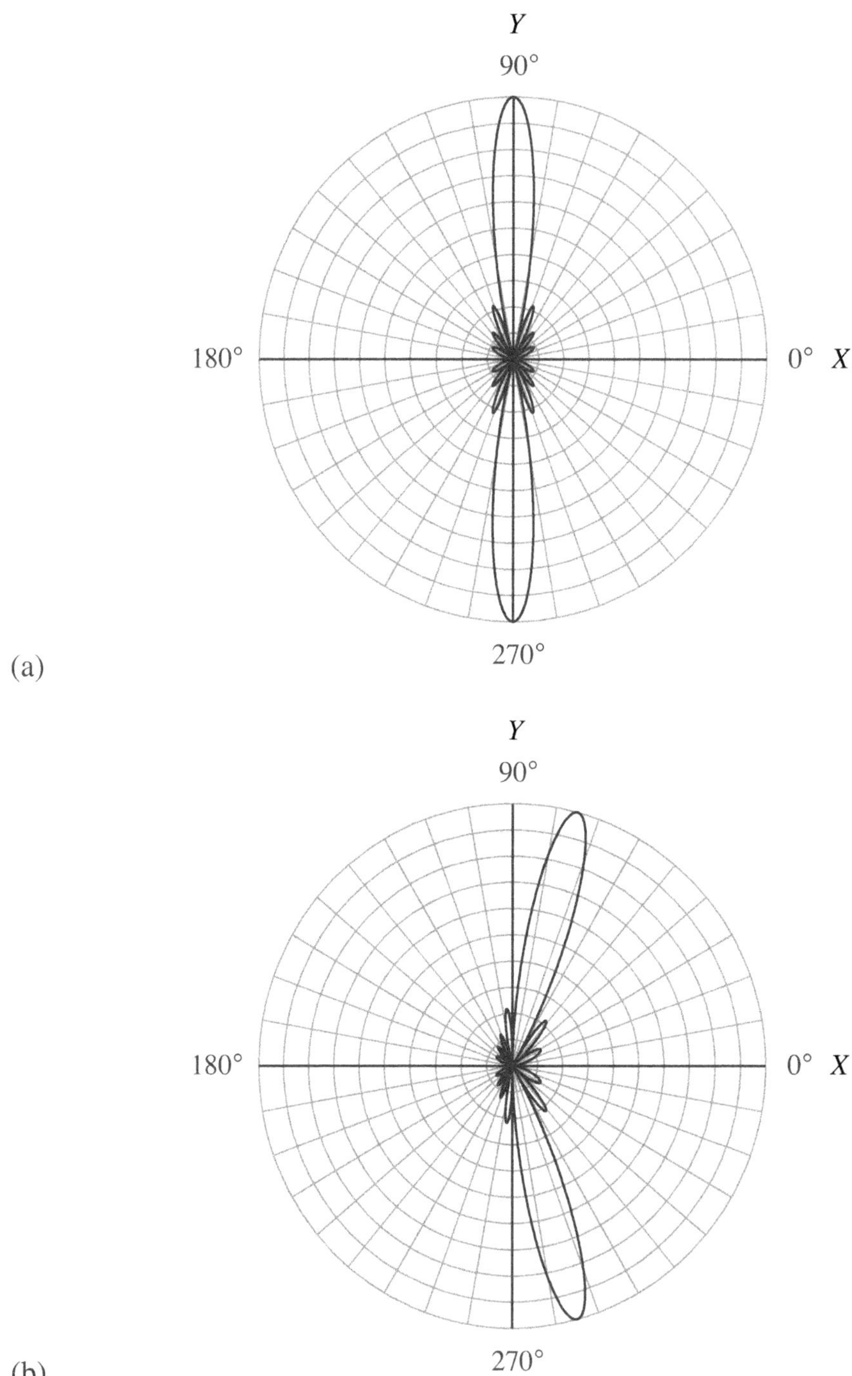

Figure 2.5-4 Polar plots, as a function of the azimuthal (bearing) angle ψ, of the magnitude of the normalized, horizontal, far-field beam pattern of the rectangular amplitude window given by (2.5-21) for $L/\lambda = 4$ and for beam-steer angle (a) $\psi' = 90°$ (broadside), (b) $\psi' = 75°$, (c) $\psi' = 60°$, (d) $\psi' = 45°$, (e) $\psi' = 30°$, and (f) $\psi' = 0°$ (end-fire).

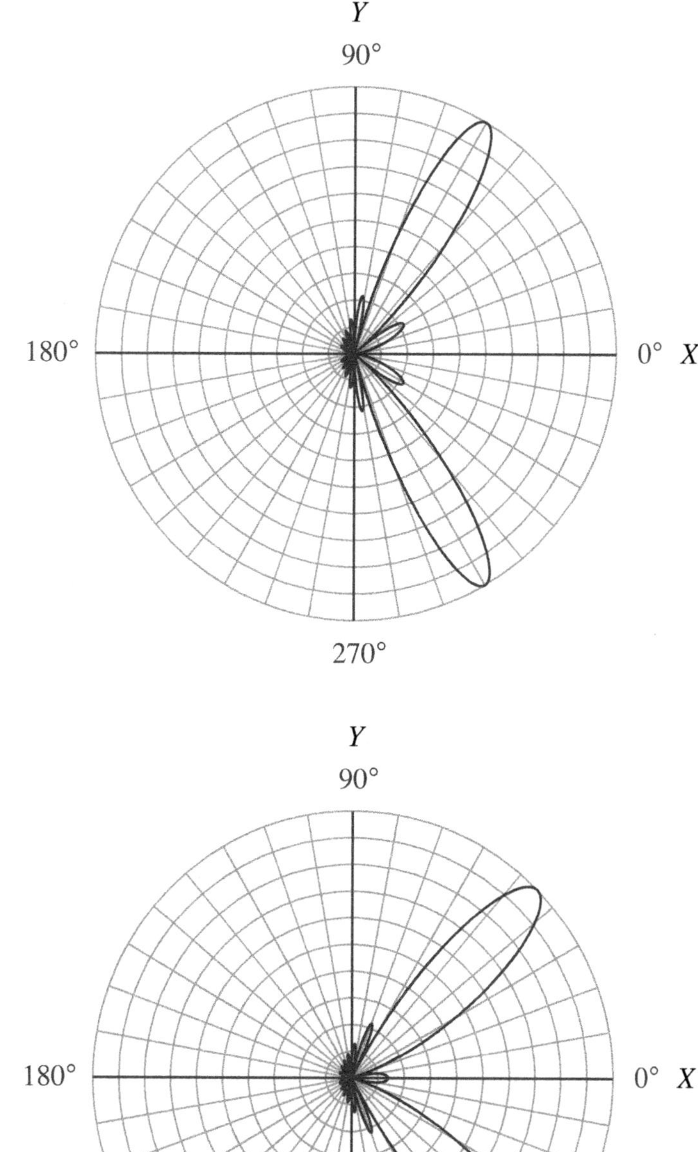

Figure 2.5-4 *continued.*

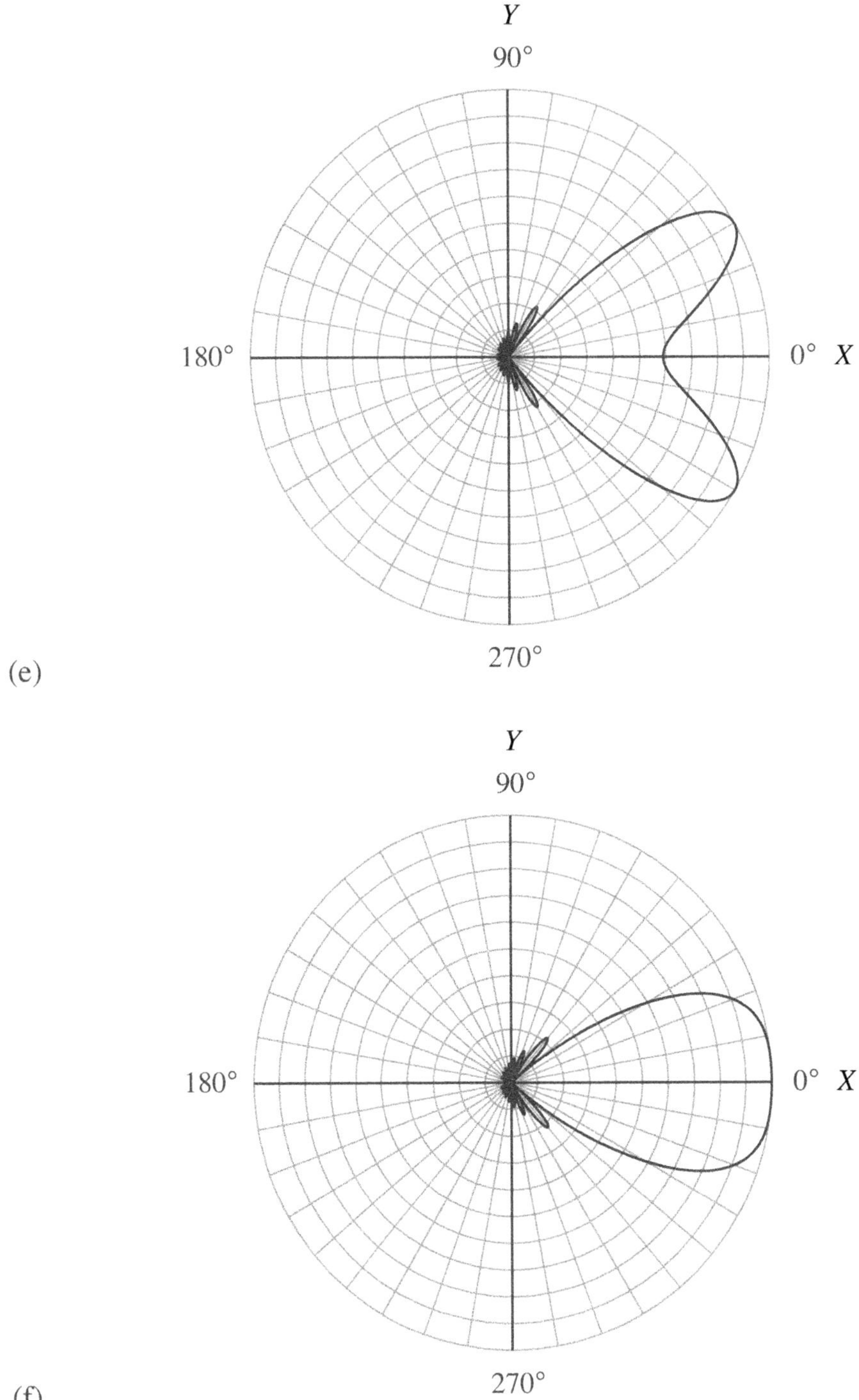

Figure 2.5-4 *continued.*

2.6 The Near-Field Beam Pattern of a Linear Aperture

In this section we shall demonstrate that the near-field beam pattern of a linear aperture is just a special case of the near-field beam pattern of a volume aperture. In Section 1.2 it was shown that the near-field beam pattern (directivity function) of a closed-surface, volume aperture – based on the Fresnel approximation of a time-independent, free-space, Green's function – is given by the following three-dimensional spatial Fresnel transform:

$$\begin{aligned}\mathcal{D}(f,r,\boldsymbol{\alpha}) &= F_{\mathbf{r}_A}\left\{A(f,\mathbf{r}_A)\exp\left(-j\frac{k}{2r}r_A^2\right)\right\} \\ &= \int_{-\infty}^{\infty} A(f,\mathbf{r}_A)\exp\left(-j\frac{k}{2r}r_A^2\right)\exp(+j2\pi\boldsymbol{\alpha}\bullet\mathbf{r}_A)\,d\mathbf{r}_A,\end{aligned} \tag{2.6-1}$$

where

$$\mathbf{r}_A = (x_A, y_A, z_A), \tag{2.6-2}$$

$$k = 2\pi f/c = 2\pi/\lambda, \tag{2.6-3}$$

$$r_A^2 = |\mathbf{r}_A|^2 = x_A^2 + y_A^2 + z_A^2, \tag{2.6-4}$$

$$r = |\mathbf{r}| = \sqrt{x^2+y^2+z^2}, \tag{2.6-5}$$

$$\boldsymbol{\alpha} = (f_X, f_Y, f_Z), \tag{2.6-6}$$

$$\boldsymbol{\alpha}\bullet\mathbf{r}_A = f_X x_A + f_Y y_A + f_Z z_A, \tag{2.6-7}$$

$$d\mathbf{r}_A = dx_A\,dy_A\,dz_A, \tag{2.6-8}$$

and $F_{\mathbf{r}_A}\{\bullet\}$ is shorthand notation for $F_{x_A}F_{y_A}F_{z_A}\{\bullet\}$. The complex aperture function $A(f,\mathbf{r}_A)$ can be expressed as

$$A(f,\mathbf{r}_A) = a(f,\mathbf{r}_A)\exp[+j\theta(f,\mathbf{r}_A)], \tag{2.6-9}$$

where $a(f,\mathbf{r}_A)$ is the *amplitude response* and $\theta(f,\mathbf{r}_A)$ is the *phase response* of the aperture at $\mathbf{r}_A$. Both $a(f,\mathbf{r}_A)$ and $\theta(f,\mathbf{r}_A)$ are *real* functions. The function $a(f,\mathbf{r}_A)$ is also known as the *amplitude window*. The units of $a(f,\mathbf{r}_A)$ and, hence, $A(f,\mathbf{r}_A)$ depend on whether the aperture is a transmit aperture (see Table 1B-1 in Appendix 1B) or a receive aperture (see Table 1B-2 in Appendix 1B).

The units of $\theta(f, \mathbf{r}_A)$ are radians.

Now consider the case of a *linear aperture* of length L meters lying along the X axis as shown in Fig. 2.1-1. As was mentioned earlier, a linear aperture can represent either a single electroacoustic transducer (e.g., a continuous line source or receiver) or a linear array of many individual electroacoustic transducers. Also shown in Fig. 2.1-1 is a field point with spherical coordinates (r, θ, ψ). The field point is in the Fresnel region of the aperture, where the Fresnel angle criterion given by either (1.2-35) or (1.2-91) is satisfied, and the range r satisfies the Fresnel range criterion given by $1.356R_A < r < \pi R_A^2/\lambda$ [see either (1.2-40) or (1.2-92)], where $R_A = L/2$ is the maximum radial extent of the linear aperture. Some typical values for the range to the near-field/far-field boundary $r_{\text{NF/FF}} = \pi R_A^2/\lambda$ for three different frequencies f and two different lengths L for a *linear towed array* (an example of a linear aperture) are shown in Table 2.6-1, along with values for $r_{\min} = 1.356R_A$.

Table 2.6-1 Values for the Range to the Near-Field/Far-Field Boundary $r_{\text{NF/FF}}$ for a *Linear Towed Array* (an Example of a Linear Aperture)

c (m/sec)	f (Hz)	λ (m)	L (m)	R_A (m)	$r_{\min}$ (m)	$r_{\text{NF/FF}}$ (km)
1500	60	25.0	100	50	67.8	0.314
1500	100	15.0	100	50	67.8	0.524
1500	1000	1.5	100	50	67.8	5.236
1500	60	25.0	200	100	135.6	1.257
1500	100	15.0	200	100	135.6	2.094
1500	1000	1.5	200	100	135.6	20.944

Since the linear aperture in Fig. 2.1-1 is lying along the X axis, the position vector to a point on the surface of the aperture is given by

$$\mathbf{r}_A = (x_A, 0, 0), \tag{2.6-10}$$

and as a result,

$$A(f, \mathbf{r}_A) = A(f, x_A, y_A, z_A) \rightarrow A(f, x_A)\delta(y_A)\delta(z_A). \tag{2.6-11}$$

Therefore, by substituting (2.6-11) into (2.6-1) and using the sifting property of impulse functions, we obtain the following one-dimensional spatial Fresnel transform for the near-field beam pattern of a linear aperture lying along the X

axis:

$$\boxed{\begin{aligned}\mathcal{D}(f,r,f_X) &= F_{x_A}\left\{A(f,x_A)\exp\left(-j\frac{k}{2r}x_A^2\right)\right\} \\ &= \int_{-L/2}^{L/2} A(f,x_A)\exp\left(-j\frac{k}{2r}x_A^2\right)\exp(+j2\pi f_X x_A)\,dx_A\end{aligned}} \tag{2.6-12}$$

where

$$A(f,x_A) = a(f,x_A)\exp[+j\theta(f,x_A)] \tag{2.6-13}$$

is the complex frequency response (complex aperture function) of the linear aperture,

$$k = 2\pi f/c = 2\pi/\lambda \tag{2.6-14}$$

is the wavenumber in radians per meter, and

$$f_X = u/\lambda = \sin\theta\cos\psi/\lambda \tag{2.6-15}$$

is a spatial frequency in the X direction with units of cycles per meter [see (1.2-43)]. Since f_X can be expressed in terms of the spherical angles θ and ψ, the near-field beam pattern can ultimately be expressed as a function of frequency f, the range r to a field point, and the spherical angles θ and ψ, that is, $\mathcal{D}(f,r,f_X) \to \mathcal{D}(f,r,\theta,\psi)$. If the linear aperture is a transmit aperture, then the aperture function $A(f,x_A)$ has units of $\left(\left(\mathrm{m}^3/\mathrm{sec}\right)/\mathrm{V}\right)/\mathrm{m}$ and the near-field beam pattern $\mathcal{D}(f,r,f_X)$ has units of $\left(\mathrm{m}^3/\mathrm{sec}\right)/\mathrm{V}$. However, if the linear aperture is a receive aperture, then $A(f,x_A)$ has units of $\left(\mathrm{V}/\left(\mathrm{m}^2/\mathrm{sec}\right)\right)/\mathrm{m}$ and $\mathcal{D}(f,r,f_X)$ has units of $\mathrm{V}/\left(\mathrm{m}^2/\mathrm{sec}\right)$. These units are summarized in Table 2.6-2. Equations (2.6-1) and (2.6-12) are most accurate in the Fresnel region of an aperture where both the Fresnel angle criterion and the Fresnel range criterion are satisfied. They are less accurate in the near-field region outside the Fresnel region where $r < r_{\mathrm{NF/FF}}$, but the Fresnel angle criterion is *not* satisfied. As was discussed in Subsection 1.2.1, the Fresnel region is only a subset of the near-field.

Finally, if a linear aperture lies along either the Y or Z axis instead of the X axis, then simply replace x_A and f_X with either y_A and f_Y, or z_A and f_Z, respectively, in (2.6-12), where spatial frequencies f_Y and f_Z are given by (1.2-44) and (1.2-45), respectively.

Table 2.6-2 Units of the Complex Aperture Function $A(f, x_A)$ and Corresponding Near-Field Beam Pattern $\mathcal{D}(f, r, f_X)$ for a Linear Aperture

Linear Aperture	$A(f, x_A)$	$\mathcal{D}(f, r, f_X)$
transmit	$\left(\left(\text{m}^3/\text{sec}\right)/\text{V}\right)/\text{m}$	$\left(\text{m}^3/\text{sec}\right)/\text{V}$
receive	$\left(\text{V}/\left(\text{m}^2/\text{sec}\right)\right)/\text{m}$	$\text{V}/\left(\text{m}^2/\text{sec}\right)$

2.6.1 Aperture Focusing

In order to discuss the concept of *aperture focusing*, let us investigate what happens to the near-field beam pattern given by (2.6-12) when the phase response $\theta(f, x_A)$ along the length of the aperture is a quadratic function of x_A, that is [see (2.4-1)],

$$\theta(f, x_A) = \theta_2(f) x_A^2, \tag{2.6-16}$$

where

$$\theta_2(f) = \frac{k}{2r'} \tag{2.6-17}$$

and

$$k = 2\pi f / c = 2\pi / \lambda . \tag{2.6-18}$$

The quadratic phase response given by (2.6-16) is an *even* function of x_A and is a *parabola* that passes through the origin. The parameter r' in (2.6-17) is referred to as the *focal range*. It is the near-field range from the aperture where the far-field beam pattern of the aperture will be in focus. If (2.6-13), (2.6-16), and (2.6-17) are substituted into (2.6-12), then

$$\mathcal{D}(f, r, f_X) = \int_{-L/2}^{L/2} a(f, x_A) \exp\left[-j\frac{k}{2}\left(\frac{1}{r} - \frac{1}{r'}\right) x_A^2\right] \exp(+j2\pi f_X x_A) dx_A , \tag{2.6-19}$$

and if (2.6-19) is evaluated at the near-field range $r = r'$, then it reduces to

$$\mathcal{D}(f, r', f_X) = \int_{-L/2}^{L/2} a(f, x_A) \exp(+j2\pi f_X x_A) dx_A , \tag{2.6-20}$$

or

$$\mathcal{D}(f, r', f_X) = F_{x_A}\{a(f, x_A)\} = D(f, f_X), \tag{2.6-21}$$

where $D(f, f_X)$ is the *far-field* beam pattern of the amplitude response (amplitude window) $a(f, x_A)$. Therefore, (2.6-21) indicates that the far-field beam pattern of the amplitude window can be *focused* at the near-field range $r = r'$ meters from the aperture if the phase response $\theta(f, x_A)$ along the length of the aperture is given by (2.6-16) through (2.6-18). The quadratic phase response accurately compensates for wavefront curvature in the Fresnel region of the aperture. The physical situation represented by (2.6-21) is known as *aperture focusing*. However, in the near-field region of the aperture outside the Fresnel region, a quadratic phase response can only approximately focus the far-field beam pattern.

A good way to think about aperture focusing is to compare it with taking a photograph. If the subject is far enough away (in the far-field), we do not concern ourselves with the actual range to the subject – we simply set the focus on the camera to infinity and take the picture. However, if the subject is nearby (in the near-field), we do concern ourselves with the actual range to the subject since we must refocus the camera whenever the subject moves.

2.6.2 Beam Steering and Aperture Focusing

Let us investigate what happens to the near-field beam pattern given by (2.6-12) when the phase response $\theta(f, x_A)$ along the length of the aperture is equal to the sum of a linear and quadratic function of x_A, that is [see (2.4-1)],

$$\theta(f, x_A) = \theta_1(f)x_A + \theta_2(f)x_A^2, \tag{2.6-22}$$

where

$$\theta_1(f) = -2\pi f_X', \tag{2.6-23}$$

$$\theta_2(f) = \frac{k}{2r'}, \tag{2.6-24}$$

$$f_X' = u'/\lambda = \sin\theta' \cos\psi'/\lambda, \tag{2.6-25}$$

and

$$k = 2\pi f/c = 2\pi/\lambda. \tag{2.6-26}$$

If (2.6-13) and (2.6-22) through (2.6-24) are substituted into (2.6-12), then

$$\mathcal{D}(f, r, f_X) = \int_{-L/2}^{L/2} a(f, x_A)\exp\left[-j\frac{k}{2}\left(\frac{1}{r} - \frac{1}{r'}\right)x_A^2\right]\exp\left[+j2\pi(f_X - f_X')x_A\right]dx_A, \tag{2.6-27}$$

and if (2.6-27) is evaluated at the near-field range $r = r'$, then it reduces to

$$\mathcal{D}(f, r', f_X) = \int_{-L/2}^{L/2} a(f, x_A) \exp[+j2\pi(f_X - f'_X)x_A] dx_A , \qquad (2.6\text{-}28)$$

or

$$\mathcal{D}(f, r', f_X) = F_{x_A}\{a(f, x_A)\exp(-j2\pi f'_X x_A)\} = D(f, f_X - f'_X) . \qquad (2.6\text{-}29)$$

Equation (2.6-29) indicates that the far-field beam pattern $D(f, f_X)$ of the amplitude response (amplitude window) $a(f, x_A)$ is *in focus* at the near-field range $r = r'$ meters from the aperture, and has been *steered* in the direction $u = u'$ in direction-cosine space, which is equivalent to steering (tilting) the beam pattern to $\theta = \theta'$ and $\psi = \psi'$ [see (2.6-25)].

Problems

Section 2.2

2-1 Show that (2.2-4) is equal to (2.2-5).

2-2 Verify (2.2-9).

2-3 Derive the *normalized*, far-field beam patterns of the following aperture functions, expressing your answers in terms of the spherical angles θ and ψ:

(a) $A(f, y_A) = A_A \text{rect}(y_A / L)$

(b) $A(f, z_A) = A_A \text{rect}(z_A / L)$

2-4 The aperture function of a linear aperture lying along the X axis is given by

$$A(f, x_A) = a(f, x_A) = \begin{cases} -A_A, & -L/2 \le x_A < 0 \\ A_A, & 0 \le x_A \le L/2. \end{cases}$$

Find the unnormalized, far-field beam pattern of the aperture. This amplitude response is an example of an *odd* function of x_A.

2-5 Derive (2.2-20) by evaluating the spatial-domain, Fourier integral.

2-6 Show that

$$F_{f_X}^{-1}\{\delta(f_X \pm f'_X)\} = \exp(\pm j2\pi f'_X x_A).$$

From this result we can write that [see (2.2-26)]

$$F_{x_A}\{\exp(\pm j2\pi f'_X x_A)\} = \delta(f_X \pm f'_X),$$

and if $f'_X = 0$, then $F_{x_A}\{1\} = \delta(f_X)$ and $F_{f_X}^{-1}\{\delta(f_X)\} = 1$.

2-7 Show that

$$g(x) \underset{x}{*} \delta(x \pm x_0) = g(x \pm x_0),$$

where the asterisk denotes convolution with respect to x, and x_0 is an arbitrary constant.

2-8 The complex frequency response (complex aperture function) of a linear aperture lying along the Y axis is modeled by the cosine amplitude window. Evaluate the *normalized*, far-field beam pattern of the aperture at $\theta = 40°$ and $\psi = 108°$. Use $L/\lambda = 2$.

2-9 If the aperture function of a linear aperture lying along the X axis is modeled by the *sine* amplitude window, that is, if

$$A(f, x_A) = a(f, x_A) = A_A \sin(\pi x_A / L)\text{rect}(x_A / L),$$

then find the unnormalized, far-field beam pattern of the aperture. The sine amplitude window is an example of an *odd* function of x_A.

2-10 Derive the *normalized*, far-field beam patterns of the Hanning, Hamming, and Blackman amplitude windows. Use the following trigonometric identity: $\sin(\alpha \pm \pi) = -\sin\alpha$.

Section 2.2 Appendix 2A

2-11 Find the transmitter sensitivity functions for the following complex frequency responses of a continuous line source lying along the X axis where $A_A(f)$ is a real, nonnegative function of frequency:

(a) rectangular amplitude window: $A_T(f, x_A) = A_A(f)\text{rect}(x_A / L)$

(b) cosine amplitude window: $A_T(f, x_A) = A_A(f)\cos(\pi x_A / L)\text{rect}(x_A / L)$

2-12 If the transmitter sensitivity level (transmitting voltage response) of a continuous line source at 16 kHz is 140 dB re 1 μPa/V at 1 m, then what is the value of the magnitude of the transmitter sensitivity function in $\left(\text{m}^3/\text{sec}\right)/\text{V}$? Use $\rho_0 = 1026 \text{ kg}/\text{m}^3$ for the ambient density of seawater. The length L of the continuous line source is less than 0.632 m or 24.9 in.

2-13 If the receiver sensitivity level (open circuit receiving response) of a continuous line receiver at 2 kHz is −195 dB re 1 V/μPa, then what is the value of the magnitude of the receiver sensitivity function in $\text{V}/\left(\text{m}^2/\text{sec}\right)$? Use $\rho_0 = 1026 \text{ kg}/\text{m}^3$ for the ambient density of seawater.

Section 2.3

2-14 Compute the 3-dB beamwidths of the horizontal, far-field beam patterns of the rectangular, triangular, cosine, Hanning, Hamming, and Blackman amplitude windows for $L/\lambda = 0.5$, 1, 2, and 4.

2-15 Consider a *linear towed array* (an example of a linear aperture) lying along the X axis. The amplitude response of the aperture is modeled by the rectangular amplitude window. How long must the aperture be in order for the 3-dB beamwidth of the horizontal, far-field beam pattern to be 10° at broadside for $f = 100$ Hz? Use $c = 1500$ m/sec.

Section 2.4

2-16 Consider a linear aperture lying along the X axis. If $f = 1$ kHz and $c = 1500$ m/sec, then

(a) what must the phase response of the aperture be in order to steer the mainlobe of the aperture's far-field beam pattern to $\theta' = 49°$ and $\psi' = 71°$.

(b) Repeat (a) for a linear aperture lying along the Y axis.

(c) Repeat (a) for a linear aperture lying along the Z axis.

Section 2.5

2-17 Compute the 3-dB beamwidths of the horizontal, far-field beam patterns of the rectangular, triangular, cosine, Hanning, Hamming, and Blackman amplitude windows for the following beam-steer angles: $\psi' = 90°$

(broadside), $75°$, $60°$, $45°$, $30°$, and $0°$ (end-fire). Use $L/\lambda = 4$.

2-18 The complex aperture function of a *linear towed array* (an example of a linear aperture) lying along the X axis is modeled by the rectangular amplitude window. The length of the aperture is 200 m. Using $f = 60$ Hz and $c = 1500$ m/sec, compute the 3-dB beamwidth of the horizontal, far-field beam pattern of this aperture at

(a) broadside

(b) a beam-steer angle of $30°$

(c) Compute *both* half-beamwidth angles at a beam-steer angle of $30°$.

(d) Repeat (a) through (c) for $f = 100$ Hz.

2-19 Consider a linear aperture lying along the X axis.

(a) Show that the 3-dB beamwidth of the vertical, far-field beam pattern in the XZ plane is given by

$$\Delta\theta = \sin^{-1}\left(\sin\theta' + \frac{\Delta u}{2}\right) - \sin^{-1}\left(\sin\theta' - \frac{\Delta u}{2}\right), \qquad \left|\sin\theta' \pm \frac{\Delta u}{2}\right| \le 1,$$

where θ' is the beam-steer angle in the XZ plane. Note that $\theta' = 0°$ corresponds to an unsteered vertical beam pattern and, in this case, $\Delta\theta$ reduces to (2.3-19) since

$$\sin^{-1}(-x) = -\sin^{-1}(x).$$

Hint: follow a procedure analogous to the one used to derive (2.5-12).

(b) Show that the 3-dB beamwidth of the vertical, far-field beam pattern in the XZ plane steered to end-fire is given by

$$\Delta\theta = 2\cos^{-1}\left(1 - \frac{\Delta u}{2}\right), \qquad \left|1 - \frac{\Delta u}{2}\right| \le 1.$$

Hint: follow a procedure analogous to the one used to derive (2.5-19).

Section 2.6

2-20 The spherical coordinates of a sound source (target) as measured from the center of a *linear towed array* (an example of a linear aperture) lying along the X axis are $r_S = 4$ km, $\theta_S = 100°$, and $\psi_S = 75°$. The length of the aperture is 100 m. If $f = 1$ kHz and $c = 1500$ m/sec,

(a) is the sound source in the aperture's Fresnel region or far-field?

(b) What must the phase response along the length of the aperture be if the aperture's far-field beam pattern is to be focused (if required) and steered to coordinates (r_S, θ_S, ψ_S)?

(c) If the range to the sound source r_S and the vertical angle θ_S remain the same, but the bearing angle changes to $\psi_S = 65°$, is the sound source in the aperture's Fresnel region or far-field?

Appendix 2A Transmitter and Receiver Sensitivity Functions of a Continuous Line Transducer

Transmitter Sensitivity Function

For the case of a continuous line source lying along the X axis, the complex, *transmitter sensitivity function* $\mathcal{S}_T(f)$ is related to the complex, transmit frequency response $A_T(f, x_A)$ of the transducer as follows (see Example 2B-1 in Appendix 2B):

$$\boxed{\mathcal{S}_T(f) = \int_{-\infty}^{\infty} A_T(f, x_A)\,dx_A} \tag{2A-1}$$

where $\mathcal{S}_T(f)$ has units of $\left(\text{m}^3/\text{sec}\right)/\text{V}$ and $A_T(f, x_A)$ has units of $\left(\left(\text{m}^3/\text{sec}\right)/\text{V}\right)/\text{m}$ (see Table 2.1-1). However, manufacturers of electroacoustic transducers usually describe the *transmitting voltage response* of a transducer by plotting the *transmitter sensitivity level* $\text{TSL}(f)$ in dB re TS_{ref} at 1 m versus frequency f in hertz. The notation "dB re TS_{ref}" means decibels relative to a reference transmitter sensitivity. In underwater acoustics, the reference transmitter sensitivity TS_{ref} is usually $1\,\mu\text{Pa/V}$. The corresponding *transmitter sensitivity* $\text{TS}(f)$ with units of Pa/V is given by

$$\boxed{\text{TS}(f) = \text{TS}_{\text{ref}}\,10^{[\text{TSL}(f)/20]}} \tag{2A-2}$$

The transmitter sensitivity $\text{TS}(f)$ is a *real*, *nonnegative* function of frequency.

Values for the magnitude of the transmitter sensitivity function, $|\mathcal{S}_T(f)|$ in $(\mathrm{m}^3/\mathrm{sec})/\mathrm{V}$, for a continuous line source of length L meters lying along the X axis and centered at the origin of the coordinate system as shown in Fig. 2.1-1, can be obtained from measured values of the transmitter sensitivity, $\mathrm{TS}(f)$ in $\mathrm{Pa/V}$, by using the following formula (see Example 2B-1 in Appendix 2B):

$$|\mathcal{S}_T(f)| = \frac{2r}{f\rho_0}\mathrm{TS}(f)\Big|_{r=1\,\mathrm{m}}, \qquad L < 0.632\ \mathrm{m} \tag{2A-3}$$

where

$$r = \sqrt{y^2 + z^2} \tag{2A-4}$$

is the distance from the center of the continuous line source to a field point at *broadside* relative to the continuous line source, ρ_0 is the constant ambient (equilibrium) density of the fluid medium in kilograms per cubic meter, and $\mathrm{TS}(f)$ is given by (2A-2). Note that (2A-3) is valid only if the length L of the continuous line source is less than 0.632 m or 24.9 in.

Receiver Sensitivity Function

For the case of a continuous line receiver lying along the X axis, the complex, *receiver sensitivity function* $\mathcal{S}_R(f)$ is related to the complex, receive frequency response $A_R(f, x_A)$ of the transducer as follows:

$$\mathcal{S}_R(f) = \int_{-\infty}^{\infty} A_R(f, x_A)\,dx_A \tag{2A-5}$$

where $\mathcal{S}_R(f)$ has units of $\mathrm{V}/(\mathrm{m}^2/\mathrm{sec})$ and $A_R(f, x_A)$ has units of $(\mathrm{V}/(\mathrm{m}^2/\mathrm{sec}))/\mathrm{m}$ (see Table 2.1-1). However, manufacturers of electroacoustic transducers usually describe the *open circuit receiving response* of a transducer by plotting the *receiver sensitivity level* $\mathrm{RSL}(f)$ in $\mathrm{dB\ re\ RS_{ref}}$ versus frequency f in hertz. The notation "$\mathrm{dB\ re\ RS_{ref}}$" means decibels relative to a reference receiver sensitivity. In underwater acoustics, the reference receiver sensitivity $\mathrm{RS_{ref}}$ is usually $1\,\mathrm{V}/\mu\mathrm{Pa}$. The corresponding *receiver sensitivity* $\mathrm{RS}(f)$ with units of $\mathrm{V/Pa}$ is given by

$$\boxed{\text{RS}(f) = \text{RS}_{\text{ref}}\, 10^{[\text{RSL}(f)/20]}} \tag{2A-6}$$

The receiver sensitivity $\text{RS}(f)$ is a *real*, *nonnegative* function of frequency. What we need to do next is to derive the conversion factor that will allow us to convert m^2/sec to Pa for a time-harmonic acoustic field.

The derivation of the desired conversion factor in this case is simple. By computing the magnitude of (2B-24), we obtain

$$\left|p_f(\mathbf{r})\right| = 2\pi f \rho_0 \left|\varphi_f(\mathbf{r})\right|, \tag{2A-7}$$

from which we obtain the conversion factor

$$\boxed{\frac{\left|p_f(\mathbf{r})\right|}{\left|\varphi_f(\mathbf{r})\right|} = 2\pi f \rho_0 \, \frac{\text{Pa}}{\text{m}^2/\text{sec}}} \tag{2A-8}$$

where $p_f(\mathbf{r}) \equiv p_f(x, y, z)$ and $\varphi_f(\mathbf{r}) \equiv \varphi_f(x, y, z)$. The conversion factor on the right-hand-side of (2A-8) is used to convert velocity potential in m^2/sec to acoustic pressure in Pa for a time-harmonic acoustic field. Therefore, values for the magnitude of the receiver sensitivity function, $\left|\mathcal{S}_R(f)\right|$ in $\text{V}/\left(\text{m}^2/\text{sec}\right)$, can be obtained from measured values of the receiver sensitivity, $\text{RS}(f)$ in V/Pa, by using the following formula:

$$\boxed{\left|\mathcal{S}_R(f)\right| = 2\pi f \rho_0 \text{RS}(f)} \tag{2A-9}$$

where ρ_0 is the constant ambient (equilibrium) density of the fluid medium in kilograms per cubic meter and $\text{RS}(f)$ is given by (2A-6).

Appendix 2B Radiation from a Linear Aperture

An exact solution of the linear wave equation given by (1.2-1) for free-space propagation in an ideal (nonviscous), homogeneous, fluid medium is given by [see (1.2-3)]

$$\varphi(t, \mathbf{r}) = \int_{-\infty}^{\infty} \int_{V_0} X(f, \mathbf{r}_0) A_T(f, \mathbf{r}_0) g_f\!\left(\mathbf{r}\,\middle|\,\mathbf{r}_0\right) dV_0 \exp(+j2\pi f t)\, df\,, \tag{2B-1}$$

where $\varphi(t, \mathbf{r})$ is the scalar velocity potential in squared meters per second,

$X(f,\mathbf{r}_0)$ is the complex frequency spectrum of the input electrical signal at location $\mathbf{r}_0$ of the transmit aperture with units of volts per hertz, $A_T(f,\mathbf{r}_0)$ is the complex frequency response of the transmit aperture at $\mathbf{r}_0$ with units of $\left(\left(\mathrm{m}^3/\mathrm{sec}\right)/\mathrm{V}\right)/\mathrm{m}^3$,

$$g_f\left(\mathbf{r}\,|\,\mathbf{r}_0\right) = -\frac{\exp\left(-jk\left|\mathbf{r}-\mathbf{r}_0\right|\right)}{4\pi\left|\mathbf{r}-\mathbf{r}_0\right|} = -\frac{\exp(-jkR)}{4\pi R} \tag{2B-2}$$

is the time-independent, free-space, Green's function of an unbounded, ideal (nonviscous), homogeneous, fluid medium with units of inverse meters,

$$k = 2\pi f/c = 2\pi/\lambda \tag{2B-3}$$

is the wavenumber in radians per meter, $c = f\lambda$ is the constant speed of sound in the fluid medium in meters per second, λ is the wavelength in meters,

$$\mathbf{r} = x\hat{x} + y\hat{y} + z\hat{z} \tag{2B-4}$$

is the position vector to a field point,

$$\mathbf{r}_0 = x_0\hat{x} + y_0\hat{y} + z_0\hat{z} \tag{2B-5}$$

is the position vector to a source point, and

$$R = \left|\mathbf{r}-\mathbf{r}_0\right| = \sqrt{(x-x_0)^2 + (y-y_0)^2 + (z-z_0)^2} \tag{2B-6}$$

is the range in meters between a source point and a field point. If an identical input electrical signal is applied at all locations $\mathbf{r}_0$ of the transmit aperture, then (see Example 1.2-1)

$$x(t,\mathbf{r}_0) = x(t) \tag{2B-7}$$

and

$$X(f,\mathbf{r}_0) = X(f), \tag{2B-8}$$

where $X(f)$ is the complex frequency spectrum of $x(t)$. Substituting (2B-8) into (2B-1) yields

$$\varphi(t,\mathbf{r}) = \int_{-\infty}^{\infty} X(f) \int_{V_0} A_T(f,\mathbf{r}_0)\, g_f\left(\mathbf{r}\,|\,\mathbf{r}_0\right) dV_0 \exp(+j2\pi f t)\, df. \tag{2B-9}$$

For the case of a linear aperture lying along the X axis,

$$\mathbf{r}_0 = (x_0, 0, 0), \tag{2B-10}$$

and as a result,

$$A_T(f, \mathbf{r}_0) = A_T(f, x_0, y_0, z_0) \rightarrow A_T(f, x_0)\delta(y_0)\delta(z_0). \tag{2B-11}$$

Substituting (2B-11) into (2B-9) yields

$$\boxed{\varphi(t, \mathbf{r}) = \int_{-\infty}^{\infty} X(f) \int_{-\infty}^{\infty} A_T(f, x_0) g_f\left(\mathbf{r} \middle| \mathbf{r}_0\right) dx_0 \exp(+j2\pi f t) df} \tag{2B-12}$$

where $g_f\left(\mathbf{r} \middle| \mathbf{r}_0\right)$ is given by (2B-2) and

$$R = \left|\mathbf{r} - \mathbf{r}_0\right| = \sqrt{(x - x_0)^2 + y^2 + z^2}\,. \tag{2B-13}$$

Equation (2B-12) is a general expression for the scalar velocity potential of the acoustic field radiated by a linear aperture lying along the X axis. The corresponding solution for the radiated acoustic pressure $p(t, \mathbf{r})$ in pascals can be obtained by substituting (2B-12) into (1.2-8). Doing so yields

$$\boxed{p(t, \mathbf{r}) = -j2\pi\rho_0 \int_{-\infty}^{\infty} f X(f) \int_{-\infty}^{\infty} A_T(f, x_0) g_f\left(\mathbf{r} \middle| \mathbf{r}_0\right) dx_0 \exp(+j2\pi f t) df} \tag{2B-14}$$

where ρ_0 is the constant ambient (equilibrium) density of the fluid medium in kilograms per cubic meter, and $g_f\left(\mathbf{r} \middle| \mathbf{r}_0\right)$ is given by (2B-2) and (2B-13).

If the input electrical signal is time-harmonic, that is, if (see Example 1.2-1)

$$x(t) = A_x \exp(+j2\pi f_0 t), \tag{2B-15}$$

where A_x is the complex amplitude with units of volts, then

$$X(f) = A_x \delta(f - f_0), \tag{2B-16}$$

where the impulse function $\delta(f - f_0)$ has units of inverse hertz. Substituting (2B-16) into (2B-12) yields

$$\varphi(t, \mathbf{r}) = A_x \int_{-\infty}^{\infty} A_T(f_0, x_0) g_{f0}\left(\mathbf{r} \middle| \mathbf{r}_0\right) dx_0 \exp(+j2\pi f_0 t), \tag{2B-17}$$

and by replacing f_0 with f,

$$\boxed{\varphi(t,\mathbf{r}) = A_x \int_{-\infty}^{\infty} A_T(f,x_0)\, g_f\left(\mathbf{r} \,\middle|\, \mathbf{r}_0\right) dx_0 \exp(+j2\pi f t)} \tag{2B-18}$$

or

$$\varphi(t,\mathbf{r}) = \varphi_f(\mathbf{r}) \exp(+j2\pi f t), \tag{2B-19}$$

where

$$\varphi_f(\mathbf{r}) = A_x \int_{-\infty}^{\infty} A_T(f,x_0)\, g_f\left(\mathbf{r} \,\middle|\, \mathbf{r}_0\right) dx_0 \tag{2B-20}$$

and $g_f\left(\mathbf{r} \,\middle|\, \mathbf{r}_0\right)$ is given by (2B-2) and (2B-13). Equation (2B-18) is the time-harmonic, scalar velocity potential of the acoustic field radiated by a linear aperture lying along the X axis. The corresponding solution for the time-harmonic, radiated acoustic pressure $p(t,\mathbf{r})$ in pascals is given by

$$\boxed{p(t,\mathbf{r}) = -j2\pi f \rho_0 A_x \int_{-\infty}^{\infty} A_T(f,x_0)\, g_f\left(\mathbf{r} \,\middle|\, \mathbf{r}_0\right) dx_0 \exp(+j2\pi f t)} \tag{2B-21}$$

or

$$p(t,\mathbf{r}) = p_f(\mathbf{r}) \exp(+j2\pi f t), \tag{2B-22}$$

where

$$p_f(\mathbf{r}) = -j2\pi f \rho_0 A_x \int_{-\infty}^{\infty} A_T(f,x_0)\, g_f\left(\mathbf{r} \,\middle|\, \mathbf{r}_0\right) dx_0 \tag{2B-23}$$

and $g_f\left(\mathbf{r} \,\middle|\, \mathbf{r}_0\right)$ is given by (2B-2) and (2B-13). Note that

$$p_f(\mathbf{r}) = -j2\pi f \rho_0 \varphi_f(\mathbf{r}), \tag{2B-24}$$

where $\varphi_f(\mathbf{r})$ is given by (2B-20).

Example 2B-1 Transmitter Sensitivity Function and Source Strength of a Continuous Line Source

Consider the case of a continuous line source (an example of a linear aperture) lying along the X axis and centered at the origin of the coordinate system as shown in Fig. 2.1-1. The transducer is L meters in length and extends from $-L/2$ to $L/2$ along the X axis. Therefore,

$$\int_{-\infty}^{\infty} A_T(f, x_0) g_f(\mathbf{r}|\mathbf{r}_0) dx_0 = \int_{-L/2}^{L/2} A_T(f, x_0) g_f(\mathbf{r}|\mathbf{r}_0) dx_0 , \tag{2B-25}$$

where $g_f(\mathbf{r}|\mathbf{r}_0)$ is given by (2B-2) and (2B-13). If the radiated acoustic field is measured at *broadside* at $\mathbf{r} = (0, y, z)$, then (2B-13) reduces to

$$R = |\mathbf{r} - \mathbf{r}_0| = r\sqrt{1 + (x_0/r)^2} , \tag{2B-26}$$

where

$$r = \sqrt{y^2 + z^2} \tag{2B-27}$$

is the range in meters to a field point. Note that $r \neq 0$. Setting x_0 equal to either the upper or lower limit of integration $\pm L/2$ in (2B-26) yields

$$R = |\mathbf{r} - \mathbf{r}_0| = r\sqrt{1 + \left(\frac{L}{2r}\right)^2} , \tag{2B-28}$$

and if

$$\left(\frac{L}{2r}\right)^2 < 0.1 , \tag{2B-29}$$

or

$$L < 0.632\, r , \tag{2B-30}$$

then

$$R = |\mathbf{r} - \mathbf{r}_0| \approx r, \qquad L < 0.632\, r , \tag{2B-31}$$

for $\mathbf{r} = (0, y, z)$ and $\mathbf{r}_0 = (x_0, 0, 0)$, where $|x_0| \leq L/2$. Therefore, substituting (2B-31) into (2B-2) yields

$$g_f(\mathbf{r}|\mathbf{r}_0) = g_f(0, y, z | x_0, 0, 0) \approx -\frac{\exp(-jkr)}{4\pi r}, \qquad L < 0.632\, r , \tag{2B-32}$$

and substituting (2B-32) into (2B-25) yields

$$\int_{-\infty}^{\infty} A_T(f, x_0) g_f(\mathbf{r}|\mathbf{r}_0) dx_0 \approx -\frac{\exp(-jkr)}{4\pi r} \mathcal{S}_T(f), \qquad L < 0.632\, r , \tag{2B-33}$$

where

$$\boxed{\mathcal{S}_T(f) = \int_{-L/2}^{L/2} A_T(f, x_0)\,dx_0} \tag{2B-34}$$

is the *transmitter sensitivity function* in $\left(\text{m}^3/\text{sec}\right)/\text{V}$. Next we shall use (2B-33) to compute the time-harmonic, radiated acoustic pressure at $\mathbf{r} = (0, y, z)$.

The time-harmonic, radiated acoustic pressure at $\mathbf{r} = (0, y, z)$ is given by

$$p(t, 0, y, z) = p_f(0, y, z)\exp(+j2\pi ft), \tag{2B-35}$$

where, by substituting (2B-33) into (2B-23),

$$p_f(0, y, z) \approx j\frac{f\rho_0}{2r} A_x \mathcal{S}_T(f)\exp(-jkr), \qquad L < 0.632r, \tag{2B-36}$$

or

$$p_f(0, y, z) \approx j\frac{f\rho_0}{2r} S_0 \exp(-jkr), \qquad L < 0.632r, \tag{2B-37}$$

where

$$\boxed{S_0 = A_x \mathcal{S}_T(f)} \tag{2B-38}$$

is the source strength of the continuous line source in cubic meters per second at frequency f hertz.

From (2B-35) it can be seen that

$$|p(t, 0, y, z)| = |p_f(0, y, z)|, \tag{2B-39}$$

where from (2B-36)

$$|p_f(0, y, z)| \approx \frac{f\rho_0}{2r} |A_x| |\mathcal{S}_T(f)|, \qquad L < 0.632r. \tag{2B-40}$$

Therefore, using (2B-40) to solve for $|\mathcal{S}_T(f)|$ yields

$$\boxed{|\mathcal{S}_T(f)| \approx \frac{2r}{f\rho_0}\text{TS}(f), \qquad L < 0.632r} \tag{2B-41}$$

where

$$\boxed{\text{TS}(f) = |p_f(0, y, z)| / |A_x|} \tag{2B-42}$$

is the *transmitter sensitivity* in pascals per volt, and r is the range to a field point with coordinates $(0, y, z)$ and is given by (2B-27). Recall that the range r is the distance from the center of the continuous line source to a field point at *broadside* relative to the continuous line source. Evaluating (2B-41) at $r = 1\text{ m}$ yields

$$\boxed{\left|\mathcal{S}_T(f)\right| \approx \frac{2r}{f\rho_0}\mathrm{TS}(f)\Big|_{r=1\text{ m}}, \qquad L < 0.632\text{ m}} \tag{2B-43}$$

Note that (2B-43) is valid only if the length L of the continuous line source is less than 0.632 m or 24.9 in.

The magnitude of the source strength can be obtained from (2B-37) as follows:

$$\boxed{\left|S_0\right| \approx \frac{2r}{f\rho_0}\left|p_f(0, y, z)\right|, \qquad L < 0.632r} \tag{2B-44}$$

where r is the range to a field point with coordinates $(0, y, z)$ and is given by (2B-27). Evaluating (2B-44) at $r = 1\text{ m}$ yields

$$\boxed{\left|S_0\right| \approx \frac{2r}{f\rho_0}\left|p_f(0, y, z)\right|\Big|_{r=1\text{ m}}, \qquad L < 0.632\text{ m}} \tag{2B-45}$$

Note that (2B-45) is valid only if the length L of the continuous line source is less than 0.632 m or 24.9 in. Also, from (2B-38),

$$\boxed{\left|S_0\right| = \left|A_x\right|\left|\mathcal{S}_T(f)\right|} \tag{2B-46}$$

where the transmitter sensitivity function $\mathcal{S}_T(f)$ is given by (2B-34). ■

Appendix 2C Symmetry Properties and Far-Field Beam Patterns

The far-field beam pattern of a linear aperture lying along the X axis is given by [see (2.1-9)]

$$D(f, f_X) = \int_{-L/2}^{L/2} A(f, x_A)\exp(+j2\pi f_X x_A)\,dx_A\,, \tag{2C-1}$$

where

$$A(f, x_A) = a(f, x_A)\exp[+j\theta(f, x_A)] \tag{2C-2}$$

is the complex frequency response (complex aperture function) of the linear aperture [see (2.1-10)]. If the phase response is zero, that is, if

$$\theta(f, x_A) = 0\,, \tag{2C-3}$$

then

$$D(f, f_X) = \int_{-L/2}^{L/2} a(f, x_A)\exp(+j2\pi f_X x_A)dx_A \tag{2C-4}$$

or

$$D(f, f_X) = \int_{-L/2}^{L/2} a(f, x_A)\cos(2\pi f_X x_A)dx_A + j\int_{-L/2}^{L/2} a(f, x_A)\sin(2\pi f_X x_A)dx_A\,. \tag{2C-5}$$

If the amplitude response (amplitude window) $a(f, x_A)$ is an *even* function of x_A, that is, if

$$a(f, -x_A) = a(f, x_A), \tag{2C-6}$$

then the product $a(f, x_A)\cos(2\pi f_X x_A)$ is an even function of x_A, and the product $a(f, x_A)\sin(2\pi f_X x_A)$ is an odd function of x_A. Therefore, the far-field beam pattern given by (2C-5) reduces to

$$D(f, f_X) = 2\int_{0}^{L/2} a(f, x_A)\cos(2\pi f_X x_A)dx_A\,. \tag{2C-7}$$

Equation (2C-7) is a *real*, *even* function of spatial frequency $f_X = u/\lambda$ and, hence, direction cosine u, since $D(f, -f_X) = D(f, f_X)$ and $D(f, -u) = D(f, u)$.

If the amplitude response (amplitude window) $a(f, x_A)$ is an *odd* function of x_A, that is, if

$$a(f, -x_A) = -a(f, x_A), \tag{2C-8}$$

then the product $a(f, x_A)\cos(2\pi f_X x_A)$ is an odd function of x_A, and the product $a(f, x_A)\sin(2\pi f_X x_A)$ is an even function of x_A. Therefore, the far-field beam pattern given by (2C-5) reduces to

$$D(f, f_X) = j2\int_{0}^{L/2} a(f, x_A)\sin(2\pi f_X x_A)dx_A\,. \tag{2C-9}$$

Equation (2C-9) is an *imaginary*, *odd* function of spatial frequency $f_X = u/\lambda$ and, hence, direction cosine u, since $D(f, -f_X) = -D(f, f_X)$ and $D(f, -u) = -D(f, u)$.

Appendix 2D Computing the Normalization Factor

The far-field beam pattern of a linear aperture lying along the X axis is given by [see (2.1-9)]

$$D(f, f_X) = \int_{-L/2}^{L/2} A(f, x_A)\exp(+j2\pi f_X x_A)\,dx_A\,, \tag{2D-1}$$

where

$$A(f, x_A) = a(f, x_A)\exp[+j\theta(f, x_A)] \tag{2D-2}$$

is the complex frequency response (complex aperture function) of the linear aperture [see (2.1-10)]. Computing the magnitude of (2D-1) yields

$$\begin{aligned} |D(f, f_X)| &= \left|\int_{-L/2}^{L/2} A(f, x_A)\exp(+j2\pi f_X x_A)\,dx_A\right| \\ &\le \int_{-L/2}^{L/2} |A(f, x_A)\exp(+j2\pi f_X x_A)|\,dx_A \\ &\le \int_{-L/2}^{L/2} |A(f, x_A)||\exp(+j2\pi f_X x_A)|\,dx_A \\ &\le \int_{-L/2}^{L/2} |a(f, x_A)|\,dx_A. \end{aligned} \tag{2D-3}$$

Therefore,

$$\max|D(f, f_X)| = \int_{-L/2}^{L/2} |a(f, x_A)|\,dx_A\,, \tag{2D-4}$$

and by following the steps in (2D-3), it can also be shown that

$$\max|\mathcal{S}(f)| = \int_{-L/2}^{L/2} |a(f, x_A)|\,dx_A\,, \tag{2D-5}$$

where $\mathcal{S}(f)$ is the element sensitivity function (see Appendix 2A). The absolute value of the real amplitude response $a(f, x_A)$ is required in (2D-4) and (2D-5) because $a(f, x_A)$ can have, in general, both positive and negative values as a function of x_A.

Equation (2D-4) is the absolute maximum value of $|D(f, f_X)|$. The actual maximum value may be less than (2D-4) as indicated by (2D-3). However, if $a(f, x_A) \ge 0$, then the normalization factor $D_{\max}$ is

$$\boxed{D_{\max} = \max|D(f, f_X)| = \int_{-L/2}^{L/2} a(f, x_A)\,dx_A} \tag{2D-6}$$

or

$$\boxed{D_{\max} = \max\left|\boldsymbol{S}(f)\right| = \int_{-L/2}^{L/2} a(f, x_A)dx_A} \tag{2D-7}$$

Note that if (2D-1) is evaluated at broadside (i.e., at $f_X = 0$) and if the phase response $\theta(f, x_A) = 0$, then

$$D(f, 0) = \int_{-L/2}^{L/2} a(f, x_A)dx_A \tag{2D-8}$$

and

$$\left|D(f, 0)\right| = \left|\int_{-L/2}^{L/2} a(f, x_A)dx_A\right| \leq \int_{-L/2}^{L/2} \left|a(f, x_A)\right| dx_A \,. \tag{2D-9}$$

If in addition $a(f, x_A) \geq 0$, then

$$\left|D(f, 0)\right| = \left|\int_{-L/2}^{L/2} a(f, x_A)dx_A\right| = \int_{-L/2}^{L/2} a(f, x_A)dx_A \,, \tag{2D-10}$$

and by comparing (2D-6) and (2D-10),

$$\boxed{D_{\max} = \max\left|D(f, f_X)\right| = \left|D(f, 0)\right|} \tag{2D-11}$$

For example, the cosine amplitude window is given by [see (2.2-30)]

$$a(f, x_A) = A_A \cos(\pi x_A / L)\text{rect}(x_A / L) \,. \tag{2D-12}$$

Since $a(f, x_A) \geq 0$ (see Fig. 2.2-8),

$$\begin{aligned} D_{\max} &= A_A \int_{-L/2}^{L/2} \cos(\pi x_A / L) dx_A \\ &= A_A \frac{L}{\pi} \sin(\pi x_A / L)\Big|_{-L/2}^{L/2} \\ &= A_A \frac{L}{\pi}\left[\sin(\pi/2) - \sin(-\pi/2)\right] \\ &= 2A_A L/\pi \,. \end{aligned} \tag{2D-13}$$

Equation (2D-13) agrees with (2.2-34).

Appendix 2E Summary of One-Dimensional Spatial Fourier Transforms

Table 2E-1 Summary of One-Dimensional Spatial Fourier Transforms

$A(f,x_A)=F_{f_X}^{-1}\{D(f,f_X)\}$	$D(f,f_X)=F_{x_A}\{A(f,x_A)\}$
$=\int_{-\infty}^{\infty} D(f,f_X)\exp(-j2\pi f_X x_A)\,df_X$	$=\int_{-\infty}^{\infty} A(f,x_A)\exp(+j2\pi f_X x_A)\,dx_A$
$f_X=\dfrac{u}{\lambda}=\dfrac{\sin\theta\cos\psi}{\lambda}$	$f_X=\dfrac{u}{\lambda}=\dfrac{\sin\theta\cos\psi}{\lambda}$

$A(f,x_A)$	$D(f,f_X)$
$A_A(f)$	$A_A(f)\delta(f_X)$
$A_A(f)\delta(x_A \pm d)$	$A_A(f)\exp(\mp j2\pi f_X d)$
$A_A(f)\exp(\pm j2\pi f_X' x_A)$	$A_A(f)\delta(f_X \pm f_X')$
$A_A(f)\mathrm{rect}(x_A/L)$	$A_A(f)L\,\mathrm{sinc}(f_X L)$
$A_A(f)\mathrm{tri}(x_A/L)$	$\left[A_A(f)L/2\right]\mathrm{sinc}^2(f_X L/2)$
$A_A(f)\cos(2\pi x_A/d)$	$\dfrac{A_A(f)}{2}\left[\delta\left(f_X+\dfrac{1}{d}\right)+\delta\left(f_X-\dfrac{1}{d}\right)\right]$
$A_A(f)\cos(2\pi x_A/d)\mathrm{rect}(x_A/L)$	$\dfrac{A_A(f)L}{2}\left\{\mathrm{sinc}\left[\left(f_X+\dfrac{1}{d}\right)L\right]+\mathrm{sinc}\left[\left(f_X-\dfrac{1}{d}\right)L\right]\right\}$
$A_A(f)\cos(\pi x_A/L)\mathrm{rect}(x_A/L)$	$\dfrac{2A_A(f)L}{\pi}\dfrac{\cos(\pi f_X L)}{1-(2f_X L)^2}$

Comments

1. The function $A_A(f)$ can also be equal to a constant A_A.

Chapter 3

Complex Aperture Theory – Planar Apertures

3.1 The Far-Field Beam Pattern of a Planar Aperture

In this section we shall demonstrate that the far-field beam pattern of a planar aperture is just a special case of the far-field beam pattern of a volume aperture. In Section 1.3 it was shown that the far-field beam pattern (directivity function) of a closed-surface, volume aperture – based on the Fraunhofer approximation of a time-independent, free-space, Green's function – is given by the following three-dimensional spatial Fourier transform:

$$D(f,\boldsymbol{\alpha}) = F_{\mathbf{r}_A}\{A(f,\mathbf{r}_A)\} = \int_{-\infty}^{\infty} A(f,\mathbf{r}_A)\exp(+j2\pi\boldsymbol{\alpha}\boldsymbol{\cdot}\mathbf{r}_A)\,d\mathbf{r}_A\,, \tag{3.1-1}$$

where

$$\mathbf{r}_A = (x_A, y_A, z_A)\,, \tag{3.1-2}$$

$$\boldsymbol{\alpha} = (f_X, f_Y, f_Z)\,, \tag{3.1-3}$$

$$\boldsymbol{\alpha}\boldsymbol{\cdot}\mathbf{r}_A = f_X x_A + f_Y y_A + f_Z z_A\,, \tag{3.1-4}$$

$$d\mathbf{r}_A = dx_A\,dy_A\,dz_A\,, \tag{3.1-5}$$

and $F_{\mathbf{r}_A}\{\bullet\}$ is shorthand notation for $F_{x_A}F_{y_A}F_{z_A}\{\bullet\}$. The complex aperture function $A(f,\mathbf{r}_A)$ can be expressed as

$$A(f,\mathbf{r}_A) = a(f,\mathbf{r}_A)\exp[+j\theta(f,\mathbf{r}_A)]\,, \tag{3.1-6}$$

where $a(f,\mathbf{r}_A)$ is the *amplitude response* and $\theta(f,\mathbf{r}_A)$ is the *phase response* of the aperture at $\mathbf{r}_A$. Both $a(f,\mathbf{r}_A)$ and $\theta(f,\mathbf{r}_A)$ are *real* functions. The function $a(f,\mathbf{r}_A)$ is also known as the *amplitude window*. The units of $a(f,\mathbf{r}_A)$ and, hence, $A(f,\mathbf{r}_A)$ depend on whether the aperture is a transmit aperture (see Table 1B-1 in Appendix 1B) or a receive aperture (see Table 1B-2 in Appendix 1B). The units of $\theta(f,\mathbf{r}_A)$ are radians.

Now consider the case of a *planar aperture* of arbitrary shape lying in the XY plane as shown in Fig. 3.1-1. A planar aperture can represent either a single electroacoustic transducer (e.g., a rectangular or circular piston) or a planar array of many individual electroacoustic transducers. Also shown in Fig. 3.1-1 is a field point with spherical coordinates (r, θ, ψ). The field point is in the far-field region of the aperture, where the range r satisfies the Fraunhofer (far-field) range criterion given by $r > \pi R_A^2/\lambda > 2.414R_A$ [see either (1.2-39) or (1.2-93)], and R_A is the maximum radial extent of the planar aperture.

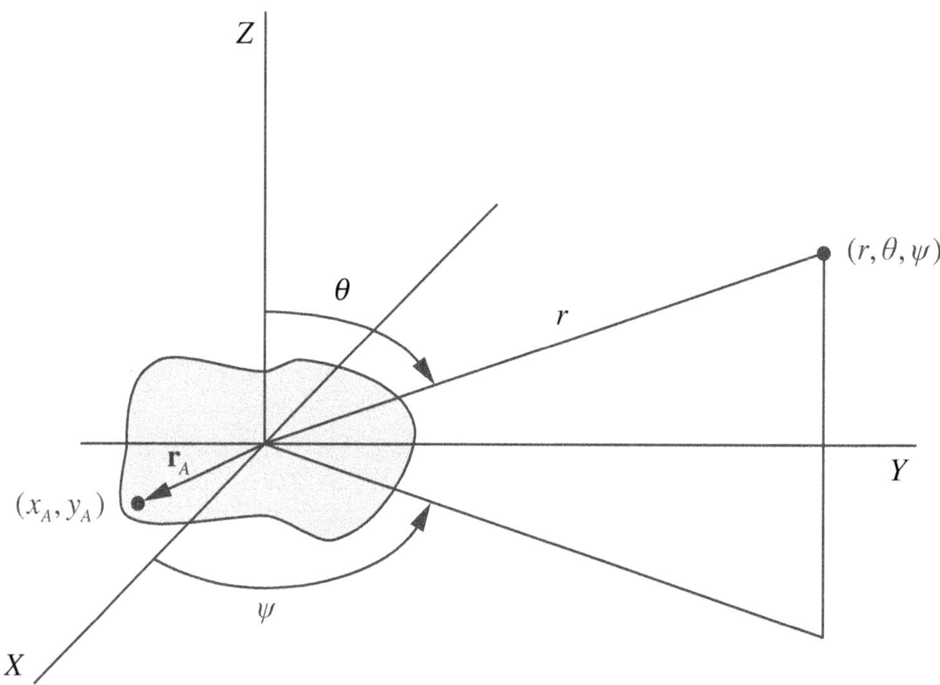

Figure 3.1-1 Planar aperture of arbitrary shape lying in the XY plane. Also shown is a field point with spherical coordinates (r, θ, ψ).

Since the planar aperture in Fig. 3.1-1 is lying in the XY plane, the position vector to a point on the surface of the aperture is given by

$$\mathbf{r}_A = (x_A, y_A, 0), \tag{3.1-7}$$

and as a result,

$$A(f, \mathbf{r}_A) = A(f, x_A, y_A, z_A) \rightarrow A(f, x_A, y_A)\delta(z_A). \tag{3.1-8}$$

Therefore, by substituting (3.1-8) into (3.1-1) and using the sifting property of impulse functions, we obtain the following two-dimensional spatial Fourier transform for the far-field beam pattern of a planar aperture lying in the XY plane:

$$\boxed{\begin{aligned} D(f, f_X, f_Y) &= F_{x_A} F_{y_A}\{A(f, x_A, y_A)\} \\ &= \int_{-\infty}^{\infty}\int_{-\infty}^{\infty} A(f, x_A, y_A)\exp[+j2\pi(f_X x_A + f_Y y_A)]dx_A\, dy_A \end{aligned}} \tag{3.1-9}$$

where

$$A(f, x_A, y_A) = a(f, x_A, y_A)\exp[+j\theta(f, x_A, y_A)] \tag{3.1-10}$$

is the complex frequency response (complex aperture function) of the planar aperture, and

$$f_X = u/\lambda = \sin\theta\cos\psi/\lambda \tag{3.1-11}$$

and

$$f_Y = v/\lambda = \sin\theta\sin\psi/\lambda \tag{3.1-12}$$

are spatial frequencies in the X and Y directions, respectively, with units of cycles per meter [see (1.2-43) and (1.2-44)]. Since f_X and f_Y can be expressed in terms of the spherical angles θ and ψ, the far-field beam pattern can ultimately be expressed as a function of frequency f and the spherical angles θ and ψ, that is, $D(f, f_X, f_Y) \to D(f, \theta, \psi)$. If the planar aperture is a transmit aperture, then the aperture function $A(f, x_A, y_A)$ has units of $\left(\left(\text{m}^3/\text{sec}\right)/\text{V}\right)/\text{m}^2$ and the far-field beam pattern $D(f, f_X, f_Y)$ has units of $\left(\text{m}^3/\text{sec}\right)/\text{V}$. However, if the planar aperture is a receive aperture, then $A(f, x_A, y_A)$ has units of $\left(\text{V}/\left(\text{m}^2/\text{sec}\right)\right)/\text{m}^2$ and $D(f, f_X, f_Y)$ has units of $\text{V}/\left(\text{m}^2/\text{sec}\right)$. These units are summarized in Table 3.1-1. In addition to describing the performance of an electroacoustic transducer by its far-field beam pattern, values of the frequency-dependent, *transmitter* and *receiver sensitivity functions* are also used. The relationships between the complex frequency response $A(f, x_A, y_A)$ of a single electroacoustic transducer and complex, transmitter and receiver sensitivity functions are discussed in Appendix 3A.

Table 3.1-1 Units of the Complex Aperture Function $A(f, x_A, y_A)$ and Corresponding Far-Field Beam Pattern $D(f, f_X, f_Y)$ for a Planar Aperture

Planar Aperture	$A(f, x_A, y_A)$	$D(f, f_X, f_Y)$
transmit	$\left(\left(\text{m}^3/\text{sec}\right)/\text{V}\right)/\text{m}^2$	$\left(\text{m}^3/\text{sec}\right)/\text{V}$
receive	$\left(\text{V}/\left(\text{m}^2/\text{sec}\right)\right)/\text{m}^2$	$\text{V}/\left(\text{m}^2/\text{sec}\right)$

Finally, if a planar aperture lies in the XZ plane instead of the XY plane, then simply replace y_A and f_Y with z_A and f_Z, respectively, in (3.1-9), where spatial frequency f_Z is given by (1.2-45). And if a planar aperture lies in the YZ plane instead of the XY plane, then simply replace x_A and f_X with z_A and f_Z, respectively, in (3.1-9). With the use of (3.1-9), we will derive the normalized, far-field beam patterns of rectangular and circular pistons in Sections 3.2 and 3.3, respectively. Although we shall be concentrating on single, planar, electroacoustic transducers in this chapter, as we shall discover in Chapter 8, some of the results obtained in this chapter are directly applicable to planar arrays. In other words, we are laying the foundation for planar array theory in this chapter as well.

3.2 The Far-Field Beam Pattern of a Rectangular Piston

One of the most common examples of a planar aperture lying in the XY plane is a single electroacoustic transducer, rectangular in shape, with sides equal to L_X and L_Y meters in length, as shown in Fig. 3.2-1. The simplest mathematical model for the complex frequency response (complex aperture function) for this rectangular-shaped transducer is given by

$$A(f, x_A, y_A) = A_A \operatorname{rect}(x_A/L_X)\operatorname{rect}(y_A/L_Y), \tag{3.2-1}$$

where the *real*, *positive* constant A_A carries the correct units of $A(f, x_A, y_A)$ (see Table 3.1-1) and the rectangle function is defined as follows:

$$\operatorname{rect}\left(\frac{x}{L}\right) \triangleq \begin{cases} 1, & |x| \leq L/2 \\ 0, & |x| > L/2. \end{cases} \tag{3.2-2}$$

According to (3.2-1), the complex frequency response of the transducer is constant across the entire surface of the transducer, regardless of the value of frequency f. It is important to note that although A_A is shown here as a constant, it is, in general, a *real*, *nonnegative* function of frequency, that is, $A_A \rightarrow A_A(f)$. A single electroacoustic transducer, rectangular in shape, with complex frequency response given by (3.2-1), is referred to as a *rectangular piston*. The maximum radial extent of a rectangular-shaped, planar aperture lying in the XY plane is given by

$$R_A = \sqrt{\left(\frac{L_X}{2}\right)^2 + \left(\frac{L_Y}{2}\right)^2}. \tag{3.2-3}$$

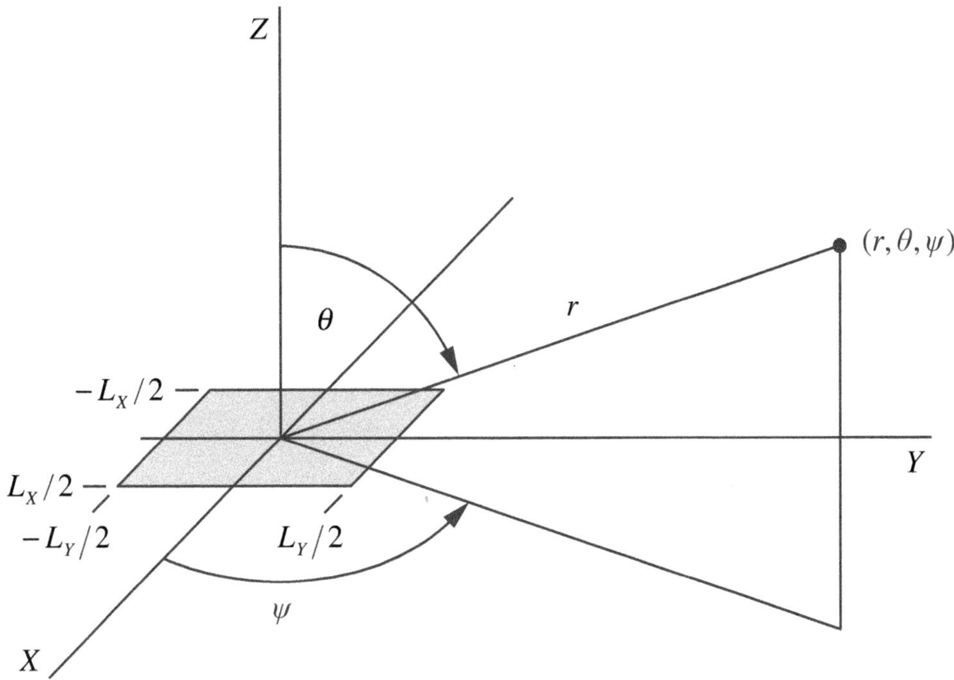

Figure 3.2-1 A single electroacoustic transducer, rectangular in shape, with sides equal to L_X and L_Y meters in length, lying in the XY plane. Also shown is a field point with spherical coordinates (r, θ, ψ).

The far-field beam pattern of the rectangular piston can be obtained by substituting (3.2-1) into (3.1-9). Doing so yields

$$\begin{aligned} D(f, f_X, f_Y) &= F_{x_A} F_{y_A}\{A(f, x_A, y_A)\} \\ &= A_A F_{x_A} F_{y_A}\{\text{rect}(x_A/L_X)\text{rect}(y_A/L_Y)\} \\ &= A_A F_{x_A}\{\text{rect}(x_A/L_X)\} F_{y_A}\{\text{rect}(y_A/L_Y)\}, \end{aligned} \tag{3.2-4}$$

or

$$\boxed{D(f, f_X, f_Y) = A_A L_X L_Y \,\text{sinc}(f_X L_X)\text{sinc}(f_Y L_Y)} \tag{3.2-5}$$

where use was made of (2.2-5) (Table 2E-1 can also be used). Equation (3.2-5) is the *unnormalized*, far-field beam pattern of a rectangular piston lying in the XY plane, where $L_X L_Y$ is the area of the rectangular piston in squared meters. Since $\text{sinc}(0) = 1$ is the maximum value of a sinc function, the normalization factor $D_{\max} = \max|D(f, f_X, f_Y)|$ – which is the *maximum* value of the *magnitude* of the unnormalized, far-field beam pattern – is given by

$$D_{\max} = |D(f,0,0)| = A_A L_X L_Y \,. \tag{3.2-6}$$

Whether or not the normalization factor is equal to the magnitude of the unnormalized, far-field beam pattern evaluated at broadside, $D_{\max}$ can also be found by using the standard calculus approach for finding the maximum value of a function, or by using the formulas in Appendix 3C. Dividing (3.2-5) by (3.2-6) yields the following expression for the *normalized*, far-field beam pattern of a rectangular piston lying in the XY plane:

$$\boxed{D_N(f, f_X, f_Y) = \operatorname{sinc}(f_X L_X)\operatorname{sinc}(f_Y L_Y)} \tag{3.2-7}$$

Note that $|D_N(f,0,0)| = 1$. Since $f_X = u/\lambda$ and $f_Y = v/\lambda$, (3.2-7) can also be expressed as

$$\boxed{D_N(f,u,v) = \operatorname{sinc}\left(\frac{L_X}{\lambda}u\right)\operatorname{sinc}\left(\frac{L_Y}{\lambda}v\right)} \tag{3.2-8}$$

The magnitude of the normalized, far-field beam pattern given by (3.2-8) is plotted as a function of direction cosines u and v in Fig. 3.2-2 for $L_Y = L_X = L$ and $L/\lambda = 2.5$. As can be seen from the equations developed in this section, the rectangular piston is just a two-dimensional generalization of a linear aperture lying along the X axis with a rectangular amplitude window for a complex frequency response (see Subsection 2.2.1).

When plotting the far-field beam pattern of a planar aperture in direction-cosine space, care must be taken when trying to solve for the spherical angles θ and ψ from (u,v) coordinates. Since $u = \sin\theta\cos\psi$ and $v = \sin\theta\sin\psi$, it can be shown that

$$\theta = \sin^{-1}\sqrt{u^2+v^2}, \qquad \sqrt{u^2+v^2} \le 1, \tag{3.2-9}$$

and

$$\psi = \tan^{-1}(v/u). \tag{3.2-10}$$

The inequality $\sqrt{u^2+v^2} \le 1$ appearing in (3.2-9) is illustrated in Fig. 3.2-3. Only those (u,v) coordinates inside or on the unit circle shown in Fig. 3.2-3 can be used for computing θ and ψ. Also, since the azimuthal (bearing) angle ψ is measured in a counter-clockwise direction from the positive u (X) axis, it takes on values in the range $0° \le \psi \le 360°$. However, the arctangent function in (3.2-10) may only give values for ψ in the range $-90° \le \psi \le 90°$ or

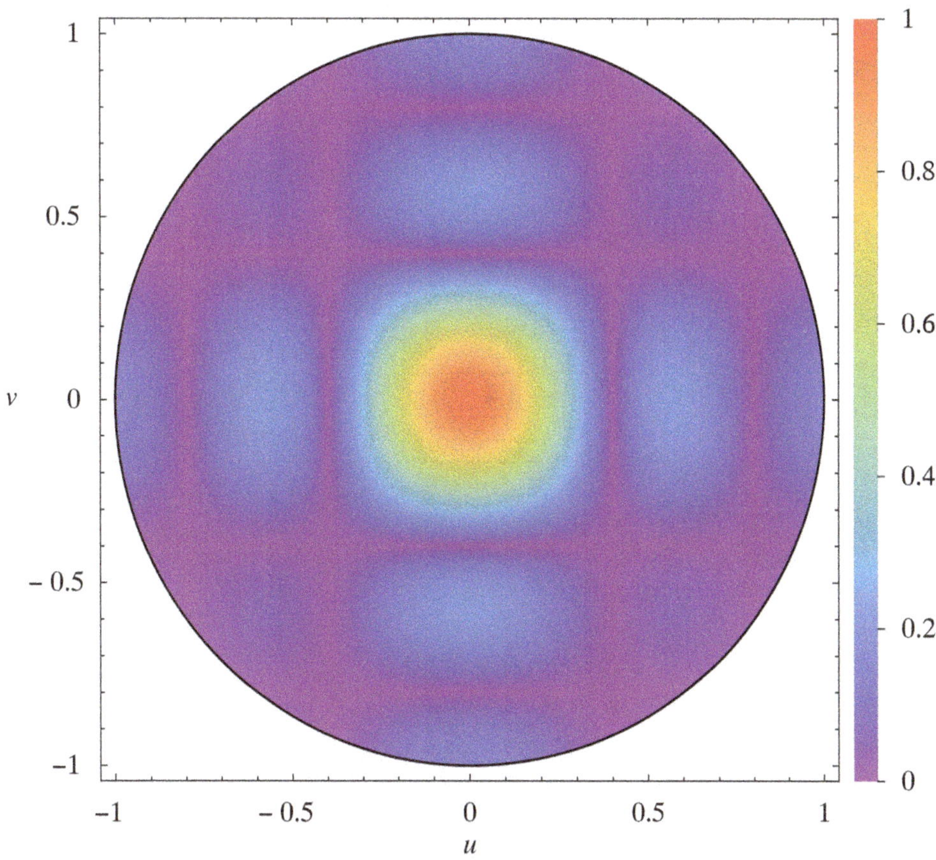

Figure 3.2-2 Magnitude of the normalized, far-field beam pattern of a rectangular piston lying in the XY plane, plotted as a function of direction cosines u and v for $L_Y = L_X = L$ and $L/\lambda = 2.5$.

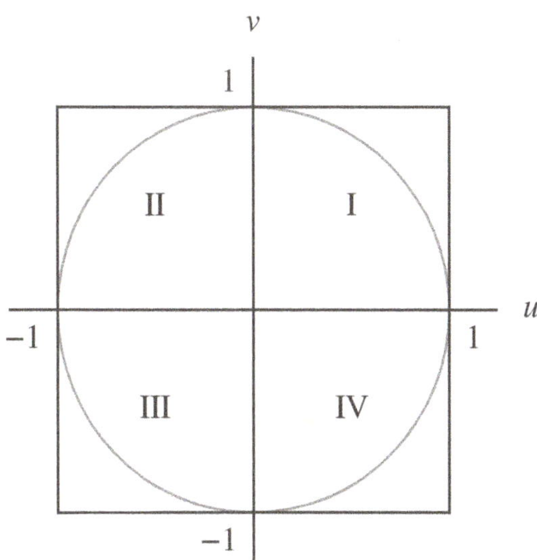

Figure 3.2-3 Unit circle with radius $\sqrt{u^2+v^2}=1$ in the uv plane. Points inside or on the unit circle satisfy the inequality $\sqrt{u^2+v^2}\le 1$.

$-180° \le \psi \le 180°$. Therefore, by taking into account the signs (quadrant) of the (u,v) coordinates, one may have to add either $180°$ or $360°$ to the value obtained from (3.2-10) so that ψ has the correct value (see Table 3.2-1).

Table 3.2-1 Range of Values for Azimuthal (Bearing) Angle ψ Versus Quadrant Number

Quadrant	u	v	ψ
I	+	+	$0° \le \psi \le 90°$
II	–	+	$90° \le \psi \le 180°$
III	–	–	$180° \le \psi \le 270°$
IV	+	–	$270° \le \psi \le 360°$

Example 3.2-1 3-dB Beamwidths of the Vertical Far-Field Beam Patterns of a Rectangular Piston

In this example we shall derive equations for the 3-dB beamwidths of the vertical far-field beam patterns in the XZ and YZ planes for a rectangular piston lying in the XY plane. Since direction cosine $v=0$ in the XZ plane, substituting $v=0$ into (3.2-8) yields the following expression for the normalized, *vertical*, far-field beam pattern in the XZ plane of a rectangular piston lying in the XY plane:

$$D_N(f,u,0) = \text{sinc}\left(\frac{L_X}{\lambda}u\right). \tag{3.2-11}$$

Equation (3.2-11) is also the normalized, far-field beam pattern of a linear aperture lying along the X axis with a rectangular amplitude window for a complex frequency response (see Subsection 2.2.1). Therefore, by combining the results from Examples 2.3-1 and 2.3-3, we obtain

$$\Delta\theta = 2\sin^{-1}\left(0.443\lambda/L_X\right), \tag{3.2-12}$$

where $\Delta\theta$ is the 3-dB beamwidth in degrees of the vertical, far-field beam pattern in the XZ plane.

Since direction cosine $u=0$ in the YZ plane, substituting $u=0$ into (3.2-8) yields the following expression for the normalized, *vertical*, far-field beam pattern in the YZ plane of a rectangular piston lying in the XY plane:

$$D_N(f,0,v) = \text{sinc}\left(\frac{L_Y}{\lambda}v\right). \tag{3.2-13}$$

Equation (3.2-13) is also the normalized, far-field beam pattern of a linear aperture lying along the Y axis with a rectangular amplitude window for a complex frequency response. Therefore,

$$\Delta\theta = 2\sin^{-1}\left(0.443\lambda/L_Y\right), \tag{3.2-14}$$

where $\Delta\theta$ is the 3-dB beamwidth in degrees of the vertical, far-field beam pattern in the YZ plane. By comparing (3.2-12) and (3.2-14), it can be seen that the 3-dB beamwidths in the XZ and YZ planes are not equal as long as $L_Y \neq L_X$. This implies that the 3-dB beamwidth in the XZ plane can be made smaller than the 3-dB beamwidth in the YZ plane ($L_X > L_Y$), or vice-versa ($L_Y > L_X$). ■

3.3 The Far-Field Beam Pattern of a Circular Piston

We know from our discussion in Section 3.1 that the far-field beam pattern of an arbitrarily shaped planar aperture lying in the XY plane is given by the following two-dimensional spatial Fourier transform [see (3.1-9)]:

$$D(f, f_X, f_Y) = \int_{-\infty}^{\infty}\int_{-\infty}^{\infty} A(f, x_A, y_A)\exp\left[+j2\pi(f_X x_A + f_Y y_A)\right]dx_A\, dy_A\,, \tag{3.3-1}$$

where

$$f_X = u/\lambda = \sin\theta\cos\psi/\lambda \tag{3.3-2}$$

and

$$f_Y = v/\lambda = \sin\theta\sin\psi/\lambda\,. \tag{3.3-3}$$

Now consider the problem of computing the far-field beam pattern of a planar aperture lying in the XY plane that is circular in shape, with radius a meters and complex frequency response $A(f, r_A, \phi_A)$ expressed in polar coordinates r_A and ϕ_A (see Fig. 3.3-1). A circular-shaped, planar aperture can represent either a single electroacoustic transducer (e.g., a circular piston) or a planar array of concentric circular arrays composed of many individual electroacoustic transducers. The maximum radial extent of a circular-shaped, planar aperture is $R_A = a$. In order to solve this problem, we need to transform (3.3-1) from rectangular coordinates to polar coordinates.

By referring to Fig. 3.3-2, we can write that

$$x_A = r_A\cos\phi_A\,, \tag{3.3-4}$$

$$y_A = r_A\sin\phi_A\,, \tag{3.3-5}$$

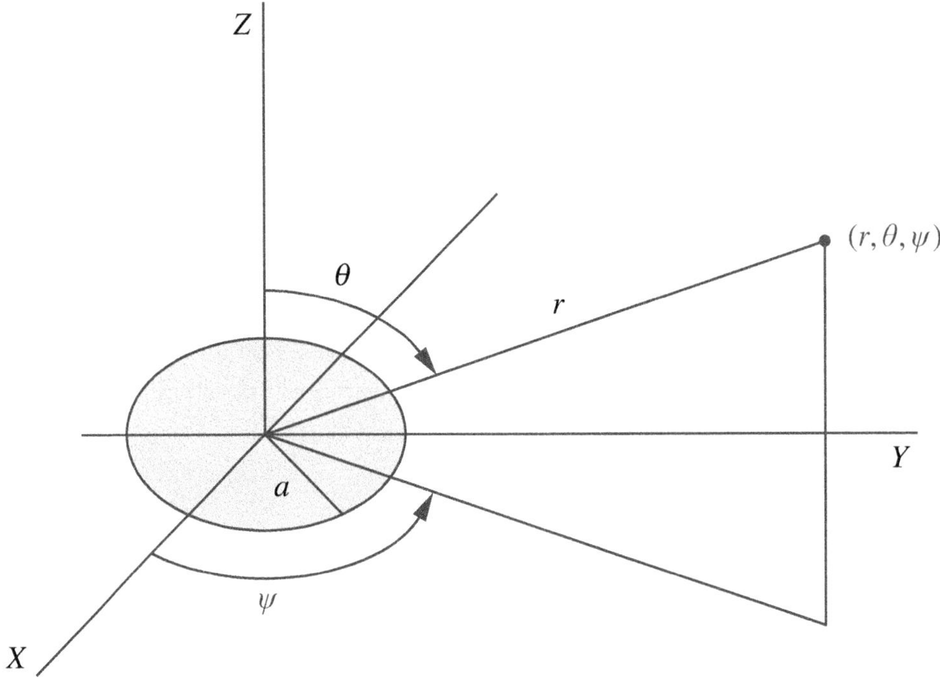

Figure 3.3-1 A single electroacoustic transducer, circular in shape, with radius a meters, lying in the XY plane. Also shown is a field point with spherical coordinates (r, θ, ψ).

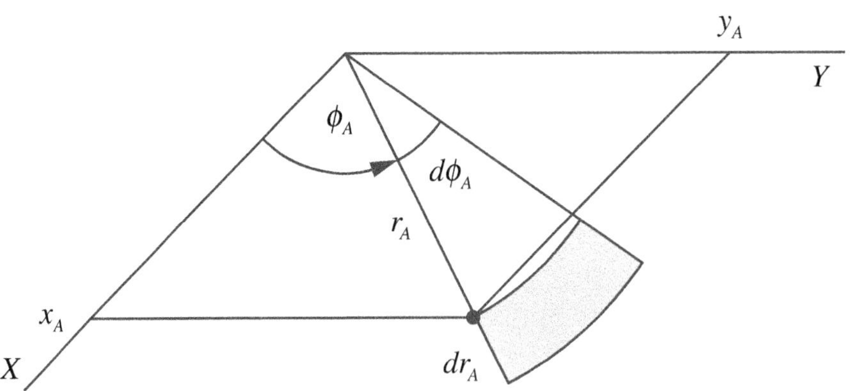

Figure 3.3-2 Relationship between the rectangular coordinates (x_A, y_A) and the polar coordinates (r_A, ϕ_A). Also shown is the infinitesimal area $r_A\,dr_A\,d\phi_A$.

$$dx_A\,dy_A \rightarrow r_A\,dr_A\,d\phi_A, \tag{3.3-6}$$

and

$$A(f, x_A, y_A) = A(f, r_A \cos\phi_A, r_A \sin\phi_A) \rightarrow A(f, r_A, \phi_A). \tag{3.3-7}$$

Substituting (3.3-2) through (3.3-7) into (3.3-1), and using the trigonometric identity

$$\cos(\alpha-\beta)=\cos\alpha\cos\beta+\sin\alpha\sin\beta\,, \tag{3.3-8}$$

yields

$$\boxed{D(f,\theta,\psi)=\int_0^{2\pi}\int_0^{a} A(f,r_A,\phi_A)\exp\left[+j\frac{2\pi r_A}{\lambda}\sin\theta\cos(\psi-\phi_A)\right]r_A\,dr_A\,d\phi_A} \tag{3.3-9}$$

where

$$A(f,r_A,\phi_A)=a(f,r_A,\phi_A)\exp[+j\theta(f,r_A,\phi_A)] \tag{3.3-10}$$

is the complex frequency response (complex aperture function) of the circular aperture. Note that $2\pi r_A$ in (3.3-9) is the circumference of a circle with radius r_A meters. Equation (3.3-9) will be used in Example 8.1-6 to derive the far-field beam pattern of a planar array of concentric circular arrays.

Circular Symmetry

Before we derive the far-field beam pattern of a circular piston, let us first consider the important special case when the complex frequency response $A(f,r_A,\phi_A)$ is *circularly symmetric*, that is,

$$A(f,r_A,\phi_A)=A_r(f,r_A)\,, \tag{3.3-11}$$

where

$$A_r(f,r_A)=a_r(f,r_A)\exp[+j\theta_r(f,r_A)]\,. \tag{3.3-12}$$

The complex frequency response given by (3.3-11) is said to be *circularly symmetric*, or to have *circular symmetry*, because its value does not depend on the polar angle ϕ_A to an aperture point. Substituting (3.3-11) into (3.3-9) and interchanging the order of integration yields

$$D(f,\theta,\psi)=\int_0^{a} A_r(f,r_A)\int_0^{2\pi}\exp\left[+j\frac{2\pi r_A}{\lambda}\sin\theta\cos(\psi-\phi_A)\right]d\phi_A\,r_A\,dr_A\,. \tag{3.3-13}$$

If we let

$$b=\frac{2\pi r_A}{\lambda}\sin\theta\,, \tag{3.3-14}$$

and since

$$\cos(\psi-\phi_A)=\sin\left(\psi-\phi_A+\frac{\pi}{2}\right), \tag{3.3-15}$$

substituting (3.3-14) and (3.3-15) into (3.3-13) yields

$$D(f,\theta,\psi)=\int_0^a A_r(f,r_A)\int_0^{2\pi}\exp(+jb\sin\alpha)\,d\phi_A\,r_A\,dr_A\,, \tag{3.3-16}$$

where

$$\alpha=\psi-\phi_A+\frac{\pi}{2}. \tag{3.3-17}$$

We next take advantage of the following identity:

$$\exp(+jb\sin\alpha)=\sum_{n=-\infty}^{\infty}J_n(b)\exp(+jn\alpha)\,, \tag{3.3-18}$$

where

$$J_n(b)=\frac{1}{2\pi}\int_{-\pi}^{\pi}\exp[\pm j(b\sin\beta-n\beta)]\,d\beta \tag{3.3-19}$$

or

$$J_n(b)=\frac{1}{\pi}\int_0^{\pi}\cos(b\sin\beta-n\beta)\,d\beta \tag{3.3-20}$$

is the *nth-order Bessel function of the first kind.* With the use of (3.3-17) and (3.3-18), the innermost integral in (3.3-16) can be rewritten as

$$\int_0^{2\pi}\exp(+jb\sin\alpha)\,d\phi_A=\sum_{n=-\infty}^{\infty}J_n(b)\exp\{+jn[\psi+(\pi/2)]\}\int_0^{2\pi}\exp(-jn\phi_A)\,d\phi_A\,. \tag{3.3-21}$$

Since

$$\int_0^{2\pi}\exp(-jn\phi_A)\,d\phi_A=\begin{cases}2\pi, & n=0\\ 0, & n\neq 0,\end{cases} \tag{3.3-22}$$

(3.3-21) reduces to

$$\int_0^{2\pi}\exp(+jb\sin\alpha)\,d\phi_A=2\pi J_0(b)\,, \tag{3.3-23}$$

and by substituting (3.3-23) into (3.3-16) and then making use of (3.3-14), we obtain

$$\boxed{D(f,\theta,\psi)=2\pi\int_0^a A_r(f,r_A)J_0\left(\frac{2\pi r_A}{\lambda}\sin\theta\right)r_A\,dr_A} \tag{3.3-24}$$

Equation (3.3-24) is the far-field beam pattern of a circular-shaped, planar aperture lying in the XY plane with radius a meters and circularly symmetric complex frequency response given by (3.3-11). The far-field beam pattern given by (3.3-24) is *axisymmetric* because it's value does not depend on the azimuthal (bearing) angle ψ to a field point. Equation (3.3-24) can also be expressed as follows:

$$D(f,f_r)=H_0\{A_r(f,r_A)\}=2\pi\int_0^a A_r(f,r_A)J_0(2\pi f_r r_A)r_A\,dr_A\,, \tag{3.3-25}$$

where

$$f_r=\sqrt{f_X^2+f_Y^2}=\sin\theta/\lambda \tag{3.3-26}$$

is the spatial frequency in the horizontal (polar) radial direction with units of cycles per meter, and $H_0\{A_r(f,r_A)\}$ is the *zeroth-order Hankel transform* of $A_r(f,r_A)$, also known as the *Fourier-Bessel transform* of $A_r(f,r_A)$. Note that $k_r=2\pi f_r$ is the propagation-vector component in the horizontal (polar) radial direction with units of radians per meter. The circular piston is considered next.

Circular Piston

Another very common example of a planar aperture lying in the XY plane (analogous to the rectangular piston discussed in Section 3.2) is a single electroacoustic transducer, circular in shape, with radius a meters (see Fig. 3.3-1). The simplest mathematical model for the complex frequency response (complex aperture function) for this circular-shaped transducer is given by

$$A(f,r_A,\phi_A)=A_r(f,r_A)=A_A\operatorname{circ}(r_A/a), \tag{3.3-27}$$

where the real, positive constant A_A carries the correct units of $A(f,r_A,\phi_A)$ (see Table 3.1-1) and the circle function is defined as follows:

$$\operatorname{circ}\left(\frac{r}{a}\right)\triangleq\begin{cases}1, & r\le a\\ 0, & r>a.\end{cases} \tag{3.3-28}$$

Therefore, according to (3.3-27), the complex frequency response of the transducer is circularly symmetric and is equal to a constant across the entire surface of the transducer, regardless of the value of frequency f. It is important to note that although A_A is shown here as a constant, it is, in general, a real, nonnegative function of frequency, that is, $A_A \rightarrow A_A(f)$. A single electroacoustic transducer, circular in shape, with complex frequency response given by (3.3-27), is referred to as a *circular piston.*

The far-field beam pattern of the circular piston can be obtained by substituting

$$A_r(f, r_A) = A_A \operatorname{circ}(r_A/a) \tag{3.3-29}$$

into (3.3-24). Doing so yields

$$D(f, \theta, \psi) = 2\pi A_A \int_0^a J_0\left(\frac{2\pi r_A}{\lambda}\sin\theta\right) r_A \, dr_A \,, \tag{3.3-30}$$

and by using the identity

$$\int_0^x J_0(\alpha)\alpha \, d\alpha = x J_1(x) \,, \tag{3.3-31}$$

(3.3-30) reduces to

$$\boxed{D(f, \theta, \psi) = 2A_A \pi a^2 \frac{J_1\left(\frac{2\pi a}{\lambda}\sin\theta\right)}{\frac{2\pi a}{\lambda}\sin\theta}} \tag{3.3-32}$$

Equation (3.3-32) is the *unnormalized*, far-field beam pattern of a circular piston with radius a meters lying in the XY plane. It is axisymmetric because its value does not depend on the azimuthal (bearing) angle ψ to a field point. The factor $2\pi a$ is the circumference of the circular piston in meters and πa^2 is the area in squared meters. Equation (3.3-32) can also be expressed as

$$D(f, f_r) = H_0\{A_A \operatorname{circ}(r_A/a)\} = 2A_A \pi a^2 \frac{J_1(2\pi f_r a)}{2\pi f_r a} \,, \tag{3.3-33}$$

where f_r is given by (3.3-26). Since the maximum value of the magnitude of (3.3-32) is at broadside where $\theta = 0°$, the normalization factor is given by

$$D_{\max} = |D(f, 0°, \psi)| = A_A \pi a^2 \,. \tag{3.3-34}$$

Whether or not the normalization factor is equal to the magnitude of the unnormalized, far-field beam pattern evaluated at broadside, $D_{\max}$ can also be found by using the standard calculus approach for finding the maximum value of a function, or by using the formulas in Appendix 3C. Dividing (3.3-32) by (3.3-34) yields the following expression for the *normalized*, far-field beam pattern of a circular piston with radius a meters lying in the XY plane:

$$\boxed{D_N(f,\theta,\psi)=2\frac{J_1\left(\frac{2\pi a}{\lambda}\sin\theta\right)}{\frac{2\pi a}{\lambda}\sin\theta}} \tag{3.3-35}$$

or

$$D_N(f,f_r)=2\frac{J_1(2\pi f_r a)}{2\pi f_r a} \tag{3.3-36}$$

where f_r is given by (3.3-26). Figure 3.3-3 is a plot of the magnitude of (3.3-35) for $ka=11$, where the level of the first sidelobe is approximately $-17.6\,\text{dB}$. Figure 3.3-4 is a plot of the magnitude of (3.3-36) for $ka=11$, where f_r was replaced by $\sqrt{u^2+v^2}/\lambda$ [see (3.3-26)].

Example 3.3-1 3-dB Beamwidth of the Vertical Far-Field Beam Pattern of a Circular Piston

In this example we shall compute the 3-dB beamwidth $\Delta\theta$ of the vertical, far-field beam pattern of a circular piston lying in the XY plane using the procedure discussed in Example 2.3-3. We shall then compare this beamwidth with the 3-dB beamwidth of the vertical, far-field beam pattern of a rectangular piston that has the same area as the circular piston. If we evaluate (3.3-35) at the location of a 3-dB-down point, then

$$D_N(f,\theta_+,\psi)=2\frac{J_1\left(\frac{2\pi a}{\lambda}\sin\theta_+\right)}{\frac{2\pi a}{\lambda}\sin\theta_+}=\frac{\sqrt{2}}{2}, \tag{3.3-37}$$

or

$$J_1(\pi x)-\frac{\sqrt{2}}{4}\pi x=0, \tag{3.3-38}$$

where

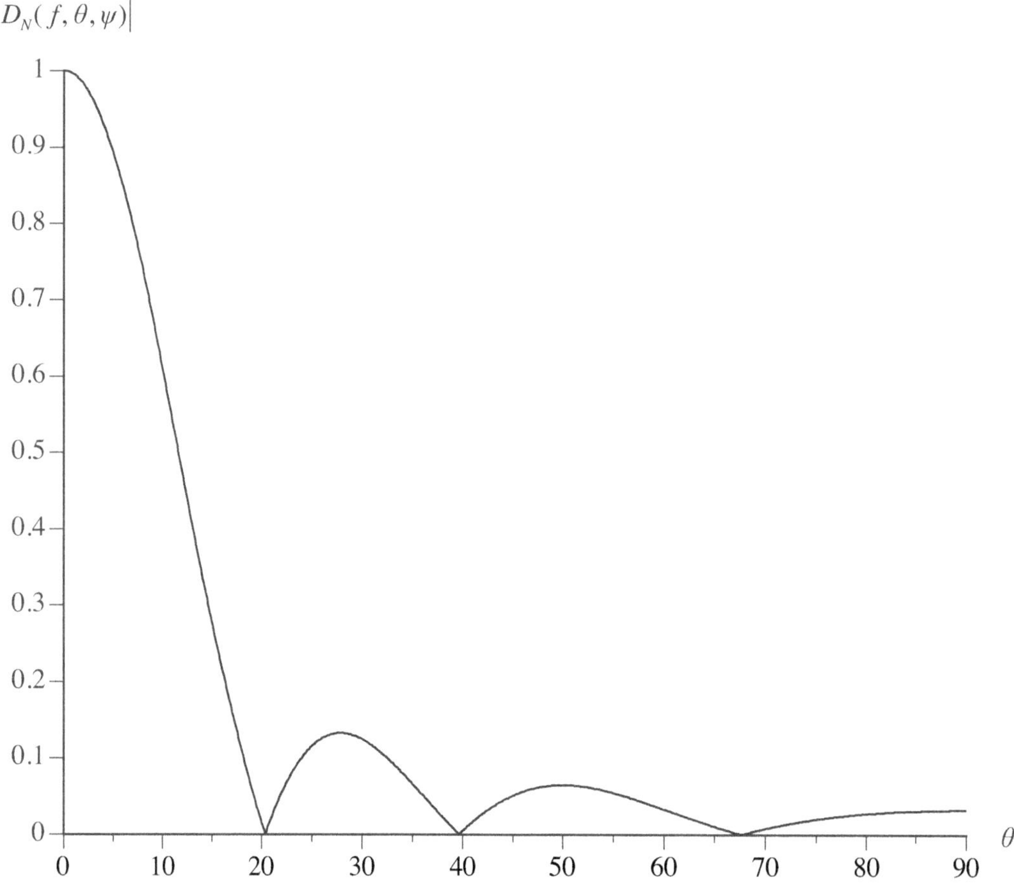

Figure 3.3-3 Magnitude of the normalized, far-field beam pattern of a circular piston lying in the XY plane given by (3.3-35) for $ka = 11$, where θ is the vertical angle in degrees.

$$x = (2a/\lambda)\sin\theta_+ . \tag{3.3-39}$$

Since

$$\theta_+ = \Delta\theta/2 , \tag{3.3-40}$$

substituting (3.3-40) into (3.3-39) yields

$$\Delta\theta = 2\sin^{-1}\left(x\frac{\lambda}{2a}\right). \tag{3.3-41}$$

The solution (root) of (3.3-38) is $x \approx 0.514$, and by substituting this result into (3.3-41), we obtain the following expression for the 3-dB beamwidth $\Delta\theta$ (in degrees) of the vertical, far-field beam pattern of a circular piston lying in the XY plane:

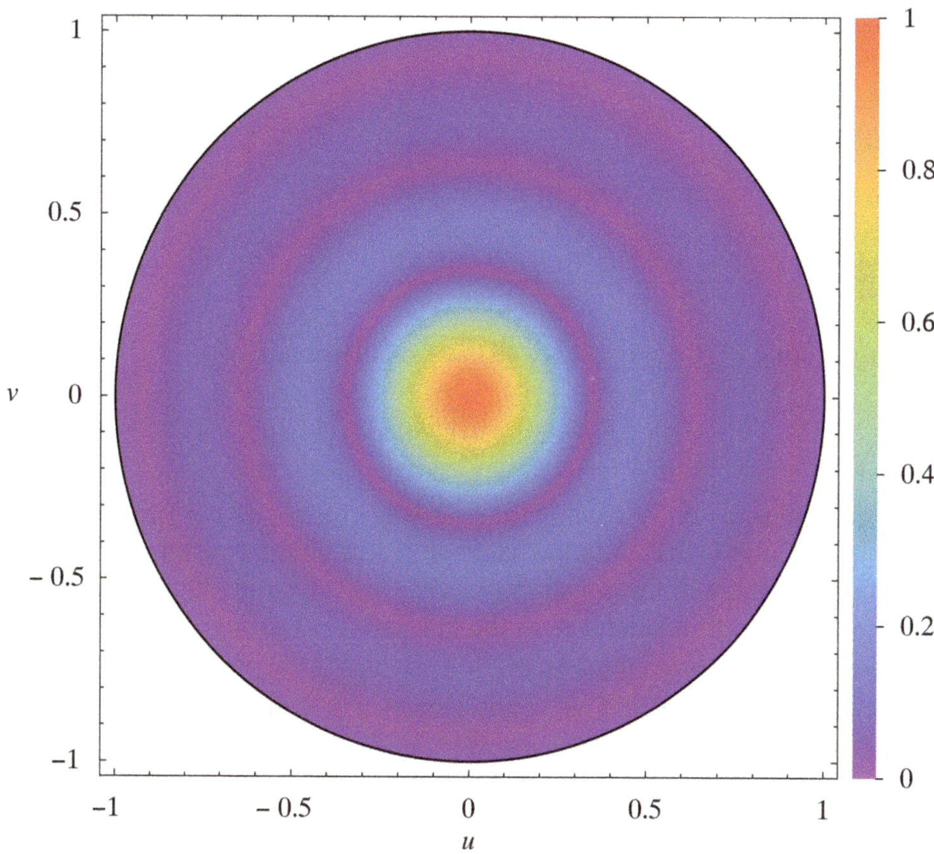

Figure 3.3-4 Magnitude of the normalized, far-field beam pattern of a circular piston lying in the XY plane given by (3.3-36) for $ka = 11$, where f_r was replaced by $\sqrt{u^2 + v^2}/\lambda$.

$$\boxed{\Delta\theta = 2\sin^{-1}(0.257\,\lambda/a)} \tag{3.3-42}$$

where a is the radius of the circular piston in meters.

Let us conclude this example by comparing (3.3-42) for a circular piston with the 3-dB beamwidth $\Delta\theta$ of the vertical, far-field beam pattern in the XZ plane of a rectangular piston lying in the XY plane given by

$$\Delta\theta = 2\sin^{-1}(0.443\lambda/L_X), \tag{3.3-43}$$

where L_X is the length (in meters) of the rectangular piston in the X direction (see Example 3.2-1). In order to make a fair comparison, consider rectangular and circular pistons of the same area. Therefore, if we let $L_Y = L_X$, then the area of the rectangular piston is L_X^2. If we then let

$$L_X^2 = \pi a^2, \tag{3.3-44}$$

where πa^2 is the area of the circular piston, then

$$L_X = \sqrt{\pi}\, a, \tag{3.3-45}$$

and by substituting (3.3-45) into (3.3-43), we finally obtain

$$\Delta\theta = 2\sin^{-1}(0.250\,\lambda/a) \tag{3.3-46}$$

for the rectangular piston. By comparing (3.3-42) for a circular piston with (3.3-46) for a rectangular piston with the same area as a circular piston, it can be seen that a circular piston will have a slightly larger 3-dB beamwidth. However, recall that the levels of the first sidelobes of the normalized, far-field beam patterns of a circular piston and a rectangular amplitude window are approximately -17.6 dB and -13.3 dB, respectively. The slight increase in beamwidth is offset by an additional, approximate, 4.3 dB sidelobe suppression provided by a circular piston. ■

3.4 Beam Steering

Following the same procedure used in Section 2.4, but generalizing it for the planar aperture problem, let $D(f, f_X, f_Y)$ be the far-field beam pattern of the amplitude window $a(f, x_A, y_A)$:

$$\begin{aligned} D(f, f_X, f_Y) &= F_{x_A} F_{y_A}\{a(f, x_A, y_A)\} \\ &= \int_{-\infty}^{\infty}\int_{-\infty}^{\infty} a(f, x_A, y_A)\exp[+j2\pi(f_X x_A + f_Y y_A)]\,dx_A\,dy_A. \end{aligned} \tag{3.4-1}$$

Let $D'(f, f_X, f_Y)$ be the far-field beam pattern of the aperture function $A(f, x_A, y_A)$ given by (3.1-10):

$$\begin{aligned} D'(f, f_X, f_Y) &= F_{x_A} F_{y_A}\{a(f, x_A, y_A)\exp[+j\theta(f, x_A, y_A)]\} \\ &= \int_{-\infty}^{\infty}\int_{-\infty}^{\infty} a(f, x_A, y_A)\exp[+j\theta(f, x_A, y_A)]\exp[+j2\pi(f_X x_A + f_Y y_A)]\,dx_A\,dy_A. \end{aligned}$$

(3.4-2)

Now let the phase response across the surface of the aperture be a linear function of both x_A and y_A, that is, let

$$\theta(f, x_A, y_A) = -2\pi f_X' x_A - 2\pi f_Y' y_A, \tag{3.4-3}$$

where

$$f_X' = u'/\lambda = \sin\theta' \cos\psi'/\lambda \tag{3.4-4}$$

and

$$f_Y' = v'/\lambda = \sin\theta' \sin\psi'/\lambda. \tag{3.4-5}$$

If (3.4-3) is substituted into (3.4-2), then

$$\begin{aligned} D'(f, f_X, f_Y) &= F_{x_A} F_{y_A}\left\{a(f, x_A, y_A)\exp\left[-j2\pi(f_X' x_A + f_Y' y_A)\right]\right\} \\ &= \int_{-\infty}^{\infty}\int_{-\infty}^{\infty} a(f, x_A, y_A)\exp\left\{+j2\pi\left[(f_X - f_X')x_A + (f_Y - f_Y')y_A\right]\right\}dx_A\, dy_A, \end{aligned} \tag{3.4-6}$$

and by comparing (3.4-6) with (3.4-1),

$$D'(f, f_X, f_Y) = D(f, f_X - f_X', f_Y - f_Y'), \tag{3.4-7}$$

or

$$F_{x_A} F_{y_A}\left\{a(f, x_A, y_A)\exp\left[-j2\pi(f_X' x_A + f_Y' y_A)\right]\right\} = D(f, f_X - f_X', f_Y - f_Y'), \tag{3.4-8}$$

where $D(f, f_X, f_Y)$ is given by (3.4-1). Equation (3.4-7) can also be expressed as

$$D'(f, u, v) = D(f, u - u', v - v'), \tag{3.4-9}$$

since $D(f, f_X, f_Y) = D(f, u/\lambda, v/\lambda) \to D(f, u, v)$. Therefore, a linear phase response across the surface of the aperture in both the X and Y directions will cause the beam pattern $D(f, u, v)$ to be steered in the direction $u = u'$ and $v = v'$ in direction-cosine space, which is equivalent to steering (tilting) the beam pattern to $\theta = \theta'$ and $\psi = \psi'$. Equation (3.4-9) indicates that the far-field beam pattern $D'(f, u, v)$ is simply a translated or shifted version of $D(f, u, v)$ in the uv plane, implying that the shape and beamwidth of the mainlobe of the beam pattern remain unchanged (see Fig. 3.4-1). However, recall from linear aperture theory, that when a beam pattern is steered, the shape and beamwidth of the mainlobe of the beam pattern *do change* when the magnitude of the beam pattern is plotted as a function of the spherical angles θ and ψ (see Section 2.5).

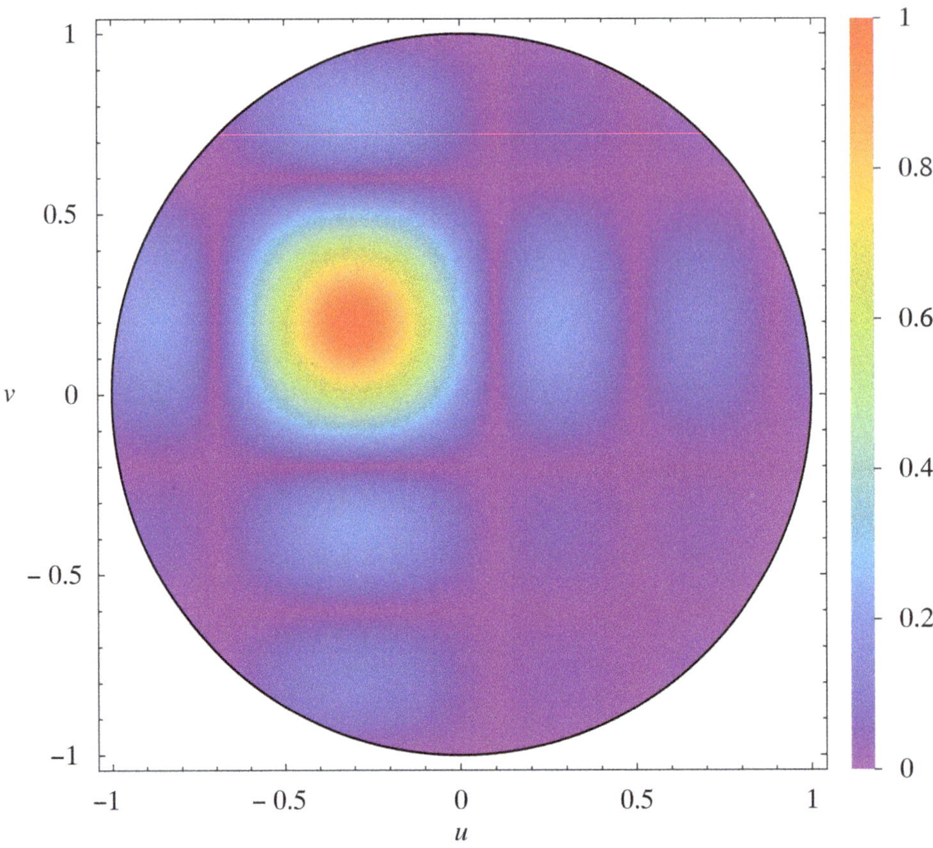

Figure 3.4-1 Magnitude of the normalized, far-field beam pattern of a rectangular piston lying in the XY plane, steered in the direction $u = u' = -0.3$ and $v = v' = 0.2$ in direction-cosine space, for $L_Y = L_X = L$ and $L/\lambda = 2.5$.

3.5 The Near-Field Beam Pattern of a Planar Aperture

In this section we shall demonstrate that the near-field beam pattern of a planar aperture is just a special case of the near-field beam pattern of a volume aperture. In Section 1.2 it was shown that the near-field beam pattern (directivity function) of a closed-surface, volume aperture – based on the Fresnel approximation of a time-independent, free-space, Green's function – is given by the following three-dimensional spatial Fresnel transform:

$$\begin{aligned}\mathcal{D}(f, r, \boldsymbol{\alpha}) &= F_{\mathbf{r}_A}\left\{ A(f, \mathbf{r}_A)\exp\left(-j\frac{k}{2r}r_A^2\right)\right\} \\ &= \int_{-\infty}^{\infty} A(f, \mathbf{r}_A)\exp\left(-j\frac{k}{2r}r_A^2\right)\exp(+j2\pi\boldsymbol{\alpha}\bullet\mathbf{r}_A)\,d\mathbf{r}_A,\end{aligned} \tag{3.5-1}$$

where

$$\mathbf{r}_A = (x_A, y_A, z_A), \tag{3.5-2}$$

$$k = 2\pi f/c = 2\pi/\lambda, \tag{3.5-3}$$

$$r_A^2 = |\mathbf{r}_A|^2 = x_A^2 + y_A^2 + z_A^2, \tag{3.5-4}$$

$$r = |\mathbf{r}| = \sqrt{x^2 + y^2 + z^2}, \tag{3.5-5}$$

$$\boldsymbol{\alpha} = (f_X, f_Y, f_Z), \tag{3.5-6}$$

$$\boldsymbol{\alpha} \cdot \mathbf{r}_A = f_X x_A + f_Y y_A + f_Z z_A, \tag{3.5-7}$$

$$d\mathbf{r}_A = dx_A\, dy_A\, dz_A, \tag{3.5-8}$$

and $F_{\mathbf{r}_A}\{\bullet\}$ is shorthand notation for $F_{x_A} F_{y_A} F_{z_A}\{\bullet\}$. The complex aperture function $A(f, \mathbf{r}_A)$ can be expressed as

$$A(f, \mathbf{r}_A) = a(f, \mathbf{r}_A)\exp[+j\theta(f, \mathbf{r}_A)], \tag{3.5-9}$$

where $a(f, \mathbf{r}_A)$ is the *amplitude response* and $\theta(f, \mathbf{r}_A)$ is the *phase response* of the aperture at $\mathbf{r}_A$. Both $a(f, \mathbf{r}_A)$ and $\theta(f, \mathbf{r}_A)$ are *real* functions. The function $a(f, \mathbf{r}_A)$ is also known as the *amplitude window*. The units of $a(f, \mathbf{r}_A)$ and, hence, $A(f, \mathbf{r}_A)$ depend on whether the aperture is a transmit aperture (see Table 1B-1 in Appendix 1B) or a receive aperture (see Table 1B-2 in Appendix 1B). The units of $\theta(f, \mathbf{r}_A)$ are radians.

Now consider the case of a *planar aperture* of arbitrary shape lying in the XY plane as shown in Fig. 3.1-1. As was mentioned earlier, a planar aperture can represent either a single electroacoustic transducer (e.g., a rectangular or circular piston) or a planar array of many individual electroacoustic transducers. Also shown in Fig. 3.1-1 is a field point with spherical coordinates (r, θ, ψ). The field point is in the Fresnel region of the aperture, where the Fresnel angle criterion given by either (1.2-35) or (1.2-91) is satisfied, and the range r satisfies the Fresnel range criterion given by $1.356R_A < r < \pi R_A^2/\lambda$ [see either (1.2-40) or (1.2-92)], where R_A is the maximum radial extent of the planar aperture.

Since the planar aperture in Fig. 3.1-1 is lying in the XY plane, the position vector to a point on the surface of the aperture is given by

$$\mathbf{r}_A = (x_A, y_A, 0), \tag{3.5-10}$$

and as a result,

$$A(f, \mathbf{r}_A) = A(f, x_A, y_A, z_A) \rightarrow A(f, x_A, y_A)\delta(z_A). \tag{3.5-11}$$

Therefore, by substituting (3.5-11) into (3.5-1) and using the sifting property of impulse functions, we obtain the following two-dimensional spatial Fresnel transform for the near-field beam pattern of a planar aperture lying in the XY plane:

$$\begin{aligned} \mathcal{D}(f, r, f_X, f_Y) &= F_{x_A} F_{y_A}\left\{ A(f, x_A, y_A)\exp\left[-j\frac{k}{2r}(x_A^2 + y_A^2)\right]\right\} \\ &= \int_{-\infty}^{\infty}\int_{-\infty}^{\infty} A(f, x_A, y_A)\exp\left[-j\frac{k}{2r}(x_A^2 + y_A^2)\right]\times \\ &\qquad \exp\left[+j2\pi(f_X x_A + f_Y y_A)\right]dx_A\, dy_A \end{aligned} \tag{3.5-12}$$

where

$$A(f, x_A, y_A) = a(f, x_A, y_A)\exp[+j\theta(f, x_A, y_A)] \tag{3.5-13}$$

is the complex frequency response (complex aperture function) of the planar aperture,

$$k = 2\pi f/c = 2\pi/\lambda \tag{3.5-14}$$

is the wavenumber in radians per meter, and

$$f_X = u/\lambda = \sin\theta\cos\psi/\lambda \tag{3.5-15}$$

and

$$f_Y = v/\lambda = \sin\theta\sin\psi/\lambda \tag{3.5-16}$$

are spatial frequencies in the X and Y directions, respectively, with units of cycles per meter [see (1.2-43) and (1.2-44)]. Since f_X and f_Y can be expressed in terms of the spherical angles θ and ψ, the near-field beam pattern can ultimately be expressed as a function of frequency f, the range r to a field point, and the spherical angles θ and ψ, that is, $\mathcal{D}(f, r, f_X, f_Y) \rightarrow \mathcal{D}(f, r, \theta, \psi)$. If the planar aperture is a transmit aperture, then the aperture function $A(f, x_A, y_A)$ has units of $\left(\left(\text{m}^3/\text{sec}\right)/\text{V}\right)/\text{m}^2$ and the near-field beam pattern $\mathcal{D}(f, r, f_X, f_Y)$ has units of $\left(\text{m}^3/\text{sec}\right)/\text{V}$. However, if the planar aperture is a receive aperture, then $A(f, x_A, y_A)$ has units of $\left(\text{V}/\left(\text{m}^2/\text{sec}\right)\right)/\text{m}^2$ and $\mathcal{D}(f, r, f_X, f_Y)$ has units of $\text{V}/\left(\text{m}^2/\text{sec}\right)$. These units are summarized in Table 3.5-1.

Table 3.5-1 Units of the Complex Aperture Function $A(f, x_A, y_A)$ and Corresponding Near-Field Beam Pattern $\mathcal{D}(f, r, f_X, f_Y)$ for a Planar Aperture

Planar Aperture	$A(f, x_A, y_A)$	$\mathcal{D}(f, r, f_X, f_Y)$
transmit	$\left(\left(\text{m}^3/\text{sec}\right)/\text{V}\right)/\text{m}^2$	$\left(\text{m}^3/\text{sec}\right)/\text{V}$
receive	$\left(\text{V}/\left(\text{m}^2/\text{sec}\right)\right)/\text{m}^2$	$\text{V}/\left(\text{m}^2/\text{sec}\right)$

Equations (3.5-1) and (3.5-12) are most accurate in the Fresnel region of an aperture where both the Fresnel angle criterion and the Fresnel range criterion are satisfied. They are less accurate in the near-field region outside the Fresnel region where $r < r_{\text{NF/FF}}$, but the Fresnel angle criterion is *not* satisfied. As was discussed in Subsection 1.2.1, the Fresnel region is only a subset of the near-field.

Finally, if a planar aperture lies in the XZ plane instead of the XY plane, then simply replace y_A and f_Y with z_A and f_Z, respectively, in (3.5-12), where spatial frequency f_Z is given by (1.2-45). And if a planar aperture lies in the YZ plane instead of the XY plane, then simply replace x_A and f_X with z_A and f_Z, respectively, in (3.5-12).

3.5.1 Beam Steering and Aperture Focusing

Following the same procedure used in Subsection 2.6.2, but generalizing it for the planar aperture problem, let the phase response across the surface of the aperture be equal to the sum of linear and quadratic functions of both x_A and y_A, that is, let

$$\theta(f, x_A, y_A) = -2\pi f'_X x_A - 2\pi f'_Y y_A + \frac{k}{2r'}(x_A^2 + y_A^2), \tag{3.5-17}$$

where

$$f'_X = u'/\lambda = \sin\theta' \cos\psi'/\lambda, \tag{3.5-18}$$

$$f'_Y = v'/\lambda = \sin\theta' \sin\psi'/\lambda, \tag{3.5-19}$$

and

$$k = 2\pi f/c = 2\pi/\lambda. \tag{3.5-20}$$

The parameter r' in (3.5-17) is referred to as the *focal range*. It is the near-field range from the aperture where the far-field beam pattern of the aperture will be in focus. Note that the third term on the right-hand side of (3.5-17) is a paraboloid that passes through the origin.

If (3.5-13) and (3.5-17) are substituted into (3.5-12), then

$$\mathcal{D}(f,r,f_X,f_Y)=\int_{-\infty}^{\infty}\int_{-\infty}^{\infty} a(f,x_A,y_A)\exp\left[-j\frac{k}{2}\left(\frac{1}{r}-\frac{1}{r'}\right)(x_A^2+y_A^2)\right]\times \exp\{+j2\pi[(f_X-f_X')x_A+(f_Y-f_Y')y_A]\}dx_A\,dy_A, \tag{3.5-21}$$

and if (3.5-21) is evaluated at the near-field range $r=r'$, then it reduces to

$$\mathcal{D}(f,r',f_X,f_Y)=\int_{-\infty}^{\infty}\int_{-\infty}^{\infty} a(f,x_A,y_A)\exp\{+j2\pi[(f_X-f_X')x_A+(f_Y-f_Y')y_A]\}dx_A\,dy_A \tag{3.5-22}$$

or

$$\begin{aligned}\mathcal{D}(f,r',f_X,f_Y)&=F_{x_A}F_{y_A}\{a(f,x_A,y_A)\exp[-j2\pi(f_X'x_A+f_Y'y_A)]\}\\&=D(f,f_X-f_X',f_Y-f_Y').\end{aligned} \tag{3.5-23}$$

Equation (3.5-23) indicates that the far-field beam pattern $D(f,f_X,f_Y)$ of the amplitude response (amplitude window) $a(f,x_A,y_A)$ is *in focus* at the near-field range $r=r'$ meters from the aperture, and has been *steered* in the direction $u=u'$ and $v=v'$ in direction-cosine space, which is equivalent to steering (tilting) the beam pattern to $\theta=\theta'$ and $\psi=\psi'$ [see (3.5-18) and (3.5-19)]. However, in the near-field region of the aperture outside the Fresnel region, a quadratic phase response can only approximately focus the far-field beam pattern.

Problems

Section 3.1

3-1 Find both the unnormalized and normalized, far-field beam patterns of the following aperture function, where $A_A(f)$ is a real, nonnegative function of frequency:

$$A(f,x_A,y_A)=A_A(f)\cos\left(\frac{\pi x_A}{L_X}\right)\text{rect}\left(\frac{x_A}{L_X}\right)\cos\left(\frac{\pi y_A}{L_Y}\right)\text{rect}\left(\frac{y_A}{L_Y}\right).$$

Section 3.1 Appendix 3A

3-2 If the aperture function given in Problem 3-1 is a *transmit* aperture function, then find the corresponding transmitter sensitivity function.

Section 3.2 Appendix 3A

3-3 Find the transmitter sensitivity function for the following complex frequency response of a rectangular piston, where $A_A(f)$ is a real, nonnegative function of frequency:

$$A_T(f, x_A, y_A) = A_A(f)\operatorname{rect}(x_A/L_X)\operatorname{rect}(y_A/L_Y).$$

Section 3.2

3-4 Find the spherical angles θ and ψ from the following (u,v) coordinates:

(a) $u=-0.3$ and $v=0.2$

(b) $u=0.5$ and $v=-0.5$

(c) $u=0.1$ and $v=0.4$

(d) $u=-0.2$ and $v=-0.7$

3-5 Consider a rectangular-shaped, planar array of electroacoustic transducers (an example of a rectangular-shaped, planar aperture) lying in the XY plane. If the operating frequency is 31 kHz and the aperture is modeled as a rectangular piston, then what should the lengths of the aperture be in the X and Y directions if the 3-dB beamwidths of the vertical, far-field beam patterns in the XZ and YZ planes are required to be $1°$ and $25°$, respectively.

Section 3.3

3-6 Verify (3.3-22).

3-7 Verify (3.3-32).

3-8 Verify (3.3-34). Note:

$$J_0(0)=1,$$

$$J_n(0)=0, \qquad n \neq 0,$$

$$2J_n'(x) = J_{n-1}(x) - J_{n+1}(x),$$

and

$$J_n'(x) = \frac{d}{dx} J_n(x) .$$

Section 3.3 Appendix 3A

3-9 Find the transmitter sensitivity function for the following complex frequency response of a circular piston, where $A_A(f)$ is a real, nonnegative function of frequency:

$$A_T(f, r_A, \phi_A) = A_A(f)\text{circ}(r_A/a) .$$

3-10 If the transmitter sensitivity level (transmitting voltage response) of a circular piston at 14 kHz is $150\text{ dB re }1\,\mu\text{Pa/V}$ at 1 m, then what is the value of the magnitude of the transmitter sensitivity function in $\left(\text{m}^3/\text{sec}\right)/\text{V}$? Use $\rho_0 = 1026\text{ kg/m}^3$ for the ambient density of seawater. The radius a of the circular piston is less than 0.316 m or 12.4 in.

3-11 If the receiver sensitivity level (open circuit receiving response) of a circular piston at 14 kHz is $-170\text{ dB re }1\text{ V}/\mu\text{Pa}$, then what is the value of the magnitude of the receiver sensitivity function in $\text{V}/\left(\text{m}^2/\text{sec}\right)$? Use $\rho_0 = 1026\text{ kg/m}^3$ for the ambient density of seawater.

Section 3.3

3-12 Consider a thin, planar electroacoustic transducer, circular in shape with radius b meters, lying in the XY plane. The complex frequency response (complex aperture function) of the transducer is circularly symmetric and is given by

$$A_r(f, r_A) = \begin{cases} +A_A, & 0 \le r_A < a, \\ -A_A, & a \le r_A \le b, \end{cases}$$

where A_A is a real, positive constant and $b > a$. Find the unnormalized, far-field beam pattern of this transducer.

3-13 Consider a rectangular and circular piston of *equal areas* lying in the XY plane. With $L_Y = L_X$ and for $a/\lambda = 0.5$, 1, 2, and 4, compute the 3-dB beamwidths in degrees of the vertical, far-field beam patterns of both pistons in the XZ plane.

Section 3.4

3-14 Repeat Problem 3-1 using the following modified aperture function, where $A_A(f)$ is a real, nonnegative function of frequency:

$$A(f, x_A, y_A) = A_A(f)\cos\left(\frac{\pi x_A}{L_X}\right)\text{rect}\left(\frac{x_A}{L_X}\right)\exp(-j2\pi f'_X x_A)\times$$
$$\cos\left(\frac{\pi y_A}{L_Y}\right)\text{rect}\left(\frac{y_A}{L_Y}\right)\exp(-j2\pi f'_Y y_A).$$

Section 3.5

3-15 The spherical coordinates of a sound source (target) as measured from the center of a rectangular-shaped, planar array of electroacoustic transducers (an example of a rectangular-shaped, planar aperture) lying in the XY plane are $r_S = 210$ m, $\theta_S = 60°$, and $\psi_S = 105°$. If $L_Y = L_X = 3$ m, $f = 25$ kHz, and $c = 1500$ m/sec,

(a) is the sound source in the aperture's Fresnel region or far-field?

(b) What must the phase response across the surface of the aperture be if the aperture's far-field beam pattern is to be focused (if required) and steered to coordinates (r_S, θ_S, ψ_S)?

Appendix 3A Transmitter and Receiver Sensitivity Functions of a Planar Transducer

Transmitter Sensitivity Function

For the case of a planar electroacoustic transducer lying in the XY plane – such as a rectangular or circular piston – that is being used as a transmitter, the complex, *transmitter sensitivity function* $\mathcal{S}_T(f)$ is related to the complex, transmit frequency response $A_T(f, x_A, y_A)$ of the transducer as follows (see Example 3B-1 in Appendix 3B):

$$\boxed{\mathcal{S}_T(f) = \int_{-\infty}^{\infty}\int_{-\infty}^{\infty} A_T(f, x_A, y_A)\,dx_A\,dy_A} \qquad \text{(3A-1)}$$

where $\mathcal{S}_T(f)$ has units of $(\text{m}^3/\text{sec})/\text{V}$ and $A_T(f, x_A, y_A)$ has units of

$\left(\left(\mathrm{m}^3/\sec\right)/\mathrm{V}\right)/\mathrm{m}^2$ (see Table 3.1-1). However, manufacturers of electroacoustic transducers usually describe the *transmitting voltage response* of a transducer by plotting the *transmitter sensitivity level* $\mathrm{TSL}(f)$ in $\mathrm{dB\ re\ TS_{ref}}$ at $1\ \mathrm{m}$ versus frequency f in hertz. The notation "$\mathrm{dB\ re\ TS_{ref}}$" means decibels relative to a reference transmitter sensitivity. In underwater acoustics, the reference transmitter sensitivity $\mathrm{TS_{ref}}$ is usually $1\ \mu\mathrm{Pa/V}$. The corresponding *transmitter sensitivity* $\mathrm{TS}(f)$ with units of $\mathrm{Pa/V}$ is given by

$$\boxed{\mathrm{TS}(f) = \mathrm{TS_{ref}}\, 10^{[\mathrm{TSL}(f)/20]}} \tag{3A-2}$$

The transmitter sensitivity $\mathrm{TS}(f)$ is a *real, nonnegative* function of frequency.

Values for the magnitude of the transmitter sensitivity function, $|\mathcal{S}_T(f)|$ in $\left(\mathrm{m}^3/\sec\right)/\mathrm{V}$, for a planar electroacoustic transducer lying in the XY plane and centered at the origin of the coordinate system as shown in Figures 3.2-1 and 3.3-1, can be obtained from measured values of the transmitter sensitivity, $\mathrm{TS}(f)$ in $\mathrm{Pa/V}$, by using the following formula (see Example 3B-1 in Appendix 3B):

$$\boxed{|\mathcal{S}_T(f)| = \frac{2z}{f\rho_0}\mathrm{TS}(f)\bigg|_{z=1\ \mathrm{m}}, \qquad R_A < 0.316\ \mathrm{m}} \tag{3A-3}$$

where the field point is located at coordinates $(0, 0, 1\ \mathrm{m})$ and is at *broadside* relative to the transducer, ρ_0 is the constant ambient (equilibrium) density of the fluid medium in kilograms per cubic meter, $\mathrm{TS}(f)$ is given by (3A-2), and R_A is the maximum *radial* extent of the transducer in meters. If the transducer is rectangular in shape, then R_A is given by (3B-28). If the transducer is circular in shape, then R_A is equal to the radius of the transducer. Note that (3A-3) is valid only if R_A is less than $0.316\ \mathrm{m}$ or $12.4\ \mathrm{in}$.

Receiver Sensitivity Function

For the case of a planar electroacoustic transducer lying in the XY plane – such as a rectangular or circular piston – that is being used as a receiver, the complex, *receiver sensitivity function* $\mathcal{S}_R(f)$ is related to the complex, receive frequency response $A_R(f, x_A, y_A)$ of the transducer as follows:

$$\boxed{\mathcal{S}_R(f) = \int_{-\infty}^{\infty}\int_{-\infty}^{\infty} A_R(f, x_A, y_A)\, dx_A\, dy_A} \tag{3A-4}$$

where $\mathcal{S}_R(f)$ has units of $\text{V}/(\text{m}^2/\text{sec})$ and $A_R(f, x_A, y_A)$ has units of $(\text{V}/(\text{m}^2/\text{sec}))/\text{m}^2$ (see Table 3.1-1). However, manufacturers of electroacoustic transducers usually describe the *open circuit receiving response* of a transducer by plotting the *receiver sensitivity level* $\text{RSL}(f)$ in $\text{dB re RS}_{\text{ref}}$ versus frequency f in hertz. The notation "$\text{dB re RS}_{\text{ref}}$" means decibels relative to a reference receiver sensitivity. In underwater acoustics, the reference receiver sensitivity RS_{ref} is usually $1\,\text{V}/\mu\text{Pa}$. The corresponding *receiver sensitivity* $\text{RS}(f)$ with units of V/Pa is given by

$$\boxed{\text{RS}(f) = \text{RS}_{\text{ref}}\, 10^{[\text{RSL}(f)/20]}} \tag{3A-5}$$

The receiver sensitivity $\text{RS}(f)$ is a *real, nonnegative* function of frequency. What we need to do next is to derive the conversion factor that will allow us to convert m^2/sec to Pa for a time-harmonic acoustic field.

The derivation of the desired conversion factor in this case is simple. By computing the magnitude of (3B-24), we obtain

$$|p_f(\mathbf{r})| = 2\pi f \rho_0 |\varphi_f(\mathbf{r})|, \tag{3A-6}$$

from which we obtain the conversion factor

$$\boxed{\frac{|p_f(\mathbf{r})|}{|\varphi_f(\mathbf{r})|} = 2\pi f \rho_0 \,\frac{\text{Pa}}{\text{m}^2/\text{sec}}} \tag{3A-7}$$

where $p_f(\mathbf{r}) \equiv p_f(x, y, z)$ and $\varphi_f(\mathbf{r}) \equiv \varphi_f(x, y, z)$. The conversion factor on the right-hand-side of (3A-7) is used to convert velocity potential in m^2/sec to acoustic pressure in Pa for a time-harmonic acoustic field. Therefore, values for the magnitude of the receiver sensitivity function, $|\mathcal{S}_R(f)|$ in $\text{V}/(\text{m}^2/\text{sec})$, can be obtained from measured values of the receiver sensitivity, $\text{RS}(f)$ in V/Pa, by using the following formula:

$$\boxed{|\mathcal{S}_R(f)| = 2\pi f \rho_0 \text{RS}(f)} \tag{3A-8}$$

where ρ_0 is the constant ambient (equilibrium) density of the fluid medium in kilograms per cubic meter and $\text{RS}(f)$ is given by (3A-5).

Appendix 3B Radiation from a Planar Aperture

An exact solution of the linear wave equation given by (1.2-1) for free-space propagation in an ideal (nonviscous), homogeneous, fluid medium is given by [see (1.2-3)]

$$\varphi(t,\mathbf{r}) = \int_{-\infty}^{\infty} \int_{V_0} X(f,\mathbf{r}_0) A_T(f,\mathbf{r}_0) g_f(\mathbf{r}|\mathbf{r}_0) dV_0 \exp(+j2\pi ft) df, \quad \text{(3B-1)}$$

where $\varphi(t,\mathbf{r})$ is the scalar velocity potential in squared meters per second, $X(f,\mathbf{r}_0)$ is the complex frequency spectrum of the input electrical signal at location $\mathbf{r}_0$ of the transmit aperture with units of volts per hertz, $A_T(f,\mathbf{r}_0)$ is the complex frequency response of the transmit aperture at $\mathbf{r}_0$ with units of $\left(\left(\mathrm{m}^3/\mathrm{sec}\right)/\mathrm{V}\right)/\mathrm{m}^3$,

$$g_f(\mathbf{r}|\mathbf{r}_0) = -\frac{\exp(-jk|\mathbf{r}-\mathbf{r}_0|)}{4\pi|\mathbf{r}-\mathbf{r}_0|} = -\frac{\exp(-jkR)}{4\pi R} \quad \text{(3B-2)}$$

is the time-independent, free-space, Green's function of an unbounded, ideal (nonviscous), homogeneous, fluid medium with units of inverse meters,

$$k = 2\pi f/c = 2\pi/\lambda \quad \text{(3B-3)}$$

is the wavenumber in radians per meter, $c = f\lambda$ is the constant speed of sound in the fluid medium in meters per second, λ is the wavelength in meters,

$$\mathbf{r} = x\hat{x} + y\hat{y} + z\hat{z} \quad \text{(3B-4)}$$

is the position vector to a field point,

$$\mathbf{r}_0 = x_0\hat{x} + y_0\hat{y} + z_0\hat{z} \quad \text{(3B-5)}$$

is the position vector to a source point, and

$$R = |\mathbf{r}-\mathbf{r}_0| = \sqrt{(x-x_0)^2 + (y-y_0)^2 + (z-z_0)^2} \quad \text{(3B-6)}$$

is the range in meters between a source point and a field point. If an identical input electrical signal is applied at all locations $\mathbf{r}_0$ of the transmit aperture, then (see Example 1.2-1)

$$x(t, \mathbf{r}_0) = x(t) \tag{3B-7}$$

and

$$X(f, \mathbf{r}_0) = X(f), \tag{3B-8}$$

where $X(f)$ is the complex frequency spectrum of $x(t)$. Substituting (3B-8) into (3B-1) yields

$$\varphi(t, \mathbf{r}) = \int_{-\infty}^{\infty} X(f) \int_{V_0} A_T(f, \mathbf{r}_0) g_f(\mathbf{r}|\mathbf{r}_0) dV_0 \exp(+j2\pi ft) df. \tag{3B-9}$$

For the case of a planar aperture lying in the XY plane,

$$\mathbf{r}_0 = (x_0, y_0, 0), \tag{3B-10}$$

and as a result,

$$A_T(f, \mathbf{r}_0) = A_T(f, x_0, y_0, z_0) \rightarrow A_T(f, x_0, y_0)\delta(z_0). \tag{3B-11}$$

Substituting (3B-11) into (3B-9) yields

$$\boxed{\varphi(t, \mathbf{r}) = \int_{-\infty}^{\infty} X(f) \int_{-\infty}^{\infty} \int_{-\infty}^{\infty} A_T(f, x_0, y_0) g_f(\mathbf{r}|\mathbf{r}_0) dx_0\, dy_0 \exp(+j2\pi ft) df} \tag{3B-12}$$

where $g_f(\mathbf{r}|\mathbf{r}_0)$ is given by (3B-2) and

$$R = |\mathbf{r} - \mathbf{r}_0| = \sqrt{(x - x_0)^2 + (y - y_0)^2 + z^2}. \tag{3B-13}$$

Equation (3B-12) is a general expression for the scalar velocity potential of the acoustic field radiated by a planar aperture lying in the XY plane. The corresponding solution for the radiated acoustic pressure $p(t, \mathbf{r})$ in pascals can be obtained by substituting (3B-12) into (1.2-8). Doing so yields

$$\boxed{p(t, \mathbf{r}) = -j2\pi\rho_0 \int_{-\infty}^{\infty} fX(f) \int_{-\infty}^{\infty} \int_{-\infty}^{\infty} A_T(f, x_0, y_0) g_f(\mathbf{r}|\mathbf{r}_0) dx_0\, dy_0 \exp(+j2\pi ft) df} \tag{3B-14}$$

where ρ_0 is the constant ambient (equilibrium) density of the fluid medium in kilograms per cubic meter, and $g_f(\mathbf{r}|\mathbf{r}_0)$ is given by (3B-2) and (3B-13).

If the input electrical signal is time-harmonic, that is, if (see Example 1.2-1)

$$x(t) = A_x \exp(+j2\pi f_0 t), \tag{3B-15}$$

where A_x is the complex amplitude with units of volts, then

$$X(f) = A_x \delta(f - f_0), \tag{3B-16}$$

where the impulse function $\delta(f - f_0)$ has units of inverse hertz. Substituting (3B-16) into (3B-12) yields

$$\varphi(t, \mathbf{r}) = A_x \int_{-\infty}^{\infty}\int_{-\infty}^{\infty} A_T(f_0, x_0, y_0) g_{f_0}\left(\mathbf{r}\,\middle|\,\mathbf{r}_0\right) dx_0\, dy_0 \exp(+j2\pi f_0 t), \tag{3B-17}$$

and by replacing f_0 with f,

$$\boxed{\varphi(t, \mathbf{r}) = A_x \int_{-\infty}^{\infty}\int_{-\infty}^{\infty} A_T(f, x_0, y_0) g_f\left(\mathbf{r}\,\middle|\,\mathbf{r}_0\right) dx_0\, dy_0 \exp(+j2\pi f t)} \tag{3B-18}$$

or

$$\varphi(t, \mathbf{r}) = \varphi_f(\mathbf{r}) \exp(+j2\pi f t), \tag{3B-19}$$

where

$$\varphi_f(\mathbf{r}) = A_x \int_{-\infty}^{\infty}\int_{-\infty}^{\infty} A_T(f, x_0, y_0) g_f\left(\mathbf{r}\,\middle|\,\mathbf{r}_0\right) dx_0\, dy_0 \tag{3B-20}$$

and $g_f\left(\mathbf{r}\,\middle|\,\mathbf{r}_0\right)$ is given by (3B-2) and (3B-13). Equation (3B-18) is the time-harmonic, scalar velocity potential of the acoustic field radiated by a planar aperture lying in the XY plane. The corresponding solution for the time-harmonic, radiated acoustic pressure $p(t, \mathbf{r})$ in pascals is given by

$$\boxed{p(t, \mathbf{r}) = -j2\pi f \rho_0 A_x \int_{-\infty}^{\infty}\int_{-\infty}^{\infty} A_T(f, x_0, y_0) g_f\left(\mathbf{r}\,\middle|\,\mathbf{r}_0\right) dx_0\, dy_0 \exp(+j2\pi f t)} \tag{3B-21}$$

or

$$p(t, \mathbf{r}) = p_f(\mathbf{r}) \exp(+j2\pi f t), \tag{3B-22}$$

where

$$p_f(\mathbf{r}) = -j2\pi f \rho_0 A_x \int_{-\infty}^{\infty}\int_{-\infty}^{\infty} A_T(f, x_0, y_0) g_f\left(\mathbf{r}\,\middle|\,\mathbf{r}_0\right) dx_0\, dy_0 \tag{3B-23}$$

and $g_f\left(\mathbf{r}\,\middle|\,\mathbf{r}_0\right)$ is given by (3B-2) and (3B-13). Note that

$$p_f(\mathbf{r}) = -j2\pi f \rho_0 \varphi_f(\mathbf{r}), \tag{3B-24}$$

where $\varphi_f(\mathbf{r})$ is given by (3B-20).

Example 3B-1 Transmitter Sensitivity Function and Source Strength of a Planar Transducer

Consider the case of a single electroacoustic transducer, rectangular in shape (an example of a planar aperture), with sides equal to L_X and L_Y meters in length, lying in the XY plane and centered at the origin of the coordinate system as shown in Fig. 3.2-1. The transducer extends from $-L_X/2$ to $L_X/2$ along the X axis, and from $-L_Y/2$ to $L_Y/2$ along the Y axis. Therefore,

$$\int_{-\infty}^{\infty}\int_{-\infty}^{\infty} A_T(f, x_0, y_0) g_f(\mathbf{r}|\mathbf{r}_0) dx_0\, dy_0 = \int_{-L_X/2}^{L_X/2}\int_{-L_Y/2}^{L_Y/2} A_T(f, x_0, y_0) g_f(\mathbf{r}|\mathbf{r}_0) dx_0\, dy_0\,, \tag{3B-25}$$

where $g_f(\mathbf{r}|\mathbf{r}_0)$ is given by (3B-2) and (3B-13). If the radiated acoustic field is measured at *broadside* at $\mathbf{r} = (0, 0, z)$, where $z \neq 0$, then (3B-13) reduces to

$$R = |\mathbf{r} - \mathbf{r}_0| = z\sqrt{1 + (r_0/z)^2}\,, \tag{3B-26}$$

where

$$r_0 = \sqrt{x_0^2 + y_0^2} \tag{3B-27}$$

is the range (polar radius) in meters to a source point in the XY plane. The maximum value of r_0, which is the maximum radial extent of the aperture R_A, is given by

$$R_A = \max r_0 = \sqrt{\left(\frac{L_X}{2}\right)^2 + \left(\frac{L_Y}{2}\right)^2}\,, \tag{3B-28}$$

and is obtained when x_0 and y_0 are equal to either the upper or lower limits of integration $\pm L_X/2$ and $\pm L_Y/2$, respectively. Setting r_0 equal to R_A in (3B-26) yields

$$R = |\mathbf{r} - \mathbf{r}_0| = z\sqrt{1 + (R_A/z)^2}\,, \tag{3B-29}$$

and if

$$\left(R_A/z\right)^2 < 0.1, \tag{3B-30}$$

or

$$R_A < 0.316z, \tag{3B-31}$$

then

$$R = \left|\mathbf{r} - \mathbf{r}_0\right| \approx z, \qquad R_A < 0.316z, \tag{3B-32}$$

for $\mathbf{r} = (0, 0, z)$ and $\mathbf{r}_0 = (x_0, y_0, 0)$, where $\left|x_0\right| \le L_X/2$ and $\left|y_0\right| \le L_Y/2$. Therefore, substituting (3B-32) into (3B-2) yields

$$g_f\left(\mathbf{r}\left|\mathbf{r}_0\right.\right) = g_f\left(0, 0, z\left|x_0, y_0, 0\right.\right) \approx -\frac{\exp(-jkz)}{4\pi z}, \qquad R_A < 0.316z, \tag{3B-33}$$

and substituting (3B-33) into (3B-25) yields

$$\int_{-\infty}^{\infty}\int_{-\infty}^{\infty} A_T(f, x_0, y_0)\, g_f\left(\mathbf{r}\left|\mathbf{r}_0\right.\right) dx_0\, dy_0 \approx -\frac{\exp(-jkz)}{4\pi z}\mathcal{S}_T(f), \qquad R_A < 0.316z, \tag{3B-34}$$

where

$$\boxed{\mathcal{S}_T(f) = \int_{-L_X/2}^{L_X/2}\int_{-L_Y/2}^{L_Y/2} A_T(f, x_0, y_0)\, dx_0\, dy_0} \tag{3B-35}$$

is the *transmitter sensitivity function* in $\left(\text{m}^3/\text{sec}\right)/\text{V}$. Next we shall use (3B-34) to compute the time-harmonic, radiated acoustic pressure at $\mathbf{r} = (0, 0, z)$.

The time-harmonic, radiated acoustic pressure at $\mathbf{r} = (0, 0, z)$ is given by

$$p(t, 0, 0, z) = p_f(0, 0, z)\exp(+j2\pi f t), \tag{3B-36}$$

where, by substituting (3B-34) into (3B-23),

$$p_f(0, 0, z) \approx j\frac{f\rho_0}{2z} A_x \mathcal{S}_T(f)\exp(-jkz), \qquad R_A < 0.316z, \tag{3B-37}$$

or

$$p_f(0, 0, z) \approx j\frac{f\rho_0}{2z} S_0 \exp(-jkz), \qquad R_A < 0.316z, \tag{3B-38}$$

where

$$\boxed{S_0 = A_x \mathcal{S}_T(f)} \tag{3B-39}$$

is the source strength of the planar transducer in cubic meters per second at frequency f hertz.

From (3B-36) it can be seen that

$$\left|p(t,0,0,z)\right| = \left|p_f(0,0,z)\right|, \tag{3B-40}$$

where from (3B-37)

$$\left|p_f(0,0,z)\right| \approx \frac{f\rho_0}{2z}\left|A_x\right|\left|\mathcal{S}_T(f)\right|, \qquad R_A < 0.316z. \tag{3B-41}$$

Therefore, using (3B-41) to solve for $\left|\mathcal{S}_T(f)\right|$ yields

$$\boxed{\left|\mathcal{S}_T(f)\right| \approx \frac{2z}{f\rho_0}\text{TS}(f), \qquad R_A < 0.316z} \tag{3B-42}$$

where

$$\boxed{\text{TS}(f) = \left|p_f(0,0,z)\right| / \left|A_x\right|} \tag{3B-43}$$

is the *transmitter sensitivity* in pascals per volt. Recall that the field point at coordinates $(0,0,z)$ is at *broadside* relative to the planar transducer. Evaluating (3B-42) at $z = 1\text{ m}$ yields

$$\boxed{\left|\mathcal{S}_T(f)\right| \approx \frac{2z}{f\rho_0}\text{TS}(f)\Big|_{z=1\text{ m}}, \qquad R_A < 0.316\text{ m}} \tag{3B-44}$$

where R_A is given by (3B-28). If the transducer is circular in shape, then R_A is equal to the radius of the transducer because r_0 given by (3B-27) is the polar radius in meters to a source point in the XY plane. Note that (3B-44) is valid only if the maximum *radial* extent of the aperture (transducer) R_A is less than 0.316 m or 12.4 in.

The magnitude of the source strength can be obtained from (3B-38) as follows:

$$\boxed{\left|S_0\right| \approx \frac{2z}{f\rho_0}\left|p_f(0,0,z)\right|, \qquad R_A < 0.316z} \tag{3B-45}$$

Evaluating (3B-45) at $z = 1\text{ m}$ yields

$$\boxed{|S_0| \approx \frac{2z}{f\rho_0}|p_f(0,0,z)|\Big|_{z=1\text{ m}}, \qquad R_A < 0.316\text{ m}} \tag{3B-46}$$

Note that (3B-46) is valid only if the maximum *radial* extent of the aperture (transducer) R_A is less than 0.316 m or 12.4 in. Also, from (3B-39),

$$\boxed{|S_0| = |A_x||\mathcal{S}_T(f)|} \tag{3B-47}$$

where the transmitter sensitivity function $\mathcal{S}_T(f)$ is given by (3B-35). ■

Appendix 3C Computing the Normalization Factor

The far-field beam pattern of a planar aperture lying in the XY plane is given by [see (3.1-9)]

$$D(f, f_X, f_Y) = \int_{-\infty}^{\infty}\int_{-\infty}^{\infty} A(f, x_A, y_A)\exp[+j2\pi(f_X x_A + f_Y y_A)]dx_A\, dy_A, \tag{3C-1}$$

where

$$A(f, x_A, y_A) = a(f, x_A, y_A)\exp[+j\theta(f, x_A, y_A)] \tag{3C-2}$$

is the complex frequency response (complex aperture function) of the planar aperture [see (3.1-10)]. Computing the magnitude of (3C-1) yields

$$\begin{aligned}
|D(f, f_X, f_Y)| &= \left|\int_{-\infty}^{\infty}\int_{-\infty}^{\infty} A(f, x_A, y_A)\exp[+j2\pi(f_X x_A + f_Y y_A)]dx_A\, dy_A\right| \\
&\le \int_{-\infty}^{\infty}\int_{-\infty}^{\infty} |A(f, x_A, y_A)\exp[+j2\pi(f_X x_A + f_Y y_A)]|dx_A\, dy_A \\
&\le \int_{-\infty}^{\infty}\int_{-\infty}^{\infty} |A(f, x_A, y_A)||\exp[+j2\pi(f_X x_A + f_Y y_A)]|dx_A\, dy_A \\
&\le \int_{-\infty}^{\infty}\int_{-\infty}^{\infty} |a(f, x_A, y_A)|dx_A\, dy_A.
\end{aligned} \tag{3C-3}$$

Therefore,

$$\max|D(f, f_X, f_Y)| = \int_{-\infty}^{\infty}\int_{-\infty}^{\infty} |a(f, x_A, y_A)|dx_A\, dy_A, \tag{3C-4}$$

and by following the steps in (3C-3), it can also be shown that

$$\max|\mathcal{S}(f)| = \int_{-\infty}^{\infty}\int_{-\infty}^{\infty} |a(f, x_A, y_A)| \, dx_A \, dy_A \,, \tag{3C-5}$$

where $\mathcal{S}(f)$ is the element sensitivity function (see Appendix 3A). The absolute value of the real amplitude response $a(f, x_A, y_A)$ is required in (3C-4) and (3C-5) because $a(f, x_A, y_A)$ can have, in general, both positive and negative values as a function of x_A and y_A.

Equation (3C-4) is the absolute maximum value of $|D(f, f_X, f_Y)|$. The actual maximum value may be less than (3C-4) as indicated by (3C-3). However, if $a(f, x_A, y_A) \geq 0$, then the normalization factor $D_{\max}$ is

$$\boxed{D_{\max} = \max|D(f, f_X, f_Y)| = \int_{-\infty}^{\infty}\int_{-\infty}^{\infty} a(f, x_A, y_A) \, dx_A \, dy_A} \tag{3C-6}$$

or

$$\boxed{D_{\max} = \max|\mathcal{S}(f)| = \int_{-\infty}^{\infty}\int_{-\infty}^{\infty} a(f, x_A, y_A) \, dx_A \, dy_A} \tag{3C-7}$$

Note that if (3C-1) is evaluated at broadside (i.e., at $f_X = 0$ and $f_Y = 0$) and if the phase response $\theta(f, x_A, y_A) = 0$, then

$$D(f, 0, 0) = \int_{-\infty}^{\infty}\int_{-\infty}^{\infty} a(f, x_A, y_A) \, dx_A \, dy_A \tag{3C-8}$$

and

$$|D(f, 0, 0)| = \left|\int_{-\infty}^{\infty}\int_{-\infty}^{\infty} a(f, x_A, y_A) \, dx_A \, dy_A\right| \leq \int_{-\infty}^{\infty}\int_{-\infty}^{\infty} |a(f, x_A, y_A)| \, dx_A \, dy_A \,. \tag{3C-9}$$

If in addition $a(f, x_A, y_A) \geq 0$, then

$$|D(f, 0, 0)| = \left|\int_{-\infty}^{\infty}\int_{-\infty}^{\infty} a(f, x_A, y_A) \, dx_A \, dy_A\right| = \int_{-\infty}^{\infty}\int_{-\infty}^{\infty} a(f, x_A, y_A) \, dx_A \, dy_A \,, \tag{3C-10}$$

and by comparing (3C-6) and (3C-10),

$$\boxed{D_{\max} = \max|D(f, f_X, f_Y)| = |D(f, 0, 0)|} \tag{3C-11}$$

If the planar aperture is circular in shape with radius a meters, then [see (3.3-9)]

$$D(f,\theta,\psi)=\int_0^{2\pi}\int_0^{a} A(f,r_A,\phi_A)\exp\left[+j\frac{2\pi r_A}{\lambda}\sin\theta\cos(\psi-\phi_A)\right]r_A\,dr_A\,d\phi_A\,, \tag{3C-12}$$

where

$$A(f,r_A,\phi_A)=a(f,r_A,\phi_A)\exp[+j\theta(f,r_A,\phi_A)] \tag{3C-13}$$

is the complex frequency response (complex aperture function) of the circular aperture, where r_A and ϕ_A are the polar coordinates of an aperture point [see (3.3-10)]. If $a(f,r_A,\phi_A)\geq 0$, then

$$\boxed{D_{\max}=\max|D(f,\theta,\psi)|=\int_0^{2\pi}\int_0^{a} a(f,r_A,\phi_A)r_A\,dr_A\,d\phi_A} \tag{3C-14}$$

or

$$\boxed{D_{\max}=\max|S(f)|=\int_0^{2\pi}\int_0^{a} a(f,r_A,\phi_A)r_A\,dr_A\,d\phi_A} \tag{3C-15}$$

Note that if (3C-12) is evaluated at broadside (i.e., at $\theta=0°$), and if the phase response $\theta(f,r_A,\phi_A)=0$ and the amplitude response $a(f,r_A,\phi_A)\geq 0$, then

$$\boxed{D_{\max}=\max|D(f,\theta,\psi)|=|D(f,0°,\psi)|} \tag{3C-16}$$

Chapter 4

Time-Average Radiated Acoustic Power

4.1 Directivity and Directivity Index

In this section we shall derive an equation for the *directivity* of an aperture, where the aperture is characterized by its far-field beam pattern. Recall that the term "aperture" can refer to either a single electroacoustic transducer or an array of electroacoustic transducers. The *directivity index* is simply the decibel equivalent of the directivity.

The directivity of an aperture, when used in the active mode as a transmitter, is a measure of its ability to concentrate the available acoustic power into a preferred direction. When used in the passive mode as a receiver, the directivity of an aperture is a measure of its ability to distinguish between several sound-sources located at different positions in the fluid medium. Therefore, directivity is basically a measure of the beamwidth and sidelobe levels of a far-field beam pattern. In the analysis that follows, the aperture is being used as a transmitter. Note that if the transducer or transducers that make up an aperture are *reversible*, that is, if they can be used as either transmitters or receivers, then the transmit and receive far-field beam patterns of the aperture are *identical*.

In order to derive an equation for the directivity of an aperture when used as a transmitter (a sound-source), we must derive equations for both the time-average intensity vector of the radiated acoustic field, and the time-average radiated acoustic power. However, in order to derive equations for the intensity and power, we must first derive equations for both the radiated acoustic pressure and the radiated acoustic fluid-velocity-vector due to the sound-source. We begin by using the following results obtained in Example 1.3-1: when an identical, time-harmonic, input electrical signal with complex amplitude A_x volts (V) is applied at all points on the surface of a *closed-surface*, transmit volume aperture, the scalar velocity potential in squared meters per second and the acoustic pressure in pascals (Pa) of the radiated acoustic field in the *far-field* region of the aperture are given by

$$\varphi(t,r,\theta,\psi) \approx -\frac{A_x}{4\pi r} D(f,\theta,\psi) \exp\left[+j2\pi f\left(t-\frac{r}{c}\right)\right] \tag{4.1-1}$$

and

$$p(t,r,\theta,\psi) \approx jk\rho_0 c \frac{A_x}{4\pi r} D(f,\theta,\psi) \exp\left[+j2\pi f\left(t-\frac{r}{c}\right)\right], \tag{4.1-2}$$

respectively, for $t \geq r/c$, where $D(f,\theta,\psi)$ is the unnormalized, far-field beam pattern of the aperture with units of $\left(\text{m}^3/\text{sec}\right)/\text{V}$,

$$k = 2\pi f/c = 2\pi/\lambda \tag{4.1-3}$$

is the wavenumber in radians per meter, ρ_0 is the constant ambient (equilibrium) density of the fluid medium in kilograms per cubic meter, and c is the constant speed of sound in the fluid medium in meters per second. Note that $1\text{ Pa} = 1\text{ N/m}^2$, where $1\text{ N} = 1\text{ kg-m/sec}^2$. Therefore, $1\text{ Pa} = 1\text{ kg}/(\text{m-sec}^2)$. Also note that the ambient density ρ_0 and the speed of sound c are constants because we are dealing with a homogeneous fluid medium. Since we already have an equation for the radiated acoustic pressure as given by (4.1-2), we need to derive an equation for the radiated acoustic fluid-velocity-vector next.

Recall from Subsection 1.2.1 that the acoustic fluid-velocity-vector (a.k.a. the acoustic particle-velocity-vector) $\mathbf{u}(t,r,\theta,\psi)$ in meters per second can be obtained from the scalar velocity potential $\varphi(t,r,\theta,\psi)$ as follows:

$$\mathbf{u}(t,r,\theta,\psi) = \nabla\varphi(t,r,\theta,\psi), \tag{4.1-4}$$

where

$$\nabla = \frac{\partial}{\partial r}\hat{r} + \frac{1}{r}\frac{\partial}{\partial\theta}\hat{\theta} + \frac{1}{r\sin\theta}\frac{\partial}{\partial\psi}\hat{\psi} \tag{4.1-5}$$

is the gradient expressed in the spherical coordinates (r,θ,ψ), and $\hat{r}$, $\hat{\theta}$, and $\hat{\psi}$ are *unit vectors* in the r, θ, and ψ directions, respectively. As will be shown later, we only need to derive an equation for the *radial* component of the acoustic fluid-velocity-vector because only the radial component contributes to the time-average radiated acoustic power. By referring to (4.1-4) and (4.1-5), the radial component of the acoustic fluid-velocity-vector is given by

$$\mathbf{u}_r(t,r,\theta,\psi) = \frac{\partial}{\partial r}\varphi(t,r,\theta,\psi)\hat{r}, \tag{4.1-6}$$

and by substituting (4.1-1) into (4.1-6), we obtain

$$\mathbf{u}_r(t,r,\theta,\psi) \approx -\frac{A_x}{4\pi}D(f,\theta,\psi)\frac{\partial}{\partial r}\left[\frac{\exp(-jkr)}{r}\right]\exp(+j2\pi ft)\hat{r}, \tag{4.1-7}$$

where k is given by (4.1-3). Since

$$\frac{\partial}{\partial r}\frac{\exp(-jkr)}{r} = -k^2\left[\frac{1}{(kr)^2} + j\frac{1}{kr}\right]\exp(-jkr), \tag{4.1-8}$$

and since the field point is in the far-field region of the aperture where it is assumed that $kr >> 1$, (4.1-8) reduces to

$$\frac{\partial}{\partial r}\frac{\exp(-jkr)}{r} \approx -jk\frac{\exp(-jkr)}{r}, \qquad kr >> 1. \tag{4.1-9}$$

Therefore, substituting (4.1-9) into (4.1-7) yields

$$\mathbf{u}_r(t,r,\theta,\psi) \approx jk\frac{A_x}{4\pi r}D(f,\theta,\psi)\exp\left[+j2\pi f\left(t-\frac{r}{c}\right)\right]\hat{r} \tag{4.1-10}$$

for $t \ge r/c$, or

$$\mathbf{u}_r(t,r,\theta,\psi) = u_r(t,r,\theta,\psi)\hat{r}, \tag{4.1-11}$$

where

$$u_r(t,r,\theta,\psi) \approx jk\frac{A_x}{4\pi r}D(f,\theta,\psi)\exp\left[+j2\pi f\left(t-\frac{r}{c}\right)\right] \tag{4.1-12}$$

is the complex acoustic *fluid-speed* (a.k.a. the complex acoustic *particle-speed*) in meters per second in the radial direction. Note that the acoustic pressure given by (4.1-2) can be expressed as

$$p(t,r,\theta,\psi) = \rho_0 c\, u_r(t,r,\theta,\psi), \tag{4.1-13}$$

where the factor $\rho_0 c$ is referred to as the characteristic impedance of the fluid medium in rayls ($1\,\text{rayl} = 1\,\text{Pa-sec/m}$). Equation (4.1-13) is a plane-wave relationship between acoustic pressure and acoustic fluid-speed. Therefore, the radiated acoustic field behaves like a plane wave in the radial direction in the far-field region of the aperture.

Now that we have equations for both the acoustic pressure and the acoustic fluid-velocity-vector in the radial direction, we can derive an equation for the time-average intensity vector in the radial direction. We begin by rewriting the acoustic pressure and acoustic fluid-velocity-vector given by (4.1-2) and (4.1-10) as follows:

$$p(t,r,\theta,\psi) = p_f(r,\theta,\psi)\exp(+j2\pi ft) \tag{4.1-14}$$

and

$$\mathbf{u}_r(t,r,\theta,\psi) = \mathbf{u}_{f,r}(r,\theta,\psi)\exp(+j2\pi ft), \tag{4.1-15}$$

where

$$p_f(r,\theta,\psi) \approx jk\rho_0 c \frac{A_x}{4\pi r} D(f,\theta,\psi)\exp(-jkr) \tag{4.1-16}$$

and

$$\mathbf{u}_{f,r}(r,\theta,\psi) \approx jk \frac{A_x}{4\pi r} D(f,\theta,\psi)\exp(-jkr)\hat{r}\,. \tag{4.1-17}$$

From (4.1-14) and (4.1-15) it is clear that the acoustic pressure and acoustic fluid-velocity-vector are time-harmonic acoustic fields. A time-harmonic acoustic field is equal to the product between a spatial-dependent part and the time-dependent part $\exp(+j2\pi ft)$. As a result, it depends on a single frequency f and is a periodic function of time. Either $\cos(2\pi ft)$ or $\sin(2\pi ft)$ can be used instead of $\exp(+j2\pi ft)$, but it is more customary to use $\exp(+j2\pi ft)$. For the special case of time-harmonic acoustic fields where

$$p(t,\mathbf{r}) = p_f(\mathbf{r})\exp(+j2\pi ft) \tag{4.1-18}$$

and

$$\mathbf{u}(t,\mathbf{r}) = \mathbf{u}_f(\mathbf{r})\exp(+j2\pi ft), \tag{4.1-19}$$

the time-average intensity vector in watts per squared meter is given by

$$\boxed{\mathbf{I}_{\text{avg}}(\mathbf{r}) = \frac{1}{2}\text{Re}\left\{p_f(\mathbf{r})\mathbf{u}_f^*(\mathbf{r})\right\}} \tag{4.1-20}$$

where the asterisk denotes complex conjugate. Therefore, in our problem, the time-average intensity vector in the radial direction can be computed by using the following equation:

$$\mathbf{I}_{\text{avg},r}(r,\theta,\psi) = \frac{1}{2}\text{Re}\left\{p_f(r,\theta,\psi)\mathbf{u}_{f,r}^*(r,\theta,\psi)\right\}. \tag{4.1-21}$$

Substituting (4.1-16) and (4.1-17) into (4.1-21) yields

$$\mathbf{I}_{\text{avg},r}(r,\theta,\psi) \approx \frac{k^2\rho_0 c}{32\pi^2 r^2}\left|A_x\right|^2 \left|D(f,\theta,\psi)\right|^2 \hat{r}\,. \tag{4.1-22}$$

We shall now use (4.1-22) to derive an equation for the time-average radiated acoustic power.

The time-average radiated acoustic power in watts can be obtained by evaluating the following closed-surface integral:

$$\boxed{P_{\text{avg}} = \oint_S \mathbf{I}_{\text{avg}}(\mathbf{r}) \bullet \mathbf{dS}} \tag{4.1-23}$$

where S is *any closed-surface* enclosing the sound-source (transmit aperture),

$$\mathbf{dS} = dS\,\hat{n}, \tag{4.1-24}$$

dS is an infinitesimal element of surface area, and $\hat{n}$ is a unit vector normal to the surface S, pointing in the conventional outward direction away from the enclosed volume and, hence, the source. Note that $1\,\text{W} = 1\,\text{J}/\text{sec}$, where $1\,\text{J} = 1\,\text{N-m}$ and $1\,\text{N} = 1\,\text{kg-m}/\text{sec}^2$. Therefore, $1\,\text{W} = 1\,\text{kg-m}^2/\text{sec}^3$. If we enclose the transmit aperture with a *sphere* of radius r meters, then

$$\mathbf{dS} = r^2 \sin\theta\, d\theta\, d\psi\, \hat{r}. \tag{4.1-25}$$

Since $\mathbf{dS}$ only has a radial component,

$$\mathbf{I}_{\text{avg}}(\mathbf{r}) \bullet \mathbf{dS} = I_{\text{avg},r}(r,\theta,\psi)\,dS \approx \frac{k^2 \rho_0 c}{32\pi^2} |A_x|^2 |D(f,\theta,\psi)|^2 \sin\theta\, d\theta\, d\psi, \tag{4.1-26}$$

where $I_{\text{avg},r}(r,\theta,\psi) = \left|\mathbf{I}_{\text{avg},r}(r,\theta,\psi)\right|$ [see (4.1-22)]. As can be seen from (4.1-26), even if we had computed all three components of the radiated acoustic fluid-velocity-vector $\mathbf{u}(t,r,\theta,\psi)$ instead of just the radial component $\mathbf{u}_r(t,r,\theta,\psi)$ so that the time-average intensity vector $\mathbf{I}_{\text{avg}}(\mathbf{r})$ would have had three components, only the magnitude of the *radial component* of $\mathbf{I}_{\text{avg}}(\mathbf{r})$ would remain after the dot product between $\mathbf{I}_{\text{avg}}(\mathbf{r})$ and $\mathbf{dS}$. And finally, by substituting (4.1-26) into (4.1-23), we obtain

$$\boxed{P_{\text{avg}} \approx \frac{k^2 \rho_0 c}{32\pi^2} |A_x|^2 \int_0^{2\pi} \int_0^{\pi} |D(f,\theta,\psi)|^2 \sin\theta\, d\theta\, d\psi} \tag{4.1-27}$$

which is the time-average acoustic power radiated by a transmit volume aperture at frequency f hertz. Equation (4.1-27) is a general result that is also applicable to linear and planar apertures since they are just special cases of a volume aperture as was discussed in Chapters 2 and 3. Therefore, (4.1-27) can be used to compute the time-average acoustic power radiated at frequency f hertz by a single electroacoustic transducer (e.g., a continuous line source or rectangular piston) or by a linear, planar, or volume array of electroacoustic transducers. All that is needed is the far-field beam pattern $D(f,\theta,\psi)$ of the aperture. We are now in a position to compute the directivity of a transmit aperture.

The *directivity* D of a transmit aperture is defined as follows:

$$\boxed{\mathrm{D} \triangleq I_{\max}/I_{\text{ref}}} \tag{4.1-28}$$

where $I_{\max}$ is the maximum value of the magnitude of the far-field, time-average intensity vector in the radial direction of the acoustic field radiated by the aperture, and I_{ref} is a reference intensity. By referring to (4.1-22),

$$I_{\max} = \frac{k^2 \rho_0 c}{32\pi^2 r^2} \left| A_x \right|^2 D_{\max}^2 , \tag{4.1-29}$$

where $D_{\max}$ is the maximum value of the magnitude of the unnormalized, far-field beam pattern, that is, $D_{\max} = \max\left| D(f, \theta, \psi) \right|$, and is referred to as the normalization factor (see Subsection 2.2.1).

The reference intensity I_{ref} is defined as the time-average intensity of the acoustic field radiated by an omnidirectional, spherical point-source that radiates the same time-average acoustic power as the aperture, that is,

$$I_{\text{ref}} \triangleq \frac{P_{\text{avg}}}{4\pi r^2} , \tag{4.1-30}$$

where P_{avg} is given by (4.1-27). Substituting (4.1-27) into (4.1-30) yields

$$I_{\text{ref}} = \frac{k^2 \rho_0 c}{32\pi^2 r^2} \left| A_x \right|^2 D_{\max}^2 \frac{1}{4\pi} \int_0^{2\pi} \int_0^{\pi} \left| D_N(f, \theta, \psi) \right|^2 \sin\theta \, d\theta \, d\psi , \tag{4.1-31}$$

where

$$D_N(f, \theta, \psi) = D(f, \theta, \psi) / D_{\max} \tag{4.1-32}$$

is the normalized, far-field beam pattern of the aperture.

Substituting (4.1-29) and (4.1-31) into (4.1-28) yields the following general expression for the directivity of a transmit volume aperture at frequency f hertz:

$$\boxed{\mathrm{D} = \frac{4\pi}{\displaystyle\int_0^{2\pi} \int_0^{\pi} \left| D_N(f, \theta, \psi) \right|^2 \sin\theta \, d\theta \, d\psi}} \tag{4.1-33}$$

Note that the directivity D is dimensionless. The *directivity index* DI is the decibel equivalent of the directivity and is defined as follows:

$$\boxed{\mathrm{DI} \triangleq 10\log_{10}\mathrm{D}\ \mathrm{dB}} \tag{4.1-34}$$

Values of the directivity range from a minimum of $\mathrm{D}=1$ $(\mathrm{DI}=0\ \mathrm{dB})$ for an *unbaffled*, omnidirectional source where $|D_N(f,\theta,\psi)|=1$ for $0\le\theta\le\pi$ and $0\le\psi\le 2\pi$, to large numbers for highly directional sources. If $|D_N(f,\theta,\psi)|=1$ for $0\le\theta\le\pi$ and $0\le\psi\le 2\pi$, then the denominator in (4.1-33) reduces to

$$\begin{aligned}\int_0^{2\pi}\int_0^{\pi}\sin\theta\, d\theta\, d\psi &= 2\pi\int_0^{\pi}\sin\theta\, d\theta \\ &= -2\pi\cos\theta\Big|_0^{\pi} \\ &= 4\pi,\end{aligned} \tag{4.1-35}$$

and as a result, the directivity $\mathrm{D}=1$ $(\mathrm{DI}=0\ \mathrm{dB})$. However, for an omnidirectional source mounted on a very large baffle in the XY plane, the range of values for θ are from 0 to $\pi/2$. Therefore, the denominator in (4.1-33) reduces to

$$\begin{aligned}\int_0^{2\pi}\int_0^{\pi/2}\sin\theta\, d\theta\, d\psi &= 2\pi\int_0^{\pi/2}\sin\theta\, d\theta \\ &= -2\pi\cos\theta\Big|_0^{\pi/2} \\ &= 2\pi,\end{aligned} \tag{4.1-36}$$

and as a result, the directivity $\mathrm{D}=2$ $(\mathrm{DI}=3\ \mathrm{dB})$. An example of an omnidirectional source is a spherical sound-source vibrating in the monopole mode of vibration. Also, as a general principle, if the size of an aperture is small compared to a wavelength, then the unnormalized, far-field beam pattern of the aperture reduces to a *constant*, that is, the aperture is omnidirectional – the magnitude of the normalized, far-field beam pattern of the aperture is equal to 1 for all angles (e.g., see Fig. 2.2-3).

The directivity of an aperture can be increased by decreasing the beamwidth of the far-field beam pattern of the aperture. The beamwidth can be decreased by keeping the size of the aperture constant while increasing the frequency, or by keeping the frequency constant while increasing the size of the aperture. And finally, it should be emphasized that both the directivity and the directivity index of either a single electroacoustic transducer or an array of electroacoustic transducers can be computed from (4.1-33) and (4.1-34), respectively, as long as the normalized, far-field beam pattern is known.

4.2 The Source Level of a Directional Sound-Source

In this section we shall derive three main equations. The first equation expresses the source level (SL) of a transmit volume aperture as a function of the time-average acoustic power P_{avg} radiated by the aperture and the directivity index (DI) of the aperture. The second equation expresses the SL of a transmit volume aperture as a function of P_0, where P_0 is the maximum value of the magnitude of the acoustic pressure measured at a distance (range) of one meter from the aperture along its acoustic axis. The third equation expresses the time-average acoustic power P_{avg} radiated by a transmit volume aperture as a function of the SL and directivity (D) of the aperture.

We begin the derivation of the first main equation by rewriting (4.1-27) as follows:

$$P_{\text{avg}} = \frac{k^2 \rho_0 c}{8\pi} |S_0|^2 \frac{1}{4\pi} \int_0^{2\pi} \int_0^{\pi} |D_N(f,\theta,\psi)|^2 \sin\theta \, d\theta \, d\psi \,, \tag{4.2-1}$$

where

$$S_0 = A_x D_{\max} \tag{4.2-2}$$

is the *maximum* value of *source strength* in cubic meters per second at frequency f hertz, and $D_N(f,\theta,\psi)$ is the normalized, far-field beam pattern of the aperture [see (4.1-32)]. Substituting (4.1-33) into (4.2-1) yields

$$P_{\text{avg}} = \frac{k^2 \rho_0 c}{8\pi} \frac{|S_0|^2}{\text{D}} \,, \tag{4.2-3}$$

and by solving for $|S_0|^2$, we obtain

$$|S_0|^2 = \text{D} \frac{8\pi}{k^2 \rho_0 c} P_{\text{avg}} \,, \tag{4.2-4}$$

where D is the directivity of the transmit volume aperture. Note that if $\text{D} = 1$, then (4.2-3) reduces to the time-average acoustic power radiated by an omnidirectional, spherical sound-source vibrating in the monopole mode of vibration with $ka \ll 1$, where a is the radius of the sphere in meters.

The *source level* (SL) is defined as follows:

$$\boxed{\text{SL} \triangleq 20 \log_{10} \left(\frac{\max\left\{ p_{\text{rms}}(\mathbf{r})\big|_{r=1\text{ m}} \right\}}{P_{\text{ref}}} \right) \text{dB re } P_{\text{ref}}} \tag{4.2-5}$$

where $\max\left\{p_{\text{rms}}(\mathbf{r})\big|_{r=1\text{ m}}\right\}$ is the maximum value of the root-mean-square (rms) acoustic pressure measured at a distance (range) of one meter from the source along its acoustic axis, that is, in the direction of the maximum response of the source, and P_{ref} is the rms reference pressure. Since we are dealing with time-harmonic acoustic fields,

$$\max\left\{p_{\text{rms}}(\mathbf{r})\big|_{r=1\text{ m}}\right\} = \frac{\sqrt{2}}{2}\max\left\{\left|p_f(r,\theta,\psi)\right|_{r=1\text{ m}}\right\}. \tag{4.2-6}$$

Computing the magnitude of the spatial-dependent part of the radiated acoustic pressure given by (4.1-16) yields

$$\left|p_f(r,\theta,\psi)\right| = k\rho_0 c\frac{|A_x|}{4\pi r}\left|D(f,\theta,\psi)\right|, \tag{4.2-7}$$

and as a result,

$$\max\left\{\left|p_f(r,\theta,\psi)\right|_{r=1\text{ m}}\right\} = k\rho_0 c\frac{|A_x|}{4\pi r}D_{\max}\Bigg|_{r=1\text{ m}}, \tag{4.2-8}$$

or

$$\max\left\{\left|p_f(r,\theta,\psi)\right|_{r=1\text{ m}}\right\} = k\rho_0 c\frac{|S_0|}{4\pi r}\Bigg|_{r=1\text{ m}}, \tag{4.2-9}$$

where $D_{\max} = \max\left|D(f,\theta,\psi)\right|$ and S_0 is the maximum value of source strength at frequency f given by (4.2-2). And by substituting (4.2-9) into (4.2-6), we obtain

$$\max\left\{p_{\text{rms}}(\mathbf{r})\big|_{r=1\text{ m}}\right\} = \frac{\sqrt{2}}{2}k\rho_0 c\frac{|S_0|}{4\pi r}\Bigg|_{r=1\text{ m}}. \tag{4.2-10}$$

With the use of (4.2-10), we shall now compute the SL according to (4.2-5). Dividing (4.2-10) by P_{ref} and then squaring yields

$$\left[\frac{\max\left\{p_{\text{rms}}(\mathbf{r})\big|_{r=1\text{ m}}\right\}}{P_{\text{ref}}}\right]^2 = k^2\frac{(\rho_0 c)^2}{P_{\text{ref}}^2}\frac{|S_0|^2}{32\pi^2 r^2}\Bigg|_{r=1\text{ m}}, \tag{4.2-11}$$

and by substituting (4.2-4) into (4.2-11), we obtain

$$\left[\frac{\max\left\{p_{\text{rms}}(\mathbf{r})\big|_{r=1\text{ m}}\right\}}{P_{\text{ref}}}\right]^2 = \mathrm{D}\frac{\rho_0 c}{P_{\text{ref}}^2}\frac{P_{\text{avg}}}{4\pi r^2}\Bigg|_{r=1\text{ m}}, \tag{4.2-12}$$

where both sides of (4.2-12) are dimensionless. Therefore,

$$10\log_{10}\left[\frac{\max\left\{p_{\text{rms}}(\mathbf{r})\big|_{r=1\text{ m}}\right\}}{P_{\text{ref}}}\right]^2 = 10\log_{10}\left[\text{D}\frac{\rho_0 c}{P_{\text{ref}}^2}\frac{P_{\text{avg}}}{4\pi r^2}\bigg|_{r=1\text{ m}}\right]$$

$$= 10\log_{10}\text{D} + 10\log_{10}(\rho_0 c) + 10\log_{10}P_{\text{avg}} - 10\log_{10}(4\pi r^2 P_{\text{ref}}^2)\big|_{r=1\text{ m}},$$

(4.2-13)

or

$$\boxed{\text{SL} = 10\log_{10}P_{\text{avg}} + \text{DI} + 10\log_{10}(\rho_0 c) - 10\log_{10}(4\pi P_{\text{ref}}^2)\text{ dB re }P_{\text{ref}}}$$

(4.2-14)

where P_{avg} is the time-average acoustic power radiated by a transmit aperture, and DI is the directivity index of the aperture given by (4.1-34). As can be seen from (4.2-14), the SL of a directional sound-source (transmit aperture) is increased by the value of the DI because the available time-average acoustic power is radiated in preferred directions by the main-lobe of the transmit aperture's far-field beam pattern. The SL of a time-harmonic, omnidirectional sound-source as a function of P_{avg} can be obtained from (4.2-14) by setting $\text{DI} = 0\text{ dB}$. For problems in underwater acoustics where $P_{\text{ref}} = 1\,\mu\text{Pa (rms)}$, (4.2-14) reduces to

$$\boxed{\text{SL} = 10\log_{10}P_{\text{avg}} + \text{DI} + 10\log_{10}(\rho_0 c) + 109.01\text{ dB re }1\,\mu\text{Pa (rms)}}$$

(4.2-15)

and if $\rho_0 = 1026\text{ kg/m}^3$ and $c = 1500\text{ m/sec}$, which are typical values for seawater, then (4.2-15) reduces to

$$\text{SL} = 10\log_{10}P_{\text{avg}} + \text{DI} + 170.88\text{ dB re }1\,\mu\text{Pa (rms)}. \tag{4.2-16}$$

The second main equation to be derived expresses the source level (SL) of a transmit volume aperture as a function of the maximum value P_0 of the magnitude of the acoustic pressure measured at a distance (range) of one meter from the aperture along its acoustic axis. From (4.2-9),

$$P_0 = \max\left\{\left|p_f(r,\theta,\psi)\right|\big|_{r=1\text{ m}}\right\} = k\rho_0 c\frac{|S_0|}{4\pi r}\bigg|_{r=1\text{ m}}. \tag{4.2-17}$$

Substituting (4.2-17) into (4.2-10) yields

$$\max\left\{p_{\text{rms}}(\mathbf{r})\Big|_{r=1\text{ m}}\right\} = \frac{\sqrt{2}}{2} P_0 , \tag{4.2-18}$$

and by substituting (4.2-18) into (4.2-5), we obtain

$$\boxed{\text{SL} = 20\log_{10}\left(\frac{\sqrt{2}P_0/2}{P_{\text{ref}}}\right) \text{dB re } P_{\text{ref}}} \tag{4.2-19}$$

Two additional useful equations shall be derived next. First, solving for P_0 from (4.2-19) yields

$$\boxed{P_0 = \sqrt{2}\, P_{\text{ref}}\, 10^{(\text{SL}/20)}} \tag{4.2-20}$$

and second, substituting wavenumber $k = 2\pi f/c$ into (4.2-17) and solving for $|S_0|$ yields

$$\boxed{|S_0| = \frac{2r}{f\rho_0} P_0 \Big|_{r=1\text{ m}}} \tag{4.2-21}$$

where $|S_0|$ is the magnitude of the maximum value of source strength in cubic meters per second at frequency f hertz [see (4.2-2)]. Therefore, for a given value of SL , P_0 can be computed using (4.2-20), and by substituting that value of P_0 into (4.2-21), the corresponding value for $|S_0|$ can be obtained.

The third and final main equation to be derived expresses the time-average acoustic power P_{avg} radiated by a transmit volume aperture as a function of the source level (SL) and directivity (D) of the aperture. From (4.2-13),

$$\text{SL} = 10\log_{10}\left[\text{D}\frac{\rho_0 c}{P_{\text{ref}}^2}\frac{P_{\text{avg}}}{4\pi r^2}\Big|_{r=1\text{ m}}\right], \tag{4.2-22}$$

and by solving for P_{avg} , we obtain

$$\boxed{P_{\text{avg}} = 4\pi r^2 \frac{P_{\text{ref}}^2}{\text{D}\rho_0 c} 10^{(\text{SL}/10)}\Big|_{r=1\text{ m}}} \tag{4.2-23}$$

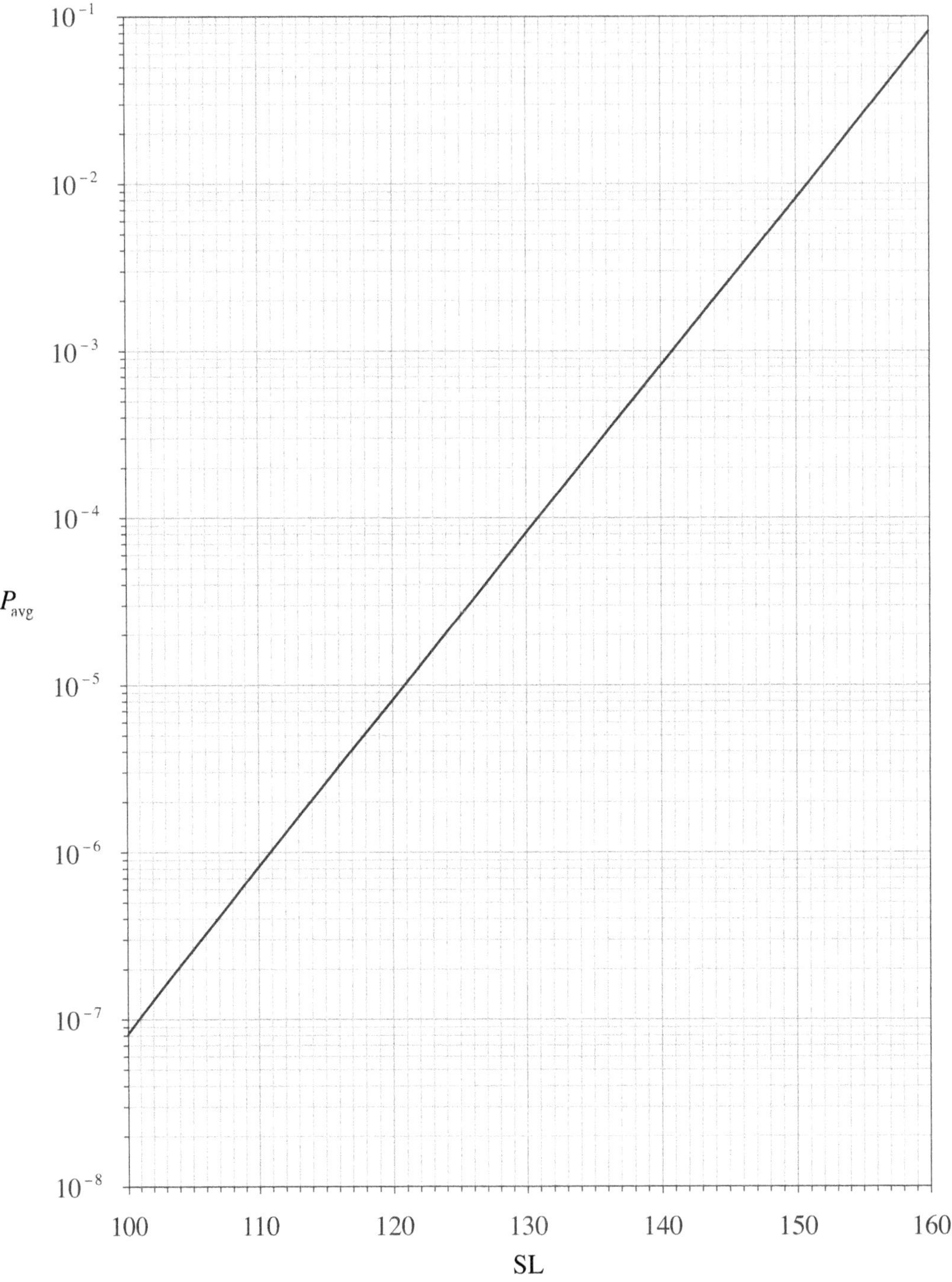

Figure 4.2-1 Time-average acoustic power P_{avg} in watts radiated by a time-harmonic, omnidirectional sound-source versus source level (SL) from 100 to 160 dB re 1 μPa (rms).

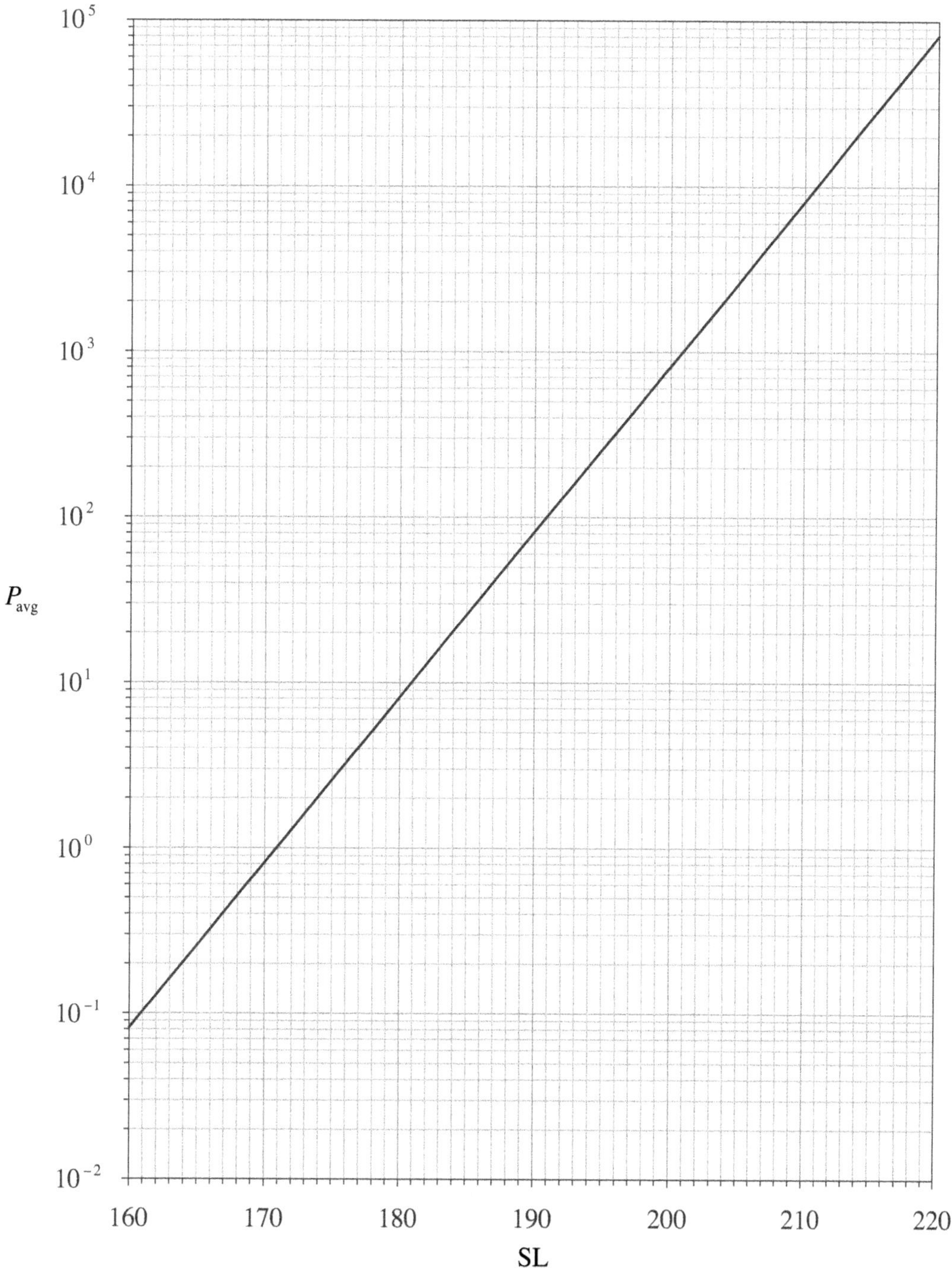

Figure 4.2-2 Time-average acoustic power P_{avg} in watts radiated by a time-harmonic, omnidirectional sound-source versus source level (SL) from 160 to 220 dB re 1 μPa (rms).

Figures 4.2-1 and 4.2-2 are plots of (4.2-23) for $\rho_0 = 1026\ \text{kg}/\text{m}^3$, $c = 1500\ \text{m/sec}$, $\text{D} = 1$ (time-harmonic, omnidirectional sound-source), and $P_{\text{ref}} = 1\ \mu\text{Pa (rms)}$ for SL values from 100 to $160\ \text{dB re}\ 1\ \mu\text{Pa (rms)}$, and from 160 to $220\ \text{dB re}\ 1\ \mu\text{Pa (rms)}$, respectively. And finally, in order to get a feeling for low and high values for P_0, if we use the same values for ρ_0, c, D, and P_{ref}, and if $P_0 = 1\ \text{Pa}$, then from (4.2-19), $\text{SL} = 116.99\ \text{dB re}\ 1\ \mu\text{Pa (rms)}$, and from (4.2-23), $P_{\text{avg}} = 4.08\ \mu\text{W}$; and if $P_0 = 1\ \text{atm} = 1.01325 \times 10^5\ \text{Pa}$, then $\text{SL} = 217.1$ dB re $1\ \mu\text{Pa (rms)}$ and $P_{\text{avg}} = 41.89\ \text{kW}$.

Problems

Section 4.1

4-1 Find the *real*, time-harmonic, radiated acoustic pressure in the far-field region of the following transmit apertures:

(a) a linear aperture lying along the Z axis with complex frequency response modeled by the cosine amplitude window

(b) a rectangular piston lying in the XY plane

4-2 Verify (4.1-8).

4-3 In Chapter 5, a side-looking sonar (SLS) lying in the YZ plane is modeled as a single rectangular piston with dimensions $L_Y = 0.1\ \text{m}$ and $L_Z = 2.5\ \text{m}$. If $f = 30\ \text{kHz}$, $\rho_0 = 1026\ \text{kg}/\text{m}^3$, and $c = 1500\ \text{m/sec}$, then find the maximum value of the magnitude of the far-field, time-average intensity vector in the radial direction of the acoustic field radiated by the SLS at a range $r = 139\ \text{m}$.

Section 4.2

4-4 If the source level of a directional transmit aperture (a sound-source) is $204\ \text{dB re}\ 1\ \mu\text{Pa (rms)}$, and the directivity index of the aperture is $3\ \text{dB}$, then what is the time-average radiated power in watts? Use $\rho_0 = 1026\ \text{kg}/\text{m}^3$, $c = 1500\ \text{m/sec}$, and $P_{\text{ref}} = 1\ \mu\text{Pa (rms)}$.

4-5 If the source level of a transmit aperture (a sound-source) is $180\ \text{dB re}\ 1\ \mu\text{Pa (rms)}$ at $25\ \text{kHz}$, then using $\rho_0 = 1026\ \text{kg}/\text{m}^3$ and $P_{\text{ref}} = 1\ \mu\text{Pa (rms)}$, find

(a) the maximum value of the magnitude of the acoustic pressure in pascals measured at a distance of one meter from the aperture along its acoustic axis.

(b) the magnitude of the maximum value of source strength in cubic meters per second.

4-6 Express P_0 as a function of P_{avg}.

Chapter 5

Side-Looking Sonar

In this chapter we shall discuss some of the fundamentals concerning the design and analysis of a *side-looking sonar* (SLS), also referred to as a *side-scan sonar*. A SLS is an active sonar system capable of producing high resolution underwater images of large areas on the ocean bottom. Side-looking sonars are commonly used to detect and identify objects lying on the ocean bottom. They are also used to perform bathymetric surveys (seafloor mapping).

A SLS generates underwater images by periodically transmitting a pulse and then receiving the scattered returns while the platform is in motion with a constant velocity vector (constant speed and direction) and at a constant height (altitude) above the ocean bottom. A SLS may be composed of a single planar aperture that is used alternately in the active (transmit) mode and passive (receive) mode, or two separate planar apertures – a transmit aperture and a receive aperture. Side-looking sonars are usually mounted on both sides of an underwater vehicle. Some SLS systems are *towed* while others are mounted on *remotely operated vehicles* (ROVs) or *autonomous underwater vehicles* (AUVs), a.k.a. *unmanned underwater vehicles* (UUVs). Understanding the fundamentals of a SLS system provides a very good foundation for the understanding of the fundamentals of a *synthetic-aperture sonar* (SAS) system (see Chapter 15).

5.1 Swath Width

We begin our discussion of side-looking sonars (SLSs) by deriving an equation for the *swath width* (SW), which is a measure of the width of the area on the ocean bottom ensonified by a SLS. We shall analyze a SLS modeled as a planar aperture lying in the YZ plane as shown in Fig. 5.1-1. The planar aperture is rectangular in shape with sides of length L_Y and L_Z meters in the Y and Z directions, respectively. The X axis is cross-range (the *cross-track direction*), the Y axis is depth, and the Z axis is down-range (the *along-track direction*). Since the planar aperture (SLS) lies in the YZ plane, its far-field beam pattern $D(f, f_Y, f_Z)$ depends on the spatial frequencies in the Y and Z directions given by

$$f_Y = v/\lambda \tag{5.1-1}$$

and

$$f_Z = w/\lambda , \tag{5.1-2}$$

respectively, with units of cycles per meter, where

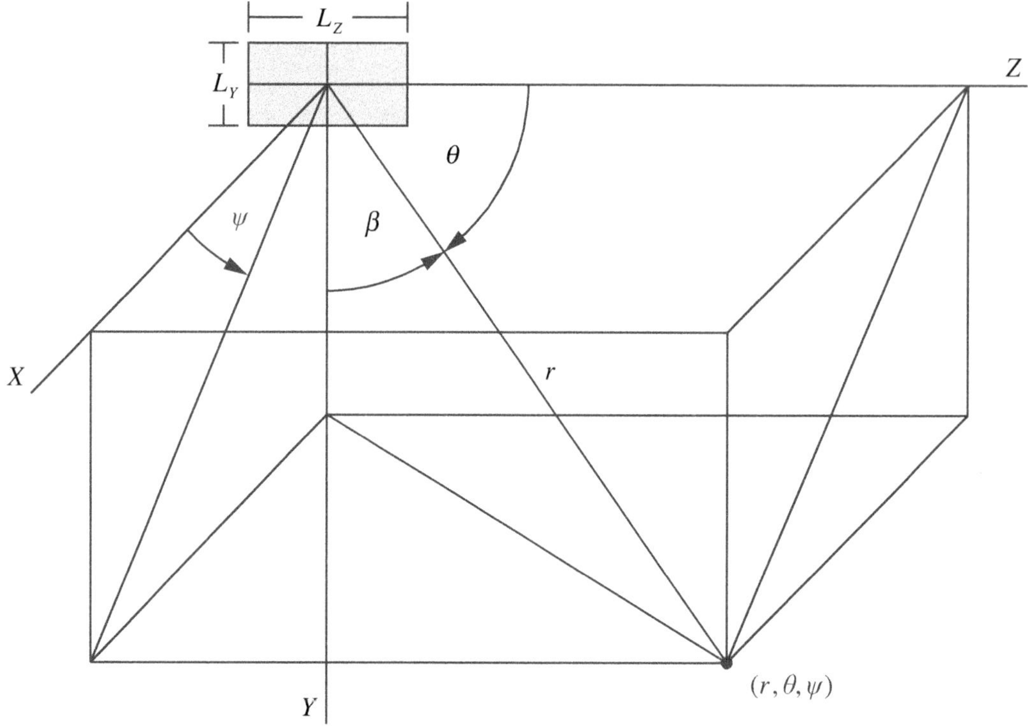

Figure 5.1-1 A SLS (planar aperture, rectangular in shape) lying in the YZ plane. Also shown is a field point in three-dimensional space with spherical coordinates (r, θ, ψ) as measured from the center of the aperture, and the angle β which is measured from the positive Y axis.

$$v = \sin\theta \sin\psi = \cos\beta \tag{5.1-3}$$

and

$$w = \cos\theta \tag{5.1-4}$$

are the dimensionless direction cosines in the Y and Z directions, respectively, λ is the wavelength in meters, and $c = f\lambda$.

In order to derive an equation for the SW, and to introduce some of the other parameters associated with a SLS and to show their interdependence, we first need to derive an equation for the 3-dB beamwidth (half-power beamwidth) of the vertical, far-field beam pattern in the XY plane. In the XY plane, $\theta = 90°$ (see Fig. 5.1-1), and as a result [see (5.1-3) and (5.1-4)],

$$v = \sin\psi = \cos\beta \tag{5.1-5}$$

and

$$w = 0. \tag{5.1-6}$$

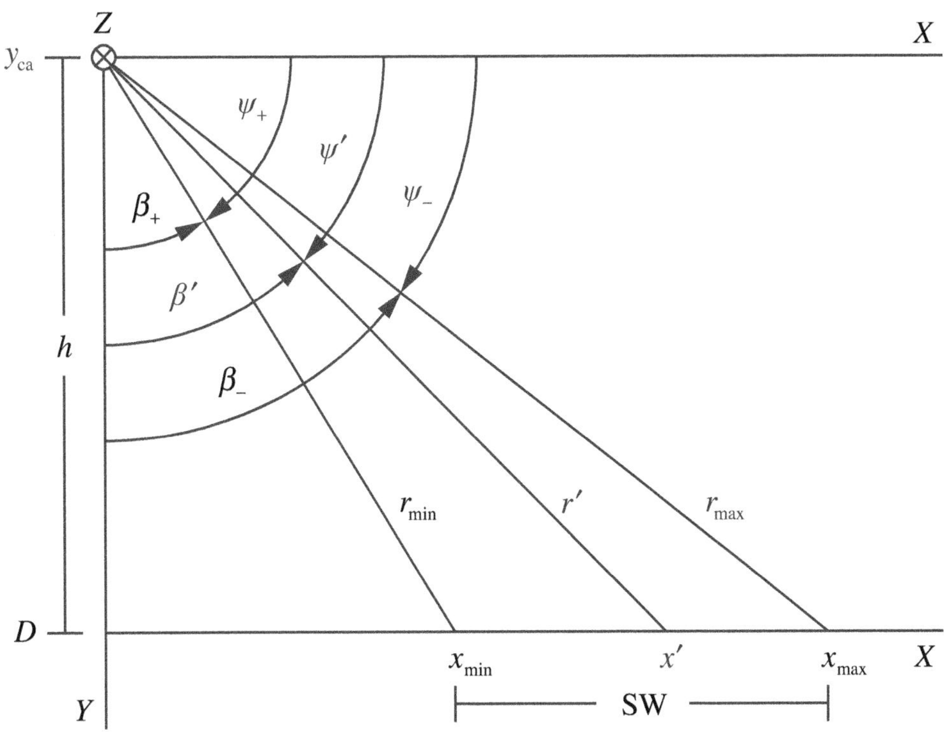

Figure 5.1-2 Angles involved in the derivation of the 3-dB beamwidth $\Delta\psi = \psi_+ - \psi_-$ (or $\Delta\beta = \beta_- - \beta_+$) of the vertical, far-field beam pattern in the XY plane. The parameters y_{ca} and D are the depths of the center of the aperture and ocean, respectively. Also shown is the swath width $\text{SW} = x_{max} - x_{min}$.

From Fig. 5.1-2 it can be seen that the 3-dB beamwidth is given by either

$$\Delta\psi = \psi_+ - \psi_- \tag{5.1-7}$$

or

$$\Delta\beta = \beta_- - \beta_+, \tag{5.1-8}$$

where ψ' (or β') is the beam-steer angle in the XY plane. For a SLS, $\psi' \neq 0°$ ($\beta' \neq 90°$) since this corresponds to steering the vertical beam pattern to broadside, and $\psi' \neq 90°$ ($\beta' \neq 0°$) since this corresponds to steering the vertical beam pattern to endfire. Note that the *half-beamwidth angles* $\psi_+ - \psi'$ and $\psi' - \psi_-$ are *not* equal for $0° < \psi' < 90°$, and that the *half-beamwidth angles* $\beta' - \beta_+$ and $\beta_- - \beta'$ are *not* equal for $0° < \beta' < 90°$. Also note that $\Delta\beta = \Delta\psi$, $\beta' - \beta_+ = \psi_+ - \psi'$, and $\beta_- - \beta' = \psi' - \psi_-$.

Following the procedure discussed in Section 2.5, let

$$v_+ = \sin\psi_+ > 0 \quad \text{or} \quad v_+ = \cos\beta_+ > 0\,, \tag{5.1-9}$$

$$v_- = \sin\psi_- > 0 \quad \text{or} \quad v_- = \cos\beta_- > 0\,, \tag{5.1-10}$$

and

$$v' = \sin\psi' > 0 \quad \text{or} \quad v' = \cos\beta' > 0\,, \tag{5.1-11}$$

where

$$v_+ > v_-\,, \tag{5.1-12}$$

$$v_+ = v' + (\Delta v/2), \tag{5.1-13}$$

$$v_- = v' - (\Delta v/2), \tag{5.1-14}$$

and Δv is the dimensionless 3-dB beamwidth of the vertical, far-field beam pattern in direction-cosine space. Therefore,

$$\psi_+ = \sin^{-1} v_+ = \sin^{-1}\left(\sin\psi' + \frac{\Delta v}{2}\right), \qquad \left|\sin\psi' + \frac{\Delta v}{2}\right| \le 1\,, \tag{5.1-15}$$

$$\psi_- = \sin^{-1} v_- = \sin^{-1}\left(\sin\psi' - \frac{\Delta v}{2}\right), \qquad \left|\sin\psi' - \frac{\Delta v}{2}\right| \le 1\,, \tag{5.1-16}$$

$$\beta_+ = \cos^{-1} v_+ = \cos^{-1}\left(\cos\beta' + \frac{\Delta v}{2}\right), \qquad \left|\cos\beta' + \frac{\Delta v}{2}\right| \le 1\,, \tag{5.1-17}$$

and

$$\beta_- = \cos^{-1} v_- = \cos^{-1}\left(\cos\beta' - \frac{\Delta v}{2}\right), \qquad \left|\cos\beta' - \frac{\Delta v}{2}\right| \le 1\,. \tag{5.1-18}$$

Substituting (5.1-15) and (5.1-16) into (5.1-7) yields

$$\boxed{\Delta\psi = \sin^{-1}\left(\sin\psi' + \frac{\Delta v}{2}\right) - \sin^{-1}\left(\sin\psi' - \frac{\Delta v}{2}\right), \qquad \left|\sin\psi' \pm \frac{\Delta v}{2}\right| \le 1} \tag{5.1-19}$$

and substituting (5.1-17) and (5.1-18) into (5.1-8) yields

$$\boxed{\Delta\beta = \cos^{-1}\left(\cos\beta' - \frac{\Delta v}{2}\right) - \cos^{-1}\left(\cos\beta' + \frac{\Delta v}{2}\right), \qquad \left|\cos\beta' \pm \frac{\Delta v}{2}\right| \le 1} \tag{5.1-20}$$

The 3-dB beamwidth (in degrees) of the vertical, far-field beam pattern in the XY plane is given by either (5.1-19) or (5.1-20).

If we let the *height* (*altitude*) h in meters of the center of the aperture above the ocean bottom be given by

$$h = D - y_{\text{ca}}, \tag{5.1-21}$$

where D and y_{ca} are the depths of the ocean and center of the aperture in meters, respectively, then from Fig. 5.1-2 it can be seen that

$$x_{\min} = h\cot\psi_{+} = h\tan\beta_{+}, \tag{5.1-22}$$

$$x' = h\cot\psi' = h\tan\beta', \tag{5.1-23}$$

$$x_{\max} = h\cot\psi_{-} = h\tan\beta_{-}, \tag{5.1-24}$$

$$r_{\min} = \sqrt{h^2 + x_{\min}^2} = \frac{h}{\sin\psi_{+}} = \frac{h}{\cos\beta_{+}}, \tag{5.1-25}$$

$$r' = \sqrt{h^2 + (x')^2} = \frac{h}{\sin\psi'} = \frac{h}{\cos\beta'}, \tag{5.1-26}$$

and

$$r_{\max} = \sqrt{h^2 + x_{\max}^2} = \frac{h}{\sin\psi_{-}} = \frac{h}{\cos\beta_{-}}, \tag{5.1-27}$$

where the angles ψ_{+}, ψ_{-}, β_{+}, and β_{-} are given by (5.1-15) through (5.1-18), respectively. Equations (5.1-25) through (5.1-27) are the *slant-ranges* corresponding to the *cross-ranges* given by (5.1-22) through (5.1-24), respectively. The cross-range $x_{\min}$ given by (5.1-22) is referred to as the *width of the blind zone* in meters because it corresponds to the distance on the ocean bottom from $x = 0$ to $x = x_{\min}$ that is *not* ensonified by the 3-dB beamwidth of the vertical, far-field beam pattern (see Fig. 5.1-2). The parameter $x_{\min}$ is also known as the width of the *one-sided* blind zone. The width of the *two-sided* blind zone is simply $2x_{\min}$.

In order to guarantee that the far-field beam pattern of the aperture (vis-à-vis its near-field beam pattern) will ensonify a point on the ocean bottom at the minimum slant-range $r_{\min}$, $r_{\min}$ must satisfy the far-field range criterion (e.g., see Section 3.1):

$$r_{\min} > \pi R_A^2/\lambda > 2.414 R_A, \tag{5.1-28}$$

where R_A is the maximum radial extent of the aperture. Since the planar aperture is rectangular in shape with sides of length L_Y and L_Z meters,

$$R_A = \sqrt{\left(\frac{L_Y}{2}\right)^2 + \left(\frac{L_Z}{2}\right)^2}. \tag{5.1-29}$$

Therefore, substituting (5.1-29) into (5.1-28) yields

$$\boxed{r_{\min} > \pi \frac{L_Y^2 + L_Z^2}{4\lambda} > 1.207\sqrt{L_Y^2 + L_Z^2}} \tag{5.1-30}$$

If (5.1-30) is satisfied, then *all* points on the ocean bottom within the swath width will be ensonified by the far-field beam pattern of the aperture vis-à-vis its near-field beam pattern. This is important because most of the parameters used to design and analyze a SLS are based on the far-field beam pattern of the SLS.

The *swath width* (SW) in meters, also referred to as the *ground-plane swath width*, is given by (see Fig. 5.1-2)

$$\boxed{\text{SW} = x_{\max} - x_{\min}} \tag{5.1-31}$$

where $x_{\min}$ and $x_{\max}$ are given by (5.1-22) and (5.1-24), respectively. The parameter SW is also known as the *one-sided* swath width. The *two-sided* swath width is simply 2SW. Note that $x' - x_{\min} \neq x_{\max} - x'$ (see Fig. 5.1-2).

An alternative expression for SW can be obtained by substituting (5.1-22) and (5.1-24) into (5.1-31). Doing so yields

$$\begin{aligned} \text{SW} &= h\left(\cot\psi_- - \cot\psi_+\right) \\ &= h\frac{\sin(\psi_+ - \psi_-)}{\sin\psi_- \sin\psi_+} = h\frac{\sin(\Delta\psi)}{v_- v_+} \\ &= h\frac{\sin(\Delta\psi)}{\left(v' - \dfrac{\Delta v}{2}\right)\left(v' + \dfrac{\Delta v}{2}\right)} = h\frac{\sin(\Delta\psi)}{(v')^2 - \left(\dfrac{\Delta v}{2}\right)^2} \end{aligned} \tag{5.1-32}$$

and finally,

$$\boxed{\text{SW} = h\frac{\sin\left(\Delta\psi(\psi', \Delta v)\right)}{\sin^2\psi' - \left(\dfrac{\Delta v}{2}\right)^2}, \qquad 0° < \psi' < 90°, \quad \left|\sin\psi' \pm \frac{\Delta v}{2}\right| \leq 1, \quad \sin\psi' > \frac{\Delta v}{2}} \tag{5.1-33}$$

where h is the height (altitude) in meters of the center of the aperture above the ocean bottom, $\Delta\psi(\psi', \Delta v)$ is the 3-dB beamwidth in degrees of the vertical, far-field beam pattern in the XY plane shown explicitly as a function of the independent variables ψ' and Δv [see (5.1-19)], ψ' is the beam-steer angle in degrees, and Δv is the dimensionless 3-dB beamwidth of the vertical, far-field beam pattern in direction-cosine space. Since $\Delta\psi = \Delta\beta$ and $\psi' = 90° - \beta'$, SW can also be expressed as

$$\boxed{\mathrm{SW} = h\frac{\sin(\Delta\beta(\beta', \Delta v))}{\cos^2\beta' - \left(\frac{\Delta v}{2}\right)^2}, \quad 0° < \beta' < 90°, \quad \left|\cos\beta' \pm \frac{\Delta v}{2}\right| \leq 1, \quad \cos\beta' > \frac{\Delta v}{2}} \tag{5.1-34}$$

where the 3-dB beamwidth $\Delta\beta(\beta', \Delta v)$ is given by (5.1-20).

The *area coverage rate* (ACR) in squared meters per second, also referred to as the *survey coverage rate*, is given by

$$\boxed{\mathrm{ACR} = \mathrm{SW} \times V} \tag{5.1-35}$$

where SW is the swath width given by either (5.1-31), (5.1-33), or (5.1-34), and V is the constant speed of the platform in meters per second. The parameter ACR is also known as the *one-sided* area coverage rate. The *two-sided* area coverage rate is simply 2 ACR. As can be seen from (5.1-35), the ACR can be increased by increasing either the SW or V or both. In general, one would like the ACR to be as big as possible.

5.2 Cross-Track (Slant-Range) Resolution

The ability of a SLS to estimate the slant-range to different objects on the ocean bottom within the SW – at the same down-range (along-track) position – is referred to as the *cross-track (slant-range) resolution.* In order to discuss this topic, we need to know how to design rectangular-envelope, CW (continuous-wave) and LFM (linear-frequency-modulated) pulses in order to provide a desired range resolution, since these two kinds of waveforms are typically transmitted by a SLS. We also need to know how to determine a minimum value for the carrier frequency. The carrier frequency of the waveform transmitted by a SLS is commonly referred to as the operating frequency of the SLS. These two waveforms are discussed in detail in Chapter 13. In this section we shall only summarize those equations from Chapter 13 that are needed to continue our discussion.

The first kind of waveform that is typically transmitted by a SLS is known as a *rectangular-envelope*, *CW* (continuous-wave) *pulse*. Specifying a value for range resolution is equivalent to specifying a desired value for the magnitude of the range estimation error of the target $|e_{r,Trgt}|$. The pulse length T (in seconds) that is required for a rectangular-envelope, CW pulse to provide a range resolution equal to $|e_{r,Trgt}|$ meters is given by

$$\boxed{T = \frac{2|e_{r,Trgt}|}{c}} \tag{5.2-1}$$

where c is the constant speed of sound in meters per second. A conservative estimate for the minimum allowed value for the carrier frequency f_c (in hertz) for a rectangular-envelope, CW pulse is given by

$$\boxed{\min f_c = 5/T} \tag{5.2-2}$$

Therefore, for a rectangular-envelope, CW pulse, f_c is chosen to satisfy the following inequality:

$$\boxed{f_c > 5/T} \tag{5.2-3}$$

The second kind of waveform that is typically transmitted by a SLS is known as a *rectangular-envelope*, *linear-frequency-modulated* (LFM) *pulse*. The *swept-bandwidth* BW_{swept} (in hertz) that is required for a rectangular-envelope, LFM pulse to provide a range resolution equal to $|e_{r,Trgt}|$ meters is given by

$$\boxed{\text{BW}_{\text{swept}} \approx \frac{c}{2|e_{r,Trgt}|}} \tag{5.2-4}$$

The pulse length T must then satisfy the inequality

$$\boxed{T \geq T_{\min}} \tag{5.2-5}$$

where

$$\boxed{T_{\min} = \frac{40}{\text{BW}_{\text{swept}}}} \tag{5.2-6}$$

is the minimum allowed value. A conservative estimate for the minimum allowed value for the carrier frequency f_c for a rectangular-envelope, LFM pulse is given by

$$\boxed{\min f_c = \text{BW}_{\text{swept}} + \frac{5}{T}} \tag{5.2-7}$$

where T must satisfy (5.2-5). Therefore, for a rectangular-envelope, LFM pulse, f_c is chosen to satisfy the following inequality:

$$\boxed{f_c > \text{BW}_{\text{swept}} + \frac{5}{T}} \tag{5.2-8}$$

As can be seen from the above equations, the minimum allowed value for the carrier frequency not only depends on the desired range resolution, but also on what kind of waveform is transmitted. In addition, the carrier frequency may need to be made much larger than the minimum allowed value in order to decrease the 3-dB beamwidth of the far-field beam pattern of the planar aperture (SLS) if the size of the aperture is small. However, keep in mind that as frequency increases, attenuation of sound in the ocean also increases. Therefore, both low and high carrier frequencies are used for short-range applications, whereas low carrier frequencies are used for long-range applications. Note that in the SLS and SAS literature, the terminology *operating frequency* f is commonly used instead of *carrier frequency* f_c. Table 5.2-1 summarizes the parameter values for both waveforms in order to obtain a cross-track (slant-range) resolution of 2.54 cm ($1 \text{ in} = 2.54 \text{ cm}$) for $c = 1500 \text{ m/sec}$. For example purposes, $T = 25 \text{ msec}$ was chosen for the rectangular-envelope, LFM pulse.

Table 5.2-1 Parameter Values for a Rectangular-Envelope, CW and LFM Pulse in Order to Obtain a Cross-Track (Slant-Range) Resolution of 2.54 cm (1 in) for $c = 1500 \text{ m/sec}$

Pulse	$\lvert e_{r,Trgt} \rvert$	BW_{swept}	$T_{\min}$	T	$\min f_c$
RE CW	2.54 cm	NA	NA	33.9 μsec	147,637.8 Hz
RE LFM	2.54 cm	29,527.6 Hz	1.355 msec	25 msec	29,727.6 Hz

5.3 Along-Track (Azimuthal) Resolution

In order to derive an equation for the *along-track (azimuthal) resolution* (see Fig. 5.3-1), we first need to derive an equation for the 3-dB beamwidth (half-power beamwidth) $\Delta\theta$ of the horizontal, far-field beam pattern in the XZ plane (see Fig. 5.3-2). In the XZ plane, $\psi = 0°$ (positive X axis) and $\beta = 90°$ (see Fig. 5.1-1), and as a result [see (5.1-3)],

$$v = 0. \tag{5.3-1}$$

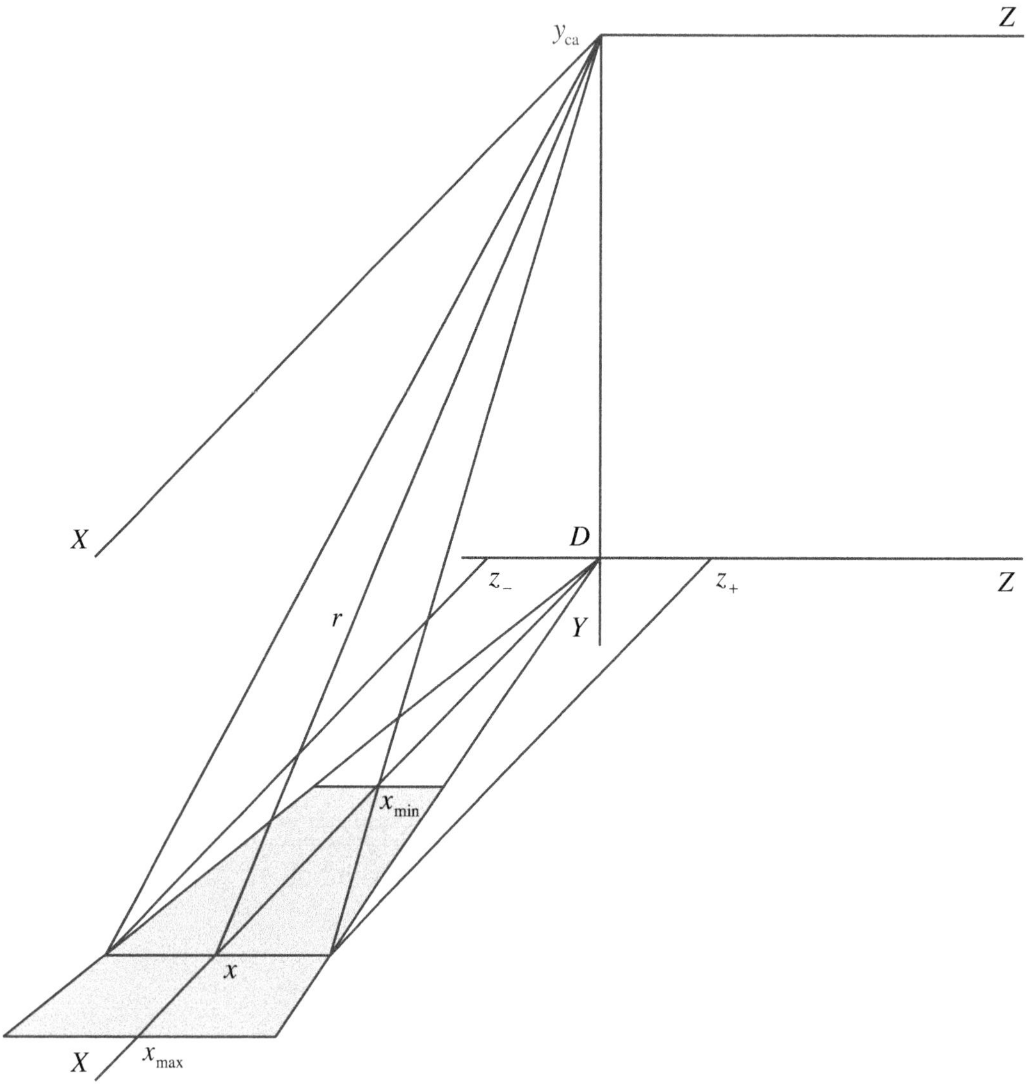

Figure 5.3-1 Along-track (azimuthal) resolution $\Delta z = z_+ - z_-$ at slant-range r and cross-range x. The shaded area represents the area on the ocean bottom within the swath width $\text{SW} = x_{max} - x_{min}$ ensonified by the 3-dB beamwidth of the horizontal, far-field beam pattern in the XZ plane.

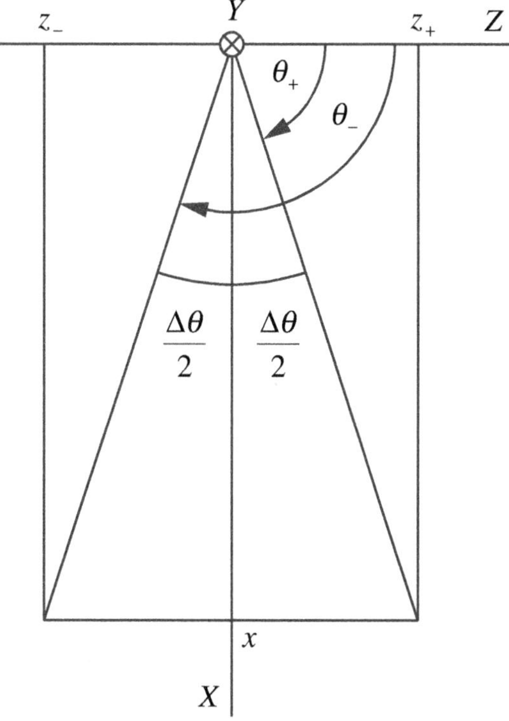

Figure 5.3-2 Angles involved in the derivation of the 3-dB beamwidth $\Delta\theta$ of the horizontal, far-field beam pattern in the XZ plane.

Following the procedure discussed in Example 2.3-2, let

$$w_+ = \cos\theta_+ > 0 \tag{5.3-2}$$

and

$$w_- = \cos\theta_- < 0 , \tag{5.3-3}$$

where from Fig. 5.3-2 it can be seen that

$$\theta_+ = \frac{\pi}{2} - \frac{\Delta\theta}{2} \tag{5.3-4}$$

and

$$\theta_- = \frac{\pi}{2} + \frac{\Delta\theta}{2} . \tag{5.3-5}$$

Substituting (5.3-4) and (5.3-5) into (5.3-2) and (5.3-3), respectively, yields

$$w_+ = \cos\left(\frac{\pi}{2} - \frac{\Delta\theta}{2}\right) = \sin\left(\frac{\Delta\theta}{2}\right) > 0 \tag{5.3-6}$$

and

$$w_- = \cos\left(\frac{\pi}{2} + \frac{\Delta\theta}{2}\right) = -\sin\left(\frac{\Delta\theta}{2}\right) < 0. \qquad (5.3\text{-}7)$$

Since the dimensionless 3-dB beamwidth Δw of the horizontal, far-field beam pattern in direction-cosine space is given by

$$\Delta w = w_+ - w_-, \qquad (5.3\text{-}8)$$

substituting (5.3-6) and (5.3-7) into (5.3-8) yields

$$\Delta w = 2\sin(\Delta\theta/2) > 0, \qquad (5.3\text{-}9)$$

or

$$\boxed{\Delta\theta = 2\sin^{-1}(\Delta w/2), \qquad \Delta w/2 \le 1} \qquad (5.3\text{-}10)$$

where $\Delta\theta$ is the 3-dB beamwidth (in degrees) of the horizontal, far-field beam pattern in the XZ plane. We are now in a position to derive equations for the along-track resolutions at the three different cross-ranges $x_{\min}$, x', and $x_{\max}$.

From Fig. 5.3-2 it can be seen that

$$z_+ = x\tan(\Delta\theta/2), \qquad (5.3\text{-}11)$$

and since

$$\Delta z = z_+ - z_- = 2z_+, \qquad (5.3\text{-}12)$$

substituting (5.3-11) into (5.3-12) yields the following expression for the along-track resolution Δz in meters at cross-range x meters:

$$\boxed{\Delta z = 2x\tan(\Delta\theta/2)} \qquad (5.3\text{-}13)$$

As can be seen from (5.3-13), for a given value of horizontal, 3-dB beamwidth $\Delta\theta$, the along-track resolution Δz increases as the cross-range x increases. As Δz *increases*, the ability of a SLS to resolve closely-spaced objects on the ocean bottom *decreases* (see Fig. 5.3-3). Therefore, in order to obtain small values for Δz for large values of x, $\Delta\theta$ must be small. The along-track (azimuthal) resolutions at cross-ranges $x_{\min}$ (the beginning or near-edge of the SW), x', and $x_{\max}$ (the end or far-edge of the SW) are given by

$$\boxed{\Delta z_{\min} = 2x_{\min}\tan(\Delta\theta/2)} \qquad (5.3\text{-}14)$$

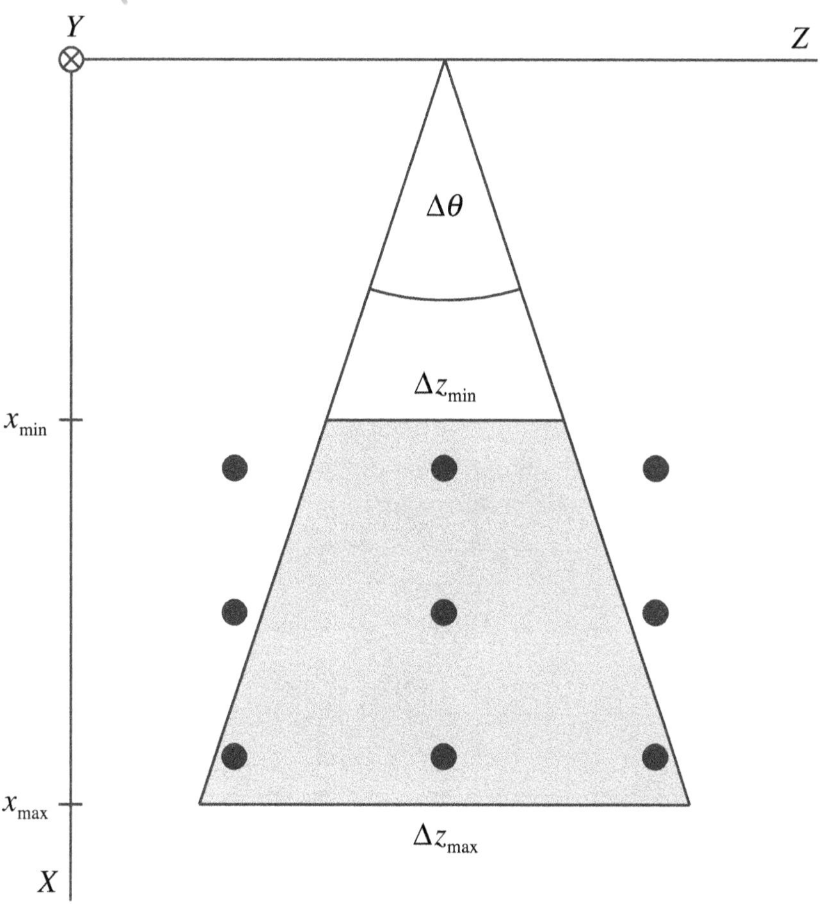

Figure 5.3-3 For a given value of horizontal, 3-dB beamwidth $\Delta\theta$, the ability of a SLS to resolve closely-spaced objects on the ocean bottom decreases as the cross-range increases.

$$\boxed{\Delta z' = 2x' \tan(\Delta\theta/2)} \tag{5.3-15}$$

and

$$\boxed{\Delta z_{\max} = 2x_{\max} \tan(\Delta\theta/2)} \tag{5.3-16}$$

respectively, where $x_{\min}$, x', and $x_{\max}$ are given by (5.1-22) through (5.1-24), respectively, and $\Delta\theta$ is the 3-dB beamwidth (in degrees) of the horizontal, far-field beam pattern in the XZ plane given by (5.3-10).

5.4 Slant-Range Ambiguity

A SLS transmits a pulse with *pulse length* T seconds every PRI seconds, where PRI is the abbreviation for *pulse repetition interval*. Sometimes the abbreviation PRP for *pulse repetition period* is used instead of PRI. The minimum and maximum *round-trip time delays* (in seconds) corresponding to the SW are given by

$$\tau_{\min} = 2r_{\min}/c \tag{5.4-1}$$

and

$$\tau_{\max} = 2r_{\max}/c, \tag{5.4-2}$$

where $r_{\min}$ and $r_{\max}$ are the slant-ranges corresponding to the cross-ranges $x_{\min}$ and $x_{\max}$, respectively (see Fig. 5.1-2), and c is the constant speed of sound in meters per second. If the relativistic effects of platform motion are ignored (i.e., time compression or time expansion), then the total duration of the received signal from the SW is $\tau_{\max} - \tau_{\min} + T$ seconds (see Fig. 5.4-1).

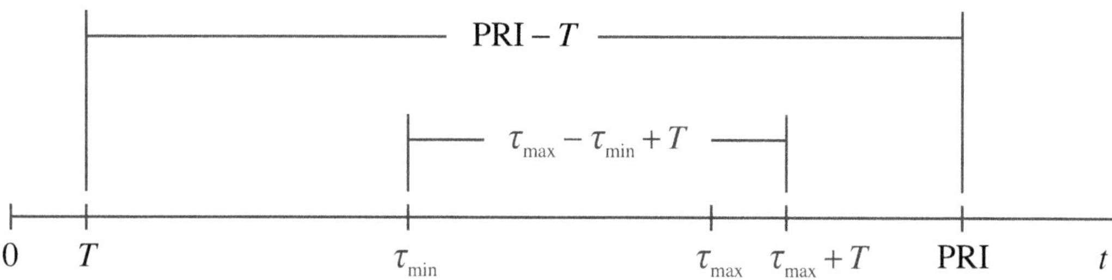

Figure 5.4-1 Time line showing the relationship between the pulse length T of the transmitted signal, the duration of the received signal $\tau_{\max} - \tau_{\min} + T$, where $\tau_{\min}$ and $\tau_{\max}$ are the minimum and maximum round-trip time delays corresponding to the swath width, and the pulse repetition interval (PRI).

In order to allow enough time for the entire received signal to arrive during the transmitter's down time before the next transmission, the following inequality must be satisfied:

$$\tau_{\max} - \tau_{\min} + T \leq \text{PRI} - T, \tag{5.4-3}$$

or

$$\text{PRI} \geq \text{PRI}_{\min}, \tag{5.4-4}$$

where

$$\boxed{\text{PRI}_{\min} = \tau_{\max} - \tau_{\min} + 2T = \frac{2\,\text{SRSW}}{c} + 2T} \tag{5.4-5}$$

and

$$\boxed{\text{SRSW} = r_{\max} - r_{\min}} \tag{5.4-6}$$

is the *slant-range swath width* (SRSW) in meters, where $r_{\min}$ and $r_{\max}$ are given by (5.1-25) and (5.1-27), respectively. Therefore, in order to avoid *slant-range ambiguities*, the PRI must be greater than or equal to the minimum value of PRI given by (5.4-5).

The minimum value of PRI given by (5.4-5) is the absolute smallest allowed value, and appears often in the SLS and SAS literature. However, from Fig. 5.4-1, it can be seen that an *operational* minimum value of PRI is given by

$$\boxed{\text{PRI}_{\min,\,\text{op}} = \tau_{\max} + T = \frac{2r_{\max}}{c} + T} \tag{5.4-7}$$

Note that

$$\text{PRI}_{\min,\,\text{op}} \geq \text{PRI}_{\min}, \tag{5.4-8}$$

and that $\text{PRI}_{\min,\,\text{op}} = \text{PRI}_{\min}$ when $\tau_{\min} = T$ [see (5.4-3)].

Since the *pulse repetition frequency* (PRF) in pulses per second (pps) – also known as the *pulse repetition rate* (PRR) – is equal to the reciprocal of the pulse repetition interval (PRI), that is, since

$$\text{PRF} = 1/\text{PRI}, \tag{5.4-9}$$

then

$$\text{PRF} \leq \text{PRF}_{\max}, \tag{5.4-10}$$

where

$$\boxed{\text{PRF}_{\max} = \frac{1}{\text{PRI}_{\min}} = \frac{1}{\tau_{\max} - \tau_{\min} + 2T} = \frac{1}{\dfrac{2\,\text{SRSW}}{c} + 2T}} \tag{5.4-11}$$

Therefore, in order to avoid slant-range ambiguities, the PRF must be less than or equal to the maximum value of PRF given by (5.4-11). The maximum value of PRF given by (5.4-11) is the absolute largest allowed value, and also appears often in the SLS and SAS literature along with $\text{PRI}_{\min}$. However, an *operational* maximum value of PRF is given by

$$\boxed{\text{PRF}_{\max,\,\text{op}} = \frac{1}{\text{PRI}_{\min,\,\text{op}}} = \frac{1}{\tau_{\max} + T} = \frac{1}{\dfrac{2r_{\max}}{c} + T}} \tag{5.4-12}$$

Note that

$$\mathrm{PRF}_{\max,\,\mathrm{op}} \le \mathrm{PRF}_{\max} \tag{5.4-13}$$

since $\mathrm{PRI}_{\min,\,\mathrm{op}} \ge \mathrm{PRI}_{\min}$, and that $\mathrm{PRF}_{\max,\,\mathrm{op}} = \mathrm{PRF}_{\max}$ when $\mathrm{PRI}_{\min,\,\mathrm{op}} = \mathrm{PRI}_{\min}$.

5.5 Azimuthal Ambiguity

The constant velocity vector of the SLS shown in Fig. 5.1-1 is given by

$$\mathbf{V} = V\hat{z}, \tag{5.5-1}$$

where V is the constant speed of the platform in meters per second and $\hat{z}$ is the unit vector in the positive Z direction. Therefore, according to (5.5-1), the SLS is traveling in the positive Z direction with a constant speed of V meters per second. A typical range of values for the speed of a SLS is from 3 to 7 knots (approximately 1.5 to 3.6 meters per second or 3.5 to 8.1 miles per hour) where $1\,\text{knot} \approx 0.514\,\text{m/sec}$ or $1\,\text{knot} \approx 1.151\,\text{mph}$. As can be seen from Fig. 5.5-1, in order to fully ensonify the ocean bottom within the SW (without leaving *gaps* in the coverage) as the SLS moves in the positive Z direction, the SLS must transmit (ping) every $\Delta z_{\min}$ meters or *less*. In other words,

$$\mathrm{PRI} \le \mathrm{PRI}_{\max}, \tag{5.5-2}$$

where

$$\boxed{\mathrm{PRI}_{\max} = \Delta z_{\min} / V} \tag{5.5-3}$$

and $\Delta z_{\min}$ is the minimum value of along-track (azimuthal) resolution at cross-range $x_{\min}$ given by (5.3-14). Therefore, in order to avoid *azimuthal ambiguities*, the PRI must be less than or equal to the maximum value of PRI given by (5.5-3).

The reciprocal of (5.5-2) is

$$\mathrm{PRF} \ge \mathrm{PRF}_{\min}, \tag{5.5-4}$$

where

$$\boxed{\mathrm{PRF}_{\min} = \frac{1}{\mathrm{PRI}_{\max}} = \frac{V}{\Delta z_{\min}}} \tag{5.5-5}$$

Therefore, in order to avoid azimuthal ambiguities, the PRF must be greater than or equal to the minimum value of PRF given by (5.5-5).

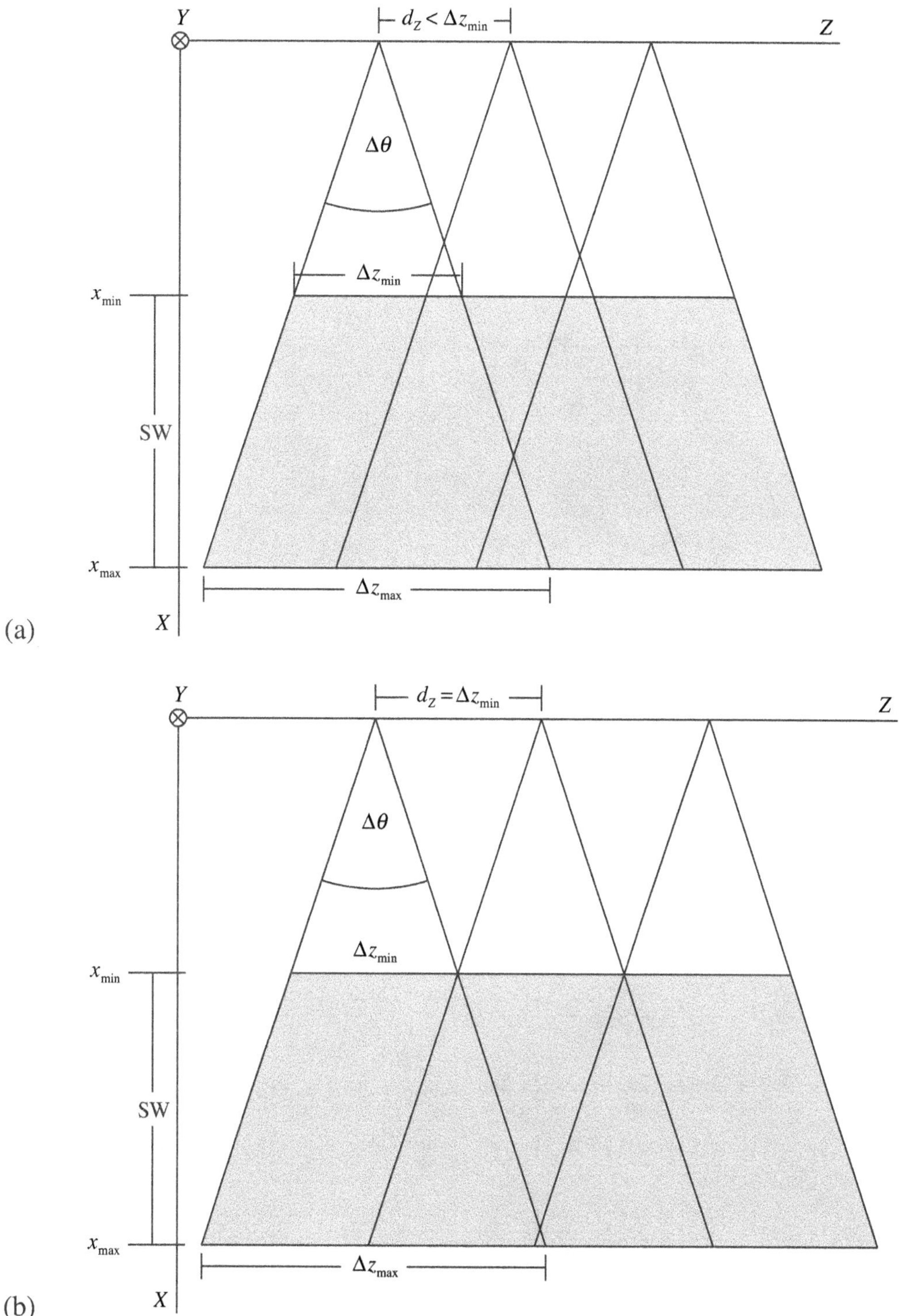

Figure 5.5-1 As the SLS moves in the positive Z direction, it transmits every d_Z meters, where (a) $d_Z < \Delta z_{min}$, (b) $d_Z = \Delta z_{min}$, and (c) $d_Z > \Delta z_{min}$. If $d_Z > \Delta z_{min}$, then there are *gaps* in the coverage within the SW as shown in (c).

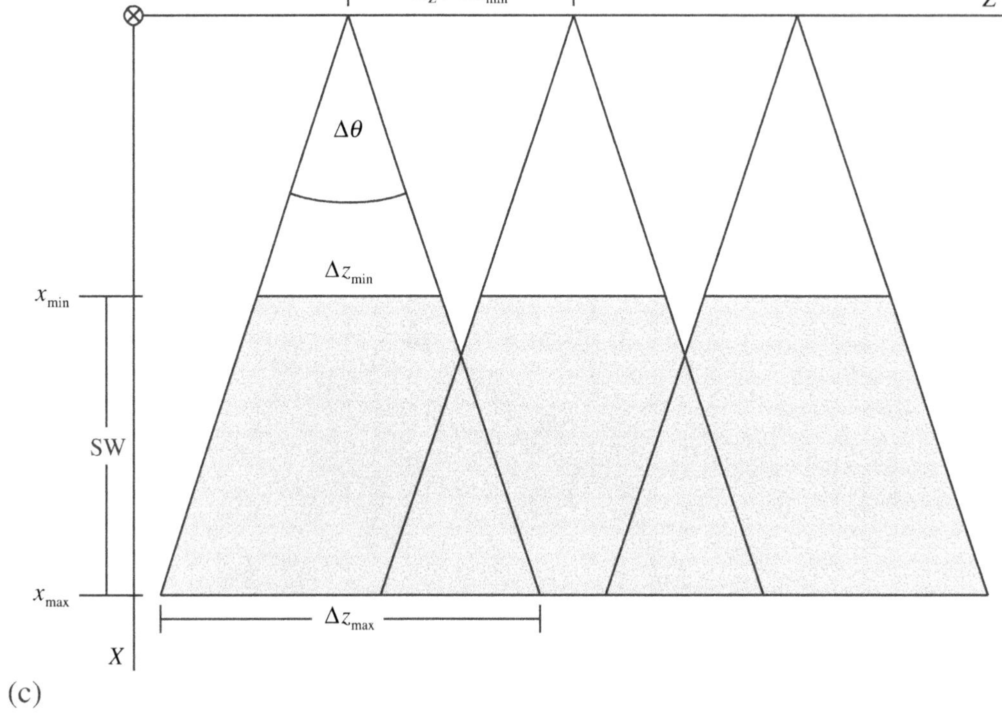

Figure 5.5-1 *continued.*

In summary, in order to avoid both slant-range and azimuthal ambiguities,

$$\boxed{\mathrm{PRI}_{\min,\,\mathrm{op}} \le \mathrm{PRI} \le \mathrm{PRI}_{\max}} \tag{5.5-6}$$

or

$$\boxed{\tau_{\max} + T \le \mathrm{PRI} \le \Delta z_{\min}/V} \tag{5.5-7}$$

where $\tau_{\max}$ is the maximum round-trip time delay corresponding to the end (far-edge) of the SW given by (5.4-2) and T is the pulse length of the transmitted signal. Equivalently,

$$\boxed{\mathrm{PRF}_{\min} \le \mathrm{PRF} \le \mathrm{PRF}_{\max,\,\mathrm{op}}} \tag{5.5-8}$$

or

$$\boxed{\frac{V}{\Delta z_{\min}} \le \mathrm{PRF} \le \frac{1}{\tau_{\max} + T}} \tag{5.5-9}$$

5.6 A Rectangular-Piston Model for a Side-Looking Sonar

In this section we shall derive approximate expressions for the dimensionless 3-dB beamwidths Δv and Δw of the vertical and horizontal, far-field beam patterns in direction-cosine space, respectively, of a SLS (a planar aperture) lying in the YZ plane as shown in Fig. 5.1-1. Having equations for Δv and Δw in terms of system parameters such as the operating (carrier) frequency and the physical dimensions of the aperture is very important because many of the SLS parameters depend on knowing the values of Δv and Δw.

As was mentioned at the beginning of this chapter, a SLS may be composed of a single planar aperture that is used alternately in the active (transmit) mode and passive (receive) mode, or two separate planar apertures – a transmit aperture and a receive aperture. A realistic SLS (planar aperture) is a *rectangular array* of identical, equally-spaced, omnidirectional, electroacoustic transducers with *rectangular amplitude weights*. A very good approximation of this kind of array is a *rectangular piston* as discussed in Section 3.2. A rectangular piston model is often used for a SLS because it is very easy to derive equations for Δv and Δw. Therefore, if the SLS (planar aperture) lying in the YZ plane is modeled as a rectangular piston with sides of length L_Y and L_Z meters in the Y and Z directions, respectively, then its normalized, far-field beam pattern is given by

$$D_N(f, f_Y, f_Z) = \operatorname{sinc}(f_Y L_Y)\operatorname{sinc}(f_Z L_Z), \tag{5.6-1}$$

or, in direction-cosine space,

$$D_N(f, v, w) = \operatorname{sinc}\left(\frac{L_Y}{\lambda}v\right)\operatorname{sinc}\left(\frac{L_Z}{\lambda}w\right), \tag{5.6-2}$$

where $c = f\lambda$.

Since direction cosine $w = 0$ in the XY plane [see (5.1-6)] and $\operatorname{sinc}(0) = 1$, the normalized, vertical, far-field beam pattern in the XY plane is given by

$$D_N(f, v, 0) = \operatorname{sinc}\left(\frac{L_Y}{\lambda}v\right), \tag{5.6-3}$$

and

$$\boxed{\Delta v/2 \approx 0.443\lambda/L_Y} \tag{5.6-4}$$

where Δv is the dimensionless 3-dB beamwidth of the vertical, far-field beam pattern in direction-cosine space (see Example 2.3-3).

Since direction cosine $v = 0$ in the XZ plane [see (5.3-1)] and $\text{sinc}(0) = 1$, the normalized, horizontal, far-field beam pattern in the XZ plane is given by

$$D_N(f, 0, w) = \text{sinc}\left(\frac{L_Z}{\lambda} w\right), \tag{5.6-5}$$

and

$$\boxed{\Delta w/2 \approx 0.443\lambda/L_Z} \tag{5.6-6}$$

where Δw is the dimensionless 3-dB beamwidth of the horizontal, far-field beam pattern in direction-cosine space (see Example 2.3-3).

5.7 Design and Analysis of a Side-Looking Sonar Mission

5.7.1 Deep Water

If a SLS is going to operate in deep water, then one can choose its height (altitude) of operation h above the ocean bottom to satisfy the far-field range criterion [see (5.1-30)]. In the case of a SLS, the far-field range criterion can be stated as follows: if $h > h_{\min}$, then *all* points on the ocean bottom within the swath width (SW) will be ensonified by the far-field beam pattern of the SLS vis-à-vis its near-field beam pattern, where $h_{\min}$ is the minimum allowed height (altitude) of operation. As was mentioned previously, this is important because most of the parameters used to design and analyze a SLS are based on the far-field beam pattern of the SLS. However, in a shallow-water environment, it may not always be possible to choose $h > h_{\min}$ (see Subsection 5.7.2). In this subsection we shall summarize those equations that need to be solved in order to determine $h_{\min}$. Solving this set of equations corresponds to following a design procedure for choosing the altitude of operation h so that $h > h_{\min}$ (see Table 5.7-1). We shall also summarize those equations that are needed to solve for the set of angles $\{\beta_+, \beta', \beta_-\}$. Solving for the required beam-steer angle β' is also part of designing the mission. Once this set of angles is known, the other parameters that describe the performance of a SLS can be evaluated in a straightforward fashion. This constitutes the analysis portion of a SLS mission.

In most SLS problems, the dimensions L_Y and L_Z, and the operating (carrier) frequency f of a SLS are given (specified). Recall from Section 5.2 that the operating frequency is determined by the waveform to be transmitted, the desired cross-track (slant-range) resolution, and the size of the aperture. The desired value for the along-track (azimuthal) resolution $\Delta z_{\min}$ at cross-range $x_{\min}$ (the beginning or near-edge of the SW) is also usually given. Since a SLS is an imaging sonar, it is not surprising that the desired cross-track and along-track resolutions are specified.

Table 5.7-1 Design Procedure for Choosing the Altitude of Operation for a SLS in Order to Satisfy the Far-Field Range Criterion in Deep Water

Given	Resulting Fixed Parameter Values
L_Y, L_Z, and f	$r_{\text{NF/FF}}$, $\Delta\psi$ ($\psi'=0°$), and $\Delta\theta$
$\Delta z_{\min}$ and $\Delta\theta$	$x_{\min}$
$r_{\text{NF/FF}}$ and $x_{\min}$ where $r_{\text{NF/FF}} > x_{\min}$	$h_{\min}$ $h > h_{\min}$ satisfies far-field range criterion

If the dimensions L_Y and L_Z, and the operating (carrier) frequency f of a SLS are given, then the values of the range to the near-field/far-field boundary $r_{\text{NF/FF}}$, the 3-dB beamwidth $\Delta\psi$ of the vertical, far-field beam pattern in the XY plane when *no* beam steering is done ($\psi'=0°$), and the 3-dB beamwidth $\Delta\theta$ of the horizontal, far-field beam pattern in the XZ plane of the SLS are *fixed*. By referring to (5.1-30), it can be seen that the range to the near-field/far-field boundary is given by

$$r_{\text{NF/FF}} = \pi \frac{L_Y^2 + L_Z^2}{4\lambda} \tag{5.7-1}$$

where $c = f\lambda$. If we let the beam-steer angles $\psi'=0°$ and $\beta'=90°$, then the 3-dB beamwidths $\Delta\psi$ and $\Delta\beta$ given by (5.1-19) and (5.1-20), respectively, reduce to

$$\Delta\psi = \Delta\beta = 2\sin^{-1}\left(\frac{\Delta v}{2}\right), \qquad \frac{\Delta v}{2} \le 1, \quad \psi'=0° \ (\beta'=90°), \tag{5.7-2}$$

since

$$\sin^{-1}(-x) = -\sin^{-1}(x) \tag{5.7-3}$$

and

$$\cos^{-1}(-x) - \cos^{-1}(x) = 2\sin^{-1}(x)\,. \tag{5.7-4}$$

And by substituting (5.6-4) into (5.7-2), we obtain

$$\Delta\psi = \Delta\beta \approx 2\sin^{-1}\left(0.443\frac{\lambda}{L_Y}\right), \qquad 0.443\frac{\lambda}{L_Y} \le 1, \quad \psi' = 0^\circ \; (\beta' = 90^\circ) \tag{5.7-5}$$

Although $\Delta\psi$ is not needed to compute h_{min}, since L_Y and f are given, the 3-dB beamwidth of the vertical, far-field beam pattern in the XY plane when *no* beam steering is done ($\psi' = 0^\circ$) is fixed and its value can be computed from (5.7-5). And if (5.6-6) is substituted into (5.3-10), then

$$\Delta\theta \approx 2\sin^{-1}\left(0.443\frac{\lambda}{L_Z}\right), \qquad 0.443\frac{\lambda}{L_Z} \le 1 \tag{5.7-6}$$

where $\Delta\theta$ is the 3-dB beamwidth of the horizontal, far-field beam pattern in the XZ plane.

Since Δz_{min} is given and $\Delta\theta$ is now known, we can use (5.3-14) to solve for x_{min}. Doing so yields

$$x_{\text{min}} = \frac{\Delta z_{\text{min}}}{2\tan(\Delta\theta/2)} \tag{5.7-7}$$

where x_{min} is the width of the one-sided blind zone, also referred to as the cross-range coordinate of the beginning (near-edge) of the SW. Now that $r_{\text{NF/FF}}$ and x_{min} are known, we are finally in a position to solve for h_{min}.

In order to satisfy the far-field range criterion, we require $r_{\text{min}} > r_{\text{NF/FF}}$. Therefore, with the use of (5.1-25), we can write that

$$r_{\text{min}} = \sqrt{h^2 + x_{\text{min}}^2} > r_{\text{NF/FF}}, \tag{5.7-8}$$

and by solving for h, we obtain

$$h > h_{\text{min}} \tag{5.7-9}$$

where

$$h_{\text{min}} = \sqrt{r_{\text{NF/FF}}^2 - x_{\text{min}}^2}, \qquad r_{\text{NF/FF}} > x_{\text{min}} \tag{5.7-10}$$

is the minimum allowed height (altitude) of operation of the SLS above the ocean bottom. If $h > h_{\min}$, then *all* points on the ocean bottom within the SW will be ensonified by the far-field beam pattern of the SLS vis-à-vis its near-field beam pattern.

Since we now have solutions for $x_{\min}$ and h, we can use them to compute the angle β_+, which will allow us to solve for the required beam-steer angle β', and then the angle β_-. Solving for β_+ from (5.1-22) yields

$$\boxed{\beta_+ = \tan^{-1}(x_{\min}/h)} \tag{5.7-11}$$

The angle β_+ must be greater than 0° (see Fig. 5.1-2). With the use of h and $x_{\min}$, or h and β_+, the slant-range $r_{\min}$ to the beginning (near-edge) of the SW can be computed from (5.1-25).

Next, solve for the beam-steer angle β' from (5.1-17). Doing so yields

$$\beta' = \cos^{-1}\left(\cos\beta_+ - \frac{\Delta v}{2}\right), \qquad \left|\cos\beta_+ - \frac{\Delta v}{2}\right| \le 1, \tag{5.7-12}$$

and by substituting (5.6-4) into (5.7-12), we obtain

$$\boxed{\beta' \approx \cos^{-1}\left(\cos\beta_+ - 0.443\frac{\lambda}{L_Y}\right), \qquad \left|\cos\beta_+ - 0.443\frac{\lambda}{L_Y}\right| \le 1} \tag{5.7-13}$$

Now that h and β' are known, the cross-range coordinate x' can be computed from (5.1-23), and with the use of h and x', or h and β', the corresponding slant-range r' can be computed from (5.1-26). And finally, substituting (5.6-4) into (5.1-18) yields

$$\boxed{\beta_- \approx \cos^{-1}\left(\cos\beta' - 0.443\frac{\lambda}{L_Y}\right), \qquad \left|\cos\beta' - 0.443\frac{\lambda}{L_Y}\right| \le 1} \tag{5.7-14}$$

The angle β_- must be less than 90° (see Fig. 5.1-2). Now that h and β_- are known, the cross-range coordinate $x_{\max}$ of the end (far-edge) of the SW can be computed from (5.1-24), and with the use of h and $x_{\max}$, or h and β_-, the corresponding slant-range $r_{\max}$ can be computed from (5.1-27).

Since $x_{\min}$ and $x_{\max}$, and $r_{\min}$ and $r_{\max}$ are now known, the SW and the slant-range swath width (SRSW) can be computed from (5.1-31) and (5.4-6), respectively. Note that the angles ψ_+, ψ', and ψ_- can be obtained from β_+, β', and β_- as follows (see Fig. 5.1-2):

$$\psi_+ = 90° - \beta_+ , \tag{5.7-15}$$

$$\psi' = 90° - \beta' , \tag{5.7-16}$$

$$\psi_- = 90° - \beta_- . \tag{5.7-17}$$

Recall that the 3-dB beamwidth of the vertical, far-field beam pattern in the XY plane is given by either [see (5.1-7) and (5.1-8)]

$$\Delta\psi = \psi_+ - \psi_- \tag{5.7-18}$$

or

$$\Delta\beta = \beta_- - \beta_+ , \tag{5.7-19}$$

where $\Delta\beta = \Delta\psi$. Using the solutions of the parameters discussed above, the remaining parameters that describe the performance of a SLS – such as the area coverage rate (ACR) and the lower and upper bounds on the pulse repetition interval (PRI) and pulse repetition frequency (PRF) that are required to avoid both slant-range and azimuthal ambiguities – can easily be evaluated.

Example 5.7-1 Deep Water Mission

The SLS (planar aperture) shown in Fig. 5.1-1 has the following specifications:

Dimensions:	$L_Y = 0.1$ m , $L_Z = 2.5$ m
Transmit Waveform:	rectangular-envelope, LFM pulse
Operating (Carrier) Frequency:	$f = 30$ kHz
Pulse Length:	$T = 25$ msec (see Table 5.2-1)
Cross-Track (Slant-Range) Resolution:	2.54 cm (1 in) (see Table 5.2-1)
Along-Track (Azimuthal) Resolution:	$\Delta z_{min} = 1$ m at cross-range x_{min}
Speed:	3 to 7 knots

In this example the SLS is meant to operate in deep water at a speed $V = 2$ m/sec (≈ 4 knots). Therefore, following the design procedure in Table 5.7-1 and using $c = 1500$ m/sec , a design and analysis of this SLS mission is summarized in Table 5.7-2. From Table 5.7-2 it can be seen that the width of the one-sided blind zone $x_{min} = 56.4$ m (this value is fixed) and the minimum altitude of operation for the SLS above the ocean bottom required to satisfy the far-field range criterion is $h_{\text{min}} = 80.5\text{ m}$. Since the SLS is operating in deep water, we have the ability to choose $h > h_{\text{min}}$. As can be seen from Table 5.7-2, $h = 85\text{ m}$ was chosen, resulting in $r_{\text{min}} = 102\text{ m}$ (which is greater than $r_{\text{NF/FF}} = 98.3\text{ m}$), a required beam-steer angle $\psi' = 37.71°$, a swath width $\text{SW} = 144.2\text{ m}$, and an area coverage rate $\text{ACR} = 288.4\text{ m}^2/\text{sec}$. As was mentioned in Section 5.1, the ACR can be increased by increasing either the SW or V or both.

Since the SLS is operating in deep water, the SW can be increased while keeping x_{min} fixed by increasing h. Note that $x_{\text{min}}/h_{\text{min}} = 0.701$ and $x_{\text{min}}/h = 0.664$ for $h = 85\text{ m}$ (see Table 5.7-2). Increasing h while keeping x_{min} fixed will increase the required beam-steer angle ψ' (see Fig. 5.7-1), which will increase the 3-dB beamwidth $\Delta\psi$ of the vertical, far-field beam pattern (see Fig. 5.7-2) and, hence, the SW (see Fig. 5.7-3). As $x_{\text{min}}/h \to 0$, $\beta_+ \to 0$ [see (5.7-11) and Fig. 5.1-2]. If a bigger value for h is chosen, then the design and analysis procedure must begin again starting with (5.7-11). Figures 5.7-1 through 5.7-3 are based on $\lambda/L_Y = 0.5$ because this is the value of the ratio in this example. ■

Table 5.7-2 Design and Analysis Parameter Values for Example 5.7-1

$c = 1500\text{ m/sec}$	$f = 30\text{ kHz}$	$\lambda = 0.05\text{ m}$	$T = 25\text{ msec}$	$V = 2\text{ m/sec}$
$L_Y = 0.1\text{ m}$ $\lambda/L_Y = 0.5$	$\Delta\psi = 25.59°\ (\psi' = 0°)$			
$L_Z = 2.5\text{ m}$ $\lambda/L_Z = 0.02$	$\Delta\theta = 1.02°$			
$r_{\text{NF/FF}} = 98.3\text{ m}$				
$h_{\text{min}} = 80.5\text{ m}$ $h = 85\text{ m}$				
$r_{\text{min}} = 102\text{ m}$ $r' = 139\text{ m}$ $r_{\text{max}} = 217.9\text{ m}$	$\psi_+ = 56.42°$ $\psi' = 37.71°$ $\psi_- = 22.96°$	$\beta_+ = 33.58°$ $\beta' = 52.29°$ $\beta_- = 67.04°$	$x_{\text{min}} = 56.4\text{ m}$ $x' = 110\text{ m}$ $x_{\text{max}} = 200.6\text{ m}$	$\Delta z_{\text{min}} = 1\text{ m}$ $\Delta z' = 1.9\text{ m}$ $\Delta z_{\text{max}} = 3.6\text{ m}$
$\text{SRSW} = 115.9\text{ m}$ $\tau_{\text{min}} = 136\text{ msec}$ $\tau_{\text{max}} = 290.5\text{ msec}$	$\psi_+ - \psi' = 18.71°$ $\psi' - \psi_- = 14.75°$ $\Delta\psi = 33.46°$	$\beta' - \beta_+ = 18.71°$ $\beta_- - \beta' = 14.75°$ $\Delta\beta = 33.46°$	$\text{SW} = 144.2\text{ m}$ $\text{ACR} = 288.4\text{ m}^2/\text{sec}$	
			$x_{\text{min}}/h_{\text{min}} = 0.701$ $x_{\text{min}}/h = 0.664$	
$\text{PRI}_{\text{min}} = 204.5\text{ msec}$ $\text{PRI}_{\text{min, op}} = 315.5\text{ msec}$ $\text{PRI}_{\text{max}} = 500\text{ msec}$	$\text{PRF}_{\text{min}} = 2\text{ pps}$ $\text{PRF}_{\text{max, op}} = 3.2\text{ pps}$ $\text{PRF}_{\text{max}} = 4.9\text{ pps}$			

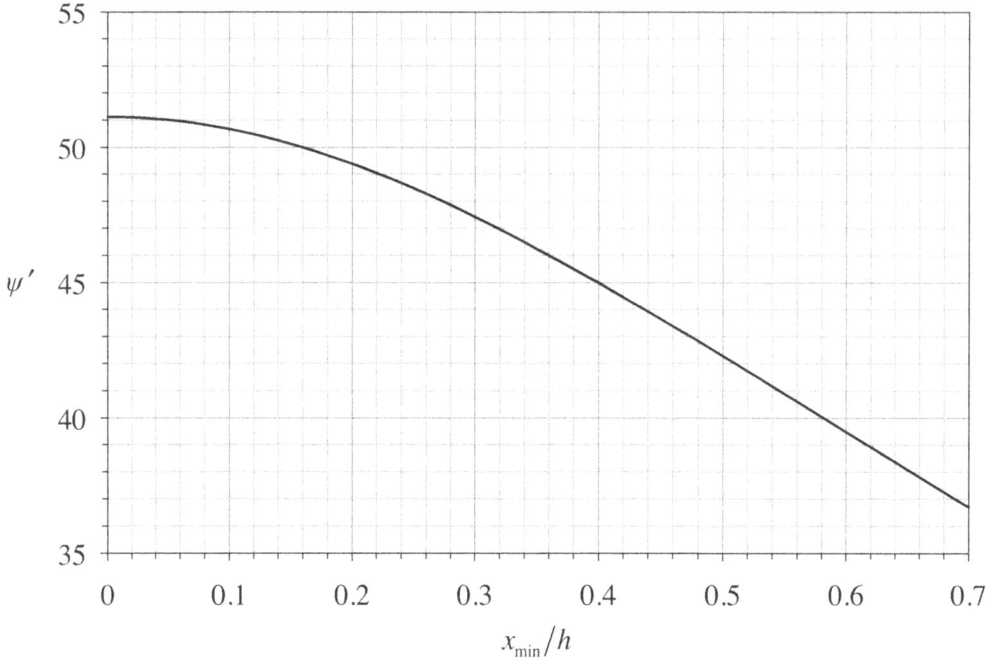

Figure 5.7-1 Beam-steer angle ψ' in degrees versus x_{min}/h for $\lambda/L_Y = 0.5$ where $h > h_{\text{min}}$ and $x_{\text{min}}/h_{\text{min}} = 0.701$.

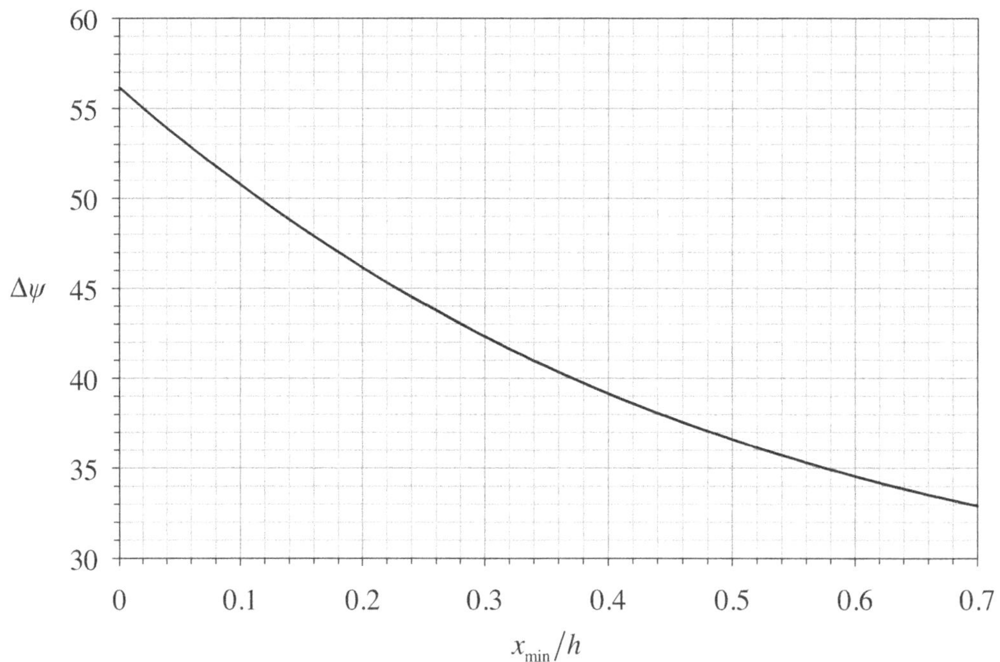

Figure 5.7-2 Three-dB beamwidth $\Delta\psi$ in degrees of the vertical, far-field beam pattern versus x_{min}/h for $\lambda/L_Y = 0.5$ where $h > h_{\text{min}}$ and $x_{\text{min}}/h_{\text{min}} = 0.701$.

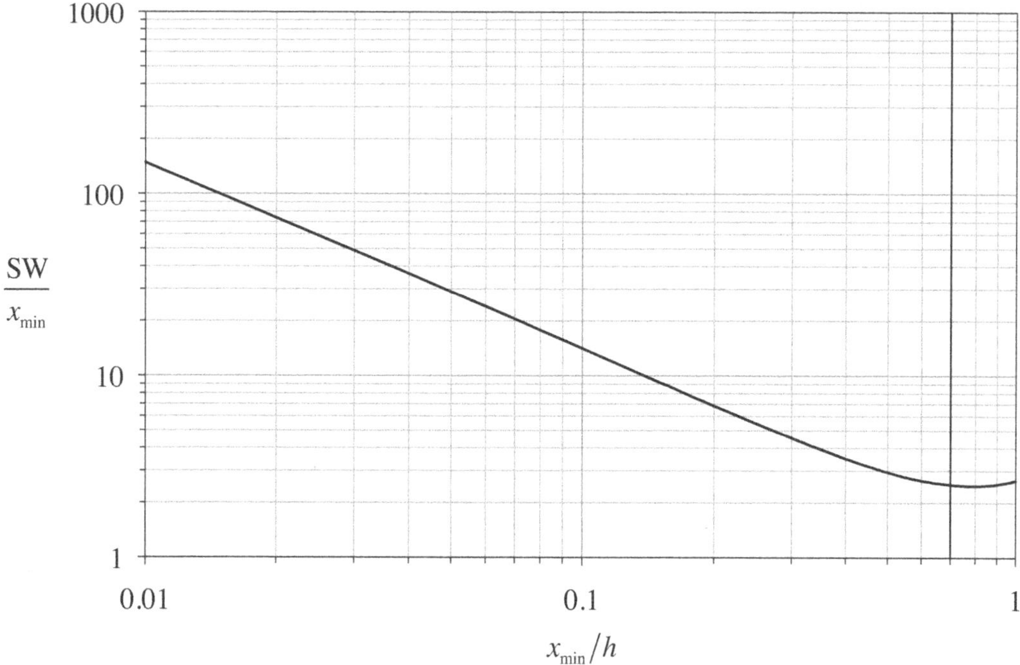

Figure 5.7-3 Ratio SW/x_{min} versus x_{min}/h for $\lambda/L_Y = 0.5$ where $h > h_{min}$ and $x_{min}/h_{min} = 0.701$.

5.7.2 Shallow Water

In shallow water, severe constraints are placed on the allowed range of values for the altitude of operation h of a SLS. In fact, the value for h may be given (specified). In this subsection, values for the following parameters are given: the operating (carrier) frequency f, the height (altitude) of operation h, the width of the one-sided blind zone x_{min} (a.k.a. the beginning or near-edge of the SW), and the along-track (azimuthal) resolution Δz_{min} at cross-range x_{min}. With the parameter values that are given, we shall determine the dimensions L_Y and L_Z of a SLS in order to guarantee that the ocean bottom will be in the far-field region of the SLS. Table 5.7-3 summarizes the design procedure for this shallow-water problem.

Since h and x_{min} are given, the slant-range r_{min} to the beginning (near-edge) of the SW can be computed from [see (5.1-25) and Fig. 5.1-2]

$$r_{min} = \sqrt{h^2 + x_{min}^2} \, . \tag{5.7-20}$$

And since x_{min} and Δz_{min} are given, using (5.3-14) to solve for $\Delta\theta$ yields

Table 5.7-3 Design Procedure for Determining the Dimensions of a SLS in Order to Satisfy the Far-Field Range Criterion in Shallow Water

Given	Resulting Fixed Parameter Values
h, $x_{\min}$, and $\Delta z_{\min}$	$r_{\min}$ and $\Delta\theta$
f and $\Delta\theta$	L_Z
f, L_Z, and $r_{\min}$	L_Y so that $r_{\min} > r_{\text{NF/FF}}$
f and L_Y	$\Delta\psi$ $(\psi' = 0°)$

$$\boxed{\Delta\theta = 2\tan^{-1}\left(\frac{\Delta z_{\min}}{2x_{\min}}\right)} \tag{5.7-21}$$

where $\Delta\theta$ is the 3-dB beamwidth of the horizontal, far-field beam pattern in the XZ plane (see Fig. 5.7-4). Now that $\Delta\theta$ is known and f is given, we can solve for L_Z by substituting (5.6-6) into (5.3-9). Doing so yields

$$\boxed{L_Z \approx 0.443\frac{\lambda}{\sin(\Delta\theta/2)}} \tag{5.7-22}$$

where L_Z is the length of the planar aperture (SLS) in the Z direction (see Fig. 5.1-1) and $c = f\lambda$ (see Fig. 5.7-5). Since f, L_Z, and $r_{\min}$ are known, we shall solve for L_Y next in order to guarantee that

$$r_{\min} > r_{\text{NF/FF}}, \tag{5.7-23}$$

where $r_{\text{NF/FF}}$ is the range to the near-field/far-field boundary given by (5.7-1). Substituting (5.7-1) into (5.7-23) yields

$$r_{\min} > \pi\frac{L_Y^2 + L_Z^2}{4\lambda}, \tag{5.7-24}$$

and by solving for L_Y, we obtain

$$\boxed{L_Y < \max L_Y} \tag{5.7-25}$$

where

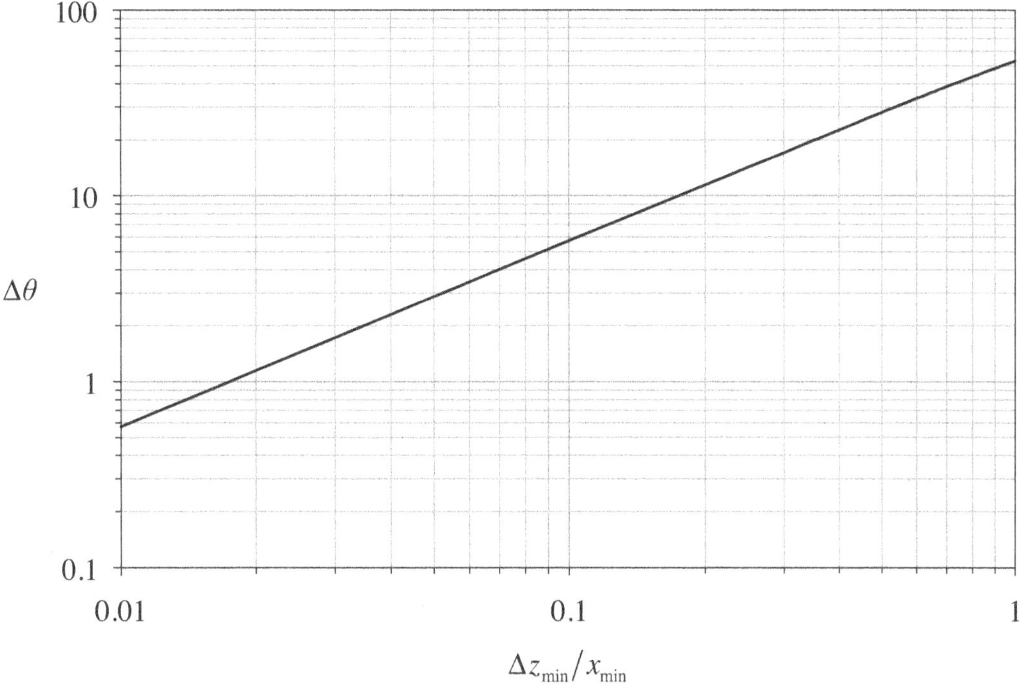

Figure 5.7-4 Three-dB beamwidth $\Delta\theta$ in degrees of the horizontal, far-field beam pattern versus $\Delta z_{\min}/x_{\min}$.

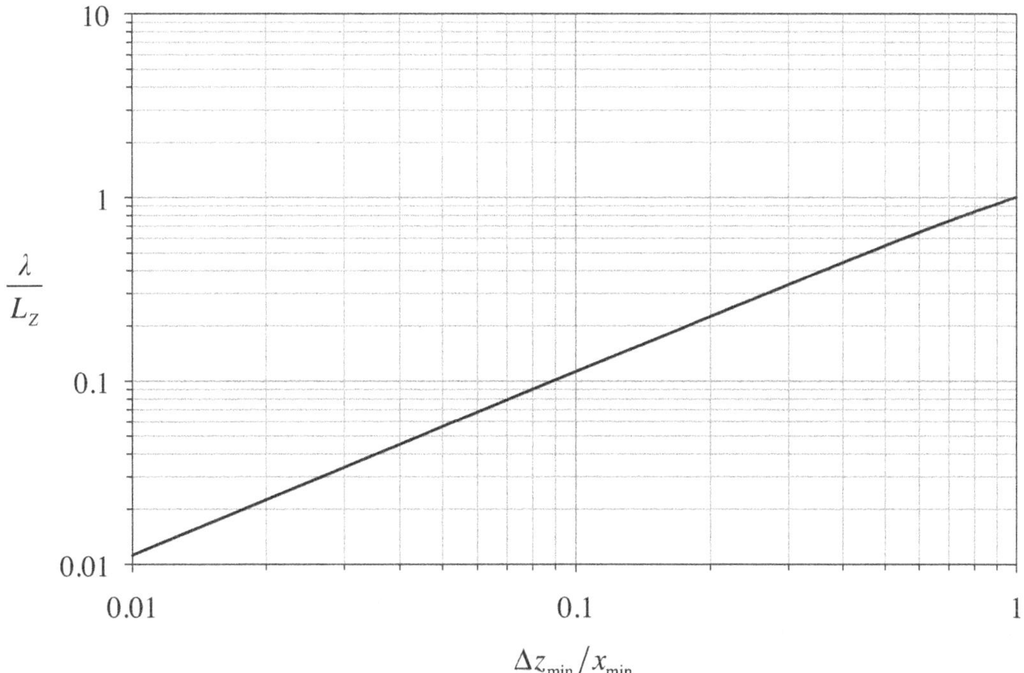

Figure 5.7-5 Ratio λ/L_Z versus $\Delta z_{\min}/x_{\min}$.

$$\boxed{\max L_Y = \sqrt{\frac{4\lambda}{\pi} r_{\min} - L_Z^2}\,, \qquad \frac{4\lambda}{\pi} r_{\min} > L_Z^2} \tag{5.7-26}$$

is the maximum allowed value for the length L_Y of the planar aperture (SLS) in the Y direction (see Fig. 5.1-1) in order to satisfy (5.7-23). For this shallow-water problem, the SW can be increased for a fixed value of the ratio $x_{\min}/h$ by decreasing L_Y since f is fixed (see Fig. 5.7-6). The SW can also be increased by increasing $x_{\min}/h$. However, if $x_{\min}/h$ is too big, then it is possible for $\beta_- \geq 90°$ ($\psi_- \leq 0°$) for a particular value of λ/L_Y, which is not allowed (see Fig. 5.7-7). In this case, the ratio $x_{\min}/h$ must be decreased.

Now that L_Y is known and f is given, we can solve for the 3-dB beamwidth $\Delta\psi$ of the vertical, far-field beam pattern in the XY plane when *no* beam steering is done ($\psi' = 0°$) by using (5.7-5). After the dimensions of the SLS have been determined, the design and analysis procedure continues with (5.7-11), as discussed in Subsection 5.7.1.

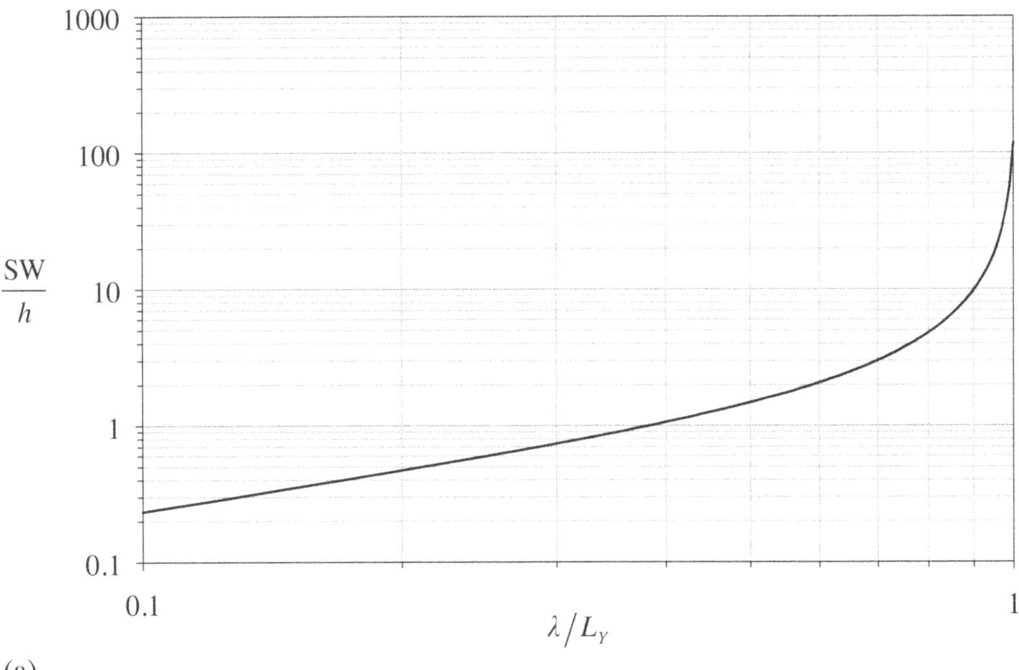

(a)

Figure 5.7-6 (a) Ratio SW/h versus λ/L_Y and (b) β_- in degrees versus λ/L_Y for $x_{\min}/h = 0.5$ where $L_Y < \max L_Y$.

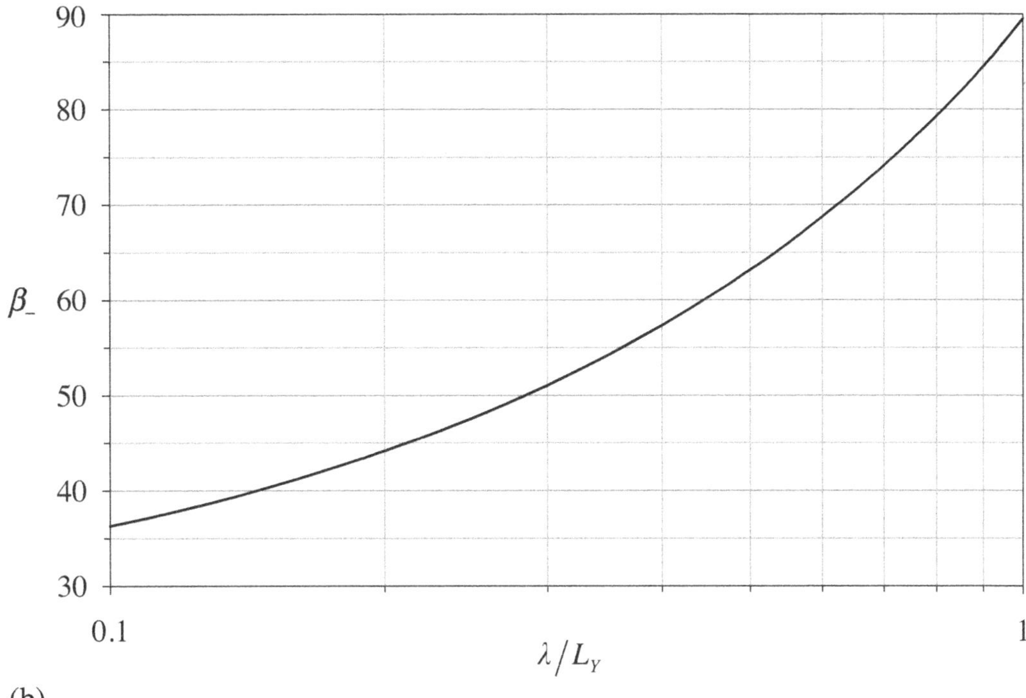

(b)

Figure 5.7-6 *continued.*

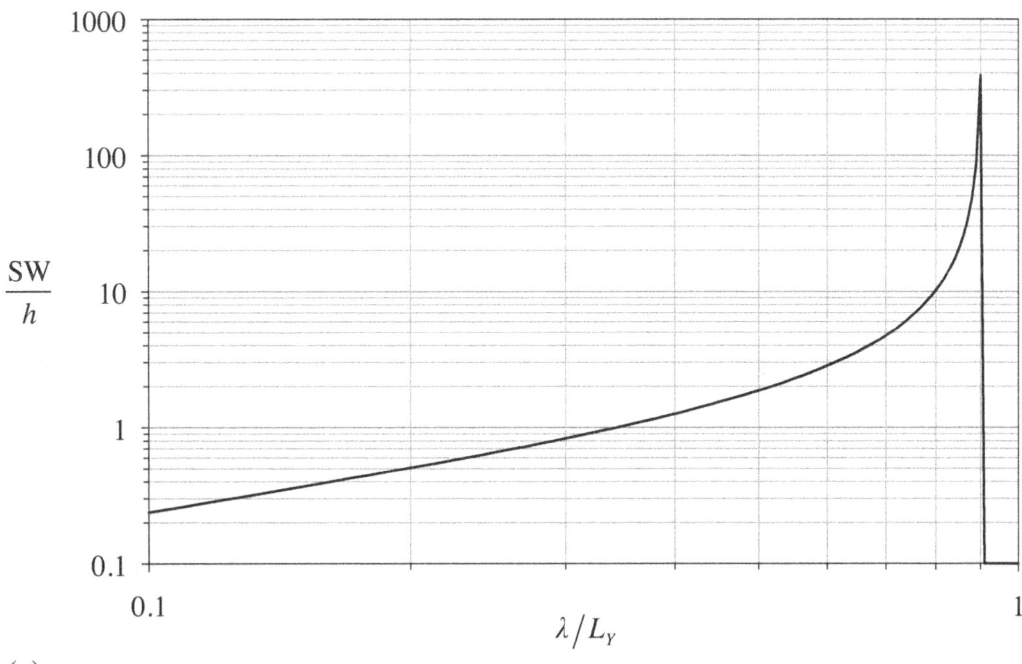

(a)

Figure 5.7-7 (a) Ratio SW/h versus λ/L_Y and (b) β_- in degrees versus λ/L_Y for $x_{\min}/h = 0.75$ where $L_Y < \max L_Y$.

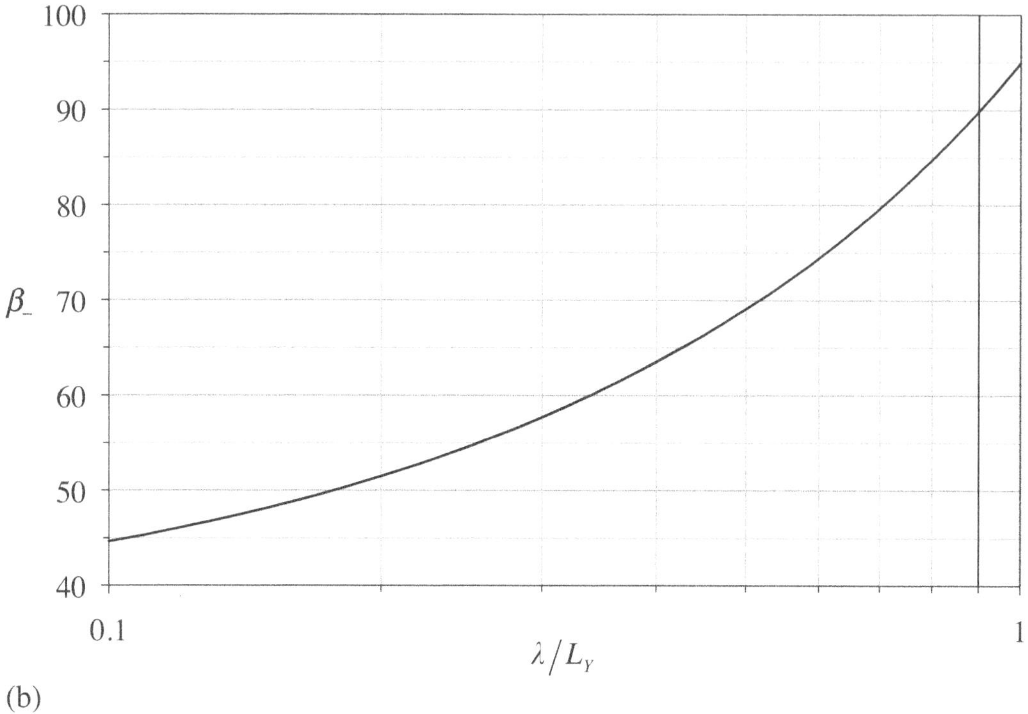

(b)

Figure 5.7-7 *continued.*

Problems

Section 5.2

5-1 A SLS transmits a rectangular-envelope, CW pulse, with a pulse length of $40\,\mu\text{sec}$. Find the cross-track (slant-range) resolution and the minimum allowed value for the operating (carrier) frequency. Use $c = 1500\text{ m/sec}$.

Section 5.3

5-2 A SLS is modeled as a rectangular piston lying in the YZ plane. If the far-edge of the swath width is at a cross-range of 200 m and the 3-dB beamwidth of the horizontal, far-field beam pattern in the XZ plane is $1°$, then what is the along-track (azimuthal) resolution in meters at the far-edge of the swath width?

Section 5.4

5-3 Show that $\text{PRI}_{\text{min, op}} \geq \text{PRI}_{\text{min}}$.

Section 5.7

5-4 Shallow-Water Problem. A SLS is modeled as a rectangular piston lying in the YZ plane with an operating frequency of $30\,\text{kHz}$. The SLS is required to operate in shallow water at an altitude of $10\,\text{m}$. The speed of sound in the ocean is $1500\,\text{m/sec}$.

(a) If you want the width of the one-sided blind zone to be $5\,\text{m}$ and the along-track resolution at the near-edge of the swath width to be $1\,\text{m}$, then compute the required value for the 3-dB beamwidth (in degrees) of the horizontal, far-field beam pattern in the XZ plane.

(b) Using your result from part (a), solve for the length L_Z of the aperture in the Z direction.

(c) Find the maximum allowed value for the length L_Y of the aperture in the Y direction in order to satisfy $r_{\text{min}} > r_{\text{NF/FF}}$.

Chapter 6

Array Theory – Linear Arrays

6.1 The Far-Field Beam Pattern of a Linear Array

An array can be thought of as a *sampled aperture*, that is, an aperture that is excited only at certain points or in certain localized areas. An array consists of individual electroacoustic transducers called *elements*. When an array is used in the *active mode*, the electroacoustic transducers are used as *transmitters*, converting electrical signals (voltages and currents) into acoustic signals (sound waves). When an array is used in the *passive mode*, the electroacoustic transducers are used as *receivers*, converting acoustic signals into electrical signals for further signal processing. In underwater acoustic applications, when an electroacoustic transducer is used as a transmitter, it is referred to as a *projector*, and when it is used as a receiver, it is referred to as a *hydrophone*. In seismic applications, an electroacoustic transducer used as a receiver is referred to as a *geophone*.

We know from complex aperture theory that an aperture function and its corresponding far-field beam pattern (directivity function) form a spatial Fourier transform pair. Therefore, since a linear array is an example of a linear aperture, the far-field beam pattern of a linear array lying along the X axis is given by the following one-dimensional spatial Fourier transform (see Section 2.1):

$$D(f, f_X) = F_{x_A}\{A(f, x_A)\} = \int_{-\infty}^{\infty} A(f, x_A)\exp(+j2\pi f_X x_A)\,dx_A\,, \qquad (6.1\text{-}1)$$

where $A(f, x_A)$ is the complex frequency response (complex aperture function) of the linear array, and

$$f_X = u/\lambda = \sin\theta\cos\psi/\lambda \qquad (6.1\text{-}2)$$

is a spatial frequency in the X direction with units of cycles per meter. Since f_X can be expressed in terms of the spherical angles θ and ψ, the far-field beam pattern can ultimately be expressed as a function of frequency f and the spherical angles θ and ψ, that is, $D(f, f_X) \rightarrow D(f, \theta, \psi)$. We shall consider a field point to be in the far-field region of an array if the range r to the field point – as measured from the center of the array – satisfies the Fraunhofer (far-field) range criterion given by

$$r > \pi R_A^2/\lambda > 2.414 R_A\,, \qquad (6.1\text{-}3)$$

where R_A is the maximum radial extent of the array. If the overall length of the linear array is L_A meters, then $R_A = L_A/2$.

If a linear array lies along either the Y or Z axis instead of the X axis, then simply replace x_A and f_X with either y_A and f_Y, or z_A and f_Z, respectively, in (6.1-1), where spatial frequencies f_Y and f_Z are given by (1.2-44) and (1.2-45), respectively. As can be seen from (6.1-1), in order to derive the far-field beam pattern of a linear array, we must first specify a mathematical model for its complex frequency response.

6.1.1 Even Number of Elements

Consider a linear array composed of an *even* number of elements lying along the X axis, where each element is a single continuous line source or receiver L meters in length (see Fig. 6.1-1). If the elements are *not* identical, that is, if each element has a different complex frequency response, and if the elements are *not* equally spaced, as shown in Fig. 6.1-1, then the complex frequency response (complex aperture function) of this array can be expressed as

$$A(f, x_A) = \sum_{n=-N/2}^{-1} c_n(f) e_n(f, x_A - x_n) + \sum_{n=1}^{N/2} c_n(f) e_n(f, x_A - x_n), \tag{6.1-4}$$

or

$$A(f, x_A) = \sum_{n=1}^{N/2} \left[c_{-n}(f) e_{-n}(f, x_A - x_{-n}) + c_n(f) e_n(f, x_A - x_n) \right], \tag{6.1-5}$$

where N is the total even number of elements,

$$\boxed{c_n(f) = a_n(f) \exp[+j\theta_n(f)]} \tag{6.1-6}$$

is the frequency-dependent, *complex weight* associated with element n, $a_n(f)$ is a real, frequency-dependent, *dimensionless*, amplitude weight, $\theta_n(f)$ is a real, frequency-dependent, phase weight in radians, $e_n(f, x_A)$ is the complex frequency response (complex aperture function) of element n, also known as the *element function*, and x_n is the x coordinate of the center of element n. Note that because the elements are not equally spaced, $x_{-n} \neq -x_n$ in general. Complex weights are implemented by using digital signal processing (digital beamforming), as will be discussed in Section 6.4. Linear arrays of planar elements, such as rectangular and circular pistons, will be discussed in Chapter 8 (see Example 8.1-6).

Equation (6.1-4) indicates that the complex frequency response of the array is modeled as the sum (superposition) of the complex-weighted frequency responses of all the elements in the array. The complex weights are used to

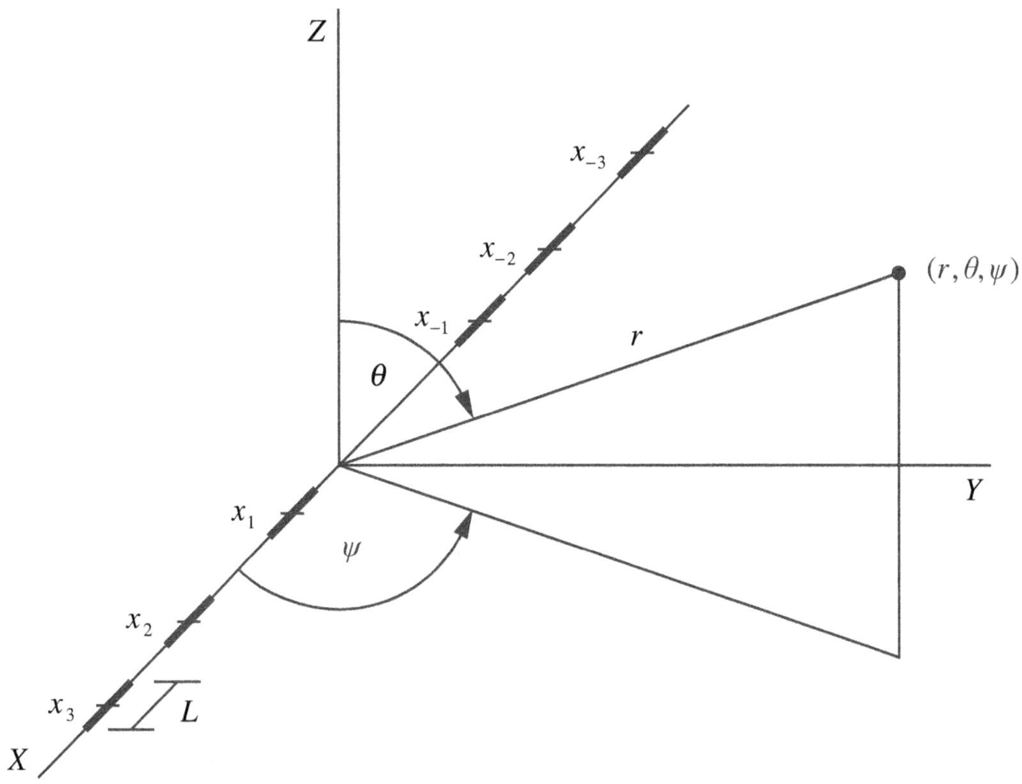

Figure 6.1-1 Linear array composed of an even number of unevenly-spaced, continuous line elements lying along the X axis. Each element is L meters in length. Also shown is a field point with spherical coordinates (r, θ, ψ).

control the complex frequency response of the array and, thus, the array's far-field beam pattern via amplitude and phase weighting, as will be discussed later in this chapter. Substituting (6.1-5) into (6.1-1) yields the following expression for the far-field beam pattern:

$$D(f, f_X) = \sum_{n=1}^{N/2} \left[c_{-n}(f) E_{-n}(f, f_X) \exp(+j2\pi f_X x_{-n}) + c_n(f) E_n(f, f_X) \exp(+j2\pi f_X x_n) \right] \tag{6.1-7}$$

where

$$F_{x_A}\{e_n(f, x_A - x_n)\} = E_n(f, f_X) \exp(+j2\pi f_X x_n) \tag{6.1-8}$$

and

$$E_n(f, f_X) = F_{x_A}\{e_n(f, x_A)\} \tag{6.1-9}$$

is the far-field beam pattern of element (electroacoustic transducer) n. The units of $e_n(f, x_A)$, $A(f, x_A)$, $E_n(f, f_X)$, and $D(f, f_X)$ are summarized in Table 6.1-1.

Table 6.1-1 Units of the Complex Frequency Responses $e_n(f, x_A)$ and $A(f, x_A)$ and Corresponding Far-Field Beam Patterns $E_n(f, f_X)$ and $D(f, f_X)$ for a Linear Array

Linear Array	$e_n(f, x_A)$, $A(f, x_A)$	$E_n(f, f_X)$, $D(f, f_X)$
active (transmit)	$\left(\left(\text{m}^3/\text{sec}\right)/\text{V}\right)/\text{m}$	$\left(\text{m}^3/\text{sec}\right)/\text{V}$
passive (receive)	$\left(\text{V}/\left(\text{m}^2/\text{sec}\right)\right)/\text{m}$	$\text{V}/\left(\text{m}^2/\text{sec}\right)$

If all the elements in the array are *identical* (i.e., all the elements have the same complex frequency response), but still *not* equally spaced, then (6.1-7) reduces to

$$D(f, f_X) = E(f, f_X)S(f, f_X) \tag{6.1-10}$$

where

$$E(f, f_X) = F_{x_A}\{e(f, x_A)\} \tag{6.1-11}$$

is the far-field beam pattern of *one* of the identical elements in the array,

$$S(f, f_X) = \sum_{n=1}^{N/2} \left[c_{-n}(f)\exp(+j2\pi f_X x_{-n}) + c_n(f)\exp(+j2\pi f_X x_n)\right] \tag{6.1-12}$$

is the *array factor*, and if $a_{-n}(f) > 0$ and $a_n(f) > 0 \;\; \forall \;\; n$, then

$$S_{\max} = \max\left|S(f, f_X)\right| = \sum_{n=1}^{N/2} \left[a_{-n}(f) + a_n(f)\right] \tag{6.1-13}$$

is the *normalization factor* for the array factor (see Appendix 6A). As can be seen from (6.1-12), $S(f, f_X)$ is *dimensionless* and depends on the set of complex weights used and the locations (the x coordinates) of all the elements along the X axis. The array factor $S(f, f_X)$ contains the spatial information of the array. If one or more elements in the array are broken (not transmitting and/or receiving), set $c_n(f) = 0$ for those elements.

Equation (6.1-10) is referred to as the *Product Theorem* for a linear array composed of an even number of identical elements. The Product Theorem states that the far-field beam pattern $D(f, f_X)$ of a linear array composed of an even number of *identical*, unevenly-spaced, complex-weighted elements is equal to the *product* of the far-field beam pattern $E(f, f_X)$ of *one* of the identical elements in the array, and the dimensionless array factor $S(f, f_X)$, which is related to the far-field beam pattern of an equivalent linear array composed of an even number of identical, unevenly-spaced, complex-weighted, *omnidirectional point-elements.* Although the Product Theorem requires identical elements, equal spacing of the elements is *not* required. The interpretation of $S(f, f_X)$ shall be discussed next.

In acoustic wave-propagation theory, an impulse function is used as a mathematical model for an omnidirectional point-source. Similarly, the mathematical model that we shall use for the complex frequency response of an omnidirectional point-element lying along the X axis is given by

$$e(f, x_A) = \mathcal{S}(f)\delta(x_A), \tag{6.1-14}$$

where $\mathcal{S}(f)$ is the complex, *element sensitivity function.* A real-world example of an omnidirectional point-element is a spherical sound-source manufactured to vibrate in the monopole mode of vibration and whose circumference is small compared to a wavelength. If the point-element is used in the active mode as a transmitter, then the element sensitivity function $\mathcal{S}(f)$ is referred to as the *transmitter sensitivity function* with units of $\left(\mathrm{m}^3/\mathrm{sec}\right)/\mathrm{V}$ (see Appendix 6B). If the point-element is used in the passive mode as a receiver, then $\mathcal{S}(f)$ is referred to as the *receiver sensitivity function* with units of $\mathrm{V}/\left(\mathrm{m}^2/\mathrm{sec}\right)$ (see Appendix 6B). Note that $\int_{-\infty}^{\infty} e(f, x_A)dx_A = \mathcal{S}(f)$ because $\int_{-\infty}^{\infty} \delta(x_A)dx_A = 1$ (see Appendix 6C). The impulse function $\delta(x_A)$ has units of inverse meters. The units of $\mathcal{S}(f)$ are summarized in Table 6.1-2.

Table 6.1-2 Units of the Element Sensitivity Function $\mathcal{S}(f)$

Omnidirectional Point-Element	$\mathcal{S}(f)$
active (transmit)	$\left(\mathrm{m}^3/\mathrm{sec}\right)/\mathrm{V}$
passive (receive)	$\mathrm{V}/\left(\mathrm{m}^2/\mathrm{sec}\right)$

Since

$$F_{x_A}\{\delta(x_A)\}=1, \tag{6.1-15}$$

taking the spatial Fourier transform of (6.1-14) yields

$$E(f, f_X)=\mathcal{S}(f). \tag{6.1-16}$$

Equation (6.1-16) is the far-field beam pattern of an omnidirectional point-element, and as expected, it is *not* a function of the spherical angles θ and ψ. The term "omnidirectional" means that an electroacoustic transducer transmits and/or receives equally well in all directions. Therefore, by substituting (6.1-16) into (6.1-10), we obtain the following expression for the far-field beam pattern of a linear array composed of an even number of identical, unevenly-spaced, complex-weighted, omnidirectional point-elements lying along the X axis:

$$\boxed{D(f, f_X)=\mathcal{S}(f)S(f, f_X)} \tag{6.1-17}$$

where $S(f, f_X)$ is given by (6.1-12). The corresponding complex frequency response (complex aperture function) of the array can be obtained by taking the inverse spatial Fourier transform of (6.1-17). Doing so yields

$$\boxed{A(f, x_A)=\mathcal{S}(f)\sum_{n=1}^{N/2}\left[c_{-n}(f)\delta(x_A-x_{-n})+c_n(f)\delta(x_A-x_n)\right]} \tag{6.1-18}$$

since

$$F_{f_X}^{-1}\{\exp(+j2\pi f_X x_n)\}=\delta(x_A-x_n). \tag{6.1-19}$$

Note that the summation in (6.1-18) is in the form of the impulse response of a finite impulse response (FIR) digital filter. In Appendix 6D, it is shown that a linear array composed of an even number of identical, complex-weighted elements can be thought of as a one-dimensional, spatial FIR filter, where $s(f, x_A)=F_{f_X}^{-1}\{S(f, f_X)\}$ is the spatial impulse response of the array at frequency f hertz.

If all the elements in the array are *equally spaced*, then

$$\boxed{x_n=(n-0.5)d, \qquad n=1, 2, \ldots, N/2} \tag{6.1-20}$$

and

$$\boxed{x_{-n}=-x_n, \qquad n=1, 2, \ldots, N/2} \tag{6.1-21}$$

where d is the *interelement spacing* in meters.

If all the elements in the array are equally spaced and the complex weights satisfy the conjugate symmetry property

$$c_{-n}(f) = c_n^*(f), \qquad n = 1, 2, \ldots, N/2, \tag{6.1-22}$$

where the asterisk denotes complex conjugate, then the array factor $S(f, f_X)$ given by (6.1-12) can be simplified. Before we begin the procedure of simplifying $S(f, f_X)$, let us first examine the consequences of (6.1-22). Substituting (6.1-6) into (6.1-22) yields

$$a_{-n}(f)\exp[+j\theta_{-n}(f)] = a_n(f)\exp[-j\theta_n(f)], \qquad n = 1, 2, \ldots, N/2, \tag{6.1-23}$$

from which we obtain the following two equations:

$$a_{-n}(f) = a_n(f), \qquad n = 1, 2, \ldots, N/2, \tag{6.1-24}$$

and

$$\theta_{-n}(f) = -\theta_n(f), \qquad n = 1, 2, \ldots, N/2. \tag{6.1-25}$$

Equations (6.1-24) and (6.1-25) indicate that if the complex weights satisfy the conjugate symmetry property given by (6.1-22), then the amplitude and phase weights are *even* and *odd* functions of index n, respectively.

We begin the procedure of simplifying $S(f, f_X)$ by substituting (6.1-20) through (6.1-22) into (6.1-12) which yields

$$\begin{aligned} S(f, f_X) &= \sum_{n=1}^{N/2} \left[c_n^*(f)\exp[-j2\pi f_X(n-0.5)d] + c_n(f)\exp[+j2\pi f_X(n-0.5)d] \right] \\ &= \sum_{n=1}^{N/2} \left\{ \left[c_n(f)\exp[+j2\pi f_X(n-0.5)d] \right]^* + \left[c_n(f)\exp[+j2\pi f_X(n-0.5)d] \right] \right\}. \end{aligned} \tag{6.1-26}$$

Since $Z^* + Z = 2\,\mathrm{Re}\{Z\}$, (6.1-26) reduces to

$$S(f, f_X) = 2\sum_{n=1}^{N/2} \mathrm{Re}\left\{ c_n(f)\exp[+j2\pi f_X(n-0.5)d] \right\}, \tag{6.1-27}$$

and by substituting (6.1-6) into (6.1-27), we finally obtain

$$\boxed{S(f, f_X) = 2\sum_{n=1}^{N/2} a_n(f)\cos[2\pi f_X(n-0.5)d + \theta_n(f)]} \tag{6.1-28}$$

Equation (6.1-28) is valid only if all the identical elements in the array are equally spaced and the complex weights satisfy the conjugate symmetry property given by (6.1-22).

Example 6.1-1 Two-Element Interferometer

A *two-element interferometer* is a linear array of two identical, complex-weighted, omnidirectional point-elements that are equally spaced about the origin of the coordinate system (the center of the array as shown in Fig. 6.1-2) where the complex weights satisfy the conjugate symmetry property given by (6.1-22). Since the elements are equally spaced, substituting $N=2$ into (6.1-20) and (6.1-21) yields the following x coordinates of the centers of the two elements:

$$x_1 = d/2 \tag{6.1-29}$$

and

$$x_{-1} = -x_1 = -d/2 \tag{6.1-30}$$

so that the interelement spacing is d meters. Since the complex weights satisfy the conjugate symmetry property given by (6.1-22), substituting $N=2$ into (6.1-24) and (6.1-25) yields

$$a_{-1}(f) = a_1(f) \tag{6.1-31}$$

and

$$\theta_{-1}(f) = -\theta_1(f). \tag{6.1-32}$$

And since the elements are identical, substituting $N=2$ into (6.1-28), and then substituting the result into (6.1-17) yields

$$\boxed{D(f, f_X) = 2a_1(f)\mathcal{S}(f)\cos[\pi f_X d + \theta_1(f)]} \tag{6.1-33}$$

where $\mathcal{S}(f)$ is the complex, element sensitivity function. Equation (6.1-33) is the *unnormalized*, far-field beam pattern of a two-element interferometer lying along the X axis.

The *dimensionless*, *normalized*, far-field beam pattern is defined as follows:

$$\boxed{D_N(f, f_X) \triangleq D(f, f_X)/D_{\max}} \tag{6.1-34}$$

where $D_{\max} = \max|D(f, f_X)|$ is the *normalization factor* – it is the *maximum* value of the *magnitude* of the unnormalized, far-field beam pattern. Therefore, by referring to (6.1-33),

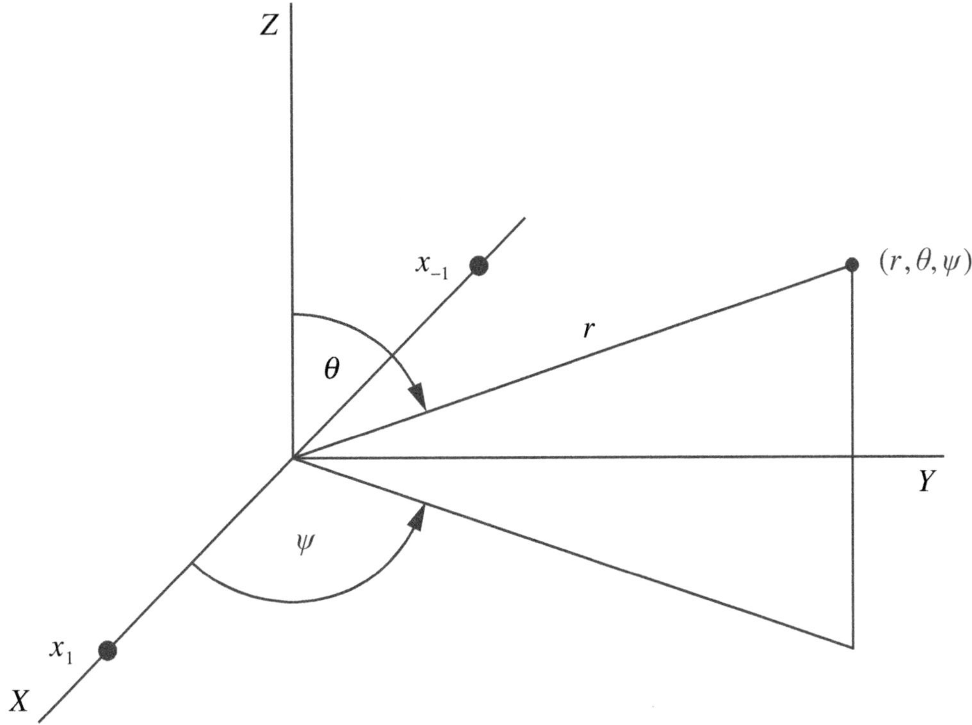

Figure 6.1-2 Linear array of two omnidirectional point-elements lying along the X axis. Also shown is a field point with spherical coordinates (r, θ, ψ).

$$|D(f, f_X)| = 2|a_1(f)||S(f)||\cos[\pi f_X d + \theta_1(f)]|, \tag{6.1-35}$$

and since the maximum value of the magnitude of a cosine function is 1,

$$D_{\max} = 2|a_1(f)||S(f)|. \tag{6.1-36}$$

Dividing (6.1-35) by (6.1-36) yields the following expression for the magnitude of the *normalized*, far-field beam pattern of a two-element interferometer lying along the X axis:

$$\boxed{|D_N(f, f_X)| = |\cos[\pi f_X d + \theta_1(f)]|} \tag{6.1-37}$$

Because we have not yet discussed phase weights in any great detail, let $\theta_1(f) = 0$. In other words, if only amplitude weighting is done (no phase weighting), then (6.1-37) reduces to

$$|D_N(f, f_X)| = |\cos(\pi f_X d)|. \tag{6.1-38}$$

Since spatial frequency

$$f_X = u/\lambda = \sin\theta\cos\psi/\lambda\,, \tag{6.1-39}$$

(6.1-38) can also be expressed as

$$\left|D_N(f,u)\right| = \left|\cos\left(\pi\frac{d}{\lambda}u\right)\right|, \tag{6.1-40}$$

or

$$\left|D_N(f,\theta,\psi)\right| = \left|\cos\left(\pi\frac{d}{\lambda}\sin\theta\cos\psi\right)\right|. \tag{6.1-41}$$

At broadside, where direction cosine $u=0$, (6.1-40) reduces to $\left|D_N(f,0)\right|=1$, and by setting $\theta=90°$ in (6.1-41), we obtain

$$\left|D_N(f,90°,\psi)\right| = \left|\cos\left(\pi\frac{d}{\lambda}\cos\psi\right)\right|, \tag{6.1-42}$$

which is the magnitude of the normalized, *horizontal*, far-field beam pattern in the XY plane of a two-element interferometer lying along the X axis (see Fig. 6.1-3).

As can be seen in Fig. 6.1-3, the mainlobe exists on both the port and starboard (left and right) sides of the array. This is an example of the *port/starboard (left/right) ambiguity problem* associated with linear arrays since a target at broadside on the port side will produce the same output electrical signal from the array as a target at broadside on the starboard side. Solutions to the port/starboard (left/right) ambiguity problem are discussed in Chapters 8 and 9, Examples 8.2-1 and 9.1-2, respectively. ■

Example 6.1-2 Dipole

A *dipole* is a linear array of two identical, complex-weighted, omnidirectional point-elements that are equally spaced about the origin of the coordinate system (the center of the array as shown in Fig. 6.1-2) where the complex weights do *not* satisfy the conjugate symmetry property given by (6.1-22). Since the elements are equally spaced, substituting $N=2$ into (6.1-20) and (6.1-21) yields the following x coordinates of the centers of the two elements:

$$x_1 = d/2 \tag{6.1-43}$$

and

$$x_{-1} = -x_1 = -d/2 \tag{6.1-44}$$

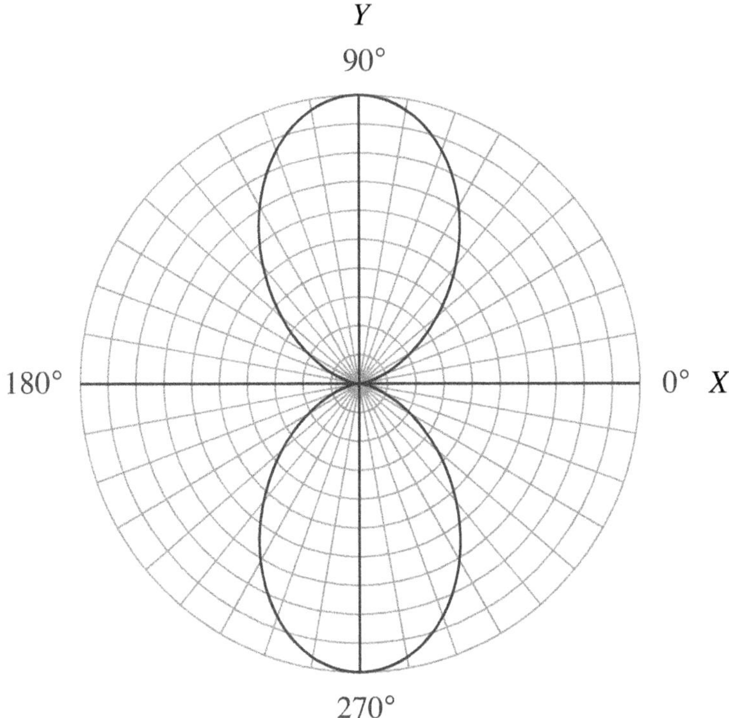

Figure 6.1-3 Polar plot, as a function of the azimuthal (bearing) angle ψ, of the magnitude of the normalized, horizontal, far-field beam pattern of a two-element interferometer lying along the X axis given by (6.1-42) for $d/\lambda = 0.5$.

so that the interelement spacing is d meters. In a dipole, *both* the amplitude and phase weights have *odd* symmetry, that is,

$$a_{-1}(f) = -a_1(f) \tag{6.1-45}$$

and

$$\theta_{-1}(f) = -\theta_1(f). \tag{6.1-46}$$

And since the elements are identical, substituting $N = 2$, (6.1-6), and (6.1-43) through (6.1-46) into (6.1-12) yields

$$S(f, f_X) = a_1(f)\left[\exp\{+j[\pi f_X d + \theta_1(f)]\} - \exp\{-j[\pi f_X d + \theta_1(f)]\}\right], \tag{6.1-47}$$

which reduces to

$$S(f, f_X) = j2a_1(f)\sin[\pi f_X d + \theta_1(f)]. \tag{6.1-48}$$

Substituting (6.1-48) into (6.1-17) yields

$$\boxed{D(f, f_X) = j2a_1(f)\mathcal{S}(f)\sin[\pi f_X d + \theta_1(f)]} \tag{6.1-49}$$

where $\mathcal{S}(f)$ is the complex, element sensitivity function. Equation (6.1-49) is the *unnormalized*, far-field beam pattern of a dipole lying along the X axis.

The magnitude of the far-field beam pattern given by (6.1-49) is

$$|D(f, f_X)| = 2|a_1(f)||\mathcal{S}(f)||\sin[\pi f_X d + \theta_1(f)]|, \tag{6.1-50}$$

and since the maximum value of the magnitude of a sine function is 1,

$$D_{\max} = 2|a_1(f)||\mathcal{S}(f)|. \tag{6.1-51}$$

Dividing (6.1-50) by (6.1-51) yields the following expression for the magnitude of the *normalized*, far-field beam pattern of a dipole lying along the X axis:

$$\boxed{|D_N(f, f_X)| = |\sin[\pi f_X d + \theta_1(f)]|} \tag{6.1-52}$$

If we let $\theta_1(f) = 0$, that is, if only amplitude weighting is done (no phase weighting), then (6.1-52) reduces to

$$|D_N(f, f_X)| = |\sin(\pi f_X d)|. \tag{6.1-53}$$

Since spatial frequency

$$f_X = u/\lambda = \sin\theta\cos\psi/\lambda, \tag{6.1-54}$$

(6.1-53) can also be expressed as

$$|D_N(f, u)| = \left|\sin\left(\pi \frac{d}{\lambda} u\right)\right|, \tag{6.1-55}$$

or

$$|D_N(f, \theta, \psi)| = \left|\sin\left(\pi \frac{d}{\lambda} \sin\theta\cos\psi\right)\right|. \tag{6.1-56}$$

At broadside, where direction cosine $u = 0$, (6.1-55) reduces to $|D_N(f, 0)| = 0$, and by setting $\theta = 90°$ in (6.1-56), we obtain

$$|D_N(f, 90°, \psi)| = \left|\sin\left(\pi \frac{d}{\lambda} \cos\psi\right)\right|, \tag{6.1-57}$$

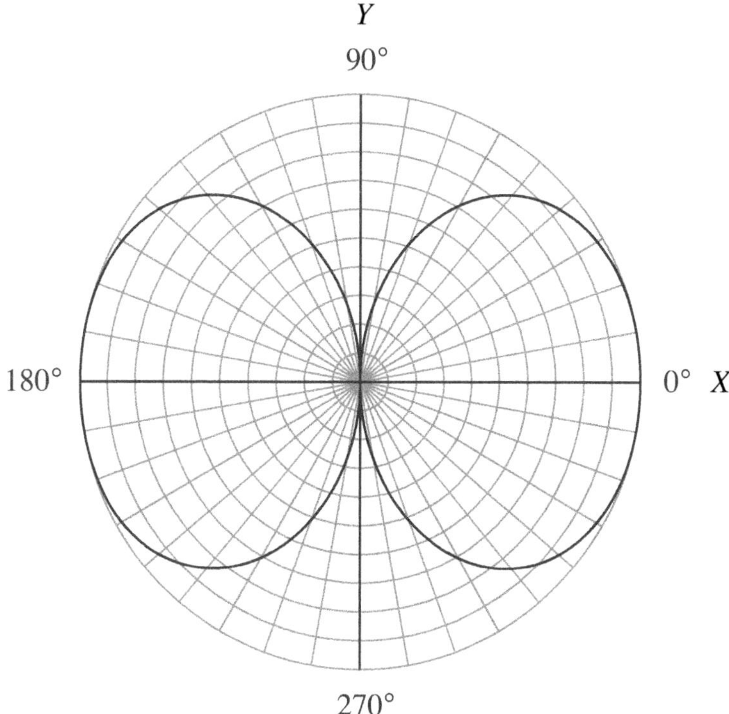

Figure 6.1-4 Polar plot, as a function of the azimuthal (bearing) angle ψ, of the magnitude of the normalized, horizontal, far-field beam pattern of a dipole lying along the X axis given by (6.1-57) for $d/\lambda = 0.5$.

which is the magnitude of the normalized, *horizontal*, far-field beam pattern in the XY plane of a dipole lying along the X axis (see Fig. 6.1-4). Unlike the far-field beam pattern of a two-element interferometer which has a maximum at broadside (see Fig. 6.1-3), the dipole has a null. ■

Example 6.1-3 Cardioid Beam Pattern

Consider a linear array of two identical, complex-weighted, omni-directional point-elements that are equally spaced about the origin of the coordinate system (the center of the array as shown in Fig. 6.1-2). Since the elements are equally spaced, substituting $N = 2$ into (6.1-20) and (6.1-21) yields the following x coordinates of the centers of the two elements:

$$x_1 = d/2 \tag{6.1-58}$$

and

$$x_{-1} = -x_1 = -d/2 \tag{6.1-59}$$

so that the interelement spacing is d meters. The amplitude weights are equal, that is,

$$a_{-1}(f) = a_1(f), \tag{6.1-60}$$

and the phase weights are

$$\theta_1(f) = 0 \tag{6.1-61}$$

and

$$\theta_{-1}(f) = \pi/2. \tag{6.1-62}$$

Substituting (6.1-60) through (6.1-62) into (6.1-6) yields the following complex weights:

$$c_1(f) = a_1(f) \tag{6.1-63}$$

and

$$c_{-1}(f) = a_1(f)\exp(+j\pi/2) = ja_1(f). \tag{6.1-64}$$

Substituting $N = 2$, (6.1-58), (6.1-59), (6.1-63), and (6.1-64) into (6.1-12) yields the array factor

$$S(f, f_X) = a_1(f)\left[j\exp(-j\pi f_X d) + \exp(+j\pi f_X d)\right]. \tag{6.1-65}$$

If the interelement spacing

$$d = \lambda/4, \tag{6.1-66}$$

and since spatial frequency

$$f_X = u/\lambda = \sin\theta\cos\psi/\lambda, \tag{6.1-67}$$

the array factor given by (6.1-65) can be rewritten as

$$S(f, u) = (1 + j)a_1(f)\left[\cos(\pi u/4) + \sin(\pi u/4)\right], \tag{6.1-68}$$

or

$$S(f, u) = \sqrt{2}\,(1 + j)a_1(f)\cos\left[\pi(u - 1)/4\right], \tag{6.1-69}$$

since

$$\cos\alpha + \sin\alpha = \sqrt{2}\cos\left[\alpha - (\pi/4)\right]. \tag{6.1-70}$$

And since the elements are identical, the *unnormalized*, far-field beam pattern of the array is given by [see (6.1-17)]

$$\boxed{D(f,u) = \sqrt{2}(1+j)a_1(f)\mathcal{S}(f)\cos[\pi(u-1)/4]} \tag{6.1-71}$$

where $\mathcal{S}(f)$ is the complex, element sensitivity function.

The magnitude of the far-field beam pattern given by (6.1-71) is

$$|D(f,u)| = 2|a_1(f)||\mathcal{S}(f)||\cos[\pi(u-1)/4]|, \tag{6.1-72}$$

and since the maximum value of $|\cos[\pi(u-1)/4]|$ occurs at $u=1$ where $|\cos(0)|=1$, the normalization factor is

$$D_{\max} = \max|D(f,u)| = 2|a_1(f)||\mathcal{S}(f)|. \tag{6.1-73}$$

Dividing (6.1-72) by (6.1-73) yields the following expression for the magnitude of the *normalized*, far-field beam pattern:

$$\boxed{|D_N(f,u)| = |\cos[\pi(u-1)/4]|} \tag{6.1-74}$$

where $|D_N(f,1)|=1$. Substituting $u=\sin\theta\cos\psi$ into (6.1-74) yields

$$|D_N(f,\theta,\psi)| = \left|\cos\left[\frac{\pi}{4}(\sin\theta\cos\psi - 1)\right]\right|, \tag{6.1-75}$$

and by setting $\theta = 90°$ in (6.1-75), we obtain

$$|D_N(f,90°,\psi)| = \left|\cos\left[\frac{\pi}{4}(\cos\psi - 1)\right]\right|, \tag{6.1-76}$$

which is the magnitude of the normalized, *horizontal*, far-field beam pattern in the XY plane (see Fig. 6.1-5). The beam pattern shown in Fig. 6.1-5 is referred to as a *cardioid* (heart-shaped) beam pattern. The mainlobe is aligned along the positive X axis since the maximum value of the normalized beam pattern occurs at $u=1$, which is one of the end-fire locations in direction-cosine space for a linear array lying along the X axis.

If the following set of phase weights are used instead,

$$\theta_1(f) = 0 \tag{6.1-77}$$

and

$$\theta_{-1}(f) = -\pi/2, \tag{6.1-78}$$

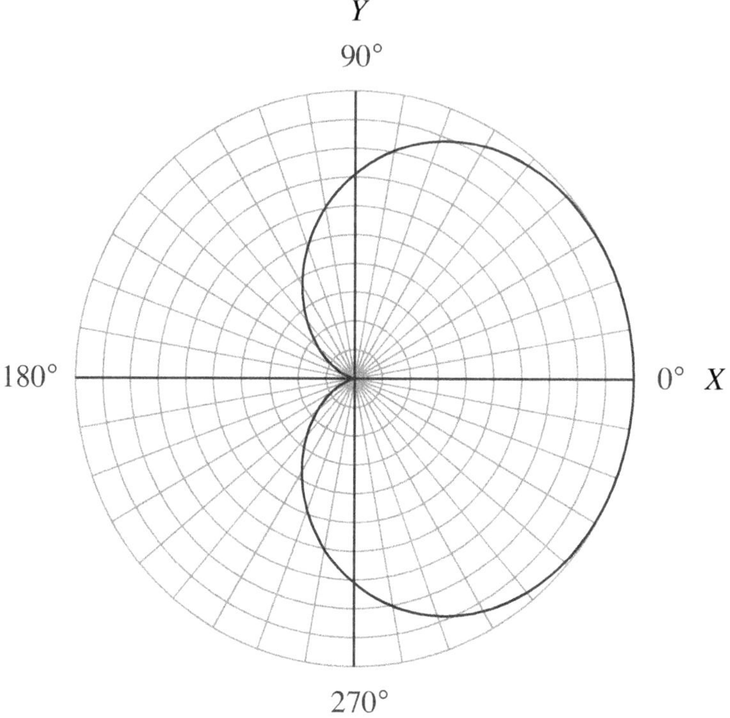

Figure 6.1-5 Polar plot, as a function of the azimuthal (bearing) angle ψ, of the magnitude of the normalized, horizontal, far-field beam pattern given by (6.1-76).

then the *unnormalized*, far-field beam pattern is given by

$$\boxed{D(f,u) = \sqrt{2}\,(1-j)a_1(f)\mathcal{S}(f)\cos[\pi(u+1)/4]} \tag{6.1-79}$$

the magnitude of the *normalized*, far-field beam pattern is given by

$$\boxed{|D_N(f,u)| = |\cos[\pi(u+1)/4]|} \tag{6.1-80}$$

or

$$|D_N(f,\theta,\psi)| = \left|\cos\left[\frac{\pi}{4}(\sin\theta\cos\psi + 1)\right]\right|, \tag{6.1-81}$$

and by setting $\theta = 90°$ in (6.1-81), we obtain

$$|D_N(f,90°,\psi)| = \left|\cos\left[\frac{\pi}{4}(\cos\psi + 1)\right]\right|, \tag{6.1-82}$$

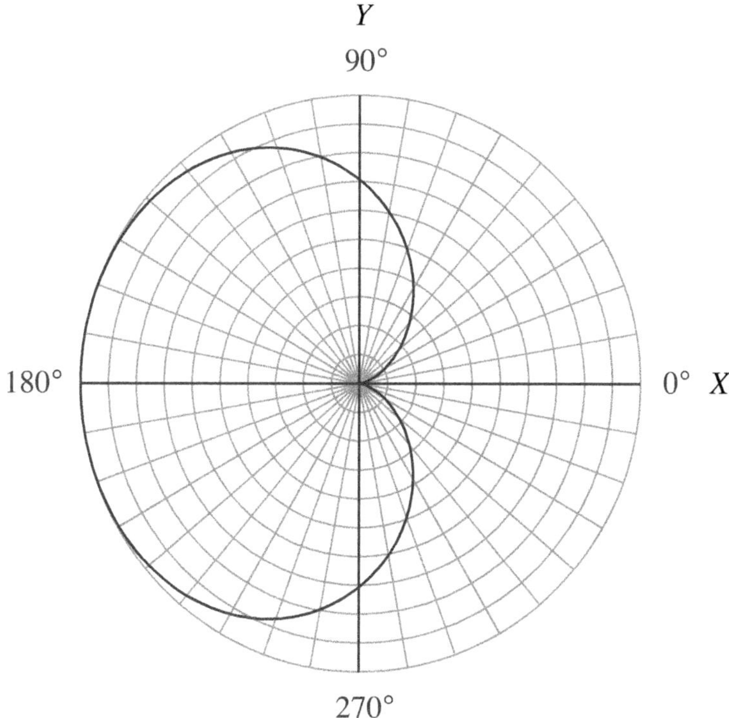

Figure 6.1-6 Polar plot, as a function of the azimuthal (bearing) angle ψ, of the magnitude of the normalized, horizontal, far-field beam pattern given by (6.1-82).

which is the magnitude of the normalized, *horizontal*, far-field beam pattern in the XY plane (see Fig. 6.1-6). The mainlobe is now aligned along the negative X axis since the maximum value of the normalized beam pattern occurs at $u = -1$, which is the other end-fire location in direction-cosine space for a linear array lying along the X axis. The *triplet circular array* discussed in Example 8.1-7 also has a cardioid far-field beam pattern. ■

6.1.2 Odd Number of Elements

Consider a linear array composed of an *odd* number of elements lying along the X axis, where each element is a single continuous line source or receiver L meters in length (see Fig. 6.1-7). If the elements are *not* identical, that is, if each element has a different complex frequency response, and if the elements are *not* equally spaced, as shown in Fig. 6.1-7, then the complex frequency response (complex aperture function) of this array can be expressed as

$$A(f, x_A) = \sum_{n=-N'}^{N'} c_n(f) e_n(f, x_A - x_n), \tag{6.1-83}$$

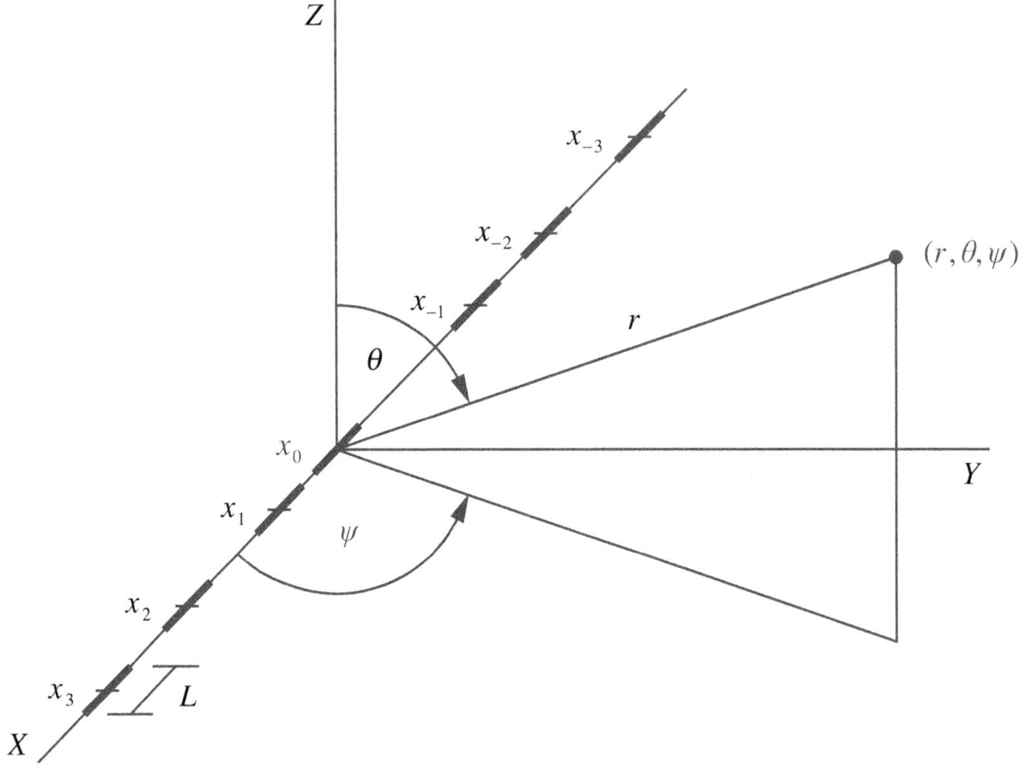

Figure 6.1-7 Linear array composed of an odd number of unevenly-spaced, continuous line elements lying along the X axis. Each element is L meters in length. Also shown is a field point with spherical coordinates (r,θ,ψ).

where

$$N' = (N-1)/2\,, \tag{6.1-84}$$

N is the total odd number of elements, $c_n(f)$ is the frequency-dependent, complex weight associated with element n [see (6.1-6)], $e_n(f, x_A)$ is the complex frequency response (complex aperture function) of element n, also known as the element function, and x_n is the x coordinate of the center of element n. Note that $x_0 = 0$ because element $n = 0$ is centered at the origin of the coordinate system, and because the elements are not equally spaced, $x_{-n} \neq -x_n$ in general. As was mentioned in Subsection 6.1.1, complex weights are implemented by using digital signal processing (digital beamforming) (see Section 6.4), and linear arrays of planar elements, such as rectangular and circular pistons, will be discussed in Chapter 8 (see Example 8.1-8).

Equation (6.1-83) indicates that the complex frequency response of the array is modeled as the sum (superposition) of the complex-weighted frequency responses of all the elements in the array. The complex weights are used to control the complex frequency response of the array and, thus, the array's far-field beam pattern via amplitude and phase weighting, as will be discussed later in this chapter. Substituting (6.1-83) into (6.1-1) yields the following expression for the far-field beam pattern:

$$\boxed{D(f, f_X) = \sum_{n=-N'}^{N'} c_n(f) E_n(f, f_X) \exp(+j2\pi f_X x_n)} \tag{6.1-85}$$

where $E_n(f, f_X)$ is the far-field beam pattern of element (electroacoustic transducer) n [see (6.1-8) and (6.1-9)]. The units of $e_n(f, x_A)$, $A(f, x_A)$, $E_n(f, f_X)$, and $D(f, f_X)$ are summarized in Table 6.1-1.

If all the elements in the array are *identical* (i.e., all the elements have the same complex frequency response), but still *not* equally spaced, then (6.1-85) reduces to

$$\boxed{D(f, f_X) = E(f, f_X) S(f, f_X)} \tag{6.1-86}$$

where $E(f, f_X)$ is the far-field beam pattern of *one* of the identical elements in the array [see (6.1-11)],

$$\boxed{S(f, f_X) = \sum_{n=-N'}^{N'} c_n(f) \exp(+j2\pi f_X x_n)} \tag{6.1-87}$$

is the *array factor*, and if $a_n(f) > 0 \;\; \forall \;\; n$, then

$$\boxed{S_{\max} = \max\left|S(f, f_X)\right| = \sum_{n=-N'}^{N'} a_n(f)} \tag{6.1-88}$$

is the *normalization factor* for the array factor (see Appendix 6A). As can be seen from (6.1-87), $S(f, f_X)$ is *dimensionless* and depends on the set of complex weights used and the locations (the x coordinates) of all the elements along the X axis. The array factor $S(f, f_X)$ contains the spatial information of the array. If one or more elements in the array are broken (not transmitting and/or receiving), set $c_n(f) = 0$ for those elements. Equation (6.1-86) is referred to as the *Product Theorem* for a linear array composed of an odd number of identical elements. It has the same form and meaning as the Product Theorem for a linear array composed of an even number of identical elements given by (6.1-10), with the exception that the equations for the array factor $S(f, f_X)$ are different – compare

(6.1-12) with (6.1-87).

If the identical elements in the array are omnidirectional point-elements, then substituting (6.1-16) into (6.1-86) yields

$$\boxed{D(f, f_X) = \mathcal{S}(f) S(f, f_X)} \tag{6.1-89}$$

where $\mathcal{S}(f)$ is the complex, element sensitivity function (see Table 6.1-2 and Appendix 6B), and $S(f, f_X)$ is given by (6.1-87). The corresponding complex frequency response (complex aperture function) of the array can be obtained by taking the inverse spatial Fourier transform of (6.1-89). Doing so yields [see (6.1-19)]

$$\boxed{A(f, x_A) = \mathcal{S}(f) \sum_{n=-N'}^{N'} c_n(f) \delta(x_A - x_n)} \tag{6.1-90}$$

Note that the summation in (6.1-90) is in the form of the impulse response of a finite impulse response (FIR) digital filter. In Appendix 6D, it is shown that a linear array composed of an odd number of identical, complex-weighted elements can be thought of as a one-dimensional, spatial FIR filter, where $s(f, x_A) = F_{f_X}^{-1}\{S(f, f_X)\}$ is the spatial impulse response of the array at frequency f hertz.

If all the elements in the array are *equally spaced*, then

$$\boxed{x_n = nd, \qquad n = -N', \ldots, 0, \ldots, N'} \tag{6.1-91}$$

and

$$\boxed{x_{-n} = -x_n, \qquad n = 1, \ldots, N'} \tag{6.1-92}$$

where d is the *interelement spacing* in meters.

If all the elements in the array are equally spaced and the complex weights satisfy the conjugate symmetry property

$$c_{-n}(f) = c_n^*(f), \qquad n = 1, 2, \ldots, N', \tag{6.1-93}$$

then the array factor $S(f, f_X)$ given by (6.1-87) can be simplified. As was discussed before [see (6.1-22) through (6.1-25)], (6.1-93) implies that the amplitude and phase weights are even and odd functions of index n, respectively. Therefore,

$$a_{-n}(f) = a_n(f), \qquad n = 1, 2, \ldots, N', \tag{6.1-94}$$

and

$$\theta_{-n}(f) = -\theta_n(f), \qquad n = 1, 2, \ldots, N'. \tag{6.1-95}$$

Since an odd function passes through the origin, the phase weight associated with element $n = 0$ – which is centered at $x_0 = 0$ – is

$$\theta_0(f) = 0. \tag{6.1-96}$$

Therefore, the complex weight for element $n = 0$ is

$$c_0(f) = a_0(f). \tag{6.1-97}$$

In order to simplify $S(f, f_X)$, first rewrite it as follows:

$$S(f, f_X) = c_0(f) + \sum_{n=1}^{N'} \left[c_{-n}(f) \exp(+j2\pi f_X x_{-n}) + c_n(f) \exp(+j2\pi f_X x_n) \right]. \tag{6.1-98}$$

Then, by substituting (6.1-91) through (6.1-93), and (6.1-97) into (6.1-98), we obtain

$$\begin{aligned} S(f, f_X) &= a_0(f) + \sum_{n=1}^{N'} \left[c_n^*(f) \exp(-j2\pi f_X nd) + c_n(f) \exp(+j2\pi f_X nd) \right] \\ &= a_0(f) + \sum_{n=1}^{N'} \left\{ \left[c_n(f) \exp(+j2\pi f_X nd) \right]^* + \left[c_n(f) \exp(+j2\pi f_X nd) \right] \right\}. \end{aligned} \tag{6.1-99}$$

Since $Z^* + Z = 2\,\mathrm{Re}\{Z\}$, (6.1-99) reduces to

$$S(f, f_X) = a_0(f) + 2\sum_{n=1}^{N'} \mathrm{Re}\{ c_n(f) \exp(+j2\pi f_X nd) \}, \tag{6.1-100}$$

and by substituting (6.1-6) into (6.1-100), we finally obtain

$$\boxed{S(f, f_X) = a_0(f) + 2\sum_{n=1}^{N'} a_n(f) \cos[2\pi f_X nd + \theta_n(f)]} \tag{6.1-101}$$

where N' is given by (6.1-84). Equation (6.1-101) is valid only if all the identical elements in the array are equally spaced and the complex weights satisfy the conjugate symmetry property given by (6.1-93).

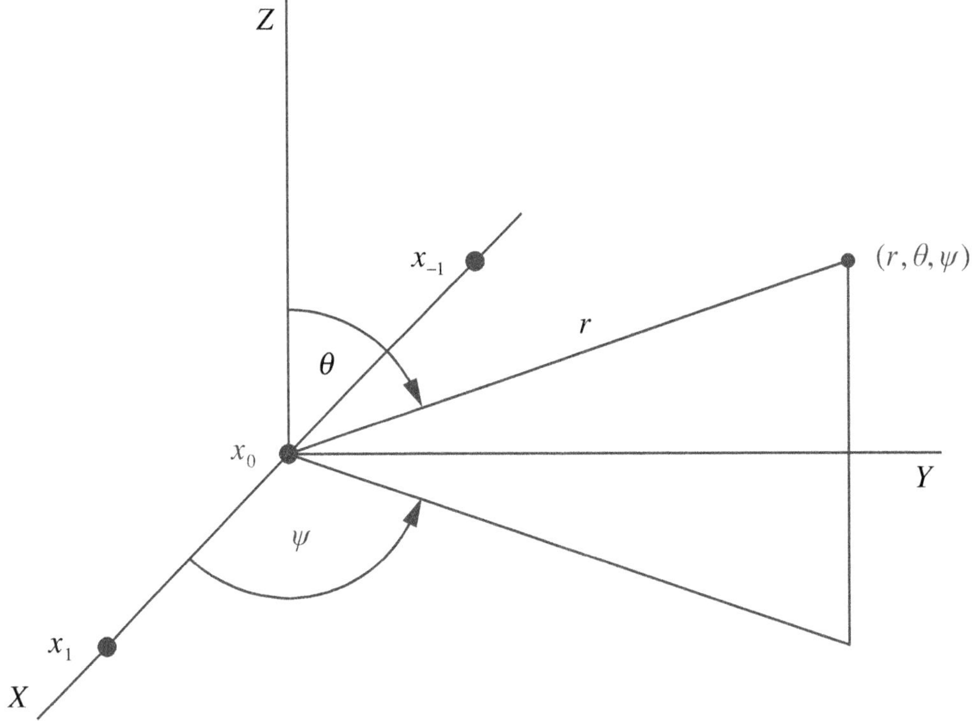

Figure 6.1-8 Linear array of three omnidirectional point-elements lying along the X axis. Also shown is a field point with spherical coordinates (r, θ, ψ).

Example 6.1-4 Axial Quadrupole

An *axial quadrupole* is a linear array of three identical, equally-spaced, complex-weighted, omnidirectional point-elements where the complex weights satisfy the conjugate symmetry property given by (6.1-93) (see Fig. 6.1-8). Substituting $N = 3$ into (6.1-84) yields $N' = 1$. Since the elements are equally spaced, substituting $N' = 1$ into (6.1-91) yields the following x coordinates of the centers of the three elements:

$$x_{-1} = -d\,, \tag{6.1-102}$$

$$x_0 = 0\,, \tag{6.1-103}$$

and

$$x_1 = d \tag{6.1-104}$$

so that the interelement spacing is d meters. Since the complex weights satisfy the conjugate symmetry property given by (6.1-93), substituting $N' = 1$ into (6.1-94) and (6.1-95) yields

$$a_{-1}(f) = a_1(f) \tag{6.1-105}$$

and

$$\theta_{-1}(f) = -\theta_1(f). \tag{6.1-106}$$

For an axial quadrupole,

$$a_0(f) = -2a_1(f). \tag{6.1-107}$$

And since the elements are identical, substituting (6.1-107) and $N' = 1$ into (6.1-101), and then substituting the result into (6.1-89) yields

$$\boxed{D(f, f_X) = 2a_1(f)\mathcal{S}(f)\left[\cos[2\pi f_X d + \theta_1(f)] - 1\right]} \tag{6.1-108}$$

where $\mathcal{S}(f)$ is the complex, element sensitivity function. Equation (6.1-108) is the *unnormalized*, far-field beam pattern of an axial quadrupole lying along the X axis.

The magnitude of the far-field beam pattern given by (6.1-108) is

$$|D(f, f_X)| = 2|a_1(f)||\mathcal{S}(f)||\cos[2\pi f_X d + \theta_1(f)] - 1|, \tag{6.1-109}$$

and since the maximum value of $|\cos[2\pi f_X d + \theta_1(f)] - 1|$ is 2,

$$D_{\max} = 4|a_1(f)||\mathcal{S}(f)|. \tag{6.1-110}$$

Dividing (6.1-109) by (6.1-110) yields the following expression for the magnitude of the *normalized*, far-field beam pattern of an axial quadrupole lying along the X axis:

$$\boxed{|D_N(f, f_X)| = 0.5|\cos[2\pi f_X d + \theta_1(f)] - 1|} \tag{6.1-111}$$

If we let $\theta_1(f) = 0$, that is, if only amplitude weighting is done (no phase weighting), then (6.1-111) reduces to

$$|D_N(f, f_X)| = 0.5|\cos(2\pi f_X d) - 1|. \tag{6.1-112}$$

Since spatial frequency

$$f_X = u/\lambda = \sin\theta\cos\psi/\lambda, \tag{6.1-113}$$

(6.1-112) can also be expressed as

$$|D_N(f,u)| = \frac{1}{2}\left|\cos\left(2\pi\frac{d}{\lambda}u\right) - 1\right|, \tag{6.1-114}$$

or

$$|D_N(f,\theta,\psi)| = \frac{1}{2}\left|\cos\left(2\pi\frac{d}{\lambda}\sin\theta\cos\psi\right) - 1\right|. \tag{6.1-115}$$

At broadside where direction cosine $u=0$, (6.1-114) reduces to $|D_N(f,0)|=0$, and by setting $\theta=90°$ in (6.1-115), we obtain

$$|D_N(f,90°,\psi)| = \frac{1}{2}\left|\cos\left(2\pi\frac{d}{\lambda}\cos\psi\right) - 1\right|, \tag{6.1-116}$$

which is the magnitude of the normalized, *horizontal*, far-field beam pattern in the XY plane of an axial quadrupole lying along the X axis (see Fig. 6.1-9). The far-field beam patterns of both the dipole (see Fig. 6.1-4) and the axial quadrupole have nulls at broadside. ■

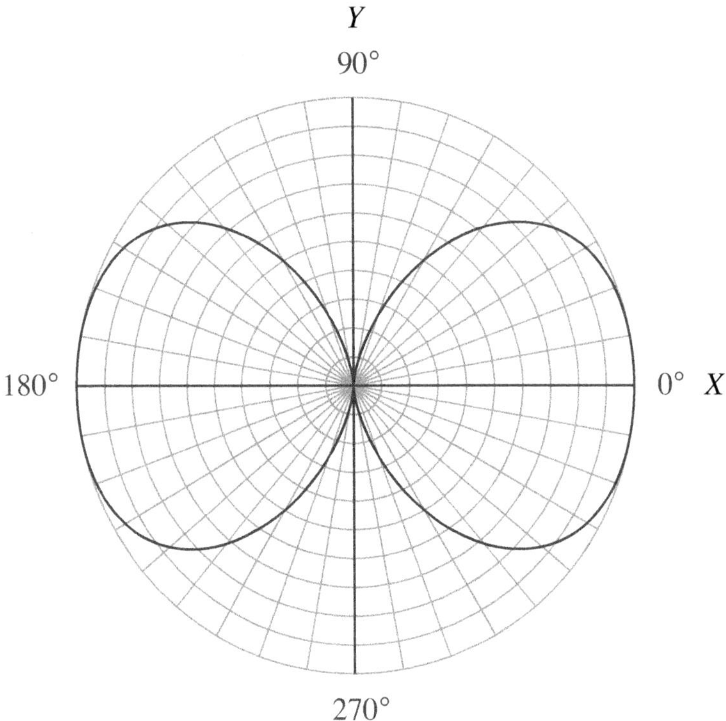

Figure 6.1-9 Polar plot, as a function of the azimuthal (bearing) angle ψ, of the magnitude of the normalized, horizontal, far-field beam pattern of an axial quadrupole lying along the X axis given by (6.1-116) for $d/\lambda = 0.5$.

6.2 Common Amplitude Weights and Corresponding Far-Field Beam Patterns

The most common set of amplitude weights are simply sampled values of the continuous amplitude windows already discussed in Section 2.2. For example, consider the following set of equations, where $n=-(N-1)/2,\ldots,0,\ldots,(N-1)/2$ and N is odd:

(1) rectangular amplitude weight:

$$a_n(f)=1 \tag{6.2-1}$$

(2) triangular amplitude weight:

$$a_n(f)=1-\frac{|n|}{(N-1)/2} \tag{6.2-2}$$

(3) cosine amplitude weight:

$$a_n(f)=\cos\left(\frac{\pi n}{N-1}\right) \tag{6.2-3}$$

(4) Hanning amplitude weight:

$$a_n(f)=0.5+0.5\cos\left(\frac{2\pi n}{N-1}\right) \tag{6.2-4}$$

(5) Hamming amplitude weight:

$$a_n(f)=0.54+0.46\cos\left(\frac{2\pi n}{N-1}\right) \tag{6.2-5}$$

(6) Blackman amplitude weight:

$$a_n(f)=0.42+0.5\cos\left(\frac{2\pi n}{N-1}\right)+0.08\cos\left(\frac{4\pi n}{N-1}\right). \tag{6.2-6}$$

Although the amplitude weight $a_n(f)$ is, in general, a function of frequency, the amplitude weights given by (6.2-1) through (6.2-6) are not functions of frequency, but they are *even* functions of index (element number) n. Amplitude weights, as well as phase weights, are implemented by using digital signal processing (digital beamforming), as will be discussed in Section 6.4. The process of amplitude weighting an array of elements is also known as amplitude shading or as applying an amplitude window to the array. Note that the amplitude weights given by (6.2-1) through (6.2-6) are also used in frequency spectrum analysis problems as well.

The far-field beam pattern $D(f, f_X)$ of a linear array composed of an odd number N of identical elements lying along the X axis, where each element is a single continuous line source or receiver L meters in length, is given by the Product Theorem [see (6.1-86)]

$$D(f, f_X) = E(f, f_X)S(f, f_X), \tag{6.2-7}$$

where $E(f, f_X)$ is the far-field beam pattern of one of the identical elements in the array, and the array factor $S(f, f_X)$ is given by (6.1-87) for N odd. If the elements are equally spaced and only amplitude weighting is done (no phase weighting), then $S(f, f_X)$ reduces to

$$S(f, f_X) = \sum_{n=-N'}^{N'} a_n(f)\exp(+j2\pi f_X nd), \tag{6.2-8}$$

where

$$N' = (N-1)/2. \tag{6.2-9}$$

Next we shall show how to obtain closed-form expressions for (6.2-8) for the amplitude weights given by (6.2-1), and (6.2-3) through (6.2-6).

Consider the following set of amplitude weights:

$$a_n(f) = A\cos\left(\frac{b\pi n}{N-1}\right), \qquad n = -N', \ldots, 0, \ldots, N', \tag{6.2-10}$$

where A and b are real constants. Note that the amplitude weights given by (6.2-1), and (6.2-3) through (6.2-6) all involve one or more terms of the form of (6.2-10), where A takes on values of 1, 0.5, 0.54, 0.46, 0.42, and 0.08; and b takes on values of 0, 1, 2, and 4. Since (6.2-10) can be rewritten as

$$a_n(f) = \frac{A}{2}\exp\left(+j\frac{b\pi n}{N-1}\right) + \frac{A}{2}\exp\left(-j\frac{b\pi n}{N-1}\right), \tag{6.2-11}$$

and since

$$\sum_{k=-K}^{K} \exp(+jk\alpha) = \frac{\sin[(2K+1)\alpha/2]}{\sin(\alpha/2)}, \tag{6.2-12}$$

substituting (6.2-11) into (6.2-8), and using (6.2-12) yields

$$S(f, f_X) = \frac{A}{2} \frac{\sin\left(\pi f_X N d + \dfrac{b\pi N}{2(N-1)}\right)}{\sin\left(\pi f_X d + \dfrac{b\pi}{2(N-1)}\right)} + \frac{A}{2} \frac{\sin\left(\pi f_X N d - \dfrac{b\pi N}{2(N-1)}\right)}{\sin\left(\pi f_X d - \dfrac{b\pi}{2(N-1)}\right)} \tag{6.2-13}$$

Because the set of amplitude weights given by (6.2-10) is analogous to the continuous aperture function given by (2.2-23), it is not surprising that (6.2-13) is analogous to (2.2-29).

Equation (6.2-13) *is a very important result because it can be used to obtain closed-form expressions for the array factor* $S(f, f_X)$ *given by* (6.2-8) *for the rectangular, cosine, Hanning, Hamming, and Blackman amplitude weights.* For example, if we let $A = 1$ and $b = 0$, then (6.2-10) reduces to

$$a_n(f) = 1, \qquad n = -N', \ldots, 0, \ldots, N', \tag{6.2-14}$$

which is the set of rectangular amplitude weights given by (6.2-1). And with $A = 1$ and $b = 0$, (6.2-13) reduces to

$$S(f, f_X) = \frac{\sin(\pi f_X N d)}{\sin(\pi f_X d)} \tag{6.2-15}$$

where the normalization factor $S_{\max} = N$ [see (6.1-88)]. Since $a_n(f) > 0 \ \forall \ n$ in this case, the array factor has its maximum value at broadside, that is, $S_{\max} = |S(f, 0)| = N$ [see (6.2-8) or use L'Hospital's rule to evaluate (6.2-15) at $f_X = 0$]. In contrast, for the dipole discussed in Example 6.1-2, and the axial quadrupole discussed in Example 6.1-4, the array factor has a null at broadside, that is, $S(f, 0) = 0$. Since spatial frequency

$$f_X = u/\lambda = \sin\theta \cos\psi / \lambda, \tag{6.2-16}$$

(6.2-15) can also be expressed as

$$S(f, u) = \frac{\sin\left(N\pi \dfrac{d}{\lambda} u\right)}{\sin\left(\pi \dfrac{d}{\lambda} u\right)}, \tag{6.2-17}$$

or

$$S(f,\theta,\psi)=\frac{\sin\left(N\pi\frac{d}{\lambda}\sin\theta\cos\psi\right)}{\sin\left(\pi\frac{d}{\lambda}\sin\theta\cos\psi\right)}. \tag{6.2-18}$$

As you may have noticed, (6.2-10) and (6.2-13) are not applicable for the set of triangular amplitude weights given by (6.2-2). Therefore, for completeness, the closed-form expression for the array factor $S(f, f_X)$ given by (6.2-8) for the set of triangular amplitude weights is given by

$$\boxed{S(f, f_X)=\frac{2}{N-1}\frac{\sin^2\left[\pi f_X(N-1)d/2\right]}{\sin^2(\pi f_X d)}} \tag{6.2-19}$$

where the normalization factor $S_{\max}=(N-1)/2$ [see (6.1-88)]. Since $a_n(f)\geq 0 \;\forall\; n$ in this case, the array factor has its maximum value at broadside, that is, $S_{\max}=|S(f,0)|=(N-1)/2$.

Because the amplitude weights given by (6.2-1) through (6.2-6) are *even* functions of index (element number) n, their array factors are *real*, *even* functions of spatial frequency f_X (direction cosine u) [e.g., see (6.2-15) and (6.2-19)]. In general, if an amplitude weight is an *even* function of index n, then the corresponding array factor will be a *real*, *even* function of the appropriate spatial frequency (direction cosine). Similarly, if an amplitude weight is an *odd* function of index n, then the corresponding array factor will be an *imaginary*, *odd* function of the appropriate spatial frequency (direction cosine).

Example 6.2-1 Application of the Product Theorem

Consider a linear array composed of an odd number N of identical, equally-spaced elements lying along the X axis, where each element is a single continuous line source or receiver L meters in length. If rectangular amplitude weights are used and no phase weighting is done, then the far-field beam pattern of this array can be obtained by substituting (6.2-15) into the Product Theorem given by (6.2-7). Doing so yields

$$D(f, f_X)=E(f, f_X)\frac{\sin(\pi f_X N d)}{\sin(\pi f_X d)}. \tag{6.2-20}$$

If the complex frequency responses (complex aperture functions) of the identical elements are modeled by the continuous rectangular amplitude window, then from Subsection 2.2.1,

$$e(f, x_A) = A_A \operatorname{rect}(x_A / L) \tag{6.2-21}$$

and

$$E(f, f_X) = A_A L \operatorname{sinc}(f_X L), \tag{6.2-22}$$

where

$$\operatorname{sinc}(x) \triangleq \frac{\sin(\pi x)}{\pi x} \tag{6.2-23}$$

and the real, positive constant A_A carries the correct units of $e(f, x_A)$ (see Table 6.1-1). Although A_A is shown here as a constant, as was previously mentioned in Subsection 2.2.1, A_A is, in general, a real, nonnegative function of frequency, that is, $A_A \rightarrow A_A(f)$. Therefore, substituting (6.2-22) into (6.2-20) yields

$$D(f, f_X) = A_A L \operatorname{sinc}(f_X L) \frac{\sin(\pi f_X N d)}{\sin(\pi f_X d)}, \tag{6.2-24}$$

where, at broadside,

$$D(f, 0) = N A_A L . \tag{6.2-25}$$

If the identical elements in the array are omnidirectional point-elements instead, then [see (6.1-14)]

$$e(f, x_A) = \mathcal{S}(f) \delta(x_A) \tag{6.2-26}$$

and [see (6.1-16)]

$$E(f, f_X) = \mathcal{S}(f), \tag{6.2-27}$$

where $\mathcal{S}(f)$ is the complex, element sensitivity function (see Table 6.1-2 and Appendix 6B). Therefore, substituting (6.2-27) into (6.2-20) yields

$$D(f, f_X) = \mathcal{S}(f) \frac{\sin(\pi f_X N d)}{\sin(\pi f_X d)}, \tag{6.2-28}$$

where, at broadside,

$$D(f, 0) = N \mathcal{S}(f). \tag{6.2-29}$$

We shall finish our discussion in this example by investigating what happens to the far-field beam pattern $E(f, f_X)$ given by (6.2-22) and, hence, the far-field beam pattern $D(f, f_X)$ given by (6.2-24), if the length L is small compared to a wavelength. Since

$$\sin\alpha \approx \alpha, \qquad \alpha \leq 0.0873 \text{ rad}, \tag{6.2-30}$$

then

$$\sin(\pi f_X L) \approx \pi f_X L \tag{6.2-31}$$

whenever

$$\pi f_X L = \pi u L / \lambda \le 0.0873 . \tag{6.2-32}$$

Setting direction cosine u equal to its maximum positive value of 1 (worst case – harder to satisfy the inequality) yields

$$L/\lambda \le 0.0278 . \tag{6.2-33}$$

Therefore, if (6.2-33) is satisfied, then the approximation given by (6.2-31) is valid and (6.2-22) reduces to

$$E(f, f_X) = A_A L \frac{\sin(\pi f_X L)}{\pi f_X L} \approx A_A L, \qquad \frac{L}{\lambda} \le 0.0278 , \tag{6.2-34}$$

and by substituting (6.2-34) into (6.2-20), we obtain

$$D(f, f_X) \approx A_A L \frac{\sin(\pi f_X N d)}{\sin(\pi f_X d)}, \qquad \frac{L}{\lambda} \le 0.0278 , \tag{6.2-35}$$

where, at broadside,

$$D(f, 0) \approx N A_A L, \qquad L/\lambda \le 0.0278 . \tag{6.2-36}$$

Compare (6.2-35) with (6.2-24) and (6.2-28). Since the far-field beam pattern given by (6.2-34) is not a function of the spherical angles θ and ψ , the identical elements are omnidirectional. In other words, the elements transmit and/or receive equally well in all directions. If we let $L_{\max} = 0.0278\lambda$, then the far-field beam patterns $E(f, f_X)$ and $D(f, f_X)$ given by (6.2-34) and (6.2-35) are valid as long as $L \le L_{\max}$ (see Table 6.2-1). ■

Table 6.2-1 Values for the Maximum Length $L_{\max} = 0.0278\lambda$ for Three Different Frequencies f. Note that $1 \text{ in} = 2.54 \text{ cm}$.

c (m/sec)	f (kHz)	λ (m)	$L_{\max}$ (cm)	$L_{\max}$ (in)
1500	0.1	15.0	41.7	16.4
1500	1.0	1.5	4.17	1.64
1500	10.0	0.15	0.417	0.164

In Example 6.2-1, we were able to obtain three different closed-form expressions for the unnormalized, far-field beam pattern $D(f, f_X)$ of a linear array, namely (6.2-24), (6.2-28), and (6.2-35). If a closed-form expression for the far-field beam pattern of a linear array can be derived, then the dimensionless 3-dB beamwidth Δu in direction-cosine space of its beam pattern can be obtained using the procedure outlined in Example 2.3-3. Once Δu has been determined, the 3-dB beamwidths $\Delta\theta$ and $\Delta\psi$ in degrees, of the vertical and horizontal, far-field beam patterns can be obtained by using (2.3-19) and (2.3-25), respectively.

Example 6.2-2 Closed-Form Expression for the Array Factor for Rectangular Amplitude Weights when *N* is Even

Consider a linear array composed of an *even* number N of identical, equally-spaced, complex-weighted elements lying along the X axis, where each element is a single continuous line source or receiver L meters in length, and the complex weights satisfy the conjugate symmetry property given by (6.1-22). The far-field beam pattern of this array is given by the Product Theorem [see (6.1-10)]

$$D(f, f_X) = E(f, f_X)S(f, f_X), \tag{6.2-37}$$

where [see (6.1-27)]

$$S(f, f_X) = 2\sum_{n=1}^{N/2} \operatorname{Re}\{c_n(f)\exp[+j2\pi f_X(n-0.5)d]\}, \tag{6.2-38}$$

or

$$S(f, f_X) = 2\operatorname{Re}\left\{\exp(-j\pi f_X d)\sum_{n=1}^{N/2} c_n(f)\exp(+j2\pi f_X nd)\right\}. \tag{6.2-39}$$

If rectangular amplitude weights are used and no phase weighting is done, then $c_n(f) = 1 \ \forall \ n$ and (6.2-39) reduces to

$$S(f, f_X) = 2\operatorname{Re}\left\{\exp(-j\pi f_X d)\sum_{n=1}^{N/2} \exp(+j2\pi f_X nd)\right\}. \tag{6.2-40}$$

In order to obtain a closed-form expression for $S(f, f_X)$ as given by (6.2-40), we need to rewrite the summation by letting $m = n - 1$. Doing so yields

$$\sum_{n=1}^{N/2} \exp(+j2\pi f_X nd) = \exp(+j2\pi f_X d)\sum_{m=0}^{M} \exp(+j2\pi f_X md) = \exp(+j2\pi f_X d)\sum_{m=0}^{M} x^m, \tag{6.2-41}$$

where

$$M = \frac{N}{2} - 1 \tag{6.2-42}$$

and

$$x = \exp(+j2\pi f_X d). \tag{6.2-43}$$

Since

$$\sum_{m=0}^{M} x^m = \frac{x^{M+1} - 1}{x - 1}, \qquad x \neq 1, \tag{6.2-44}$$

$$\sum_{n=1}^{N/2} \exp(+j2\pi f_X nd) = \exp(+j\pi f_X d)\exp(+j\pi f_X Nd/2)\frac{\sin(\pi f_X Nd/2)}{\sin(\pi f_X d)}. \tag{6.2-45}$$

Substituting (6.2-45) into (6.2-40) yields the desired result:

$$\boxed{S(f, f_X) = 2\cos(\pi f_X Nd/2)\frac{\sin(\pi f_X Nd/2)}{\sin(\pi f_X d)}} \tag{6.2-46}$$

where the normalization factor $S_{\max} = N$ [see (6.1-13)]. Note that in this case, $|S(f, 0)| = N$. Compare (6.2-46) for N *even* with (6.2-15) for N *odd.* ■

6.3 Dolph-Chebyshev Amplitude Weights

The far-field beam pattern of a linear array of N identical, complex-weighted elements lying along the X axis, where each element is a single continuous line source or receiver L meters in length, is given by the Product Theorem

$$D(f, f_X) = E(f, f_X)S(f, f_X). \tag{6.3-1}$$

If the elements are equally-spaced, the complex weights satisfy the conjugate symmetry property given by (6.1-22) for N even, and (6.1-93) for N odd, and no phase weighting is done [i.e., $\theta_n(f) = 0 \;\; \forall \;\; n$], then the array factor given by (6.1-28) for N *even* reduces to

$$S(f, f_X) = 2\sum_{n=1}^{N/2} a_n(f)\cos[2\pi f_X(n - 0.5)d], \tag{6.3-2}$$

and the array factor given by (6.1-101) for N *odd* reduces to

$$S(f, f_X) = a_0(f) + 2\sum_{n=1}^{N'} a_n(f)\cos(2\pi f_X nd), \tag{6.3-3}$$

where

$$N' = (N-1)/2. \tag{6.3-4}$$

We know from our discussion of linear apertures in Section 2.2 that different amplitude windows yield far-field beam patterns with different sidelobe levels and beamwidths, and as sidelobe levels decrease, beamwidth generally increases. This principle also applies to the far-field beam patterns of amplitude-weighted arrays. However, the Dolph-Chebyshev method of computing amplitude weights makes it possible to *optimize* the array factor $S(f, f_X)$ so that for any specified sidelobe level relative to the level of the mainlobe, the *smallest* possible mainlobe beamwidth for $S(f, f_X)$ is achieved, or for any specified mainlobe beamwidth, the *lowest* possible sidelobe level for $S(f, f_X)$ is achieved. The number of elements N can be either even or odd, and the resulting Dolph-Chebyshev amplitude weights are *even* functions of element number n. Although the sidelobe level and mainlobe beamwidth of $S(f, f_X)$ can be optimized, the overall far-field beam pattern of the array $D(f, f_X)$ is still equal to $S(f, f_X)$ multiplied by the far-field beam pattern $E(f, f_X)$ of one of the identical elements in the array. The far-field beam pattern $E(f, f_X)$ has its own sidelobe levels and mainlobe beamwidth, unless it is omnidirectional. The one disadvantage of the Dolph-Chebyshev method is that *all* of the sidelobes of $S(f, f_X)$ are at the *same* level – they do not fall off in value.

We begin our discussion of the Dolph-Chebyshev method of computing amplitude weights by letting

$$\alpha = \pi f_X d. \tag{6.3-5}$$

Substituting (6.3-5) into (6.3-2) and (6.3-3) yields

$$\begin{aligned} S(f, f_X) &= 2\sum_{n=1}^{N/2} a_n(f)\cos[(2n-1)\alpha] \\ &= 2a_1(f)\cos(\alpha) + 2a_2(f)\cos(3\alpha) + \ldots + 2a_{N/2}(f)\cos[(N-1)\alpha] \end{aligned} \tag{6.3-6}$$

and

$$\begin{aligned} S(f, f_X) &= a_0(f) + 2\sum_{n=1}^{N'} a_n(f)\cos(2n\alpha) \\ &= a_0(f) + 2a_1(f)\cos(2\alpha) + 2a_2(f)\cos(4\alpha) + \ldots + 2a_{N'}(f)\cos(2N'\alpha), \end{aligned} \tag{6.3-7}$$

respectively. Since the m th-degree Chebyshev polynomial is given by

$$T_m(x) = \begin{cases} \cos[m\cos^{-1}x], & |x| \le 1 \\ \cosh[m\cosh^{-1}x], & x \ge 1 \\ (-1)^m \cosh[m\cosh^{-1}(-x)], & x \le -1 \end{cases} \tag{6.3-8}$$

if we also let [see (6.3-5)]

$$\alpha = \cos^{-1}x \tag{6.3-9}$$

so that

$$x = \cos\alpha\,, \tag{6.3-10}$$

then

$$T_m(x) = \cos(m\alpha), \qquad |x| \le 1. \tag{6.3-11}$$

By comparing (6.3-11) with (6.3-6) and (6.3-7), it can be seen that $S(f, f_X)$ can be expressed as a sum of Chebyshev polynomials. The degree of the polynomial $S(f, f_X)$, which is called the *array polynomial*, is

$$\boxed{i = N - 1} \tag{6.3-12}$$

or one less than the total number of elements in the array. Because the Dolph-Chebyshev method depends on certain properties of Chebyshev polynomials, we must digress for a moment to discuss these properties before continuing further.

All Chebyshev polynomials possess the following two important properties:

(1) $T_m(x)$ has unit-magnitude maxima and minima that occur alternately at

$$x_k = \cos\left(k\pi/m\right), \qquad k = 1, 2, \ldots, m-1, \tag{6.3-13}$$

where $-1 < x_k < 1$ and $|T_m(x_k)| = 1$. Therefore, Chebyshev polynomials have an "equal ripple" characteristic in the interval $-1 < x < 1$.

(2) At $x = \pm 1$, $|T_m(\pm 1)| = 1$. However, $|T_m(x)| > 1$ for $|x| > 1$ and

$$\left|\frac{d}{dx}T_m(x)\right|_{x=\pm 1} = m^2. \tag{6.3-14}$$

Chebyshev polynomials possess additional properties, but they are not pertinent to the present problem.

In the Dolph-Chebyshev method, the magnitude of the level of the mainlobe of $S(f, f_X)$ corresponds to the value of $T_i(x_0)$, where i is the degree of the array polynomial given by (6.3-12) and $x_0 > 1$. The magnitude of the level of the sidelobes of $S(f, f_X)$ is numerically equal to the absolute value of the interior maxima and minima, that is, 1. Note that it is the "equal ripple" characteristic of Chebyshev polynomials, as discussed in property (1), that is responsible for generating sidelobes all at the same level in the Dolph-Chebyshev method of computing amplitude weights. Therefore, the ratio

$$\frac{|\text{mainlobe level}|}{|\text{sidelobe level}|} = |\text{mainlobe level}| = T_i(x_0) = r > 1, \tag{6.3-15}$$

is used to determine the value of x_0, that is, we solve for x_0 using the equation $T_i(x_0) = r$. But in order to compute a value for x_0, we need a value for r.

A value for the ratio r is determined as follows: If you want the magnitude of the level of the mainlobe to be R dB above the magnitude of the level of the sidelobes, then

$$20 \log_{10} r = \text{R dB}, \tag{6.3-16}$$

from which we obtain

$$\boxed{r = 10^{\text{R}/20}} \tag{6.3-17}$$

Since the ratio $r > 1$, in order for $T_i(x_0)$ to be greater than 1, x_0 needs to be greater than 1 (see Property 2). Therefore, by referring to (6.3-8),

$$T_i(x_0) = \cosh[i \cosh^{-1} x_0] = r, \qquad x_0 > 1. \tag{6.3-18}$$

Solving for x_0 yields

$$\boxed{x_0 = \cosh\left(\frac{1}{i}\cosh^{-1} r\right), \qquad r > 1} \tag{6.3-19}$$

where i is given by (6.3-12) and r is given by (6.3-17). Now that we know how to compute a value for x_0, we can use it to compute the amplitude weights.

If the total number of elements N is *even*, then the Dolph-Chebyshev amplitude weights are given by[1]

$$a_n(f) = \sum_{j=n}^{N/2} \left[(-1)^{(N/2)-j} \right] \frac{N-1}{N+2(j-1)} \binom{(N/2)+j-1}{2j-1} \binom{2j-1}{j-n} x_0^{2j-1}, \quad n = 1, 2, \ldots, \frac{N}{2} \tag{6.3-20}$$

where

$$\binom{a}{b} = \frac{a!}{b!(a-b)!} \tag{6.3-21}$$

is the binomial coefficient, x_0 is given by (6.3-19), and

$$a_{-n}(f) = a_n(f), \qquad n = 1, 2, \ldots, N/2\,. \tag{6.3-22}$$

If the total number of elements N is *odd*, then the Dolph-Chebyshev amplitude weights are given by[1]

$$a_n(f) = \sum_{j=n}^{N'} \left[(-1)^{N'-j} \right] \frac{N'}{N'+j} \binom{N'+j}{2j} \binom{2j}{j-n} x_0^{2j}, \qquad n = 0, 1, \ldots, N' \tag{6.3-23}$$

where N' is given by (6.3-4),

$$\binom{a}{b} = \frac{a!}{b!(a-b)!} \tag{6.3-24}$$

is the binomial coefficient, x_0 is given by (6.3-19), and

$$a_{-n}(f) = a_n(f), \qquad n = 1, 2, \ldots, N'\,. \tag{6.3-25}$$

In both cases, after the amplitude weights have been computed, they can be normalized by dividing each amplitude weight by the largest value.

Next we shall derive an equation for the 3-dB beamwidth $\Delta\psi$ of the horizontal, angular profile of the array factor $S(f, f_X)$. Recall that $S(f, f_X)$ is dimensionless – it is not a far-field beam pattern by itself since $D(f, f_X) = E(f, f_X) S(f, f_X)$. Since the right-hand sides of (6.3-5) and (6.3-9)

[1] R. S. Elliott, *Antenna Theory and Design*, Prentice-Hall, Englewood Cliffs, NJ, 1981, pg. 147.

must be equal,

$$\pi f_X d = \pi \frac{d}{\lambda} u = \cos^{-1} x\,, \tag{6.3-26}$$

or

$$u = \sin\theta \cos\psi = \frac{\lambda}{\pi d} \cos^{-1} x\,. \tag{6.3-27}$$

Also, since the linear array lies along the X axis, the horizontal, angular profile of $S(f, f_X)$ lies in the XY plane (see Example 2.3-2). In the XY plane, $u = \cos\psi$ since $\theta = 90°$. If we let

$$u_+ = \cos\psi_+ > 0, \qquad \theta_+ = 90°\,, \tag{6.3-28}$$

then from (6.3-27),

$$u_+ = \cos\psi_+ = \frac{\lambda}{\pi d} \cos^{-1} x_+\,, \tag{6.3-29}$$

or

$$\psi_+ = \cos^{-1}\left(\frac{\lambda}{\pi d} \cos^{-1} x_+ \right), \tag{6.3-30}$$

where x_+ is the value of x that corresponds to the bearing angle ψ_+ to a 3-dB down point.

In order to evaluate $\cos^{-1} x_+$ in (6.3-30), it is required that $|x_+| \le 1$. Since the magnitude of the level of the mainlobe is equal to r [see (6.3-15)], the magnitude of the 3-dB down-point is $\sqrt{2}\,r/2$. However, since $\sqrt{2}\,r/2 > 1$ and $|T_i(x)| > 1$ only if $|x| > 1$, x_+ has to be scaled by x_0 so that $x = x_0 x_+ > 1$. Therefore, solving for x_+ from [see (6.3-8) for $x \ge 1$]

$$T_i(x_0 x_+) = \cosh[i \cosh^{-1}(x_0 x_+)] = \sqrt{2}\,r/2 \tag{6.3-31}$$

yields

$$\boxed{x_+ = \frac{1}{x_0} \cosh\left(\frac{1}{i} \cosh^{-1}\left(\sqrt{2}\,r/2\right) \right), \qquad r > \sqrt{2}} \tag{6.3-32}$$

where i is given by (6.3-12), r is given by (6.3-17), and x_0 is given by (6.3-19). And since (see Example 2.3-2),

$$\Delta\psi = 180° - 2\psi_+\,, \tag{6.3-33}$$

substituting (6.3-30) into (6.3-33) yields

$$\boxed{\Delta\psi = 180° - 2\cos^{-1}\left(\frac{\lambda}{\pi d}\cos^{-1}x_+\right), \qquad \left|\frac{\lambda}{\pi d}\cos^{-1}x_+\right| \le 1} \tag{6.3-34}$$

which is the 3-dB beamwidth in degrees of the horizontal, angular profile of the array factor $S(f, f_X)$ in the XY plane, where x_+ is given by (6.3-32). Note that $\cos^{-1}x_+$ must be expressed in radians.

Example 6.3-1

Consider a linear array of 7 identical, equally-spaced elements lying along the X axis, where each element is a single continuous line source or receiver L meters in length. The interelement spacing $d = \lambda/2$ meters. Later, in Section 6.5, we shall show that if $d < \lambda/2$, then grating lobes (extraneous mainlobes) can be avoided for all possible directions of beam steering. Find the set of *normalized*, Dolph-Chebyshev amplitude weights so that the magnitude of the level of the mainlobe of the array factor $S(f, f_X)$ is 30 dB above the magnitude of the level of its sidelobes. Also, compute the 3-dB beamwidth in degrees of the horizontal, angular profile of $S(f, f_X)$ in the XY plane.

Since $N = 7$ [see (6.3-12)],

$$i = 6, \tag{6.3-35}$$

and since $\mathrm{R} = 30$ dB [see (6.3-17)],

$$r = 31.6227766. \tag{6.3-36}$$

Substituting (6.3-35) and (6.3-36) into (6.3-19) yields

$$x_0 = 1.2484890. \tag{6.3-37}$$

Since $N = 7$ [see (6.3-4)],

$$N' = 3. \tag{6.3-38}$$

And finally, substituting (6.3-37) and (6.3-38) into (6.3-23), *normalizing* the Dolph-Chebyshev amplitude weights, and using (6.3-25) yields

$$a_0(f) = 1 \tag{6.3-39}$$

and

$$a_{-1}(f) = a_1(f) = 0.873814 \tag{6.3-40}$$

$$a_{-2}(f) = a_2(f) = 0.568269 \tag{6.3-41}$$

$$a_{-3}(f) = a_3(f) = 0.264225 \tag{6.3-42}$$

We shall compute the 3-dB beamwidth next. Substituting (6.3-35) through (6.3-37) into (6.3-32) yields

$$x_{+} = 0.9670442\,, \tag{6.3-43}$$

and by substituting (6.3-43) and $d = \lambda/2$ into (6.3-34), we obtain

$$\Delta\psi = 18.9°\,, \tag{6.3-44}$$

which is the 3-dB beamwidth of the horizontal, angular profile of the array factor $S(f, f_X)$ in the XY plane for $N = 7$, $d = \lambda/2$, and $\text{R} = 30$ dB.

In order to verify that the magnitude of the level of the mainlobe of $S(f, f_X)$ is 30 dB above the magnitude of the level of its sidelobes, we shall plot $|S_N(f, u)|$ in decibels versus direction cosine u. But first, we need to derive an equation for the normalized array factor $S_N(f, f_X)$, where $S(f, f_X)$ is given by (6.3-3).

The normalized array factor $S_N(f, f_X)$ is defined as follows:

$$\boxed{S_N(f, f_X) \triangleq S(f, f_X)/S_{\max}} \tag{6.3-45}$$

where $S_{\max} = \max|S(f, f_X)|$ is the normalization factor. Since the Dolph-Chebyshev amplitude weights are *positive* in value for all n, the maximum value of the magnitude of $S(f, f_X)$ given by (6.3-3) occurs at broadside, that is, when spatial frequency $f_X = 0$. Therefore, the normalization factor $S_{\max} = |S(f, 0)|$. Evaluating (6.3-3) at $f_X = 0$ yields

$$S_{\max} = a_0(f) + 2\sum_{n=1}^{N'} a_n(f)\,, \tag{6.3-46}$$

and by substituting (6.3-3) and (6.3-46) into (6.3-45), we obtain

$$S_N(f, f_X) = \frac{a_0(f) + 2\sum_{n=1}^{N'} a_n(f)\cos(2\pi f_X n d)}{a_0(f) + 2\sum_{n=1}^{N'} a_n(f)}\,. \tag{6.3-47}$$

The normalization factor given by (6.1-88) for N odd can also be used.

Since

$$f_X = u/\lambda = \sin\theta\cos\psi/\lambda\,, \tag{6.3-48}$$

substituting (6.3-48) into (6.3-47) yields

$$S_N(f,u) = \frac{a_0(f) + 2\sum_{n=1}^{N'} a_n(f)\cos\left(2n\pi\frac{d}{\lambda}u\right)}{a_0(f) + 2\sum_{n=1}^{N'} a_n(f)}. \tag{6.3-49}$$

And since $d = \lambda/2$ in this example, (6.3-49) reduces to

$$S_N(f,u) = \frac{a_0(f) + 2\sum_{n=1}^{N'} a_n(f)\cos(n\pi u)}{a_0(f) + 2\sum_{n=1}^{N'} a_n(f)}. \tag{6.3-50}$$

Figure 6.3-1 is a plot of the magnitude of (6.3-50) in decibels versus direction cosine u for $N' = 3$ and the set of Dolph-Chebyshev amplitude weights given by (6.3-39) through (6.3-42). As can be seen from Fig. 6.3-1, the level of the mainlobe is 30 dB above the level of the sidelobes. ■

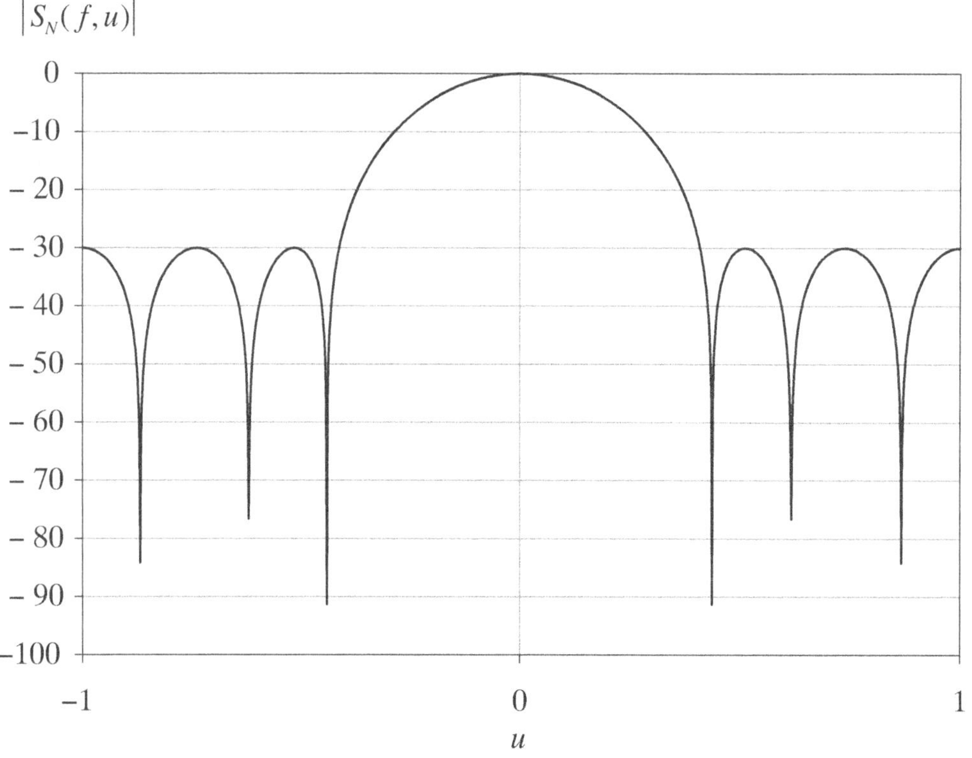

Figure 6.3-1 Plot of the magnitude of (6.3-50) in decibels versus direction cosine u for $N' = 3$ and the set of Dolph-Chebyshev amplitude weights given by (6.3-39) through (6.3-42).

6.4 The Phased Array – Beam Steering

Consider a linear array composed of an odd number N of identical, unevenly-spaced, complex-weighted, omnidirectional point-elements lying along the X axis. From the Product Theorem for an odd number of elements, the far-field beam pattern of this array is given by [see (6.1-89)]

$$D(f, f_X) = \mathcal{S}(f)S(f, f_X), \tag{6.4-1}$$

where $\mathcal{S}(f)$ is the complex, element sensitivity function (see Table 6.1-2 and Appendix 6B),

$$S(f, f_X) = \sum_{n=-N'}^{N'} c_n(f)\exp(+j2\pi f_X x_n), \tag{6.4-2}$$

$$c_n(f) = a_n(f)\exp[+j\theta_n(f)], \tag{6.4-3}$$

and

$$N' = (N-1)/2. \tag{6.4-4}$$

Let $D(f, f_X)$ be the far-field beam pattern of the array when it is only amplitude weighted (i.e., $\theta_n(f) = 0 \ \forall \ n$), and let $D'(f, f_X)$ be the far-field beam pattern of the array when it is complex weighted. Therefore,

$$D(f, f_X) = \mathcal{S}(f) \sum_{n=-N'}^{N'} a_n(f)\exp(+j2\pi f_X x_n) \tag{6.4-5}$$

and

$$D'(f, f_X) = \mathcal{S}(f) \sum_{n=-N'}^{N'} a_n(f)\exp\{+j[2\pi f_X x_n + \theta_n(f)]\}. \tag{6.4-6}$$

We already know from linear aperture theory that in order to do beam steering, there must be a linear phase response along the length of the aperture (see Section 2.4). A set of linear phase weights that will produce a linear phase response along the length of the linear array under discussion is given by

$$\boxed{\theta_n(f) = -2\pi f'_X x_n, \qquad n = -N', \ldots, 0, \ldots, N'} \tag{6.4-7}$$

where

$$\boxed{f'_X = u'/\lambda = \sin\theta' \cos\psi' / \lambda} \tag{6.4-8}$$

If the elements are *equally spaced*, then $x_n = nd$ [see (6.1-91)]. Compare (6.4-7) and (6.4-8) with (2.4-4) through (2.4-7). If (6.4-7) is substituted into (6.4-6), then

$$D'(f, f_X) = S(f) \sum_{n=-N'}^{N'} a_n(f) \exp\left[+j2\pi(f_X - f'_X)x_n\right], \tag{6.4-9}$$

and by comparing (6.4-9) with (6.4-5),

$$D'(f, f_X) = D(f, f_X - f'_X). \tag{6.4-10}$$

Since spatial frequency

$$f_X = u/\lambda = \sin\theta\cos\psi/\lambda, \tag{6.4-11}$$

(6.4-10) can also be expressed as

$$D'(f, u) = D(f, u - u'). \tag{6.4-12}$$

Equation (6.4-12) indicates that a set of linear phase weights applied along the length of the array will steer the far-field beam pattern $D(f, u)$ of a set of amplitude weights in the direction $u = u'$ in direction-cosine space, which is equivalent to steering (tilting) the beam pattern to $\theta = \theta'$ and $\psi = \psi'$ [see (6.4-8)]. Equal spacing of the elements is *not* required in order to do beam steering. Equation (6.4-12) also indicates that the far-field beam pattern $D'(f, u)$ is simply a translated or shifted version of $D(f, u)$ along the u axis, implying that the shape and beamwidth of the mainlobe of the beam pattern remain unchanged. However, recall from linear aperture theory, that when a beam pattern is steered, the shape and beamwidth of the mainlobe of the beam pattern do change when the magnitude of the beam pattern is plotted as a function of the spherical angles θ and ψ (see Section 2.5).

It is also important to note that a phase weight $\theta_n(f)$ in radians is equivalent to a *time delay* τ'_n in seconds, that is,

$$\theta_n(f) \triangleq -2\pi f \tau'_n, \tag{6.4-13}$$

or

$$\tau'_n = -\theta_n(f)/(2\pi f). \tag{6.4-14}$$

Substituting (6.4-7) and (6.4-8) into (6.4-14) yields

$$\boxed{\tau'_n = \frac{u'}{c} x_n, \qquad n = -N', \ldots, 0, \ldots, N'} \tag{6.4-15}$$

where $c = f\lambda$.

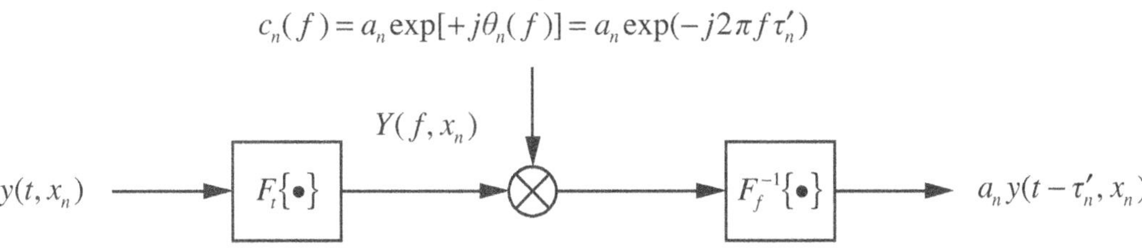

Figure 6.4-1 Beamforming. Implementing a complex weight (amplitude and phase weights) using forward and inverse Fourier transforms.

Figure 6.4-1 illustrates how to implement the complex weight

$$c_n(f) = a_n \exp[+j\theta_n(f)] = a_n \exp(-j2\pi f \tau_n') \tag{6.4-16}$$

using forward and inverse Fourier transforms. This procedure is referred to as *beamforming*. Note that the amplitude weight a_n is *not* a function of frequency. In Fig. 6.4-1, $y(t, x_n)$ is the output electrical signal from element n in a linear array. The complex weight $c_n(f)$ is just a complex number in a computer program. The expression $F_t\{\bullet\}$ represents a forward Fourier transform with respect to time t, and $F_f^{-1}\{\bullet\}$ represents an inverse Fourier transform with respect to frequency f. In practical applications, the forward and inverse Fourier transforms are computed using forward and inverse *discrete Fourier transforms* (DFTs), which can be evaluated very quickly by using forward and inverse *fast-Fourier-transform* (FFT) algorithms. This is known as *digital beamforming* or *FFT beamforming*. Since the phase weight $\theta_n(f)$ is a function of frequency f, an appropriate phase weight must be applied to *each* frequency component contained in the complex frequency spectrum $Y(f, x_n)$ shown in Fig. 6.4-1 in order to produce the correct time delay τ_n'. This must be done at each element in a linear array.

Example 6.4-1 Steering the Null of a Dipole

The magnitude of the normalized, far-field beam pattern of a dipole lying along the X axis is given by (see Example 6.1-2)

$$|D_N(f, f_X)| = |\sin[\pi f_X d + \theta_1(f)]|. \tag{6.4-17}$$

Since $N = 2$ (an even number) for a dipole, (6.4-7) cannot be used for beam steering. However, all we need to do is to find the equation for the phase weight $\theta_1(f)$ that will generate the factor $f_X - f_X'$ on the right-hand side of (6.4-17).

Therefore, if we let

$$\theta_1(f) = -\pi f'_X d, \tag{6.4-18}$$

then substituting (6.4-18) into (6.4-17) yields

$$|D_N(f, f_X)| = |\sin[\pi(f_X - f'_X)d]|, \tag{6.4-19}$$

where

$$f'_X = u'/\lambda = \sin\theta' \cos\psi'/\lambda. \tag{6.4-20}$$

For completeness, the phase weight at element $n = -1$ is [see (6.1-46)]

$$\theta_{-1}(f) = -\theta_1(f) = \pi f'_X d. \tag{6.4-21}$$

Since spatial frequency

$$f_X = u/\lambda = \sin\theta \cos\psi/\lambda, \tag{6.4-22}$$

(6.4-19) can also be expressed as

$$|D_N(f, u)| = \left|\sin\left[\pi \frac{d}{\lambda}(u - u')\right]\right|, \tag{6.4-23}$$

or

$$|D_N(f, \theta, \psi)| = \left|\sin\left[\pi \frac{d}{\lambda}(\sin\theta\cos\psi - \sin\theta'\cos\psi')\right]\right|. \tag{6.4-24}$$

Setting $\theta = 90°$ and $\theta' = 90°$ in (6.4-24) yields

$$|D_N(f, 90°, \psi)| = \left|\sin\left[\pi \frac{d}{\lambda}(\cos\psi - \cos\psi')\right]\right|, \qquad \theta' = 90°, \tag{6.4-25}$$

which is the magnitude of the normalized, *horizontal*, far-field beam pattern in the XY plane of a dipole lying along the X axis, steered to bearing angle ψ' (see Fig. 6.4-2).

Let us end our discussion in this example by deriving the equivalent set of time delays required to steer the far-field beam pattern of a dipole. We cannot use τ'_n given by (6.4-15) because (6.4-15) is only applicable for N odd. However, we can use the general expression for τ'_n given by (6.4-14). Therefore, substituting (6.4-18), (6.4-20), (6.4-21), and $c = f\lambda$ into (6.4-14) yields

$$\tau_1' = \frac{u'}{c}\frac{d}{2} \tag{6.4-26}$$

and

$$\tau_{-1}' = -\frac{u'}{c}\frac{d}{2} \tag{6.4-27}$$

where $u' = \sin\theta' \cos\psi'$. ■

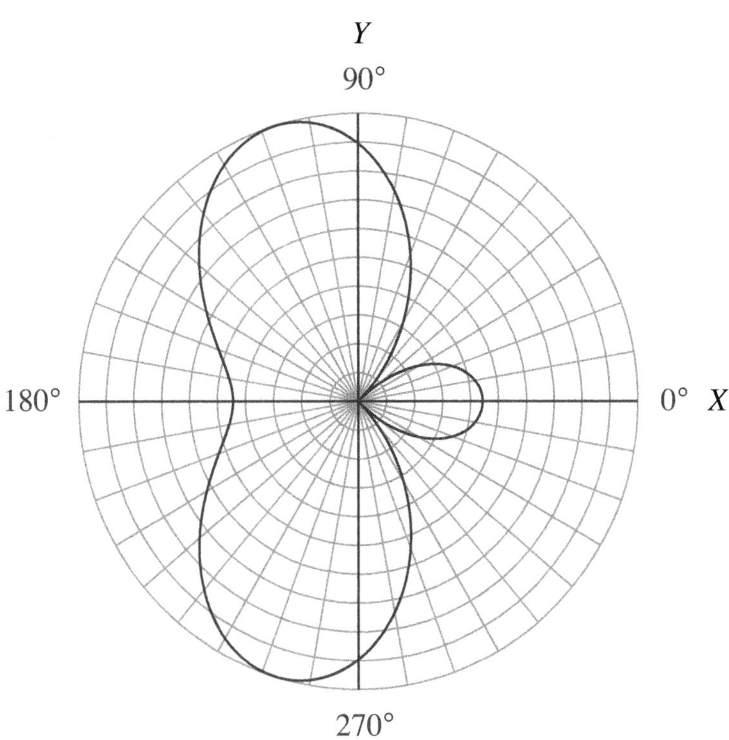

Figure 6.4-2 Polar plot, as a function of the azimuthal (bearing) angle ψ, of the magnitude of the normalized, horizontal, far-field beam pattern of a dipole lying along the X axis given by (6.4-25) for $d/\lambda = 0.5$ and beam-steer angle $\psi' = 45°$.

6.5 Far-Field Beam Patterns and the Spatial Discrete Fourier Transform

In this section we shall show that the far-field beam pattern of a linear array composed of an odd number N of identical, equally-spaced, complex-weighted, omnidirectional point-elements lying along the X axis can be computed using a *spatial* discrete Fourier transform. The far-field beam pattern of the aforementioned array can be obtained by substituting (6.1-87) and (6.1-91) into (6.1-89). Doing so yields

$$D(f, f_X) = \mathcal{S}(f) \sum_{n=-N'}^{N'} c_n(f) \exp(+j2\pi f_X nd), \tag{6.5-1}$$

where $\mathcal{S}(f)$ is the complex, element sensitivity function (see Table 6.1-2 and Appendix 6B),

$$c_n(f) = a_n(f) \exp[+j\theta_n(f)], \tag{6.5-2}$$

and

$$N' = (N-1)/2 . \tag{6.5-3}$$

According to (6.5-1), $D(f, f_X)$ is evaluated at continuous values of spatial frequency f_X. A more efficient way to compute $D(f, f_X)$ is to evaluate it at the following set of *discrete* values of f_X:

$$f_X = m\Delta f_X, \qquad m = -N'', \ldots, 0, \ldots, N'', \tag{6.5-4}$$

where

$$\Delta f_X = \frac{1}{(N+Z)d} \le \frac{1}{Nd} \tag{6.5-5}$$

is the spatial-frequency spacing with units of cycles per meter, Z is a *positive integer* ($Z \ge 0$), d is the interelement spacing in meters, and

$$N'' = \begin{cases} (N+Z)/2, & N+Z \text{ even} \\ (N+Z-1)/2, & N+Z \text{ odd.} \end{cases} \tag{6.5-6}$$

The ratio $(N+Z)/2$ is referred to as the *folding-bin* whether $N+Z$ is even or odd. Substituting (6.5-4) and (6.5-5) into (6.5-1) yields

$$\boxed{D(f, m) = \mathcal{S}(f) \sum_{n=-N'}^{N'} c_n(f) W_{N+Z}^{mn}, \qquad m = -N'', \ldots, 0, \ldots, N''} \tag{6.5-7}$$

where the summation on the right-hand side of (6.5-7) is the *spatial discrete Fourier transform* (DFT) of the set of N complex weights $c_n(f)$, and

$$W_{N+Z} = \exp\left(+j\frac{2\pi}{N+Z}\right). \tag{6.5-8}$$

Several comments are in order regarding (6.5-7). First, the index m is referred to as the DFT *bin number*. Evaluating the far-field beam pattern $D(f, m)$ at bin number m corresponds to evaluating the beam pattern at spatial frequency $f_X = m\Delta f_X$. Second, when the integer $Z=0$, the spatial-frequency spacing (bin spacing) Δf_X is equal to the *maximum* allowed value $1/(Nd)$ [see (6.5-5)], which is the spatial analog of the maximum allowed value of frequency spacing (bin spacing) in hertz in order to avoid aliasing when performing a time-domain DFT. Choosing $Z>0$ decreases Δf_X and increases N''. Increasing N'' means that $D(f, m)$ is evaluated at more bins yielding a smoother curve for plotting purposes. For example, if we want to evaluate $D(f, m)$ at twice as many bins by reducing the bin spacing Δf_X by a factor of two, then set $Z=N$ so that $N+Z=2N$. This is equivalent to adding $Z=N$ zeros to the original data sequence so that $N+Z=2N$. Choosing $Z>0$ is referred to as *padding-with-zeros*. Third, since the spatial DFT is *periodic* with *period* $N+Z$, the far-field beam pattern given by (6.5-7) is *periodic* with *period* $N+Z$, that is,

$$D(f, m+N+Z) = D(f, m). \tag{6.5-9}$$

It is the equal spacing of the elements that is responsible for the periodicity. The periodicity of the DFT is the reason why the range of values for bin number m need only cover one period, as shown in (6.5-7). The reason that negative as well as positive values for bin number m are used is to cover the entire visible region, as shall be discussed later. The fourth and final comment is that a DFT can be evaluated very quickly by using a *fast-Fourier-transform* (FFT) algorithm. Therefore, the far-field beam pattern $D(f, m)$ given by (6.5-7) can be evaluated by using either a DFT or a FFT algorithm. See Appendix 6E for an equation analogous to (6.5-7) for the case when the number of elements N is *even*.

Most commonly available FFT algorithms assume that the total number of data points to be processed is equal to an integer power of two, that is,

$$N = 2^b, \tag{6.5-10}$$

where b is a nonzero, positive integer. If N does not satisfy (6.5-10), then simply pick a value for Z so that $N+Z=2^b$. Since in this section the number of elements N and, hence, complex weights is odd, Z must also be odd in order for $N+Z$ to be equal to an integer power of two, which is an even number. For example, if $N=7$ and $Z=9$, then $N+Z=16=2^4$. It is very important to remember that even if $N+Z \neq 2^b$ so that a FFT algorithm cannot be used to evaluate the DFT in (6.5-7), (6.5-7) can still be used to compute the far-field beam pattern $D(f, m)$.

In order to relate a bin number to values of the spherical angles θ and ψ, we first need to derive a relationship between bin number m and the corresponding value of direction cosine u_m. Since [see (6.5-4)]

$$f_X = \frac{u}{\lambda} = m\Delta f_X, \qquad m = -N'', \ldots, 0, \ldots, N'', \tag{6.5-11}$$

solving for direction cosine u yields

$$u = u_m = m\lambda\Delta f_X, \qquad m = -N'', \ldots, 0, \ldots, N'', \tag{6.5-12}$$

and by substituting (6.5-5) into (6.5-12), we obtain

$$\boxed{u_m = \frac{m}{N+Z}\frac{\lambda}{d}, \qquad m = -N'', \ldots, 0, \ldots, N''} \tag{6.5-13}$$

where

$$u_m = \sin\theta_m \cos\psi_m \tag{6.5-14}$$

and N'' is given by (6.5-6). If we restrict ourselves to the vertical beam pattern in the XZ plane, then $u_m = \sin\theta_m$ when $\psi_m = 0°$, and $u_m = -\sin\theta_m$ when $\psi_m = 180°$. Similarly, if we restrict ourselves to the horizontal beam pattern in the XY plane, then $u_m = \cos\psi_m$ since $\theta_m = 90°$. Also, since the visible region corresponds to $-1 \le u_m \le 1$, and since u_m will take on negative values only for negative values of m, it is important to evaluate the far-field beam pattern given by (6.5-7) at negative and positive bin numbers m.

If we let δu be the direction-cosine bin spacing, then by using (6.5-13)

$$\delta u = u_m - u_{m-1} = \frac{1}{N+Z}\frac{\lambda}{d}. \tag{6.5-15}$$

Solving for Z from (6.5-15) yields

$$\boxed{Z = \frac{1}{\delta u}\frac{\lambda}{d} - N} \tag{6.5-16}$$

Equation (6.5-16) is used to solve for the value of Z that is required in order to obtain a desired direction-cosine bin spacing δu.

The *normalized*, far-field beam pattern $D_N(f, m)$ is given by

$$D_N(f, m) = D(f, m)/D_{\max}, \tag{6.5-17}$$

where $D(f, m)$ is the unnormalized, far-field beam pattern given by (6.5-7), and if $a_n(f) > 0 \;\; \forall \;\; n$, then

$$\boxed{D_{\max} = \max|D(f,m)| = |S(f)| \sum_{n=-N'}^{N'} a_n(f)} \tag{6.5-18}$$

is the normalization factor [use the procedure in Appendix 6A for N odd to obtain (6.5-18)]. Therefore, dividing the magnitude of (6.5-7) by (6.5-18) yields

$$\boxed{|D_N(f,m)| = \frac{\left|\sum_{n=-N'}^{N'} c_n(f) W_{N+Z}^{mn}\right|}{\sum_{n=-N'}^{N'} a_n(f)}, \qquad m = -N'', \ldots, 0, \ldots, N''} \tag{6.5-19}$$

Figure 6.5-1 is a plot of the magnitudes of the normalized, far-field beam patterns of the rectangular, triangular, and cosine amplitude weights given by (6.2-1) through (6.2-3). Compare Fig. 6.5-1 with Fig. 2.2-9. Figure 6.5-2 is a plot of the magnitudes of the normalized, far-field beam patterns of the Hanning, Hamming, and Blackman amplitude weights given by (6.2-4) through (6.2-6). Compare Fig. 6.5-2 with Fig. 2.2-11. Figures 6.5-1 and 6.5-2 were obtained by evaluating (6.5-19) for $N = 63$ and $Z = 193$ so that $N + Z = 256 = 2^8$. The plots are only shown for positive values of m since the magnitudes of the far-field beam patterns are symmetric about $m = 0$.

6.5.1 Grating Lobes

Since the far-field beam pattern given by (6.5-1) or, equivalently, (6.5-7) is *periodic* [see (6.5-9)] because the elements are equally spaced, there is a potential problem with *grating lobes*. Grating lobes are *extraneous, unwanted* mainlobes that exist within the visible region. One way to avoid grating lobes is to use unevenly spaced elements. However, since the elements in our linear array are equally spaced, in order to guarantee only one mainlobe within the visible region, steered in the desired direction, the interelement spacing d must be chosen properly.

The criterion that d must satisfy in order to avoid grating lobes can be derived by first substituting (6.5-2) and (6.4-7) into (6.5-1). Doing so yields

$$D(f, f_X) = S(f) \sum_{n=-N'}^{N'} a_n(f) \exp[+j2\pi(f_X - f_X')nd]. \tag{6.5-20}$$

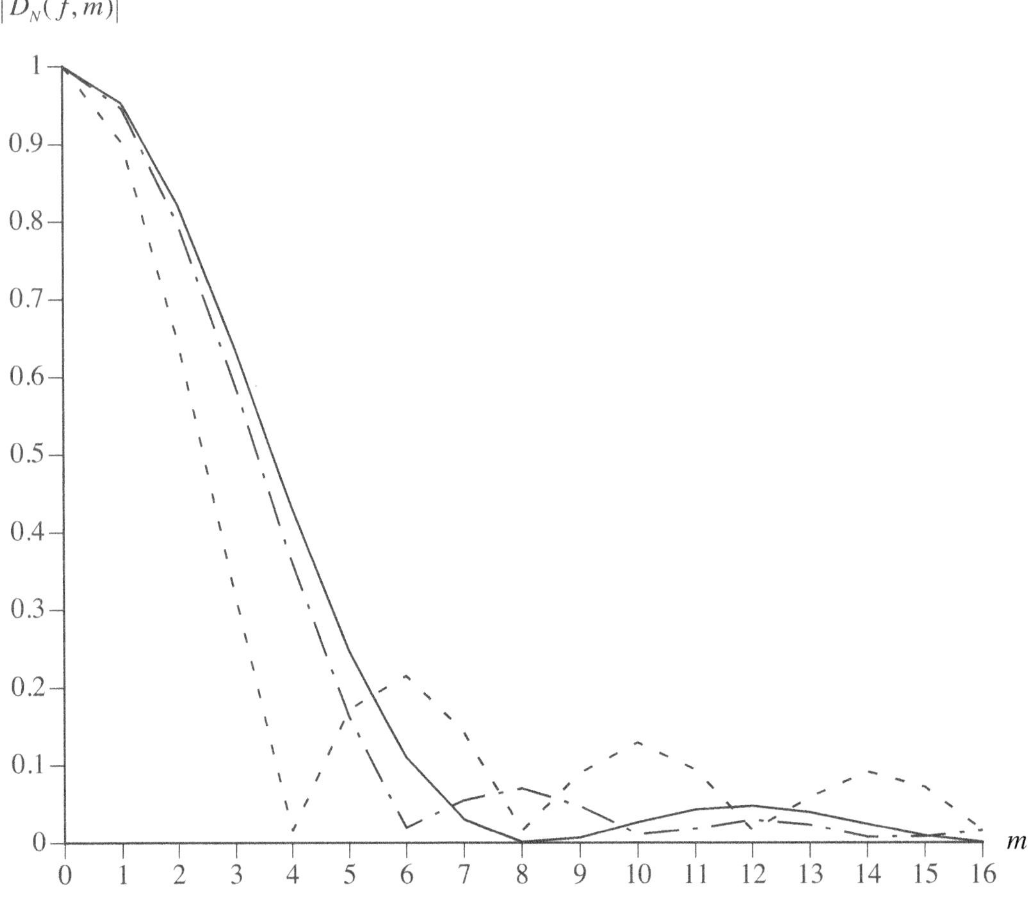

Figure 6.5-1 Magnitudes of the normalized, far-field beam patterns of the rectangular amplitude weights (dashed curve), triangular amplitude weights (solid curve), and cosine amplitude weights (dot-dash curve) plotted as a function of bin number m.

If the amplitude weight $a_n(f) > 0 \;\; \forall \;\; n$, then the maximum value of the magnitude of (6.5-20) occurs at $f_X = f_X'$ and is given by

$$\begin{aligned} D_{\max} &= \max|D(f, f_X)| \\ &= |D(f, f_X')| \\ &= |S(f)| \sum_{n=-N'}^{N'} a_n(f). \end{aligned} \tag{6.5-21}$$

Equation (6.5-21) indicates that the mainlobe of the unnormalized, far-field beam pattern $D(f, f_X)$ has been steered to $f_X = f_X'$. However, the maximum value $D_{\max}$ also occurs at

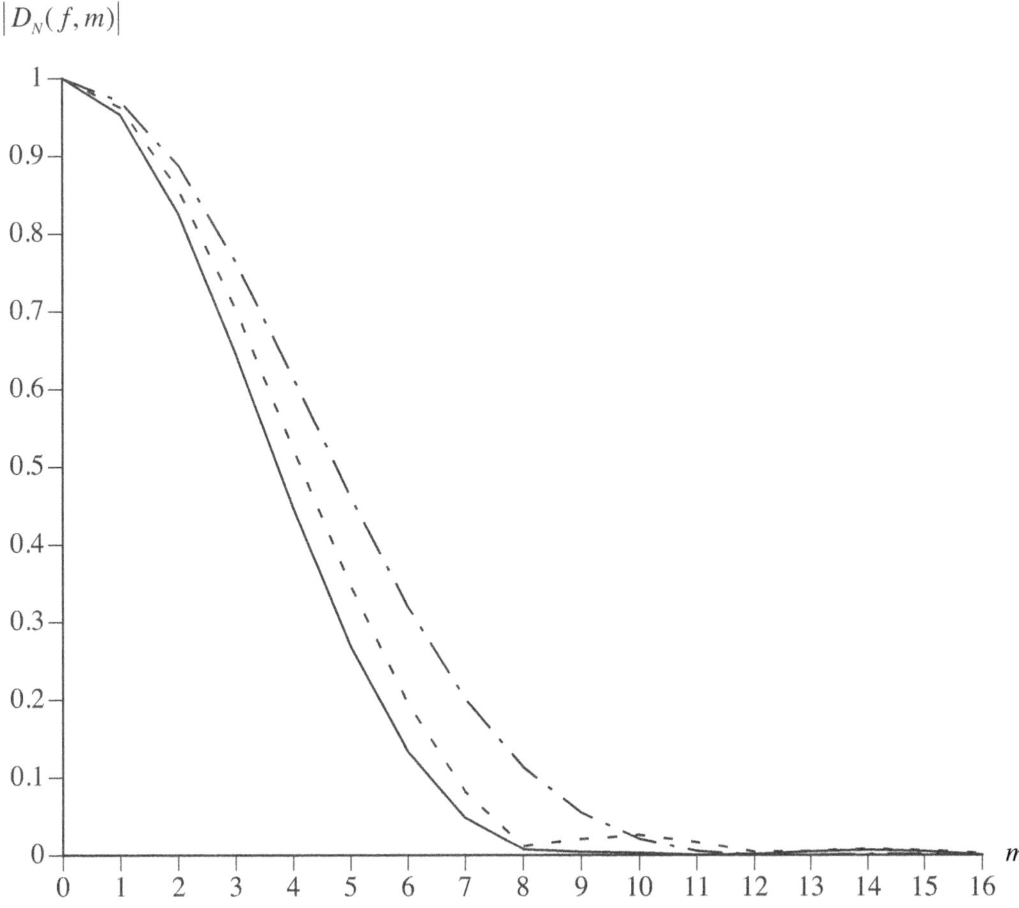

Figure 6.5-2 Magnitudes of the normalized, far-field beam patterns of the Hanning amplitude weights (dashed curve), Hamming amplitude weights (solid curve), and Blackman amplitude weights (dot-dash curve) plotted as a function of bin number m.

$$f_X = f_X' + \frac{i}{d}, \qquad i = \pm 1, \pm 2, \pm 3, \ldots, \tag{6.5-22}$$

because substituting (6.5-22) into the magnitude of (6.5-20) yields

$$\begin{aligned}
\left| D\left(f, f_X' + \frac{i}{d} \right) \right| &= |S(f)| \left| \sum_{n=-N'}^{N'} a_n(f) \exp(+j2\pi i n) \right| \\
&= |S(f)| \left| \sum_{n=-N'}^{N'} a_n(f) \right| \\
&= |S(f)| \sum_{n=-N'}^{N'} a_n(f) \\
&= D_{\max}, \qquad i = \pm 1, \pm 2, \pm 3, \ldots
\end{aligned} \tag{6.5-23}$$

Therefore, if we express (6.5-22) in terms of direction cosines u and u', then we can state that grating lobes will exist at $u = u_g$, where

$$\boxed{u_g = u' + i\frac{\lambda}{d}, \qquad i = \pm 1, \pm 2, \pm 3, \ldots} \tag{6.5-24}$$

if $|u_g| \le 1$, that is, if u_g is in the *visible region.*

If grating lobes can be avoided when a beam pattern is steered to end-fire ($u' = \pm 1$), then grating lobes can be avoided for all possible directions of beam steering. Therefore, if a beam pattern is steered to $u' = 1$, then from (6.5-24), the location of the first possible grating lobe within the visible region occurs when $i = -1$ and is given by

$$u_g = 1 - \frac{\lambda}{d}, \qquad i = -1. \tag{6.5-25}$$

If $d < \lambda/2$, then $|u_g| > 1$, which means that *no* grating lobes will exist in the visible region, even if a beam pattern is steered to end-fire. However, since an array must be able to operate over a range of frequencies in general, if

$$\boxed{d < \lambda_{\min}/2} \tag{6.5-26}$$

where $\lambda_{\min}$ is the minimum wavelength associated with the maximum frequency $f_{\max} = c/\lambda_{\min}$ in hertz, then *grating lobes will be avoided for all possible directions of beam steering*. If *no* beam steering is done, that is, if $u' = 0$, then grating lobes will be avoided if $d < \lambda_{\min}$.

In order to illustrate grating lobes, consider the magnitude of the normalized, horizontal, far-field beam pattern in the XY plane of a linear array of 7 identical, equally-spaced, rectangular-amplitude-weighted, omnidirectional point-elements lying along the X axis steered to bearing angle ψ':

$$|D_N(f, 90°, \psi)| = \frac{1}{7}\left|\frac{\sin\left(7\pi\frac{d}{\lambda}(\cos\psi - \cos\psi')\right)}{\sin\left(\pi\frac{d}{\lambda}(\cos\psi - \cos\psi')\right)}\right|, \qquad \theta' = 90°. \tag{6.5-27}$$

Figures 6.5-3 and 6.5-4 are polar plots of (6.5-27) for beam-steer angles $\psi' = 45°$ and $\psi' = 0°$ (end-fire), respectively, for $d/\lambda = 1$, 0.5, and 0.45. Figure 6.5-3 (b) shows that for $\psi' = 45°$, $d = \lambda/2$ is sufficient to avoid grating lobes. However, for $\psi'=0°$ (end-fire), Fig. 6.5-4 (b) shows that a grating lobe does exist for $d=\lambda/2$, and only when $d<\lambda/2$, does the grating lobe disappear [see Fig. 6.5-4 (c)]. Figures 6.5-3 and 6.5-4 also show that as the ratio d/λ *decreases*, mainlobe beamwidth *increases*. Recall from linear aperture theory that mainlobe beamwidth will increase when the frequency (wavelength) is held constant and the length of the aperture is decreased, or when the length of the aperture is held constant and the frequency is decreased (wavelength is increased).

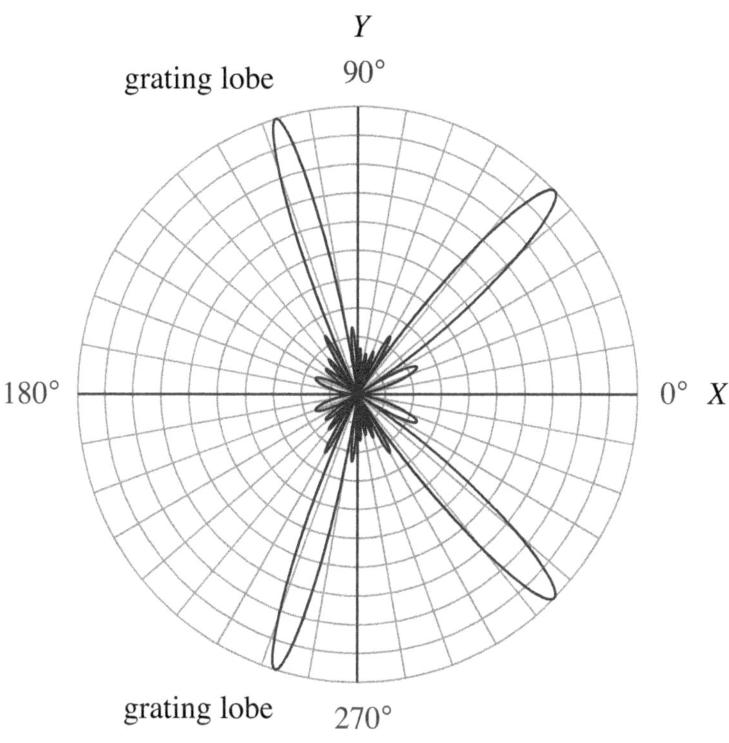

Figure 6.5-3 Polar plots of the magnitudes of the normalized, horizontal, far-field beam patterns of a linear array of $N=7$ identical, equally-spaced, rectangular amplitude-weighted, omnidirectional point-elements lying along the X axis for beam-steer angle $\psi'=45°$ and (a) $d/\lambda=1$, (b) $d/\lambda=0.5$, and (c) $d/\lambda=0.45$.

Example 6.5-1 Spatial-Domain Sampling Theorem

In this example we shall use the *spatial version* of the sampling theorem from time-domain signal processing to derive the criterion that the interelement spacing d must satisfy in order to avoid grating lobes for all possible directions of beam steering. The time-domain sampling theorem states that in order to avoid aliasing, a time-domain signal must be sampled at a rate

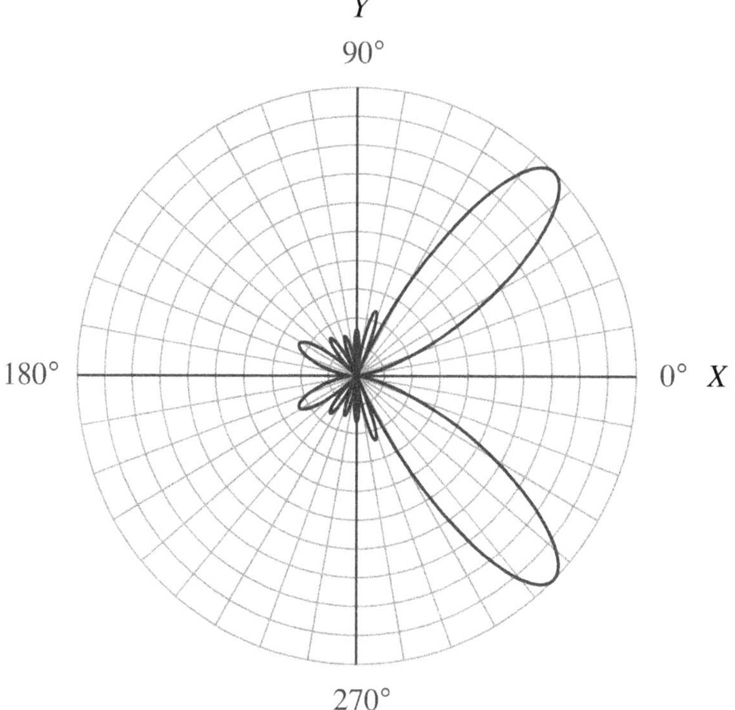

(b)

Y

90°

180° 0° X

270°

(c)

Figure 6.5-3 *continued.*

$$f_S = \frac{1}{T_S} \geq 2 f_{\max}, \tag{6.5-28}$$

where f_S is the sampling frequency in hertz (or sampling rate in samples per second), T_S is the sampling period in seconds, and $f_{\max}$ is the highest frequency component in hertz contained in the frequency spectrum of the signal. The minimum sampling frequency, $2 f_{\max}$, is known as the Nyquist rate. In practice, it is always recommended to sample at a rate greater than the Nyquist rate.

The *spatial-domain sampling theorem* based on (6.5-28) is given by

$$\boxed{f_{SX} = \frac{1}{d} \geq 2 \max f_X} \tag{6.5-29}$$

where f_{SX} is the spatial sampling frequency in the X direction in cycles per meter

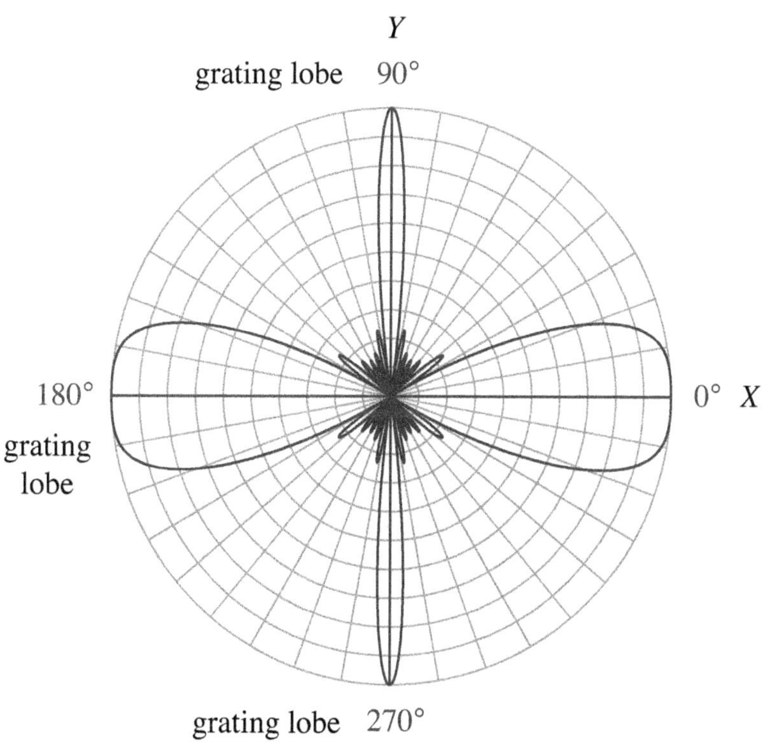

Figure 6.5-4 Polar plots of the magnitudes of the normalized, horizontal, far-field beam patterns of a linear array of $N = 7$ identical, equally-spaced, rectangular amplitude-weighted, omnidirectional point-elements lying along the X axis for beam-steer angle $\psi' = 0°$ (end-fire) and (a) $d/\lambda = 1$, (b) $d/\lambda = 0.5$, and (c) $d/\lambda = 0.45$.

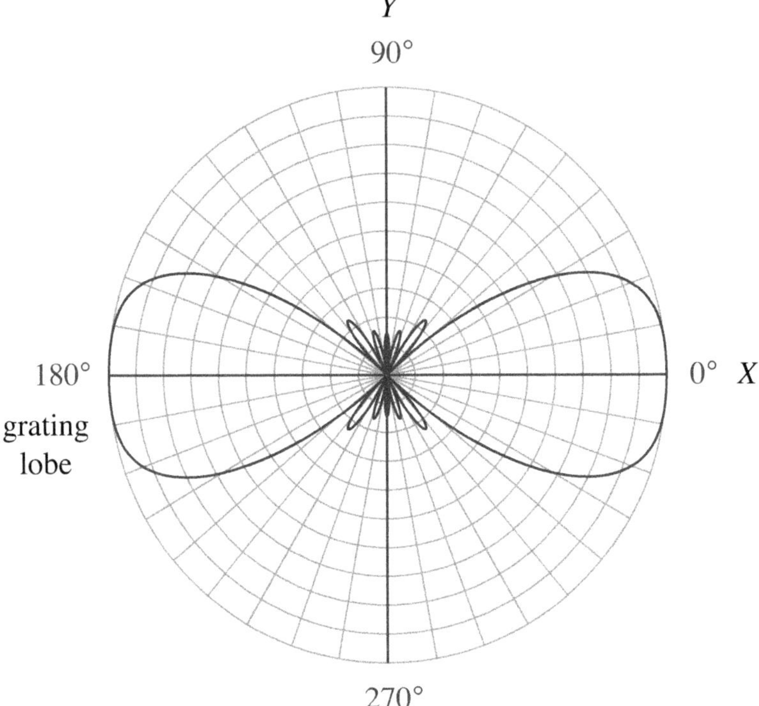

(b)

Y
90°
180°
0° X
270°

(c)

Figure 6.5-4 *continued.*

(a.k.a. the spatial sampling rate in samples per meter), d is the sampling period (interelement spacing) in meters, and $\max f_X$ is the highest spatial-frequency component in the X direction in cycles per meter contained in the spatial Fourier transform (angular spectrum) of a signal that is a function of both time t and spatial coordinate x. For example, a linear array of equally-spaced elements lying along the X axis and being operated in the passive mode, samples (measures) acoustic fields incident upon the array every d meters. Therefore, the output electrical signals from the elements are functions of both time and space. Since

$$\max f_X = u_{\max}/\lambda_{\min}, \tag{6.5-30}$$

substituting (6.5-30) into (6.5-29) and setting $u_{\max} = 1$ yields

$$d \le \lambda_{\min}/2, \tag{6.5-31}$$

or

$$\boxed{d < \lambda_{\min}/2} \tag{6.5-32}$$

where $\lambda_{\min}$ corresponds to $f_{\max} = c/\lambda_{\min}$, the highest frequency component in hertz contained in the frequency spectrum of the signal. Therefore, if the interelement spacing satisfies (6.5-32) [which is identical to (6.5-26)], then grating lobes will be avoided for all possible directions of beam steering. ■

6.6 The Near-Field Beam Pattern of a Linear Array

As was mentioned at the beginning of Section 6.1, a linear array is an example of a linear aperture. Therefore, the near-field beam pattern (directivity function) of a linear array lying along the X axis is given by the following one-dimensional spatial Fresnel transform (see Section 2.6):

$$\begin{aligned} \mathcal{D}(f, r, f_X) &= F_{x_A}\left\{A(f, x_A)\exp\left(-j\frac{k}{2r}x_A^2\right)\right\} \\ &= \int_{-\infty}^{\infty} A(f, x_A)\exp\left(-j\frac{k}{2r}x_A^2\right)\exp(+j2\pi f_X x_A)\,dx_A, \end{aligned} \tag{6.6-1}$$

where $A(f, x_A)$ is the complex frequency response (complex aperture function) of the linear array,

$$k = 2\pi f/c = 2\pi/\lambda \tag{6.6-2}$$

is the wavenumber in radians per meter, and

$$f_X = u/\lambda = \sin\theta\cos\psi/\lambda \tag{6.6-3}$$

is a spatial frequency in the X direction with units of cycles per meter. Since f_X can be expressed in terms of the spherical angles θ and ψ, the near-field beam pattern can ultimately be expressed as a function of frequency f, the range r to a field point, and the spherical angles θ and ψ, that is, $\mathcal{D}(f,r,f_X) \rightarrow \mathcal{D}(f,r,\theta,\psi)$. We shall consider a field point to be in the Fresnel (near-field) region of an array if the Fresnel angle criterion given by either (1.2-35) or (1.2-91) is satisfied, and the range r to the field point – as measured from the center of the array – satisfies the Fresnel range criterion given by

$$1.356R_A < r < \pi R_A^2/\lambda\,, \tag{6.6-4}$$

where R_A is the maximum radial extent of the array. If the overall length of the linear array is L_A meters, then $R_A = L_A/2$. Some typical values for the range to the near-field/far-field boundary $r_{\text{NF/FF}} = \pi R_A^2/\lambda$ for three different frequencies f and two different lengths L_A for a linear towed array are shown in Table 6.6-1, along with values for $r_{\min} = 1.356R_A$.

Table 6.6-1 Values for the Range to the Near-Field/Far-Field Boundary $r_{\text{NF/FF}}$ for a Linear Towed Array

c (m/sec)	f (Hz)	λ (m)	L_A (m)	R_A (m)	$r_{\min}$ (m)	$r_{\text{NF/FF}}$ (km)
1500	60	25.0	100	50	67.8	0.314
1500	100	15.0	100	50	67.8	0.524
1500	1000	1.5	100	50	67.8	5.236
1500	60	25.0	200	100	135.6	1.257
1500	100	15.0	200	100	135.6	2.094
1500	1000	1.5	200	100	135.6	20.944

If a linear array lies along either the Y or Z axis instead of the X axis, then simply replace x_A and f_X with either y_A and f_Y, or z_A and f_Z, respectively, in (6.6-1), where spatial frequencies f_Y and f_Z are given by (1.2-44) and (1.2-45), respectively. As can be seen from (6.6-1), in order to derive the near-field beam pattern of a linear array, we must first specify a mathematical model for its complex frequency response.

Consider a linear array composed of an odd number N of identical, unevenly-spaced, complex-weighted, omnidirectional point-elements lying along the X axis. The complex frequency response of this array is given by [see (6.1-90)]

$$A(f, x_A) = \mathcal{S}(f) \sum_{n=-N'}^{N'} c_n(f)\delta(x_A - x_n), \tag{6.6-5}$$

where $\mathcal{S}(f)$ is the complex, element sensitivity function (see Table 6.1-2 and Appendix 6B),

$$c_n(f) = a_n(f)\exp[+j\theta_n(f)], \tag{6.6-6}$$

$$N' = (N-1)/2, \tag{6.6-7}$$

and the impulse function $\delta(x_A - x_n)$ has units of inverse meters. Substituting (6.6-5) into (6.6-1) and making use of the sifting property of impulse functions yields the following expression for the near-field beam pattern:

$$\boxed{\mathcal{D}(f, r, f_X) = \mathcal{S}(f) \sum_{n=-N'}^{N'} c_n(f)\exp\left(-j\frac{k}{2r}x_n^2\right)\exp(+j2\pi f_X x_n)} \tag{6.6-8}$$

If the elements are *equally spaced*, then $x_n = nd$ [see (6.1-91)]. The units of $A(f, x_A)$ and $\mathcal{D}(f, r, f_X)$ are summarized in Table 6.6-2. Equation (6.6-8) is most accurate in the Fresnel region of the linear array where both the Fresnel angle criterion and the Fresnel range criterion are satisfied. It is less accurate in the near-field region outside the Fresnel region where $r < r_{\text{NF/FF}}$, but the Fresnel angle criterion is *not* satisfied. As was discussed in Subsection 1.2.1, the Fresnel region is only a subset of the near-field.

Table 6.6-2 Units of the Complex Frequency Response $A(f, x_A)$ and Corresponding Near-Field Beam Pattern $\mathcal{D}(f, r, f_X)$ for a Linear Array

Linear Array	$A(f, x_A)$	$\mathcal{D}(f, r, f_X)$
active (transmit)	$\left(\left(\text{m}^3/\text{sec}\right)/\text{V}\right)/\text{m}$	$\left(\text{m}^3/\text{sec}\right)/\text{V}$
passive (receive)	$\left(\text{V}/\left(\text{m}^2/\text{sec}\right)\right)/\text{m}$	$\text{V}/\left(\text{m}^2/\text{sec}\right)$

6.6.1 Beam Steering and Array Focusing

We already know from linear aperture theory that in order to do beam steering and aperture focusing in the Fresnel region of a linear aperture, there must be a linear plus quadratic phase response along the length of the aperture (see Subsection 2.6.2). A set of phase weights that will produce a linear plus quadratic phase response along the length of the linear array under discussion is given by

$$\boxed{\theta_n(f) = -2\pi f_X' x_n + \frac{k}{2r'} x_n^2, \qquad n = -N', \ldots, 0, \ldots, N'} \tag{6.6-9}$$

where

$$\boxed{f_X' = u'/\lambda = \sin\theta' \cos\psi' / \lambda} \tag{6.6-10}$$

and

$$k = 2\pi f / c = 2\pi/\lambda. \tag{6.6-11}$$

The parameter r' in (6.6-9) is the *focal range*. It is the near-field range from the array where the far-field beam pattern of the array will be in focus. Compare (6.6-9) through (6.6-11) with (2.6-22) through (2.6-26).

If (6.6-6) and (6.6-9) are substituted into (6.6-8), then

$$\mathcal{D}(f, r, f_X) = \mathcal{S}(f) \sum_{n=-N'}^{N'} a_n(f) \exp\left\{-j\frac{k}{2}\left[\frac{1}{r} - \frac{1}{r'}\right] x_n^2\right\} \exp\left[+j2\pi(f_X - f_X')x_n\right], \tag{6.6-12}$$

and if (6.6-12) is evaluated at the near-field range $r = r'$, then it reduces to

$$\mathcal{D}(f, r', f_X) = D(f, f_X - f_X'), \tag{6.6-13}$$

where

$$D(f, f_X) = \mathcal{S}(f) \sum_{n=-N'}^{N'} a_n(f) \exp(+j2\pi f_X x_n) \tag{6.6-14}$$

is the far-field beam pattern of the array when it is only amplitude weighted. Equation (6.6-13) indicates that the far-field beam pattern $D(f, f_X)$ of the amplitude weights $a_n(f)$ is *in focus* at the near-field range $r = r'$ meters from the array, and has been *steered* in the direction $u = u'$ in direction-cosine space, which is equivalent to steering (tilting) the beam pattern to $\theta = \theta'$ and $\psi = \psi'$

[see (6.6-10)]. However, in the near-field region of the array outside the Fresnel region, a set of quadratic phase weights can only approximately focus the far-field beam pattern. Equal spacing of the elements is *not* required in order to do beam steering and array focusing.

As was discussed in Section 6.4, a phase weight $\theta_n(f)$ in radians is equivalent to a time delay τ'_n in seconds, that is,

$$\theta_n(f) \triangleq -2\pi f \tau'_n, \tag{6.6-15}$$

or

$$\tau'_n = -\theta_n(f)/(2\pi f). \tag{6.6-16}$$

Substituting (6.6-9) through (6.6-11) into (6.6-16) yields

$$\boxed{\tau'_n = \frac{u'}{c} x_n - \frac{1}{2r'c} x_n^2, \qquad n = -N', \ldots, 0, \ldots, N'} \tag{6.6-17}$$

where $c = f\lambda$. The phase weights given by (6.6-9) or, equivalently, the time delays given by (6.6-17) are implemented by using the beamforming procedure shown in Fig. 6.4-1.

Example 6.6-1 Beam Steering and Focusing in the Fresnel (Near-Field) Region

Consider a linear array of 11 identical, equally-spaced, complex-weighted, omnidirectional point-elements lying along the X axis. The array is operated at a frequency of $1\,\text{kHz}$, rectangular amplitude weights are used, the speed of sound $c = 1500\,\text{m/sec}$, and half-wavelength interelement spacing is used. With the information that is given, we can make the following calculations:

$$N' = \frac{N-1}{2} = \frac{11-1}{2} = 5, \tag{6.6-18}$$

$$\lambda = \frac{c}{f} = \frac{1500\,\text{m/sec}}{1000\,\text{Hz}} = 1.5\,\text{m}, \tag{6.6-19}$$

$$d = \frac{\lambda}{2} = \frac{1.5\,\text{m}}{2} = 0.75\,\text{m}, \tag{6.6-20}$$

the overall length of the array is

$$L_A = (N-1)d = (11-1) \times 0.75 \text{ m} = 7.5 \text{ m}, \tag{6.6-21}$$

and the maximum radial extent of the array is

$$R_A = \frac{L_A}{2} = \frac{7.5 \text{ m}}{2} = 3.75 \text{ m}. \tag{6.6-22}$$

Therefore, the Fresnel region begins at a range of

$$r_{\min} = 1.356 R_A = 1.356 \times 3.75 \text{ m} = 5.085 \text{ m}, \tag{6.6-23}$$

and the range to the near-field/far-field boundary is

$$r_{\text{NF/FF}} = \frac{\pi R_A^2}{\lambda} = \frac{\pi (3.75 \text{ m})^2}{1.5 \text{ m}} = 29.5 \text{ m}. \tag{6.6-24}$$

Figure 1.2-4 illustrates the Fresnel region in the XY plane – in terms of both range and angle – for a linear aperture (linear array) lying along the X axis.

Since

$$f_X = u/\lambda = \sin\theta \cos\psi/\lambda, \tag{6.6-25}$$

$$f_X' = u'/\lambda = \sin\theta' \cos\psi'/\lambda, \tag{6.6-26}$$

and the elements are equally spaced, substituting (6.1-91), (6.6-25), and (6.6-26) into the near-field beam pattern given by (6.6-12) yields

$$\mathcal{D}(f,r,u) = \mathcal{S}(f) \sum_{n=-N'}^{N'} a_n(f) \exp\left\{-j\frac{k}{2}\left[\frac{1}{r} - \frac{1}{r'}\right](nd)^2\right\} \exp\left[+j2n\pi\frac{d}{\lambda}(u-u')\right], \tag{6.6-27}$$

or

$$\mathcal{D}(f,r,\theta,\psi) = \mathcal{S}(f) \sum_{n=-N'}^{N'} a_n(f) \exp\left\{-j\frac{k}{2}\left[\frac{1}{r} - \frac{1}{r'}\right](nd)^2\right\} \times \exp\left[+j2n\pi\frac{d}{\lambda}(\sin\theta\cos\psi - \sin\theta'\cos\psi')\right]. \tag{6.6-28}$$

Setting $\theta = 90°$ and $\theta' = 90°$ in (6.6-28) yields

$$\mathcal{D}(f,r,90°,\psi)=\mathcal{S}(f)\sum_{n=-N'}^{N'} a_n(f)\exp\left\{-j\frac{k}{2}\left[\frac{1}{r}-\frac{1}{r'}\right](nd)^2\right\}\times$$
$$\exp\left[+j2n\pi\frac{d}{\lambda}(\cos\psi-\cos\psi')\right], \qquad \theta'=90°, \tag{6.6-29}$$

which is the unnormalized, *horizontal*, near-field beam pattern in the XY plane of a linear array composed of an odd number N of identical, equally-spaced, complex-weighted, omnidirectional point-elements lying along the X axis, with focal range r' and steered to bearing angle ψ'. Equation (6.6-29) is applicable to the array being considered in this example. If quadratic phase weights are *not* used (*no* focusing), then (6.6-29) reduces to

$$\mathcal{D}(f,r,90°,\psi)=\mathcal{S}(f)\sum_{n=-N'}^{N'} a_n(f)\exp\left[-j\frac{k}{2r}(nd)^2\right]\times$$
$$\exp\left[+j2n\pi\frac{d}{\lambda}(\cos\psi-\cos\psi')\right], \qquad \theta'=90°. \tag{6.6-30}$$

Since rectangular amplitude weights are used in this example, $a_n(f)=1\ \forall\ n$.

Figures 6.6-1 (a), 6.6-2 (a), and 6.6-3 (a) are polar plots of the *normalized* magnitudes of the *unfocused*, horizontal, near-field beam patterns of the array obtained by evaluating (6.6-30) at $r=8$ m for beam-steer angles $\psi'=90°$ (broadside), $81°$, and $72°$ (Fresnel angular limit of $18°$ off broadside), respectively. The magnitude of (6.6-30) was normalized by setting the element sensitivity function $\mathcal{S}(f)$ equal to one and then determining the normalization factor $\mathcal{D}_{\max}$ numerically. Similarly, Figs. 6.6-1 (b), 6.6-2 (b), and 6.6-3 (b) are polar plots of the *normalized* magnitudes of the *focused*, horizontal, near-field beam patterns of the array obtained by evaluating (6.6-29) at $r=8$ m for $r'=8$ m and $\psi'=90°$, $81°$, and $72°$, respectively. Note that these figures show that the far-field beam pattern of the array has been focused at the near-field range $r=8$ m and steered. ■

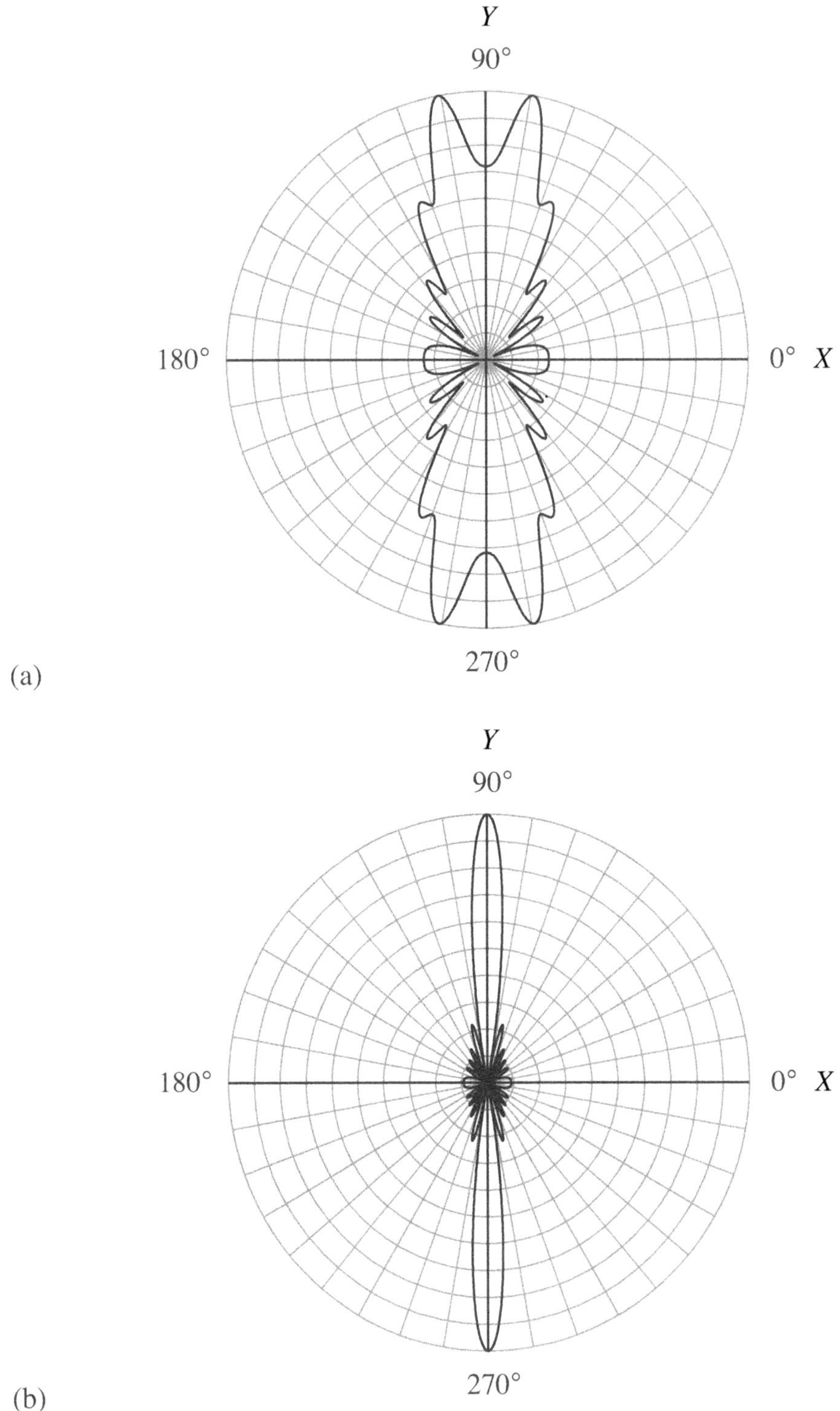

Figure 6.6-1 Polar plots of the normalized magnitudes of the (a) *unfocused* and (b) *focused* horizontal, near-field beam patterns of the array discussed in Example 6.6-1 for beam-steer angle $\psi' = 90°$ (broadside).

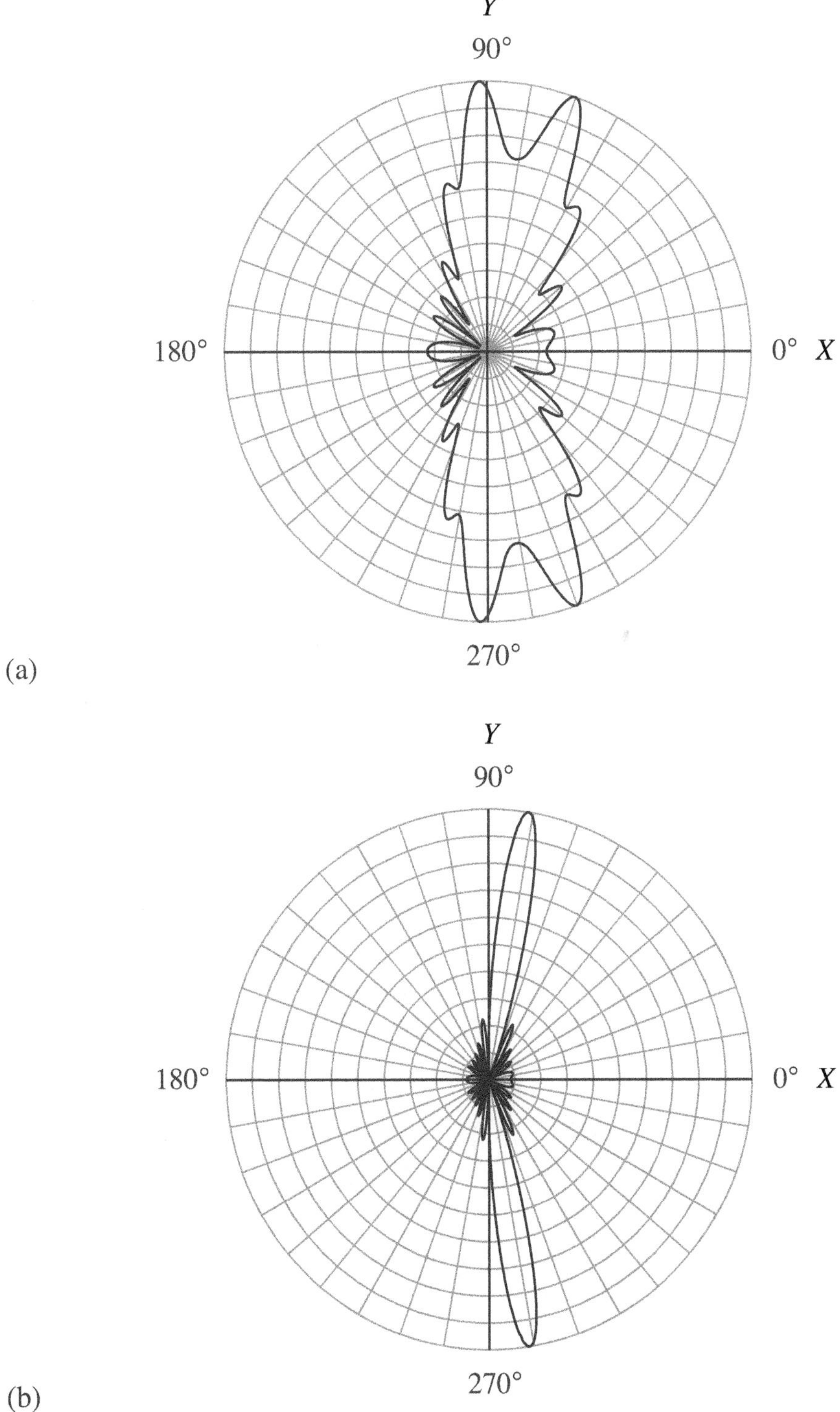

Figure 6.6-2 Polar plots of the normalized magnitudes of the (a) *unfocused* and (b) *focused* horizontal, near-field beam patterns of the array discussed in Example 6.6-1 for beam-steer angle $\psi' = 81°$.

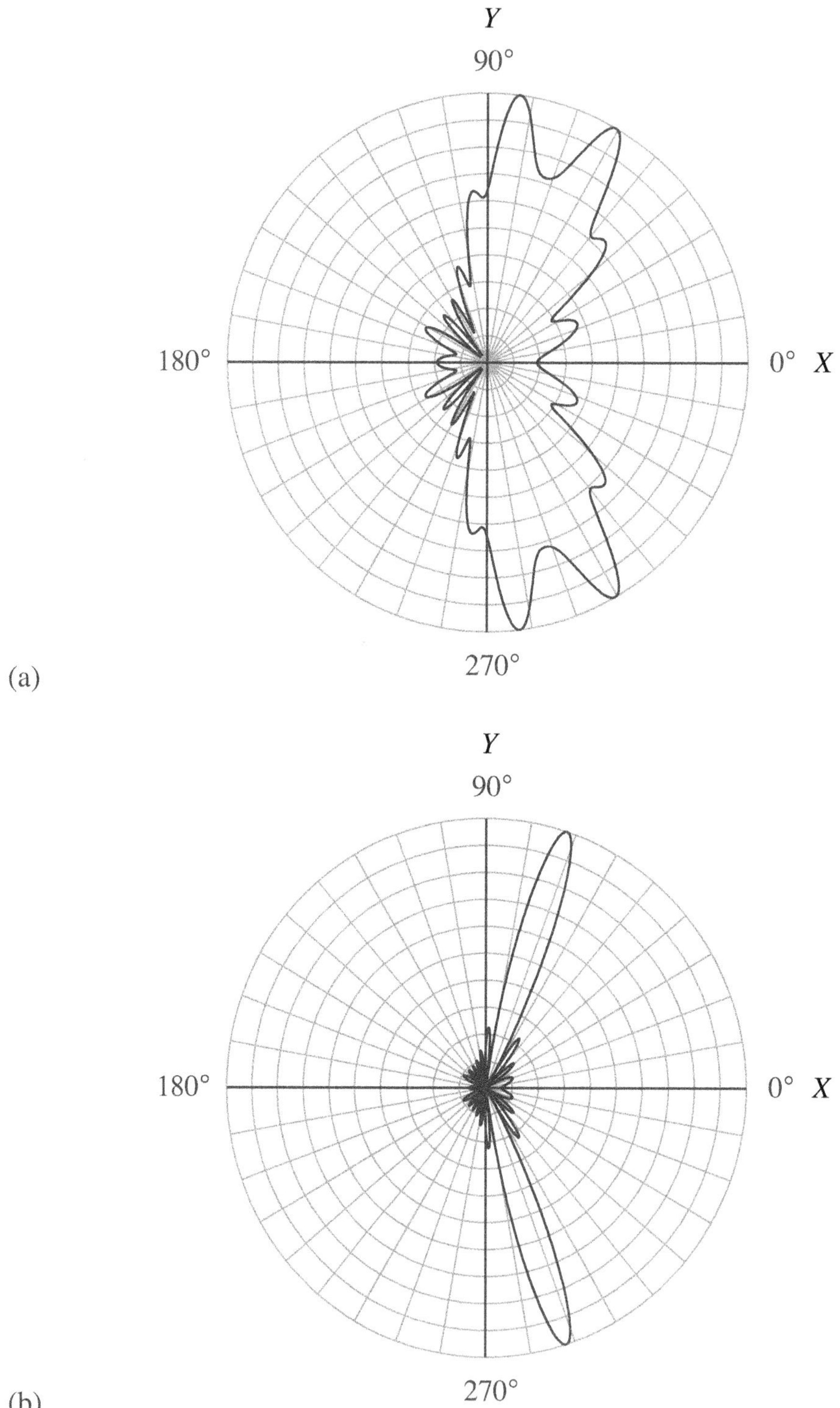

Figure 6.6-3 Polar plots of the normalized magnitudes of the (a) *unfocused* and (b) *focused* horizontal, near-field beam patterns of the array discussed in Example 6.6-1 for beam-steer angle $\psi' = 72°$.

Problems

Section 6.1 Appendix 6B

6-1 If the transmitter sensitivity level (transmitting voltage response) of an omnidirectional point-element at 18 kHz is $150\ \text{dB re}\ 1\ \mu\text{Pa/V}$ at 1 m, then what is the value of the magnitude of the transmitter sensitivity function in $\left(\text{m}^3/\text{sec}\right)/\text{V}$? Use $\rho_0 = 1026\ \text{kg/m}^3$ for the ambient density of seawater.

6-2 If the receiver sensitivity level (open circuit receiving response) of an omnidirectional point-element at 18 kHz is $-190\ \text{dB re}\ 1\ \text{V}/\mu\text{Pa}$, then what is the value of the magnitude of the receiver sensitivity function in $\text{V}/\left(\text{m}^2/\text{sec}\right)$? Use $\rho_0 = 1026\ \text{kg/m}^3$ for the ambient density of seawater.

Section 6.1

6-3 Find the unnormalized, far-field beam pattern of a two-element interferometer lying along the (a) Y axis and (b) Z axis as a function of frequency f and spherical angles θ and ψ.

6-4 Find the unnormalized, far-field beam pattern of a dipole lying along the (a) Y axis and (b) Z axis as a function of frequency f and spherical angles θ and ψ.

6-5 Find the unnormalized, far-field beam pattern of an axial quadrupole lying along the (a) Y axis and (b) Z axis as a function of frequency f and spherical angles θ and ψ.

6-6 A linear array composed of an even number N of identical, equally-spaced, complex-weighted, omnidirectional point-elements is lying along the X axis. The complex weights satisfy the conjugate symmetry property given by (6.1-22). The following set of amplitude weights are used: $a_{-n}(f) = a_n(f) = a\ \forall\ n$. With the amplitude weights as given, this array can be thought of as a *generalized, two-element interferometer*. Find the unnormalized, far-field beam pattern of this array as a function of frequency f and spatial frequency f_X.

6-7 A linear array composed of an even number N of identical, equally-spaced, complex-weighted, omnidirectional point-elements is lying along the X axis. The following set of amplitude and phase weights are used:

$a_n(f) = a$, $a_{-n}(f) = -a_n(f) = -a$, and $\theta_{-n}(f) = -\theta_n(f)$ $\forall$ n. Since both the amplitude and phase weights are odd functions of index n, this array can be thought of as a *generalized dipole*. Find the unnormalized, far-field beam pattern of this array as a function of frequency f and spatial frequency f_X.

6-8 A linear array composed of an odd number N of identical, equally-spaced, complex-weighted, omnidirectional point-elements is lying along the X axis. The complex weights satisfy the conjugate symmetry property given by (6.1-93). The following set of amplitude weights are used:

$$a_{-n}(f) = a_n(f) = a, \qquad n = 1, 2, \ldots, (N-1)/2$$
$$a_0(f) = -(N-1)a.$$

With the amplitude weights as given, this array can be thought of as a *generalized axial quadrupole*. Find the unnormalized, far-field beam pattern of this array as a function of frequency f and spatial frequency f_X.

Section 6.2

6-9 Show that $S(f, f_X)$ given by (6.2-15) is equal to N at $f_X = 0$ (broadside).

6-10 Using the procedure outlined in Example 2.3-3, (a) find the transcendental equation that must be solved in order to compute the dimensionless 3-dB beamwidth Δu in direction-cosine space of the far-field beam pattern given by (6.2-28). (b) If $N = 7$ and $d = \lambda/2$, compute the 3-dB beamwidth $\Delta\psi$ in degrees of the horizontal beam pattern.

Section 6.3

6-11 Consider a linear array of 6 identical, equally-spaced elements lying along the X axis, where each element is a single continuous line source or receiver L meters in length. The interelement spacing $d = \lambda/2$ meters.

(a) Find the set of *normalized* Dolph-Chebyshev amplitude weights so that the magnitude of the level of the mainlobe of the array factor $S(f, f_X)$ is 30 dB above the magnitude of the level of its sidelobes.

(b) Compute the 3-dB beamwidth $\Delta\psi$ in degrees of the horizontal, angular profile of the array factor $S(f, f_X)$.

(c) Compute the 3-dB beamwidth $\Delta\theta$ in degrees of the *vertical*, angular profile of the array factor $S(f, f_X)$. Compare $\Delta\theta$ with $\Delta\psi$ from Part (b).

Section 6.4

6-12 Consider a linear array composed of an odd number N of identical, equally-spaced, complex-weighted, omnidirectional point-elements. Find the equations for the phase weights and equivalent time delays that will steer the far-field beam pattern of this array to $\theta = 50°$ and $\psi = 112°$ if the array is lying along the

(a) X axis.

(b) Y axis.

(c) Find the equations for the phase weights and equivalent time delays that will steer the far-field beam pattern of this array to $\theta = 50°$ if the array is lying along the Z axis.

6-13 Consider a linear array composed of an *even* number N of identical, *unevenly-spaced*, complex-weighted, omnidirectional point-elements lying along the X axis. Find the equations for the phase weights and equivalent time delays that must be used in order to do beam steering. Also consider the special case of equally-spaced elements.

6-14 Find the equations for the phase weights and equivalent time delays that are required in order to steer the mainlobe of the far-field beam pattern of the two-element interferometer discussed in Example 6.1-1.

6-15 Find the equations for the phase weights and equivalent time delays that are required in order to steer the far-field beam pattern of the axial quadrupole discussed in Example 6.1-4.

Section 6.5

6-16 Verify (6.5-9).

6-17 The far-field beam pattern of a linear array of 15 identical, equally-spaced, complex-weighted, omnidirectional point-elements lying along the X axis is computed using a spatial DFT. Half-wavelength interelement spacing is used.

(a) What is the direction-cosine bin spacing if padding-with-zeros is not done?

(b) Solve for the number of zeros that is required in order to obtain a direction-cosine bin spacing of 0.05 .

6-18 Assume that an array must be designed to operate at any one of the following three frequencies: $f = 100$, 300, and 1000 Hz. What criterion must the interelement spacing satisfy in order to avoid grating lobes for all possible directions of beam steering regardless of which frequency is used? Use $c = 1500$ m/sec.

6-19 Consider a linear array composed of an odd number N of identical, equally-spaced, complex-weighted, omnidirectional point-elements lying along the Z axis. The interelement spacing is d meters (unspecified) and the operating frequency is also unspecified. If you only plan to steer the far-field beam pattern of the array to $\theta' = 45°$ and never to end-fire ($\theta = 0°$ or $\theta = 180°$), then

(a) what is the location of the *first* possible grating lobe in direction-cosine space?

(b) What inequality must d satisfy in order to avoid the first and, thus, all grating lobes?

(c) If the operating frequency is 1 kHz and $c = 1500$ m/sec, then what inequality must d satisfy?

(d) Repeat Parts (a) and (b) for $\theta' = 135°$.

Section 6.6

6-20 Consider a linear towed array composed of an odd number N of identical, equally-spaced, complex-weighted, omnidirectional point-elements lying along the X axis. The overall length of the array is 200 m, the operating frequency is 300 Hz, the interelement spacing $d = \lambda/2$ meters, and $c = 1500$ m/sec. The spherical coordinates of a sound source (target) as measured from the center of the array are $r_S = 5$ km, $\theta_S = 100°$, and $\psi_S = 75°$.

(a) How many elements are in the array?

(b) Is the sound source in the array's Fresnel region or far-field?

(c) What is the equation for the phase weights that must be applied across the array if the array's far-field beam pattern is to be focused (if required) and steered to coordinates (r_S, θ_S, ψ_S)?

(d) What is the equation for the equivalent time delays?

6-21 Consider a linear array composed of an *even* number N of identical, *unevenly-spaced*, complex-weighted, omnidirectional point-elements lying along the X axis.

(a) Find the equation for the near-field beam pattern of this array. Also consider the special case of equally-spaced elements.

(b) What is the equation for the phase weights that must be used in order to do beam steering and array focusing?

(c) What is the equation for the equivalent time delays?

Appendix 6A Normalization Factor for the Array Factor for N Even and Odd

N Even

The general expression for the array factor $S(f, f_X)$ for N even is given by [see (6.1-12)]

$$S(f, f_X) = \sum_{n=1}^{N/2} \left[c_{-n}(f) \exp(+j2\pi f_X x_{-n}) + c_n(f) \exp(+j2\pi f_X x_n) \right]. \quad \text{(6A-1)}$$

Computing the magnitude of $S(f, f_X)$ yields

$$\begin{aligned} |S(f, f_X)| &= \left| \sum_{n=1}^{N/2} \left[c_{-n}(f) \exp(+j2\pi f_X x_{-n}) + c_n(f) \exp(+j2\pi f_X x_n) \right] \right| \\ &\le \sum_{n=1}^{N/2} \left| \left[c_{-n}(f) \exp(+j2\pi f_X x_{-n}) + c_n(f) \exp(+j2\pi f_X x_n) \right] \right| \\ &\le \sum_{n=1}^{N/2} \left[\left| c_{-n}(f) \exp(+j2\pi f_X x_{-n}) \right| + \left| c_n(f) \exp(+j2\pi f_X x_n) \right| \right] \\ &\le \sum_{n=1}^{N/2} \left[\left| c_{-n}(f) \right| \left| \exp(+j2\pi f_X x_{-n}) \right| + \left| c_n(f) \right| \left| \exp(+j2\pi f_X x_n) \right| \right], \end{aligned} \quad \text{(6A-2)}$$

which further reduces to

$$\left|S(f,f_X)\right| \le \sum_{n=1}^{N/2}\left[\left|a_{-n}(f)\right| + \left|a_n(f)\right|\right]. \tag{6A-3}$$

Therefore,

$$\max\left|S(f,f_X)\right| = \sum_{n=1}^{N/2}\left[\left|a_{-n}(f)\right| + \left|a_n(f)\right|\right]. \tag{6A-4}$$

The absolute value of the real amplitude weights is required because amplitude weights can have both positive and negative values in general.

Equation (6A-4) is the absolute maximum value of $\left|S(f,f_X)\right|$. The actual maximum value may be less than (6A-4) as indicated by (6A-3). However, if $a_{-n}(f) > 0$ and $a_n(f) > 0 \ \forall \ n$, then the normalization factor $S_{\max}$ for N even is

$$\boxed{S_{\max} = \max\left|S(f,f_X)\right| = \sum_{n=1}^{N/2}\left[a_{-n}(f) + a_n(f)\right]} \tag{6A-5}$$

Note that if (6A-1) is evaluated at broadside (i.e., at $f_X = 0$) and if no phase weighting is done, then

$$S(f,0) = \sum_{n=1}^{N/2}\left[a_{-n}(f) + a_n(f)\right] \tag{6A-6}$$

and

$$\left|S(f,0)\right| = \left|\sum_{n=1}^{N/2}\left[a_{-n}(f) + a_n(f)\right]\right| \le \sum_{n=1}^{N/2}\left[\left|a_{-n}(f)\right| + \left|a_n(f)\right|\right]. \tag{6A-7}$$

If $a_{-n}(f) > 0$ and $a_n(f) > 0 \ \forall \ n$, then

$$\left|S(f,0)\right| = \left|\sum_{n=1}^{N/2}\left[a_{-n}(f) + a_n(f)\right]\right| = \sum_{n=1}^{N/2}\left[a_{-n}(f) + a_n(f)\right], \tag{6A-8}$$

and by comparing (6A-5) and (6A-8),

$$\boxed{S_{\max} = \max\left|S(f,f_X)\right| = \left|S(f,0)\right|} \tag{6A-9}$$

N Odd

The general expression for the array factor $S(f,f_X)$ for N odd is given by [see (6.1-87)]

$$S(f, f_X) = \sum_{n=-N'}^{N'} c_n(f)\exp(+j2\pi f_X x_n). \tag{6A-10}$$

Computing the magnitude of $S(f, f_X)$ yields

$$\begin{aligned} |S(f, f_X)| &= \left|\sum_{n=-N'}^{N'} c_n(f)\exp(+j2\pi f_X x_n)\right| \\ &\le \sum_{n=-N'}^{N'} |c_n(f)\exp(+j2\pi f_X x_n)| \\ &\le \sum_{n=-N'}^{N'} |c_n(f)||\exp(+j2\pi f_X x_n)|, \end{aligned} \tag{6A-11}$$

which further reduces to

$$|S(f, f_X)| \le \sum_{n=-N'}^{N'} |a_n(f)|. \tag{6A-12}$$

Therefore,

$$\max|S(f, f_X)| = \sum_{n=-N'}^{N'} |a_n(f)|. \tag{6A-13}$$

The absolute value of the real amplitude weights is required because amplitude weights can have both positive and negative values in general.

Equation (6A-13) is the absolute maximum value of $|S(f, f_X)|$. The actual maximum value may be less than (6A-13) as indicated by (6A-12). However, if $a_n(f) > 0 \ \forall \ n$, then the normalization factor $S_{\max}$ for N odd is

$$\boxed{S_{\max} = \max|S(f, f_X)| = \sum_{n=-N'}^{N'} a_n(f)} \tag{6A-14}$$

Note that if (6A-10) is evaluated at broadside (i.e., at $f_X = 0$) and if no phase weighting is done, then

$$S(f, 0) = \sum_{n=-N'}^{N'} a_n(f) \tag{6A-15}$$

and

$$|S(f, 0)| = \left|\sum_{n=-N'}^{N'} a_n(f)\right| \le \sum_{n=-N'}^{N'} |a_n(f)|. \tag{6A-16}$$

If $a_n(f) > 0 \;\; \forall \;\; n$, then

$$|S(f,0)| = \left| \sum_{n=-N'}^{N'} a_n(f) \right| = \sum_{n=-N'}^{N'} a_n(f), \tag{6A-17}$$

and by comparing (6A-14) and (6A-17),

$$\boxed{S_{\max} = \max|S(f, f_X)| = |S(f,0)|} \tag{6A-18}$$

Appendix 6B Transmitter and Receiver Sensitivity Functions of an Omnidirectional Point-Element

Transmitter Sensitivity Function

Manufacturers of electroacoustic transducers usually describe the *transmitting voltage response* of a transducer by plotting the *transmitter sensitivity level* $\text{TSL}(f)$ in $\text{dB re TS}_{\text{ref}}$ at 1 m versus frequency f in hertz. The notation "$\text{dB re TS}_{\text{ref}}$" means decibels relative to a reference transmitter sensitivity. In underwater acoustics, the reference transmitter sensitivity TS_{ref} is usually $1\,\mu\text{Pa/V}$. The corresponding *transmitter sensitivity* $\text{TS}(f)$ with units of Pa/V is given by

$$\boxed{\text{TS}(f) = \text{TS}_{\text{ref}}\, 10^{[\text{TSL}(f)/20]}} \tag{6B-1}$$

The transmitter sensitivity $\text{TS}(f)$ is a *real*, *nonnegative* function of frequency.

Values for the magnitude of the transmitter sensitivity function, $|\mathcal{S}_T(f)|$ in $\left(\text{m}^3/\text{sec}\right)/\text{V}$, for an omnidirectional point-source centered at the origin of a coordinate system, can be obtained from measured values of the transmitter sensitivity, $\text{TS}(f)$ in Pa/V, by using the following formula (see Appendix 6C):

$$\boxed{|\mathcal{S}_T(f)| = \left.\frac{2r}{f\rho_0}\text{TS}(f)\right|_{r=1\text{ m}}} \tag{6B-2}$$

where

$$r = \sqrt{x^2 + y^2 + z^2} \tag{6B-3}$$

is the distance from the center of the omnidirectional point-source to a field point,

ρ_0 is the constant ambient (equilibrium) density of the fluid medium in kilograms per cubic meter, and $\text{TS}(f)$ is given by (6B-1).

Receiver Sensitivity Function

Manufacturers of electroacoustic transducers usually describe the *open circuit receiving response* of a transducer by plotting the *receiver sensitivity level* $\text{RSL}(f)$ in dB re RS_{ref} versus frequency f in hertz. The notation "dB re RS_{ref}" means decibels relative to a reference receiver sensitivity. In underwater acoustics, the reference receiver sensitivity RS_{ref} is usually $1\,\text{V}/\mu\text{Pa}$. The corresponding *receiver sensitivity* $\text{RS}(f)$ with units of V/Pa is given by

$$\boxed{\text{RS}(f) = \text{RS}_{\text{ref}}\, 10^{[\text{RSL}(f)/20]}} \tag{6B-4}$$

The receiver sensitivity $\text{RS}(f)$ is a *real, nonnegative* function of frequency. What we need to do next is to derive the conversion factor that will allow us to convert m^2/sec to Pa for a time-harmonic acoustic field so that values for the magnitude of the receiver sensitivity function, $|\mathcal{S}_R(f)|$ in $\text{V}/(\text{m}^2/\text{sec})$, can be obtained from measured values of the receiver sensitivity $\text{RS}(f)$ in V/Pa.

The derivation of the desired conversion factor in this case is simple. By computing the magnitude of (6C-27), we obtain

$$|p_f(\mathbf{r})| = 2\pi f \rho_0 |\varphi_f(\mathbf{r})|, \tag{6B-5}$$

from which we obtain the conversion factor

$$\boxed{\frac{|p_f(\mathbf{r})|}{|\varphi_f(\mathbf{r})|} = 2\pi f \rho_0 \frac{\text{Pa}}{\text{m}^2/\text{sec}}} \tag{6B-6}$$

where $p_f(\mathbf{r}) \equiv p_f(x,y,z)$ and $\varphi_f(\mathbf{r}) \equiv \varphi_f(x,y,z)$. The conversion factor on the right-hand side of (6B-6) is used to convert velocity potential in m^2/sec to acoustic pressure in Pa for a time-harmonic acoustic field. Therefore, values for the magnitude of the receiver sensitivity function, $|\mathcal{S}_R(f)|$ in $\text{V}/(\text{m}^2/\text{sec})$, can be obtained from measured values of the receiver sensitivity, $\text{RS}(f)$ in V/Pa, by using the following formula:

$$\boxed{|\mathcal{S}_R(f)| = 2\pi f \rho_0 \text{RS}(f)} \tag{6B-7}$$

where ρ_0 is the constant ambient (equilibrium) density of the fluid medium in kilograms per cubic meter and $\mathrm{RS}(f)$ is given by (6B-4).

Appendix 6C Radiation from an Omnidirectional Point-Source

An exact solution of the linear wave equation given by (1.2-1) for free-space propagation in an ideal (nonviscous), homogeneous, fluid medium is given by [see (1.2-3)]

$$\varphi(t,\mathbf{r}) = \int_{-\infty}^{\infty} \int_{V_0} X(f,\mathbf{r}_0) A_T(f,\mathbf{r}_0) g_f\left(\mathbf{r}\,\middle|\,\mathbf{r}_0\right) dV_0 \exp(+j2\pi f t)\, df\,, \quad \text{(6C-1)}$$

where $\varphi(t,\mathbf{r})$ is the scalar velocity potential in squared meters per second, $X(f,\mathbf{r}_0)$ is the complex frequency spectrum of the input electrical signal at location $\mathbf{r}_0$ of the transmit aperture with units of volts per hertz, $A_T(f,\mathbf{r}_0)$ is the complex frequency response of the transmit aperture at $\mathbf{r}_0$ with units of $\left(\left(\mathrm{m}^3/\mathrm{sec}\right)/\mathrm{V}\right)/\mathrm{m}^3$,

$$g_f\left(\mathbf{r}\,\middle|\,\mathbf{r}_0\right) = -\frac{\exp\left(-jk\left|\mathbf{r}-\mathbf{r}_0\right|\right)}{4\pi\left|\mathbf{r}-\mathbf{r}_0\right|} = -\frac{\exp(-jkR)}{4\pi R} \quad \text{(6C-2)}$$

is the time-independent, free-space, Green's function of an unbounded, ideal (nonviscous), homogeneous, fluid medium with units of inverse meters,

$$k = 2\pi f/c = 2\pi/\lambda \quad \text{(6C-3)}$$

is the wavenumber in radians per meter, $c = f\lambda$ is the constant speed of sound in the fluid medium in meters per second, λ is the wavelength in meters,

$$\mathbf{r} = x\hat{x} + y\hat{y} + z\hat{z} \quad \text{(6C-4)}$$

is the position vector to a field point,

$$\mathbf{r}_0 = x_0\hat{x} + y_0\hat{y} + z_0\hat{z} \quad \text{(6C-5)}$$

is the position vector to a source point, and

$$R = \left|\mathbf{r}-\mathbf{r}_0\right| = \sqrt{(x-x_0)^2 + (y-y_0)^2 + (z-z_0)^2} \quad \text{(6C-6)}$$

is the range in meters between a source point and a field point. If an identical input electrical signal is applied at all locations $\mathbf{r}_0$ of the transmit aperture, then (see Example 1.2-1)

$$x(t, \mathbf{r}_0) = x(t) \tag{6C-7}$$

and

$$X(f, \mathbf{r}_0) = X(f), \tag{6C-8}$$

where $X(f)$ is the complex frequency spectrum of $x(t)$. Substituting (6C-8) into (6C-1) yields

$$\varphi(t, \mathbf{r}) = \int_{-\infty}^{\infty} X(f) \int_{V_0} A_T(f, \mathbf{r}_0) g_f(\mathbf{r}|\mathbf{r}_0) dV_0 \exp(+j2\pi ft) df. \tag{6C-9}$$

If the transmit aperture is an omnidirectional point-source, then the complex frequency response of the transmit aperture is given by

$$A_T(f, \mathbf{r}_0) = \mathcal{S}_T(f)\delta(\mathbf{r}_0 - \mathbf{r}_S), \tag{6C-10}$$

where $\mathcal{S}_T(f)$ is the complex, *transmitter sensitivity function* of the transducer in $(\text{m}^3/\text{sec})/\text{V}$, and the impulse function $\delta(\mathbf{r}_0 - \mathbf{r}_S)$, with units of inverse cubic meters, represents a unit-amplitude, omnidirectional point-source at $\mathbf{r}_S = (x_S, y_S, z_S)$. Note that $\int_{V_0} A_T(f, \mathbf{r}_0) dV_0 = \mathcal{S}_T(f)$ because $\int_{V_0} \delta(\mathbf{r}_0 - \mathbf{r}_S) dV_0 = 1$. This result is consistent with the results obtained in Examples 2B-1 and 3B-1, that is, integrating the complex frequency response of a transducer over its spatial coordinates yields the complex sensitivity function of the transducer. Substituting (6C-10) into (6C-9) yields

$$\boxed{\varphi(t, \mathbf{r}) = \int_{-\infty}^{\infty} X(f) \mathcal{S}_T(f) g_f(\mathbf{r}|\mathbf{r}_S) \exp(+j2\pi ft) df} \tag{6C-11}$$

where

$$g_f(\mathbf{r}|\mathbf{r}_S) = -\frac{\exp(-jk|\mathbf{r} - \mathbf{r}_S|)}{4\pi|\mathbf{r} - \mathbf{r}_S|} = -\frac{\exp(-jkR)}{4\pi R}, \tag{6C-12}$$

$$\mathbf{r}_S = x_S\hat{x} + y_S\hat{y} + z_S\hat{z} \tag{6C-13}$$

is the position vector to the point-source, and

$$R = |\mathbf{r} - \mathbf{r}_S| = \sqrt{(x - x_S)^2 + (y - y_S)^2 + (z - z_S)^2} \tag{6C-14}$$

is the range in meters between the point-source and a field point. Equation (6C-11) is a general expression for the scalar velocity potential of the acoustic field radiated by an omnidirectional point-source at $\mathbf{r}_S = (x_S, y_S, z_S)$. The corresponding solution for the radiated acoustic pressure $p(t, \mathbf{r})$ in pascals can be obtained by substituting (6C-11) into (1.2-8). Doing so yields

$$\boxed{p(t,\mathbf{r}) = -j2\pi\rho_0 \int_{-\infty}^{\infty} f X(f) \mathcal{S}_T(f) g_f(\mathbf{r}\,|\,\mathbf{r}_S) \exp(+j2\pi f t)\, df} \tag{6C-15}$$

where ρ_0 is the constant ambient (equilibrium) density of the fluid medium in kilograms per cubic meter.

If the input electrical signal is time-harmonic, that is, if (see Example 1.2-1)

$$x(t) = A_x \exp(+j2\pi f_0 t), \tag{6C-16}$$

where A_x is a complex amplitude with units of volts, then

$$X(f) = A_x \delta(f - f_0), \tag{6C-17}$$

where the impulse function $\delta(f - f_0)$ has units of inverse hertz. Substituting (6C-17) into (6C-11) yields

$$\varphi(t,\mathbf{r}) = A_x \mathcal{S}_T(f_0) g_{f_0}(\mathbf{r}\,|\,\mathbf{r}_S) \exp(+j2\pi f_0 t), \tag{6C-18}$$

and by replacing f_0 with f,

$$\boxed{\varphi(t,\mathbf{r}) = A_x \mathcal{S}_T(f) g_f(\mathbf{r}\,|\,\mathbf{r}_S) \exp(+j2\pi f t)} \tag{6C-19}$$

or

$$\varphi(t,\mathbf{r}) = \varphi_f(\mathbf{r}) \exp(+j2\pi f t), \tag{6C-20}$$

where

$$\varphi_f(\mathbf{r}) = A_x \mathcal{S}_T(f) g_f(\mathbf{r}\,|\,\mathbf{r}_S). \tag{6C-21}$$

Equation (6C-19) is the time-harmonic, scalar velocity potential of the acoustic field radiated by an omnidirectional point-source at $\mathbf{r}_S = (x_S, y_S, z_S)$. The corresponding solution for the time-harmonic, radiated acoustic pressure $p(t, \mathbf{r})$ in pascals is given by

$$\boxed{p(t,\mathbf{r}) = -j2\pi f \rho_0 A_x \mathcal{S}_T(f) g_f(\mathbf{r}\,|\,\mathbf{r}_S) \exp(+j2\pi f t)} \tag{6C-22}$$

or

$$p(t,\mathbf{r}) = p_f(\mathbf{r})\exp(+j2\pi ft), \tag{6C-23}$$

where

$$p_f(\mathbf{r}) = -j2\pi f\rho_0 A_x \mathcal{S}_T(f) g_f(\mathbf{r}|\mathbf{r}_S), \tag{6C-24}$$

or

$$p_f(\mathbf{r}) = -j2\pi f\rho_0 S_0 g_f(\mathbf{r}|\mathbf{r}_S) \tag{6C-25}$$

where

$$\boxed{S_0 = A_x \mathcal{S}_T(f)} \tag{6C-26}$$

is the source strength of the omnidirectional point-source in cubic meters per second at frequency f hertz. Note that

$$p_f(\mathbf{r}) = -j2\pi f\rho_0 \varphi_f(\mathbf{r}), \tag{6C-27}$$

where $\varphi_f(\mathbf{r})$ is given by (6C-21).

From (6C-23) it can be seen that

$$|p(t,\mathbf{r})| = |p_f(\mathbf{r})|, \tag{6C-28}$$

where from (6C-24)

$$|p_f(\mathbf{r})| = \frac{f\rho_0}{2R}|A_x||\mathcal{S}_T(f)|. \tag{6C-29}$$

Therefore, using (6C-29) to solve for $|\mathcal{S}_T(f)|$ yields

$$\boxed{|\mathcal{S}_T(f)| = \frac{2R}{f\rho_0}\text{TS}(f)} \tag{6C-30}$$

where $\mathcal{S}_T(f)$ is the transmitter sensitivity function in $\left(\text{m}^3/\text{sec}\right)/\text{V}$,

$$\boxed{\text{TS}(f) = |p_f(x,y,z)|/|A_x|} \tag{6C-31}$$

is the *transmitter sensitivity* in Pa/V, and the range R is given by (6C-14). If the omnidirectional point-source is placed at the origin of the coordinate system, that is, if $\mathbf{r}_S = \mathbf{0} = (0,0,0)$, then R given by (6C-14) reduces to

$$R = |\mathbf{r}| = r, \tag{6C-32}$$

where

$$r = \sqrt{x^2 + y^2 + z^2}. \tag{6C-33}$$

Evaluating (6C-30) at $R = r = 1\text{ m}$ yields

$$\boxed{|\mathcal{S}_T(f)| = \frac{2r}{f\rho_0}\mathrm{TS}(f)\Big|_{r=1\text{ m}}} \tag{6C-34}$$

where the range r is given by (6C-33). Note that the range R in (6C-30) and the range r in (6C-34) can be measured in any direction from the center of the omnidirectional point-source.

The magnitude of the source strength can be obtained from (6C-25) as follows:

$$\boxed{|S_0| = \frac{2R}{f\rho_0}|p_f(x,y,z)|} \tag{6C-35}$$

where the range R is given by (6C-14). If the omnidirectional point-source is placed at the origin of the coordinate system so that $R = r$ [see (6C-32)], then evaluating (6C-35) at $R = r = 1\text{ m}$ yields

$$\boxed{|S_0| = \frac{2r}{f\rho_0}|p_f(x,y,z)|\Big|_{r=1\text{ m}}} \tag{6C-36}$$

where the range r is given by (6C-33). Also, from (6C-26),

$$\boxed{|S_0| = |A_x||\mathcal{S}_T(f)|} \tag{6C-37}$$

Let us conclude our discussion in this Appendix by deriving the equation for the source distribution $x_M(t,\mathbf{r})$ for a time-harmonic, omnidirectional point-source. The source distribution $x_M(t,\mathbf{r})$, with units of inverse seconds, appears on the right-hand side of the linear wave equation given by (1.2-1). Since the complex frequency spectrum of the source distribution is given by (see Subsection 1.1.1)

$$X_M(f, \mathbf{r}_0) = X(f, \mathbf{r}_0) A_T(f, \mathbf{r}_0), \tag{6C-38}$$

substituting (6C-8), (6C-10), and (6C-17) into (6C-38) yields

$$X_M(f, \mathbf{r}_0) = A_x \mathcal{S}_T(f) \delta(f - f_0) \delta(\mathbf{r}_0 - \mathbf{r}_S). \tag{6C-39}$$

Taking the inverse Fourier transform of (6C-39) yields

$$x_M(t, \mathbf{r}_0) = A_x \mathcal{S}_T(f_0) \delta(\mathbf{r}_0 - \mathbf{r}_S) \exp(+j2\pi f_0 t), \tag{6C-40}$$

and by replacing $\mathbf{r}_0$ with $\mathbf{r}$, and f_0 with f, we obtain

$$\boxed{x_M(t, \mathbf{r}) = A_x \mathcal{S}_T(f) \delta(\mathbf{r} - \mathbf{r}_S) \exp(+j2\pi f t)} \tag{6C-41}$$

or

$$\boxed{x_M(t, \mathbf{r}) = S_0 \delta(\mathbf{r} - \mathbf{r}_S) \exp(+j2\pi f t)} \tag{6C-42}$$

where the source strength S_0 is given by (6C-26). Therefore, (6C-19) is an exact solution of the linear wave equation given by (1.2-1) when the source distribution is given by (6C-41), or equivalently, by (6C-42).

Appendix 6D One-Dimensional Spatial FIR Filters

N Even

The Product Theorem for a linear array composed of an even number of identical elements is given by [see (6.1-10)]

$$D(f, f_X) = E(f, f_X) S(f, f_X), \tag{6D-1}$$

where $D(f, f_X)$ is the far-field beam pattern of the array, $E(f, f_X)$ is the far-field beam pattern of *one* of the identical elements in the array, and

$$S(f, f_X) = \sum_{n=1}^{N/2} \left[c_{-n}(f) \exp(+j2\pi f_X x_{-n}) + c_n(f) \exp(+j2\pi f_X x_n) \right] \tag{6D-2}$$

is the dimensionless array factor [see (6.1-12)]. Taking the inverse spatial Fourier transform of the Product Theorem given by (6D-1) yields

$$A(f, x_A) = e(f, x_A) \underset{x_A}{*} s(f, x_A), \tag{6D-3}$$

where

$$s(f,x_A)=F_{f_X}^{-1}\{S(f,f_X)\}=\sum_{n=1}^{N/2}\left[c_{-n}(f)\delta(x_A-x_{-n})+c_n(f)\delta(x_A-x_n)\right] \quad \text{(6D-4)}$$

since

$$F_{f_X}^{-1}\{\exp(+j2\pi f_X x_n)\}=\delta(x_A-x_n). \quad \text{(6D-5)}$$

If the elements are omnidirectional point-elements, then the element function $e(f,x_A)$ is given by [see (6.1-14)]

$$e(f,x_A)=\mathcal{S}(f)\delta(x_A), \quad \text{(6D-6)}$$

where $\mathcal{S}(f)$ is the element sensitivity function. If we set $\mathcal{S}(f)=1\left(\text{m}^3/\text{sec}\right)/\text{V}$ or $\mathcal{S}(f)=1\,\text{V}/\left(\text{m}^2/\text{sec}\right)$, then (6D-6) reduces to

$$e(f,x_A)=\delta(x_A). \quad \text{(6D-7)}$$

Substituting the "input signal" given by (6D-7) into (6D-3) yields the "output signal"

$$A(f,x_A)=\delta(x_A)\underset{x_A}{*}s(f,x_A)=s(f,x_A), \quad \text{(6D-8)}$$

where $s(f,x_A)$ given by (6D-4) can be thought of as the spatial impulse response of the array at frequency f hertz. Since (6D-4) is in the form of the impulse response of a finite impulse response (FIR) digital filter, a linear array composed of an even number of identical, complex-weighted elements can be thought of as a one-dimensional, spatial FIR filter.

N Odd

The Product Theorem for a linear array composed of an odd number of identical elements is given by [see (6.1-86)]

$$D(f,f_X)=E(f,f_X)S(f,f_X), \quad \text{(6D-9)}$$

where $D(f,f_X)$ is the far-field beam pattern of the array, $E(f,f_X)$ is the far-field beam pattern of *one* of the identical elements in the array, and

$$S(f,f_X)=\sum_{n=-N'}^{N'}c_n(f)\exp(+j2\pi f_X x_n) \quad \text{(6D-10)}$$

is the dimensionless array factor [see (6.1-87)]. Taking the inverse spatial Fourier transform of the Product Theorem given by (6D-9) yields

$$A(f, x_A) = e(f, x_A) \underset{x_A}{*} s(f, x_A), \tag{6D-11}$$

where, with the use of (6D-5),

$$s(f, x_A) = F_{f_X}^{-1}\{S(f, f_X)\} = \sum_{n=-N'}^{N'} c_n(f)\delta(x_A - x_n). \tag{6D-12}$$

Substituting the "input signal" given by (6D-7) into (6D-11) yields the "output signal"

$$A(f, x_A) = \delta(x_A) \underset{x_A}{*} s(f, x_A) = s(f, x_A), \tag{6D-13}$$

where $s(f, x_A)$ given by (6D-12) can be thought of as the spatial impulse response of the array at frequency f hertz. Since (6D-12) is in the form of the impulse response of a finite impulse response (FIR) digital filter, a linear array composed of an odd number of identical, complex-weighted elements can be thought of as a one-dimensional, spatial FIR filter.

Appendix 6E Far-Field Beam Patterns and the Spatial Discrete Fourier Transform for N Even

The far-field beam pattern of a linear array composed of an even number N of identical, equally-spaced, complex-weighted, omnidirectional point-elements lying along the X axis is given by [see (6.1-17)]

$$D(f, f_X) = \mathcal{S}(f) S(f, f_X), \tag{6E-1}$$

where $\mathcal{S}(f)$ is the complex, element sensitivity function, and by substituting (6.1-20) and (6.1-21) into (6.1-12),

$$S(f, f_X) = \sum_{n=1}^{N/2} \left[c_{-n}(f) \exp[-j2\pi f_X (n-0.5)d] + c_n(f) \exp[+j2\pi f_X (n-0.5)d] \right], \tag{6E-2}$$

or

$$\begin{aligned} S(f, f_X) = {} & \exp(+j\pi f_X d) \sum_{n=1}^{N/2} c_{-n}(f) \exp(-j2\pi f_X nd) + \\ & \exp(-j\pi f_X d) \sum_{n=1}^{N/2} c_n(f) \exp(+j2\pi f_X nd), \end{aligned} \tag{6E-3}$$

where d is the interelement spacing in meters. Therefore, substituting (6E-3) into (6E-1) yields

$$D(f, f_X) = \mathcal{S}(f)\exp(+j\pi f_X d)\sum_{n=1}^{N/2} c_{-n}(f)\exp(-j2\pi f_X nd) + \mathcal{S}(f)\exp(-j\pi f_X d)\sum_{n=1}^{N/2} c_n(f)\exp(+j2\pi f_X nd). \tag{6E-4}$$

The far-field beam pattern given by (6E-4) can be expressed in terms of two spatial discrete Fourier transforms (DFTs) by appropriately discretizing spatial frequency f_X. Since both summations on the right-hand side of (6E-4) involve only $N/2$ complex weights, f_X is discretized as follows:

$$f_X = m\Delta f_X, \qquad m = -N'', \ldots, 0, \ldots, N'', \tag{6E-5}$$

where

$$\Delta f_X = \frac{1}{(N'+Z)d} \le \frac{1}{N'd}, \tag{6E-6}$$

$$N' = N/2, \tag{6E-7}$$

and

$$N'' = \begin{cases} (N'+Z)/2, & N'+Z \text{ even}, \\ (N'+Z-1)/2, & N'+Z \text{ odd}. \end{cases} \tag{6E-8}$$

Compare (6E-6) and (6E-8) with (6.5-5) and (6.5-6), respectively, for N odd. Substituting (6E-5) through (6E-7) into (6E-4) yields

$$\boxed{D(f, m) = \mathcal{S}(f)\left[W_{N'+Z}^{m/2} \sum_{n=1}^{N'} c_{-n}(f) W_{N'+Z}^{-mn} + W_{N'+Z}^{-m/2} \sum_{n=1}^{N'} c_n(f) W_{N'+Z}^{mn} \right], \quad m = -N'', \ldots, 0, \ldots, N''} \tag{6E-9}$$

where [see (6.5-8)]

$$W_{N'+Z} = \exp\left(+j\frac{2\pi}{N'+Z} \right). \tag{6E-10}$$

The first summation on the right-hand side of (6E-9) is the spatial DFT of the set of $N/2$ complex weights along the negative X axis, whereas the second summation is the spatial DFT of the set of $N/2$ complex weights along the positive X axis. Both spatial DFTs are periodic with period $N'+Z$ because

$$
\begin{aligned}
W_{N'+Z}^{\pm(m+N'+Z)n} &= W_{N'+Z}^{\pm mn} W_{N'+Z}^{\pm(N'+Z)n} \\
&= W_{N'+Z}^{\pm mn} \exp(\pm j2n\pi) \\
&= W_{N'+Z}^{\pm mn}.
\end{aligned}
\tag{6E-11}
$$

The relationship between bin number m and the corresponding value of direction cosine u_m is given by [see (6.5-13)]

$$
u_m = \frac{m}{N'+Z}\frac{\lambda}{d}, \qquad m = -N'', \ldots, 0, \ldots, N'' \tag{6E-12}
$$

where N' is given by (6E-7) and N'' is given by (6E-8). If we let δu be the direction-cosine bin spacing, then by using (6E-12)

$$
\delta u = u_m - u_{m-1} = \frac{1}{N'+Z}\frac{\lambda}{d}. \tag{6E-13}
$$

Solving for Z from (6E-13) yields

$$
Z = \frac{1}{\delta u}\frac{\lambda}{d} - N' \tag{6E-14}
$$

Equation (6E-14) is used to solve for the number of zeros Z that are required in order to obtain a desired direction-cosine bin spacing δu.

And finally, the normalized, far-field beam pattern $D_N(f, m)$ is given by

$$
D_N(f, m) = D(f, m)/D_{\max}, \tag{6E-15}
$$

where $D(f, m)$ is the unnormalized, far-field beam pattern given by (6E-9), and if $a_{-n}(f) > 0$ and $a_n(f) > 0 \;\; \forall \; n$, then

$$
D_{\max} = \max|D(f, m)| = |S(f)| \sum_{n=1}^{N/2} \left[a_{-n}(f) + a_n(f)\right] \tag{6E-16}
$$

is the normalization factor [use the procedure in Appendix 6A for N even to obtain (6E-16)]. Figure 6E-1 is a plot of the magnitude of (6E-15) using rectangular amplitude weights, $N = 6$, $Z = 125$, and (6E-12) with $d = \lambda/2$. With $N = 6$ and $Z = 125$, $N' = 3$ and $N' + Z = 128 = 2^7$.

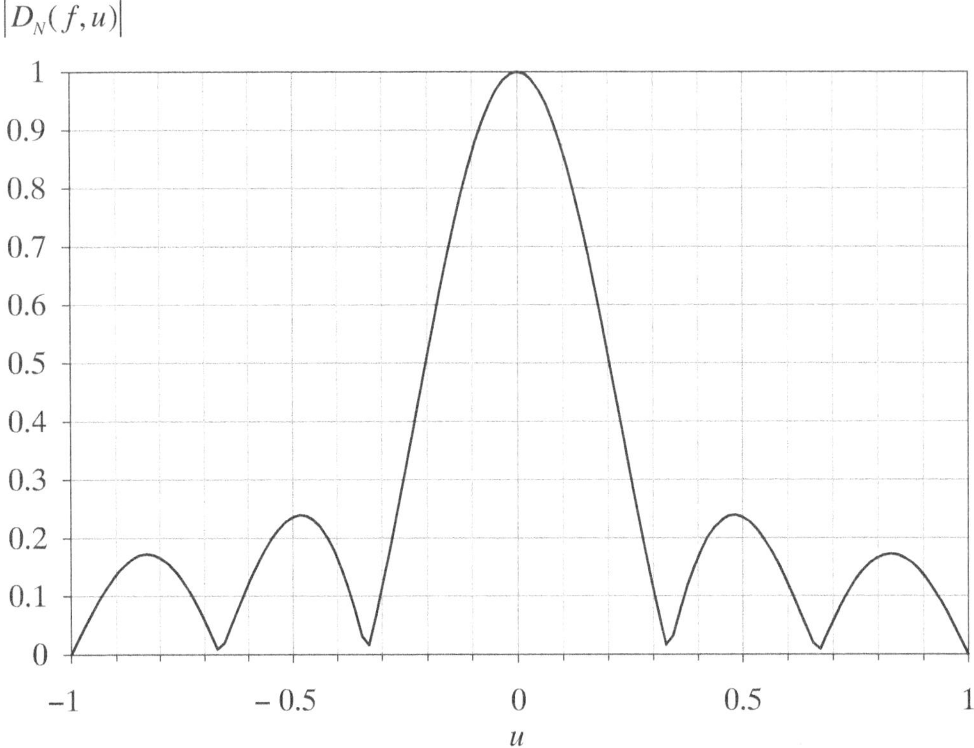

Figure 6E-1 Magnitude of the normalized, far-field beam pattern of a linear array of 6 identical, equally-spaced, rectangular amplitude-weighted, omni-directional point-elements lying along the X axis using a spatial DFT and plotted as a function of direction cosine u.

Chapter 7

Array Gain

Using an array of electroacoustic transducers (elements) in the passive mode as a receiver can increase the probability of detecting very weak signals corrupted by additive noise. This is especially important in underwater acoustics when trying to detect sound fields radiated by potential targets of interest such as submarines, surface ships, and marine mammals (e.g., whales and dolphins). Sound radiated by marine life is referred to as biologics. Besides detection, it is also very important to be able to classify (identify) sound-sources (targets).

An array has its own output signal-to-noise ratio (SNR) that can be very large compared to the SNR at the output from a single element in an array. Array gain (AG) is simply a decibel measure of the increase in the output SNR that is obtained by using an array of elements versus a single element. Increasing the AG increases the chances of both detection and classification.

7.1 General Definition of Array Gain for a Linear Array

In this section we shall define array gain (AG) for a linear array composed of an odd number N of identical, equally-spaced, complex-weighted, omnidirectional point-elements lying along the X axis. The array is being used in the passive mode as a receiver. The received signal at the output of each element in the array is a *random process* and is equal to the sum of the output signals due to the acoustic field radiated by a target, the ambient (background) noise in the ocean, and receiver noise. Therefore, the received electrical signal (in volts) at the output of element i in the array *before* and *after* complex weighting can be expressed as

$$r'(t, x_i) = y'_{Trgt}(t, x_i) + y'_{n_a}(t, x_i) + n'_r(t, x_i) \tag{7.1-1}$$

and

$$r(t, x_i) = y_{Trgt}(t, x_i) + y_{n_a}(t, x_i) + n_r(t, x_i), \tag{7.1-2}$$

respectively, where $y'_{Trgt}(t, x_i)$, $y'_{n_a}(t, x_i)$, and $n'_r(t, x_i)$ are the output electrical signals from element i due to the target, ambient noise, and receiver noise, respectively, *before* complex weighting; and $y_{Trgt}(t, x_i)$, $y_{n_a}(t, x_i)$, and $n_r(t, x_i)$ are the output electrical signals from element i due to the target, ambient noise, and receiver noise, respectively, *after* complex weighting, where x_i is the x coordinate of the center of element i (see Fig. 7.1-1). As was discussed in Section 6.4, the procedure illustrated in Fig. 7.1-1 is known as beamforming. In practical

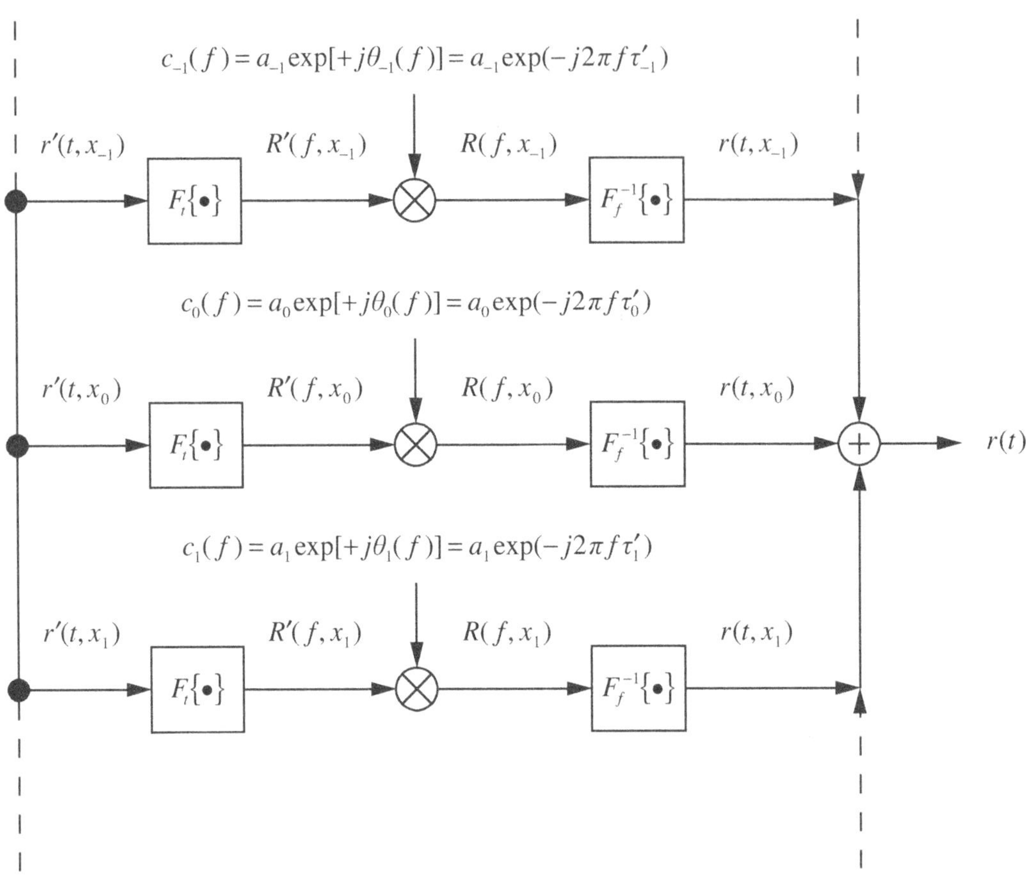

Figure 7.1-1 Beamforming for a linear array. Implementing complex weights (amplitude and phase weights) using forward and inverse Fourier transforms.

applications, the forward and inverse Fourier transforms are computed using forward and inverse *discrete Fourier transforms* (DFTs), which can be evaluated very quickly by using forward and inverse *fast-Fourier-transform* (FFT) algorithms. This is known as digital beamforming or FFT beamforming.

The *total* output electrical signal from the array (in volts) is given by

$$r(t) = \sum_{i=-N'}^{N'} r(t, x_i), \tag{7.1-3}$$

and by substituting (7.1-2) into (7.1-3), we obtain

$$r(t) = y_{Trgt}(t) + y_{n_a}(t) + n_r(t), \tag{7.1-4}$$

where

$$y_{Trgt}(t) = \sum_{i=-N'}^{N'} y_{Trgt}(t, x_i), \tag{7.1-5}$$

$$y_{n_a}(t) = \sum_{i=-N'}^{N'} y_{n_a}(t, x_i), \tag{7.1-6}$$

and

$$n_r(t) = \sum_{i=-N'}^{N'} n_r(t, x_i) \tag{7.1-7}$$

are the *total* output electrical signals from the array due to the target, ambient noise, and receiver noise, respectively, where

$$N' = (N-1)/2. \tag{7.1-8}$$

If we let

$$z(t) = y_{n_a}(t) + n_r(t), \tag{7.1-9}$$

then the *output signal-to-noise ratio of the array* is defined as follows:

$$\boxed{\mathrm{SNR}_\mathrm{A} \triangleq \frac{E\left\{\left|y_{Trgt}(t)\right|^2\right\}}{E\left\{\left|z(t)\right|^2\right\}}} \tag{7.1-10}$$

where the second moment (mean-squared value)

$$E\left\{\left|y_{Trgt}(t)\right|^2\right\} = \sum_{i=-N'}^{N'} \sum_{j=-N'}^{N'} E\left\{y_{Trgt}(t, x_i)\, y_{Trgt}^*(t, x_j)\right\} \tag{7.1-11}$$

is the average signal power at the output of the array, and

$$E\left\{\left|z(t)\right|^2\right\} = E\left\{\left|y_{n_a}(t)\right|^2\right\} + 2\,\mathrm{Re}\left\{E\left\{y_{n_a}(t)\, n_r^*(t)\right\}\right\} + E\left\{\left|n_r(t)\right|^2\right\} \tag{7.1-12}$$

is the average noise power at the output of the array, where

$$E\left\{\left|y_{n_a}(t)\right|^2\right\} = \sum_{i=-N'}^{N'} \sum_{j=-N'}^{N'} E\left\{y_{n_a}(t, x_i)\, y_{n_a}^*(t, x_j)\right\} \tag{7.1-13}$$

and

$$E\left\{\left|n_r(t)\right|^2\right\} = \sum_{i=-N'}^{N'} \sum_{j=-N'}^{N'} E\left\{n_r(t, x_i)\, n_r^*(t, x_j)\right\} \tag{7.1-14}$$

are the average noise powers at the output of the array due to ambient noise and receiver noise, respectively. Assuming that $y_{n_a}(t)$ and $n_r(t)$ are *statistically independent*, *zero-mean* random processes,

$$E\left\{y_{n_a}(t)n_r^*(t)\right\}=E\left\{y_{n_a}(t)\right\}E\left\{n_r^*(t)\right\}=0\,, \tag{7.1-15}$$

and as a result, (7.1-12) reduces to

$$E\left\{|z(t)|^2\right\}=E\left\{|y_{n_a}(t)|^2\right\}+E\left\{|n_r(t)|^2\right\}. \tag{7.1-16}$$

The average powers are functions of time in general. However, if $y_{Trgt}(t)$, $y_{n_a}(t)$, and $n_r(t)$ are *wide-sense stationary* (WSS) random processes, then the average powers are *constants*. The expected value (ensemble average) $E\left\{y_{Trgt}(t,x_i)y_{Trgt}^*(t,x_j)\right\}$ is the *cross-correlation* of $y_{Trgt}(t,x_i)$ and $y_{Trgt}(t,x_j)$, $E\left\{y_{n_a}(t,x_i)y_{n_a}^*(t,x_j)\right\}$ is the *cross-correlation* of $y_{n_a}(t,x_i)$ and $y_{n_a}(t,x_j)$, and $E\left\{n_r(t,x_i)n_r^*(t,x_j)\right\}$ is the *cross-correlation* of $n_r(t,x_i)$ and $n_r(t,x_j)$. The SNR_A given by (7.1-10) is dimensionless.

In a similar manner, the *output signal-to-noise ratio at the center element in the array* (element $i=0$) is defined as follows:

$$\boxed{\text{SNR}_0 \triangleq \frac{E\left\{|y_{Trgt}(t,x_0)|^2\right\}}{E\left\{|z(t,x_0)|^2\right\}}} \tag{7.1-17}$$

where $E\left\{|y_{Trgt}(t,x_0)|^2\right\}$ is the average signal power, and assuming that $y_{n_a}(t,x_0)$ and $n_r(t,x_0)$ are statistically independent, zero-mean random processes,

$$E\left\{|z(t,x_0)|^2\right\}=E\left\{|y_{n_a}(t,x_0)|^2\right\}+E\left\{|n_r(t,x_0)|^2\right\} \tag{7.1-18}$$

is the average noise power, where $E\left\{|y_{n_a}(t,x_0)|^2\right\}$ and $E\left\{|n_r(t,x_0)|^2\right\}$ are the average noise powers due to ambient noise and receiver noise, respectively, all at the output of element $i=0$ *after* complex weighting. The average powers are functions of time in general. However, if $y_{Trgt}(t,x_0)$, $y_{n_a}(t,x_0)$, and $n_r(t,x_0)$ are WSS in time, then the average powers are constants. The SNR_0 given by (7.1-17) is dimensionless.

Array gain (AG) is defined as follows:

$$\boxed{\text{AG} \triangleq 10\log_{10}\frac{\text{SNR}_\text{A}}{\text{SNR}_0}\ \text{dB}} \tag{7.1-19}$$

where SNR_A is given by (7.1-10) and SNR_0 is given by (7.1-17). As can be seen from (7.1-19), AG is a decibel measure of the increase in the output signal-to-noise ratio that is obtained by using an array of elements versus a single element. Array gain depends on the statistical properties of the signal (the acoustic field radiated by a target), ambient noise, and receiver noise. Therefore, in general, the same array will have different AG values when placed in different signal and ambient noise sound fields. In order to evaluate the equation for AG as given by (7.1-19) (see Section 7.5), we need mathematical models for the acoustic field radiated by a target (see Section 7.2), the total output signal from a linear array due to the target (see Section 7.3), and the total output signal from a linear array due to ambient noise in the ocean and receiver noise (see Section 7.4).

7.2 Acoustic Field Radiated by a Target

A mathematical model for the acoustic field radiated by a sound-source (target) depends on the models used for both the target and the fluid medium. The model that we shall use for the target is the same as that used in Subsection 1.2.2 – the target is modeled as an omnidirectional point-source located at $\mathbf{r}_S = (x_S, y_S, z_S)$ with arbitrary time dependence $s_0(t)$, where $s_0(t)$, $t \geq 0$, is the source strength of the target in cubic meters per second (see Fig. 1.2-5). In this chapter, $s_0(t)$ is assumed to be a *wide-sense stationary* (WSS) *random process*. The model that we shall *initially* use for the fluid medium is also the same as that used in Subsection 1.2.2 – the fluid medium is unbounded, ideal (nonviscous), and homogeneous (constant speed of sound). However, later in this section we shall generalize the fluid-medium model to include viscosity (resistance to fluid flow).

The receive aperture in Subsection 1.2.2 was a closed-surface, volume aperture. The receive aperture in this chapter is a linear array, as described in Section 7.1. Since a linear array is an example of a linear aperture, we shall first obtain results for a receive linear aperture lying along the X axis. By applying the receive volume aperture results obtained in Subsection 1.2.2 to a receive linear aperture lying along the X axis, the position vector to an aperture point, $\mathbf{r}_R$, is given by

$$\mathbf{r}_R = x_R \hat{x}, \tag{7.2-1}$$

and the velocity potential in squared meters per second of the output acoustic signal from the fluid medium due to the target and incident upon the aperture at x_R is given by

$$y_{M,Trgt}(t, x_R) = -\frac{1}{4\pi R} s_0(t-\tau), \tag{7.2-2}$$

for $t \geq \tau$, where

$$\begin{aligned} R &= |\mathbf{r}_R - \mathbf{r}_S| \\ &= \sqrt{(x_R - x_S)^2 + (y_R - y_S)^2 + (z_R - z_S)^2} \\ &= \sqrt{(x_R - x_S)^2 + y_S^2 + z_S^2} \end{aligned} \tag{7.2-3}$$

is the range in meters between the target and an aperture point,

$$\tau = R/c \tag{7.2-4}$$

is the corresponding *one-way* travel time or *one-way* time delay in seconds, and c is the constant speed of sound in the fluid medium in meters per second. Equation (7.2-2) is the velocity potential of a spherical wave with arbitrary time dependence propagating in an unbounded, ideal (nonviscous), homogeneous, fluid medium.

The position vector to the sound-source (target) is given by either

$$\mathbf{r}_S = x_S \hat{x} + y_S \hat{y} + z_S \hat{z} \tag{7.2-5}$$

or

$$\mathbf{r}_S = r_S \hat{r}_S, \tag{7.2-6}$$

where

$$r_S = |\mathbf{r}_S| = \sqrt{x_S^2 + y_S^2 + z_S^2} \tag{7.2-7}$$

is the range to the target in meters as measured from the origin of the coordinate system (the center of the array) (see Fig. 1.2-5),

$$\hat{r}_S = u_S \hat{x} + v_S \hat{y} + w_S \hat{z} \tag{7.2-8}$$

is the unit vector in the direction of $\mathbf{r}_S$, and

$$u_S = \sin\theta_S \cos\psi_S, \tag{7.2-9}$$

$$v_S = \sin\theta_S \sin\psi_S, \tag{7.2-10}$$

$$w_S = \cos\theta_S \tag{7.2-11}$$

are dimensionless direction cosines with respect to the X, Y, and Z axes, respectively (see Fig. 1.2-6). The spherical coordinates of the target's location

(r_S, θ_S, ψ_S) are *unknown* a priori.

All real fluids possess viscosity (resistance to fluid flow). For example, in ocean acoustics, seawater is correctly modeled as a viscous fluid with a frequency-dependent attenuation coefficient. Because of viscosity, sound is attenuated by conversion of its acoustic energy into heat. In order to take viscosity into account, we need to work in the frequency domain. Since both the target and aperture are not in motion, output (received) frequencies η are equal to input (transmitted) frequencies f since there is no Doppler shift nor bandwidth compression or expansion (see Subsection 1.1.2). Therefore, taking the Fourier transform of (7.2-2) with respect to time t yields

$$Y_{M,Trgt}(f, x_R) = -\frac{1}{4\pi R} S_0(f) \exp(-j2\pi f \tau), \tag{7.2-12}$$

and by substituting (7.2-3) and (7.2-4) into (7.2-12), we obtain

$$Y_{M,Trgt}(f, x_R) = S_0(f) g_f(\mathbf{r}_R | \mathbf{r}_S), \tag{7.2-13}$$

where $Y_{M,Trgt}(f, x_R)$ is the complex frequency spectrum of the acoustic field radiated by the target and incident upon the receive aperture at x_R, with units of $(\mathrm{m}^2/\mathrm{sec})/\mathrm{Hz}$; $S_0(f)$ is the complex frequency spectrum of the target's source strength in $(\mathrm{m}^3/\mathrm{sec})/\mathrm{Hz}$,

$$g_f(\mathbf{r}_R | \mathbf{r}_S) = -\frac{\exp(-jk|\mathbf{r}_R - \mathbf{r}_S|)}{4\pi|\mathbf{r}_R - \mathbf{r}_S|} \tag{7.2-14}$$

is the time-independent, free-space, Green's function of an unbounded, ideal (nonviscous), homogeneous, fluid medium with units of inverse meters,

$$k = 2\pi f / c = 2\pi / \lambda \tag{7.2-15}$$

is the wavenumber in radians per meter, and $c = f\lambda$.

If we next replace the real wavenumber k in (7.2-14) with the *complex wavenumber*

$$\boxed{K = k - j\alpha(f)} \tag{7.2-16}$$

where K has units of inverse meters and $\alpha(f)$ is the real, nonnegative, frequency-dependent, *attenuation coefficient* of the fluid medium in nepers (Np) per meter, then

$$g_f(\mathbf{r}_R|\mathbf{r}_S) = -\frac{\exp\left[-\alpha(f)|\mathbf{r}_R - \mathbf{r}_S|\right]}{4\pi|\mathbf{r}_R - \mathbf{r}_S|}\exp(-jk|\mathbf{r}_R - \mathbf{r}_S|) \tag{7.2-17}$$

Equation (7.2-17) is the time-independent, free-space, Green's function of an unbounded, *viscous*, homogeneous, fluid medium with units of inverse meters. The Green's function given by (7.2-17) is the complex frequency response of the fluid medium at frequency f hertz and location $\mathbf{r}_R$ due to the application of a unit-amplitude impulse (i.e., a unit-amplitude, omnidirectional point-source) at location $\mathbf{r}_S$. Substituting (7.2-17) into (7.2-13) yields

$$Y_{M,Trgt}(f, x_R) = -\frac{1}{4\pi|\mathbf{r}_R - \mathbf{r}_S|}S_0(f)\exp\left[-\alpha(f)|\mathbf{r}_R - \mathbf{r}_S|\right]\exp(-jk|\mathbf{r}_R - \mathbf{r}_S|), \tag{7.2-18}$$

or

$$Y_{M,Trgt}(f, x_R) = -\frac{1}{4\pi R}S_0(f)\exp[-\alpha(f)R]\exp(-j2\pi f\tau), \tag{7.2-19}$$

where the range R and time delay τ are given by (7.2-3) and (7.2-4), respectively. If we let

$$S_0'(f, x_R) = S_0(f)\exp[-\alpha(f)R] \tag{7.2-20}$$

then (7.2-19) can be rewritten as

$$Y_{M,Trgt}(f, x_R) = -\frac{1}{4\pi R}S_0'(f, x_R)\exp(-j2\pi f\tau). \tag{7.2-21}$$

Taking the inverse Fourier transform of (7.2-21) yields

$$y_{M,Trgt}(t, x_R) = -\frac{1}{4\pi R}s_0'(t - \tau, x_R) \tag{7.2-22}$$

for $t \geq \tau$, where

$$s_0'(t, x_R) = F_f^{-1}\{S_0'(f, x_R)\} \tag{7.2-23}$$

Note that $S_0'(f, x_R)$ and, hence, $s_0'(t, x_R)$ are functions of x_R because the range R is a function of x_R [see (7.2-3)]. Equation (7.2-22) is a mathematical model of the velocity potential (in squared meters per second) of the output acoustic signal from the fluid medium due to the target and incident upon the receive aperture at x_R that takes into account the viscosity of the fluid medium. It is the velocity potential of a spherical wave with arbitrary time dependence propagating in an unbounded, *viscous*, homogeneous, fluid medium. If there is no viscosity, that is, if $\alpha(f)=0$, then $s_0'(t, x_R) = s_0(t)$ and (7.2-22) reduces to (7.2-2). The mathematical model for the incident acoustic field given by (7.2-22), although simple, will allow us to discuss both near-field and far-field problems. As was discussed in Section 6.6, with the advent of very long, linear towed arrays, near-field issues cannot be ignored.

In the ocean acoustics literature, equations and values for the attenuation coefficient of seawater are usually given in decibels (dB) per meter or decibels per kilometer. For example, an accurate equation for the attenuation coefficient of seawater in decibels per kilometer is given in Appendix 7A. However, in order to evaluate decaying exponentials such as $\exp[-\alpha(f)R]$, only $\alpha(f)$ in nepers per meter can be used. Therefore, in order to convert the attenuation coefficient $\alpha'(f)$ in decibels per meter to $\alpha(f)$ in nepers per meter, use

$$\boxed{\alpha(f) = \frac{\alpha'(f)}{8.686} \frac{\text{Np}}{\text{m}}} \tag{7.2-24}$$

and to convert $\alpha(f)$ in nepers per meter to $\alpha'(f)$ in decibels per meter, use

$$\boxed{\alpha'(f) = 8.686\,\alpha(f) \frac{\text{dB}}{\text{m}}} \tag{7.2-25}$$

Figure 7.2-1 is a plot of the attenuation coefficient of seawater $\alpha'(f)$ in decibels per kilometer (dB/km) given by (7A-1) versus frequency f in kilohertz (kHz) for temperatures $T = 14°\,\text{C}$ ($57.2°\,\text{F}$) and $T = 20°\,\text{C}$ ($68°\,\text{F}$) at a depth of $50\,\text{m}$. A salinity of 35 ppt and an acidity $p\text{H} = 8$ was used.

With the use of (7.2-7), the range R between the target and an aperture point [see (7.2-3)]

$$R = \sqrt{(x_R - x_S)^2 + y_S^2 + z_S^2} \tag{7.2-26}$$

can be rewritten as

$$R = \sqrt{r_S^2 - 2x_S x_R + x_R^2}\,. \tag{7.2-27}$$

Since

$$x_S = r_S u_S\,, \tag{7.2-28}$$

substituting (7.2-28) into (7.2-27) yields

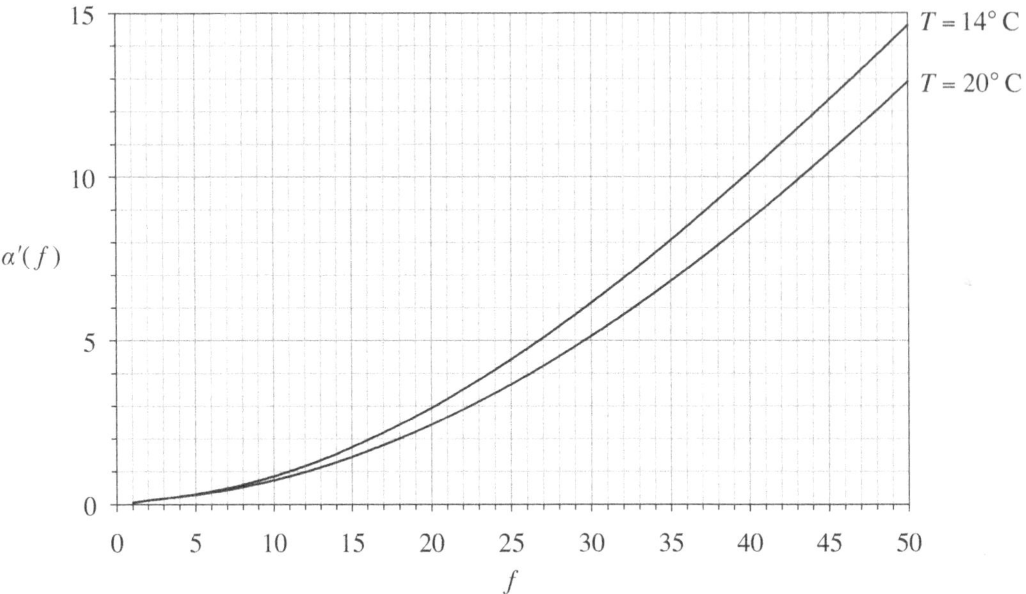

Figure 7.2-1 Attenuation coefficient of seawater $\alpha'(f)$ in decibels per kilometer (dB/km) given by (7A-1) versus frequency f in kilohertz (kHz) for temperatures $T = 14° \text{C}$ (57.2° F) and $T = 20° \text{C}$ (68° F) at a depth of 50 m. A salinity of 35 ppt and an acidity $p\text{H} = 8$ was used.

$$R = \sqrt{r_S^2 - 2r_S u_S x_R + x_R^2}\,, \tag{7.2-29}$$

where r_S is given by (7.2-7) and u_S is given by (7.2-9). Equations (7.2-26), (7.2-27), and (7.2-29) are all *exact* expressions for the range R that can be used to compute both the amplitude factor $1/R$ and time delay $\tau = R/c$ in (7.2-22). For example, if (7.2-29) is used, then

$$\frac{1}{R} = \frac{1}{\sqrt{r_S^2 - 2r_S u_S x_R + x_R^2}} \tag{7.2-30}$$

and

$$\tau = \frac{1}{c}\sqrt{r_S^2 - 2r_S u_S x_R + x_R^2}\,. \tag{7.2-31}$$

If the target is in the Fresnel region of the aperture, then

$$\frac{1}{R} \approx \frac{1}{r_S} \tag{7.2-32}$$

is a good approximation of the amplitude factor in (7.2-22) (see Subsection 1.2.2).

However, in order to obtain a good approximation of the time delay τ in (7.2-22), use (see Subsection 1.2.2)

$$R \approx r_S - \hat{r}_S \bullet \mathbf{r}_R + \frac{1}{2r_S} r_R^2 . \tag{7.2-33}$$

Substituting (7.2-1) and (7.2-8) into (7.2-33), and since $r_R^2 = |\mathbf{r}_R|^2 = x_R^2$, (7.2-33) reduces to

$$R \approx r_S - u_S x_R + \frac{1}{2r_S} x_R^2 , \tag{7.2-34}$$

so that

$$\tau \approx \tau_S - \frac{u_S}{c} x_R + \frac{1}{2 r_S c} x_R^2 , \tag{7.2-35}$$

where

$$\tau_S = r_S / c \tag{7.2-36}$$

is the *one-way* time delay in seconds associated with the path length between the sound-source (target) and the center of the aperture. The time delay τ_S is also *unknown* a priori because the range r_S is unknown a priori.

If the target is in the far-field region of the aperture, then (7.2-32) is also a good approximation of the amplitude factor in (7.2-22) (see Subsection 1.3.2). However, in order to obtain a good approximation of the time delay τ in (7.2-22), use (see Subsection 1.3.2)

$$R \approx r_S - \hat{r}_S \bullet \mathbf{r}_R , \tag{7.2-37}$$

or

$$R \approx r_S - u_S x_R , \tag{7.2-38}$$

so that

$$\tau \approx \tau_S - \frac{u_S}{c} x_R , \tag{7.2-39}$$

where τ_S is given by (7.2-36).

7.3 Total Output Signal from a Linear Array Due to the Target

By applying the receive volume aperture results obtained in Subsection 1.1.2 to a receive linear aperture lying along the X axis, and since the target and aperture are not in motion (output (received) frequencies η are equal to input (transmitted) frequencies f since there is no Doppler shift and no bandwidth compression or expansion), we can write that

$$Y_{Trgt}(f, x_R) = Y_{M,Trgt}(f, x_R) A_R(f, x_R), \tag{7.3-1}$$

where $Y_{Trgt}(f, x_R)$ is the complex frequency spectrum of the output electrical signal from the receive aperture at x_R due to the target, with units of $(\mathrm{V/Hz})/\mathrm{m}$; $Y_{M,Trgt}(f, x_R)$ is the complex frequency spectrum of the acoustic field radiated by the target and incident upon the receive aperture at x_R, with units of $(\mathrm{m}^2/\mathrm{sec})/\mathrm{Hz}$, and is given by (7.2-21); and $A_R(f, x_R)$ is the complex frequency response of the receive aperture at x_R (a.k.a. the complex receive aperture function), with units of $(\mathrm{V}/(\mathrm{m}^2/\mathrm{sec}))/\mathrm{m}$, where f represents frequencies in hertz transmitted by the target.

Since the receive linear aperture is a linear array composed of an odd number N of identical, equally-spaced, complex-weighted, omnidirectional point-elements lying along the X axis, the complex frequency response of the array is given by (see Subsection 6.1.2)

$$A_R(f, x_R) = \mathcal{S}_R(f) \sum_{i=-N'}^{N'} c_i(f) \delta(x_R - x_i), \tag{7.3-2}$$

where $\mathcal{S}_R(f)$ is the complex, receiver sensitivity function with units of $\mathrm{V}/(\mathrm{m}^2/\mathrm{sec})$ (see Table 6.1-2 and Appendix 6B),

$$c_i(f) = a_i \exp[+j\theta_i(f)] = a_i \exp(-j2\pi f \tau_i') \tag{7.3-3}$$

is the dimensionless complex weight associated with element i in the array, the impulse function $\delta(x_R - x_i)$ has units of inverse meters,

$$x_i = id, \qquad i = -N', \ldots, 0, \ldots, N', \tag{7.3-4}$$

is the x coordinate of the center of element i, d is the interelement spacing in meters, and

$$N' = (N-1)/2. \tag{7.3-5}$$

Note that the amplitude weight a_i is not a function of frequency. The different equations that can be used for the time delay τ_i' for purposes of beam steering and array focusing shall be discussed later in this section.

Substituting (7.3-2) into (7.3-1) yields the following expression for the complex frequency spectrum of the output electrical signal from the array at x_R due to the target, with units of $(\mathrm{V/Hz})/\mathrm{m}$:

$$Y_{Trgt}(f, x_R) = Y_{M,Trgt}(f, x_R)\,S_R(f) \sum_{i=-N'}^{N'} c_i(f)\,\delta(x_R - x_i). \tag{7.3-6}$$

The complex frequency spectrum (in volts per hertz) of the *total* output electrical signal from the array due to the target is given by [see (1.2-99)]

$$Y_{Trgt}(f) = \int_{-\infty}^{\infty} Y_{Trgt}(f, x_R)\,dx_R\,. \tag{7.3-7}$$

Substituting (7.3-6) into (7.3-7) yields

$$Y_{Trgt}(f) = S_R(f) \sum_{i=-N'}^{N'} c_i(f) \int_{-\infty}^{\infty} Y_{M,Trgt}(f, x_R)\,\delta(x_R - x_i)\,dx_R\,, \tag{7.3-8}$$

and since the impulse function will be nonzero only when $x_R = x_i$, by using the sifting property of impulse functions, (7.3-8) reduces to

$$Y_{Trgt}(f) = S_R(f) \sum_{i=-N'}^{N'} c_i(f)\,Y_{M,Trgt}(f, x_i)\,, \tag{7.3-9}$$

or

$$Y_{Trgt}(f) = \sum_{i=-N'}^{N'} Y_{M,Trgt}(f, x_i)\,S_R(f)\,c_i(f)\,. \tag{7.3-10}$$

Before proceeding further with our analysis, let us summarize our results up to this point.

Since $x_R = x_i = id$ [see (7.3-4)], the position vector to an aperture point, $\mathbf{r}_R$, given by (7.2-1) becomes

$$\mathbf{r}_R = \mathbf{r}_i = x_i\hat{x} = id\hat{x}, \qquad i = -N', \ldots, 0, \ldots, N'\,, \tag{7.3-11}$$

where $\mathbf{r}_i$ is the position vector to element i in the array. Therefore, the velocity potential in squared meters per second of the output acoustic signal from the fluid medium due to the target and incident upon element i in the array is given by [see (7.2-22)]

$$\boxed{y_{M,Trgt}(t, x_i) = -\frac{1}{4\pi R_i}\,s_0'(t - \tau_i, x_i)} \tag{7.3-12}$$

for $t \geq \tau_i$, where

$$\boxed{s_0'(t, x_i) = F_f^{-1}\{S_0'(f, x_i)\}} \tag{7.3-13}$$

has units of cubic meters per second [see (7.2-23)],

$$\boxed{S_0'(f, x_i) = S_0(f)\exp[-\alpha(f)R_i]} \tag{7.3-14}$$

has units of $(\mathrm{m}^3/\mathrm{sec})/\mathrm{Hz}$ [see (7.2-20)], $S_0(f)$ is the complex frequency spectrum of the target's source strength in $(\mathrm{m}^3/\mathrm{sec})/\mathrm{Hz}$, $\alpha(f)$ is the frequency-dependent attenuation coefficient in nepers per meter,

$$R_i = |\mathbf{r}_i - \mathbf{r}_S| = \sqrt{r_S^2 - 2r_S u_S x_i + x_i^2} \tag{7.3-15}$$

is the range in meters between the target and the center of element i [see (7.2-29)], and

$$\tau_i = R_i/c \tag{7.3-16}$$

is the corresponding *one-way* travel time or *one-way* time delay in seconds [see (7.2-4)]. The exact equations for the amplitude factor $1/R_i$ and time delay τ_i are [see (7.2-30) and (7.2-31)]

$$\frac{1}{R_i} = \frac{1}{\sqrt{r_S^2 - 2r_S u_S x_i + x_i^2}} \tag{7.3-17}$$

and

$$\tau_i = \frac{1}{c}\sqrt{r_S^2 - 2r_S u_S x_i + x_i^2} \, . \tag{7.3-18}$$

If the target is in the Fresnel region of the array, then [see (7.2-32) and (7.2-35)]

$$\frac{1}{R_i} \approx \frac{1}{r_S} \tag{7.3-19}$$

and

$$\tau_i \approx \tau_S - \frac{u_S}{c}x_i + \frac{1}{2r_S c}x_i^2 , \tag{7.3-20}$$

where τ_S is given by (7.2-36). If the target is in the far-field region of the array, then the amplitude factor given by (7.3-19) can also be used and [see (7.2-39)]

$$\tau_i \approx \tau_S - \frac{u_S}{c} x_i, \tag{7.3-21}$$

where τ_S is given by (7.2-36). And finally, the complex frequency spectrum of the acoustic field radiated by the target and incident upon element i in the array, with units of $\left(\mathrm{m}^2/\mathrm{sec}\right)/\mathrm{Hz}$, is given by [see (7.2-21)]

$$Y_{M,Trgt}(f, x_i) = -\frac{1}{4\pi R_i} S_0'(f, x_i) \exp(-j2\pi f \tau_i), \tag{7.3-22}$$

or

$$Y_{M,Trgt}(f, x_i) = S_0(f) g_f\left(\mathbf{r}_i \middle| \mathbf{r}_S\right), \tag{7.3-23}$$

where

$$\boxed{g_f\left(\mathbf{r}_i \middle| \mathbf{r}_S\right) = -\frac{\exp\left[-\alpha(f)\left|\mathbf{r}_i - \mathbf{r}_S\right|\right]}{4\pi\left|\mathbf{r}_i - \mathbf{r}_S\right|} \exp\left(-jk\left|\mathbf{r}_i - \mathbf{r}_S\right|\right)} \tag{7.3-24}$$

is the time-independent, free-space, Green's function of an unbounded, viscous, homogeneous, fluid medium with units of inverse meters, and k is the wavenumber in radians per meter given by (7.2-15). The Green's function given by (7.3-24) is the complex frequency response of the fluid medium at frequency f hertz and location $\mathbf{r}_i$ due to the application of a unit-amplitude impulse (i.e., a unit-amplitude, omnidirectional point-source) at location $\mathbf{r}_S$.

Let us now continue with the derivation of $Y_{Trgt}(f)$ by rewriting (7.3-10) as follows:

$$Y_{Trgt}(f) = \sum_{i=-N'}^{N'} Y_{Trgt}(f, x_i), \tag{7.3-25}$$

where

$$Y_{Trgt}(f, x_i) = Y'_{Trgt}(f, x_i) c_i(f) \tag{7.3-26}$$

and

$$Y'_{Trgt}(f, x_i) = Y_{M,Trgt}(f, x_i) \mathcal{S}_R(f) \tag{7.3-27}$$

are the complex frequency spectra (in volts per hertz) of the output electrical signal from element i in the array due to the target *after* and *before* the application of the complex weight $c_i(f)$, respectively (see Fig. 7.3-1). Substituting (7.3-22) and (7.3-14) into (7.3-27) yields

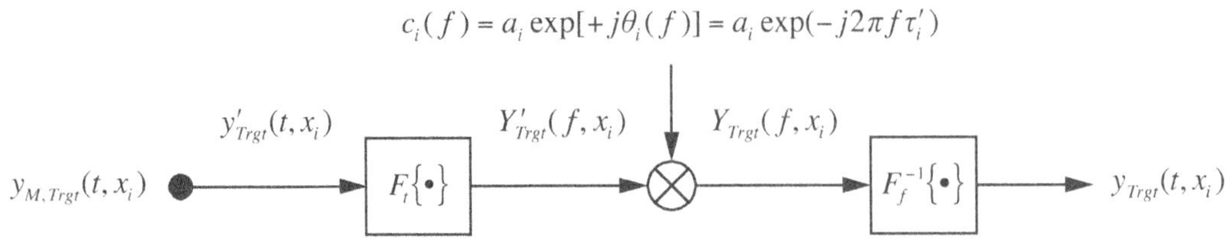

$$c_i(f) = a_i \exp[+j\theta_i(f)] = a_i \exp(-j2\pi f \tau_i')$$

$$y_{M,Trgt}(t,x_i) \to y'_{Trgt}(t,x_i) \to F_t\{\bullet\} \to Y'_{Trgt}(f,x_i) \to \otimes \to Y_{Trgt}(f,x_i) \to F_f^{-1}\{\bullet\} \to y_{Trgt}(t,x_i)$$

Figure 7.3-1 Beamforming. Implementing a complex weight (amplitude and phase weights) using forward and inverse Fourier transforms.

$$Y'_{Trgt}(f,x_i) = S(f,x_i)\exp(-j2\pi f\tau_i), \tag{7.3-28}$$

where

$$\boxed{S(f,x_i) = -\frac{1}{4\pi R_i} S_0(f)\exp[-\alpha(f)R_i]\mathcal{S}_R(f)} \tag{7.3-29}$$

has units of volts per hertz. Note that $S(f,x_i)$ is a function of x_i because the exact range R_i between the target and the center of each element in the array is a function of x_i [see (7.3-15)]. Substituting (7.3-3) and (7.3-28) into (7.3-26) yields

$$Y_{Trgt}(f,x_i) = a_i S(f,x_i)\exp\left[-j2\pi f(\tau_i + \tau_i')\right], \tag{7.3-30}$$

and by substituting (7.3-30) into (7.3-25), we finally obtain

$$Y_{Trgt}(f) = \sum_{i=-N'}^{N'} a_i S(f,x_i)\exp\left[-j2\pi f(\tau_i + \tau_i')\right]. \tag{7.3-31}$$

Equation (7.3-31) is the complex frequency spectrum (in volts per hertz) of the *total* output electrical signal from the array due to the target. We shall obtain a time-domain model next.

Taking the inverse Fourier transform of (7.3-28) yields

$$y'_{Trgt}(t,x_i) = s(t-\tau_i, x_i), \tag{7.3-32}$$

which is the output electrical signal in volts from element i in the array due to the target, *before* complex weighting (see Fig. 7.3-1), for $t \geq \tau_i$, where

$$\boxed{s(t,x_i) = F_f^{-1}\{S(f,x_i)\} = -\frac{1}{4\pi R_i} F_f^{-1}\{S_0(f)\exp[-\alpha(f)R_i]\mathcal{S}_R(f)\}} \tag{7.3-33}$$

The function $s(t, x_i)$ is a *time-domain, WSS random process* because it is the result of linear, time-invariant, space-invariant, deterministic filtering of the target's source strength $s_0(t)$, which was assumed to be WSS. Although we are working with a simple spherical-wave model for the acoustic field incident upon the array [see (7.3-12)], (7.3-33) provides some insight as to why it is so difficult to detect quiet submarines. If a target is radiating a very weak acoustic field, then the magnitudes of the frequency components contained in the magnitude spectrum of the target's source strength, $|S_0(f)|$, will be very small. By the time the radiated acoustic field reaches the array, the magnitude spectrum $|S_0(f)|$ is attenuated by the decaying exponential $\exp[-\alpha(f)R_i]$ and the amplitude factor $1/R_i$. The magnitude spectrum $|S_0(f)|$ is further attenuated by the filtering action of the elements in the array as modeled by the receiver sensitivity function $\mathcal{S}_R(f)$ and, in particular, its magnitude response $|\mathcal{S}_R(f)|$.

The next step is to inverse Fourier transform (7.3-25) and (7.3-30). Doing so yields

$$y_{Trgt}(t) = \sum_{i=-N'}^{N'} y_{Trgt}(t, x_i) \tag{7.3-34}$$

and

$$\boxed{y_{Trgt}(t, x_i) = a_i s(t - [\tau_i + \tau_i'], x_i)} \tag{7.3-35}$$

respectively, and by substituting (7.3-35) into (7.3-34), we obtain

$$\boxed{y_{Trgt}(t) = \sum_{i=-N'}^{N'} a_i s(t - [\tau_i + \tau_i'], x_i)} \tag{7.3-36}$$

where $s(t, x_i)$ is given by (7.3-33). Equation (7.3-35) is the output electrical signal (in volts) from element i in the array due to the target, *after* complex weighting (see Fig. 7.3-1), for $t \geq \tau_i + \tau_i'$. Equation (7.3-36) is the *total* output electrical signal from the array (in volts) due to the target. Both $y_{Trgt}(t, x_i)$ and $y_{Trgt}(t)$ are WSS in time because $s(t, x_i)$ is WSS in time.

From (7.3-35) it can be seen that the individual output electrical signals $y_{Trgt}(t, x_i)$, $i = -N', \ldots, 0, \ldots, N'$, are *out-of-phase* with each other because $\tau_i + \tau_i'$ is different at each element i in the array. As a result, the time-average power of the total output electrical signal from the array, $y_{Trgt}(t)$, given by (7.3-36) will *not* be maximized because out-of-phase signals add destructively (destructive interference). Therefore, in order to maximize the time-average power of $y_{Trgt}(t)$, all the individual signals $y_{Trgt}(t, x_i)$, $i=-N',\ldots,0,\ldots,N'$, must be *in-phase*, which requires that $\tau_i+\tau'_i$ be equal to either *zero* or a *constant* $\forall\ i$.

In order to cophase all the individual signals $y_{Trgt}(t, x_i)$, $i=-N',\ldots,0,\ldots,N'$, the correct equation for the time delay τ'_i must be used, which depends on the model used for τ_i. For example, if you want to be able to cophase $y_{Trgt}(t, x_i)$ $\forall\ i$ regardless of where the target is relative to the array – near-field, Fresnel, or far-field region – then [see (7.3-18)]

$$\tau_i=\frac{1}{c}\sqrt{r_S^2-2r_S u_S x_i+x_i^2}\ , \tag{7.3-37}$$

where [see (7.2-9)]

$$u_S=\sin\theta_S\cos\psi_S\ . \tag{7.3-38}$$

However, since the spherical coordinates of the target's location (r_S, θ_S, ψ_S) are unknown a priori,

$$\boxed{\tau'_i=-\frac{1}{c}\sqrt{(r')^2-2r'u'x_i+x_i^2}} \tag{7.3-39}$$

where

$$u'=\sin\theta'\cos\psi'\ . \tag{7.3-40}$$

Adding (7.3-37) and (7.3-39) yields

$$\tau_i+\tau'_i=\frac{1}{c}\left[\sqrt{r_S^2-2r_S u_S x_i+x_i^2}-\sqrt{(r')^2-2r'u'x_i+x_i^2}\ \right]. \tag{7.3-41}$$

Therefore, if the far-field beam pattern of the array is *focused* at the range to the target ($r'=r_S$) and *steered* in the direction of the acoustic field incident upon the array (in this case, $\theta'=\theta_S$ and $\psi'=\psi_S$ so that $u'=u_S$), then (7.3-41) reduces to

$$\tau_i+\tau'_i=0, \qquad i=-N',\ldots,0,\ldots,N'\ , \tag{7.3-42}$$

and, as a result, all the individual output electrical signals due to the target will be in-phase. Substituting (7.3-42) into (7.3-35) and (7.3-36) yields

$$\boxed{y_{Trgt}(t, x_i)=a_i\, s(t, x_i)} \tag{7.3-43}$$

and

$$\boxed{y_{Trgt}(t)=\sum_{i=-N'}^{N'} a_i\, s(t, x_i)} \tag{7.3-44}$$

respectively, where $s(t, x_i)$ is given by (7.3-33). If rectangular amplitude weights are used, then $a_i = 1 \;\; \forall \;\; i$, and (7.3-44) reduces to

$$\boxed{y_{Trgt}(t) = \sum_{i=-N'}^{N'} s(t, x_i)} \tag{7.3-45}$$

Equations (7.3-44) and (7.3-45) both indicate that $y_{Trgt}(t)$ is equal to the sum of N signals that are in-phase (constructive interference).

If the target is in the Fresnel region of the array, then a good approximation of τ_i is given by [see (7.3-20)]

$$\tau_i \approx \tau_S - \frac{u_S}{c} x_i + \frac{1}{2 r_S c} x_i^2 , \tag{7.3-46}$$

where the time delay τ_S is given by (7.2-36). Based on the model for τ_i given by (7.3-46), and ignoring τ_S (see Subsection 6.6.1),

$$\boxed{\tau_i' = \frac{u'}{c} x_i - \frac{1}{2r'c} x_i^2} \tag{7.3-47}$$

Adding (7.3-46) and (7.3-47) yields

$$\tau_i + \tau_i' = \tau_S + \frac{1}{c}(u' - u_S) x_i + \frac{1}{2c} \frac{r' - r_S}{r' r_S} x_i^2 . \tag{7.3-48}$$

If the far-field beam pattern of the array is *focused* at the range to the target ($r' = r_S$) and *steered* in the direction of the acoustic field incident upon the array (in this case, $\theta' = \theta_S$ and $\psi' = \psi_S$ so that $u' = u_S$), then (7.3-48) reduces to

$$\tau_i + \tau_i' = \tau_S , \qquad i = -N', \ldots, 0, \ldots, N' . \tag{7.3-49}$$

Since τ_S is a constant, all the individual output electrical signals due to the target will be in-phase. Substituting (7.3-49) into (7.3-35) and (7.3-36), and setting $R_i = r_S$ in (7.3-29) and (7.3-33) yields

$$\boxed{y_{Trgt}(t, x_i) = a_i s(t - \tau_S)} \tag{7.3-50}$$

and

$$\boxed{y_{Trgt}(t) = s(t - \tau_S) \sum_{i=-N'}^{N'} a_i} \tag{7.3-51}$$

respectively, for $t \geq \tau_S$, $S(f, x_i) \approx S(f)$ where [see (7.3-29)]

$$\boxed{S(f) = -\frac{1}{4\pi r_S} S_0(f) \exp[-\alpha(f) r_S] \boldsymbol{S}_R(f)} \tag{7.3-52}$$

and $s(t, x_i) \approx s(t)$ where [see (7.3-33)]

$$\boxed{s(t) = F_f^{-1}\{S(f)\} = -\frac{1}{4\pi r_S} F_f^{-1}\{S_0(f) \exp[-\alpha(f) r_S] \boldsymbol{S}_R(f)\}} \tag{7.3-53}$$

The function $s(t)$ is a WSS random process because it is the result of linear, time-invariant, space-invariant, deterministic filtering of the target's source strength $s_0(t)$, which was assumed to be WSS. If rectangular amplitude weights are used, then $a_i = 1 \;\; \forall \; i$, and (7.3-51) reduces to

$$\boxed{y_{Trgt}(t) = N s(t - \tau_S)} \tag{7.3-54}$$

for $t \geq \tau_S$. Equation (7.3-54) is the result of N identical signals adding in-phase (constructive interference).

If the target is in the far-field region of the array, then a good approximation of τ_i is given by [see (7.3-21)]

$$\tau_i \approx \tau_S - \frac{u_S}{c} x_i, \tag{7.3-55}$$

where the time delay τ_S is given by (7.2-36). Based on the model for τ_i given by (7.3-55), and ignoring τ_S (see Section 6.4),

$$\boxed{\tau_i' = \frac{u'}{c} x_i} \tag{7.3-56}$$

Adding (7.3-55) and (7.3-56) yields

$$\tau_i + \tau_i' = \tau_S + \frac{1}{c}(u' - u_S) x_i. \tag{7.3-57}$$

If the far-field beam pattern of the array is *steered* in the direction of the acoustic field incident upon the array (in this case, $\theta' = \theta_S$ and $\psi' = \psi_S$ so that $u' = u_S$), then (7.3-57) reduces to

$$\tau_i + \tau_i' = \tau_S, \qquad i = -N', \ldots, 0, \ldots, N'. \tag{7.3-58}$$

Since τ_S is a constant, all the individual output electrical signals due to the target will be in-phase. And since (7.3-58) is identical to (7.3-49), (7.3-50) through (7.3-54) are also applicable to the far-field case.

At this point in our discussion, we need to answer the following practical question: How do we determine when all the individual output electrical signals due to the target are in-phase when using the beamformer shown in Fig. 7.1-1? The answer is to compute the time-average power of the total output electrical signal from the array, $r(t)$, given by (7.1-4), for every set of phase weights used. One set of phase weights corresponds to evaluating

$$\theta_i(f) = -2\pi f \tau_i', \qquad i = -N', \ldots, 0, \ldots, N', \tag{7.3-59}$$

$\forall\ i$, using one set of values for (r', θ', ψ') to evaluate τ_i' given by (7.3-39) or (7.3-47), or using one set of values for (θ', ψ') to evaluate τ_i' given by (7.3-56). The phase weights given by (7.3-59) are then used to form the set of complex weights [see (7.3-3) and Fig. 7.1-1]

$$c_i(f) = a_i \exp[+j\theta_i(f)] = a_i \exp(-j2\pi f \tau_i'), \qquad i = -N', \ldots, 0, \ldots, N'. \tag{7.3-60}$$

Using one set of values for (r', θ', ψ') or (θ', ψ') corresponds to forming *one beam*. Forming *multiple beams* with multiple "look directions" means using multiple sets of values for (r', θ', ψ') or (θ', ψ'). Note that the set of complex frequency spectra $R'(f, x_i)$, $i = -N', \ldots, 0, \ldots, N'$, only needs to be computed *once* in order to form multiple beams (see Fig. 7.1-1). By using *parallel signal processing*, multiple beams can be formed *simultaneously*. The set of values (r', θ', ψ') or (θ', ψ') that maximizes the time-average power of $r(t)$ is the best *estimate* of (r_S, θ_S, ψ_S) or (θ_S, ψ_S) for an omnidirectional, point-source target in an unbounded, viscous, homogeneous, fluid medium. A signal processing algorithm that can be used for computing the time-average power of $r(t)$ is discussed in Appendix 7B.

After the time-average power of $r(t)$ has been maximized, a *target signature analysis* can be performed by computing the power spectrum of $r(t)$ (see Appendix 7B). Estimating the frequency components (spectral lines) of the sound-field radiated by a target via a power spectrum calculation is referred to as performing a target signature analysis because different sound-sources (targets) such as submarines, surface ships, and marine mammals are known to radiate energy at different sets of frequencies. A target's unique set of radiated frequencies is known as a *target signature*. The information obtained from a target signature analysis can then be used to classify (identify) the target. This is known as *target classification*.

7.3.1 FFT Beamforming for Linear Arrays

In this subsection we shall show how to implement complex weights in a linear array using the method of *FFT beamforming*. We shall also show how to compute the beam pattern of the linear array by using frequency-domain data already obtained from the method of FFT beamforming. Although the output electrical signals from the individual elements in the array are equal to the sum of the output electrical signals due to the target, ambient noise, and receiver noise; for example purposes, we shall only work with the output electrical signals due to the target.

The first step is to sample $y'_{Trgt}(t, x_i)$, which is the output electrical signal from element i in the array due to the target *before* complex weighting given by [see (7.3-32) and Fig. 7.3-1]

$$y'_{Trgt}(t, x_i) = s(t - \tau_i, x_i), \tag{7.3-61}$$

for $t \geq \tau_i$, where $s(t, x_i)$ is given by (7.3-33) and τ_i is given by (7.3-16). If we sample $y'_{Trgt}(t, x_i)$ at a rate of f_S samples per second, then

$$y'_{Trgt}(t_l, x_i) = s(t_l - \tau_i, x_i), \tag{7.3-62}$$

or

$$y'_{Trgt}(lT_S, id) = s(lT_S - \tau_i, id), \tag{7.3-63}$$

where

$$t_l = lT_S, \qquad l = 0, 1, \ldots, L-1, \tag{7.3-64}$$

is the sampling time instant in seconds, $T_S = 1/f_S$ is the sampling period in seconds, L is the total number of time samples taken,

$$x_i = id, \qquad i = -N', \ldots, 0, \ldots, N', \tag{7.3-65}$$

is the x coordinate of the center of element i [see (7.3-4)], d is the interelement spacing in meters, and [see (7.3-5)]

$$N' = (N-1)/2, \tag{7.3-66}$$

where N is the total odd number of identical, equally-spaced, complex-weighted, omnidirectional point-elements in the linear array. Note that

$$y'_{Trgt}(l,i) \equiv y'_{Trgt}(t_l, x_i) = y'_{Trgt}(lT_S, id). \tag{7.3-67}$$

By referring to Fig. 7.3-1, the second step is to compute the time-domain Fourier transform of $y'_{Trgt}(l,i)$. Using a modified version of the algorithm discussed in Appendix 7B for computing time-domain Fourier transforms yields

$$\hat{Y}'_{Trgt}(q,i) = T_S\, DFT_l\{y'_{Trgt}(l,i)\}, \qquad q = -L'', \ldots, 0, \ldots, L'', \tag{7.3-68}$$

or

$$\hat{Y}'_{Trgt}(q,i) = T_S \sum_{l=0}^{L-1} y'_{Trgt}(l,i) W_{L+Z}^{-ql}, \qquad q = -L'', \ldots, 0, \ldots, L'', \tag{7.3-69}$$

where

$$\hat{Y}'_{Trgt}(q,i) \equiv \hat{Y}'_{Trgt}(qf_0, id) = \hat{Y}'_{Trgt}(f, x_i)\Big|_{f=qf_0} \tag{7.3-70}$$

is an estimate of the theoretical frequency spectrum $Y'_{Trgt}(f, x_i)$ given by (7.3-28) at discrete frequencies $f = qf_0$,

$$W_{L+Z} = \exp\left(+j\frac{2\pi}{L+Z}\right), \tag{7.3-71}$$

$$f_0 = \frac{1}{T_0} = \frac{1}{(L+Z)T_S} = \frac{f_S}{L+Z} \tag{7.3-72}$$

is the frequency-domain DFT bin spacing [the fundamental frequency of $y'_{Trgt}(t, x_i)$] in hertz, T_0 is the fundamental period of $y'_{Trgt}(t, x_i)$ in seconds,

$$L'' = \begin{cases} (L+Z)/2, & L+Z \text{ even} \\ (L+Z-1)/2, & L+Z \text{ odd}, \end{cases} \tag{7.3-73}$$

and Z is the total number of zeros used for padding-with-zeros. See Section 6.5 or Appendix 7B for an explanation of padding-with-zeros. Substituting (7.3-67) and (7.3-63) into (7.3-68) and using the time-shifting property of Fourier transforms yields

$$\begin{aligned}\hat{Y}'_{Trgt}(q,i) &= T_S\, DFT_l\{s(lT_S - \tau_i, id)\} \\ &= T_S\, DFT_l\{s(l,i)\}\exp(-j2\pi q f_0 \tau_i), \qquad q = -L'', \ldots, 0, \ldots, L'',\end{aligned} \tag{7.3-74}$$

or

$$\hat{Y}'_{Trgt}(q,i) = \hat{S}(q,i)\exp(-j2\pi q f_0 \tau_i), \tag{7.3-75}$$

where

$$\hat{S}(q,i) = T_S\, DFT_l\{s(l,i)\}, \qquad q = -L'', \ldots, 0, \ldots, L'', \tag{7.3-76}$$

is an estimate of the theoretical frequency spectrum $S(f, x_i)$ given by (7.3-29) at discrete frequencies $f = qf_0$, and

$$s(l,i) \equiv s(t_l, x_i) = s(lT_S, id). \tag{7.3-77}$$

As can be seen from (7.3-29), $S(f, x_i)$ is related to the complex frequency spectrum of the target's source strength $S_0(f)$.

By referring to Fig. 7.3-1, the third step is to multiply $\hat{Y}'_{Trgt}(q,i)$ by the complex weight [see (7.3-3)]

$$c_i(qf_0) = a_i \exp[+j\theta_i(qf_0)] = a_i \exp(-j2\pi q f_0 \tau'_i). \tag{7.3-78}$$

Doing so yields

$$\hat{Y}_{Trgt}(q,i) = \hat{Y}'_{Trgt}(q,i)\, c_i(qf_0), \tag{7.3-79}$$

and by substituting (7.3-75) and (7.3-78) into (7.3-79), we obtain

$$\hat{Y}_{Trgt}(q,i) = a_i \hat{S}(q,i)\exp\left[-j2\pi q f_0 (\tau_i + \tau'_i)\right], \tag{7.3-80}$$

which is an estimate of the theoretical frequency spectrum $Y_{Trgt}(f, x_i)$ given by (7.3-30) at discrete frequencies $f = qf_0$.

By referring to Fig. 7.3-1, the fourth and final step is to compute the inverse Fourier transform of $\hat{Y}_{Trgt}(q,i)$. The inverse Fourier transform of $\hat{Y}_{Trgt}(q,i)$ is computed as follows:

$$\hat{y}_{Trgt}(l,i) = \frac{1}{T_S} IDFT_q\left\{\hat{Y}_{Trgt}(q,i)\right\}, \qquad l = 0, 1, \ldots, L+Z-1, \tag{7.3-81}$$

or

$$\hat{y}_{Trgt}(l,i) = \frac{1}{T_S}\frac{1}{L+Z}\sum_{q=-L''}^{L''} \hat{Y}_{Trgt}(q,i) W_{L+Z}^{ql}, \qquad l = 0, 1, \ldots, L+Z-1, \tag{7.3-82}$$

where IDFT is the abbreviation for inverse discrete Fourier transform.

Substituting (7.3-80) into (7.3-81) yields

$$\hat{y}_{Trgt}(l,i) = a_i \frac{1}{T_S} IDFT_q\left\{\hat{S}(q,i)\exp[-j2\pi q f_0(\tau_i + \tau_i')]\right\}, \qquad l = 0,1,\ldots, L+Z-1, \tag{7.3-83}$$

and by using the time-shifting property of Fourier transforms, we obtain

$$\hat{y}_{Trgt}(l,i) = a_i \hat{s}(lT_S - [\tau_i + \tau_i'], i), \qquad l = 0,1,\ldots, L+Z-1, \tag{7.3-84}$$

which is an estimate of time samples of the output electrical signal $y_{Trgt}(t, x_i)$ from element i in the array due to the target *after* complex weighting [see (7.3-35) and Fig. 7.3-1], where

$$\hat{s}(l,i) = \frac{1}{T_S} IDFT_q\left\{\hat{S}(q,i)\right\} \tag{7.3-85}$$

is an *estimate* of time samples of $s(t, x_i)$ because $\hat{Y}_{Trgt}(q,i)$ was obtained by multiplying the *estimate* $\hat{Y}'_{Trgt}(q,i)$ by the complex weight $c_i(qf_0)$ [see (7.3-79)].

In order to compute the beam pattern of the linear array, compute the *spatial* DFT of the frequency-domain data $\hat{Y}_{Trgt}(q,i)$ already obtained from the method of FFT beamforming as follows (see Section 6.5):

$$\hat{\mathrm{Y}}_{Trgt}(q,m) = DFT_i\left\{\hat{Y}_{Trgt}(q,i)\right\}, \qquad m = -N'',\ldots,0,\ldots,N'', \tag{7.3-86}$$

or

$$\hat{\mathrm{Y}}_{Trgt}(q,m) = \sum_{i=-N'}^{N'} \hat{Y}_{Trgt}(q,i) W_{N+Z_X}^{mi}, \qquad m = -N'',\ldots,0,\ldots,N'', \tag{7.3-87}$$

where

$$\hat{\mathrm{Y}}_{Trgt}(q,m) \equiv \hat{\mathrm{Y}}_{Trgt}(qf_0, m\Delta f_X) = \hat{\mathrm{Y}}_{Trgt}(f, f_X)\Big|_{\substack{f=qf_0 \\ f_X = m\Delta f_X}} \tag{7.3-88}$$

is an estimate of the frequency-and-angular spectrum of $y_{Trgt}(t, x_i)$ at discrete frequencies $f = qf_0$, and discrete spatial frequencies $f_X = m\Delta f_X$, or equivalently, direction cosine values u_m;

$$W_{N+Z_X} = \exp\left(+j\frac{2\pi}{N+Z_X}\right), \tag{7.3-89}$$

$$\Delta f_X = \frac{1}{(N+Z_X)d} \tag{7.3-90}$$

is the spatial-frequency spacing in the X direction with units of cycles per meter,

$$N'' = \begin{cases} (N+Z_X)/2, & N+Z_X \text{ even} \\ (N+Z_X-1)/2, & N+Z_X \text{ odd}, \end{cases} \tag{7.3-91}$$

$$u_m = \frac{m}{N+Z_X}\frac{\lambda}{d}, \qquad m = -N'', \ldots, 0, \ldots, N'', \tag{7.3-92}$$

is the value of direction cosine u at DFT bin m, and Z_X is the total number of zeros used for padding-with-zeros in the X direction. Next we shall show that $\hat{Y}_{Trgt}(q,m)$ is related to the beam pattern of the linear array.

For example, if the target is in the far-field region of the array, then substituting (7.3-57) and (7.3-65) into (7.3-80), and setting $R_i = r_S$ in (7.3-29) yields

$$\hat{Y}_{Trgt}(q,i) = \hat{S}(q)\exp(-j2\pi q f_0 \tau_S) a_i \exp\left[-j2\pi q f_0 \frac{(u'-u_S)}{c} id\right], \tag{7.3-93}$$

where $\hat{S}(q)$ is an estimate of the theoretical frequency spectrum $S(f)$ given by (7.3-52) at discrete frequencies $f = qf_0$, and τ_S is the one-way time delay in seconds associated with the path length between the sound-source (target) and the center of the array [see (7.2-36)]. As can be seen from (7.3-52), $S(f)$ is related to the complex frequency spectrum of the target's source strength $S_0(f)$. Note that when $u' \neq u_S$, the *phase* of $\hat{Y}_{Trgt}(q,i)$ given by (7.3-93) is *different* at each element i. As a result, the output electrical signals from each element in the array will be *out-of-phase*.

If the far-field beam pattern of the array is steered in the direction of the acoustic field incident upon the array (in this case, $\theta' = \theta_S$ and $\psi' = \psi_S$ so that $u' = u_S$), then (7.3-93) reduces to

$$\hat{Y}_{Trgt}(q,i) = \hat{S}(q)\exp(-j2\pi q f_0 \tau_S) a_i . \tag{7.3-94}$$

The *phase* of $\hat{Y}_{Trgt}(q,i)$ given by (7.3-94) is now the *same* at each element i. Therefore, the output electrical signals from each element in the array will be *in-phase* and, as a result, the array gain will be *maximized*. Substituting (7.3-94) into (7.3-87) yields

$$\hat{Y}_{Trgt}(q,m) = \hat{S}(q)\exp(-j2\pi q f_0 \tau_S) \sum_{i=-N'}^{N'} a_i W_{N+Z_X}^{mi}, \qquad m = -N'',\ldots,0,\ldots,N''.$$

(7.3-95)

The summation in (7.3-95) is the *array factor* for the set of amplitude weights a_i. The array factor is directly proportional to the far-field beam pattern of the array. At broadside, that is, at $m=0$, (7.3-95) reduces to

$$\hat{Y}_{Trgt}(q,0) = \hat{S}(q)\exp(-j2\pi q f_0 \tau_S) \sum_{i=-N'}^{N'} a_i . \tag{7.3-96}$$

When $u' \neq u_S$, substituting (7.3-93) into (7.3-87) yields

$$\hat{Y}_{Trgt}(q,m) = \hat{S}(q)\exp(-j2\pi q f_0 \tau_S) \times$$
$$\sum_{i=-N'}^{N'} a_i \exp\left[-j2\pi q f_0 \frac{(u'-u_S)}{c} id\right] W_{N+Z_X}^{mi}, \qquad m = -N'',\ldots,0,\ldots,N''.$$

(7.3-97)

The summation in (7.3-97) is the array factor for the set of amplitude weights a_i steered in the direction $u = u' - u_S$ in direction-cosine space.

In order to prove the above statement, we need to find the value of DFT bin m, or equivalently, direction cosine $u = u_m$, so that (7.3-97) reduces to (7.3-96). Start by substituting (7.3-89) into (7.3-97). Doing so yields

$$\hat{Y}_{Trgt}(q,m) = \hat{S}(q)\exp(-j2\pi q f_0 \tau_S) \sum_{i=-N'}^{N'} a_i \exp\left\{-j2\pi\left[q f_0 \frac{(u'-u_S)}{c} d - \frac{m}{N+Z_X}\right] i\right\}.$$

(7.3-98)

As can be seen from (7.3-98), the value of m that will reduce (7.3-98) to (7.3-96) is given by

$$m = m' = (N+Z_X) q f_0 \frac{(u'-u_S)}{c} d \tag{7.3-99}$$

since this value of m will cause the exponent inside the summation to be equal to 0. Although the parameter values on the right-hand side of (7.3-99) are such that m' is not an integer in general, for example purposes, we shall assume it is. Therefore, substituting (7.3-99) into (7.3-92), and since

$$c = f\lambda = q f_0 \lambda, \tag{7.3-100}$$

$$u_{m'} = u' - u_S . \tag{7.3-101}$$

In other words, when $u' \neq u_S$, the value of the array factor at broadside will be located at $u = u' - u_S$ in direction-cosine space, that is, the array factor will be steered to $u = u' - u_S$.

If *no* beam steering is done, that is, if $u' = 0$, then $u_{m'} = -u_S$, or $-u_{m'} = u_S$. If the sound-source (target) lies in the XY plane, then $u_S = \cos\psi_S$ since $\theta_S = 90°$. Therefore,

$$-u_{m'} = \cos\psi_S , \tag{7.3-102}$$

or

$$\psi_S = \cos^{-1}(-u_{m'}) , \tag{7.3-103}$$

which is the azimuthal (bearing) angle in the direction of the acoustic field incident upon the array. In order to obtain the correct value for ψ_S, the *port/starboard (left/right) ambiguity problem* associated with linear apertures (linear arrays) has to be resolved (see Examples 8.2-1 and 9.1-2).

7.4 Total Output Signal from a Linear Array Due to Ambient Noise and Receiver Noise

In addition to an output electrical signal from each element in the array due to the sound-field radiated by the target, there is also present an output electrical signal due to ambient (background) noise in the ocean. We shall assume that the ambient-noise sound-field is *isotropic* (omnidirectional). Sources of ambient noise in the ocean are, for example, shipping, seismic activity, waves, rain, wind, ocean thermal noise, and biological.

Let $y_{M,n_a}(t, x_i)$ be the velocity potential in squared meters per second of the output acoustic signal from the fluid medium due to ambient noise and incident upon element i in the array. The velocity potential $y_{M,n_a}(t, x_i)$ is assumed to be a *zero-mean, time-domain* and *spatial-domain, wide-sense stationary* (WSS) *random process* $\forall\ i$. A spatial-domain WSS random process is also referred to as being *statistically homogeneous.* It is also assumed that $y_{M,n_a}(t, x_i)$ and $y_{M,n_a}(t, x_j)$ are *uncorrelated* for $i \neq j$. The complex frequency spectrum (in volts per hertz) of the *total* output electrical signal from the array due to ambient noise, $Y_{n_a}(f)$, can be expressed as follows [see (7.3-25) through (7.3-27)]:

$$Y_{n_a}(f) = \sum_{i=-N'}^{N'} Y_{n_a}(f, x_i) , \tag{7.4-1}$$

where

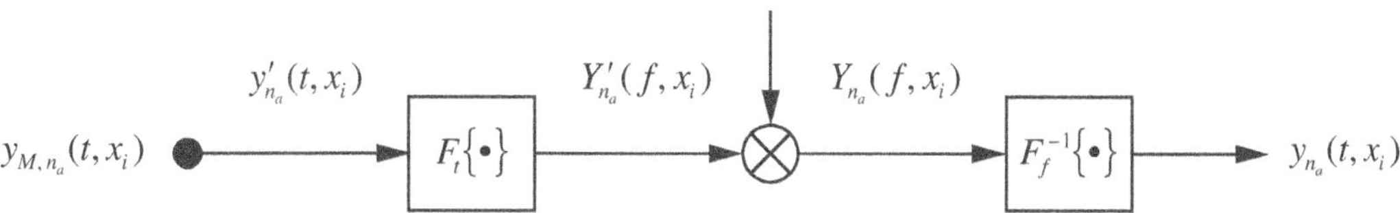

Figure 7.4-1 Beamforming. Implementing a complex weight (amplitude and phase weights) using forward and inverse Fourier transforms.

$$Y_{n_a}(f, x_i) = Y'_{n_a}(f, x_i) c_i(f) \tag{7.4-2}$$

and

$$Y'_{n_a}(f, x_i) = N_a(f, x_i) = Y_{M,n_a}(f, x_i) \mathcal{S}_R(f) \tag{7.4-3}$$

are the complex frequency spectra (in volts per hertz) of the output electrical signal from element i in the array due to ambient noise, *after* and *before* the application of the complex weight $c_i(f)$, respectively (see Fig. 7.4-1), and

$$Y_{M,n_a}(f, x_i) = F_t\{y_{M,n_a}(t, x_i)\} \tag{7.4-4}$$

is the complex frequency spectrum of the ambient noise incident upon element i in the array with units of $(\mathrm{m}^2/\mathrm{sec})/\mathrm{Hz}$. Substituting $Y'_{n_a}(f, x_i) = N_a(f, x_i)$ [see (7.4-3)] and (7.3-3) into (7.4-2) yields

$$Y_{n_a}(f, x_i) = a_i N_a(f, x_i) \exp(-j2\pi f \tau_i'), \tag{7.4-5}$$

and by substituting (7.4-5) into (7.4-1), we obtain

$$Y_{n_a}(f) = \sum_{i=-N'}^{N'} a_i N_a(f, x_i) \exp(-j2\pi f \tau_i'). \tag{7.4-6}$$

Equation (7.4-6) is the complex frequency spectrum (in volts per hertz) of the *total* output electrical signal from the array due to ambient noise, where $N_a(f, x_i)$ is given (7.4-3). We shall obtain a time-domain model next.

Taking the inverse Fourier transform of (7.4-3) yields

$$y'_{n_a}(t, x_i) = n_a(t, x_i) = F_f^{-1}\{Y_{M,n_a}(f, x_i) \mathcal{S}_R(f)\}, \tag{7.4-7}$$

which is the output electrical signal in volts from element i in the array due to ambient noise, *before* complex weighting (see Fig. 7.4-1). The signal $n_a(t, x_i)$ is a zero-mean, time-domain and spatial-domain, WSS random process $\forall\ i$, and $n_a(t, x_i)$ and $n_a(t, x_j)$ are uncorrelated for $i \neq j$, because $n_a(t, x_i)$ is the result of linear, time-invariant, space-invariant, deterministic filtering of the velocity potential $y_{M,n_a}(t, x_i)\ \forall\ i$. Taking the inverse Fourier transform of (7.4-1) and (7.4-5) yields

$$y_{n_a}(t) = \sum_{i=-N'}^{N'} y_{n_a}(t, x_i) \tag{7.4-8}$$

and

$$\boxed{y_{n_a}(t, x_i) = a_i n_a(t - \tau_i', x_i)} \tag{7.4-9}$$

respectively, and by substituting (7.4-9) into (7.4-8), we obtain

$$\boxed{y_{n_a}(t) = \sum_{i=-N'}^{N'} a_i n_a(t - \tau_i', x_i)} \tag{7.4-10}$$

where $n_a(t, x_i)$ is given by (7.4-7). Equation (7.4-9) is the output electrical signal (in volts) from element i in the array due to ambient noise, *after* complex weighting (see Fig. 7.4-1). The signal $y_{n_a}(t, x_i)$ is a zero-mean, time-domain and spatial-domain, WSS random process $\forall\ i$, and $y_{n_a}(t, x_i)$ and $y_{n_a}(t, x_j)$ are uncorrelated for $i \neq j$. Equation (7.4-10) is the *total* output electrical signal from the array (in volts) due to ambient noise, and is a zero-mean, WSS random process.

In addition to ambient noise, there is also receiver noise present at the output of each element in the array. The output electrical signals in volts from element i in the array due to receiver noise, *before* and *after* complex weighting (see Fig. 7.4-2), are given by

$$n_r'(t, x_i) = n(t, x_i) \tag{7.4-11}$$

and

$$\boxed{n_r(t, x_i) = a_i n(t - \tau_i', x_i)} \tag{7.4-12}$$

respectively, where $n(t, x_i)$ is assumed to be a *zero-mean, time-domain* and *spatial-domain, WSS random process* $\forall\ i$, and $n(t, x_i)$ and $n(t, x_j)$ are *uncorrelated* for $i \neq j$. As a result, $n_r(t, x_i)$ is a zero-mean, time-domain and spatial-domain, WSS random process $\forall\ i$, and $n_r(t, x_i)$ and $n_r(t, x_j)$ are uncorrelated for $i \neq j$. The *total* output electrical signal from the array (in volts) due to receiver noise is given by

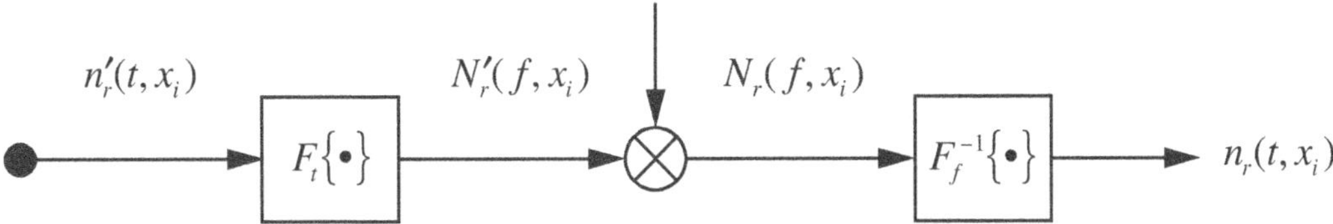

Figure 7.4-2 Beamforming. Implementing a complex weight (amplitude and phase weights) using forward and inverse Fourier transforms.

$$n_r(t) = \sum_{i=-N'}^{N'} n_r(t, x_i) \tag{7.4-13}$$

or

$$\boxed{n_r(t) = \sum_{i=-N'}^{N'} a_i n(t - \tau_i', x_i)} \tag{7.4-14}$$

where $n_r(t)$ is a zero-mean, WSS random process.

7.5 Evaluation of the Equation for Array Gain

In order to evaluate the equation for AG as given by (7.1-19), we need to decide on which model to use for $y_{Trgt}(t, x_i)$ – the output electrical signal from element i in the array due to the target, after complex weighting. There are three different models to choose from, namely, (7.3-35), (7.3-43), and (7.3-50). We shall use (7.3-50) in this section. In Section 7.3 it was shown that if the target is in the Fresnel or far-field region of the array, and if the far-field beam pattern of the array is focused at the range to the target and steered in the direction of the acoustic field incident upon the array (Fresnel region case), or if the far-field beam pattern is simply steered in the direction of the incident acoustic field (far-field region case), then [see (7.3-50)]

$$y_{Trgt}(t, x_i) = a_i s(t - \tau_S), \qquad i = -N', \ldots, 0, \ldots, N', \tag{7.5-1}$$

for $t \geq \tau_S$, where $s(t)$ is given by (7.3-53). As was discussed in Section 7.3, since all the individual signals $y_{Trgt}(t, x_i)$ given by (7.5-1) are *in-phase*, the time-average power of $y_{Trgt}(t)$ given by (7.1-5) will be maximized. As a result, the time-average power of $r(t)$ given by (7.1-4) will also be maximized.

We begin the evaluation of the equation for AG by first computing the output signal-to-noise ratio of the array SNR_A given by (7.1-10). With the use of (7.5-1), the cross-correlation of $y_{Trgt}(t, x_i)$ and $y_{Trgt}(t, x_j)$ is

$$E\left\{y_{Trgt}(t, x_i)\, y^*_{Trgt}(t, x_j)\right\} = a_i a_j E\left\{\left|s(t-\tau_S)\right|^2\right\} = a_i a_j E\left\{\left|s(t)\right|^2\right\}, \tag{7.5-2}$$

or

$$E\left\{y_{Trgt}(t, x_i)\, y^*_{Trgt}(t, x_j)\right\} = a_i a_j E\left\{\left|s(t)\right|^2\right\}, \qquad i, j = -N', \ldots, 0, \ldots, N', \tag{7.5-3}$$

since $s(t)$ is WSS. The second moment (mean-squared value) $E\left\{\left|s(t)\right|^2\right\}$ is the *constant* average power of $s(t)$, where $s(t)$ is given by (7.3-53). Substituting (7.5-3) into (7.1-11) yields

$$E\left\{\left|y_{Trgt}(t)\right|^2\right\} = E\left\{\left|s(t)\right|^2\right\} \sum_{i=-N'}^{N'} a_i \sum_{j=-N'}^{N'} a_j = E\left\{\left|s(t)\right|^2\right\} \left[\sum_{i=-N'}^{N'} a_i\right]^2, \tag{7.5-4}$$

which is the *constant* average power of the signal $y_{Trgt}(t)$ from the array due to the target. Equation (7.5-4) is the numerator of SNR_A.

The denominator of SNR_A is given by (7.1-16). In order to compute the average power of the ambient noise from the array, first rewrite (7.1-13) as follows:

$$E\left\{\left|y_{n_a}(t)\right|^2\right\} = \sum_{i=-N'}^{N'} E\left\{\left|y_{n_a}(t, x_i)\right|^2\right\} + \sum_{\substack{i=-N' \\ i \neq j}}^{N'} \sum_{j=-N'}^{N'} E\left\{y_{n_a}(t, x_i)\, y^*_{n_a}(t, x_j)\right\}. \tag{7.5-5}$$

Since the output electrical signal from element i in the array due to ambient noise after complex weighting is given by [see (7.4-9)]

$$y_{n_a}(t, x_i) = a_i n_a(t - \tau'_i, x_i), \qquad i = -N', \ldots, 0, \ldots, N', \tag{7.5-6}$$

where $n_a(t, x_i)$ is given by (7.4-7), the average power of $y_{n_a}(t, x_i)$ is

$$E\left\{\left|y_{n_a}(t, x_i)\right|^2\right\} = a_i^2 E\left\{\left|n_a(t - \tau'_i, x_i)\right|^2\right\} = a_i^2 E\left\{\left|n_a(t, x_0)\right|^2\right\}, \tag{7.5-7}$$

or

$$E\left\{\left|y_{n_a}(t, x_i)\right|^2\right\} = a_i^2 E\left\{\left|n_a(t, x_0)\right|^2\right\}, \qquad i = -N', \ldots, 0, \ldots, N', \tag{7.5-8}$$

since $n_a(t, x_i)$ is WSS in time and space $\forall\ i$. The second moment (mean-squared value) $E\{|n_a(t, x_0)|^2\}$ is the *constant* average power of the ambient noise at the output of element $i = 0$ in the array after complex weighting. Since $n_a(t, x_i)$ is zero-mean $\forall\ i$, and $n_a(t, x_i)$ and $n_a(t, x_j)$ are uncorrelated for $i \neq j$, the cross-correlation

$$\begin{aligned} E\{y_{n_a}(t, x_i)y^*_{n_a}(t, x_j)\} &= a_i a_j E\{n_a(t - \tau'_i, x_i)n^*_a(t - \tau'_j, x_j)\} \\ &= a_i a_j E\{n_a(t - \tau'_i, x_i)\}E\{n^*_a(t - \tau'_j, x_j)\} = 0, \qquad i \neq j. \end{aligned} \tag{7.5-9}$$

Recall that a cross-correlation function is equal to a cross-covariance function whenever one or both random processes are zero-mean. Since $y_{n_a}(t, x_i)$ is zero-mean $\forall\ i$, (7.5-9) agrees with the statement made in Section 7.4 that $y_{n_a}(t, x_i)$ and $y_{n_a}(t, x_j)$ are uncorrelated for $i \neq j$. Therefore, by substituting (7.5-8) and (7.5-9) into (7.5-5), we obtain

$$E\{|y_{n_a}(t)|^2\} = E\{|n_a(t, x_0)|^2\} \sum_{i=-N'}^{N'} a_i^2, \tag{7.5-10}$$

which is the *constant* average power of the ambient noise $y_{n_a}(t)$ from the array. Also recall that a mean-squared value is equal to a variance for a zero-mean random process.

In order to compute the average power of the receiver noise from the array, first rewrite (7.1-14) as follows:

$$E\{|n_r(t)|^2\} = \sum_{i=-N'}^{N'} E\{|n_r(t, x_i)|^2\} + \sum_{\substack{i=-N' \\ i\neq j}}^{N'} \sum_{j=-N'}^{N'} E\{n_r(t, x_i)n^*_r(t, x_j)\}. \tag{7.5-11}$$

Since the output electrical signal from element i in the array due to receiver noise after complex weighting is given by [see (7.4-12)]

$$n_r(t, x_i) = a_i n(t - \tau'_i, x_i), \qquad i = -N', \ldots, 0, \ldots, N', \tag{7.5-12}$$

the average power of $n_r(t, x_i)$ is

$$E\{|n_r(t, x_i)|^2\} = a_i^2 E\{|n(t - \tau'_i, x_i)|^2\} = a_i^2 E\{|n(t, x_0)|^2\}, \tag{7.5-13}$$

or

$$E\left\{\left|n_r(t,x_i)\right|^2\right\}=a_i^2 E\left\{\left|n(t,x_0)\right|^2\right\}, \qquad i=-N',\ldots,0,\ldots,N', \tag{7.5-14}$$

since $n(t,x_i)$ is WSS in time and space $\forall\ i$. The second moment (mean-squared value) $E\left\{\left|n(t,x_0)\right|^2\right\}$ is the *constant* average power of the receiver noise at the output of element $i=0$ in the array after complex weighting. Since $n(t,x_i)$ is zero-mean $\forall\ i$, and $n(t,x_i)$ and $n(t,x_j)$ are uncorrelated for $i\neq j$, the cross-correlation

$$\begin{aligned}E\left\{n_r(t,x_i)n_r^*(t,x_j)\right\}&=a_ia_jE\left\{n(t-\tau_i',x_i)n^*(t-\tau_j',x_j)\right\}\\&=a_ia_jE\left\{n(t-\tau_i',x_i)\right\}E\left\{n^*(t-\tau_j',x_j)\right\}=0, \qquad i\neq j.\end{aligned} \tag{7.5-15}$$

Since a cross-correlation function is equal to a cross-covariance function whenever one or both random processes are zero-mean, and since $n_r(t,x_i)$ is zero-mean $\forall\ i$, (7.5-15) agrees with the statement made in Section 7.4 that $n_r(t,x_i)$ and $n_r(t,x_j)$ are uncorrelated for $i\neq j$. Therefore, by substituting (7.5-14) and (7.5-15) into (7.5-11), we obtain

$$E\left\{\left|n_r(t)\right|^2\right\}=E\left\{\left|n(t,x_0)\right|^2\right\}\sum_{i=-N'}^{N'}a_i^2, \tag{7.5-16}$$

which is the *constant* average power of the receiver noise $n_r(t)$ from the array. Substituting (7.5-10) and (7.5-16) into (7.1-16) yields

$$E\left\{\left|z(t)\right|^2\right\}=\left[E\left\{\left|n_a(t,x_0)\right|^2\right\}+E\left\{\left|n(t,x_0)\right|^2\right\}\right]\sum_{i=-N'}^{N'}a_i^2. \tag{7.5-17}$$

Equation (7.5-17) is the denominator of $\mathrm{SNR_A}$. The output signal-to-noise ratio of the array can now be obtained by substituting (7.5-4) and (7.5-17) into (7.1-10). Doing so yields

$$\mathrm{SNR_A}=\frac{E\left\{\left|s(t)\right|^2\right\}}{E\left\{\left|n_a(t,x_0)\right|^2\right\}+E\left\{\left|n(t,x_0)\right|^2\right\}}\frac{\left[\sum_{i=-N'}^{N'}a_i\right]^2}{\sum_{i=-N'}^{N'}a_i^2}. \tag{7.5-18}$$

Next, compute the output signal-to-noise ratio at element $i=0$ in the array, SNR_0, given by (7.1-17). Evaluating (7.5-3) at $i=0$ and $j=0$, (7.5-8) at $i=0$, and (7.5-14) at $i=0$ yields

$$E\left\{\left|y_{Trgt}(t,x_0)\right|^2\right\}=a_0^2 E\left\{\left|s(t)\right|^2\right\}, \tag{7.5-19}$$

$$E\left\{\left|y_{n_a}(t,x_0)\right|^2\right\}=a_0^2 E\left\{\left|n_a(t,x_0)\right|^2\right\}, \tag{7.5-20}$$

and

$$E\left\{\left|n_r(t,x_0)\right|^2\right\}=a_0^2 E\left\{\left|n(t,x_0)\right|^2\right\}, \tag{7.5-21}$$

respectively. Substituting (7.5-20) and (7.5-21) into (7.1-18) yields

$$E\left\{\left|z(t,x_0)\right|^2\right\}=a_0^2\left[E\left\{\left|n_a(t,x_0)\right|^2\right\}+E\left\{\left|n(t,x_0)\right|^2\right\}\right], \tag{7.5-22}$$

and by substituting (7.5-19) and (7.5-22) into (7.1-17), we obtain

$$\text{SNR}_0=\frac{E\left\{\left|s(t)\right|^2\right\}}{E\left\{\left|n_a(t,x_0)\right|^2\right\}+E\left\{\left|n(t,x_0)\right|^2\right\}}. \tag{7.5-23}$$

And finally, substituting (7.5-18) and (7.5-23) into (7.1-19) yields the following expression for array gain:

$$\boxed{\text{AG}=10\log_{10}\frac{\left[\sum\limits_{i=-N'}^{N'} a_i\right]^2}{\sum\limits_{i=-N'}^{N'} a_i^2}\ \text{dB}} \tag{7.5-24}$$

If rectangular amplitude weights are used, then $a_i=1\ \forall\ i$, and (7.5-24) reduces to

$$\boxed{\text{AG}=10\log_{10} N\ \text{dB}} \tag{7.5-25}$$

where N is the total number of elements in the array. For example, if the number of elements is doubled and if (7.5-25) is applicable, then the AG is increased by 3 dB. Later, in Chapter 12, we will show how AG can increase the probability of detecting a very weak signal with a low SNR at the output of a single element in an array.

In summary, the AG formula given by (7.5-25) is based on the following major assumptions:

1) The target is modeled as an *omnidirectional point-source with arbitrary time dependence* $s_0(t)$, where $s_0(t)$ is the source strength of the target.

2) The source strength of the target $s_0(t)$ is a *wide-sense stationary* (WSS) *random process*.

3) The fluid medium is *unbounded*, *viscous*, and *homogeneous* (constant speed of sound).

4) The ambient noise is *isotropic* and the velocity potential of the output acoustic signal from the fluid medium due to ambient noise and incident upon element i in the array, $y_{M,n_a}(t,x_i)$, is a *zero-mean*, *time-domain* and *spatial-domain*, *WSS random process* $\forall\ i$. Also, $y_{M,n_a}(t,x_i)$ and $y_{M,n_a}(t,x_j)$ are *uncorrelated* for $i \neq j$.

5) The receiver noise $n(t,x_i)$ is a *zero-mean*, *time-domain* and *spatial-domain*, *WSS random process* $\forall\ i$, and $n(t,x_i)$ and $n(t,x_j)$ are *uncorrelated* for $i \neq j$. Also, the receiver noise and ambient noise are *statistically independent*.

6) The target is in either the *Fresnel* or *far-field* region of the array.

7) All the individual output electrical signals due to the target after complex weighting, $y_{Trgt}(t,x_i)$, $i=-N',\ldots,0,\ldots,N'$, are *in-phase*.

8) *Rectangular amplitude weights* are used.

Problems

Section 7.2

7-1 If the attenuation coefficient of seawater is $4.425\ \text{dB/km}$ at a frequency of $25\ \text{kHz}$, then what is its value in nepers per meter?

Section 7.3

7-2 Compute the autocorrelation function

$$R_{y_{Trgt}}(t_1,t_2)=E\{y_{Trgt}(t_1)y^*_{Trgt}(t_2)\}$$

of $y_{Trgt}(t)$ given by (7.3-51). At the end of the derivation, let $t_1=t$ and $t_2=t-\tau$.

Section 7.4

7-3 Compute the autocorrelation function

$$R_{y_{n_a}}(t_1,t_2)=E\{y_{n_a}(t_1)y^*_{n_a}(t_2)\}$$

of $y_{n_a}(t)$ given by (7.4-10). At the end of the derivation, let $t_1=t$ and $t_2=t-\tau$.

7-4 Compute the autocorrelation function

$$R_{n_r}(t_1,t_2)=E\{n_r(t_1)n^*_r(t_2)\}$$

of $n_r(t)$ given by (7.4-14). At the end of the derivation, let $t_1=t$ and $t_2=t-\tau$.

Section 7.5

7-5 For $N=7$,

(a) use (7.5-25) to compute the array gain for the set of rectangular amplitude weights.

(b) use (7.5-24) to compute the array gain for the set of Hamming amplitude weights and the normalized, Dolph-Chebyshev amplitude weights computed in Example 6.3-1.

Appendix 7A Attenuation Coefficient of Seawater

An accurate equation for the frequency-dependent attenuation coefficient of sound in seawater, valid for all oceanic conditions and frequencies from 200 Hz to 1 MHz, is given by[1]

---

[1] R. E. Francois and G. R. Garrison, "Sound Absorption Based on Ocean Measurements. Part II: Boric Acid Contribution and Equation for Total Absorption," *J. Acoust. Soc. Am.*, vol. 72, pp. 1879-1890, 1982.

$$\alpha'(f)=\frac{A_1P_1f_1}{f^2+f_1^2}f^2+\frac{A_2P_2f_2}{f^2+f_2^2}f^2+A_3P_3f^2\,, \tag{7A-1}$$

where $\alpha'(f)$ has units of decibels per kilometer (dB/km) and frequency f is in kilohertz (kHz). The various parameters appearing in (7A-1) are defined next.

Boric Acid Contribution

$$A_1=\frac{8.86}{c}\times 10^{(0.78\,p\mathrm{H}-5)} \tag{7A-2}$$

has units of dB/(km-kHz), where pH is the acidity (dimensionless) of seawater,

$$c=1412+3.21T+1.19S+0.0167y \tag{7A-3}$$

is the speed of sound in meters per second (m/sec), where T is the temperature in degrees Celsius (°C), S is the salinity in parts per thousand (ppt), and y is the depth in meters (m),

$$P_1=1\,, \tag{7A-4}$$

where P_1 is dimensionless,

$$f_1=2.8\sqrt{\frac{S}{35}}\times 10^{[4-(1245/\theta)]} \tag{7A-5}$$

is the relaxation frequency of boric acid in kHz , and

$$\theta=273+T \tag{7A-6}$$

is the absolute temperature in degrees Kelvin (°K).

Magnesium Sulfate Contribution

$$A_2=21.44\frac{S}{c}(1+0.025T) \tag{7A-7}$$

has units of dB/(km-kHz),

$$P_2=1-(1.37\times 10^{-4})y+(6.2\times 10^{-9})y^2\,, \tag{7A-8}$$

where P_2 is dimensionless, and

$$f_2 = \frac{8.17 \times 10^{[8-(1990/\theta)]}}{1 + 0.0018(S - 35)} \tag{7A-9}$$

is the relaxation frequency of magnesium sulfate in kHz.

Pure Water Contribution

$$P_3 = 1 - (3.83 \times 10^{-5})y + (4.9 \times 10^{-10})y^2, \tag{7A-10}$$

where P_3 is dimensionless.

For $T \leq 20°\,\text{C}$:

$$A_3 = 4.937 \times 10^{-4} - (2.59 \times 10^{-5})T + (9.11 \times 10^{-7})T^2 - (1.50 \times 10^{-8})T^3 \tag{7A-11}$$

has units of $\text{dB}/(\text{km-kHz}^2)$.

For $T > 20°\,\text{C}$:

$$A_3 = 3.964 \times 10^{-4} - (1.146 \times 10^{-5})T + (1.45 \times 10^{-7})T^2 - (6.5 \times 10^{-10})T^3 \tag{7A-12}$$

has units of $\text{dB}/(\text{km-kHz}^2)$.

Appendix 7B Fourier Transform, Fourier Series Coefficients, Time-Average Power, and Power Spectrum via the DFT

Fourier Transform

An estimate $\hat{X}(f)$ of the Fourier transform $X(f)$ of $x(t)$ can be computed numerically as follows:

$$\boxed{\hat{X}(q) = T_S \mathcal{X}(q), \qquad q = 0, 1, \ldots, N + Z - 1} \tag{7B-1}$$

where

$$\hat{X}(q) \equiv \hat{X}(qf_0) = \hat{X}(f)\Big|_{f=qf_0} \tag{7B-2}$$

and

$$\mathcal{X}(q) \equiv \mathcal{X}(qf_0) = \mathcal{X}(f)\big|_{f=qf_0}, \tag{7B-3}$$

where

$$\mathcal{X}(q) = DFT\{x(i)\} = \sum_{i=0}^{N-1} x(i) W_{N+Z}^{-qi}, \qquad q = 0, 1, \ldots, N+Z-1, \tag{7B-4}$$

is the *discrete Fourier transform* (DFT) of $x(t)$, $x(i) \equiv x(t_i) = x(iT_S)$, and

$$W_{N+Z} = \exp\left(+j\frac{2\pi}{N+Z}\right), \tag{7B-5}$$

where T_S is the sampling period in seconds, N is the total number of samples taken of $x(t)$, and Z is a *positive integer* ($Z \geq 0$). The waveform $x(t)$ is defined for $0 \leq t \leq T$, where T is the duration of $x(t)$ in seconds. If $x(t)$ has units of volts, then $\hat{X}(q)$ has units of volts per hertz, and if $x(t)$ has units of amperes, then $\hat{X}(q)$ has units of amperes per hertz.

The frequency components f in hertz are given by

$$f = qf_0, \qquad q = 0, 1, \ldots, N+Z-1, \tag{7B-6}$$

where

$$\boxed{f_0 = \frac{1}{T_0} = \frac{1}{(N+Z)T_S} = \frac{f_S}{N+Z}} \tag{7B-7}$$

is the DFT *bin spacing* [the fundamental frequency of $x(t)$] in hertz,

$$\boxed{T_0 = (N+Z)T_S} \tag{7B-8}$$

is the fundamental period of $x(t)$ in seconds, and $f_S = 1/T_S$ is the sampling frequency in hertz (a.k.a. the sampling rate in samples per second). As can be seen from (7B-7), when integer $Z = 0$, the DFT bin spacing $f_0 = 1/(NT_S)$, which is the *maximum* allowed value in hertz in order to avoid aliasing when performing a time-domain DFT. Choosing $Z > 0$ decreases f_0 and increases $N+Z$. Increasing $N+Z$ means that $\hat{X}(q)$ is evaluated at more bins yielding a smoother curve for plotting purposes. For example, if we want to evaluate $\hat{X}(q)$ at twice as many bins by reducing the bin spacing f_0 by a factor of two, then set $Z = N$ so that $N+Z = 2N$. This is equivalent to adding $Z = N$ zeros to the original data sequence so that $N+Z = 2N$. Choosing $Z > 0$ is referred to as padding-with-zeros.

In order to determine the value for N to be used in the DFT, set $Z=0$ and $T_0=T$, where T is the duration of $x(t)$ in seconds. Doing so yields

$$T_0=T=NT_S, \tag{7B-9}$$

or

$$\boxed{N=\frac{T}{T_S}=f_S T} \tag{7B-10}$$

The corresponding DFT bin spacing for $Z=0$ is given by

$$f_0=\frac{1}{T_0}=\frac{1}{T}=\frac{1}{NT_S}=\frac{f_S}{N}. \tag{7B-11}$$

However, if we want $f_0<1/T$, then the required value for Z is given by

$$\boxed{Z=\frac{f_S}{f_0}-N} \tag{7B-12}$$

where $f_0<1/T$ is the desired DFT bin spacing. Note that $T\le T_0$.

Fourier Series Coefficients

An estimate of the set of complex Fourier series coefficients of $x(t)$ can be computed numerically as follows:

$$\boxed{\hat{c}_q=\frac{1}{N+Z}\mathcal{X}(q), \qquad q=0,1,\ldots,N+Z-1} \tag{7B-13}$$

where $\hat{c}_q$ is an estimate of the complex Fourier series coefficient c_q of $x(t)$ at harmonic q (frequency $f=qf_0$ Hz), and $\mathcal{X}(q)$ is the DFT of $x(t)$. If $x(t)$ has units of volts, then $\hat{c}_q$ has units of volts, and if $x(t)$ has units of amperes, then $\hat{c}_q$ has units of amperes.

Time-Average Power

The time-average power of $x(t)$ in the time interval $[0,T]$ can be computed numerically as follows:

$$P_{\text{avg},x,T} = \frac{1}{T}\int_0^T |x(t)|^2 dt = \frac{T_0}{T} P_{\text{avg},x,T_0} \tag{7B-14}$$

where

$$P_{\text{avg},x,T_0} = \frac{1}{T_0}\int_0^{T_0} |x(t)|^2 dt = \frac{1}{T_0}\int_0^T |x(t)|^2 dt = \sum_{q=0}^{N+Z-1} |\hat{c}_q|^2 \tag{7B-15}$$

is the time-average power of $x(t)$ in the time interval $[0, T_0]$, $T_0 = (N+Z)T_S$ is the fundamental period of $x(t)$ in seconds, $T = NT_S$ is the duration of $x(t)$ in seconds, and $\hat{c}_q$ is an estimate of the complex Fourier series coefficient c_q of $x(t)$ at harmonic q (frequency $f = qf_0$ Hz). Equation (7B-15) is Parseval's Relation from Fourier series theory. Note that $T \le T_0$. If $x(t)$ has units of volts, then P_{avg,x,T_0} and $P_{\text{avg},x,T}$ have units of watts-ohms, and if $x(t)$ has units of amperes, then P_{avg,x,T_0} and $P_{\text{avg},x,T}$ have units of watts per ohm (see Subsection 11.1.1).

Power Spectrum

An estimate $\hat{S}_x(f)$ of the power spectrum $S_x(f)$ of $x(t)$ can be computed numerically as follows:

$$\hat{S}_x(q) = \frac{1}{T}\left|\hat{X}(q)\right|^2, \qquad q = 0, 1, \ldots, N+Z-1 \tag{7B-16}$$

where

$$\hat{S}_x(q) \equiv \hat{S}_x(qf_0) = \hat{S}_x(f)\Big|_{f=qf_0}, \tag{7B-17}$$

$\hat{X}(q)$ is the numerical Fourier transform of $x(t)$ given by (7B-1), and T is the duration of $x(t)$ in seconds. If $x(t)$ has units of volts, then $\hat{S}_x(q)$ has units of $(\text{W-}\Omega)/\text{Hz}$, and if $x(t)$ has units of amperes, then $\hat{S}_x(q)$ has units of $(\text{W}/\Omega)/\text{Hz}$.

Chapter 8

Array Theory – Planar Arrays

8.1 The Far-Field Beam Pattern of a Planar Array

Since a planar array is an example of a planar aperture, the far-field beam pattern (directivity function) of a planar array lying in the XY plane is given by the following two-dimensional spatial Fourier transform (see Section 3.1):

$$\begin{aligned} D(f, f_X, f_Y) &= F_{x_A} F_{y_A}\{A(f, x_A, y_A)\} \\ &= \int_{-\infty}^{\infty}\int_{-\infty}^{\infty} A(f, x_A, y_A)\exp[+j2\pi(f_X x_A + f_Y y_A)]dx_A\, dy_A, \end{aligned} \tag{8.1-1}$$

where $A(f, x_A, y_A)$ is the complex frequency response (complex aperture function) of the planar array, and

$$f_X = u/\lambda = \sin\theta\cos\psi/\lambda \tag{8.1-2}$$

and

$$f_Y = v/\lambda = \sin\theta\sin\psi/\lambda \tag{8.1-3}$$

are spatial frequencies in the X and Y directions, respectively, with units of cycles per meter. Since f_X and f_Y can be expressed in terms of the spherical angles θ and ψ, the far-field beam pattern can ultimately be expressed as a function of frequency f and the spherical angles θ and ψ, that is, $D(f, f_X, f_Y) \rightarrow D(f, \theta, \psi)$. We shall consider a field point to be in the far-field region of an array if the range r to the field point – as measured from the center of the array – satisfies the Fraunhofer (far-field) range criterion given by

$$r > \pi R_A^2/\lambda > 2.414 R_A, \tag{8.1-4}$$

where R_A is the maximum radial extent of the array.

If a planar array lies in the XZ plane instead of the XY plane, then simply replace y_A and f_Y with z_A and f_Z, respectively, in (8.1-1), where spatial frequency f_Z is given by (1.2-45). And if a planar array lies in the YZ plane instead of the XY plane, then simply replace x_A and f_X with z_A and f_Z, respectively, in (8.1-1). As can be seen from (8.1-1), in order to derive the far-field beam pattern of a planar array, we must first specify a mathematical model for its complex frequency response.

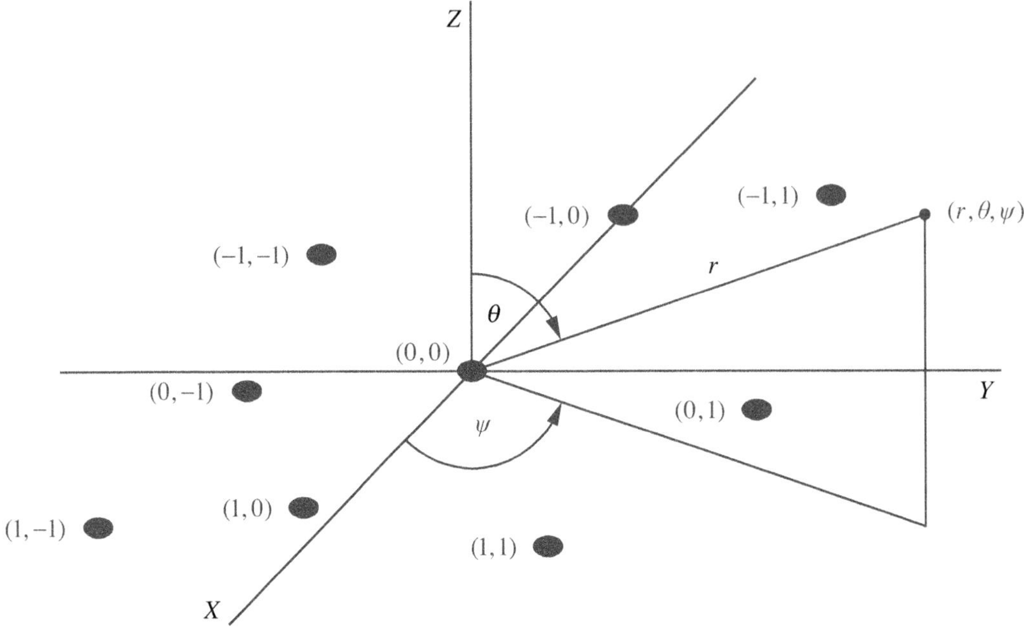

Figure 8.1-1 Planar array composed of an odd number of unevenly-spaced elements lying in the XY plane. Also shown are the element numbers (m,n) and a field point with spherical coordinates (r,θ,ψ).

Consider a planar array composed of an *odd* number $M \times N$ of elements (electroacoustic transducers) lying in the XY plane, where M and N are the total odd number of elements in the X and Y directions, respectively (see Fig. 8.1-1). For example, the elements can be omnidirectional point-elements, rectangular pistons, or circular pistons. If the elements are *not* identical, that is, if each element has a different complex frequency response, and if the elements are *not* equally spaced, as shown in Fig. 8.1-1, then the complex frequency response (complex aperture function) of this array can be expressed as

$$A(f, x_A, y_A) = \sum_{m=-M'}^{M'} \sum_{n=-N'}^{N'} c_{mn}(f) e_{mn}(f, x_A - x_{mn}, y_A - y_{mn}), \qquad (8.1\text{-}5)$$

where

$$M' = (M-1)/2, \qquad (8.1\text{-}6)$$

$$N' = (N-1)/2, \qquad (8.1\text{-}7)$$

$$\boxed{c_{mn}(f) = a_{mn}(f)\exp[+j\theta_{mn}(f)]} \qquad (8.1\text{-}8)$$

is the frequency-dependent, *complex weight* associated with element (m,n), $a_{mn}(f)$ is a real, frequency-dependent, *dimensionless*, amplitude weight, $\theta_{mn}(f)$ is a real, frequency-dependent, phase weight in radians, $e_{mn}(f, x_A, y_A)$ is the complex frequency response (complex aperture function) of element (m,n), also known as the *element function*, and x_{mn} and y_{mn} are the x and y coordinates of the center of element (m,n). Note that $x_{00}=0$ and $y_{00}=0$ because element $(0,0)$ is centered at the origin of the coordinate system. Also note that, in general, $M \neq N$, and because the elements are not equally spaced, $x_{-m\,-n} \neq -x_{mn}$ and $y_{-m\,-n} \neq -y_{mn}$. Complex weights are implemented by using digital signal processing (digital beamforming) (see Section 8.2).

Equation (8.1-5) indicates that the complex frequency response of the array is modeled as the sum (superposition) of the complex-weighted frequency responses of all the elements in the array. The complex weights are used to control the complex frequency response of the array and, thus, the array's far-field beam pattern via amplitude and phase weighting. Substituting (8.1-5) into (8.1-1) yields the following expression for the far-field beam pattern:

$$\boxed{D(f, f_X, f_Y) = \sum_{m=-M'}^{M'} \sum_{n=-N'}^{N'} c_{mn}(f) E_{mn}(f, f_X, f_Y) \exp\left[+j2\pi(f_X x_{mn} + f_Y y_{mn})\right]} \tag{8.1-9}$$

where

$$F_{x_A} F_{y_A}\left\{e_{mn}(f, x_A - x_{mn}, y_A - y_{mn})\right\} = E_{mn}(f, f_X, f_Y)\exp\left[+j2\pi(f_X x_{mn} + f_Y y_{mn})\right] \tag{8.1-10}$$

and

$$E_{mn}(f, f_X, f_Y) = F_{x_A} F_{y_A}\left\{e_{mn}(f, x_A, y_A)\right\} \tag{8.1-11}$$

is the far-field beam pattern of element (electroacoustic transducer) (m,n). The units of $e_{mn}(f, x_A, y_A)$, $A(f, x_A, y_A)$, $E_{mn}(f, f_X, f_Y)$, and $D(f, f_X, f_Y)$ are summarized in Table 8.1-1.

If all the elements in the array are *identical* (i.e., all the elements have the same complex frequency response), but still *not* equally spaced, then (8.1-9) reduces to

$$\boxed{D(f, f_X, f_Y) = E(f, f_X, f_Y) S(f, f_X, f_Y)} \tag{8.1-12}$$

where

Table 8.1-1 Units of the Complex Frequency Responses $e_{mn}(f, x_A, y_A)$ and $A(f, x_A, y_A)$ and Corresponding Far-Field Beam Patterns $E_{mn}(f, f_X, f_Y)$ and $D(f, f_X, f_Y)$ for a Planar Array

Planar Array	$e_{mn}(f, x_A, y_A)$, $A(f, x_A, y_A)$	$E_{mn}(f, f_X, f_Y)$, $D(f, f_X, f_Y)$
active (transmit)	$\left(\left(\text{m}^3/\text{sec}\right)/\text{V}\right)/\text{m}^2$	$\left(\text{m}^3/\text{sec}\right)/\text{V}$
passive (receive)	$\left(\text{V}/\left(\text{m}^2/\text{sec}\right)\right)/\text{m}^2$	$\text{V}/\left(\text{m}^2/\text{sec}\right)$

$$E(f, f_X, f_Y) = F_{x_A} F_{y_A}\{e(f, x_A, y_A)\} \tag{8.1-13}$$

is the far-field beam pattern of *one* of the identical elements in the array,

$$S(f, f_X, f_Y) = \sum_{m=-M'}^{M'} \sum_{n=-N'}^{N'} c_{mn}(f) \exp\left[+j2\pi(f_X x_{mn} + f_Y y_{mn})\right] \tag{8.1-14}$$

is the *array factor*, and if $a_{mn}(f) > 0 \;\forall\; m$ and n, then

$$S_{\max} = \max\left|S(f, f_X, f_Y)\right| = \sum_{m=-M'}^{M'} \sum_{n=-N'}^{N'} a_{mn}(f) \tag{8.1-15}$$

is the *normalization factor* for the array factor (see Appendix 8B). As can be seen from (8.1-14), $S(f, f_X, f_Y)$ is *dimensionless* and depends on the set of complex weights used and the locations (the x and y coordinates) of all the elements in the XY plane. The array factor $S(f, f_X, f_Y)$ contains the spatial information of the array. If one or more elements in the array are broken (not transmitting and/or receiving), set $c_{mn}(f) = 0$ for those elements. Equation (8.1-12) is referred to as the *Product Theorem* for a planar array composed of an odd number of identical elements. The Product Theorem states that the far-field beam pattern $D(f, f_X, f_Y)$ of a planar array composed of an odd number of *identical*, unevenly-spaced, complex-weighted elements is equal to the *product* of the far-field beam pattern $E(f, f_X, f_Y)$ of *one* of the identical elements in the array, and the dimensionless array factor $S(f, f_X, f_Y)$, which is related to the far-field beam pattern of an equivalent planar array composed of an odd number of identical, unevenly-spaced, complex-weighted, *omnidirectional point-elements*. Although the Product Theorem requires identical elements, equal spacing of the elements is *not* required. The interpretation of $S(f, f_X, f_Y)$ shall be discussed next.

Since an impulse function is used as a mathematical model for an omnidirectional point-source in acoustic wave propagation theory, the mathematical model that we shall use for the complex frequency response of an omnidirectional point-element lying in the XY plane is given by

$$e(f, x_A, y_A) = \mathcal{S}(f)\delta(x_A, y_A) = \mathcal{S}(f)\delta(x_A)\delta(y_A), \tag{8.1-16}$$

where $\mathcal{S}(f)$ is the complex, *element sensitivity function.* If the point-element is used in the active mode as a transmitter, then the element sensitivity function $\mathcal{S}(f)$ is referred to as the *transmitter sensitivity function* with units of $\left(\text{m}^3/\text{sec}\right)/\text{V}$ (see Appendix 6B). If the point-element is used in the passive mode as a receiver, then $\mathcal{S}(f)$ is referred to as the *receiver sensitivity function* with units of $\text{V}/\left(\text{m}^2/\text{sec}\right)$ (see Appendix 6B). Note that $\int_{-\infty}^{\infty}\int_{-\infty}^{\infty} e(f, x_A, y_A)\,dx_A\,dy_A = \mathcal{S}(f)$ because $\int_{-\infty}^{\infty} \delta(x_A)\,dx_A \int_{-\infty}^{\infty} \delta(y_A)\,dy_A = 1$. The impulse function $\delta(x_A, y_A) = \delta(x_A)\delta(y_A)$ has units of inverse squared-meters (m^{-2}). The units of $\mathcal{S}(f)$ are summarized in Table 8.1-2.

Table 8.1-2 Units of the Element Sensitivity Function $\mathcal{S}(f)$

Omnidirectional Point-Element	$\mathcal{S}(f)$
active (transmit)	$\left(\text{m}^3/\text{sec}\right)/\text{V}$
passive (receive)	$\text{V}/\left(\text{m}^2/\text{sec}\right)$

Since

$$F_{x_A} F_{y_A}\{\delta(x_A, y_A)\} = F_{x_A}\{\delta(x_A)\} F_{y_A}\{\delta(y_A)\} = 1, \tag{8.1-17}$$

taking the two-dimensional spatial Fourier transform of (8.1-16) yields

$$E(f, f_X, f_Y) = \mathcal{S}(f). \tag{8.1-18}$$

Equation (8.1-18) is the far-field beam pattern of an omnidirectional point-element and as expected, it is *not* a function of the spherical angles θ and ψ. The term "omnidirectional" means that an electroacoustic transducer transmits and/or receives equally well in all directions. Therefore, by substituting (8.1-18) into (8.1-12), we obtain the following expression for the far-field beam pattern of a planar array composed of an odd number $M \times N$ of identical, unevenly-spaced, complex-weighted, *omnidirectional point-elements* lying in the XY plane, where M and N are the total odd number of elements in the X and Y directions, respectively:

$$D(f, f_X, f_Y) = \mathcal{S}(f) S(f, f_X, f_Y) \tag{8.1-19}$$

where $S(f, f_X, f_Y)$ is given by (8.1-14). The corresponding complex frequency response (complex aperture function) of the array can be obtained by taking the inverse two-dimensional spatial Fourier transform of (8.1-19). Doing so yields

$$A(f, x_A, y_A) = \mathcal{S}(f) \sum_{m=-M'}^{M'} \sum_{n=-N'}^{N'} c_{mn}(f) \delta(x_A - x_{mn}) \delta(y_A - y_{mn}) \tag{8.1-20}$$

since

$$\begin{aligned} F_{f_X}^{-1} F_{f_Y}^{-1} \left\{ \exp\left[+j2\pi(f_X x_{mn} + f_Y y_{mn}) \right] \right\} &= F_{f_X}^{-1} \left\{ \exp(+j2\pi f_X x_{mn}) \right\} F_{f_Y}^{-1} \left\{ \exp(+j2\pi f_Y y_{mn}) \right\} \\ &= \delta(x_A - x_{mn}) \delta(y_A - y_{mn}). \end{aligned} \tag{8.1-21}$$

Note that the double summation in (8.1-20) is in the form of the impulse response of a two-dimensional, finite impulse response (FIR) digital filter. In Appendix 8A, it is shown that a planar array composed of an odd number of identical, complex-weighted elements can be thought of as a two-dimensional, spatial FIR filter, where $s(f, x_A, y_A) = F_{f_X}^{-1} F_{f_Y}^{-1} \{ S(f, f_X, f_Y) \}$ is the spatial impulse response of the array at frequency f hertz.

Finally, if all the elements in the array are *equally spaced* in the X direction, and *equally spaced* in the Y direction, then

$$x_{mn} = x_m = m d_X, \qquad m = -M', \ldots, 0, \ldots, M' \tag{8.1-22}$$

and

$$y_{mn} = y_n = n d_Y, \qquad n = -N', \ldots, 0, \ldots, N' \tag{8.1-23}$$

where d_X and d_Y are the *interelement spacings* in meters in the X and Y directions, respectively. Note that, in general, $d_X \neq d_Y$.

Example 8.1-1 Planar Array of Rectangular Pistons

Consider a planar array composed of an odd number $M \times N$ of identical, unevenly-spaced, complex-weighted, rectangular pistons lying in the XY plane, where M and N are the total odd number of elements in the X and Y directions, respectively. Since the elements are identical, the Product Theorem given by (8.1-12) shall be used to obtain the far-field beam pattern of the array. The element function is given by (see Section 3.2)

$$e(f, x_A, y_A) = A_A \operatorname{rect}(x_A/L_X)\operatorname{rect}(y_A/L_Y), \tag{8.1-24}$$

with corresponding unnormalized, far-field beam pattern

$$E(f, f_X, f_Y) = A_A L_X L_Y \operatorname{sinc}(f_X L_X)\operatorname{sinc}(f_Y L_Y), \tag{8.1-25}$$

where the real, positive constant A_A carries the correct units of $e(f, x_A, y_A)$ (see Table 8.1-1), and $L_X L_Y$ is the area of an individual rectangular piston in squared meters. Although A_A is shown here as a constant, as was previously mentioned in Section 3.2, A_A is, in general, a real, nonnegative function of frequency, that is, $A_A \to A_A(f)$. Therefore, substituting (8.1-25) into (8.1-12) yields the following expression for the unnormalized, far-field beam pattern of the array:

$$D(f, f_X, f_Y) = A_A L_X L_Y \operatorname{sinc}(f_X L_X)\operatorname{sinc}(f_Y L_Y) S(f, f_X, f_Y), \tag{8.1-26}$$

where $S(f, f_X, f_Y)$ is given by (8.1-14). If all the elements in the array are equally spaced in the X direction, and equally spaced in the Y direction, then substitute (8.1-22) and (8.1-23) into (8.1-14).

If L_X and L_Y are small compared to a wavelength so that

$$L_X/\lambda \le 0.0278 \tag{8.1-27}$$

and

$$L_Y/\lambda \le 0.0278, \tag{8.1-28}$$

then (8.1-25) reduces to (see Example 6.2-1)

$$E(f, f_X, f_Y) \approx A_A L_X L_Y, \quad L_X/\lambda \le 0.0278, \quad L_Y/\lambda \le 0.0278. \tag{8.1-29}$$

Therefore, the rectangular pistons act like omnidirectional point-elements because the far-field beam pattern $E(f, f_X, f_Y)$ is not a function of the spherical angles θ and ψ. Substituting (8.1-29) into (8.1-12) yields the following expression for the unnormalized, far-field beam pattern of the array:

$$D(f, f_X, f_Y) \approx A_A L_X L_Y S(f, f_X, f_Y), \quad L_X/\lambda \le 0.0278, \quad L_Y/\lambda \le 0.0278. \tag{8.1-30}$$

Compare (8.1-30) with (8.1-26) and (8.1-19). If we let $L_{\max} = 0.0278\lambda$, then the far-field beam patterns $E(f, f_X, f_Y)$ and $D(f, f_X, f_Y)$ given by (8.1-29) and (8.1-30) are valid as long as $L_X \le L_{\max}$ and $L_Y \le L_{\max}$ (see Table 8.1-3). ■

Table 8.1-3 Values for the Maximum Length $L_{\max} = 0.0278\lambda$ for Three Different Frequencies f. Note that 1 in = 2.54 cm.

c (m/sec)	f (kHz)	λ (m)	$L_{\max}$ (cm)	$L_{\max}$ (in)
1500	0.1	15.0	41.7	16.4
1500	1.0	1.5	4.17	1.64
1500	10.0	0.15	0.417	0.164

Example 8.1-2 Planar Array of Circular Pistons

Consider a planar array composed of an odd number $M \times N$ of identical, unevenly-spaced, complex-weighted, circular pistons lying in the XY plane, where M and N are the total odd number of elements in the X and Y directions, respectively. Since the elements are identical, the Product Theorem given by (8.1-12) shall be used to obtain the far-field beam pattern of the array. From Section 3.3, the element function is given by

$$e(f, r_A, \phi_A) = A_A \operatorname{circ}(r_A/a), \tag{8.1-31}$$

with corresponding unnormalized, far-field beam pattern

$$E(f, \theta, \psi) = 2A_A \pi a^2 \frac{J_1\left(\dfrac{2\pi a}{\lambda}\sin\theta\right)}{\dfrac{2\pi a}{\lambda}\sin\theta}, \tag{8.1-32}$$

where the real, positive constant A_A carries the correct units of $e(f, r_A, \phi_A)$ (see Table 8.1-1), a is the radius of an individual circular piston in meters, and πa^2 is the area of an individual circular piston in squared meters. Although A_A is shown here as a constant, as was previously mentioned in Section 3.3, A_A is, in general, a real, nonnegative function of frequency, that is, $A_A \to A_A(f)$. Therefore, substituting (8.1-32) into (8.1-12) yields the following expression for the unnormalized, far-field beam pattern of the array:

$$D(f,\theta,\psi) = 2A_A\pi a^2 \frac{J_1\left(\frac{2\pi a}{\lambda}\sin\theta\right)}{\frac{2\pi a}{\lambda}\sin\theta} S(f,\theta,\psi), \tag{8.1-33}$$

where $S(f,\theta,\psi)$ is obtained by substituting f_X and f_Y given by (8.1-2) and (8.1-3), respectively, into (8.1-14). If all the elements in the array are equally spaced in the X direction, and equally spaced in the Y direction, then substitute (8.1-22) and (8.1-23) into (8.1-14).

If the circumference $2\pi a$ of a circular piston is small compared to a wavelength so that

$$2\pi a/\lambda < 1, \tag{8.1-34}$$

then the magnitude of the argument of the Bessel function will also be less than 1 because $|\sin\theta| \le 1$. Since

$$J_1(x) = \frac{x}{2} - \frac{2x^3}{2\times 4^2} + \frac{3x^5}{2\times 4^2\times 6^2} - \cdots, \tag{8.1-35}$$

by using only the first term in the series expansion given by (8.1-35),

$$J_1\left(\frac{2\pi a}{\lambda}\sin\theta\right) \approx \frac{\pi a}{\lambda}\sin\theta, \qquad \frac{2\pi a}{\lambda} < 1. \tag{8.1-36}$$

Substituting (8.1-36) into (8.1-32) yields

$$E(f,\theta,\psi) \approx A_A\pi a^2, \qquad 2\pi a/\lambda < 1. \tag{8.1-37}$$

Therefore, the circular pistons act like omnidirectional point-elements because the far-field beam pattern $E(f,\theta,\psi)$ is not a function of the spherical angles θ and ψ. Substituting (8.1-37) into (8.1-12) yields the following expression for the unnormalized, far-field beam pattern of the array:

$$D(f,\theta,\psi) \approx A_A\pi a^2 S(f,\theta,\psi), \qquad 2\pi a/\lambda < 1, \tag{8.1-38}$$

or

$$D(f,f_X,f_Y) \approx A_A\pi a^2 S(f,f_X,f_Y), \qquad 2\pi a/\lambda < 1, \tag{8.1-39}$$

where $S(f,f_X,f_Y)$ is given by (8.1-14). Compare (8.1-38) with (8.1-33), and compare (8.1-39) with (8.1-19). If we let $a_{\max} = \lambda/(2\pi)$, then the far-field beam patterns $E(f,\theta,\psi)$ and $D(f,\theta,\psi)$ given by (8.1-37) and (8.1-38) are valid as long as $a < a_{\max}$ (see Table 8.1-4). By comparing Tables 8.1-3 and 8.1-4, it can be seen that for the same value of frequency, a circular piston with a bigger area than a rectangular piston can still act like an omnidirectional point-element. In other words, it is easier for a circular piston to act like an omnidirectional point-element than a rectangular piston. ■

Table 8.1-4 Values for the Maximum Radius $a_{\max} = \lambda/(2\pi)$ for Three Different Frequencies f. Note that $1 \text{ in} = 2.54 \text{ cm}$.

c (m/sec)	f (kHz)	λ (m)	$a_{\max}$ (cm)	$a_{\max}$ (in)
1500	0.1	15.0	238.7	94.0
1500	1.0	1.5	23.87	9.4
1500	10.0	0.15	2.387	0.94

Example 8.1-3 Separable Complex Weights

The product theorem for planar arrays given by (8.1-12) is a general expression. A somewhat simplified version of (8.1-12) can be obtained if the identical elements are *equally spaced* in the X direction, and *equally spaced* in the Y direction, and the complex weights $c_{mn}(f)$ are *separable*. If the complex weights are separable, then

$$c_{mn}(f) = c_m(f)w_n(f), \tag{8.1-40}$$

where

$$c_m(f) = a_m(f)\exp[+j\theta_m(f)] \tag{8.1-41}$$

is a frequency-dependent, complex weight in the X direction, and

$$w_n(f) = b_n(f)\exp[+j\phi_n(f)] \tag{8.1-42}$$

is a frequency-dependent, complex weight in the Y direction, where $a_m(f)$ and $b_n(f)$ are real, frequency-dependent, dimensionless, amplitude weights, and $\theta_m(f)$ and $\phi_n(f)$ are real, frequency-dependent, phase weights in radians. Equation (8.1-40) indicates that the set of complex weights in the X direction is independent of the set of complex weights in the Y direction.

Therefore, substituting (8.1-22), (8.1-23), and (8.1-40) into (8.1-14) yields

$$\boxed{S(f, f_X, f_Y) = S_X(f, f_X)S_Y(f, f_Y)} \tag{8.1-43}$$

where

$$S_X(f, f_X) = \sum_{m=-M'}^{M'} c_m(f) \exp(+j2\pi f_X m d_X) \tag{8.1-44}$$

and

$$S_Y(f, f_Y) = \sum_{n=-N'}^{N'} w_n(f) \exp(+j2\pi f_Y n d_Y) \tag{8.1-45}$$

and by substituting (8.1-43) into (8.1-12), we obtain the following version of the Product Theorem:

$$D(f, f_X, f_Y) = E(f, f_X, f_Y) S_X(f, f_X) S_Y(f, f_Y) \tag{8.1-46}$$

■

Example 8.1-4 Tesseral Quadrupole

Consider a planar array of four identical, complex-weighted, omnidirectional point-elements lying in the XY plane as shown in Fig. 8.1-2. Let the interelement spacings in both the X and Y directions be equal to d meters, that is, $d_Y = d_X = d$. The planar array shown in Fig. 8.1-2 with $d_Y = d_X = d$, and the amplitude and phase weights given in Table 8.1-5 are together known as a *tesseral quadrupole*.

Table 8.1-5 The x and y Coordinates of the Centers of the Elements, and the Associated Amplitude and Phase Weights of a Tesseral Quadrupole Lying in the XY Plane

(m,n)	x_{mn}	y_{mn}	$a_{mn}(f)$	$\theta_{mn}(f)$
$(-1,-1)$	$-d/2$	$-d/2$	$a(f)$	$-\theta_{11}(f)$
$(-1,1)$	$-d/2$	$d/2$	$-a(f)$	$-\theta_{1-1}(f)$
$(1,-1)$	$d/2$	$-d/2$	$-a(f)$	$\theta_{1-1}(f)$
$(1,1)$	$d/2$	$d/2$	$a(f)$	$\theta_{11}(f)$

In order to derive the far-field beam pattern of the tesseral quadrupole, we shall first model its complex frequency response. By using the mathematical model for the complex frequency response of an omnidirectional point-element as given by (8.1-16) and the principle of superposition, the complex frequency

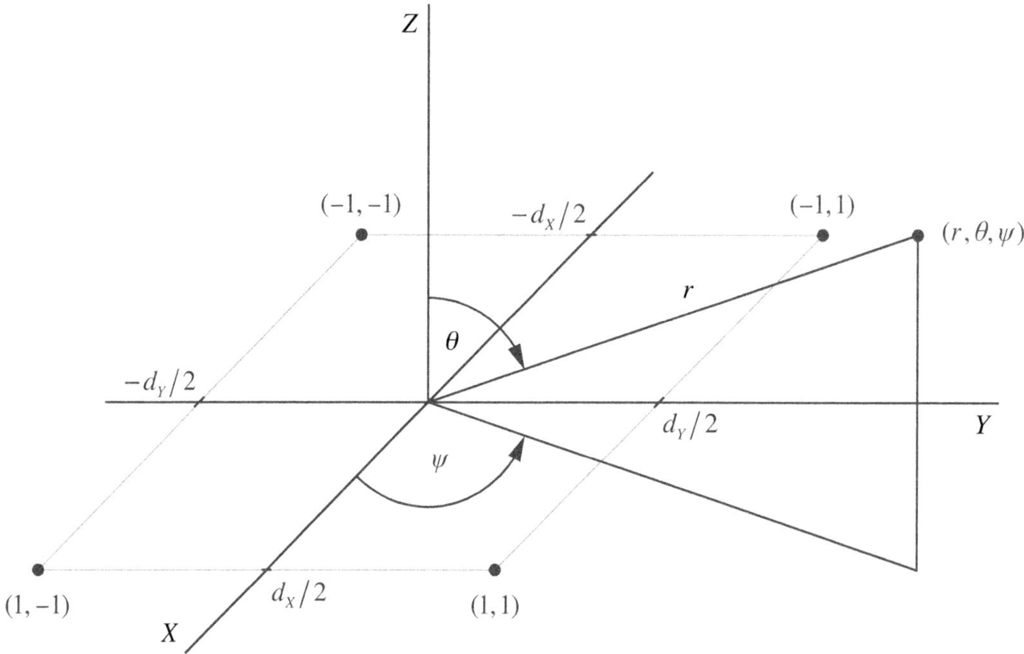

Figure 8.1-2 Planar array of four identical, omnidirectional point-elements lying in the XY plane. Also shown are the element numbers (m,n) and a field point with spherical coordinates (r, θ, ψ).

response (complex aperture function) of the tesseral quadrupole can be expressed as the sum of the complex-weighted frequency responses of all the elements in the array as follows:

$$\begin{aligned} A(f, x_A, y_A) = {} & c_{-1-1}(f)\mathcal{S}(f)\delta(x_A - x_{-1-1})\delta(y_A - y_{-1-1}) + \\ & c_{-11}(f)\mathcal{S}(f)\delta(x_A - x_{-11})\delta(y_A - y_{-11}) + \\ & c_{1-1}(f)\mathcal{S}(f)\delta(x_A - x_{1-1})\delta(y_A - y_{1-1}) + \\ & c_{11}(f)\mathcal{S}(f)\delta(x_A - x_{11})\delta(y_A - y_{11}), \end{aligned} \tag{8.1-47}$$

where $\mathcal{S}(f)$ is the complex, element sensitivity function. Since [see (8.1-21)]

$$\begin{aligned} F_{x_A} F_{y_A}\left\{\delta(x_A - x_{mn})\delta(y_A - y_{mn})\right\} &= F_{x_A}\left\{\delta(x_A - x_{mn})\right\} F_{y_A}\left\{\delta(y_A - y_{mn})\right\} \\ &= \exp\left[+j2\pi(f_X x_{mn} + f_Y y_{mn})\right], \end{aligned} \tag{8.1-48}$$

taking the two-dimensional spatial Fourier transform of (8.1-47) with respect to x_A and y_A yields

$$\begin{aligned} D(f, f_X, f_Y) = {} & c_{-1-1}(f)\mathcal{S}(f)\exp\left[+j2\pi(f_X x_{-1-1} + f_Y y_{-1-1})\right] + \\ & c_{-11}(f)\mathcal{S}(f)\exp\left[+j2\pi(f_X x_{-11} + f_Y y_{-11})\right] + \\ & c_{1-1}(f)\mathcal{S}(f)\exp\left[+j2\pi(f_X x_{1-1} + f_Y y_{1-1})\right] + \\ & c_{11}(f)\mathcal{S}(f)\exp\left[+j2\pi(f_X x_{11} + f_Y y_{11})\right]. \end{aligned} \tag{8.1-49}$$

Substituting (8.1-8) and the specifications given in Table 8.1-5 into (8.1-49) and rearranging terms yields

$$\begin{aligned} D(f, f_X, f_Y) = {} & a(f)\mathcal{S}(f)\times \\ & \Big\{\Big[\exp\{+j[\pi(f_X + f_Y)d + \theta_{11}(f)]\} + \exp\{-j[\pi(f_X + f_Y)d + \theta_{11}(f)]\}\Big] - \\ & \Big[\exp\{+j[\pi(f_X - f_Y)d + \theta_{1-1}(f)]\} + \exp\{-j[\pi(f_X - f_Y)d + \theta_{1-1}(f)]\}\Big]\Big\}, \end{aligned} \tag{8.1-50}$$

or

$$\begin{aligned} D(f, f_X, f_Y) = {} & 2a(f)\mathcal{S}(f)\times \\ & \Big\{\cos\big[\pi(f_X + f_Y)d + \theta_{11}(f)\big] - \cos\big[\pi(f_X - f_Y)d + \theta_{1-1}(f)\big]\Big\}. \end{aligned} \tag{8.1-51}$$

Because we have not yet discussed phase weights for planar arrays in any great detail, let $\theta_{11}(f) = 0$ and $\theta_{1-1}(f) = 0$. In other words, if only amplitude weighting is done (no phase weighting), then (8.1-51) reduces to

$$D(f, f_X, f_Y) = 2a(f)\mathcal{S}(f)\left[\cos(\pi f_X d + \pi f_Y d) - \cos(\pi f_X d - \pi f_Y d)\right], \tag{8.1-52}$$

and by using the trigonometric identity

$$\cos(\alpha + \beta) - \cos(\alpha - \beta) = -2\sin\alpha\sin\beta, \tag{8.1-53}$$

(8.1-52) can be rewritten as

$$\boxed{D(f, f_X, f_Y) = -4a(f)\mathcal{S}(f)\sin(\pi f_X d)\sin(\pi f_Y d)} \tag{8.1-54}$$

where $\mathcal{S}(f)$ is the complex, element sensitivity function. Equation (8.1-54) is the *unnormalized*, far-field beam pattern of a tesseral quadrupole lying in the XY

plane with no phase weighting.

The magnitude of the far-field beam pattern given by (8.1-54) is

$$|D(f, f_X, f_Y)| = 4|a(f)||S(f)||\sin(\pi f_X d)||\sin(\pi f_Y d)|, \tag{8.1-55}$$

and since the maximum value of the magnitude of a sine function is 1, the normalization factor is

$$D_{\max} = \max|D(f, f_X, f_Y)| = 4|a(f)||S(f)|. \tag{8.1-56}$$

Dividing (8.1-55) by (8.1-56) yields the following expression for the magnitude of the *normalized*, far-field beam pattern of a tesseral quadrupole lying in the XY plane with no phase weighting:

$$\boxed{|D_N(f, f_X, f_Y)| = |\sin(\pi f_X d)||\sin(\pi f_Y d)|} \tag{8.1-57}$$

Since spatial frequencies

$$f_X = u/\lambda = \sin\theta\cos\psi/\lambda \tag{8.1-58}$$

and

$$f_Y = v/\lambda = \sin\theta\sin\psi/\lambda, \tag{8.1-59}$$

(8.1-57) can also be expressed as

$$|D_N(f, u, v)| = \left|\sin\left(\pi\frac{d}{\lambda}u\right)\right|\left|\sin\left(\pi\frac{d}{\lambda}v\right)\right|, \tag{8.1-60}$$

or

$$|D_N(f, \theta, \psi)| = \left|\sin\left(\pi\frac{d}{\lambda}\sin\theta\cos\psi\right)\right|\left|\sin\left(\pi\frac{d}{\lambda}\sin\theta\sin\psi\right)\right|. \tag{8.1-61}$$

At broadside ($\theta = 0°$) where direction cosines $u = 0$ and $v = 0$, (8.1-60) and (8.1-61) reduce to $|D_N(f, 0, 0)| = 0$ and $|D_N(f, 0°, \psi)| = 0$, respectively. Furthermore, in the XZ plane where $\psi = 0°$ or $\psi = 180°$, $v = 0$. As a result, (8.1-60) reduces to $|D_N(f, u, 0)| = 0$, and (8.1-61) reduces to $|D_N(f, \theta, 0°)| = 0$ and $|D_N(f, \theta, 180°)| = 0$. Similarly, in the YZ plane where $\psi = 90°$ or $\psi = 270°$, $u = 0$. As a result, (8.1-60) reduces to $|D_N(f, 0, v)| = 0$ and (8.1-61) reduces to $|D_N(f, \theta, 90°)| = 0$ and $|D_N(f, \theta, 270°)| = 0$. Therefore, the far-field beam pattern of the tesseral quadrupole not only has a null at broadside like the far-field beam patterns of both the dipole (see Fig. 6.1-4) and the axial quadrupole (see Fig. 6.1-9), it also has nulls in the XZ and YZ planes. And finally, if $d/\lambda = 0.5$, then

$|D_N(f,u,v)|=1$ when $(u,v)=(-1,-1)$, $(-1,1)$, $(1,-1)$, and $(1,1)$. However, these (u,v) coordinates do not lie inside or on the unit circle (see Fig. 8.1-3). ■

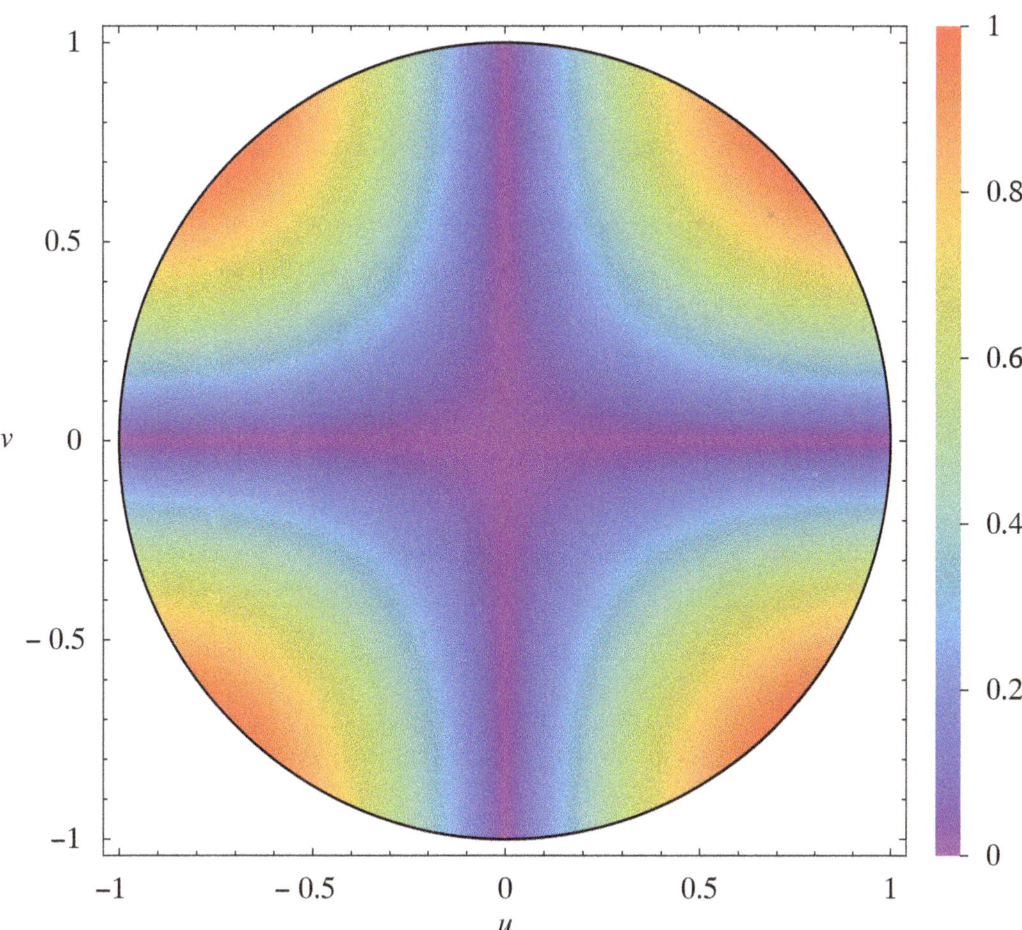

Figure 8.1-3 Magnitude of the normalized, far-field beam pattern of a tesseral quadrupole lying in the XY plane, plotted as a function of direction cosines u and v for $d/\lambda = 0.5$.

Example 8.1-5 Mainlobe in a Half-Space

Consider a planar array of four identical, complex-weighted, omnidirectional point-elements lying in the XY plane as shown in Fig. 8.1-2. The elements are equally spaced in the X direction, and equally spaced in the Y direction. If the complex weights $c_{mn}(f)$ are separable [see (8.1-40)], and since $E(f, f_X, f_Y) = \mathcal{S}(f)$, where $\mathcal{S}(f)$ is the complex, element sensitivity function, then the unnormalized, far-field beam pattern of this array is given by the Product

Theorem [see (8.1-46)]

$$D(f, f_X, f_Y) = \mathcal{S}(f) S_X(f, f_X) S_Y(f, f_Y), \tag{8.1-62}$$

where

$$S_X(f, f_X) = c_{-1}(f)\exp(+j2\pi f_X x_{-1}) + c_1(f)\exp(+j2\pi f_X x_1) \tag{8.1-63}$$

and

$$S_Y(f, f_Y) = w_{-1}(f)\exp(+j2\pi f_Y y_{-1}) + w_1(f)\exp(+j2\pi f_Y y_1) \tag{8.1-64}$$

since there are two elements in both the X and Y directions [see (6.1-12)]. The x and y coordinates of the centers of the elements, and the associated complex weights are given in Table 8.1-6.

Table 8.1-6 The x and y Coordinates of the Centers of the Elements, and the Associated Complex Weights of a Four-Element Planar Array Lying in the XY Plane

x_1	x_{-1}	$c_1(f)$	$c_{-1}(f)$	y_1	y_{-1}	$w_1(f)$	$w_{-1}(f)$
$d_X/2$	$-d_X/2$	$a(f)$	$a(f)\exp(\pm j\pi/2)$	$d_Y/2$	$-d_Y/2$	$b(f)$	$b(f)$

Substituting the specifications given in Table 8.1-6 into (8.1-63) and (8.1-64) yields

$$S_X(f, f_X) = 2a(f)\cos\left[\pi f_X d_X \mp (\pi/4)\right]\exp(\pm j\pi/4) \tag{8.1-65}$$

and

$$S_Y(f, f_Y) = 2b(f)\cos(\pi f_Y d_Y), \tag{8.1-66}$$

respectively, and by substituting (8.1-65) and (8.1-66) into (8.1-62), we obtain

$$D(f, f_X, f_Y) = 4a(f)b(f)\mathcal{S}(f)\cos\left[\pi f_X d_X \mp (\pi/4)\right]\cos(\pi f_Y d_Y)\exp(\pm j\pi/4). \tag{8.1-67}$$

If the interelement spacings

$$d_X = \lambda/4 \tag{8.1-68}$$

and

$$d_Y = \lambda/2, \tag{8.1-69}$$

and by using (8.1-58) and (8.1-59), (8.1-67) can be rewritten as

$$\boxed{D(f,u,v) = 4a(f)b(f)\mathcal{S}(f)\cos[\pi(u \mp 1)/4]\cos(\pi v/2)\exp(\pm j\pi/4)} \tag{8.1-70}$$

Equation (8.1-70) is the *unnormalized*, far-field beam pattern of the array with interelement spacings given by (8.1-68) and (8.1-69), where $\mathcal{S}(f)$ is the complex, element sensitivity function.

The magnitude of the far-field beam pattern given by (8.1-70) is

$$|D(f,u,v)| = 4|a(f)||b(f)||\mathcal{S}(f)||\cos[\pi(u \mp 1)/4]||\cos(\pi v/2)|, \tag{8.1-71}$$

and since the maximum value of $|\cos[\pi(u \mp 1)/4]||\cos(\pi v/2)|$ occurs at $u = \pm 1$ and $v = 0$ where $|\cos(0)||\cos(0)| = 1$, the normalization factor is

$$D_{\max} = \max|D(f,u,v)| = 4|a(f)||b(f)||\mathcal{S}(f)|. \tag{8.1-72}$$

Note that $u = \pm 1$ are the end-fire locations in direction-cosine space in the X direction. Dividing (8.1-71) by (8.1-72) yields the following expression for the magnitude of the *normalized*, far-field beam pattern of the array:

$$\boxed{|D_N(f,u,v)| = |\cos[\pi(u \mp 1)/4]||\cos(\pi v/2)|} \tag{8.1-73}$$

where $|D_N(f,\pm 1,0)| = 1$. Substituting $u = \sin\theta\cos\psi$ and $v = \sin\theta\sin\psi$ into (8.1-73) yields

$$|D_N(f,\theta,\psi)| = \left|\cos\left[\frac{\pi}{4}(\sin\theta\cos\psi \mp 1)\right]\right|\left|\cos\left(\frac{\pi}{2}\sin\theta\sin\psi\right)\right|, \tag{8.1-74}$$

and by setting $\theta = 90°$ in (8.1-74), we obtain

$$|D_N(f,90°,\psi)| = \left|\cos\left[\frac{\pi}{4}(\cos\psi \mp 1)\right]\right|\left|\cos\left(\frac{\pi}{2}\sin\psi\right)\right|, \tag{8.1-75}$$

which is the magnitude of the normalized, *horizontal*, far-field beam pattern in the XY plane. The $\mp$ sign on the right-hand side of (8.1-75) corresponds to the $\pm\pi/2$ phase weight in Table 8.1-6 (see Figs. 8.1-4 and 8.1-5). Like the cardioid beam patterns in Example 6.1-3, the mainlobes of the beam patterns in Figs. 8.1-4 and 8.1-5 exist in a half-space. A *twin-line planar array*, which is one solution to the port/starboard (left/right) ambiguity problem associated with linear arrays, and is a generalization of the planar array shown in Fig. 8.1-2, is discussed in Example 8.2-1. ■

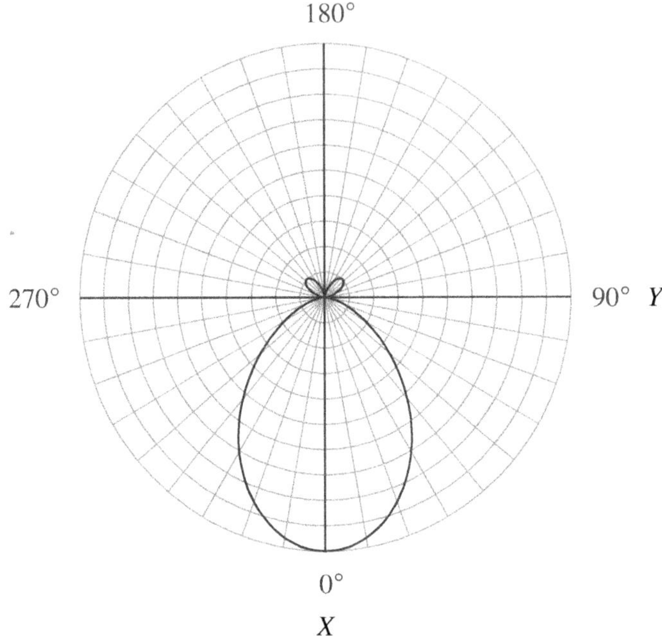

Figure 8.1-4 Polar plot, as a function of the azimuthal (bearing) angle ψ , of the magnitude of the normalized, horizontal, far-field beam pattern given by (8.1-75) for the phase weight $\pi/2$ in Table 8.1-6.

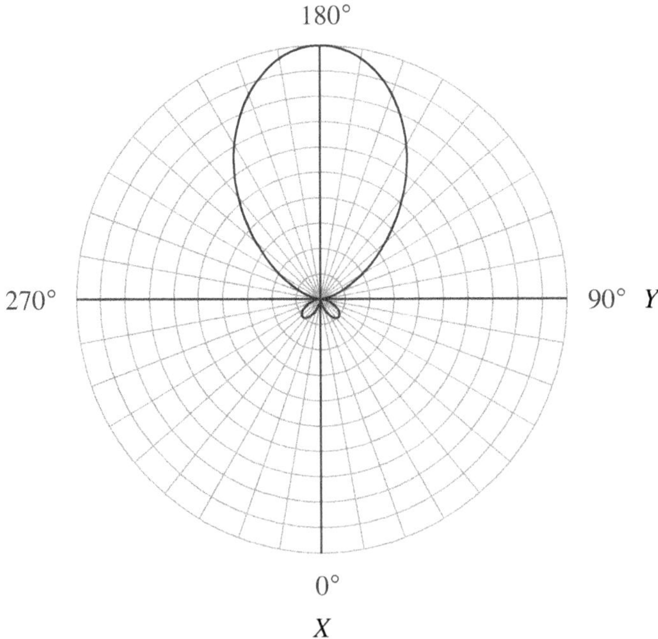

Figure 8.1-5 Polar plot, as a function of the azimuthal (bearing) angle ψ , of the magnitude of the normalized, horizontal, far-field beam pattern given by (8.1-75) for the phase weight $-\pi/2$ in Table 8.1-6.

Example 8.1-6 Concentric Circular Arrays

In Section 3.3 it was shown that the far-field beam pattern of a planar, circular aperture lying in the XY plane with radius a meters and with complex frequency response (complex aperture function) expressed in terms of the polar coordinates r_A and ϕ_A is given by (see Figures 3.3-1 and 3.3-2)

$$D(f,\theta,\psi)=\int_0^{2\pi}\int_0^{a} A(f,r_A,\phi_A)\exp\left[+j\frac{2\pi r_A}{\lambda}\sin\theta\cos(\psi-\phi_A)\right]r_A\,dr_A\,d\phi_A\,. \tag{8.1-76}$$

In this example we shall use (8.1-76) to derive the far-field beam pattern of a planar array of *concentric circular arrays* of identical, equally-spaced, complex-weighted, omnidirectional point-elements lying in the XY plane (see Fig. 8.1-6). However, in order to use (8.1-76), we first need to model the complex frequency response $A(f,r_A,\phi_A)$ of the planar array.

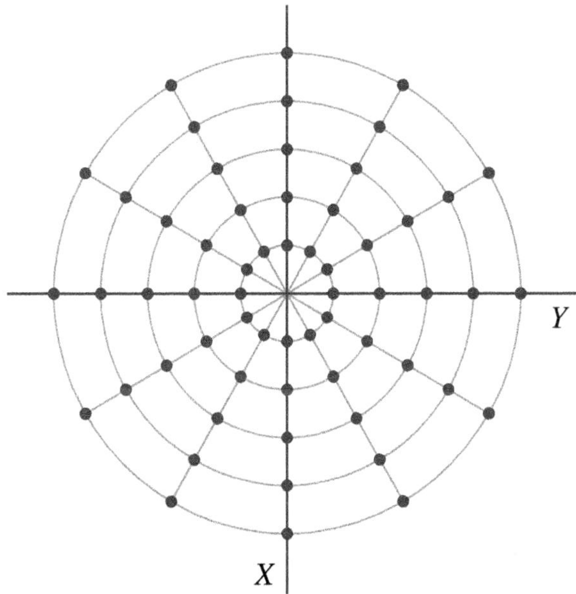

Figure 8.1-6 Planar array of concentric circular arrays of identical, equally-spaced, omnidirectional point-elements lying in the XY plane.

The mathematical model for the complex frequency response of an omnidirectional point-element can be expressed in polar coordinates as

$$e(f,r_A-r_0,\phi_A-\phi_0)=\mathcal{S}(f)\frac{\delta(r_A-r_0)}{r_A}\delta(\phi_A-\phi_0)\,, \tag{8.1-77}$$

where $\mathcal{S}(f)$ is the complex, element sensitivity function, and (r_0, ϕ_0) are the polar coordinates of the center of the element. Note that $\int_0^{2\pi}\int_0^{\infty} e(f, r_A - r_0, \phi_A - \phi_0) r_A \, dr_A \, d\phi_A = \mathcal{S}(f)$ because $\int_0^{\infty} \delta(r_A - r_0) \, dr_A \int_0^{2\pi} \delta(\phi_A - \phi_0) \, d\phi_A = 1$. The impulse function $\delta(r_A - r_0)$ has units of inverse meters and $\delta(\phi_A - \phi_0)$ has units of inverse radians. Therefore, with the use of (8.1-77) and the principle of superposition, the complex frequency response (complex aperture function) of the planar array shown in Fig. 8.1-6 can be expressed as the sum of the complex-weighted frequency responses of all the elements in the array as follows:

$$A(f, r_A, \phi_A) = \frac{\mathcal{S}(f)}{r_A} \sum_{m=1}^{M} \sum_{n=1}^{N} c_{mn}(f) \delta(r_A - r_m) \delta(\phi_A - \phi_n), \tag{8.1-78}$$

where M is the total number (even or odd) of concentric circular arrays, N is the total number (even or odd) of omnidirectional point-elements per circular array,

$$c_{mn}(f) = a_{mn}(f) \exp[+j\theta_{mn}(f)] \tag{8.1-79}$$

is the frequency-dependent, complex weight associated with element (m,n), where $a_{mn}(f)$ is a real, frequency-dependent, dimensionless, amplitude weight, and $\theta_{mn}(f)$ is a real, frequency-dependent, phase weight in radians, and (r_m, ϕ_n) are the polar coordinates of the center of element (m,n). Since the elements are equally spaced in the r and ϕ directions,

$$r_m = m\Delta r, \tag{8.1-80}$$

where

$$\Delta r = a/M \tag{8.1-81}$$

and a is the maximum radial extent of the array in meters, and

$$\phi_n = (n-1)\Delta\phi, \tag{8.1-82}$$

where

$$\Delta\phi = 360°/N. \tag{8.1-83}$$

Substituting (8.1-78) into (8.1-76) yields the following expression for the far-field beam pattern of a planar array of concentric circular arrays of identical, equally-spaced, complex-weighted, omnidirectional point-elements lying in the XY plane:

$$\boxed{D(f, \theta, \psi) = \mathcal{S}(f) \sum_{m=1}^{M} \sum_{n=1}^{N} c_{mn}(f) \exp\left[+j\frac{2\pi r_m}{\lambda} \sin\theta \cos(\psi - \phi_n)\right]} \tag{8.1-84}$$

where $\mathcal{S}(f)$ is the complex, element sensitivity function, and the double summation is the dimensionless array factor. Since

$$\cos(\psi-\phi_n)=\cos\psi\cos\phi_n+\sin\psi\sin\phi_n\,, \tag{8.1-85}$$

(8.1-84) can also be expressed as

$$D(f,u,v)=\mathcal{S}(f)\sum_{m=1}^{M}\sum_{n=1}^{N}c_{mn}(f)\exp\left[+j\frac{2\pi}{\lambda}(ux_{mn}+vy_{mn})\right], \tag{8.1-86}$$

or

$$D(f,f_X,f_Y)=\mathcal{S}(f)\sum_{m=1}^{M}\sum_{n=1}^{N}c_{mn}(f)\exp\left[+j2\pi(f_X x_{mn}+f_Y y_{mn})\right], \tag{8.1-87}$$

where

$$x_{mn}=r_m\cos\phi_n \tag{8.1-88}$$

and

$$y_{mn}=r_m\sin\phi_n \tag{8.1-89}$$

are the rectangular coordinates of element (m,n), and spatial frequencies

$$f_X=u/\lambda=\sin\theta\cos\psi/\lambda \tag{8.1-90}$$

and

$$f_Y=v/\lambda=\sin\theta\sin\psi/\lambda\,. \tag{8.1-91}$$

For the special case of a *single circular array*, $M=1$, and as a result, $m=1$ and $r_1=\Delta r=a$. Therefore, the far-field beam patterns given by (8.1-84), (8.1-86), and (8.1-87) reduce as follows:

$$\boxed{D(f,\theta,\psi)=\mathcal{S}(f)\sum_{n=1}^{N}c_n(f)\exp\left[+j\frac{2\pi a}{\lambda}\sin\theta\cos(\psi-\phi_n)\right]} \tag{8.1-92}$$

$$D(f,u,v)=\mathcal{S}(f)\sum_{n=1}^{N}c_n(f)\exp\left[+j\frac{2\pi}{\lambda}(ux_n+vy_n)\right], \tag{8.1-93}$$

and

$$D(f,f_X,f_Y)=\mathcal{S}(f)\sum_{n=1}^{N}c_n(f)\exp\left[+j2\pi(f_X x_n+f_Y y_n)\right], \tag{8.1-94}$$

respectively, where

$$x_n=a\cos\phi_n \tag{8.1-95}$$

and

$$y_n=a\sin\phi_n\,. \tag{8.1-96}$$

■

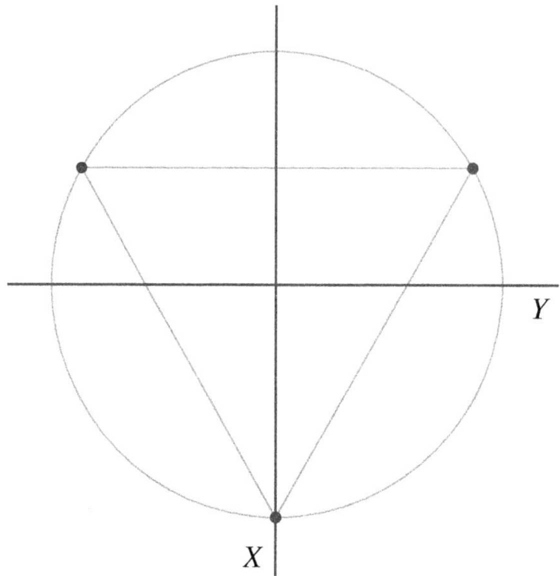

Figure 8.1-7 A triplet circular array.

Example 8.1-7 Triplet – Cardioid Beam Pattern

A special case of a single circular array is a *triplet*. A triplet is composed of three identical, equally-spaced, complex-weighted, omnidirectional point-elements that form an equilateral triangle (see Fig. 8.1-7). Substituting (8.1-95), (8.1-96), and $N = 3$ into (8.1-93) yields the following *unnormalized*, far-field beam pattern of a triplet lying in the XY plane:

$$D(f,u,v) = \mathcal{S}(f)\sum_{n=1}^{3} c_n(f)\exp\left[+jka(u\cos\phi_n + v\sin\phi_n)\right], \tag{8.1-97}$$

where

$$ka = 2\pi a/\lambda, \tag{8.1-98}$$

a is the radius of the circular array in meters, and by substituting (8.1-83) and $N = 3$ into (8.1-82),

$$\phi_n = (n-1)120°,\ n = 1, 2, 3. \tag{8.1-99}$$

Evaluating (8.1-99) yields $\phi_1 = 0°$, $\phi_2 = 120°$, and $\phi_3 = 240°$. Therefore, (8.1-97) can be rewritten as

$$D(f,u,v) = \mathcal{S}(f)\left[c_1(f)\exp(+jkau) + c_2(f)\exp\left[-j\frac{ka}{2}\left(u - \sqrt{3}\,v\right)\right] + c_3(f)\exp\left[-j\frac{ka}{2}\left(u + \sqrt{3}\,v\right)\right]\right], \tag{8.1-100}$$

where $\mathcal{S}(f)$ is the complex, element sensitivity function.

In order to create a *cardioid* (heart-shaped) beam pattern from a triplet, set complex weight

$$c_1(f) = -[c_2(f) + c_3(f)]\exp\left[+j\frac{3}{2}ka\right]. \tag{8.1-101}$$

If we let

$$c_3(f) = c_2(f) = \mathrm{a}(f), \tag{8.1-102}$$

where $\mathrm{a}(f)$ is a real, frequency-dependent, dimensionless, amplitude weight; and

$$ka = \pi/3, \tag{8.1-103}$$

then

$$c_1(f) = -2\,\mathrm{a}(f)\exp(+j\pi/2) = -j2\,\mathrm{a}(f), \tag{8.1-104}$$

or

$$c_1(f) = 2\,\mathrm{a}(f)\exp(-j\pi/2). \tag{8.1-105}$$

Substituting (8.1-102), (8.1-103), and (8.1-105) into (8.1-100) yields the *unnormalized, cardioid*, far-field beam pattern of a triplet lying in the XY plane:

$$\boxed{D(f,u,v) = 2\,\mathrm{a}(f)\mathcal{S}(f)\exp\left(+j\frac{\pi}{3}u\right)\left[\exp\left(-j\frac{\pi}{2}u\right)\cos\left(\frac{\sqrt{3}}{6}\pi v\right) - j\right]} \tag{8.1-106}$$

The magnitude of the beam pattern given by (8.1-106) is

$$|D(f,u,v)| = 2|\mathrm{a}(f)||\mathcal{S}(f)|\left|\exp\left(-j\frac{\pi}{2}u\right)\cos\left(\frac{\sqrt{3}}{6}\pi v\right) - j\right|, \tag{8.1-107}$$

and since the maximum value of

$$\left|\exp\left(-j\frac{\pi}{2}u\right)\cos\left(\frac{\sqrt{3}}{6}\pi v\right) - j\right|$$

occurs at $u = 1$ and $v = 0$ where

$$\left|\exp\left(-j\frac{\pi}{2}\right) - j\right| = |-j2| = 2, \tag{8.1-108}$$

the normalization factor is

$$D_{\max} = \max|D(f,u,v)| = 4|\mathrm{a}(f)||S(f)|. \tag{8.1-109}$$

Dividing (8.1-107) by (8.1-109) yields the following expression for the magnitude of the *normalized*, *cardioid*, far-field beam pattern of a triplet lying in the XY plane:

$$\boxed{|D_N(f,u,v)| = \frac{1}{2}\left|\exp\left(-j\frac{\pi}{2}u\right)\cos\left(\frac{\sqrt{3}}{6}\pi v\right) - j\right|} \tag{8.1-110}$$

where $|D_N(f,1,0)| = 1$. Substituting $u = \sin\theta\cos\psi$ and $v = \sin\theta\sin\psi$ into (8.1-110) yields

$$|D_N(f,\theta,\psi)| = \frac{1}{2}\left|\exp\left(-j\frac{\pi}{2}\sin\theta\cos\psi\right)\cos\left(\frac{\sqrt{3}}{6}\pi\sin\theta\sin\psi\right) - j\right|, \tag{8.1-111}$$

and by setting $\theta = 90°$ in (8.1-111), we obtain

$$|D_N(f,90°,\psi)| = \frac{1}{2}\left|\exp\left(-j\frac{\pi}{2}\cos\psi\right)\cos\left(\frac{\sqrt{3}}{6}\pi\sin\psi\right) - j\right|, \tag{8.1-112}$$

which is the magnitude of the normalized, *horizontal*, far-field beam pattern in the XY plane (see Fig. 8.1-8).

In the XZ plane, $\psi = 0°$ (positive X axis) or $\psi = 180°$ (negative X axis). Therefore, the magnitude of the normalized, *vertical*, far-field beam pattern in the XZ plane can be obtained by evaluating (8.1-111) at $\psi = 0°$ and $\psi = 180°$ for $0° \le \theta \le 180°$, that is,

$$|D_N(f,\theta,0°)| = \frac{1}{2}\left|\exp\left(-j\frac{\pi}{2}\sin\theta\right) - j\right|, \qquad 0° \le \theta \le 180°, \tag{8.1-113}$$

and

$$|D_N(f,\theta,180°)| = \frac{1}{2}\left|\exp\left(+j\frac{\pi}{2}\sin\theta\right) - j\right|, \qquad 0° \le \theta \le 180°. \tag{8.1-114}$$

Combining (8.1-113) and (8.1-114) yields Fig. 8.1-9.

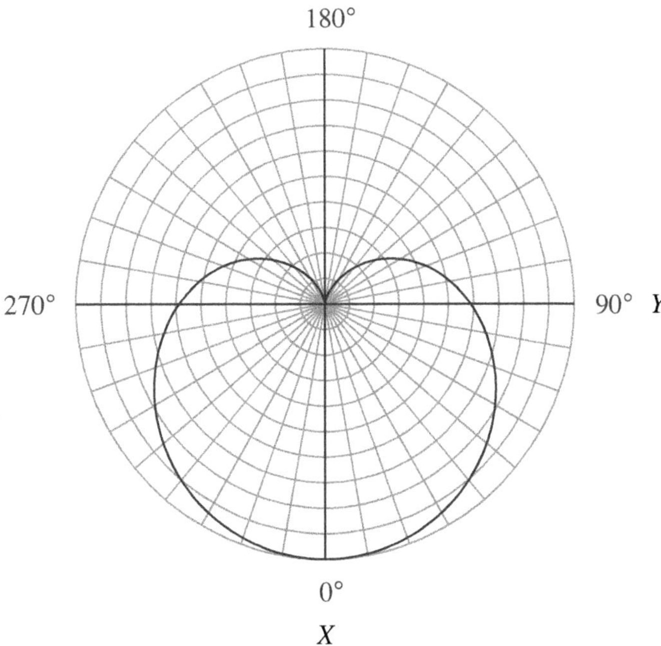

Figure 8.1-8 Polar plot, as a function of the azimuthal (bearing) angle ψ, of the magnitude of the normalized, horizontal, far-field beam pattern of a triplet given by (8.1-112).

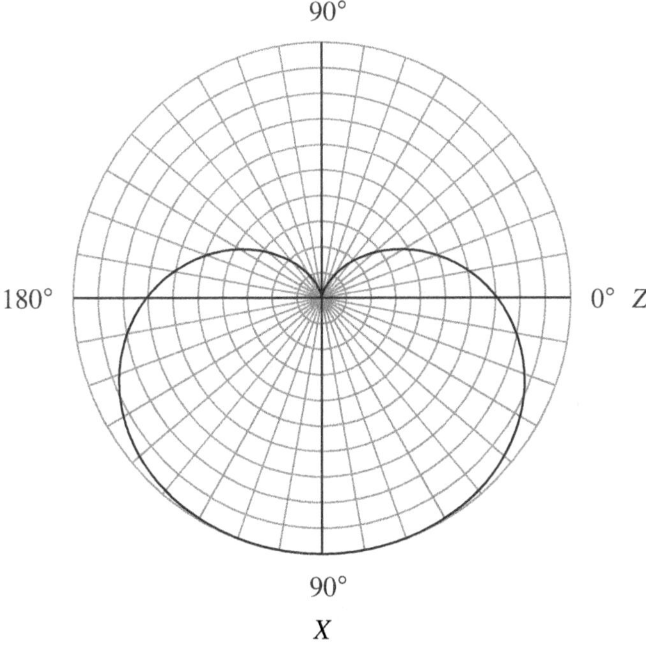

Figure 8.1-9 Polar plot, as a function of the vertical angle θ, of the magnitude of the normalized, vertical, far-field beam pattern of a triplet given by combining (8.1-113) and (8.1-114).

If the complex weight

$$c_1(f) = 2\,\mathrm{a}(f)\exp(+j\pi/2), \tag{8.1-115}$$

then the *unnormalized*, *cardioid*, far-field beam pattern of a triplet lying in the XY plane is given by

$$\boxed{D(f,u,v) = 2\,\mathrm{a}(f)\,\mathcal{S}(f)\exp\left(+j\frac{\pi}{3}u\right)\left[\exp\left(-j\frac{\pi}{2}u\right)\cos\left(\frac{\sqrt{3}}{6}\pi v\right) + j\right]} \tag{8.1-116}$$

and the mainlobes of the horizontal and vertical beam patterns are aligned along the negative X axis instead of the positive X axis (see Figs. 8.1-10 and 8.1-11). One solution to the port/starboard (left/right) ambiguity problem associated with linear arrays is a linear array composed of triplet elements perpendicular to the array axis (see Example 9.1-2). ■

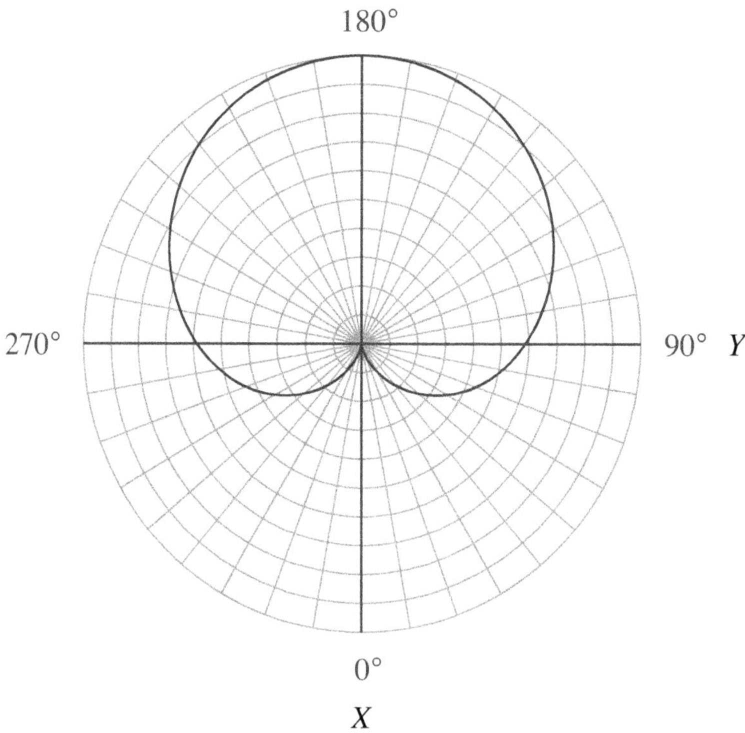

Figure 8.1-10 Polar plot, as a function of the azimuthal (bearing) angle ψ, of the magnitude of the normalized, horizontal, far-field beam pattern of (8.1-116) for a triplet.

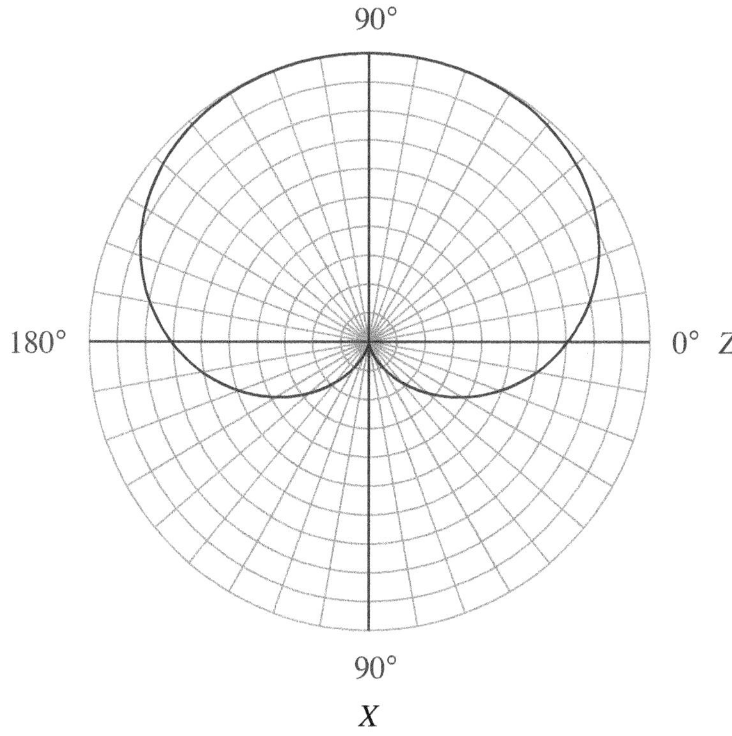

Figure 8.1-11 Polar plot, as a function of the vertical angle θ, of the magnitude of the normalized, vertical, far-field beam pattern of (8.1-116) for a triplet.

Example 8.1-8 Linear Array in a Plane

In this example we shall derive the far-field beam pattern of a linear array composed of an odd number M of unevenly-spaced, complex-weighted elements lying in the XY plane (see Fig. 8.1-12). For example, the elements can be omnidirectional point-elements, rectangular pistons, or circular pistons. If the elements are *not* identical, that is, if each element has a different complex frequency response, and if the elements are *not* equally spaced, as shown in Fig. 8.1-12, then by using the principle of superposition, the complex frequency response (complex aperture function) of this array can be expressed as the sum of the complex-weighted frequency responses of all the elements in the array as follows:

$$A(f, x_A, y_A) = \sum_{m=-M'}^{M'} c_m(f) e_m(f, x_A - x_m, y_A - y_m), \tag{8.1-117}$$

where

$$M' = (M-1)/2, \tag{8.1-118}$$

$$c_m(f) = a_m(f) \exp[+j\theta_m(f)] \tag{8.1-119}$$

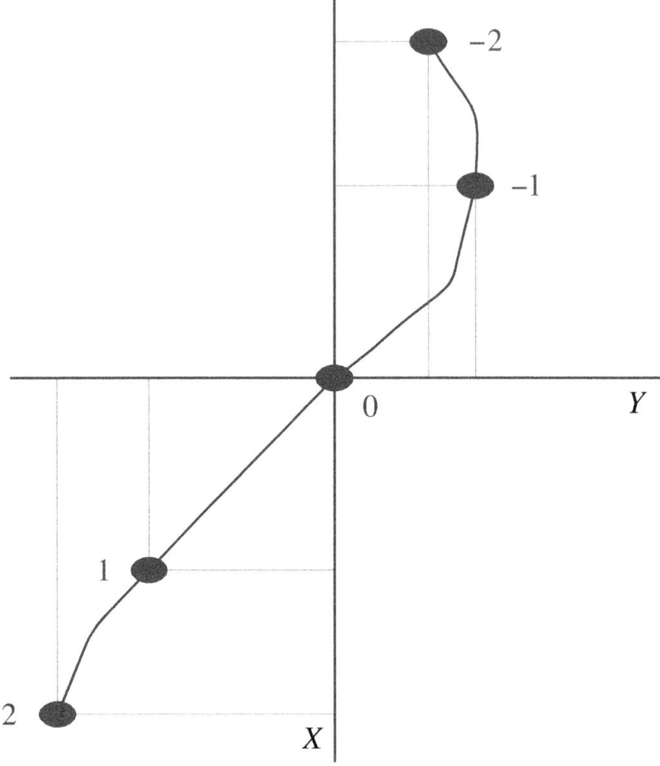

Figure 8.1-12 Linear array composed of an odd number of unevenly-spaced elements lying in the XY plane. Also shown are the element numbers.

is the frequency-dependent, complex weight associated with element m, $a_m(f)$ is a real, frequency-dependent, dimensionless, amplitude weight, $\theta_m(f)$ is a real, frequency-dependent, phase weight in radians, $e_m(f, x_A, y_A)$ is the complex frequency response (complex aperture function) of element m, also known as the element function, and x_m and y_m are the x and y coordinates of the center of element m. Note that $x_0 = 0$ and $y_0 = 0$ because element $m = 0$ is centered at the origin of the coordinate system.

Taking the two-dimensional spatial Fourier transform of (8.1-117) with respect to x_A and y_A yields the following expression for the far-field beam pattern:

$$\boxed{D(f, f_X, f_Y) = \sum_{m=-M'}^{M'} c_m(f) E_m(f, f_X, f_Y) \exp\left[+j2\pi(f_X x_m + f_Y y_m)\right]} \tag{8.1-120}$$

where

$$F_{x_A} F_{y_A}\{e_m(f, x_A - x_m, y_A - y_m)\} = E_m(f, f_X, f_Y)\exp[+j2\pi(f_X x_m + f_Y y_m)] \tag{8.1-121}$$

and

$$E_m(f, f_X, f_Y) = F_{x_A} F_{y_A}\{e_m(f, x_A, y_A)\} \tag{8.1-122}$$

is the far-field beam pattern of element (electroacoustic transducer) m. If all the elements in the array are *identical* (i.e., all the elements have the same complex frequency response), but still *not* equally spaced, then (8.1-120) reduces to the Product Theorem [see (8.1-12)]

$$\boxed{D(f, f_X, f_Y) = E(f, f_X, f_Y) S(f, f_X, f_Y)} \tag{8.1-123}$$

where

$$E(f, f_X, f_Y) = F_{x_A} F_{y_A}\{e(f, x_A, y_A)\} \tag{8.1-124}$$

is the far-field beam pattern of *one* of the identical elements in the array, but now

$$\boxed{S(f, f_X, f_Y) = \sum_{m=-M'}^{M'} c_m(f)\exp[+j2\pi(f_X x_m + f_Y y_m)]} \tag{8.1-125}$$

Note that the dimensionless array factor $S(f, f_X, f_Y)$ given by (8.1-125) for a *linear* array in the XY plane is different than $S(f, f_X, f_Y)$ given by (8.1-14) for a planar array in the XY plane. ■

8.2 The Phased Array – Beam Steering

Consider a planar array composed of an odd number $M \times N$ of identical, unevenly-spaced, complex-weighted, omnidirectional point-elements lying in the XY plane, where M and N are the total odd number of elements in the X and Y directions, respectively. From the Product Theorem for an odd number of elements, the far-field beam pattern of this array is given by [see (8.1-19)]

$$D(f, f_X, f_Y) = \mathcal{S}(f) S(f, f_X, f_Y), \tag{8.2-1}$$

where $\mathcal{S}(f)$ is the complex, element sensitivity function (see Table 8.1-2 and Appendix 6B),

$$S(f, f_X, f_Y) = \sum_{m=-M'}^{M'} \sum_{n=-N'}^{N'} c_{mn}(f)\exp\left[+j2\pi(f_X x_{mn} + f_Y y_{mn})\right], \tag{8.2-2}$$

$$c_{mn}(f) = a_{mn}(f)\exp[+j\theta_{mn}(f)], \tag{8.2-3}$$

$$M' = (M-1)/2, \tag{8.2-4}$$

and

$$N' = (N-1)/2. \tag{8.2-5}$$

Let $D(f, f_X, f_Y)$ be the far-field beam pattern of the array when it is only amplitude weighted (i.e., $\theta_{mn}(f) = 0 \;\; \forall \;\; m$ and n), and let $D'(f, f_X, f_Y)$ be the far-field beam pattern of the array when it is complex weighted. Therefore,

$$D(f, f_X, f_Y) = \mathcal{S}(f) \sum_{m=-M'}^{M'} \sum_{n=-N'}^{N'} a_{mn}(f)\exp\left[+j2\pi(f_X x_{mn} + f_Y y_{mn})\right] \tag{8.2-6}$$

and

$$D'(f, f_X, f_Y) = \mathcal{S}(f) \sum_{m=-M'}^{M'} \sum_{n=-N'}^{N'} a_{mn}(f)\exp\left\{+j\left[2\pi f_X x_{mn} + 2\pi f_Y y_{mn} + \theta_{mn}(f)\right]\right\}. \tag{8.2-7}$$

We already know from planar aperture theory that in order to do beam steering, there must be a linear phase response across the surface of the aperture in both the X and Y directions (see Section 3.4). A set of linear phase weights that will produce a linear phase response across the surface of the planar array under discussion in both the X and Y directions is given by

$$\boxed{\theta_{mn}(f) = -2\pi f'_X x_{mn} - 2\pi f'_Y y_{mn}, \qquad \begin{array}{l} m = -M', \ldots, 0, \ldots, M' \\ n = -N', \ldots, 0, \ldots, N' \end{array}} \tag{8.2-8}$$

where

$$\boxed{f'_X = u'/\lambda = \sin\theta' \cos\psi' / \lambda} \tag{8.2-9}$$

and

$$\boxed{f'_Y = v'/\lambda = \sin\theta' \sin\psi' / \lambda} \tag{8.2-10}$$

If the elements are *equally spaced*, then $x_{mn} = md_X$ and $y_{mn} = nd_Y$ [see (8.1-22) and (8.1-23)], and in order to avoid grating lobes for all possible directions of beam steering, $d_X < \lambda_{\min}/2$ and $d_Y < \lambda_{\min}/2$, where $\lambda_{\min}$ is the minimum wavelength associated with the maximum frequency $f_{\max} = c/\lambda_{\min}$ in hertz (see Subsection 6.5.1). Compare (8.2-8) through (8.2-10) with (3.4-3) through (3.4-5). If (8.2-8) is substituted into (8.2-7), then

$$D'(f,f_X,f_Y)=S(f)\sum_{m=-M'}^{M'}\sum_{n=-N'}^{N'}a_{mn}(f)\exp\left\{+j2\pi\left[(f_X-f'_X)x_{mn}+(f_Y-f'_Y)y_{mn}\right]\right\},\tag{8.2-11}$$

and by comparing (8.2-11) with (8.2-6),

$$D'(f,f_X,f_Y)=D(f,f_X-f'_X,f_Y-f'_Y).\tag{8.2-12}$$

Since spatial frequencies

$$f_X=u/\lambda=\sin\theta\cos\psi/\lambda\tag{8.2-13}$$

and

$$f_Y=v/\lambda=\sin\theta\sin\psi/\lambda,\tag{8.2-14}$$

(8.2-12) can also be expressed as

$$D'(f,u,v)=D(f,u-u',v-v').\tag{8.2-15}$$

Equation (8.2-15) indicates that a set of linear phase weights applied across the surface of the array in both the X and Y directions will steer the far-field beam pattern $D(f,u,v)$ of a set of amplitude weights in the direction $u=u'$ and $v=v'$ in direction-cosine space, which is equivalent to steering (tilting) the beam pattern to $\theta=\theta'$ and $\psi=\psi'$ [see (8.2-9) and (8.2-10)]. Equal spacing of the elements is *not* required in order to do beam steering. Equation (8.2-15) also indicates that the far-field beam pattern $D'(f,u,v)$ is simply a translated or shifted version of $D(f,u,v)$ in the uv plane, implying that the shape and beamwidth of the mainlobe of the beam pattern remain unchanged. However, recall from linear aperture theory, that when a beam pattern is steered, the shape and beamwidth of the mainlobe of the beam pattern do change when the magnitude of the beam pattern is plotted as a function of the spherical angles θ and ψ (see Section 2.5).

Finally, note that a phase weight $\theta_{mn}(f)$ in radians is equivalent to a *time delay* τ'_{mn} in seconds, that is,

$$\theta_{mn}(f)\triangleq-2\pi f\tau'_{mn},\tag{8.2-16}$$

or

$$\tau'_{mn}=-\theta_{mn}(f)/(2\pi f).\tag{8.2-17}$$

Substituting (8.2-8) through (8.2-10) into (8.2-17) yields

$$\boxed{\tau'_{mn}=\frac{u'}{c}x_{mn}+\frac{v'}{c}y_{mn},\qquad \begin{array}{l}m=-M',\ldots,0,\ldots,M'\\ n=-N',\ldots,0,\ldots,N'\end{array}}\tag{8.2-18}$$

where $c = f\lambda$. The phase weights given by (8.2-8) or, equivalently, the time delays given by (8.2-18) are implemented by using the beamforming procedure shown in Fig. 8.2-1. Note that the amplitude weight a_{mn} in Fig. 8.2-1 is *not* a function of frequency.

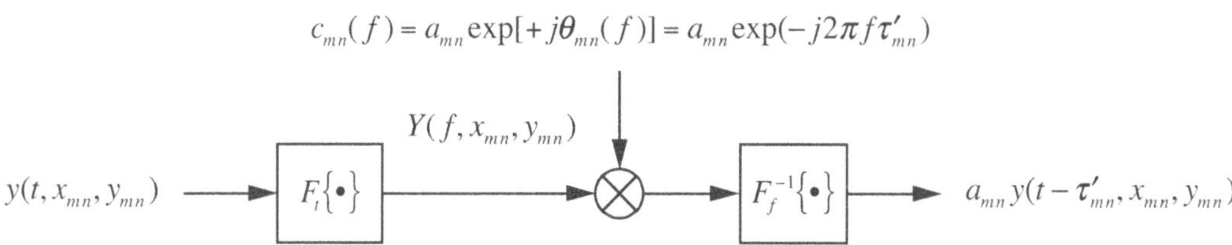

Figure 8.2-1 Beamforming. Implementing a complex weight (amplitude and phase weights) using forward and inverse Fourier transforms.

In Fig. 8.2-1, $y(t, x_{mn}, y_{mn})$ is the output electrical signal from element (m,n) in a planar array. The complex weight $c_{mn}(f)$ is just a complex number in a computer program. The expression $F_t\{\bullet\}$ represents a forward Fourier transform with respect to time t, and $F_f^{-1}\{\bullet\}$ represents an inverse Fourier transform with respect to frequency f. In practical applications, the forward and inverse Fourier transforms are computed using forward and inverse *discrete Fourier transforms* (DFTs), which can be evaluated very quickly by using forward and inverse *fast-Fourier-transform* (FFT) algorithms. This is known as *digital beamforming* or *FFT beamforming*. Since the phase weight $\theta_{mn}(f)$ is a function of frequency f, an appropriate phase weight must be applied to *each* frequency component contained in the complex frequency spectrum $Y(f, x_{mn}, y_{mn})$ shown in Fig. 8.2-1 in order to produce the correct time delay τ'_{mn}. This must be done at each element in a planar array.

Example 8.2-1 Twin-Line Planar Array

One solution to the port/starboard (left/right) ambiguity problem associated with linear arrays is a twin-line planar array. Consider a twin-line planar array of identical, complex-weighted, omnidirectional point-elements lying in the XY plane as shown in Fig. 8.2-2. The elements are equally spaced in the X direction, and equally spaced in the Y direction. If the complex weights $c_{mn}(f)$ are separable [see (8.1-40)], and since $E(f, f_X, f_Y) = \mathcal{S}(f)$, where $\mathcal{S}(f)$ is the complex, element sensitivity function, then the unnormalized, far-field beam pattern of this array is given by the Product Theorem [see (8.1-46)]

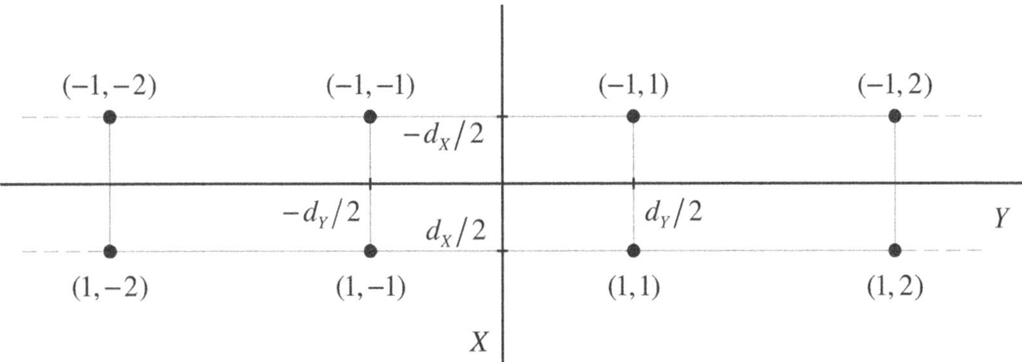

Figure 8.2-2 Top view of a twin-line planar array of identical, omnidirectional point-elements lying in the XY plane. The positive Z axis points toward the surface. Also shown are the element numbers (m,n).

$$D(f, f_X, f_Y) = \mathcal{S}(f) S_X(f, f_X) S_Y(f, f_Y), \tag{8.2-19}$$

where

$$S_X(f, f_X) = c_{-1}(f)\exp(+j2\pi f_X x_{-1}) + c_1(f)\exp(+j2\pi f_X x_1) \tag{8.2-20}$$

since there are two elements in the X direction, and

$$S_Y(f, f_Y) = \sum_{n=1}^{N/2} \left[w_{-n}(f)\exp(+j2\pi f_Y y_{-n}) + w_n(f)\exp(+j2\pi f_Y y_n) \right], \tag{8.2-21}$$

where N is the total *even* number of elements *per line* in the Y direction [see (6.1-12)].

If

$$c_{-1}(f) = c_1^*(f), \tag{8.2-22}$$

and since

$$x_1 = d_X/2 \tag{8.2-23}$$

and

$$x_{-1} = -x_1, \tag{8.2-24}$$

then (8.2-20) reduces to [see (6.1-28)]

$$S_X(f, f_X) = 2a_1(f)\cos[\pi f_X d_X + \theta_1(f)], \tag{8.2-25}$$

where the amplitude weights in the X direction

$$a_{-1}(f) = a_1(f), \tag{8.2-26}$$

and the phase weights in the X direction

$$\theta_{-1}(f) = -\theta_1(f). \tag{8.2-27}$$

In order to do beam steering, let

$$\theta_1(f) = -\pi f'_X d_X, \tag{8.2-28}$$

where f'_X is given by (8.2-9). Substituting (8.2-28) into (8.2-25) yields

$$S_X(f, f_X) = 2a_1(f)\cos\left[\pi(f_X - f'_X)d_X\right]. \tag{8.2-29}$$

Similarly, if

$$w_{-n}(f) = w_n^*(f), \qquad n = 1, 2, \ldots, N/2, \tag{8.2-30}$$

and since

$$y_n = (n - 0.5)d_Y, \qquad n = 1, 2, \ldots, N/2, \tag{8.2-31}$$

and

$$y_{-n} = -y_n, \qquad n = 1, 2, \ldots, N/2, \tag{8.2-32}$$

then (8.2-21) reduces to [see (6.1-28)]

$$S_Y(f, f_Y) = 2\sum_{n=1}^{N/2} b_n(f)\cos\left[2\pi f_Y(n - 0.5)d_Y + \phi_n(f)\right], \tag{8.2-33}$$

where the amplitude weights in the Y direction

$$b_{-n}(f) = b_n(f), \qquad n = 1, 2, \ldots, N/2, \tag{8.2-34}$$

and the phase weights in the Y direction

$$\phi_{-n}(f) = -\phi_n(f), \qquad n = 1, 2, \ldots, N/2. \tag{8.2-35}$$

In order to do beam steering, let

$$\phi_n(f) = -2\pi f'_Y(n - 0.5)d_Y, \tag{8.2-36}$$

where f'_Y is given by (8.2-10). Substituting (8.2-36) into (8.2-33) yields

$$S_Y(f, f_Y) = 2\sum_{n=1}^{N/2} b_n(f)\cos\left[2\pi(f_Y - f'_Y)(n - 0.5)d_Y\right]. \tag{8.2-37}$$

Substituting the array factors given by (8.2-29) and (8.2-37) into (8.2-19) yields

$$D(f, f_X, f_Y) = 4 a_1(f) \mathcal{S}(f) \cos\left[\pi(f_X - f'_X) d_X\right] \times \sum_{n=1}^{N/2} b_n(f) \cos\left[2\pi(f_Y - f'_Y)(n-0.5) d_Y\right], \tag{8.2-38}$$

and by substituting (8.2-9), (8.2-10), (8.2-13), and (8.2-14) into (8.2-38), we obtain

$$D(f, u, v) = 4 a_1(f) \mathcal{S}(f) \cos\left[\pi(u - u') d_X/\lambda\right] \times \sum_{n=1}^{N/2} b_n(f) \cos\left[2\pi(v - v')(n-0.5) d_Y/\lambda\right]. \tag{8.2-39}$$

If the interelement spacings

$$d_X = \lambda/4 \tag{8.2-40}$$

and

$$d_Y = \lambda/2, \tag{8.2-41}$$

then (8.2-39) reduces to

$$\boxed{D(f, u, v) = 4 a_1(f) \mathcal{S}(f) \cos\left[\pi(u - u')/4\right] \sum_{n=1}^{N/2} b_n(f) \cos\left[\pi(v - v')(n-0.5)\right]} \tag{8.2-42}$$

Equation (8.2-42) is the *unnormalized*, far-field beam pattern of a twin-line planar array lying in the XY plane with interelement spacings given by (8.2-40) and (8.2-41), where $\mathcal{S}(f)$ is the complex, element sensitivity function.

The magnitude of the far-field beam pattern given by (8.2-42) is

$$|D(f, u, v)| = 4|a_1(f)||\mathcal{S}(f)||\cos[\pi(u - u')/4]|\left|\sum_{n=1}^{N/2} b_n(f) \cos[\pi(v - v')(n-0.5)]\right|, \tag{8.2-43}$$

and since

$$|D(f, u', v')| = 4|a_1(f)||\mathcal{S}(f)|\left|\sum_{n=1}^{N/2} b_n(f)\right|, \tag{8.2-44}$$

if the amplitude weights

$$b_n(f) > 0, \qquad n = 1, 2, \ldots, N/2, \tag{8.2-45}$$

then the normalization factor is

$$D_{\max} = \max\left|D(f,u,v)\right| = 4\left|a_1(f)\right|\left|S(f)\right|\sum_{n=1}^{N/2} b_n(f). \tag{8.2-46}$$

Dividing (8.2-43) by (8.2-46) yields the following expression for the magnitude of the *normalized*, far-field beam pattern of a twin-line planar array lying in the XY plane:

$$\boxed{\left|D_N(f,u,v)\right| = \left|\cos[\pi(u-u')/4]\right| \frac{\left|\sum_{n=1}^{N/2} b_n(f)\cos[\pi(v-v')(n-0.5)]\right|}{\sum_{n=1}^{N/2} b_n(f)}} \tag{8.2-47}$$

where $\left|D_N(f,u',v')\right| = 1$. Equation (8.2-47) can also be expressed as

$$\left|D_N(f,\theta,\psi)\right| = \left|\cos[\pi(\sin\theta\cos\psi - \sin\theta'\cos\psi')/4]\right| \times \frac{\left|\sum_{n=1}^{N/2} b_n(f)\cos[\pi(\sin\theta\sin\psi - \sin\theta'\sin\psi')(n-0.5)]\right|}{\sum_{n=1}^{N/2} b_n(f)}, \tag{8.2-48}$$

and by setting $\theta = 90°$ and $\theta' = 90°$ in (8.2-48), we obtain

$$\left|D_N(f,90°,\psi)\right| = \left|\cos[\pi(\cos\psi - \cos\psi')/4]\right| \times \frac{\left|\sum_{n=1}^{N/2} b_n(f)\cos[\pi(\sin\psi - \sin\psi')(n-0.5)]\right|}{\sum_{n=1}^{N/2} b_n(f)}, \tag{8.2-49}$$

which is the magnitude of the normalized, *horizontal*, far-field beam pattern in the XY plane (see Figs. 8.2-3 and 8.2-4). If the array is towed in the positive Y direction, then $\psi = 0°$ is broadside on the starboard (right) side of the array, and $\psi = 180°$ is broadside on the port (left) side of the array. Like the cardioid beam patterns in Examples 6.1-3 and 8.1-7, the mainlobes of the beam patterns in Figs. 8.2-3 and 8.2-4 exist in a half-space. Another solution to the port/starboard (left/right) ambiguity problem is a linear array composed of triplet elements perpendicular to the array axis (see Example 9.1-2). ■

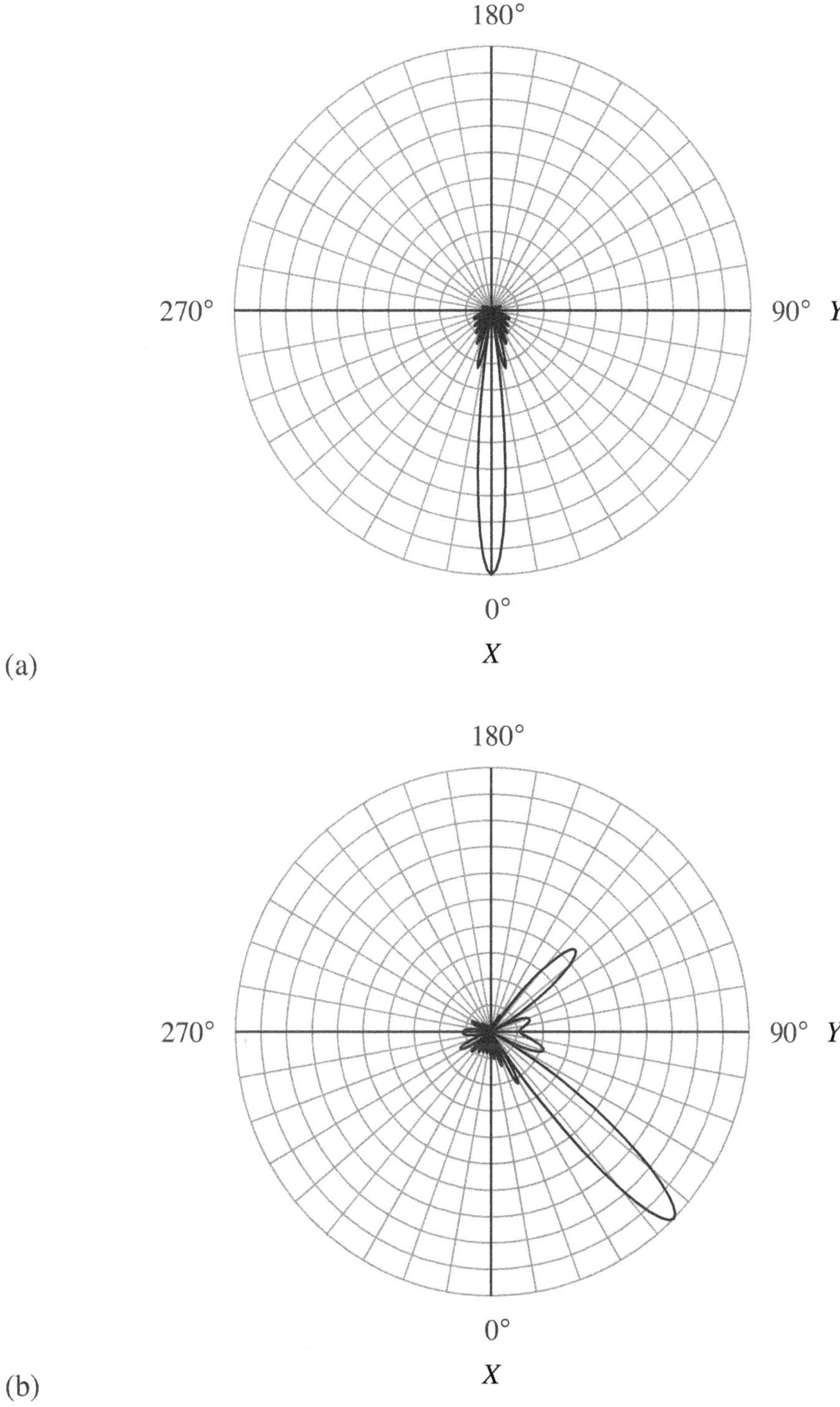

Figure 8.2-3 Polar plot, as a function of the azimuthal (bearing) angle ψ, of the magnitude of the normalized, horizontal, far-field beam pattern of a twin-line planar array given by (8.2-49) for $N = 12$, rectangular amplitude weights, and beam-steer angles (a) $\psi' = 0°$ and (b) $\psi' = 45°$.

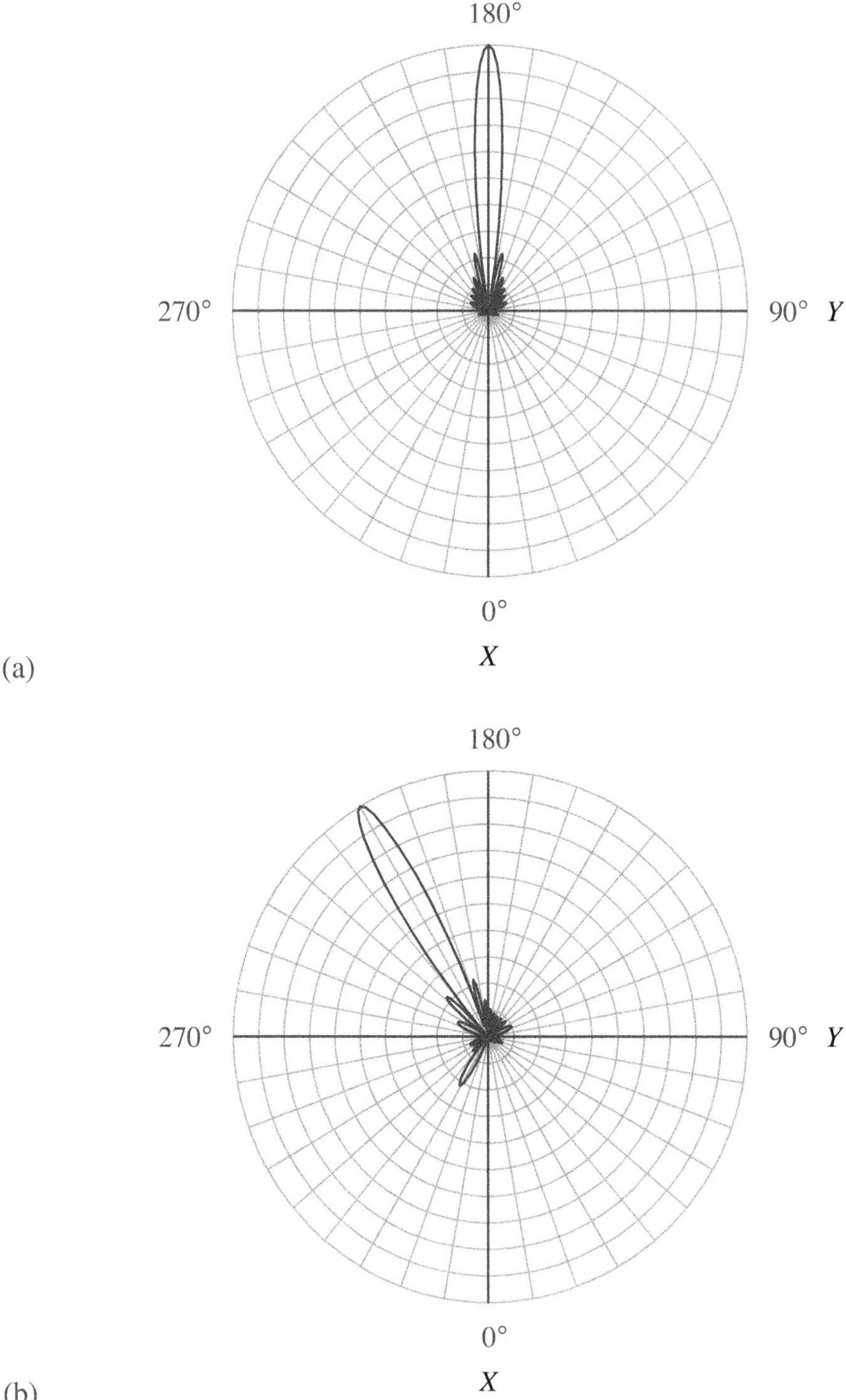

Figure 8.2-4 Polar plot, as a function of the azimuthal (bearing) angle ψ, of the magnitude of the normalized, horizontal, far-field beam pattern of a twin-line planar array given by (8.2-49) for $N = 12$, rectangular amplitude weights, and beam-steer angles (a) $\psi' = 180°$ and (b) $\psi' = 210°$.

8.3 Far-Field Beam Patterns and the Two-Dimensional Spatial Discrete Fourier Transform

In Section 6.5 it was shown that the far-field beam pattern of a linear array composed of an odd number N of identical, equally-spaced, complex-weighted, omnidirectional point-elements lying along the X axis can be computed using a one-dimensional, spatial discrete Fourier transform (DFT). In this section we shall generalize those one-dimensional results and show that the far-field beam pattern of a planar array composed of an odd number $M \times N$ of identical, equally-spaced, complex-weighted, omnidirectional point-elements lying in the XY plane, where M and N are the total odd number of elements in the X and Y directions, respectively, can be computed using a two-dimensional, spatial DFT or, if separable complex weights are used, the product of two, one-dimensional, spatial DFTs. The far-field beam pattern of the aforementioned planar array can be obtained by substituting (8.1-14), (8.1-22), and (8.1-23) into (8.1-19). Doing so yields

$$D(f, f_X, f_Y) = \mathcal{S}(f) \sum_{m=-M'}^{M'} \sum_{n=-N'}^{N'} c_{mn}(f) \exp\left[+j2\pi(f_X m d_X + f_Y n d_Y)\right], \quad (8.3\text{-}1)$$

where $\mathcal{S}(f)$ is the complex, element sensitivity function (see Table 8.1-2 and Appendix 6B),

$$c_{mn}(f) = a_{mn}(f) \exp[+j\theta_{mn}(f)], \quad (8.3\text{-}2)$$

$$M' = (M-1)/2, \quad (8.3\text{-}3)$$

and

$$N' = (N-1)/2. \quad (8.3\text{-}4)$$

According to (8.3-1), $D(f, f_X, f_Y)$ is evaluated at continuous values of spatial frequencies f_X and f_Y. Following the procedure in Section 6.5, a more efficient way to compute $D(f, f_X, f_Y)$ is to evaluate it at the following set of discrete values of f_X and f_Y:

$$f_X = r\Delta f_X, \qquad r = -M'', \ldots, 0, \ldots, M'', \quad (8.3\text{-}5)$$

and

$$f_Y = s\Delta f_Y, \qquad s = -N'', \ldots, 0, \ldots, N'', \quad (8.3\text{-}6)$$

where

$$\Delta f_X = \frac{1}{(M + Z_X) d_X} \le \frac{1}{M d_X} \quad (8.3\text{-}7)$$

and

$$\Delta f_Y = \frac{1}{(N + Z_Y)d_Y} \leq \frac{1}{Nd_Y} \tag{8.3-8}$$

are the spatial-frequency spacings (in cycles per meter) in the X and Y directions, respectively, $Z_X \geq 0$ and $Z_Y \geq 0$ are *positive integers* used for padding-with-zeros, d_X and d_Y are the interelement spacings (in meters) in the X and Y directions, respectively,

$$M'' = \begin{cases} (M + Z_X)/2, & M + Z_X \text{ even} \\ (M + Z_X - 1)/2, & M + Z_X \text{ odd} \end{cases} \tag{8.3-9}$$

and

$$N'' = \begin{cases} (N + Z_Y)/2, & N + Z_Y \text{ even} \\ (N + Z_Y - 1)/2, & N + Z_Y \text{ odd.} \end{cases} \tag{8.3-10}$$

Substituting (8.3-5) through (8.3-8) into (8.3-1) yields

$$\boxed{D(f, r, s) = \mathcal{S}(f) \sum_{m=-M'}^{M'} \sum_{n=-N'}^{N'} c_{mn}(f) W_{M+Z_X}^{rm} W_{N+Z_Y}^{sn}, \qquad \begin{array}{l} r = -M'', \ldots, 0, \ldots, M'' \\ s = -N'', \ldots, 0, \ldots, N'' \end{array}} \tag{8.3-11}$$

where the double summation on the right-hand side of (8.3-11) is the *two-dimensional*, spatial DFT of the set of $M \times N$ complex weights $c_{mn}(f)$,

$$W_{M+Z_X} = \exp\left(+j\frac{2\pi}{M + Z_X} \right), \tag{8.3-12}$$

and

$$W_{N+Z_Y} = \exp\left(+j\frac{2\pi}{N + Z_Y} \right). \tag{8.3-13}$$

Equation (8.3-11) can also be expressed as a sequence of two, one-dimensional spatial DFTs as follows:

$$\boxed{D(f, r, s) = \mathcal{S}(f) \sum_{n=-N'}^{N'} S_X(f, r, n) W_{N+Z_Y}^{sn}, \qquad \begin{array}{l} r = -M'', \ldots, 0, \ldots, M'' \\ s = -N'', \ldots, 0, \ldots, N'' \end{array}} \tag{8.3-14}$$

where

$$S_X(f,r,n)=\sum_{m=-M'}^{M'} c_{mn}(f)W_{M+Z_X}^{rm}, \qquad r=-M'',\ldots,0,\ldots,M'' \quad n=-N',\ldots,0,\ldots,N' \tag{8.3-15}$$

is directly proportional to the far-field beam pattern of a linear array composed of an odd number M of identical, equally-spaced, complex-weighted, omnidirectional point-elements lying along a line with y coordinate nd_Y that is parallel to the X axis.

If separable complex weights are used, then substituting (8.1-55) into (8.3-11) yields

$$D(f,r,s)=\mathcal{S}(f)S_X(f,r)S_Y(f,s), \qquad r=-M'',\ldots,0,\ldots,M'' \quad s=-N'',\ldots,0,\ldots,N'' \tag{8.3-16}$$

where

$$S_X(f,r)=\sum_{m=-M'}^{M'} c_m(f)W_{M+Z_X}^{rm}, \qquad r=-M'',\ldots,0,\ldots,M'' \tag{8.3-17}$$

and

$$S_Y(f,s)=\sum_{n=-N'}^{N'} w_n(f)W_{N+Z_Y}^{sn}, \qquad s=-N'',\ldots,0,\ldots,N'' \tag{8.3-18}$$

are one-dimensional, spatial DFTs in the X and Y directions, respectively, where

$$c_m(f)=a_m(f)\exp[+j\theta_m(f)] \tag{8.3-19}$$

is the complex weight in the X direction, and

$$w_n(f)=b_n(f)\exp[+j\phi_n(f)] \tag{8.3-20}$$

is the complex weight in the Y direction. Note that $S_X(f,r,0)=S_X(f,r)$ if $c_{m0}(f)=c_m(f)$ [see (8.3-15)].

In order to relate the DFT bin numbers r and s to values of the spherical angles θ and ψ, we first need to derive relationships between bin numbers r and s and the corresponding values of direction cosines u_r and v_s, respectively. Since [see (8.3-5)]

$$f_X = \frac{u}{\lambda} = r\Delta f_X, \qquad r = -M'', \ldots, 0, \ldots, M'', \tag{8.3-21}$$

solving for direction cosine u yields

$$u = u_r = r\lambda\Delta f_X, \qquad r = -M'', \ldots, 0, \ldots, M'', \tag{8.3-22}$$

and by substituting (8.3-7) into (8.3-22), we obtain

$$\boxed{u_r = \frac{r}{M + Z_X}\frac{\lambda}{d_X}, \qquad r = -M'', \ldots, 0, \ldots, M''} \tag{8.3-23}$$

where M'' is given by (8.3-9). Similarly, since [see (8.3-6)],

$$f_Y = \frac{v}{\lambda} = s\Delta f_Y, \qquad s = -N'', \ldots, 0, \ldots, N'', \tag{8.3-24}$$

solving for direction cosine v yields

$$v = v_s = s\lambda\Delta f_Y, \qquad s = -N'', \ldots, 0, \ldots, N'', \tag{8.3-25}$$

and by substituting (8.3-8) into (8.3-25), we obtain

$$\boxed{v_s = \frac{s}{N + Z_Y}\frac{\lambda}{d_Y}, \qquad s = -N'', \ldots, 0, \ldots, N''} \tag{8.3-26}$$

where N'' is given by (8.3-10). Substituting $u = u_r$ and $v = v_s$ into (3.2-9) and (3.2-10) yields

$$\boxed{\theta_{rs} = \sin^{-1}\sqrt{u_r^2 + v_s^2}, \qquad \sqrt{u_r^2 + v_s^2} \le 1} \tag{8.3-27}$$

and

$$\boxed{\psi_{rs} = \tan^{-1}(v_s/u_r)} \tag{8.3-28}$$

respectively. Equations (8.3-27) and (8.3-28) are used to compute values for the spherical angles θ and ψ at DFT bins r and s. As discussed in Section 3.2, only those (u_r, v_s) coordinates inside or on the unit circle shown in Fig. 3.2-3 can be used for computing θ_{rs} and ψ_{rs}. Also, by taking into account the signs (quadrant) of the (u_r, v_s) coordinates, one may have to add either 180° or 360° to the value obtained from (8.3-28) so that ψ_{rs} has the correct value (see Table 3.2-1).

If we let δu be the direction cosine u bin spacing, then by using (8.3-23)

$$\delta u = u_r - u_{r-1} = \frac{1}{M + Z_X} \frac{\lambda}{d_X}. \tag{8.3-29}$$

Solving for Z_X using (8.3-29) yields

$$\boxed{Z_X = \frac{1}{\delta u} \frac{\lambda}{d_X} - M} \tag{8.3-30}$$

Equation (8.3-30) is used to solve for the value of Z_X that is required in order to obtain a desired direction-cosine bin spacing δu. Similarly, if we let δv be the direction cosine v bin spacing, then by using (8.3-26)

$$\delta v = v_s - v_{s-1} = \frac{1}{N + Z_Y} \frac{\lambda}{d_Y}. \tag{8.3-31}$$

Solving for Z_Y using (8.3-31) yields

$$\boxed{Z_Y = \frac{1}{\delta v} \frac{\lambda}{d_Y} - N} \tag{8.3-32}$$

Equation (8.3-32) is used to solve for the value of Z_Y that is required in order to obtain a desired direction-cosine bin spacing δv. See Section 6.5 for more discussion on *folding bins*, *padding-with-zeros*, and the *periodicity* of the DFT.

The *normalized*, far-field beam pattern $D_N(f, r, s)$ is given by

$$D_N(f, r, s) = D(f, r, s)/D_{\max}, \tag{8.3-33}$$

where $D(f, r, s)$ is the unnormalized, far-field beam pattern given by (8.3-11), and if $a_{mn}(f) > 0 \; \forall \; m$ and n, then

$$\boxed{D_{\max} = \max\left|D(f, r, s)\right| = \left|S(f)\right| \sum_{m=-M'}^{M'} \sum_{n=-N'}^{N'} a_{mn}(f)} \tag{8.3-34}$$

is the normalization factor [use the procedure in Appendix 6A for N odd to obtain (8.3-34)]. Therefore, dividing the magnitude of (8.3-11) by (8.3-34) yields

$$\left|D_N(f,r,s)\right| = \frac{\left|\sum_{m=-M'}^{M'}\sum_{n=-N'}^{N'} c_{mn}(f) W_{M+Z_X}^{rm} W_{N+Z_Y}^{sn}\right|}{\sum_{m=-M'}^{M'}\sum_{n=-N'}^{N'} a_{mn}(f)}, \qquad \begin{array}{l} r=-M'',\ldots,0,\ldots,M'' \\ s=-N'',\ldots,0,\ldots,N'' \end{array}$$

(8.3-35)

8.4 The Near-Field Beam Pattern of a Planar Array

As was mentioned at the beginning of Section 8.1, a planar array is an example of a planar aperture. Therefore, the near-field beam pattern (directivity function) of a planar array lying in the XY plane is given by the following two-dimensional spatial Fresnel transform (see Section 3.5):

$$\begin{aligned} \mathcal{D}(f,r,f_X,f_Y) &= F_{x_A}F_{y_A}\left\{A(f,x_A,y_A)\exp\left[-j\frac{k}{2r}(x_A^2+y_A^2)\right]\right\} \\ &= \int_{-\infty}^{\infty}\int_{-\infty}^{\infty} A(f,x_A,y_A)\exp\left[-j\frac{k}{2r}(x_A^2+y_A^2)\right]\times \\ &\qquad \exp\left[+j2\pi(f_X x_A+f_Y y_A)\right]dx_A\,dy_A, \end{aligned} \tag{8.4-1}$$

where $A(f,x_A,y_A)$ is the complex frequency response (complex aperture function) of the planar array,

$$k = 2\pi f/c = 2\pi/\lambda \tag{8.4-2}$$

is the wavenumber in radians per meter, and

$$f_X = u/\lambda = \sin\theta\cos\psi/\lambda \tag{8.4-3}$$

and

$$f_Y = v/\lambda = \sin\theta\sin\psi/\lambda \tag{8.4-4}$$

are spatial frequencies in the X and Y directions, respectively, with units of cycles per meter. Since f_X and f_Y can be expressed in terms of the spherical angles θ and ψ, the near-field beam pattern can ultimately be expressed as a function of frequency f, the range r to a field point, and the spherical angles θ and ψ, that is, $\mathcal{D}(f,r,f_X,f_Y)\rightarrow\mathcal{D}(f,r,\theta,\psi)$. We shall consider a field point to be in the Fresnel (near-field) region of an array if the Fresnel angle criterion given by either (1.2-35) or (1.2-91) is satisfied, and the range r to the field point – as measured from the center of the array – satisfies the Fresnel range criterion given by

$$1.356 R_A < r < \pi R_A^2 / \lambda, \tag{8.4-5}$$

where R_A is the maximum radial extent of the array.

If a planar array lies in the XZ plane instead of the XY plane, then simply replace y_A and f_Y with z_A and f_Z, respectively, in (8.4-1), where spatial frequency f_Z is given by (1.2-45). And if a planar array lies in the YZ plane instead of the XY plane, then simply replace x_A and f_X with z_A and f_Z, respectively, in (8.4-1). As can be seen from (8.4-1), in order to derive the near-field beam pattern of a planar array, we must first specify a mathematical model for its complex frequency response.

Consider a planar array composed of an odd number $M \times N$ of identical, unevenly-spaced, complex-weighted, omnidirectional point-elements lying in the XY plane, where M and N are the total odd number of elements in the X and Y directions, respectively. The complex frequency response of this array is given by [see (8.1-20)]

$$A(f, x_A, y_A) = \mathcal{S}(f) \sum_{m=-M'}^{M'} \sum_{n=-N'}^{N'} c_{mn}(f)\,\delta(x_A - x_{mn})\,\delta(y_A - y_{mn}), \tag{8.4-6}$$

where $\mathcal{S}(f)$ is the complex, element sensitivity function (see Table 8.1-2 and Appendix 6B),

$$c_{mn}(f) = a_{mn}(f) \exp[+j\theta_{mn}(f)], \tag{8.4-7}$$

$$M' = (M-1)/2, \tag{8.4-8}$$

$$N' = (N-1)/2, \tag{8.4-9}$$

and the impulse functions $\delta(x_A - x_{mn})$ and $\delta(y_A - y_{mn})$ have units of inverse meters. Substituting (8.4-6) into (8.4-1) and making use of the sifting property of impulse functions yields the following expression for the near-field beam pattern:

$$\boxed{\begin{aligned}\mathcal{D}(f, r, f_X, f_Y) = \mathcal{S}(f) \sum_{m=-M'}^{M'} \sum_{n=-N'}^{N'} c_{mn}(f) \exp\left[-j\frac{k}{2r}(x_{mn}^2 + y_{mn}^2)\right] \times \\ \exp\left[+j2\pi(f_X x_{mn} + f_Y y_{mn})\right]\end{aligned}} \tag{8.4-10}$$

If the elements are *equally spaced*, then $x_{mn} = m d_X$ and $y_{mn} = n d_Y$ [see (8.1-22) and (8.1-23)]. The units of $A(f, x_A, y_A)$ and $\mathcal{D}(f, r, f_X, f_Y)$ are summarized in Table 8.4-1. Equation (8.4-10) is most accurate in the Fresnel region of the planar array where both the Fresnel angle criterion and the Fresnel range criterion are satisfied. It is less accurate in the near-field region outside the Fresnel region where $r < r_{\text{NF/FF}}$, but the Fresnel angle criterion is *not* satisfied. As was discussed in Subsection 1.2.1, the Fresnel region is only a subset of the near-field.

Table 8.4-1 Units of the Complex Aperture Function $A(f, x_A, y_A)$ and Corresponding Near-Field Beam Pattern $\mathcal{D}(f, r, f_X, f_Y)$ for a Planar Array

Planar Array	$A(f, x_A, y_A)$	$\mathcal{D}(f, r, f_X, f_Y)$
active (transmit)	$\left(\left(\text{m}^3/\text{sec}\right)/\text{V}\right)/\text{m}^2$	$\left(\text{m}^3/\text{sec}\right)/\text{V}$
passive (receive)	$\left(\text{V}/\left(\text{m}^2/\text{sec}\right)\right)/\text{m}^2$	$\text{V}/\left(\text{m}^2/\text{sec}\right)$

8.4.1 Beam Steering and Array Focusing

We already know from planar aperture theory that in order to do beam steering and aperture focusing in the Fresnel region of a planar aperture, there must be a linear plus quadratic phase response across the surface of the aperture in both the X and Y directions (see Subsection 3.5.1). A set of phase weights that will produce a linear plus quadratic phase response across the surface of the planar array under discussion in both the X and Y directions is given by

$$\theta_{mn}(f) = -2\pi f_X' x_{mn} - 2\pi f_Y' y_{mn} + \frac{k}{2r'}(x_{mn}^2 + y_{mn}^2), \qquad m = -M', \ldots, 0, \ldots, M' \quad n = -N', \ldots, 0, \ldots, N' \tag{8.4-11}$$

where

$$f_X' = u'/\lambda = \sin\theta' \cos\psi'/\lambda \tag{8.4-12}$$

$$f_Y' = v'/\lambda = \sin\theta' \sin\psi'/\lambda \tag{8.4-13}$$

and

$$k = 2\pi f/c = 2\pi/\lambda. \tag{8.4-14}$$

The parameter r' in (8.4-11) is the *focal range*. It is the near-field range from the array where the far-field beam pattern of the array will be in focus. Compare (8.4-11) through (8.4-14) with (3.5-17) through (3.5-20).

If (8.4-7) and (8.4-11) are substituted into (8.4-10), then

$$\mathcal{D}(f,r,f_X,f_Y) = \mathcal{S}(f) \sum_{m=-M'}^{M'} \sum_{n=-N'}^{N'} a_{mn}(f) \exp\left[-j\frac{k}{2}\left(\frac{1}{r}-\frac{1}{r'}\right)(x_{mn}^2+y_{mn}^2)\right] \times \exp\left\{+j2\pi\left[(f_X-f_X')x_{mn}+(f_Y-f_Y')y_{mn}\right]\right\}, \tag{8.4-15}$$

and if (8.4-15) is evaluated at the near-field range $r=r'$, then it reduces to

$$\mathcal{D}(f,r',f_X,f_Y) = D(f,f_X-f_X',f_Y-f_Y'), \tag{8.4-16}$$

where

$$D(f,f_X,f_Y) = \mathcal{S}(f) \sum_{m=-M'}^{M'} \sum_{n=-N'}^{N'} a_{mn}(f) \exp\left[+j2\pi(f_X x_{mn}+f_Y y_{mn})\right] \tag{8.4-17}$$

is the far-field beam pattern of the array when it is only amplitude weighted. Equation (8.4-16) indicates that the far-field beam pattern $D(f,f_X,f_Y)$ of the amplitude weights $a_{mn}(f)$ is *in focus* at the near-field range $r=r'$ meters from the array, and has been *steered* in the direction $u=u'$ and $v=v'$ in direction-cosine space, which is equivalent to steering (tilting) the beam pattern to $\theta=\theta'$ and $\psi=\psi'$ [see (8.4-12) and (8.4-13)]. However, in the near-field region of the array outside the Fresnel region, a set of quadratic phase weights can only approximately focus the far-field beam pattern. Equal spacing of the elements is *not* required in order to do beam steering and array focusing.

As was discussed in Section 8.2, a phase weight $\theta_{mn}(f)$ in radians is equivalent to a time delay τ'_{mn} in seconds, that is,

$$\theta_{mn}(f) \triangleq -2\pi f \tau'_{mn}, \tag{8.4-18}$$

or

$$\tau'_{mn} = -\theta_{mn}(f)/(2\pi f). \tag{8.4-19}$$

Substituting (8.4-11) through (8.4-14) into (8.4-19) yields

$$\boxed{\tau'_{mn} = \frac{u'}{c}x_{mn} + \frac{v'}{c}y_{mn} - \frac{1}{2r'c}(x_{mn}^2+y_{mn}^2), \qquad \begin{array}{l} m=-M',\ldots,0,\ldots,M' \\ n=-N',\ldots,0,\ldots,N' \end{array}} \tag{8.4-20}$$

where $c = f\lambda$. The phase weights given by (8.4-11) or, equivalently, the time delays given by (8.4-20) are implemented by using the beamforming procedure shown in Fig. 8.2-1.

8.5 FFT Beamforming for Planar Arrays

In this section we shall show how to implement complex weights in a planar array using the method of *FFT beamforming*. We shall also show how to compute the beam pattern of a planar array by using frequency-domain data already obtained from the method of FFT beamforming. The model that we shall use for the target is the same as that used in Subsection 1.2.2 and Section 7.2 – the target is modeled as an omnidirectional point-source located at $\mathbf{r}_S = (x_S, y_S, z_S)$ with arbitrary time dependence $s_0(t)$, where $s_0(t)$ is the source strength of the target in cubic meters per second (see Fig. 1.2-5). The model that we shall use for the fluid medium is the same as that used in Section 7.2 – the fluid medium is unbounded, viscous, and homogeneous (constant speed of sound).

The planar array in this section is being used in the passive mode, and is composed of an odd number $M \times N$ of identical, complex-weighted, omnidirectional point-elements lying in the XY plane, where M and N are the total odd number of elements in the X and Y directions, respectively. The elements are *equally spaced* in the X direction, and *equally spaced* in the Y direction. As a result, the x and y coordinates of the center of element (m,n) are given by $x_{mn} = x_m$ and $y_{mn} = y_n$ [see (8.1-22) and (8.1-23)]. Although the output electrical signals from the individual elements in the array are equal to the sum of the output electrical signals due to the target, ambient noise, and receiver noise; for example purposes, we shall only work with the output electrical signals due to the target. The procedure that we shall follow in this section is a two-dimensional generalization of the procedures used in Sections 7.2 and 7.3, and Subsection 7.3.1.

The output electrical signal in volts from element (m,n) in the array due to a sound-source (target) *before* complex weighting is given by

$$y'_{Trgt}(t, x_m, y_n) = s(t - \tau_{mn}, x_m, y_n), \tag{8.5-1}$$

for $t \geq \tau_{mn}$, where

$$s(t, x_m, y_n) = F_f^{-1}\{S(f, x_m, y_n)\} = -\frac{1}{4\pi R_{mn}} F_f^{-1}\{S_0(f)\exp[-\alpha(f)R_{mn}]\mathcal{S}_R(f)\}, \tag{8.5-2}$$

$$R_{mn} = \sqrt{r_S^2 - 2r_S(u_S x_m + v_S y_n) + x_m^2 + y_n^2} \tag{8.5-3}$$

is the range in meters between the target and the center of element (m,n),

$$r_S = |\mathbf{r}_S| = \sqrt{x_S^2 + y_S^2 + z_S^2} \tag{8.5-4}$$

is the range to the target in meters as measured from the origin of the coordinate system (the center of the array) (see Fig. 1.2-5),

$$u_S = \sin\theta_S \cos\psi_S \tag{8.5-5}$$

and

$$v_S = \sin\theta_S \sin\psi_S \tag{8.5-6}$$

are dimensionless direction cosines with respect to the X and Y axes, respectively, associated with the target's location (see Fig. 1.2-6),

$$x_m = md_X, \qquad m = -M', \ldots, 0, \ldots, M', \tag{8.5-7}$$

and

$$y_n = nd_Y, \qquad n = -N', \ldots, 0, \ldots, N', \tag{8.5-8}$$

where

$$M' = (M-1)/2 \tag{8.5-9}$$

and

$$N' = (N-1)/2, \tag{8.5-10}$$

$S_0(f)$ is the complex frequency spectrum of the target's source strength in $(\mathrm{m}^3/\mathrm{sec})/\mathrm{Hz}$, $\alpha(f)$ is the real, nonnegative, frequency-dependent, attenuation coefficient of the fluid medium in nepers (Np) per meter, $\mathcal{S}_R(f)$ is the complex, receiver sensitivity function with units of $\mathrm{V}/(\mathrm{m}^2/\mathrm{sec})$ (see Table 8.1-2 and Appendix 6B), and

$$\tau_{mn} = R_{mn}/c \tag{8.5-11}$$

is the *one-way* time delay in seconds associated with the path length between the sound-source (target) and the center of element (m,n). The spherical coordinates of the target's location (r_S, θ_S, ψ_S) are *unknown* a priori. Since element (m,n) is located at $\mathbf{r}_R = \mathbf{r}_{mn} = (x_m, y_n, 0)$, the range R_{mn} given by (8.5-3) was obtained by substituting $x_R = x_m$, $y_R = y_n$, $z_R = 0$, $x_S = r_S u_S$, and $y_S = r_S v_S$ into $R = |\mathbf{r}_R - \mathbf{r}_S| = \sqrt{(x_R - x_S)^2 + (y_R - y_S)^2 + (z_R - z_S)^2}$ and using (8.5-4).

The first step is to sample $y'_{Trgt}(t, x_m, y_n)$. If we sample $y'_{Trgt}(t, x_m, y_n)$ at a rate of f_S samples per second, then

$$y'_{Trgt}(t_l, x_m, y_n) = s(t_l - \tau_{mn}, x_m, y_n), \tag{8.5-12}$$

or

$$y'_{Trgt}(lT_S, md_X, nd_Y) = s(lT_S - \tau_{mn}, md_X, nd_Y), \tag{8.5-13}$$

where

$$t_l = lT_S, \qquad l = 0, 1, \ldots, L-1, \tag{8.5-14}$$

is the sampling time instant in seconds, $T_S = 1/f_S$ is the sampling period in seconds, and L is the total number of time samples taken. Note that

$$y'_{Trgt}(l,m,n) \equiv y'_{Trgt}(t_l, x_m, y_n) = y'_{Trgt}(lT_S, md_X, nd_Y). \tag{8.5-15}$$

By following the procedure shown in Fig. 7.3-1 for linear arrays, the second step is to compute the time-domain Fourier transform of $y'_{Trgt}(l,m,n)$. Using a modified version of the algorithm discussed in Appendix 7B for computing time-domain Fourier transforms yields

$$\hat{Y}'_{Trgt}(q,m,n) = T_S \, DFT_l\left\{y'_{Trgt}(l,m,n)\right\}, \qquad q = -L'', \ldots, 0, \ldots, L'', \tag{8.5-16}$$

or

$$\hat{Y}'_{Trgt}(q,m,n) = T_S \sum_{l=0}^{L-1} y'_{Trgt}(l,m,n) W_{L+Z}^{-ql}, \qquad q = -L'', \ldots, 0, \ldots, L'', \tag{8.5-17}$$

where

$$\hat{Y}'_{Trgt}(q,m,n) \equiv \hat{Y}'_{Trgt}(qf_0, md_X, nd_Y) = \hat{Y}'_{Trgt}(f, x_m, y_n)\Big|_{f=qf_0} \tag{8.5-18}$$

is an estimate of the frequency spectrum of $y'_{Trgt}(t, x_m, y_n)$ at discrete frequencies $f = qf_0$,

$$W_{L+Z} = \exp\left(+j\frac{2\pi}{L+Z}\right), \tag{8.5-19}$$

$$f_0 = \frac{1}{T_0} = \frac{1}{(L+Z)T_S} = \frac{f_S}{L+Z} \tag{8.5-20}$$

is the frequency-domain DFT bin spacing [the fundamental frequency of $y'_{Trgt}(t, x_m, y_n)$] in hertz, T_0 is the fundamental period of $y'_{Trgt}(t, x_m, y_n)$ in seconds,

$$L'' = \begin{cases} (L+Z)/2, & L+Z \text{ even} \\ (L+Z-1)/2, & L+Z \text{ odd,} \end{cases} \tag{8.5-21}$$

and Z is the total number of zeros used for padding-with-zeros. See Section 6.5 or Appendix 7B for an explanation of padding-with-zeros. Substituting (8.5-15) and (8.5-13) into (8.5-16) and using the time-shifting property of Fourier transforms yields

$$\hat{Y}'_{Trgt}(q,m,n) = T_S\, DFT_l\{s(lT_S - \tau_{mn}, md_X, nd_Y)\}$$
$$= T_S\, DFT_l\{s(l,m,n)\}\exp(-j2\pi q f_0 \tau_{mn}), \qquad q = -L'', \ldots, 0, \ldots, L'', \tag{8.5-22}$$

or

$$\hat{Y}'_{Trgt}(q,m,n) = \hat{S}(q,m,n)\exp(-j2\pi q f_0 \tau_{mn}), \tag{8.5-23}$$

where

$$\hat{S}(q,m,n) = T_S\, DFT_l\{s(l,m,n)\}, \qquad q = -L'', \ldots, 0, \ldots, L'', \tag{8.5-24}$$

is an estimate of the theoretical frequency spectrum [see (8.5-2)]

$$S(f, x_m, y_n) = -\frac{1}{4\pi R_{mn}} S_0(f)\exp[-\alpha(f)R_{mn}]\mathcal{S}_R(f) \tag{8.5-25}$$

at discrete frequencies $f = qf_0$, and

$$s(l,m,n) \equiv s(t_l, x_m, y_n) = s(lT_S, md_X, nd_Y). \tag{8.5-26}$$

As can be seen from (8.5-25), $S(f, x_m, y_n)$ is related to the complex frequency spectrum of the target's source strength $S_0(f)$.

The third step is to multiply $\hat{Y}'_{Trgt}(q,m,n)$ by the complex weight

$$c_{mn}(qf_0) = a_{mn}\exp[+j\theta_{mn}(qf_0)] = a_{mn}\exp(-j2\pi q f_0 \tau'_{mn}). \tag{8.5-27}$$

Doing so yields

$$\hat{Y}_{Trgt}(q,m,n) = \hat{Y}'_{Trgt}(q,m,n)\, c_{mn}(qf_0), \tag{8.5-28}$$

and by substituting (8.5-23) and (8.5-27) into (8.5-28), we obtain

$$\hat{Y}_{Trgt}(q,m,n) = a_{mn}\hat{S}(q,m,n)\exp[-j2\pi q f_0(\tau_{mn} + \tau'_{mn})]. \tag{8.5-29}$$

The fourth and final step is to compute the inverse Fourier transform of $\hat{Y}_{Trgt}(q,m,n)$. The inverse Fourier transform of $\hat{Y}_{Trgt}(q,m,n)$ is computed as follows:

$$\hat{y}_{Trgt}(l,m,n) = \frac{1}{T_S} IDFT_q\{\hat{Y}_{Trgt}(q,m,n)\}, \qquad l = 0, 1, \ldots, L + Z - 1, \tag{8.5-30}$$

or

$$\hat{y}_{Trgt}(l,m,n) = \frac{1}{T_S}\frac{1}{L+Z}\sum_{q=-L''}^{L''}\hat{Y}_{Trgt}(q,m,n)W_{L+Z}^{ql}, \qquad l = 0,1,\ldots,L+Z-1, \tag{8.5-31}$$

where IDFT is the abbreviation for inverse discrete Fourier transform. Substituting (8.5-29) into (8.5-30) yields

$$\hat{y}_{Trgt}(l,m,n) = a_{mn}\frac{1}{T_S}IDFT_q\left\{\hat{S}(q,m,n)\exp[-j2\pi qf_0(\tau_{mn}+\tau'_{mn})]\right\}, \qquad l = 0,1,\ldots,L+Z-1, \tag{8.5-32}$$

and by using the time-shifting property of Fourier transforms, we obtain

$$\hat{y}_{Trgt}(l,m,n) = a_{mn}\hat{s}(lT_S - [\tau_{mn}+\tau'_{mn}],m,n), \qquad l = 0,1,\ldots,L+Z-1, \tag{8.5-33}$$

which is an estimate of time samples of the output electrical signal $y_{Trgt}(t,x_m,y_n)$ from element (m,n) in the array due to the target *after* complex weighting, where

$$\hat{s}(l,m,n) = \frac{1}{T_S}IDFT_q\left\{\hat{S}(q,m,n)\right\} \tag{8.5-34}$$

is an *estimate* of time samples of $s(t,x_m,y_n)$ given by (8.5-2) because $\hat{Y}_{Trgt}(q,m,n)$ was obtained by multiplying the *estimate* $\hat{Y}'_{Trgt}(q,m,n)$ by the complex weight $c_{mn}(qf_0)$ [see (8.5-28)].

In order to compute the beam pattern of the planar array, compute the *two-dimensional spatial* DFT of the frequency-domain data $\hat{Y}_{Trgt}(q,m,n)$ already obtained from the method of FFT beamforming. As discussed in Section 8.3, a two-dimensional spatial DFT can be computed as a sequence of two, one-dimensional spatial DFTs. First compute the spatial DFT of $\hat{Y}_{Trgt}(q,m,n)$ in the X direction as follows:

$$\hat{Y}_{Trgt}(q,r,n) = DFT_m\left\{\hat{Y}_{Trgt}(q,m,n)\right\}, \qquad r = -M'',\ldots,0,\ldots,M'', \tag{8.5-35}$$

or (see Section 8.3)

$$\hat{Y}_{Trgt}(q,r,n) = \sum_{m=-M'}^{M'}\hat{Y}_{Trgt}(q,m,n)W_{M+Z_X}^{rm}, \qquad r = -M'',\ldots,0,\ldots,M'', \tag{8.5-36}$$

where W_{M+Z_X} is given by (8.3-12), M'' is given by (8.3-9), the DFT bin number r corresponds to discrete spatial frequency $f_X = r\Delta f_X$, where Δf_X is the spatial-frequency spacing in the X direction (with units of cycles per meter) given by (8.3-7), and Z_X is the total number of zeros used for padding-with-zeros in the X direction. The value of direction cosine u at DFT bin r is u_r, which is given by (8.3-23).

Next, compute the spatial DFT of $\hat{Y}_{Trgt}(q,r,n)$ in the Y direction as follows:

$$\hat{\mathsf{Y}}_{Trgt}(q,r,s) = DFT_n\left\{\hat{Y}_{Trgt}(q,r,n)\right\}, \qquad s = -N'',\ldots,0,\ldots,N'', \quad (8.5\text{-}37)$$

or (see Section 8.3)

$$\hat{\mathsf{Y}}_{Trgt}(q,r,s) = \sum_{n=-N'}^{N'} \hat{Y}_{Trgt}(q,r,n) W_{N+Z_Y}^{sn}, \qquad s = -N'',\ldots,0,\ldots,N'', \quad (8.5\text{-}38)$$

where W_{N+Z_Y} is given by (8.3-13), N'' is given by (8.3-10), the DFT bin number s corresponds to discrete spatial frequency $f_Y = s\Delta f_Y$, where Δf_Y is the spatial-frequency spacing in the Y direction (with units of cycles per meter) given by (8.3-8), and Z_Y is the total number of zeros used for padding-with-zeros in the Y direction. The value of direction cosine v at DFT bin s is v_s given by (8.3-26). The function $\hat{\mathsf{Y}}_{Trgt}(q,r,s)$ is an estimate of the frequency-and-angular spectrum of $y_{Trgt}(t,x_m,y_n)$ at discrete frequencies $f = qf_0$, and discrete spatial frequencies $f_X = r\Delta f_X$ and $f_Y = s\Delta f_Y$, or equivalently, direction cosine values u_r and v_s. The function $y_{Trgt}(t,x_m,y_n)$ is the output electrical signal (in volts) from element (m,n) in the array due to the target *after* complex weighting. Next we shall show that $\hat{\mathsf{Y}}_{Trgt}(q,r,s)$ is related to the beam pattern of the planar array.

For example, if the sound-source (target) is in the far-field region of the array, then set $R_{mn} = r_S$ in (8.5-25) so that $S(f,x_m,y_n) \approx S(f)$ where

$$S(f) = -\frac{1}{4\pi r_S} S_0(f)\exp[-\alpha(f)r_S]\mathcal{S}_R(f). \quad (8.5\text{-}39)$$

Also, in order to obtain a good approximation of the time delay τ_{mn}, start by substituting (7.2-8) and $\mathbf{r}_R = md_X\hat{x} + nd_Y\hat{y}$ into (7.2-37). Doing so yields

$$R_{mn} \approx r_S - u_S md_X - v_S nd_Y, \quad (8.5\text{-}40)$$

and by substituting (8.5-40) into (8.5-11), we obtain

$$\tau_{mn} \approx \tau_S - \frac{u_S}{c} m d_X - \frac{v_S}{c} n d_Y , \tag{8.5-41}$$

where

$$\tau_S = r_S / c \tag{8.5-42}$$

is the *one-way* time delay in seconds associated with the path length between the sound-source (target) and the center of the array. The time delay τ_S is also *unknown* a priori because the range r_S is unknown a priori. Based on the model for τ_{mn} given by (8.5-41), ignoring τ_S and since u_S and v_S are unknown a priori (see Section 8.2),

$$\tau'_{mn} = \frac{u'}{c} m d_X + \frac{v'}{c} n d_Y . \tag{8.5-43}$$

Adding (8.5-41) and (8.5-43) yields

$$\tau_{mn} + \tau'_{mn} = \tau_S + \frac{1}{c}(u' - u_S) m d_X + \frac{1}{c}(v' - v_S) n d_Y . \tag{8.5-44}$$

Substituting (8.5-44) into (8.5-29), and replacing $\hat{S}(q,m,n)$ with $\hat{S}(q)$ yields

$$\hat{Y}_{Trgt}(q,m,n) = \hat{S}(q) \exp(-j2\pi q f_0 \tau_S) \times \\ a_{mn} \exp\left\{-j2\pi q f_0 \left[\frac{(u' - u_S)}{c} m d_X + \frac{(v' - v_S)}{c} n d_Y \right]\right\}, \tag{8.5-45}$$

where $\hat{S}(q)$ is an estimate of the theoretical frequency spectrum $S(f)$ given by (8.5-39) at discrete frequencies $f = q f_0$. As can be seen from (8.5-39), $S(f)$ is related to the complex frequency spectrum of the target's source strength $S_0(f)$. Note that when $u' \neq u_S$ and $v' \neq v_S$, the *phase* of $\hat{Y}_{Trgt}(q,m,n)$ given by (8.5-45) is *different* at each element (m,n). As a result, the output electrical signals from each element in the array will be *out-of-phase*.

If the far-field beam pattern of the array is steered in the direction of the acoustic field incident upon the array (in this case, $\theta' = \theta_S$ and $\psi' = \psi_S$ so that $u' = u_S$ and $v' = v_S$), then (8.5-45) reduces to

$$\hat{Y}_{Trgt}(q,m,n) = \hat{S}(q) \exp(-j2\pi q f_0 \tau_S) a_{mn} . \tag{8.5-46}$$

The *phase* of $\hat{Y}_{Trgt}(q,m,n)$ given by (8.5-46) is now the *same* at each element (m,n). Therefore, the output electrical signals from each element in the array will be *in-phase* and, as a result, the array gain will be *maximized.* Substituting (8.5-46) into (8.5-36) yields

$$\hat{Y}_{Trgt}(q,r,n) = \hat{S}(q)\exp(-j2\pi q f_0 \tau_S) \sum_{m=-M'}^{M'} a_{mn} W_{M+Z_X}^{rm}, \qquad r = -M'', \ldots, 0, \ldots, M'', \tag{8.5-47}$$

and substituting (8.5-47) into (8.5-38) yields

$$\hat{Y}_{Trgt}(q,r,s) = \hat{S}(q)\exp(-j2\pi q f_0 \tau_S) \sum_{m=-M'}^{M'} \sum_{n=-N'}^{N'} a_{mn} W_{M+Z_X}^{rm} W_{N+Z_Y}^{sn}, \tag{8.5-48}$$

$$r = -M'', \ldots, 0, \ldots, M'', \quad s = -N'', \ldots, 0, \ldots, N''.$$

The double summation in (8.5-48) is the *array factor* for the set of amplitude weights a_{mn}. The array factor is directly proportional to the far-field beam pattern of the array. At broadside, that is, at $r = 0$ and $s = 0$, (8.5-48) reduces to

$$\hat{Y}_{Trgt}(q,0,0) = \hat{S}(q)\exp(-j2\pi q f_0 \tau_S) \sum_{m=-M'}^{M'} \sum_{n=-N'}^{N'} a_{mn}. \tag{8.5-49}$$

When $u' \neq u_S$ and $v' \neq v_S$, substituting (8.5-45) into (8.5-36) yields

$$\hat{Y}_{Trgt}(q,r,n) = \hat{S}(q)\exp(-j2\pi q f_0 \tau_S)\exp\left[-j2\pi q f_0 \frac{(v'-v_S)}{c} n d_Y\right] \times$$
$$\sum_{m=-M'}^{M'} a_{mn} \exp\left[-j2\pi q f_0 \frac{(u'-u_S)}{c} m d_X\right] W_{M+Z_X}^{rm},$$
$$r = -M'', \ldots, 0, \ldots, M'', \tag{8.5-50}$$

and substituting (8.5-50) into (8.5-38) yields

$$\hat{Y}_{Trgt}(q,r,s) = \hat{S}(q)\exp(-j2\pi q f_0 \tau_S) \times$$
$$\sum_{m=-M'}^{M'} \sum_{n=-N'}^{N'} a_{mn} \exp\left\{-j2\pi q f_0 \left[\frac{(u'-u_S)}{c} m d_X + \frac{(v'-v_S)}{c} n d_Y\right]\right\} W_{M+Z_X}^{rm} W_{N+Z_Y}^{sn},$$
$$r = -M'', \ldots, 0, \ldots, M'', \quad s = -N'', \ldots, 0, \ldots, N''. \tag{8.5-51}$$

The double summation in (8.5-51) is the array factor for the set of amplitude weights a_{mn} steered in the direction $u = u' - u_S$ and $v = v' - v_S$ in direction-cosine space.

In order to prove the above statement, we need to find the values of DFT bins r and s, or equivalently, direction cosines $u = u_r$ and $v = v_s$, so that (8.5-51) reduces to (8.5-49). Start by substituting (8.3-12) and (8.3-13) into (8.5-51). Doing so yields

$$\hat{Y}_{Trgt}(q,r,s) = \hat{S}(q)\exp(-j2\pi q f_0 \tau_S) \times \\ \sum_{m=-M'}^{M'} \sum_{n=-N'}^{N'} a_{mn} \exp\left\{-j2\pi\left[q f_0 \frac{(u'-u_S)}{c} d_X - \frac{r}{M+Z_X}\right] m\right\} \times \\ \exp\left\{-j2\pi\left[q f_0 \frac{(v'-v_S)}{c} d_Y - \frac{s}{N+Z_Y}\right] n\right\}. \tag{8.5-52}$$

As can be seen from (8.5-52), the values of r and s that will reduce (8.5-52) to (8.5-49) are

$$r = r' = (M + Z_X) q f_0 \frac{(u'-u_S)}{c} d_X \tag{8.5-53}$$

and

$$s = s' = (N + Z_Y) q f_0 \frac{(v'-v_S)}{c} d_Y \tag{8.5-54}$$

since these values of r and s will cause the exponents inside the double summation to be equal to 0. Although the parameter values on the right-hand sides of (8.5-53) and (8.5-54) are such that r' and s' are not integers in general, for example purposes, we shall assume they are. Therefore, substituting (8.5-53) and (8.5-54) into (8.3-23) and (8.3-26), respectively, and since

$$c = f\lambda = q f_0 \lambda, \tag{8.5-55}$$

then

$$u_{r'} = u' - u_S \tag{8.5-56}$$

and

$$v_{s'} = v' - v_S. \tag{8.5-57}$$

In other words, when $u' \neq u_S$ and $v' \neq v_S$, the value of the array factor at broadside will be located at $u = u' - u_S$ and $v = v' - v_S$ in direction-cosine space, that is, the array factor will be steered to $u = u' - u_S$ and $v = v' - v_S$.

If *no* beam steering is done, that is, if $u' = 0$ and $v' = 0$, then $u_{r'} = -u_S$ and

$v_{s'} = -v_S$, or $-u_{r'} = u_S$ and $-v_{s'} = v_S$. Substituting

$$u = -u_{r'} \tag{8.5-58}$$

and

$$v = -v_{s'} \tag{8.5-59}$$

into (3.2-9) and (3.2-10) yields

$$\theta_S = \sin^{-1}\sqrt{u_{r'}^2 + v_{s'}^2}, \qquad \sqrt{u_{r'}^2 + v_{s'}^2} \le 1 \tag{8.5-60}$$

and

$$\psi_S = \tan^{-1}\left(v_{s'}/u_{r'}\right), \tag{8.5-61}$$

which are the angles in the direction of the acoustic field incident upon the array. In order to obtain the correct value for the azimuthal (bearing) angle ψ_S, one must take into account the signs (quadrant) of the $(-u_{r'}, -v_{s'}) = (u_S, v_S)$ coordinates (see Table 3.2-1).

Problems

Section 8.1

8-1 Consider a planar array composed of an odd number $M \times N$ of identical, equally-spaced, complex-weighted elements lying in the XY plane, where M and N are the total odd number of elements in the X and Y directions, respectively. If the complex weight $c_{mn}(f)$ satisfies the conjugate symmetry property $c_{-m-n}(f) = c_{mn}^*(f)$, then what does (8.1-14) reduce to?

8-2 Consider a planar array composed of an odd number $M \times N$ of identical, equally-spaced, complex-weighted elements lying in the XY plane, where M and N are the total odd number of elements in the X and Y directions, respectively. The complex weights are separable, rectangular amplitude weights are used in both the X and Y directions, and *no* phase weighting is done.

(a) If the elements are omnidirectional point-elements, then find the unnormalized, far-field beam pattern of this array as a function of frequency f and direction cosines u and v.

(b) If the elements are rectangular pistons whose size is small compared to a wavelength so that (8.1-27) and (8.1-28) are satisfied, then find the unnormalized, far-field beam pattern of this array as a function of frequency f and direction cosines u and v.

(c) If the elements are circular pistons whose circumference is small compared to a wavelength so that (8.1-34) is satisfied, then find the unnormalized, far-field beam pattern of this array as a function of frequency f and direction cosines u and v.

8-3 A side-looking sonar (SLS) is modeled as a planar array composed of an odd number $M \times N$ of identical, equally-spaced, complex-weighted, omnidirectional point-elements lying in the YZ plane (see Fig. 5.1-1), where M and N are the total odd number of elements in the Y and Z directions, respectively, and d_Y and d_Z are the interelement spacings in meters.

(a) If the complex weights are separable, rectangular amplitude weights are used in both the Y and Z directions, and *no* phase weighting is done, then find the unnormalized, far-field beam pattern of this array as a function of frequency f and direction cosines v and w.

Using the specifications of the SLS from Example 5.7-1, if the lengths of the array in the Y and Z directions are $L_{AY} = 0.1\text{ m}$ and $L_{AZ} = 2.5\text{ m}$, respectively, the operating frequency is 30 kHz, half-wavelength interelement spacing is used in both the Y and Z directions, and $c = 1500\text{ m/sec}$, then

(b) how many elements are there in the Y direction?

(c) how many elements are there in the Z direction?

(d) what is the maximum possible value of array gain?

8-4 If the length of the sides of the equilateral triangle formed by a triplet circular array is d meters, then

(a) find the relationship between the radius a of the triplet and d.

(b) If $ka = \pi/3$ as in Example 8.1-7, then find the value of the ratio d/λ.

(c) If $ka = \pi/3$ and $c = 1500\text{ m/sec}$, then find the radius a of the triplet and the interelement spacing d for the following frequencies: $f = 100\text{ Hz}$, 500 Hz, and 1 kHz.

8-5 Consider a linear array composed of an odd number M of identical, unevenly-spaced, complex-weighted elements lying in the XY plane. Find the unnormalized, far-field beam pattern of this array if

(a) the linear array is lying along the X axis.

(b) the linear array is lying along the Y axis.

8-6 Consider a linear array composed of an odd number M of identical, unevenly-spaced, complex-weighted elements lying in the XZ plane. Find the unnormalized, far-field beam pattern of this array if

(a) the linear array is lying along the Z axis.

(b) the elements are equally-spaced rectangular pistons, rectangular amplitude weights are used, *no* phase weighting is done, and the linear array is lying along the Z axis.

Section 8.2

8-7 (a) Repeat Problem 8-3 (a) using rectangular amplitude weights *and* linear phase weights in both the Y and Z directions.

(b) Find the equations for the phase weights and equivalent time delays that will steer the far-field beam pattern of this array to $\theta' = 90°$ and $\psi' = 37.71°$ (see Example 5.7-1 and Table 5.7-2).

8-8 Find the equations for the phase weights and equivalent time delays that are required in order to steer the far-field beam pattern of the tesseral quadrupole discussed in Example 8.1-4.

8-9 Find the equations for the phase weights and equivalent time delays that will steer the far-field beam pattern of a planar array of concentric circular arrays of identical, equally-spaced, complex-weighted, omnidirectional point-elements lying in the XY plane.

8-10 Find the equations for the phase weights and equivalent time delays that are required in order to steer the mainlobe of the far-field beam pattern of the linear array in the XY plane discussed in Example 8.1-8 for the case of identical, unevenly-spaced, omnidirectional point-elements.

Section 8.3

8-11 A planar array of 505 identical, equally-spaced, complex-weighted, omnidirectional point-elements is lying in the XY plane, where the total number of elements in the X and Y directions are 101 and 5, respectively. Half-wavelength interelement spacing is used in both the X and Y directions. A two-dimensional, spatial DFT is used to compute the far-field beam pattern of the array.

(a) What are the direction-cosine bin spacings δu and δv if padding-with-zeros is not done?

(b) Solve for the number of zeros that are required in order to obtain direction-cosine bin spacings $\delta u = 0.01$ and $\delta v = 0.01$.

(c) Using your answers from Part (b), find the values of direction cosines u and v at DFT bins $r = -50$ and $s = 75$.

(d) Using your answers from Part (c), find the values of the vertical angle θ and the azimuthal (bearing) angle ψ at DFT bins $r = -50$ and $s = 75$.

Appendix 8A Two-Dimensional Spatial FIR Filters

The Product Theorem for a planar array composed of an odd number of identical elements is given by [see (8.1-12)]

$$D(f, f_X, f_Y) = E(f, f_X, f_Y)S(f, f_X, f_Y), \tag{8A-1}$$

where $D(f, f_X, f_Y)$ is the far-field beam pattern of the array, $E(f, f_X, f_Y)$ is the far-field beam pattern of *one* of the identical elements in the array, and

$$S(f, f_X, f_Y) = \sum_{m=-M'}^{M'} \sum_{n=-N'}^{N'} c_{mn}(f)\exp\left[+j2\pi(f_X x_{mn} + f_Y y_{mn})\right] \tag{8A-2}$$

is the dimensionless array factor [see (8.1-14)]. Taking the inverse two-dimensional spatial Fourier transform of the Product Theorem given by (8A-1) yields

$$A(f, x_A, y_A) = e(f, x_A, y_A) \underset{x_A,\, y_A}{**} s(f, x_A, y_A), \tag{8A-3}$$

where

$$s(f, x_A, y_A) = F_{f_X}^{-1}F_{f_Y}^{-1}\{S(f, f_X, f_Y)\} = \sum_{m=-M'}^{M'} \sum_{n=-N'}^{N'} c_{mn}(f)\delta(x_A - x_{mn})\delta(y_A - y_{mn}) \tag{8A-4}$$

since

$$\begin{aligned} F_{f_X}^{-1}F_{f_Y}^{-1}\left\{\exp\left[+j2\pi(f_X x_{mn} + f_Y y_{mn})\right]\right\} &= F_{f_X}^{-1}\left\{\exp(+j2\pi f_X x_{mn})\right\}F_{f_Y}^{-1}\left\{\exp(+j2\pi f_Y y_{mn})\right\} \\ &= \delta(x_A - x_{mn})\delta(y_A - y_{mn}). \end{aligned} \tag{8A-5}$$

If the elements are omnidirectional point-elements, then the element function $e(f, x_A, y_A)$ is given by [see (8.1-16)]

$$e(f, x_A, y_A) = \mathcal{S}(f)\delta(x_A, y_A) = \mathcal{S}(f)\delta(x_A)\delta(y_A), \tag{8A-6}$$

where $\mathcal{S}(f)$ is the element sensitivity function. If we set $\mathcal{S}(f) = 1\left(\text{m}^3/\text{sec}\right)/\text{V}$ or $\mathcal{S}(f) = 1\,\text{V}/\left(\text{m}^2/\text{sec}\right)$, then (8A-6) reduces to

$$e(f, x_A, y_A) = \delta(x_A)\delta(y_A). \tag{8A-7}$$

Substituting the "input signal" given by (8A-7) into (8A-3) yields the "output signal"

$$A(f, x_A, y_A) = \delta(x_A)\delta(y_A) \underset{x_A, y_A}{**} s(f, x_A, y_A) = s(f, x_A, y_A), \tag{8A-8}$$

where $s(f, x_A, y_A)$ given by (8A-4) can be thought of as the two-dimensional, spatial impulse response of the array at frequency f hertz. Since (8A-4) is in the form of the impulse response of a two-dimensional, finite impulse response (FIR) digital filter, a planar array composed of an odd number of identical, complex-weighted elements can be thought of as a two-dimensional, spatial FIR filter.

Appendix 8B Normalization Factor for the Array Factor

The general expression for the array factor $S(f, f_X, f_Y)$ for M and N odd is given by [see (8.1-14)]

$$S(f, f_X, f_Y) = \sum_{m=-M'}^{M'} \sum_{n=-N'}^{N'} c_{mn}(f)\exp\left[+j2\pi(f_X x_{mn} + f_Y y_{mn})\right]. \tag{8B-1}$$

Computing the magnitude of $S(f, f_X, f_Y)$ yields

$$\begin{aligned} |S(f, f_X, f_Y)| &= \left|\sum_{m=-M'}^{M'} \sum_{n=-N'}^{N'} c_{mn}(f)\exp\left[+j2\pi(f_X x_{mn} + f_Y y_{mn})\right]\right| \\ &\le \sum_{m=-M'}^{M'} \sum_{n=-N'}^{N'} \left|c_{mn}(f)\exp\left[+j2\pi(f_X x_{mn} + f_Y y_{mn})\right]\right| \qquad \text{(8B-2)} \\ &\le \sum_{m=-M'}^{M'} \sum_{n=-N'}^{N'} \left|c_{mn}(f)\right|\left|\exp\left[+j2\pi(f_X x_{mn} + f_Y y_{mn})\right]\right|, \end{aligned}$$

which further reduces to

$$\left|S(f, f_X, f_Y)\right| \leq \sum_{m=-M'}^{M'} \sum_{n=-N'}^{N'} \left|a_{mn}(f)\right|. \tag{8B-3}$$

Therefore,

$$\max\left|S(f, f_X, f_Y)\right| = \sum_{m=-M'}^{M'} \sum_{n=-N'}^{N'} \left|a_{mn}(f)\right|. \tag{8B-4}$$

The absolute value of the real amplitude weights is required because amplitude weights can have both positive and negative values in general.

Equation (8B-4) is the absolute maximum value of $\left|S(f, f_X, f_Y)\right|$. The actual maximum value may be less than (8B-4) as indicated by (8B-3). However, if $a_{mn}(f) > 0 \;\; \forall \;\; m$ and n, then the normalization factor $S_{\max}$ for M and N odd is

$$\boxed{S_{\max} = \max\left|S(f, f_X, f_Y)\right| = \sum_{m=-M'}^{M'} \sum_{n=-N'}^{N'} a_{mn}(f)} \tag{8B-5}$$

Note that if (8B-1) is evaluated at broadside (i.e., at $f_X = 0$ and $f_Y = 0$) and if no phase weighting is done, then

$$S(f, 0, 0) = \sum_{m=-M'}^{M'} \sum_{n=-N'}^{N'} a_{mn}(f) \tag{8B-6}$$

and

$$\left|S(f, 0, 0)\right| = \left|\sum_{m=-M'}^{M'} \sum_{n=-N'}^{N'} a_{mn}(f)\right| \leq \sum_{m=-M'}^{M'} \sum_{n=-N'}^{N'} \left|a_{mn}(f)\right|. \tag{8B-7}$$

If $a_{mn}(f) \geq 0 \;\; \forall \;\; m$ and n, then

$$\left|S(f, 0, 0)\right| = \left|\sum_{m=-M'}^{M'} \sum_{n=-N'}^{N'} a_{mn}(f)\right| = \sum_{m=-M'}^{M'} \sum_{n=-N'}^{N'} a_{mn}(f), \tag{8B-8}$$

and by comparing (8B-5) and (8B-8),

$$\boxed{S_{\max} = \max\left|S(f, f_X, f_Y)\right| = \left|S(f, 0, 0)\right|} \tag{8B-9}$$

Chapter 9

Array Theory – Volume Arrays

9.1 The Far-Field Beam Pattern of a Cylindrical Array

Since a volume array is an example of a volume aperture, the far-field beam pattern (directivity function) of a volume array is given by the following three-dimensional spatial Fourier transform (see Section 1.3):

$$D(f, f_X, f_Y, f_Z) = F_{x_A} F_{y_A} F_{z_A}\{A(f, x_A, y_A, z_A)\}$$
$$= \int_{-\infty}^{\infty}\int_{-\infty}^{\infty}\int_{-\infty}^{\infty} A(f, x_A, y_A, z_A) \times$$
$$\exp[+j2\pi(f_X x_A + f_Y y_A + f_Z z_A)]\,dx_A\,dy_A\,dz_A, \tag{9.1-1}$$

where $A(f, x_A, y_A, z_A)$ is the complex frequency response (complex aperture function) of the volume array, and

$$f_X = u/\lambda = \sin\theta\cos\psi/\lambda, \tag{9.1-2}$$

$$f_Y = v/\lambda = \sin\theta\sin\psi/\lambda, \tag{9.1-3}$$

and

$$f_Z = w/\lambda = \cos\theta/\lambda \tag{9.1-4}$$

are spatial frequencies in the X, Y, and Z directions, respectively, with units of cycles per meter. Since f_X, f_Y, and f_Z can be expressed in terms of the spherical angles θ and ψ, the far-field beam pattern can ultimately be expressed as a function of frequency f and the spherical angles θ and ψ, that is, $D(f, f_X, f_Y, f_Z) \rightarrow D(f, \theta, \psi)$. We shall consider a field point to be in the far-field region of an array if the range r to the field point – as measured from the center of the array – satisfies the Fraunhofer (far-field) range criterion given by

$$r > \pi R_A^2/\lambda > 2.414 R_A, \tag{9.1-5}$$

where R_A is the maximum radial extent of the array. The units of $A(f, x_A, y_A, z_A)$ and $D(f, f_X, f_Y, f_Z)$ are summarized in Table 9.1-1 (Also see Tables 1B-1 and 1B-2).

Equation (9.1-1) can also be expressed in terms of the cylindrical coordinates (r_A, ϕ_A, z_A), as shown in Fig. 9.1-1, by noting that

Table 9.1-1 Units of the Complex Frequency Response $A(f, x_A, y_A, z_A)$ and Corresponding Far-Field Beam Pattern $D(f, f_X, f_Y, f_Z)$ for a Volume Array

Volume Array	$A(f, x_A, y_A, z_A)$	$D(f, f_X, f_Y, f_Z)$
active (transmit)	$\left(\left(\text{m}^3/\text{sec}\right)/\text{V}\right)/\text{m}^3$	$\left(\text{m}^3/\text{sec}\right)/\text{V}$
passive (receive)	$\left(\text{V}/\left(\text{m}^2/\text{sec}\right)\right)/\text{m}^3$	$\text{V}/\left(\text{m}^2/\text{sec}\right)$

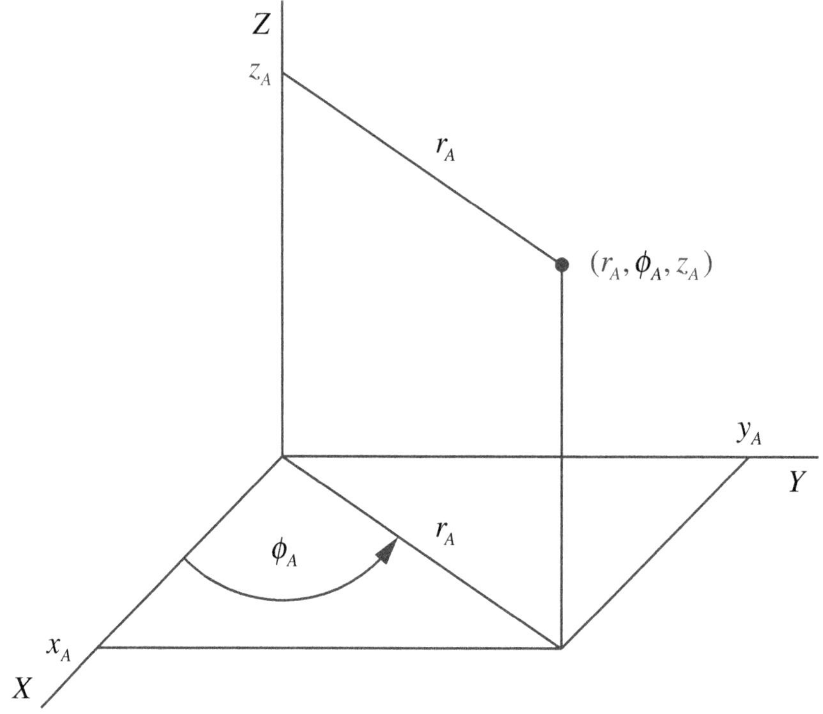

Figure 9.1-1 The cylindrical coordinates (r_A, ϕ_A, z_A).

$$x_A = r_A \cos\phi_A, \tag{9.1-6}$$

$$y_A = r_A \sin\phi_A, \tag{9.1-7}$$

$$z_A = z_A, \tag{9.1-8}$$

$$dx_A\, dy_A\, dz_A \rightarrow r_A\, dr_A\, d\phi_A\, dz_A, \tag{9.1-9}$$

and

$$A(f, x_A, y_A, z_A) = A(f, r_A \cos\phi_A, r_A \sin\phi_A, z_A) \rightarrow A(f, r_A, \phi_A, z_A). \tag{9.1-10}$$

Substituting (9.1-2) through (9.1-4), and (9.1-6) through (9.1-10) into (9.1-1), and using the trigonometric identity

$$\cos(\alpha-\beta)=\cos\alpha\cos\beta+\sin\alpha\sin\beta, \tag{9.1-11}$$

yields

$$D(f,\theta,\psi)=\int_{-\infty}^{\infty}\int_{0}^{2\pi}\int_{0}^{\infty} A(f,r_A,\phi_A,z_A)\times \exp\left\{+j\frac{2\pi}{\lambda}\left[r_A\sin\theta\cos(\psi-\phi_A)+z_A\cos\theta\right]\right\}r_A\,dr_A\,d\phi_A\,dz_A \tag{9.1-12}$$

Equation (9.1-12) shall be used to derive the far-field beam pattern of a cylindrical array of omnidirectional point-elements. A cylindrical array is an example of a *conformal array*. A conformal array is an array of elements that lie on a *curved* surface. In other words, the array conforms to the shape of the curved surface.

Consider an array of identical, equally-spaced, complex-weighted, omnidirectional point-elements lying on the surface of a cylinder with radius a meters and length L meters (see Fig. 9.1-2). The maximum radial extent of the cylindrical array is

$$R_A=\sqrt{(L/2)^2+a^2}\,. \tag{9.1-13}$$

Any single vertical column of elements or any set of two or more adjacent vertical columns of elements (linear arrays parallel to the Z axis) is known as a *stave*. Since an impulse function is used as a mathematical model for an omnidirectional point-source in acoustic wave propagation theory, the mathematical model that we shall use for the complex frequency response of an omnidirectional point-element lying on the surface of a cylinder is given by

$$e(f,r_A-r_0,\phi_A-\phi_0,z_A-z_0)=\mathcal{S}(f)\frac{\delta(r_A-r_0)}{r_A}\delta(\phi_A-\phi_0)\delta(z_A-z_0), \tag{9.1-14}$$

where $\mathcal{S}(f)$ is the complex, *element sensitivity function*, and (r_0,ϕ_0,z_0) are the cylindrical coordinates of the center of the element. If the point-element is used in the active mode as a transmitter, then the element sensitivity function $\mathcal{S}(f)$ is referred to as the *transmitter sensitivity function* with units of $\left(\text{m}^3/\text{sec}\right)/\text{V}$ (see Appendix 6B). If the point-element is used in the passive mode as a receiver, then

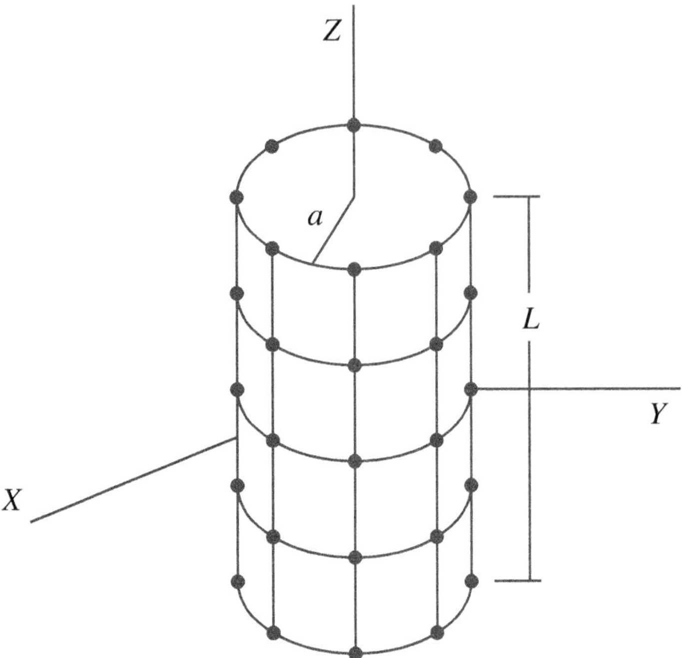

Figure 9.1-2 Cylindrical array of identical, equally-spaced, omnidirectional point-elements.

$\mathcal{S}(f)$ is referred to as the *receiver sensitivity function* with units of $\mathrm{V}/\left(\mathrm{m}^2/\mathrm{sec}\right)$ (see Appendix 6B). Note that

$$\begin{aligned}\int_{-\infty}^{\infty}\int_{0}^{2\pi}\int_{0}^{\infty} e(f, r_A - r_0, \phi_A - \phi_0, z_A - z_0) r_A dr_A d\phi_A dz_A = \mathcal{S}(f)\int_{0}^{\infty} \delta(r_A - r_0) dr_A \times \\ \int_{0}^{2\pi} \delta(\phi_A - \phi_0) d\phi_A \times \\ \int_{-\infty}^{\infty} \delta(z_A - z_0) dz_A \\ = \mathcal{S}(f).\end{aligned} \tag{9.1-15}$$

The impulse function $\delta(r_A - r_0)$ has units of inverse meters, $\delta(\phi_A - \phi_0)$ has units of inverse radians, and $\delta(z_A - z_0)$ has units of inverse meters. The units of $\mathcal{S}(f)$ are summarized in Table 9.1-2.

Therefore, with the use of (9.1-14) and the principle of superposition, the complex frequency response (complex aperture function) of the cylindrical array shown in Fig. 9.1-2 can be expressed as the sum of the complex-weighted frequency responses of all the elements in the array as follows:

Table 9.1-2 Units of the Element Sensitivity Function $\mathcal{S}(f)$

Omnidirectional Point-Element	$\mathcal{S}(f)$
active (transmit)	$(\mathrm{m}^3/\mathrm{sec})/\mathrm{V}$
passive (receive)	$\mathrm{V}/(\mathrm{m}^2/\mathrm{sec})$

$$A(f, r_A, \phi_A, z_A) = \mathcal{S}(f)\frac{\delta(r_A - a)}{r_A}\sum_{m=1}^{M}\sum_{n=-N'}^{N'} c_{mn}(f)\delta(\phi_A - \phi_m)\delta(z_A - z_n), \tag{9.1-16}$$

where M is the total number (even or odd) of omnidirectional point-elements per circular array,

$$N' = (N-1)/2, \tag{9.1-17}$$

N is the total *odd* number of circular arrays,

$$\boxed{c_{mn}(f) = a_{mn}(f)\exp[+j\theta_{mn}(f)]} \tag{9.1-18}$$

is the frequency-dependent, *complex weight* associated with element (m,n), where $a_{mn}(f)$ is a real, frequency-dependent, *dimensionless*, amplitude weight, and $\theta_{mn}(f)$ is a real, frequency-dependent, phase weight in radians, and (a, ϕ_m, z_n) are the cylindrical coordinates of the center of element (m,n). Complex weights are implemented by using digital signal processing (digital beamforming) (see Section 6.4). The complex weights are used to control the complex frequency response of the array and, thus, the array's far-field beam pattern via amplitude and phase weighting. Since the elements are equally spaced in the ϕ and Z directions,

$$\phi_m = (m-1)\Delta\phi, \tag{9.1-19}$$

where

$$\Delta\phi = 360°/M, \tag{9.1-20}$$

and

$$z_n = nd_Z, \tag{9.1-21}$$

where the interelement spacing in the Z direction (in meters) is given by

$$d_Z = \frac{L}{N-1}, \quad N \neq 1. \tag{9.1-22}$$

Substituting (9.1-16) into (9.1-12) yields the following expression for the far-field beam pattern of a cylindrical array of identical, equally-spaced, complex-weighted, omnidirectional point-elements:

$$D(f,\theta,\psi) = \mathcal{S}(f) \sum_{m=1}^{M} \sum_{n=-N'}^{N'} c_{mn}(f) \exp\left[+j\frac{2\pi a}{\lambda}\sin\theta\cos(\psi-\phi_m)\right] \times \exp\left[+j\frac{2\pi}{\lambda}z_n\cos\theta\right] \tag{9.1-23}$$

where $\mathcal{S}(f)$ is the complex, element sensitivity function, and the double summation is the dimensionless array factor. If one or more elements in the array are broken (not transmitting and/or receiving), set $c_{mn}(f)=0$ for those elements. Similarly, if only one stave is in operation at any given instant of time, set $c_{mn}(f)=0$ for those elements not in operation.

If the number of circular arrays $N=1$, then $N'=0$ [see (9.1-17)] and, as a result, $n=0$. And since $z_0=0$ [see (9.1-21)], (9.1-23) reduces to

$$D(f,\theta,\psi) = \mathcal{S}(f) \sum_{m=1}^{M} c_m(f) \exp\left[+j\frac{2\pi a}{\lambda}\sin\theta\cos(\psi-\phi_m)\right]. \tag{9.1-24}$$

Equation (9.1-24) is the far-field beam pattern of a single circular array with radius a meters lying in the XY plane [see (8.1-92)].

If the complex weights are separable, then

$$c_{mn}(f) = c_m(f)w_n(f), \tag{9.1-25}$$

where

$$c_m(f) = a_m(f)\exp[+j\theta_m(f)] \tag{9.1-26}$$

is the frequency-dependent, complex weight in the ϕ direction, and

$$w_n(f) = b_n(f)\exp[+j\phi_n(f)] \tag{9.1-27}$$

is the frequency-dependent, complex weight in the Z direction, where $a_m(f)$ and $b_n(f)$ are real, frequency-dependent, dimensionless, amplitude weights, and

$\theta_m(f)$ and $\phi_n(f)$ are real, frequency-dependent, phase weights in radians. Substituting (9.1-25) into (9.1-23) yields

$$\boxed{D(f,\theta,\psi) = \mathcal{S}(f) S_\phi(f,\theta,\psi) S_Z(f,\theta,\psi)} \tag{9.1-28}$$

where

$$\boxed{S_\phi(f,\theta,\psi) = \sum_{m=1}^{M} c_m(f) \exp\left[+j\frac{2\pi a}{\lambda}\sin\theta\cos(\psi - \phi_m)\right]} \tag{9.1-29}$$

and

$$\boxed{S_Z(f,\theta,\psi) = \sum_{n=-N'}^{N'} w_n(f) \exp\left[+j\frac{2\pi}{\lambda} z_n \cos\theta\right]} \tag{9.1-30}$$

are dimensionless array factors. Since

$$\cos(\psi - \phi_m) = \cos\psi\cos\phi_m + \sin\psi\sin\phi_m , \tag{9.1-31}$$

(9.1-29) can be rewritten as follows:

$$\boxed{S_\phi(f,\theta,\psi) = \sum_{m=1}^{M} c_m(f) \exp\left[+j\frac{2\pi}{\lambda}(x_m \sin\theta\cos\psi + y_m \sin\theta\sin\psi)\right]} \tag{9.1-32}$$

where

$$x_m = a\cos\phi_m \tag{9.1-33}$$

and

$$y_m = a\sin\phi_m \tag{9.1-34}$$

are the rectangular coordinates of element m in a circular array with radius a meters, and ϕ_m is given by (9.1-19).

9.1.1 The Phased Array – Beam Steering

Beam steering can be accomplished by operating staves in a clockwise or counterclockwise fashion. This is known as *scanning*. When a particular stave is in operation, the far-field beam pattern of the stave can be steered to $\theta = \theta'$ and $\psi = \psi'$ by using separable complex weights with the following phase weights:

$$\theta_m(f) = -\frac{2\pi}{\lambda}(x_m \sin\theta' \cos\psi' + y_m \sin\theta' \sin\psi'), \qquad m = 1, 2, \ldots, M \tag{9.1-35}$$

where x_m and y_m are given by (9.1-33) and (9.1-34), respectively, and

$$\phi_n(f) = -\frac{2\pi}{\lambda} z_n \cos\theta', \qquad n = -N', \ldots, 0, \ldots, N' \tag{9.1-36}$$

where z_n is given by (9.1-21). Substituting (9.1-26) and (9.1-35) into (9.1-32), and (9.1-27) and (9.1-36) into (9.1-30) yields

$$S_\phi(f, \theta, \psi) = \sum_{m=1}^{M} a_m(f) \exp\left[+j\frac{2\pi}{\lambda}(\sin\theta\cos\psi - \sin\theta'\cos\psi')x_m\right] \times \exp\left[+j\frac{2\pi}{\lambda}(\sin\theta\sin\psi - \sin\theta'\sin\psi')y_m\right] \tag{9.1-37}$$

and

$$S_Z(f, \theta, \psi) = \sum_{n=-N'}^{N'} b_n(f) \exp\left[+j\frac{2\pi}{\lambda}(\cos\theta - \cos\theta')z_n\right] \tag{9.1-38}$$

respectively. Therefore, the far-field beam pattern of a particular stave of the cylindrical array when beam steering is done is obtained by substituting (9.1-37) and (9.1-38) into (9.1-28), and by setting $a_m(f) = 0$ in (9.1-37) for those elements not in operation. By operating two or more staves at the same time and using *parallel signal processing*, multiple beams with multiple "look directions" can be formed *simultaneously*.

Finally, note that a phase weight $\theta_{mn}(f)$ in radians is equivalent to a *time delay* τ'_{mn} in seconds, that is,

$$\theta_{mn}(f) \triangleq -2\pi f \tau'_{mn}, \tag{9.1-39}$$

or

$$\tau'_{mn} = -\theta_{mn}(f)/(2\pi f). \tag{9.1-40}$$

In our case,

$$\theta_{mn}(f) = \theta_m(f) + \phi_n(f), \tag{9.1-41}$$

where $\theta_m(f)$ and $\phi_n(f)$ are given by (9.1-35) and (9.1-36), respectively. Substituting (9.1-41), (9.1-35), and (9.1-36) into (9.1-40) yields

$$\boxed{\tau'_{mn} = \frac{u'}{c}x_m + \frac{v'}{c}y_m + \frac{w'}{c}z_n, \qquad \begin{array}{l} m = 1, 2, \ldots, M \\ n = -N', \ldots, 0, \ldots, N' \end{array}} \tag{9.1-42}$$

where $c = f\lambda$, and

$$u' = \sin\theta' \cos\psi', \tag{9.1-43}$$

$$v' = \sin\theta' \sin\psi', \tag{9.1-44}$$

and

$$w' = \cos\theta'. \tag{9.1-45}$$

Example 9.1-1 Beam Steering the Far-Field Beam Pattern of a Stave

Let $M = 30$, where M is the total number of omnidirectional point-elements per circular array. As a result, the angular spacing in the ϕ direction $\Delta\phi = 12°$ [see (9.1-20)], and angle $\phi_m = (m-1)12°$, $m = 1, 2, \ldots, 30$ [see (9.1-19)]. Also, let a single stave be composed of five adjacent vertical columns of elements, and let the range of values for index m for stave 1 be $m = 1, 2, 3, 4, 5$; for stave 2, $m = 2, 3, 4, 5, 6$; … ; for stave 26, $m = 26, 27, 28, 29, 30$; for stave 27, $m = 27, 28, 29, 30, 1$; … ; and for stave 30, $m = 30, 1, 2, 3, 4$. Therefore, for stave 1, $\phi_1 = 0°$, $\phi_2 = 12°$, $\phi_3 = 24°$, $\phi_4 = 36°$, $\phi_5 = 48°$, beam-steer angle $\psi' = \phi_3 = 24°$, and amplitude weight $a_m(f) = 0$ for $m = 6, 7, \ldots, 30$; for stave 2, $\phi_2 = 12°$, $\phi_3 = 24°$, $\phi_4 = 36°$, $\phi_5 = 48°$, $\phi_6 = 60°$, $\psi' = \phi_4 = 36°$, and $a_m(f) = 0$ for $m = 1$ and $m = 7, 8, \ldots, 30$; etc. Note that for stave 29, $\psi' = \phi_1 = 0°$; and for stave 30, $\psi' = \phi_2 = 12°$. If stave 2 is currently in operation, then substituting (9.1-33) and (9.1-34) into (9.1-37), and summing over index m from 2 to 6 yields

$$S_\phi(f,\theta,\psi) = \sum_{m=2}^{6} a_m(f) \exp\left[+j\frac{2\pi a}{\lambda}(\sin\theta\cos\psi - \sin\theta'\cos\psi')\cos\phi_m\right] \times \exp\left[+j\frac{2\pi a}{\lambda}(\sin\theta\sin\psi - \sin\theta'\sin\psi')\sin\phi_m\right] \tag{9.1-46}$$

where $\phi_2 = 12°$, $\phi_3 = 24°$, $\phi_4 = 36°$, $\phi_5 = 48°$, $\phi_6 = 60°$, and $\psi' = \phi_4 = 36°$. Note that ψ' does not have to be equal to ϕ_4. The array factor $S_Z(f,\theta,\psi)$ given by (9.1-38) remains unchanged. However, by substituting (9.1-21) and (9.1-22) into (9.1-38), it can be rewritten as

$$S_Z(f,\theta,\psi) = \sum_{n=-N'}^{N'} b_n(f) \exp\left[+j2\pi \frac{L}{\lambda}(\cos\theta - \cos\theta') \frac{n}{N-1}\right]. \tag{9.1-47}$$

Substituting (9.1-46) and (9.1-47) into (9.1-28) yields the far-field beam pattern of stave 2. At $\theta = \theta'$ and $\psi = \psi'$,

$$\begin{aligned} D(f,\theta',\psi') &= \mathcal{S}(f) S_\phi(f,\theta',\psi') S_Z(f,\theta',\psi') \\ &= \mathcal{S}(f) \sum_{m=2}^{6} a_m(f) \sum_{n=-N'}^{N'} b_n(f). \end{aligned} \tag{9.1-48}$$

If both sets of amplitude weights are rectangular amplitude weights, that is, if $a_m(f) = 1$, $m = 2, 3, \ldots, 6$, and $b_n(f) = 1 \ \forall \ n$, then

$$D(f,\theta',\psi') = 5N\mathcal{S}(f), \tag{9.1-49}$$

where N is the total odd number of circular arrays (or elements in a linear array parallel to the Z axis).

Figure 9.1-3 is a polar plot, as a function of the azimuthal (bearing) angle ψ, of the magnitude of the normalized, horizontal, far-field beam pattern of stave 2 in the first quadrant for $M = 30$, $N = 9$, rectangular amplitude weights, $ka = 2\pi a/\lambda = 15$, $L/\lambda = 4$, and for (a) no beam steering, and for beam-steer angles (b) $\theta' = 90°$ and $\psi' = \phi_4 = 36°$, and (c) $\theta' = 90°$ and $\psi' = 30°$. Note that the axis of the mainlobe in Fig. 9.1-3 (a) is at $\psi = \phi_4 = 36°$ even though no beam steering was done. However, as can be seen in Fig. 9.1-3 (b), when beam steering is done using $\theta' = 90°$ and $\psi' = \phi_4 = 36°$, the mainlobe in Fig. 9.1-3 (b) is well defined compared to the mainlobe in Fig. 9.1-3 (a), and the sidelobes in Fig. 9.1-3 (b) are lower than the sidelobes in Fig. 9.1-3 (a). Figure 9.1-4 is a polar plot, as a function of the vertical angle θ, of the magnitude of the normalized, vertical, far-field beam pattern of stave 2 in the plane with bearing angle $\psi = \psi'$ for $M = 30$, $N = 9$, rectangular amplitude weights, $ka = 2\pi a/\lambda = 15$, $L/\lambda = 4$, and for beam-steer angles (a) $\theta' = 135°$ and $\psi' = \phi_4 = 36°$, and (b) $\theta' = 135°$ and $\psi' = 30°$. ■

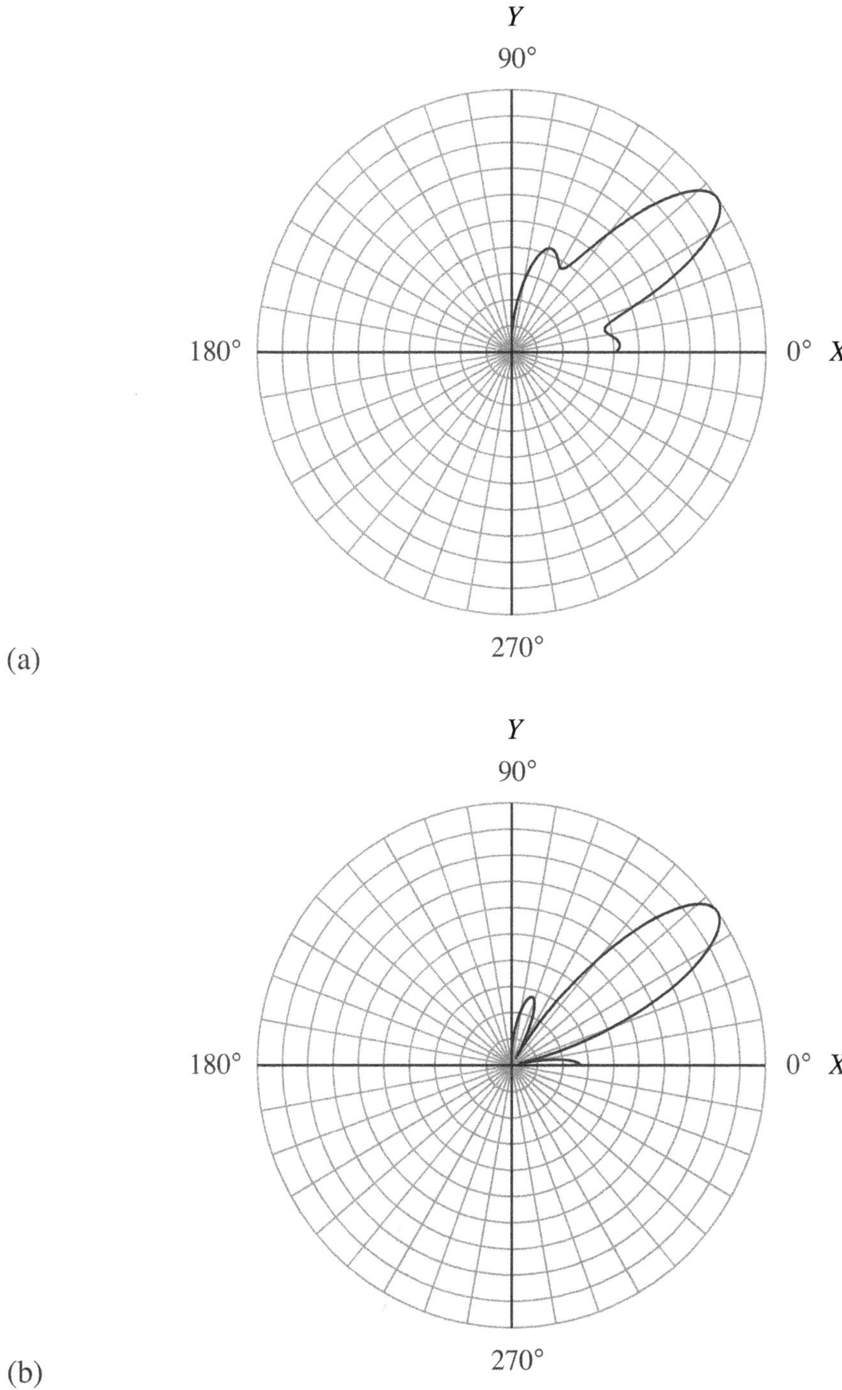

Figure 9.1-3 Polar plot, as a function of the azimuthal (bearing) angle ψ, of the magnitude of the normalized, horizontal, far-field beam pattern of stave 2 in the first quadrant for the cylindrical array discussed in Example 9.1-1 for (a) no beam steering, and for beam-steer angles (b) $\theta' = 90°$ and $\psi' = \phi_4 = 36°$, and (c) $\theta' = 90°$ and $\psi' = 30°$.

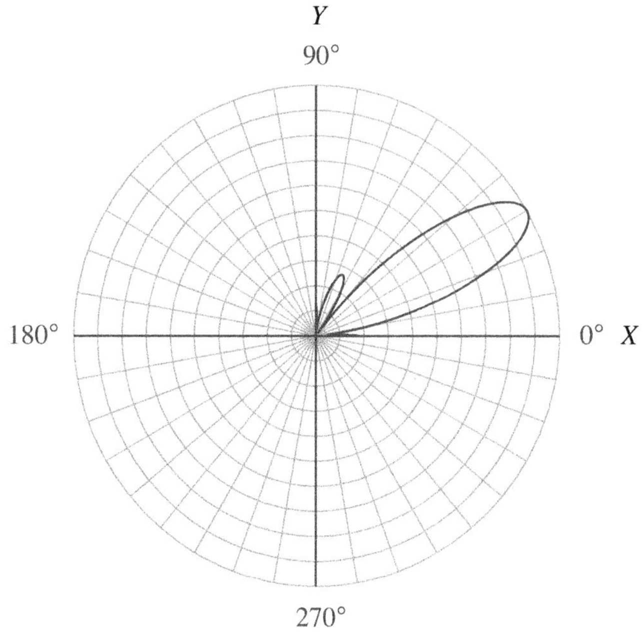

Figure 9.1-3 *continued.*

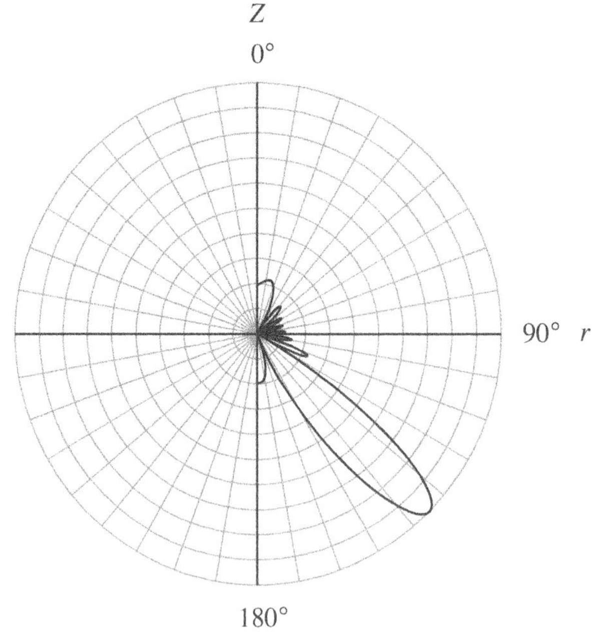

Figure 9.1-4 Polar plot, as a function of the vertical angle θ, of the magnitude of the normalized, vertical, far-field beam pattern of stave 2 in the plane with bearing angle $\psi = \psi'$ for the cylindrical array discussed in Example 9.1-1 for beam-steer angles (a) $\theta' = 135°$ and $\psi' = \phi_4 = 36°$, and (b) $\theta' = 135°$ and $\psi' = 30°$, where r is horizontal range.

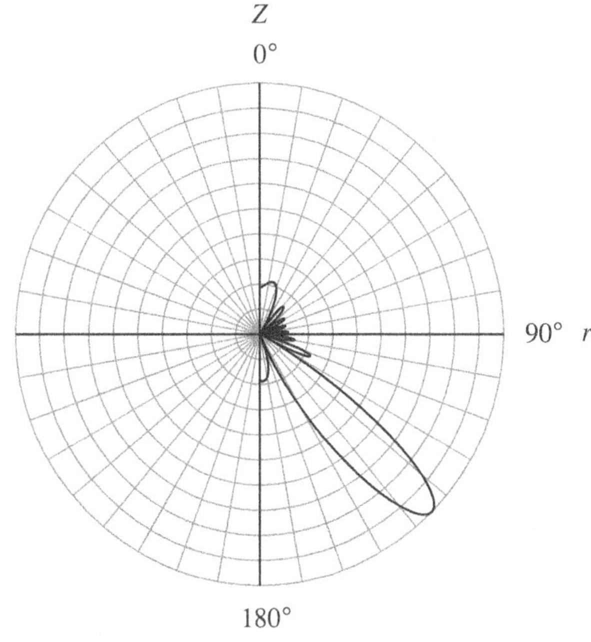

Figure 9.1-4 *continued.*

Example 9.1-2 Linear Array of Triplets

One solution to the port/starboard (left/right) ambiguity problem associated with linear arrays is a twin-line planar array (see Example 8.2-1). Another solution is a linear array of *triplets*. A single triplet is discussed in Example 8.1-7. In this example we shall consider a linear array composed of an odd number N of identical, equally-spaced, complex-weighted, triplets perpendicular to the array axis as shown in Fig. 9.1-5. Each triplet has a radius of a meters. The array shown in Fig. 9.1-5 is analogous to a cylindrical array of identical, equally-spaced, complex-weighted, omnidirectional point-elements that is lying on its side.

Since the elements (triplets) are identical, the far-field beam pattern of the array shown in Fig. 9.1-5 can be obtained by using the Product Theorem. Doing so yields

$$D(f, f_X, f_Y, f_Z) = E(f, f_X, f_Z) S_Y(f, f_Y), \tag{9.1-50}$$

or

$$D(f, u, v, w) = E(f, u, w) S_Y(f, v), \tag{9.1-51}$$

where $E(f, u, w)$ is the far-field beam pattern of a single triplet in the XZ plane,

$$S_Y(f, v) = \sum_{n=-N'}^{N'} w_n(f) \exp\left(+j2\pi v n d_Y / \lambda\right) \tag{9.1-52}$$

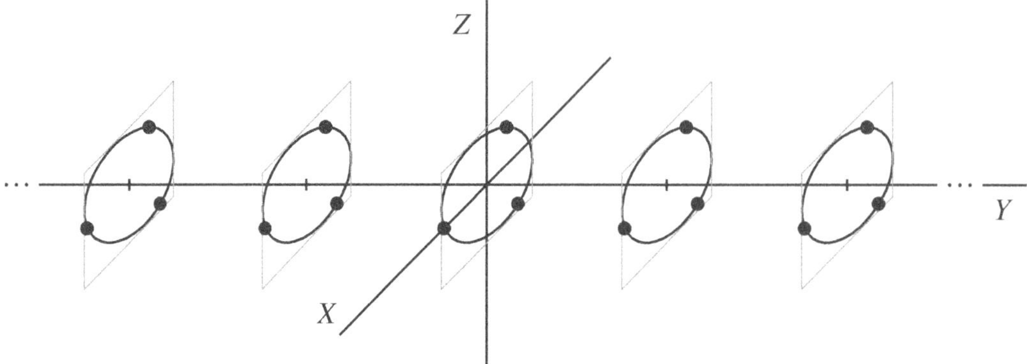

Figure 9.1-5 Linear array of identical, equally-spaced, triplets perpendicular to the array axis.

is the dimensionless array factor in the Y direction,

$$w_n(f) = b_n(f)\exp[+j\phi_n(f)] \tag{9.1-53}$$

is a frequency-dependent, complex weight associated with the nth triplet, $b_n(f)$ is a real, frequency-dependent, dimensionless, amplitude weight, $\phi_n(f)$ is a real, frequency-dependent, phase weight in radians, d_Y is the interelement spacing in meters between the triplets along the Y axis, and

$$N' = (N-1)/2\,. \tag{9.1-54}$$

Since the triplets are lying in the XZ plane, the far-field beam pattern $E(f,u,w)$ can be obtained by replacing direction cosine v in (8.1-100) with direction cosine w yielding

$$E(f,u,w) = \mathcal{S}(f)\left[c_1(f)\exp(+jkau) + c_2(f)\exp\left[-j\frac{ka}{2}\left(u-\sqrt{3}\,w\right)\right] + c_3(f)\exp\left[-j\frac{ka}{2}\left(u+\sqrt{3}\,w\right)\right]\right], \tag{9.1-55}$$

where $\mathcal{S}(f)$ is the complex, element sensitivity function of the omnidirectional point-elements in a triplet.

From Example 8.1-7, in order to create a *cardioid* (heart-shaped) beam pattern from a triplet,

$$ka = \pi/3, \tag{9.1-56}$$

$$c_1(f) = 2\,\mathrm{a}(f)\exp(-j\pi/2), \tag{9.1-57}$$

and

$$c_3(f) = c_2(f) = a(f), \tag{9.1-58}$$

where $a(f)$ is a real, frequency-dependent, dimensionless, amplitude weight. Substituting (9.1-56) through (9.1-58) into (9.1-55) yields the *unnormalized*, *cardioid*, far-field beam pattern of a triplet lying in the XZ plane:

$$E(f,u,w) = 2a(f)S(f)\exp\left(+j\frac{\pi}{3}u\right)\left[\exp\left(-j\frac{\pi}{2}u\right)\cos\left(\frac{\sqrt{3}}{6}\pi w\right) - j\right] \tag{9.1-59}$$

In order to do beam steering, let the phase weight

$$\phi_n(f) = -2\pi v' n d_Y/\lambda, \tag{9.1-60}$$

where

$$v' = \sin\theta' \sin\psi'. \tag{9.1-61}$$

Substituting (9.1-53) and (9.1-60) into (9.1-52) yields

$$S_Y(f,v) = \sum_{n=-N'}^{N'} b_n(f)\exp[+j2\pi(v - v')nd_Y/\lambda]. \tag{9.1-62}$$

Substituting (9.1-59) and (9.1-62) into (9.1-51) yields

$$D(f,u,v,w) = 2a(f)S(f)\exp\left(+j\frac{\pi}{3}u\right)\left[\exp\left(-j\frac{\pi}{2}u\right)\cos\left(\frac{\sqrt{3}}{6}\pi w\right) - j\right]\times \sum_{n=-N'}^{N'} b_n(f)\exp[+j2\pi(v - v')nd_Y/\lambda], \tag{9.1-63}$$

or

$$\begin{aligned} \dot{D}(f,\theta,\psi) = {} & 2a(f)S(f)\exp\left(+j\frac{\pi}{3}\sin\theta\cos\psi\right)\times \\ & \left[\exp\left(-j\frac{\pi}{2}\sin\theta\cos\psi\right)\cos\left(\frac{\sqrt{3}}{6}\pi\cos\theta\right) - j\right]\times \\ & \sum_{n=-N'}^{N'} b_n(f)\exp[+j2\pi(\sin\theta\sin\psi - \sin\theta'\sin\psi')nd_Y/\lambda] \end{aligned} \tag{9.1-64}$$

Equation (9.1-64) is the *unnormalized*, far-field beam pattern of a linear array of triplets with magnitude

$$|D(f,\theta,\psi)| = 2|\mathrm{a}(f)||\mathcal{S}(f)|\left|\exp\left(-j\frac{\pi}{2}\sin\theta\cos\psi\right)\cos\left(\frac{\sqrt{3}}{6}\pi\cos\theta\right) - j\right| \times$$
$$\left|\sum_{n=-N'}^{N'} b_n(f)\exp\left[+j2\pi(\sin\theta\sin\psi - \sin\theta'\sin\psi')nd_Y/\lambda\right]\right|.$$

(9.1-65)

Setting $\theta = 90°$ and $\theta' = 90°$ in (9.1-65) yields

$$|D(f,90°,\psi)| = 2|\mathrm{a}(f)||\mathcal{S}(f)|\left|\exp\left(-j\frac{\pi}{2}\cos\psi\right) - j\right| \times$$
$$\left|\sum_{n=-N'}^{N'} b_n(f)\exp\left[+j2\pi(\sin\psi - \sin\psi')nd_Y/\lambda\right]\right|, \tag{9.1-66}$$

which is the magnitude of the unnormalized, *horizontal*, far-field beam pattern in the XY plane. Figure 9.1-6 is a plot of (9.1-66) for $|\mathrm{a}(f)| = 1$ and $|\mathcal{S}(f)| = 1$. Figure 9.1-6 (b) shows that the magnitude of the mainlobe at $\psi = 45°$ is less than the maximum value at $\psi = 0°$ (broadside) shown in Fig. 9.1-6 (a). If the array is towed in the positive Y direction, then $\psi = 0°$ is broadside on the starboard (right) side of the array, and $\psi = 180°$ is broadside on the port (left) side of the array.

If the complex weight

$$c_1(f) = 2\mathrm{a}(f)\exp(+j\pi/2), \tag{9.1-67}$$

then the *unnormalized, cardioid*, far-field beam pattern of a triplet lying in the XZ plane is given by (see Example 8.1-7)

$$\boxed{E(f,u,w) = 2\mathrm{a}(f)\mathcal{S}(f)\exp\left(+j\frac{\pi}{3}u\right)\left[\exp\left(-j\frac{\pi}{2}u\right)\cos\left(\frac{\sqrt{3}}{6}\pi w\right) + j\right]}$$

(9.1-68)

and the *unnormalized*, far-field beam pattern of a linear array of triplets is given by

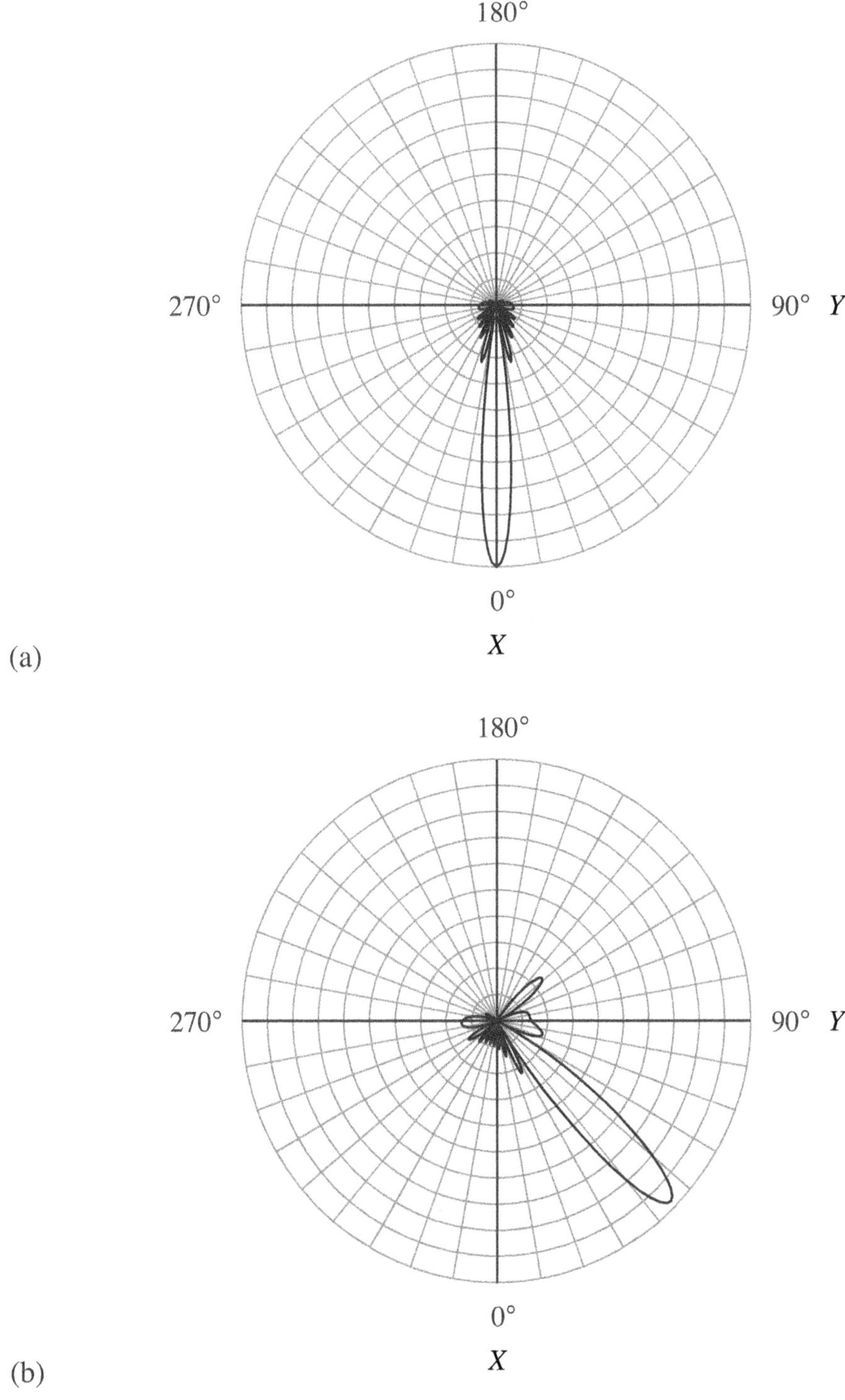

Figure 9.1-6 Polar plot, as a function of the azimuthal (bearing) angle ψ, of the magnitude of the unnormalized, horizontal, far-field beam pattern of a linear array of triplets given by (9.1-66) for $N = 11$, $d_Y/\lambda = 0.5$, rectangular amplitude weights, and beam-steer angles (a) $\psi' = 0°$ (broadside) and (b) $\psi' = 45°$.

$$\boxed{\begin{aligned} D(f,\theta,\psi) &= 2\,\mathrm{a}(f)\mathcal{S}(f)\exp\left(+j\frac{\pi}{3}\sin\theta\cos\psi\right)\times \\ &\left[\exp\left(-j\frac{\pi}{2}\sin\theta\cos\psi\right)\cos\left(\frac{\sqrt{3}}{6}\pi\cos\theta\right)+j\right]\times \\ &\sum_{n=-N'}^{N'} b_n(f)\exp\left[+j2\pi(\sin\theta\sin\psi-\sin\theta'\sin\psi')nd_Y/\lambda\right] \end{aligned}}$$

(9.1-69)

with magnitude

$$\begin{aligned} |D(f,\theta,\psi)| &= 2|\mathrm{a}(f)||\mathcal{S}(f)|\left|\exp\left(-j\frac{\pi}{2}\sin\theta\cos\psi\right)\cos\left(\frac{\sqrt{3}}{6}\pi\cos\theta\right)+j\right|\times \\ &\left|\sum_{n=-N'}^{N'} b_n(f)\exp\left[+j2\pi(\sin\theta\sin\psi-\sin\theta'\sin\psi')nd_Y/\lambda\right]\right|. \end{aligned}$$

(9.1-70)

Setting $\theta = 90°$ and $\theta' = 90°$ in (9.1-70) yields

$$\begin{aligned} |D(f,90°,\psi)| &= 2|\mathrm{a}(f)||\mathcal{S}(f)|\left|\exp\left(-j\frac{\pi}{2}\cos\psi\right)+j\right|\times \\ &\left|\sum_{n=-N'}^{N'} b_n(f)\exp\left[+j2\pi(\sin\psi-\sin\psi')nd_Y/\lambda\right]\right|, \end{aligned} \qquad (9.1\text{-}71)$$

which is the magnitude of the unnormalized, *horizontal*, far-field beam pattern in the XY plane. Figure 9.1-7 is a plot of (9.1-71) for $|\mathrm{a}(f)| = 1$ and $|\mathcal{S}(f)| = 1$. Although not very visible from the polar plot shown in Fig. 9.1-7 (b), the data shows that the magnitude of the mainlobe at $\psi = 210°$ is slightly less than the maximum value at $\psi = 180°$ (broadside) shown in Fig. 9.1-7 (a).

Like the mainlobes of the horizontal, far-field beam patterns of a twin-line planar array shown in Figs. 8.2-3 and 8.2-4, the mainlobes of the horizontal, far-field beam patterns shown in Figs. 9.1-6 and 9.1-7 exist in a half-space. Compare Figs. 8.2-3 and 8.2-4 with Figs. 9.1-6 and 9.1-7, respectively. Note that the sidelobe levels are lower in Figs. 9.1-6 and 9.1-7 compared to Figs. 8.2-3 and 8.2-4. Although the horizontal beam patterns in Figs. 8.2-3 and 8.2-4 are normalized, the unnormalized, horizontal beam patterns also have their maximum values at the beam-steer angles $\psi' = 45°$ and $\psi' = 210°$ unlike the unnormalized, horizontal beam patterns of a linear array of triplets. ■

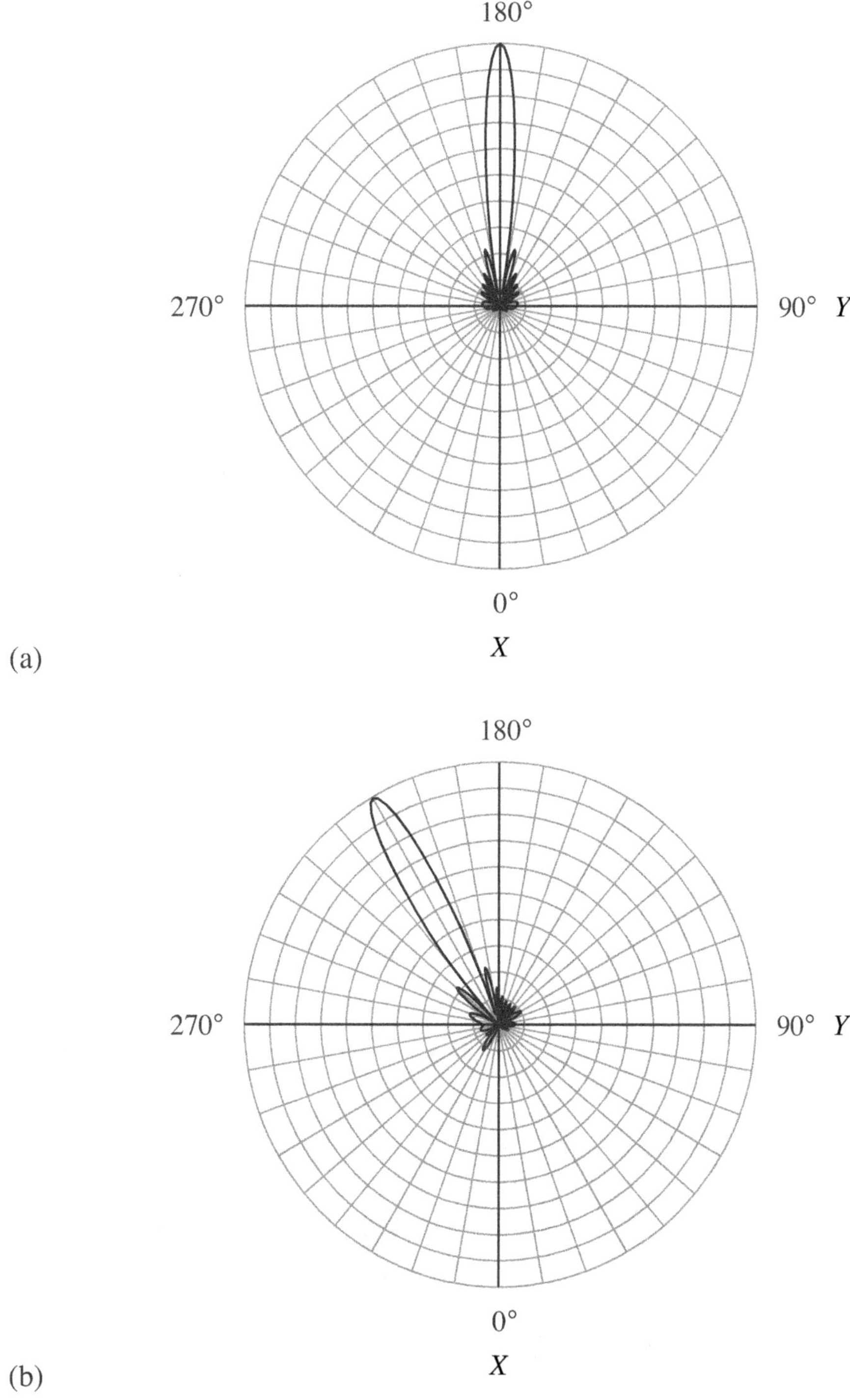

Figure 9.1-7 Polar plot, as a function of the azimuthal (bearing) angle ψ, of the magnitude of the unnormalized, horizontal, far-field beam pattern of a linear array of triplets given by (9.1-71) for $N=11$, $d_Y/\lambda=0.5$, rectangular amplitude weights, and beam-steer angles (a) $\psi'=180°$ (broadside) and (b) $\psi'=210°$.

9.2 The Far-Field Beam Pattern of a Spherical Array

Equation (9.1-1) can also be expressed in terms of the spherical coordinates (r_A, γ_A, ϕ_A), as shown in Fig. 9.2-1, by noting that

$$x_A = r_A \sin\gamma_A \cos\phi_A , \tag{9.2-1}$$

$$y_A = r_A \sin\gamma_A \sin\phi_A , \tag{9.2-2}$$

$$z_A = r_A \cos\gamma_A , \tag{9.2-3}$$

$$dx_A dy_A dz_A \rightarrow r_A^2 \sin\gamma_A dr_A d\gamma_A d\phi_A , \tag{9.2-4}$$

and

$$A(f, x_A, y_A, z_A) = A(f, r_A \sin\gamma_A \cos\phi_A, r_A \sin\gamma_A \sin\phi_A, r_A \cos\gamma_A) \rightarrow A(f, r_A, \gamma_A, \phi_A) . \tag{9.2-5}$$

Substituting (9.1-2) through (9.1-4), and (9.2-1) through (9.2-5) into (9.1-1), and using the trigonometric identity

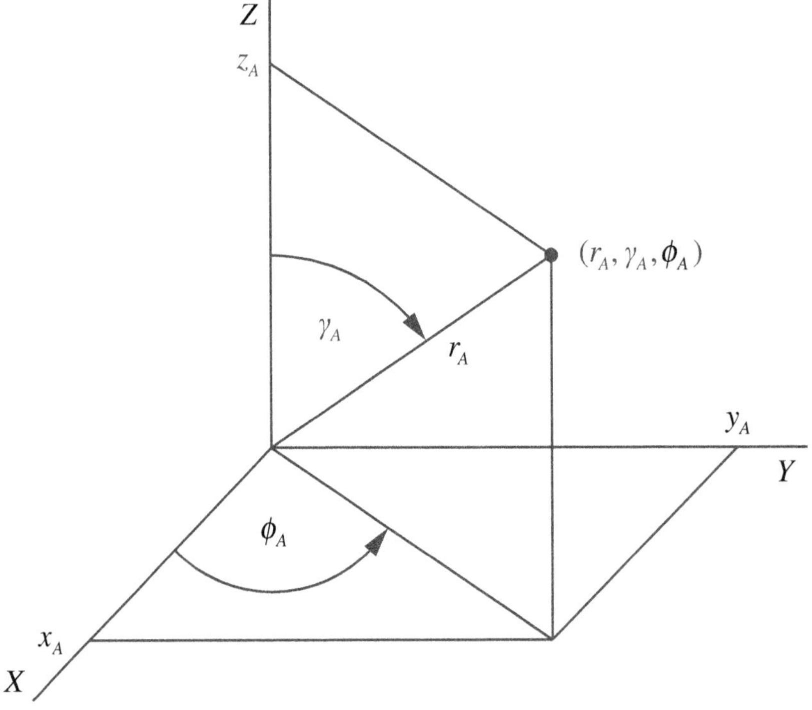

Figure 9.2-1 The spherical coordinates (r_A, γ_A, ϕ_A).

$$\cos(\alpha-\beta)=\cos\alpha\cos\beta+\sin\alpha\sin\beta, \tag{9.2-6}$$

yields

$$D(f,\theta,\psi)=\int_0^{2\pi}\int_0^{\pi}\int_0^{\infty} A(f,r_A,\gamma_A,\phi_A)\times \exp\left\{+j\frac{2\pi r_A}{\lambda}\left[\sin\gamma_A\sin\theta\cos(\psi-\phi_A)+\cos\gamma_A\cos\theta\right]\right\}\times r_A^2\sin\gamma_A\,dr_A\,d\gamma_A\,d\phi_A \tag{9.2-7}$$

Equation (9.2-7) shall be used to derive the far-field beam pattern of a spherical array of omnidirectional point-elements. Like the cylindrical array discussed in Section 9.1, a spherical array is another example of a conformal array.

Consider an array of identical, equally-spaced, complex-weighted, omnidirectional point-elements lying on the surface of a sphere with radius a meters (see Fig. 9.2-2). The maximum radial extent of the spherical array is $R_A=a$ meters. Analogous to a cylindrical array, let us consider a stave to be any single vertical line array of elements or any set of two or more adjacent vertical line arrays of elements running from north to south on the surface of the sphere (lines of longitude). Since an impulse function is used as a mathematical model for an omnidirectional point-source in acoustic wave propagation theory, the mathematical model that we shall use for the complex frequency response of an omnidirectional point-element lying on the surface of a sphere is given by

$$e(f,r_A-r_0,\gamma_A-\gamma_0,\phi_A-\phi_0)=\mathcal{S}(f)\frac{\delta(r_A-r_0)}{r_A^2}\frac{\delta(\gamma_A-\gamma_0)}{\sin\gamma_A}\delta(\phi_A-\phi_0), \tag{9.2-8}$$

where $\mathcal{S}(f)$ is the complex, element sensitivity function (see Table 9.1-2 and Appendix 6B), and (r_0,γ_0,ϕ_0) are the spherical coordinates of the center of the element. Note that

$$\begin{aligned}\int_0^{2\pi}\int_0^{\pi}\int_0^{\infty} e(f,r_A-r_0,\gamma_A-\gamma_0,\phi_A-\phi_0)r_A^2\sin\gamma_A\,dr_A\,d\gamma_A\,d\phi_A &= \mathcal{S}(f)\int_0^{\infty}\delta(r_A-r_0)\,dr_A\times\\ &\quad\int_0^{\pi}\delta(\gamma_A-\gamma_0)\,d\gamma_A\times\\ &\quad\int_0^{2\pi}\delta(\phi_A-\phi_0)\,d\phi_A\\ &= \mathcal{S}(f).\end{aligned} \tag{9.2-9}$$

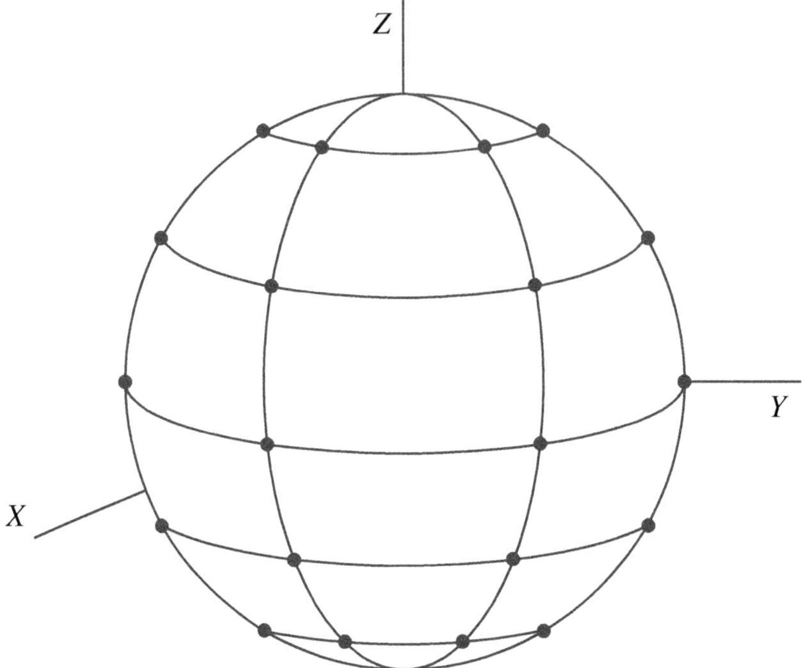

Figure 9.2-2 Spherical array of identical, equally-spaced, omnidirectional point-elements.

The impulse function $\delta(r_A - r_0)$ has units of inverse meters, and the impulse functions $\delta(\gamma_A - \gamma_0)$ and $\delta(\phi_A - \phi_0)$ have units of inverse radians.

Therefore, with the use of (9.2-8) and the principle of superposition, the complex frequency response (complex aperture function) of the spherical array shown in Fig. 9.2-2 can be expressed as the sum of the complex-weighted frequency responses of all the elements in the array as follows:

$$A(f, r_A, \gamma_A, \phi_A) = \mathcal{S}(f) \frac{\delta(r_A - a)}{r_A^2 \sin\gamma_A} \sum_{m=1}^{M} \sum_{n=1}^{N} c_{mn}(f) \delta(\gamma_A - \gamma_m) \delta(\phi_A - \phi_n), \tag{9.2-10}$$

where M is the total *odd* number of circular arrays, N is the total number (even or odd) of omnidirectional point-elements per circular array,

$$c_{mn}(f) = a_{mn}(f) \exp[+j\theta_{mn}(f)] \tag{9.2-11}$$

is the frequency-dependent, complex weight associated with element (m,n), where $a_{mn}(f)$ is a real, frequency-dependent, dimensionless, amplitude weight, and $\theta_{mn}(f)$ is a real, frequency-dependent, phase weight in radians, and

(a, γ_m, ϕ_n) are the spherical coordinates of the center of element (m,n). Complex weights are implemented by using digital signal processing (digital beamforming) (see Section 6.4). The complex weights are used to control the complex frequency response of the array and, thus, the array's far-field beam pattern via amplitude and phase weighting. Since the elements are equally spaced in the γ and ϕ directions,

$$\gamma_m = m\Delta\gamma , \tag{9.2-12}$$

where

$$\Delta\gamma = \frac{180°}{M+1}, \tag{9.2-13}$$

and

$$\phi_n = (n-1)\Delta\phi , \tag{9.2-14}$$

where

$$\Delta\phi = 360°/N . \tag{9.2-15}$$

Substituting (9.2-10) into (9.2-7) yields the following expression for the far-field beam pattern of a spherical array of identical, equally-spaced, complex-weighted, omnidirectional point-elements:

$$D(f,\theta,\psi) = \mathcal{S}(f)\sum_{m=1}^{M}\sum_{n=1}^{N} c_{mn}(f)\exp\left\{+j\frac{2\pi a}{\lambda}\left[\sin\gamma_m \sin\theta\cos(\psi-\phi_n)\right]\right\}\times \exp\left\{+j\frac{2\pi a}{\lambda}\cos\gamma_m \cos\theta\right\} \tag{9.2-16}$$

where $\mathcal{S}(f)$ is the complex, element sensitivity function, and the double summation is the dimensionless array factor. If one or more elements in the array are broken (not transmitting and/or receiving), set $c_{mn}(f)=0$ for those elements. Similarly, if only one stave is in operation at any given instant of time, set $c_{mn}(f)=0$ for those elements not in operation.

If the number of circular arrays $M=1$, then $\Delta\gamma = 90°$ [see (9.2-13)] and $m=1$. And since $\gamma_1 = 90°$ [see (9.2-12)], (9.2-16) reduces to

$$D(f,\theta,\psi) = \mathcal{S}(f)\sum_{n=1}^{N} c_n(f)\exp\left[+j\frac{2\pi a}{\lambda}\sin\theta\cos(\psi-\phi_n)\right]. \tag{9.2-17}$$

Equation (9.2-17) is the far-field beam pattern of a single circular array with radius a meters lying in the XY plane [see (8.1-92)].

Since

$$\cos(\psi - \phi_n) = \cos\psi\cos\phi_n + \sin\psi\sin\phi_n\,, \tag{9.2-18}$$

(9.2-16) can be rewritten as follows:

$$D(f,\theta,\psi) = \mathcal{S}(f)\sum_{m=1}^{M}\sum_{n=1}^{N} c_{mn}(f)\exp\left\{+j\frac{2\pi}{\lambda}\left[x_{mn}\sin\theta\cos\psi + y_{mn}\sin\theta\sin\psi\right]\right\}\times \exp\left\{+j\frac{2\pi}{\lambda}z_m\cos\theta\right\} \tag{9.2-19}$$

where

$$x_{mn} = a\sin\gamma_m\cos\phi_n\,, \tag{9.2-20}$$

$$y_{mn} = a\sin\gamma_m\sin\phi_n\,, \tag{9.2-21}$$

and

$$z_m = a\cos\gamma_m \tag{9.2-22}$$

are the rectangular coordinates of element (m,n), γ_m is given by (9.2-12), and ϕ_n is given by (9.2-14).

9.2.1 The Phased Array – Beam Steering

Beam steering can be accomplished by operating staves in a clockwise or counterclockwise fashion, analogous to the scanning procedure that was discussed for a cylindrical array in Subsection 9.1.1 (see Example 9.1-1). When a particular stave is in operation, the far-field beam pattern of the stave can be steered to $\theta = \theta'$ and $\psi = \psi'$ by using the following phase weights:

$$\theta_{mn}(f) = -\frac{2\pi}{\lambda}(x_{mn}\sin\theta'\cos\psi' + y_{mn}\sin\theta'\sin\psi' + z_m\cos\theta'), \qquad m = 1, 2, \ldots, M \quad n = 1, 2, \ldots, N \tag{9.2-23}$$

where x_{mn}, y_{mn}, and z_m are given by (9.2-20) through (9.2-22), respectively. Substituting (9.2-11) and (9.2-23) into (9.2-19) yields

$$D(f,\theta,\psi)=S(f)\sum_{m=1}^{M}\sum_{n=1}^{N}a_{mn}(f)\exp\left[+j\frac{2\pi}{\lambda}(\sin\theta\cos\psi-\sin\theta'\cos\psi')x_{mn}\right]\times$$
$$\exp\left[+j\frac{2\pi}{\lambda}(\sin\theta\sin\psi-\sin\theta'\sin\psi')y_{mn}\right]\times$$
$$\exp\left[+j\frac{2\pi}{\lambda}(\cos\theta-\cos\theta')z_{m}\right] \tag{9.2-24}$$

Set $a_{mn}(f)=0$ in (9.2-24) for those elements not in operation. As was previously mentioned in Subsection 9.1.1 for a cylindrical array, by operating two or more staves at the same time and using parallel signal processing, multiple beams with multiple "look directions" can be formed simultaneously.

The equivalent time delay τ'_{mn} in seconds at element (m,n) is given by [see (9.1-40)]

$$\tau'_{mn}=-\theta_{mn}(f)/(2\pi f). \tag{9.2-25}$$

Substituting (9.2-23) into (9.2-25) yields

$$\tau'_{mn}=\frac{u'}{c}x_{mn}+\frac{v'}{c}y_{mn}+\frac{w'}{c}z_{m}, \qquad m=1,2,\ldots,M \quad n=1,2,\ldots,N \tag{9.2-26}$$

where $c=f\lambda$, and u', v', and w' are given by (9.1-43) through (9.1-45), respectively.

Problems

Section 9.1

9-1 Consider a linear array composed of an odd number M of identical, unevenly-spaced, complex-weighted, omnidirectional point-elements in three-dimensional space (see Fig. P9-1).

(a) Find the unnormalized, far-field beam pattern of this array as a function of frequency f and spherical angles θ and ψ. **Hint:** Generalize the results in Example 8.1-8.

(b) Find the equation for the phase weights that must be used in order to do beam steering. **Hint:** Generalize the results in Example 8.2-1.

(c) Find the equation for the equivalent time delays. **Hint:** Generalize the results in Example 8.2-1.

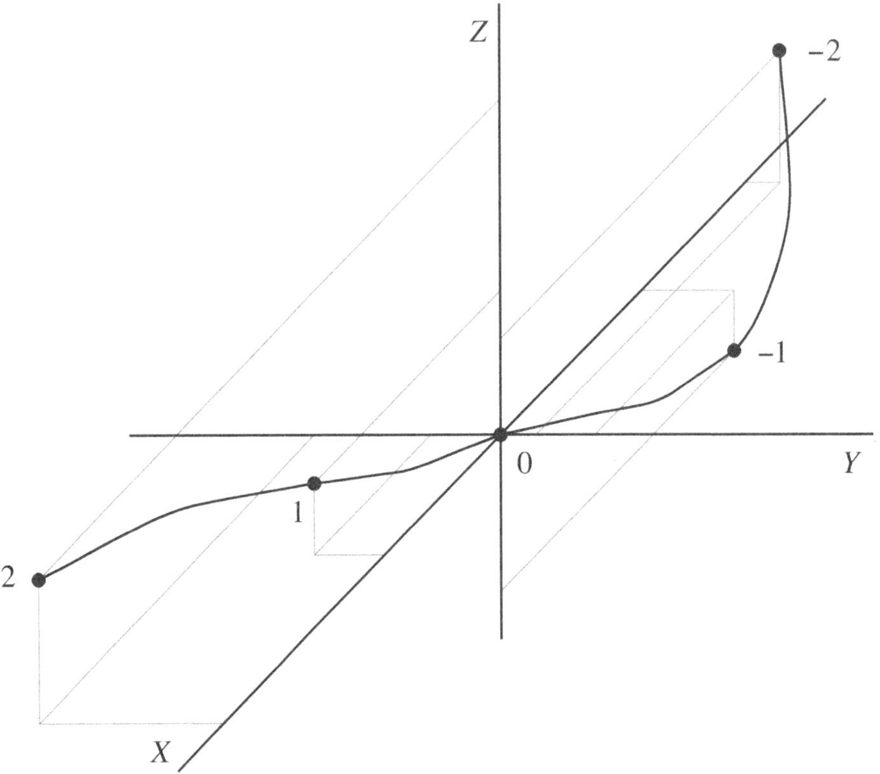

Figure P9-1

9-2 The arc length s in meters between two adjacent omnidirectional point-elements on a circular array on the surface of the cylinder shown in Fig. 9.1-2 is $s = a\Delta\phi$, where a is the radius of the cylinder in meters and $\Delta\phi$ is the angular spacing in *radians*.

(a) How many omnidirectional point-elements per circular array M must be used so that $s \le \lambda/2$?

(b) How many circular arrays N must be used so that the spacing $d_Z \le \lambda/2$?

9-3 For the cylindrical array described in Example 9.1-1,

(a) find the magnitude of the normalized, horizontal, far-field beam pattern of stave 2 for rectangular amplitude weights and beam-steer angles $\theta' = 90°$ and ψ'.

(b) find the magnitude of the normalized, vertical, far-field beam pattern of stave 2 in the plane with bearing angle $\psi = \psi'$ for rectangular amplitude weights and beam-steer angles θ' and ψ'.

9-4 Using (9.1-65), find the magnitude of the unnormalized, vertical, far-field beam pattern in the XZ plane of the linear array of triplets discussed in Example 9.1-2 when no beam steering is done.

Chapter 10

Bistatic Scattering

10.1 Target Strength

In this section we shall analyze the *bistatic scattering* problem shown in Fig. 10.1-1 for the purpose of deriving an equation that defines the *target strength* of an object. Objects of interest can be, for example, gas bubbles in water, fish, marine mammals (dolphins and whales), submarines, mines, etc. As will be shown later, the target strength of an object can be shown to be directly proportional to either the *differential scattering cross-section* or *scattering function* of an object, or to the ratio between the scattered and incident time-average intensities. Target strength is one of those important parameters that determines the signal-to-noise ratio (SNR) at a receiver and whether or not an object (target) can be detected.

A bistatic scattering problem is one in which the transmitter and receiver are *not* at the same location. Referring to Fig. 10.1-1, the transmitter (sound-source) is located at $\mathbf{r}_T = (x_T, y_T, z_T)$, the object (target or scatterer) is located at $\mathbf{r}_S = (x_S, y_S, z_S)$, where the subscript S denotes scatterer, and the receiver is located at $\mathbf{r}_R = (x_R, y_R, z_R)$. For our purposes, the transmitter is a time-harmonic, omnidirectional point-source, and the receiver is an omnidirectional point-element. The corresponding equation for the source distribution $x_M(t, \mathbf{r})$, with units of inverse seconds, is given by (see Appendix 6C)

$$x_M(t, \mathbf{r}) = S_0 \delta(\mathbf{r} - \mathbf{r}_T) \exp(+j2\pi f t), \tag{10.1-1}$$

where S_0 is the source strength in cubic meters per second at frequency f hertz, the impulse function $\delta(\mathbf{r} - \mathbf{r}_T)$, with units of inverse cubic meters, represents a unit-amplitude, omnidirectional point-source at $\mathbf{r}_T$, and f is the source frequency. An exact solution of the linear wave equation

$$\nabla^2 \varphi(t, \mathbf{r}) - \frac{1}{c^2} \frac{\partial^2}{\partial t^2} \varphi(t, \mathbf{r}) = x_M(t, \mathbf{r}), \tag{10.1-2}$$

where $\varphi(t, \mathbf{r})$ is the scalar velocity potential in squared meters per second, c is the constant speed of sound in the fluid medium in meters per second, and $x_M(t, \mathbf{r})$ is the source distribution in inverse seconds given by (10.1-1), is discussed in Appendix 10A, along with other equations that we shall need that describe the radiated acoustic field.

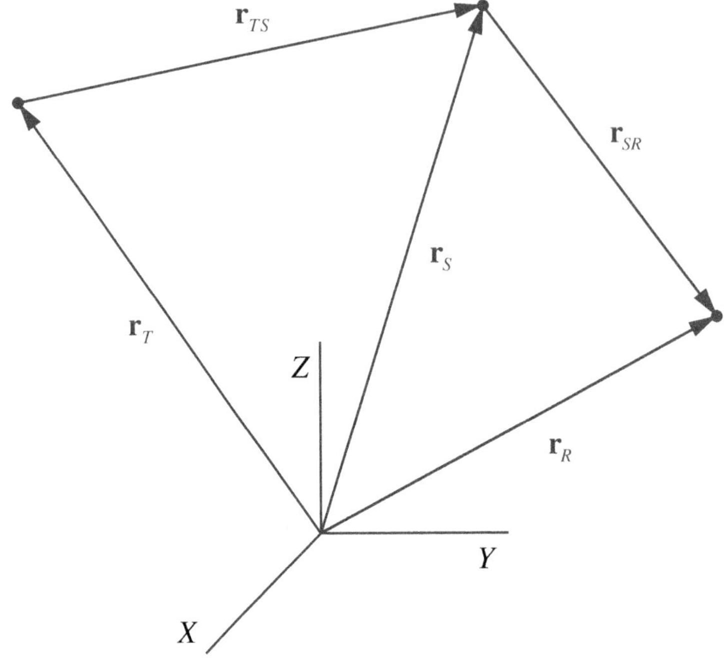

Figure 10.1-1 Bistatic scattering geometry. The transmitter (sound-source) is located at $\mathbf{r}_T = (x_T, y_T, z_T)$, the object (target or scatterer) is located at $\mathbf{r}_S = (x_S, y_S, z_S)$, and the receiver is located at $\mathbf{r}_R = (x_R, y_R, z_R)$.

We begin the analysis of the bistatic scattering problem by computing the acoustic pressure incident upon the object due to a time-harmonic, omnidirectional point-source. The time-harmonic acoustic pressure (in pascals) incident upon the object at $\mathbf{r}_S$ is given by (see Appendix 10A)

$$p_i(t, \mathbf{r}_S) = p_{f,i}(\mathbf{r}_S)\exp(+j2\pi ft), \tag{10.1-3}$$

where

$$p_{f,i}(\mathbf{r}_S) = -jk\rho_0 c\varphi_{f,i}(\mathbf{r}_S), \tag{10.1-4}$$

$$\varphi_{f,i}(\mathbf{r}_S) = S_0\, g_f\left(\mathbf{r}_S \middle| \mathbf{r}_T\right), \tag{10.1-5}$$

$$S_0 = A_x \mathcal{S}_T(f), \tag{10.1-6}$$

$$\begin{aligned} g_f\left(\mathbf{r}_S \middle| \mathbf{r}_T\right) &= -\frac{\exp\left[-\alpha(f)\left|\mathbf{r}_S - \mathbf{r}_T\right|\right]}{4\pi\left|\mathbf{r}_S - \mathbf{r}_T\right|}\exp\left(-jk\left|\mathbf{r}_S - \mathbf{r}_T\right|\right) \\ &= -\frac{\exp[-\alpha(f)r_{TS}]}{4\pi r_{TS}}\exp(-jkr_{TS}), \end{aligned} \tag{10.1-7}$$

$$\mathbf{r}_{TS} = \mathbf{r}_S - \mathbf{r}_T = (x_S - x_T)\hat{x} + (y_S - y_T)\hat{y} + (z_S - z_T)\hat{z}, \tag{10.1-8}$$

$$r_{TS} = \left|\mathbf{r}_{TS}\right| = \sqrt{(x_S - x_T)^2 + (y_S - y_T)^2 + (z_S - z_T)^2}, \tag{10.1-9}$$

and

$$\mathbf{r}_{TS} = r_{TS}\hat{r}_{TS}, \tag{10.1-10}$$

where $k = 2\pi f/c = 2\pi/\lambda$ is the real wavenumber in radians per meter, $c = f\lambda$ is the constant speed of sound in the fluid medium in meters per second, λ is the wavelength in meters, ρ_0 is the constant ambient (equilibrium) density of the fluid medium in kilograms per cubic meter, S_0 is the source strength in cubic meters per second at frequency f hertz, A_x is the complex amplitude in volts of the time-harmonic, input electrical signal applied to the omnidirectional point-source, $\mathcal{S}_T(f)$ is the complex, transmitter sensitivity function of the omnidirectional point-source in $\left(\mathrm{m}^3/\mathrm{sec}\right)/\mathrm{V}$ (see Table 6.1-2 and Appendix 6B), $\alpha(f)$ is the real, nonnegative, frequency-dependent, attenuation coefficient of seawater in nepers per meter, and $\hat{r}_{TS}$ is the dimensionless unit vector in the direction of the vector $\mathbf{r}_{TS}$.

The time-average intensity vector (in watts per squared meter) incident upon the object at $\mathbf{r}_S$ is given by (see Appendix 10A)

$$\mathbf{I}_{\mathrm{avg},i}(\mathbf{r}_S) = I_{\mathrm{avg},i}(\mathbf{r}_S)\hat{r}_{TS}, \tag{10.1-11}$$

where

$$I_{\mathrm{avg},i}(\mathbf{r}_S) = \frac{1}{2}k^2\rho_0 c\left(\frac{|S_0|}{4\pi r_{TS}}\right)^2 \exp[-2\alpha(f)r_{TS}]. \tag{10.1-12}$$

If the origin of a Cartesian coordinate system (with the same orientation as the one shown in Fig. 10.1-1) is placed at $\mathbf{r}_T$, then the spherical angles θ_{TS} and ψ_{TS} that describe the direction of the unit vector $\hat{r}_{TS}$ are measured as shown in Fig. 1.2-2 and are given by (see Appendix 10A)

$$\boxed{\theta_{TS} = \cos^{-1}\left(\frac{z_S - z_T}{r_{TS}}\right)} \tag{10.1-13}$$

and

$$\boxed{\psi_{TS} = \tan^{-1}\left(\frac{y_S - y_T}{x_S - x_T}\right)} \tag{10.1-14}$$

Equations (10.1-13) and (10.1-14) are the *angles of incidence* at the object.

If we enclose the omnidirectional point-source at $\mathbf{r}_T$ with a sphere of radius r_{TS} meters, then the time-average radiated power (in watts) is given by (see Appendix 10A)

$$P_{\text{avg},i} = \frac{1}{2} k^2 \rho_0 c \frac{|S_0|^2}{4\pi} \exp[-2\alpha(f) r_{TS}] = 4\pi r_{TS}^2 I_{\text{avg},i}(\mathbf{r}_S), \qquad (10.1\text{-}15)$$

where the source strength S_0 is given by (10.1-6), or

$$P_{\text{avg},i} = 4\pi R^2 \frac{P_0^2}{2\rho_0 c} \exp\left[-2\alpha(f)(r_{TS} - R)\right]\Big|_{R=1\text{ m}}, \qquad (10.1\text{-}16)$$

where P_0 is the magnitude of the time-harmonic, radiated acoustic pressure measured at a range $R = 1\text{ m}$ from the omnidirectional point-source at $\mathbf{r}_T$. The source level (SL) and P_0 are related as follows (see Section 4.2):

$$\text{SL} = 20\log_{10}\left(\frac{\sqrt{2}P_0/2}{P_{\text{ref}}}\right) \text{dB re } P_{\text{ref}}, \qquad (10.1\text{-}17)$$

or

$$P_0 = \sqrt{2}\, P_{\text{ref}}\, 10^{(\text{SL}/20)}, \qquad (10.1\text{-}18)$$

where $P_{\text{ref}} = 1\,\mu\text{Pa}$ (rms) is the root-mean-square (rms) reference pressure used in underwater acoustics.

In order to compute the acoustic field scattered by the object centered at $\mathbf{r}_S$, we shall treat the object as a discrete point-scatterer that acts like a time-harmonic point-source with source strength S_0' cubic meters per second at frequency f hertz, where

$$S_0' = \varphi_{f,i}(\mathbf{r}_S) S_S(f, \theta_{TS}, \psi_{TS}, \theta_{SR}, \psi_{SR}), \qquad (10.1\text{-}19)$$

$\varphi_{f,i}(\mathbf{r}_S)$ is the spatial-dependent part of the time-harmonic, scalar velocity potential (in squared meters per second) incident upon the object at $\mathbf{r}_S$ [see (10.1-5)], and $S_S(f, \theta_{TS}, \psi_{TS}, \theta_{SR}, \psi_{SR})$ is the *scattering function* (a.k.a. the scattering amplitude[1] or complex acoustical scattering length[2]) of the object (scatterer) with

---

[1] A. Ishimaru, *Wave Propagation and Scattering in Random Media*, Vol. 1, Academic Press, New York, 1978, pp. 9-12, 39-40.

[2] H. Medwin and C. S. Clay, *Fundamentals of Acoustical Oceanography*, Academic Press, Boston, 1998, pp. 235-241.

units of meters, where θ_{TS} and ψ_{TS} are the angles of incidence at the object given by (10.1-13) and (10.1-14), respectively, and θ_{SR} and ψ_{SR} are the angles of scatter at the receiver to be given later. The scattering function is a complex function (magnitude and phase) and is, in general, a function of frequency and the directions of wave propagation from the source to the object, and from the object to the receiver. For a given object and fixed scattering geometry, the magnitude of the scattering function can be relatively large for certain resonant or characteristic frequencies of the object, and relatively small for other frequencies. The resonant or characteristic frequencies of an object can be used to identify or classify an object. The magnitude of the scattering function versus frequency is referred to as the *acoustic color* of an object. Substituting (10.1-5) into (10.1-19) yields

$$S_0' = S_0 g_f\left(\mathbf{r}_S \middle| \mathbf{r}_T\right) S_S(f, \theta_{TS}, \psi_{TS}, \theta_{SR}, \psi_{SR}), \qquad (10.1\text{-}20)$$

where S_0 is given by (10.1-6) and $g_f\left(\mathbf{r}_S \middle| \mathbf{r}_T\right)$ is given by (10.1-7). For a fixed frequency and geometry, S_0' is a constant.

The time-harmonic acoustic pressure (in pascals) of the scattered acoustic field incident upon the receiver at $\mathbf{r}_R$ is given by (see Appendix 10A)

$$p_s(t, \mathbf{r}_R) = p_{f,s}(\mathbf{r}_R) \exp(+j2\pi f t), \qquad (10.1\text{-}21)$$

where

$$p_{f,s}(\mathbf{r}_R) = -jk\rho_0 c\, \varphi_{f,s}(\mathbf{r}_R), \qquad (10.1\text{-}22)$$

$$\varphi_{f,s}(\mathbf{r}_R) = S_0' g_f\left(\mathbf{r}_R \middle| \mathbf{r}_S\right), \qquad (10.1\text{-}23)$$

$$\begin{aligned} g_f\left(\mathbf{r}_R \middle| \mathbf{r}_S\right) &= -\frac{\exp\left[-\alpha(f)\left|\mathbf{r}_R - \mathbf{r}_S\right|\right]}{4\pi\left|\mathbf{r}_R - \mathbf{r}_S\right|} \exp\left(-jk\left|\mathbf{r}_R - \mathbf{r}_S\right|\right) \\ &= -\frac{\exp[-\alpha(f) r_{SR}]}{4\pi r_{SR}} \exp(-jk r_{SR}), \end{aligned} \qquad (10.1\text{-}24)$$

$$\mathbf{r}_{SR} = \mathbf{r}_R - \mathbf{r}_S = (x_R - x_S)\hat{x} + (y_R - y_S)\hat{y} + (z_R - z_S)\hat{z}, \qquad (10.1\text{-}25)$$

$$r_{SR} = \left|\mathbf{r}_{SR}\right| = \sqrt{(x_R - x_S)^2 + (y_R - y_S)^2 + (z_R - z_S)^2}, \qquad (10.1\text{-}26)$$

and

$$\mathbf{r}_{SR} = r_{SR}\hat{r}_{SR}, \qquad (10.1\text{-}27)$$

where $\hat{r}_{SR}$ is the dimensionless unit vector in the direction of the vector $\mathbf{r}_{SR}$.

The time-average intensity vector (in watts per squared meter) of the scattered acoustic field incident upon the receiver at $\mathbf{r}_R$ is given by (see Appendix 10A)

$$\mathbf{I}_{\text{avg},s}(\mathbf{r}_R) = I_{\text{avg},s}(\mathbf{r}_R)\hat{r}_{SR}, \tag{10.1-28}$$

where

$$I_{\text{avg},s}(\mathbf{r}_R) = \frac{1}{2}k^2\rho_0 c\left(\frac{|S_0'|}{4\pi r_{SR}}\right)^2 \exp[-2\alpha(f)r_{SR}]. \tag{10.1-29}$$

Substituting (10.1-20) into (10.1-29) yields

$$I_{\text{avg},s}(\mathbf{r}_R) = I_{\text{avg},i}(\mathbf{r}_S)\left[\frac{|S_S(f,\theta_{TS},\psi_{TS},\theta_{SR},\psi_{SR})|}{4\pi r_{SR}}\right]^2 \exp[-2\alpha(f)r_{SR}], \tag{10.1-30}$$

where $I_{\text{avg},i}(\mathbf{r}_S)$ is given by (10.1-12). If the origin of a Cartesian coordinate system (with the same orientation as the one shown in Fig. 10.1-1) is placed at $\mathbf{r}_S$, then the spherical angles θ_{SR} and ψ_{SR} that describe the direction of the unit vector $\hat{r}_{SR}$ are measured as shown in Fig. 1.2-2 and are given by (see Appendix 10A)

$$\boxed{\theta_{SR} = \cos^{-1}\left(\frac{z_R - z_S}{r_{SR}}\right)} \tag{10.1-31}$$

and

$$\boxed{\psi_{SR} = \tan^{-1}\left(\frac{y_R - y_S}{x_R - x_S}\right)} \tag{10.1-32}$$

Equations (10.1-31) and (10.1-32) are the *angles of scatter* at the receiver.

If we enclose the object centered at $\mathbf{r}_S$ with a sphere of radius r_{SR} meters, then the *differential* time-average scattered power (in watts) incident upon the receiver at $\mathbf{r}_R$ is given by (see Appendix 10A)

$$dP_{\text{avg},s} = \mathbf{I}_{\text{avg},s}(\mathbf{r}_R)\cdot\mathbf{dS}, \tag{10.1-33}$$

where

$$\mathbf{dS} = dS\,\hat{r}_{SR} = r_{SR}^2 \sin\theta_{SR}\, d\theta_{SR}\, d\psi_{SR}\,\hat{r}_{SR}. \tag{10.1-34}$$

Substituting (10.1-28), (10.1-30), and (10.1-34) into (10.1-33) yields

$$dP_{\text{avg},s} = I_{\text{avg},i}(\mathbf{r}_S)\sigma_{\text{d}}(f,\theta_{TS},\psi_{TS},\theta_{SR},\psi_{SR})\exp[-2\alpha(f)r_{SR}]\sin\theta_{SR}\,d\theta_{SR}\,d\psi_{SR}, \tag{10.1-35}$$

or

$$dP_{\text{avg},s} = I_{\text{avg},i}(\mathbf{r}_S)\sigma_{\text{d}}(f,\theta_{TS},\psi_{TS},\theta_{SR},\psi_{SR})\exp[-2\alpha(f)r_{SR}]d\Omega, \tag{10.1-36}$$

where

$$\boxed{\begin{aligned}\sigma_{\text{d}}(f,\theta_{TS},\psi_{TS},\theta_{SR},\psi_{SR}) &= \left[\frac{|S_S(f,\theta_{TS},\psi_{TS},\theta_{SR},\psi_{SR})|}{4\pi}\right]^2 \\ &= r_{SR}^2\frac{I_{\text{avg},s}(\mathbf{r}_R)}{I_{\text{avg},i}(\mathbf{r}_S)}\exp[+2\alpha(f)r_{SR}]\end{aligned}} \tag{10.1-37}$$

is the *differential scattering cross-section* (in squared meters) of the object and

$$d\Omega = dS/r_{SR}^2 = \sin\theta_{SR}\,d\theta_{SR}\,d\psi_{SR} \tag{10.1-38}$$

is the differential of solid angle. The right-hand side of (10.1-37) was obtained by using (10.1-30). The factor r_{SR}^2 on the right-hand side of (10.1-37) is the surface area on a sphere of radius r_{SR} meters subtended by a solid angle of one steradian. Note that the differential scattering cross-section is directly proportional to the magnitude-squared of the scattering function.[1,2] The *total* time-average scattered power (in watts) is obtained by integrating either (10.1-35) or (10.1-36). Doing so yields

$$\boxed{P_{\text{avg},s} = I_{\text{avg},i}(\mathbf{r}_S)\sigma_{\text{s}}(f,\theta_{TS},\psi_{TS})\exp[-2\alpha(f)r_{SR}]} \tag{10.1-39}$$

where

$$\boxed{\begin{aligned}\sigma_{\text{s}}(f,\theta_{TS},\psi_{TS}) &= \int_0^{2\pi}\int_0^{\pi}\sigma_{\text{d}}(f,\theta_{TS},\psi_{TS},\theta_{SR},\psi_{SR})\sin\theta_{SR}\,d\theta_{SR}\,d\psi_{SR} \\ &= \int_0^{4\pi}\sigma_{\text{d}}(f,\theta_{TS},\psi_{TS},\theta_{SR},\psi_{SR})\,d\Omega\end{aligned}} \tag{10.1-40}$$

is the *scattering cross-section*[1] (in squared meters) of the object, also known as the *total scattering cross-section*[2,3]. For a given object and fixed scattering

[3] C. S. Clay and H. Medwin, *Acoustical Oceanography*, Wiley, New York, 1977, pp. 180-184.

geometry, the scattering cross-section may be less than, equal to, or greater than the actual geometrical cross-sectional area of the object depending on the frequency.

If the scatter from the object is *omnidirectional*, that is, if the scattering function and, hence, the differential scattering cross-section are *not* functions of the angles of incidence and scatter, then $S_S(f,\theta_{TS},\psi_{TS},\theta_{SR},\psi_{SR})=S_S(f)$ and $\sigma_{\mathrm{d}}(f,\theta_{TS},\psi_{TS},\theta_{SR},\psi_{SR})=\sigma_{\mathrm{d}}(f)$. Therefore, $\sigma_{\mathrm{s}}(f,\theta_{TS},\psi_{TS})=\sigma_{\mathrm{s}}(f)$, where

$$\sigma_{\mathrm{s}}(f)=4\pi\sigma_{\mathrm{d}}(f) \tag{10.1-41}$$

and

$$\sigma_{\mathrm{d}}(f)=\left[\frac{|S_S(f)|}{4\pi}\right]^2. \tag{10.1-42}$$

Substituting (10.1-41) and (10.1-42) into (10.1-39) yields

$$P_{\mathrm{avg},s}=4\pi I_{\mathrm{avg},i}(\mathbf{r}_S)\left[\frac{|S_S(f)|}{4\pi}\right]^2\exp[-2\alpha(f)r_{SR}], \tag{10.1-43}$$

or

$$P_{\mathrm{avg},s}=4\pi r_{SR}^2 I_{\mathrm{avg},s}(\mathbf{r}_R), \tag{10.1-44}$$

where $I_{\mathrm{avg},s}(\mathbf{r}_R)$ is given by (10.1-30). Equation (10.1-44) is the total time-average scattered power (in watts) from an omnidirectional scatterer and is analogous to the time-average power radiated by an omnidirectional point-source.

Another popular scattering cross-section is the *bistatic scattering cross-section* (a.k.a. the *bistatic radar cross-section*) given by[1]

$$\sigma_{\mathrm{bi}}(f,\theta_{TS},\psi_{TS},\theta_{SR},\psi_{SR})=4\pi\sigma_{\mathrm{d}}(f,\theta_{TS},\psi_{TS},\theta_{SR},\psi_{SR}). \tag{10.1-45}$$

Substituting (10.1-37) into (10.1-45) yields

$$\boxed{\begin{aligned}\sigma_{\mathrm{bi}}(f,\theta_{TS},\psi_{TS},\theta_{SR},\psi_{SR}) &= \frac{|S_S(f,\theta_{TS},\psi_{TS},\theta_{SR},\psi_{SR})|^2}{4\pi} \\ &= 4\pi r_{SR}^2\frac{I_{\mathrm{avg},s}(\mathbf{r}_R)}{I_{\mathrm{avg},i}(\mathbf{r}_S)}\exp[+2\alpha(f)r_{SR}]\end{aligned}} \tag{10.1-46}$$

where $4\pi r_{SR}^2$ is the surface area of a sphere with radius r_{SR} meters. If we let

$$\sigma_{\mathrm{bi}}(f,\theta_{TS},\psi_{TS},\theta_{SR},\psi_{SR})=\frac{\left|S_S(f,\theta_{TS},\psi_{TS},\theta_{SR},\psi_{SR})\right|^2}{4\pi}=\pi a_{\mathrm{eff}}^2, \quad (10.1\text{-}47)$$

where πa_{eff}^2 is an *effective cross-sectional area* (in squared meters) of the object, and a_{eff} is an *effective radius* (in meters), then

$$\left|S_S(f,\theta_{TS},\psi_{TS},\theta_{SR},\psi_{SR})\right|=2\pi a_{\mathrm{eff}}, \quad (10.1\text{-}48)$$

that is, the magnitude of the scattering function can be interpreted as an *effective circumference* (in meters) of the object based on the bistatic scattering cross-section. Therefore,

$$\sigma_{\mathrm{d}}(f,\theta_{TS},\psi_{TS},\theta_{SR},\psi_{SR})=\frac{\sigma_{\mathrm{bi}}(f,\theta_{TS},\psi_{TS},\theta_{SR},\psi_{SR})}{4\pi}=\frac{a_{\mathrm{eff}}^2}{4}. \quad (10.1\text{-}49)$$

We are now in a position to define target strength.

We shall define *target strength* (TS) as follows:[2,3]

$$\boxed{\mathrm{TS}\triangleq 10\log_{10}\left[\frac{\sigma_{\mathrm{d}}(f,\theta_{TS},\psi_{TS},\theta_{SR},\psi_{SR})}{A_{\mathrm{ref}}}\right]\ \mathrm{dB\ re}\ A_{\mathrm{ref}}} \quad (10.1\text{-}50)$$

and by substituting (10.1-37) into (10.1-50), we obtain

$$\boxed{\mathrm{TS}=10\log_{10}\left[\frac{r_{SR}^2}{A_{\mathrm{ref}}}\frac{I_{\mathrm{avg},s}(\mathbf{r}_R)}{I_{\mathrm{avg},i}(\mathbf{r}_S)}\exp[+2\alpha(f)r_{SR}]\right]\mathrm{dB\ re}\ A_{\mathrm{ref}}} \quad (10.1\text{-}51)$$

where A_{ref} is a *reference cross-sectional area* commonly chosen to be equal to $1\,\mathrm{m}^2$. Since acoustic measurements are usually taken in terms of acoustic pressure, what we need to do next is to rewrite (10.1-37) and (10.1-51) so that $\sigma_{\mathrm{d}}(f,\theta_{TS},\psi_{TS},\theta_{SR},\psi_{SR})$ and TS can be related to source level (SL) and the scattered acoustic pressure incident upon the receiver. One of the most common parameters used to describe a sound-source is SL .

Since

$$I_{\mathrm{avg},i}(\mathbf{r}_S)=\frac{\left|p_{f,i}(\mathbf{r}_S)\right|^2}{2\rho_0 c} \quad (10.1\text{-}52)$$

and

$$I_{\text{avg},s}(\mathbf{r}_R) = \frac{\left|p_{f,s}(\mathbf{r}_R)\right|^2}{2\rho_0 c}, \tag{10.1-53}$$

$$\frac{I_{\text{avg},s}(\mathbf{r}_R)}{I_{\text{avg},i}(\mathbf{r}_S)} = \frac{\left|p_{f,s}(\mathbf{r}_R)\right|^2}{\left|p_{f,i}(\mathbf{r}_S)\right|^2}, \tag{10.1-54}$$

where

$$\left|p_{f,i}(\mathbf{r}_S)\right|^2 = k^2(\rho_0 c)^2 \left(\frac{|S_0|}{4\pi r_{TS}}\right)^2 \exp[-2\alpha(f) r_{TS}] \tag{10.1-55}$$

and the source strength S_0 is given by (10.1-6). Equation (10.1-55) relates the incident acoustic pressure to source strength. However, the incident acoustic pressure can also be related to source level (SL). Since (see Appendix 10A)

$$|S_0| = \frac{4\pi R}{k\rho_0 c} \exp[+\alpha(f)R] P_0 \Big|_{R=1\text{ m}}, \tag{10.1-56}$$

substituting (10.1-56) into (10.1-55) yields

$$\left|p_{f,i}(\mathbf{r}_S)\right|^2 = \left[P_0 \frac{R}{r_{TS}} \exp\left[-\alpha(f)(r_{TS} - R)\right]\right]^2 \Bigg|_{R=1\text{ m}}, \tag{10.1-57}$$

where the acoustic pressure magnitude P_0 and SL are related by (10.1-18). By substituting (10.1-57) into (10.1-54), we obtain

$$\frac{I_{\text{avg},s}(\mathbf{r}_R)}{I_{\text{avg},i}(\mathbf{r}_S)} = \left[\frac{\left|p_{f,s}(\mathbf{r}_R)\right|}{P_0} \frac{r_{TS}}{R} \exp\left[+\alpha(f)(r_{TS} - R)\right]\right]^2 \Bigg|_{R=1\text{ m}}. \tag{10.1-58}$$

Therefore, substituting (10.1-58) into (10.1-37) and (10.1-51) yields the differential scattering cross-section

$$\begin{aligned} \sigma_{\text{d}}(f, \theta_{TS}, \psi_{TS}, \theta_{SR}, \psi_{SR}) &= \left[\frac{\left|S_S(f, \theta_{TS}, \psi_{TS}, \theta_{SR}, \psi_{SR})\right|}{4\pi}\right]^2 \\ &= r_{SR}^2 \left[\frac{\left|p_{f,s}(\mathbf{r}_R)\right|}{P_0} \frac{r_{TS}}{R} \exp\left[+\alpha(f)(r_{TS} + r_{SR} - R)\right]\right]^2 \Bigg|_{R=1\text{ m}} \end{aligned}$$

(10.1-59)

and target strength

$$\mathrm{TS} = 10\log_{10}\left[\frac{r_{SR}^2}{A_{\mathrm{ref}}}\left[\frac{|p_{f,s}(\mathbf{r}_R)|}{P_0}\frac{r_{TS}}{R}\exp\left[+\alpha(f)(r_{TS}+r_{SR}-R)\right]\right]^2\Bigg|_{R=1\,\mathrm{m}}\right] \text{ dB re } A_{\mathrm{ref}} \tag{10.1-60}$$

of the object, respectively, where P_0 and SL are related by (10.1-18), and $A_{\mathrm{ref}} = 1\ \mathrm{m}^2$.

Since $p_{f,s}(\mathbf{r}_R)$ is complex, it can be expressed in polar form as follows:

$$p_{f,s}(\mathbf{r}_R) = |p_{f,s}(\mathbf{r}_R)|\exp\left[+j\angle p_{f,s}(\mathbf{r}_R)\right]. \tag{10.1-61}$$

Substituting (10.1-61) into (10.1-21) yields

$$p_s(t,\mathbf{r}_R) = |p_{f,s}(\mathbf{r}_R)|\exp\left\{+j\left[2\pi ft + \angle p_{f,s}(\mathbf{r}_R)\right]\right\}, \tag{10.1-62}$$

or, equivalently,

$$\mathrm{Re}\{p_s(t,\mathbf{r}_R)\} = |p_{f,s}(\mathbf{r}_R)|\cos\left[2\pi ft + \angle p_{f,s}(\mathbf{r}_R)\right]. \tag{10.1-63}$$

Therefore, $|p_{f,s}(\mathbf{r}_R)|$ and $\angle p_{f,s}(\mathbf{r}_R)$ are the amplitude (in pascals) and phase (in radians) of the time-harmonic, scattered acoustic pressure incident upon the receiver at $\mathbf{r}_R$.

The amplitude factor $|p_{f,s}(\mathbf{r}_R)|$ can be related to the output electrical signal from an omnidirectional point-element (the receiver) at $\mathbf{r}_R$ as follows. Since the scattered acoustic pressure incident upon the omnidirectional point-element is time-harmonic, the output electrical signal in volts is time-harmonic, that is,

$$y_s(t,\mathbf{r}_R) = y_{f,s}(\mathbf{r}_R)\exp(+j2\pi ft), \tag{10.1-64}$$

where

$$y_{f,s}(\mathbf{r}_R) = |y_{f,s}(\mathbf{r}_R)|\exp\left[+j\angle y_{f,s}(\mathbf{r}_R)\right]. \tag{10.1-65}$$

Substituting (10.1-65) into (10.1-64) yields

$$y_s(t,\mathbf{r}_R) = |y_{f,s}(\mathbf{r}_R)|\exp\left\{+j\left[2\pi ft + \angle y_{f,s}(\mathbf{r}_R)\right]\right\}, \tag{10.1-66}$$

or, equivalently,

$$\operatorname{Re}\{y_s(t,\mathbf{r}_R)\} = \left|y_{f,s}(\mathbf{r}_R)\right| \cos\left[2\pi f t + \angle y_{f,s}(\mathbf{r}_R)\right], \tag{10.1-67}$$

where $\left|y_{f,s}(\mathbf{r}_R)\right|$ and $\angle y_{f,s}(\mathbf{r}_R)$ are the amplitude (in volts) and phase (in radians) of the time-harmonic, output electrical signal. Furthermore, since $F_t\{\varphi_s(t,\mathbf{r}_R)\} = \varphi_{f,s}(\mathbf{r}_R)\delta(\eta - f)$ and $F_t\{y_s(t,\mathbf{r}_R)\} = y_{f,s}(\mathbf{r}_R)\delta(\eta - f)$, in the frequency domain at frequency $\eta = f$ Hz [see (7.3-27)],

$$y_{f,s}(\mathbf{r}_R) = \varphi_{f,s}(\mathbf{r}_R)\mathcal{S}_R(f), \tag{10.1-68}$$

where $\varphi_{f,s}(\mathbf{r}_R)$ is the spatial-dependent part of the time-harmonic velocity potential (in $\mathrm{m}^2/\mathrm{sec}$) of the scattered acoustic field incident upon the omnidirectional point-element (the receiver) at $\mathbf{r}_R$, and $\mathcal{S}_R(f)$ is the complex, receiver sensitivity function with units of $\mathrm{V}/\left(\mathrm{m}^2/\mathrm{sec}\right)$ (see Table 6.1-2 and Appendix 6B). Since [see (10.1-22)]

$$p_{f,s}(\mathbf{r}_R) = -j2\pi f \rho_0 \varphi_{f,s}(\mathbf{r}_R), \tag{10.1-69}$$

$$\left|p_{f,s}(\mathbf{r}_R)\right| = 2\pi f \rho_0 \left|\varphi_{f,s}(\mathbf{r}_R)\right|, \tag{10.1-70}$$

and from (10.1-68),

$$\left|\varphi_{f,s}(\mathbf{r}_R)\right| = \frac{\left|y_{f,s}(\mathbf{r}_R)\right|}{\left|\mathcal{S}_R(f)\right|}. \tag{10.1-71}$$

Substituting (10.1-71) into (10.1-70) yields

$$\left|p_{f,s}(\mathbf{r}_R)\right| = 2\pi f \rho_0 \frac{\left|y_{f,s}(\mathbf{r}_R)\right|}{\left|\mathcal{S}_R(f)\right|}, \tag{10.1-72}$$

and because (see Appendix 6B)

$$\left|\mathcal{S}_R(f)\right| = 2\pi f \rho_0 \mathrm{RS}(f), \tag{10.1-73}$$

where $\mathrm{RS}(f)$ is the receiver sensitivity in V/Pa, substituting (10.1-73) into (10.1-72) yields

$$\boxed{\left|p_{f,s}(\mathbf{r}_R)\right| = \frac{\left|y_{f,s}(\mathbf{r}_R)\right|}{\mathrm{RS}(f)}} \tag{10.1-74}$$

Therefore, by substituting (10.1-74) into (10.1-59) and (10.1-60), the differential scattering cross-section and target strength of an object can be computed using the measured amplitude $\left|y_{f,s}(\mathbf{r}_R)\right|$ of the time-harmonic, output electrical signal [see (10.1-67)].

In the case of a monostatic (backscatter) geometry, where the transmitter (sound-source) and receiver are at the same location,

$$\mathbf{r}_R = \mathbf{r}_T \,, \tag{10.1-75}$$

$$\mathbf{r}_{SR} = -\mathbf{r}_{TS} \,, \tag{10.1-76}$$

$$r_{SR} = r_{TS} \,, \tag{10.1-77}$$

and

$$\hat{r}_{SR} = -\hat{r}_{TS} \,. \tag{10.1-78}$$

Therefore (see Figs. 10.1-2 and 10.1-3),

$$\theta_{SR} = 180° - \theta_{TS} \tag{10.1-79}$$

and

$$\psi_{SR} = 180° + \psi_{TS} \,. \tag{10.1-80}$$

In addition, the bistatic scattering cross-section $\sigma_{\text{bi}}(f, \theta_{TS}, \psi_{TS}, \theta_{SR}, \psi_{SR})$ given by (10.1-45) is replaced by the *backscattering cross-section* (a.k.a. the *radar cross-section*[1]) $\sigma_{\text{bs}}(f, \theta_{TS}, \psi_{TS}, \theta_{SR}, \psi_{SR})$, where θ_{SR} and ψ_{SR} are given by (10.1-79) and (10.1-80), respectively.

10.2 Computing the Scattering Function of an Object

By referring to (10.1-59) and (10.1-60), it can be seen that the differential scattering cross-section and target strength of an object can be computed without knowing its scattering function. In this section we shall derive an equation for the complex scattering function that will enable us to compute its value using the scattered acoustic pressure incident upon a receiver.

We begin the derivation by expanding the equation for the spatial-dependent part of the time-harmonic, scattered acoustic pressure given by (10.1-22). Substituting (10.1-23) and (10.1-19) into (10.1-22) yields

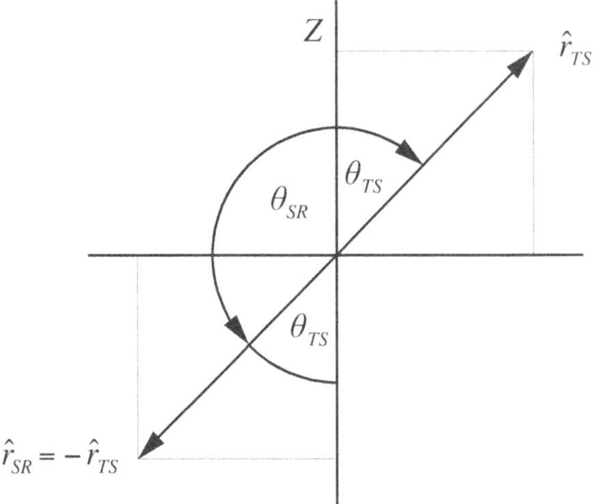

Figure 10.1-2 Measurement of the vertical angles θ_{TS} and θ_{SR} for a backscatter geometry. The vertical angles are measured from the positive Z axis to the unit vectors $\hat{r}_{TS}$ and $\hat{r}_{SR} = -\hat{r}_{TS}$ in the plane containing the unit vectors and the Z axis.

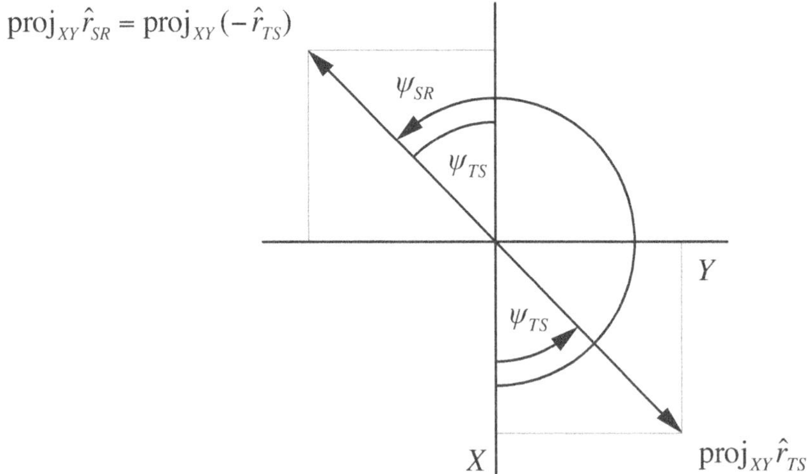

Figure 10.1-3 Measurement of the azimuthal (bearing) angles ψ_{TS} and ψ_{SR} for a backscatter geometry. The bearing angles are measured in a counter-clockwise direction from the positive X axis to the orthogonal projections $\text{proj}_{XY}\hat{r}_{TS}$ and $\text{proj}_{XY}\hat{r}_{SR} = \text{proj}_{XY}(-\hat{r}_{TS})$ of the unit vectors $\hat{r}_{TS}$ and $\hat{r}_{SR} = -\hat{r}_{TS}$ in the XY plane.

$$p_{f,s}(\mathbf{r}_R) = p_{f,i}(\mathbf{r}_S) S_S(f, \theta_{TS}, \psi_{TS}, \theta_{SR}, \psi_{SR}) g_f\left(\mathbf{r}_R \middle| \mathbf{r}_S\right), \tag{10.2-1}$$

where [see (10.1-4) and (10.1-5)]

$$p_{f,i}(\mathbf{r}_S) = -jk\rho_0 c S_0 g_f\left(\mathbf{r}_S \middle| \mathbf{r}_T\right) \tag{10.2-2}$$

is the spatial-dependent part of the time-harmonic, incident acoustic pressure. Solving for the scattering function using (10.2-1) yields

$$S_S(f, \theta_{TS}, \psi_{TS}, \theta_{SR}, \psi_{SR}) = \frac{p_{f,s}(\mathbf{r}_R)}{p_{f,i}(\mathbf{r}_S) g_f\left(\mathbf{r}_R \middle| \mathbf{r}_S\right)}, \tag{10.2-3}$$

and by substituting (10.2-2) into (10.2-3), we obtain

$$S_S(f, \theta_{TS}, \psi_{TS}, \theta_{SR}, \psi_{SR}) = j \frac{p_{f,s}(\mathbf{r}_R)}{k\rho_0 c S_0 g_f\left(\mathbf{r}_S \middle| \mathbf{r}_T\right) g_f\left(\mathbf{r}_R \middle| \mathbf{r}_S\right)}. \tag{10.2-4}$$

Further substituting (10.1-7) and (10.1-24) into (10.2-4) yields

$$S_S(f, \theta_{TS}, \psi_{TS}, \theta_{SR}, \psi_{SR}) = j \frac{(4\pi)^2 r_{TS} r_{SR}}{k\rho_0 c S_0} p_{f,s}(\mathbf{r}_R) \exp\left[+\alpha(f)(r_{TS} + r_{SR})\right] \times \exp\left[+jk(r_{TS} + r_{SR})\right], \tag{10.2-5}$$

or, since wavenumber $k = 2\pi f / c$,

$$\boxed{S_S(f, \theta_{TS}, \psi_{TS}, \theta_{SR}, \psi_{SR}) = j \frac{8\pi r_{TS} r_{SR}}{f \rho_0 S_0} p_{f,s}(\mathbf{r}_R) \exp\left[+\alpha(f)(r_{TS} + r_{SR})\right] \exp(+j2\pi f \tau)} \tag{10.2-6}$$

where the source strength S_0 is given by (10.1-6), $p_{f,s}(\mathbf{r}_R)$ is the spatial-dependent part of the time-harmonic, scattered acoustic pressure incident upon the receiver at $\mathbf{r}_R$ [see (10.1-61) through (10.1-63)], and

$$\tau = (r_{TS} + r_{SR}) / c \tag{10.2-7}$$

is the *bistatic* time delay in seconds.

The acoustic pressure $p_{f,s}(\mathbf{r}_R)$ can be related to the time-harmonic, output electrical signal from an omnidirectional point-element (the receiver) at $\mathbf{r}_R$ as follows. With the use of (10.1-65), (10.1-68), and (10.1-69),

$$\boxed{p_{f,s}(\mathbf{r}_R) = -jk\rho_0 c \frac{\left|y_{f,s}(\mathbf{r}_R)\right|}{\mathcal{S}_R(f)} \exp\left[+j\angle y_{f,s}(\mathbf{r}_R)\right]} \tag{10.2-8}$$

where $\left|y_{f,s}(\mathbf{r}_R)\right|$ and $\angle y_{f,s}(\mathbf{r}_R)$ are the amplitude (in volts) and phase (in radians) of the time-harmonic, output electrical signal $y_s(t,\mathbf{r}_R)$ [see (10.1-67)], and $\mathcal{S}_R(f)$ is the complex, receiver sensitivity function with units of $\mathrm{V}/\left(\mathrm{m}^2/\mathrm{sec}\right)$ (see Table 6.1-2 and Appendix 6B). Substituting (10.2-8) into (10.2-5) yields

$$\boxed{\begin{aligned} S_S(f,\theta_{TS},\psi_{TS},\theta_{SR},\psi_{SR}) = {} & \frac{(4\pi)^2 r_{TS} r_{SR}}{S_0 \mathcal{S}_R(f)} \left|y_{f,s}(\mathbf{r}_R)\right| \exp\left[+\alpha(f)(r_{TS}+r_{SR})\right] \times \\ & \exp\left\{+j\left[2\pi f\tau + \angle y_{f,s}(\mathbf{r}_R)\right]\right\} \end{aligned}} \tag{10.2-9}$$

where τ is given by (10.2-7). Note that the scattering function $S_S(f,\theta_{TS},\psi_{TS},\theta_{SR},\psi_{SR})$ has units of meters.

10.3 Direct Path

In this section we shall complete the analysis of the bistatic scattering problem by analyzing the *direct path* between the transmitter (sound-source) located at $\mathbf{r}_T = (x_T, y_T, z_T)$ and the receiver located at $\mathbf{r}_R = (x_R, y_R, z_R)$, where the transmitter is a time-harmonic, omnidirectional point-source, and the receiver is an omnidirectional point-element (see Fig. 10.3-1). The equations that describe the direct path are very similar to the equations that describe the acoustic field incident upon the object. The time-harmonic acoustic pressure (in pascals) incident upon the receiver at $\mathbf{r}_R$ is given by (see Appendix 10A)

$$p_d(t,\mathbf{r}_R) = p_{f,d}(\mathbf{r}_R)\exp(+j2\pi ft), \tag{10.3-1}$$

where

$$p_{f,d}(\mathbf{r}_R) = -jk\rho_0 c\,\varphi_{f,d}(\mathbf{r}_R), \tag{10.3-2}$$

$$\varphi_{f,d}(\mathbf{r}_R) = S_0\, g_f\left(\mathbf{r}_R \,\middle|\, \mathbf{r}_T\right), \tag{10.3-3}$$

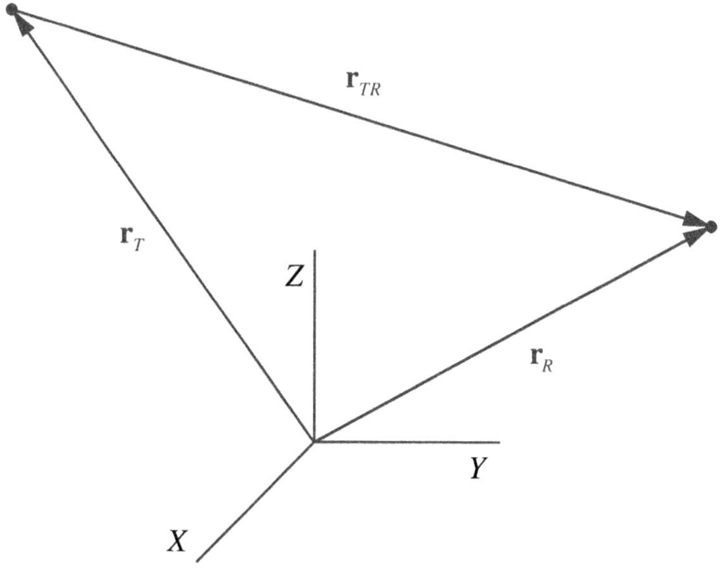

Figure 10.3-1 Direct path between the transmitter (sound-source) located at $\mathbf{r}_T = (x_T, y_T, z_T)$ and the receiver located at $\mathbf{r}_R = (x_R, y_R, z_R)$.

$$S_0 = A_x \mathcal{S}_T(f), \tag{10.3-4}$$

$$\begin{aligned} g_f(\mathbf{r}_R|\mathbf{r}_T) &= -\frac{\exp\left[-\alpha(f)\left|\mathbf{r}_R - \mathbf{r}_T\right|\right]}{4\pi\left|\mathbf{r}_R - \mathbf{r}_T\right|}\exp\left(-jk\left|\mathbf{r}_R - \mathbf{r}_T\right|\right) \\ &= -\frac{\exp[-\alpha(f)r_{TR}]}{4\pi r_{TR}}\exp(-jkr_{TR}), \end{aligned} \tag{10.3-5}$$

$$\mathbf{r}_{TR} = \mathbf{r}_R - \mathbf{r}_T = (x_R - x_T)\hat{x} + (y_R - y_T)\hat{y} + (z_R - z_T)\hat{z}, \tag{10.3-6}$$

$$r_{TR} = \left|\mathbf{r}_{TR}\right| = \sqrt{(x_R - x_T)^2 + (y_R - y_T)^2 + (z_R - z_T)^2}, \tag{10.3-7}$$

and

$$\mathbf{r}_{TR} = r_{TR}\hat{r}_{TR}, \tag{10.3-8}$$

where $\hat{r}_{TR}$ is the dimensionless unit vector in the direction of the vector $\mathbf{r}_{TR}$.

The time-average intensity vector (in watts per squared meter) incident upon the receiver at $\mathbf{r}_R$ is given by (see Appendix 10A)

$$\mathbf{I}_{\text{avg},d}(\mathbf{r}_R) = I_{\text{avg},d}(\mathbf{r}_R)\hat{r}_{TR}, \tag{10.3-9}$$

where

$$I_{\text{avg},d}(\mathbf{r}_R)=\frac{1}{2}k^2\rho_0 c\left(\frac{|S_0|}{4\pi r_{TR}}\right)^2\exp[-2\alpha(f)r_{TR}]. \tag{10.3-10}$$

If the origin of a Cartesian coordinate system (with the same orientation as the one shown in Fig. 10.1-1) is placed at $\mathbf{r}_T$, then the spherical angles θ_{TR} and ψ_{TR} that describe the direction of the unit vector $\hat{r}_{TR}$ are measured as shown in Fig. 1.2-2 and are given by (see Appendix 10A)

$$\boxed{\theta_{TR}=\cos^{-1}\left(\frac{z_R-z_T}{r_{TR}}\right)} \tag{10.3-11}$$

and

$$\boxed{\psi_{TR}=\tan^{-1}\left(\frac{y_R-y_T}{x_R-x_T}\right)} \tag{10.3-12}$$

Equations (10.3-11) and (10.3-12) are the *angles of incidence* at the receiver.

And finally, if we enclose the omnidirectional point-source at $\mathbf{r}_T$ with a sphere of radius r_{TR} meters, then the time-average radiated power (in watts) is given by (see Appendix 10A)

$$P_{\text{avg},d}=\frac{1}{2}k^2\rho_0 c\frac{|S_0|^2}{4\pi}\exp[-2\alpha(f)r_{TR}]=4\pi r_{TR}^2 I_{\text{avg},d}(\mathbf{r}_R), \tag{10.3-13}$$

where the source strength S_0 is given by (10.1-6), or

$$P_{\text{avg},d}=4\pi R^2\frac{P_0^2}{2\rho_0 c}\exp\left[-2\alpha(f)(r_{TR}-R)\right]\Bigg|_{R=1\,\text{m}}, \tag{10.3-14}$$

where P_0 is the magnitude of the time-harmonic, radiated acoustic pressure measured at a range $R=1\,\text{m}$ from the omnidirectional point-source at $\mathbf{r}_T$. The acoustic pressure magnitude P_0 and source level (SL) are related by (10.1-18).

10.4 Sonar Equations

10.4.1 Scattered Path

In this subsection we shall derive equations for the sound-pressure level

(SPL), transmission loss (TL), and acoustic signal-to-noise ratio at the location of the receiver for the bistatic scattering problem shown in Fig. 10.1-1. The receiver is an omnidirectional point-element at $\mathbf{r}_R$. In order to derive equations for the SPL and TL, we first have to find the root-mean-square (rms) scattered acoustic pressure.

The time-harmonic, scattered acoustic pressure (in pascals) incident upon the receiver at $\mathbf{r}_R$ is given by

$$p_s(t, \mathbf{r}_R) = p_{f,s}(\mathbf{r}_R)\exp(+j2\pi ft), \tag{10.4-1}$$

where, by substituting (10.2-2) into (10.2-1),

$$p_{f,s}(\mathbf{r}_R) = -jk\rho_0 c S_0 g_f\left(\mathbf{r}_S \middle| \mathbf{r}_T\right) S_S(f, \theta_{TS}, \psi_{TS}, \theta_{SR}, \psi_{SR}) g_f\left(\mathbf{r}_R \middle| \mathbf{r}_S\right). \tag{10.4-2}$$

Further substituting (10.1-7) and (10.1-24) into (10.4-2) yields

$$p_{f,s}(\mathbf{r}_R) = -j\frac{k\rho_0 c S_0}{(4\pi)^2 r_{TS} r_{SR}} S_S(f, \theta_{TS}, \psi_{TS}, \theta_{SR}, \psi_{SR}) \exp\left[-\alpha(f)(r_{TS} + r_{SR})\right] \times \exp\left[-jk(r_{TS} + r_{SR})\right]. \tag{10.4-3}$$

Since the scattered acoustic pressure $p_s(t, \mathbf{r}_R)$ is a time-harmonic acoustic field, its rms value is given by

$$p_{\text{rms},s}(\mathbf{r}_R) = \frac{\sqrt{2}}{2}\left|p_{f,s}(\mathbf{r}_R)\right|. \tag{10.4-4}$$

Computing the magnitude of (10.4-3) yields

$$\left|p_{f,s}(\mathbf{r}_R)\right| = \frac{k\rho_0 c|S_0|}{(4\pi)^2 r_{TS} r_{SR}}\left|S_S(f, \theta_{TS}, \psi_{TS}, \theta_{SR}, \psi_{SR})\right| \exp\left[-\alpha(f)(r_{TS} + r_{SR})\right], \tag{10.4-5}$$

and by substituting (10.1-56) into (10.4-5), we obtain

$$\left|p_{f,s}(\mathbf{r}_R)\right| = P_0 \frac{R}{r_{TS} r_{SR}} \frac{\left|S_S(f, \theta_{TS}, \psi_{TS}, \theta_{SR}, \psi_{SR})\right|}{4\pi} \exp\left[-\alpha(f)(r_{TS} + r_{SR} - R)\right]\Bigg|_{R=1\text{ m}}. \tag{10.4-6}$$

Therefore, substituting (10.4-6) into (10.4-4) yields the rms scattered acoustic pressure

$$p_{\text{rms},s}(\mathbf{r}_R)=\frac{\sqrt{2}}{2}P_0\frac{R}{r_{TS}r_{SR}}\frac{\left|S_S(f,\theta_{TS},\psi_{TS},\theta_{SR},\psi_{SR})\right|}{4\pi}\exp\left[-\alpha(f)(r_{TS}+r_{SR}-R)\right]\Bigg|_{R=1\text{ m}}. \tag{10.4-7}$$

Now that we have an equation for the rms scattered acoustic pressure, we can begin the SPL calculation by 1) dividing both sides of (10.4-7) by the rms reference pressure P_{ref}, 2) squaring both sides of the resulting equation, and 3) multiplying the right-hand side of the resulting equation by $A_{\text{ref}}/A_{\text{ref}}=1$, where A_{ref} is a reference cross-sectional area. Doing steps one through three yields

$$\left[\frac{p_{\text{rms},s}(\mathbf{r}_R)}{P_{\text{ref}}}\right]^2=\left(\frac{\sqrt{2}P_0/2}{P_{\text{ref}}}\right)^2\left(\frac{R\sqrt{A_{\text{ref}}}}{r_{TS}r_{SR}}\right)^2\frac{\sigma_{\text{d}}(f,\theta_{TS},\psi_{TS},\theta_{SR},\psi_{SR})}{A_{\text{ref}}}\times \left\{\exp\left[-\alpha(f)(r_{TS}+r_{SR}-R)\right]\right\}^2\Bigg|_{R=1\text{ m}}, \tag{10.4-8}$$

where

$$\sigma_{\text{d}}(f,\theta_{TS},\psi_{TS},\theta_{SR},\psi_{SR})=\left[\frac{\left|S_S(f,\theta_{TS},\psi_{TS},\theta_{SR},\psi_{SR})\right|}{4\pi}\right]^2 \tag{10.4-9}$$

is the differential scattering cross-section in squared meters [see (10.1-37)]. Note that both sides of (10.4-8) are *dimensionless*. Taking $10\log_{10}$ of (10.4-8) and setting $R=1\text{ m}$ and $A_{\text{ref}}=1\text{ m}^2$ in the radicand yields

$$\text{SPL}=\text{SL}-20\log_{10}(r_{TS}r_{SR})+\text{TS}+20\log_{10}\left\{\exp\left[-\alpha(f)(r_{TS}+r_{SR}-1)\right]\right\}, \tag{10.4-10}$$

where

$$\text{SPL}=20\log_{10}\left[\frac{p_{\text{rms},s}(\mathbf{r}_R)}{P_{\text{ref}}}\right]\text{ dB re }P_{\text{ref}} \tag{10.4-11}$$

is the sound-pressure level at the receiver,

$$\text{SL}=20\log_{10}\left(\frac{\sqrt{2}P_0/2}{P_{\text{ref}}}\right)\text{ dB re }P_{\text{ref}} \tag{10.4-12}$$

is the source level [see (10.1-17)], where P_0 is the magnitude of the time-harmonic, radiated acoustic pressure measured at a range $R = 1\text{ m}$ from the omnidirectional point-source at $\mathbf{r}_T$,

$$\text{TS} = 10\log_{10}\left[\frac{\sigma_{\text{d}}(f,\theta_{TS},\psi_{TS},\theta_{SR},\psi_{SR})}{A_{\text{ref}}}\right] \text{dB re } A_{\text{ref}} \tag{10.4-13}$$

is the target strength [see (10.1-50)], $\alpha(f)$ is the real, nonnegative, frequency-dependent, attenuation coefficient of seawater in nepers per meter, $P_{\text{ref}} = 1\,\mu\text{Pa}$ (rms) is the rms reference pressure used in underwater acoustics, and $A_{\text{ref}} = 1\text{ m}^2$ is the reference cross-sectional area. Since

$$\begin{aligned} 20\log_{10}\left\{\exp\left[-\alpha(f)(r_{TS}+r_{SR}-1)\right]\right\} &= -20\,\alpha(f)(r_{TS}+r_{SR}-1)\log_{10}e \\ &= -8.686\,\alpha(f)(r_{TS}+r_{SR}-1) \\ &= -\alpha'(f)(r_{TS}+r_{SR}-1), \end{aligned} \tag{10.4-14}$$

where

$$\boxed{\alpha'(f) = 8.686\,\alpha(f)} \tag{10.4-15}$$

is the attenuation coefficient in decibels per meter (see Section 7.2), substituting (10.4-14) into (10.4-10) and rearranging terms yields the following equation for the sound-pressure level at the location of the receiver for the bistatic scattering problem shown in Fig. 10.1-1:

$$\boxed{\text{SPL} = \text{SL} - 20\log_{10}(r_{TS}r_{SR}) - \alpha'(f)(r_{TS}+r_{SR}-1) + \text{TS}} \tag{10.4-16}$$

Equation (10.4-16) is an example of what is commonly referred to as an *active sonar equation* because of the intentional transmission of sound and the appearance of the TS term. Now that we have an equation for the SPL, we can compute the transmission loss (TL) given by

$$\text{TL} = \text{SL} - \text{SPL}\,. \tag{10.4-17}$$

Substituting (10.4-16) into (10.4-17) yields

$$\boxed{\text{TL} = 20\log_{10}(r_{TS}r_{SR}) + \alpha'(f)(r_{TS}+r_{SR}-1) - \text{TS}} \tag{10.4-18}$$

For the case of a monostatic (backscatter) geometry, where the transmitter (sound-source) and receiver are at the same location, $r_{SR} = r_{TS}$ [see (10.1-77)]. Therefore, (10.4-16) and (10.4-18) reduce as follows:

$$\boxed{\text{SPL} = \text{SL} - 40\log_{10}(r_{TS}) - \alpha'(f)(2r_{TS} - 1) + \text{TS}} \tag{10.4-19}$$

and

$$\boxed{\text{TL} = 40\log_{10}(r_{TS}) + \alpha'(f)(2r_{TS} - 1) - \text{TS}} \tag{10.4-20}$$

Acoustic Signal-to-Noise Ratio

In the development of the sonar equations in this subsection, ambient noise in the ocean was not taken into account. In order to develop sonar equations that include ambient noise, we shall use the following definition for the *acoustic signal-to-noise ratio* (SNR_a) as a starting point:

$$\boxed{\text{SNR}_\text{a} \triangleq \frac{\left\langle \left| p_s(t, \mathbf{r}_R) \right|^2 \right\rangle}{E\left\{ \left| p_{n_a}(t, \mathbf{r}_R) \right|^2 \right\}}} \tag{10.4-21}$$

where $\left\langle \left| p_s(t, \mathbf{r}_R) \right|^2 \right\rangle$ is the time-average second moment (time-average power) of the *deterministic* (nonrandom), scattered acoustic pressure $p_s(t, \mathbf{r}_R)$ incident upon the receiver at $\mathbf{r}_R$, and $E\left\{ \left| p_{n_a}(t, \mathbf{r}_R) \right|^2 \right\}$ is the statistical second moment (statistical average power) of the acoustic pressure due to ambient noise $p_{n_a}(t, \mathbf{r}_R)$ incident upon the receiver at $\mathbf{r}_R$, where $E\{\bullet\}$ is the ensemble-average (expectation) operator. The case where $p_s(t, \mathbf{r}_R)$ is a *random process* shall be considered later in this subsection. For an *aperiodic*, *finite duration* signal defined in the time interval $[0, T]$,

$$\left\langle \left| p_s(t, \mathbf{r}_R) \right|^2 \right\rangle = \frac{1}{T} \int_0^T \left| p_s(t, \mathbf{r}_R) \right|^2 dt = p^2_{\text{rms},\, s}(\mathbf{r}_R), \tag{10.4-22}$$

and for a *periodic* signal with *fundamental period* T_0 seconds,

$$\left\langle \left| p_s(t, \mathbf{r}_R) \right|^2 \right\rangle = \frac{1}{T_0} \int_0^{T_0} \left| p_s(t, \mathbf{r}_R) \right|^2 dt = p^2_{\text{rms},\, s}(\mathbf{r}_R), \tag{10.4-23}$$

where $p_{\text{rms},s}(\mathbf{r}_R)$ is the root-mean-square (rms) value of the scattered acoustic pressure $p_s(t,\mathbf{r}_R)$.

Although we know from (10.4-22) and (10.4-23) that the numerator of (10.4-21) is equal to $p^2_{\text{rms},s}(\mathbf{r}_R)$, we shall compute $\left\langle |p_s(t,\mathbf{r}_R)|^2 \right\rangle$ using our model for $p_s(t,\mathbf{r}_R)$ as a check. In order to evaluate the numerator, we shall treat $p_s(t,\mathbf{r}_R)$ as being a *real* function by rewriting (10.4-1) as

$$p_s(t,\mathbf{r}_R) = |p_{f,s}(\mathbf{r}_R)| \cos\left[2\pi f t + \angle p_{f,s}(\mathbf{r}_R)\right], \tag{10.4-24}$$

where $|p_{f,s}(\mathbf{r}_R)|$ and $\angle p_{f,s}(\mathbf{r}_R)$ are the magnitude and phase of $p_{f,s}(\mathbf{r}_R)$, respectively. Since $p_s(t,\mathbf{r}_R)$ is a periodic function of time with fundamental period $T_0 = 1/f$ seconds, the time-average second moment (time-average power) of $p_s(t,\mathbf{r}_R)$ is given by [see (10.4-23)]

$$\begin{aligned}
\left\langle |p_s(t,\mathbf{r}_R)|^2 \right\rangle &= \frac{1}{T_0}\int_0^{T_0} |p_s(t,\mathbf{r}_R)|^2 \, dt \\
&= \frac{1}{T_0}|p_{f,s}(\mathbf{r}_R)|^2 \int_0^{T_0} \cos^2\left[2\pi f t + \angle p_{f,s}(\mathbf{r}_R)\right] dt \\
&= \frac{1}{2T_0}|p_{f,s}(\mathbf{r}_R)|^2 \int_0^{T_0} \left\{1 + \cos\left[4\pi f t + 2\angle p_{f,s}(\mathbf{r}_R)\right]\right\} dt \\
&= \frac{1}{2}|p_{f,s}(\mathbf{r}_R)|^2,
\end{aligned} \tag{10.4-25}$$

or

$$\left\langle |p_s(t,\mathbf{r}_R)|^2 \right\rangle = p^2_{\text{rms},s}(\mathbf{r}_R) \tag{10.4-26}$$

as expected, where $p_{\text{rms},s}(\mathbf{r}_R)$ is given by (10.4-7).

In order to evaluate the denominator of (10.4-21), we shall assume that the random process $p_{n_a}(t,\mathbf{r}_R)$ is *wide-sense stationary* (WSS) *in time*. Therefore, the autocorrelation function of $p_{n_a}(t,\mathbf{r}_R)$ at $\mathbf{r}_R$, given by

$$R_{p_{n_a}}(t,t',\mathbf{r}_R,\mathbf{r}_R) = E\left\{p_{n_a}(t,\mathbf{r}_R)\, p^*_{n_a}(t',\mathbf{r}_R)\right\}, \tag{10.4-27}$$

is a function of *time difference*, that is,

$$R_{p_{n_a}}(\Delta t,\mathbf{r}_R,\mathbf{r}_R) = E\left\{p_{n_a}(t,\mathbf{r}_R)\, p^*_{n_a}(t',\mathbf{r}_R)\right\}, \tag{10.4-28}$$

where

$$\Delta t = t - t' . \tag{10.4-29}$$

Since $p_{n_a}(t, \mathbf{r}_R)$ is WSS in time, its autocorrelation function $R_{p_{n_a}}(\Delta t, \mathbf{r}_R, \mathbf{r}_R)$ forms a Fourier transform pair with its power spectrum $S_{p_{n_a}}(\eta, \mathbf{r}_R)$ with units of pascals-squared per hertz:

$$R_{p_{n_a}}(\Delta t, \mathbf{r}_R, \mathbf{r}_R) \leftrightarrow S_{p_{n_a}}(\eta, \mathbf{r}_R) , \tag{10.4-30}$$

where

$$S_{p_{n_a}}(\eta, \mathbf{r}_R) = F_{\Delta t}\left\{R_{p_{n_a}}(\Delta t, \mathbf{r}_R, \mathbf{r}_R)\right\} = \int_{-\infty}^{\infty} R_{p_{n_a}}(\Delta t, \mathbf{r}_R, \mathbf{r}_R)\exp(-j2\pi\eta\Delta t)\,d\Delta t \tag{10.4-31}$$

and

$$R_{p_{n_a}}(\Delta t, \mathbf{r}_R, \mathbf{r}_R) = F_{\eta}^{-1}\left\{S_{p_{n_a}}(\eta, \mathbf{r}_R)\right\} = \int_{-\infty}^{\infty} S_{p_{n_a}}(\eta, \mathbf{r}_R)\exp(+j2\pi\eta\Delta t)\,d\eta , \tag{10.4-32}$$

where η is frequency in hertz.

If we let $t' = t$, then $\Delta t = 0$, and as a result, the average power of $p_{n_a}(t, \mathbf{r}_R)$ is given by

$$P_{\text{avg}, p_{n_a}}(\mathbf{r}_R) = R_{p_{n_a}}(0, \mathbf{r}_R, \mathbf{r}_R) = E\left\{\left|p_{n_a}(t, \mathbf{r}_R)\right|^2\right\}, \tag{10.4-33}$$

or

$$P_{\text{avg}, p_{n_a}}(\mathbf{r}_R) = \int_{-\infty}^{\infty} S_{p_{n_a}}(\eta, \mathbf{r}_R)\,d\eta , \tag{10.4-34}$$

where $P_{\text{avg}, p_{n_a}}(\mathbf{r}_R)$ has units of pascals-squared. If we also treat $p_{n_a}(t, \mathbf{r}_R)$ as a *real* function of time and space, its power spectrum $S_{p_{n_a}}(\eta, \mathbf{r}_R)$ is an *even* function of frequency η and (10.4-34) reduces to

$$P_{\text{avg}, p_{n_a}}(\mathbf{r}_R) = 2\int_{0}^{\infty} S_{p_{n_a}}(\eta, \mathbf{r}_R)\,d\eta . \tag{10.4-35}$$

Therefore, if the acoustic pressure due to ambient noise $p_{n_a}(t, \mathbf{r}_R)$ is WSS in time, and is treated as a real function of time and space, the SNR_a given by (10.4-21) can be rewritten as

$$\boxed{\text{SNR}_\text{a} = \frac{p^2_{\text{rms}, s}(\mathbf{r}_R)}{P_{\text{avg}, p_{n_a}}(\mathbf{r}_R)}} \tag{10.4-36}$$

where $p_{\text{rms}, s}(\mathbf{r}_R)$ is given by (10.4-7) and $P_{\text{avg}, p_{n_a}}(\mathbf{r}_R)$ is given by (10.4-35).

Since the scattered acoustic pressure $p_s(t, \mathbf{r}_R)$ is time-harmonic in this analysis, the *acoustic signal-to-noise ratio at frequency f hertz* can be defined as

$$\boxed{\mathrm{SNR}_{\mathrm{a},\, f} \triangleq \frac{p^2_{\mathrm{rms},\, s}(\mathbf{r}_R)}{P_{\mathrm{avg},\, p_{n_a},\, f}(\mathbf{r}_R)}} \tag{10.4-37}$$

where from (10.4-35), the average power of the ambient noise $p_{n_a}(t, \mathbf{r}_R)$ at frequency f hertz is

$$P_{\mathrm{avg},\, p_{n_a},\, f}(\mathbf{r}_R) = 2\int_{f-(\Delta f/2)}^{f+(\Delta f/2)} S_{p_{n_a}}(\eta, \mathbf{r}_R)\, d\eta\,, \tag{10.4-38}$$

or

$$P_{\mathrm{avg},\, p_{n_a},\, f}(\mathbf{r}_R) = 2S_{p_{n_a}}(f, \mathbf{r}_R)\Delta f\,, \tag{10.4-39}$$

where $\Delta f = 1\,\mathrm{Hz}$. Therefore, substituting (10.4-39) into (10.4-37) yields

$$\boxed{\mathrm{SNR}_{\mathrm{a},\, f} = \frac{p^2_{\mathrm{rms},\, s}(\mathbf{r}_R)}{2S_{p_{n_a}}(f, \mathbf{r}_R)\Delta f}} \tag{10.4-40}$$

If (10.4-40) is rewritten as

$$\mathrm{SNR}_{\mathrm{a},\, f} = \frac{\left[p_{\mathrm{rms},\, s}(\mathbf{r}_R)/P_{\mathrm{ref}}\right]^2}{2S_{p_{n_a}}(f, \mathbf{r}_R)\Delta f / P^2_{\mathrm{ref}}}\,, \tag{10.4-41}$$

then the acoustic signal-to-noise ratio (in decibels) at frequency f hertz is given by

$$\boxed{\mathrm{SNR}_{\mathrm{a},\, f}\,(\mathrm{dB}) = \mathrm{SPL} - \mathrm{NL}} \tag{10.4-42}$$

where

$$\mathrm{SNR}_{\mathrm{a},\, f}\,(\mathrm{dB}) = 10\log_{10}\mathrm{SNR}_{\mathrm{a},\, f}\,, \tag{10.4-43}$$

$$\mathrm{SPL} = 20\log_{10}\left[\frac{p_{\mathrm{rms},\, s}(\mathbf{r}_R)}{P_{\mathrm{ref}}}\right] \mathrm{dB\ re\ } P_{\mathrm{ref}} \tag{10.4-44}$$

is the sound-pressure level given by (10.4-16) or (10.4-19),

$$\boxed{\text{NL} = 10\log_{10}\left[\frac{2S_{p_{n_a}}(f,\mathbf{r}_R)\Delta f}{P_{\text{ref}}^2}\right] \text{dB re } P_{\text{ref}}^2} \tag{10.4-45}$$

is the *noise level*[4] where $\Delta f = 1\,\text{Hz}$, and $P_{\text{ref}} = 1\,\mu\text{Pa (rms)}$ is the rms reference pressure used in underwater acoustics. The NL can be expressed as a *noise spectrum level*[5] (NSL) as follows:

$$\text{NSL} = 10\log_{10}\left[\frac{2S_{p_{n_a}}(f,\mathbf{r}_R)}{P_{\text{ref}}^2/\Delta f}\right], \tag{10.4-46}$$

or

$$\boxed{\text{NSL} = 10\log_{10}\left[\frac{2S_{p_{n_a}}(f,\mathbf{r}_R)}{S_{\text{ref}}}\right] \text{dB re } S_{\text{ref}}} \tag{10.4-47}$$

where $S_{\text{ref}} = P_{\text{ref}}^2/\text{Hz}$ since $\Delta f = 1\,\text{Hz}$. For a bistatic scattering problem, SPL is given by (10.4-16), and for a monostatic (backscatter) geometry, SPL is given by (10.4-19).

If the scattering function $S_S(f,\theta_{TS},\psi_{TS},\theta_{SR},\psi_{SR})$ of the object (scatterer) is treated as a *random variable* (see Section 10.6), then the scattered acoustic pressure $p_s(t,\mathbf{r}_R)$ is a *random process*. Therefore, replacing the time-average $\left\langle |p_s(t,\mathbf{r}_R)|^2 \right\rangle$ in the numerator of (10.4-21) with the ensemble-average $E\left\{|p_s(t,\mathbf{r}_R)|^2\right\}$ yields the following general expression for the *acoustic signal-to-noise ratio*:

$$\boxed{\text{SNR}_\text{a} \triangleq \frac{E\left\{|p_s(t,\mathbf{r}_R)|^2\right\}}{E\left\{|p_{n_a}(t,\mathbf{r}_R)|^2\right\}}} \tag{10.4-48}$$

With the use of (10.4-24),

$$E\left\{|p_s(t,\mathbf{r}_R)|^2\right\} = \frac{1}{2}E\left\{|p_{f,s}(\mathbf{r}_R)|^2\right\} + \frac{1}{2}E\left\{|p_{f,s}(\mathbf{r}_R)|^2\cos\left[4\pi f t + 2\angle p_{f,s}(\mathbf{r}_R)\right]\right\}. \tag{10.4-49}$$

[4] C. S. Clay and H. Medwin, *Acoustical Oceanography*, Wiley, New York, 1977, pp. 150-151.
[5] *Ibid.*, pp. 119.

If the *random variables* $\left|p_{f,s}(\mathbf{r}_R)\right|$ and $\angle p_{f,s}(\mathbf{r}_R)$ are *uncorrelated*, then

$$E\left\{\left|p_{f,s}(\mathbf{r}_R)\right|^2 \cos\left[4\pi ft + 2\angle p_{f,s}(\mathbf{r}_R)\right]\right\} = E\left\{\left|p_{f,s}(\mathbf{r}_R)\right|^2\right\} \times E\left\{\cos\left[4\pi ft + 2\angle p_{f,s}(\mathbf{r}_R)\right]\right\}, \tag{10.4-50}$$

and if the random variable $\angle p_{f,s}(\mathbf{r}_R)$ is *uniformly distributed in the interval* $[0, 2\pi]$, then

$$E\left\{\cos\left[4\pi ft + 2\angle p_{f,s}(\mathbf{r}_R)\right]\right\} = 0, \tag{10.4-51}$$

and as a result,

$$E\left\{\left|p_{f,s}(\mathbf{r}_R)\right|^2 \cos\left[4\pi ft + 2\angle p_{f,s}(\mathbf{r}_R)\right]\right\} = 0. \tag{10.4-52}$$

Substituting (10.4-52) into (10.4-49) yields

$$E\left\{\left|p_s(t,\mathbf{r}_R)\right|^2\right\} = \frac{1}{2} E\left\{\left|p_{f,s}(\mathbf{r}_R)\right|^2\right\}. \tag{10.4-53}$$

Because $p_s(t,\mathbf{r}_R)$ is time-harmonic in this analysis, (10.4-53) can be rewritten as

$$E\left\{\left|p_s(t,\mathbf{r}_R)\right|^2\right\} = E\left\{p^2_{\mathrm{rms},s}(\mathbf{r}_R)\right\}, \tag{10.4-54}$$

where $p_{\mathrm{rms},s}(\mathbf{r}_R)$ is given by (10.4-7).

Since the acoustic pressure due to ambient noise $p_{n_a}(t,\mathbf{r}_R)$ is assumed to be WSS in time, and is being treated as a real function of time and space, then from (10.4-33)

$$E\left\{\left|p_{n_a}(t,\mathbf{r}_R)\right|^2\right\} = P_{\mathrm{avg},\,p_{n_a}}(\mathbf{r}_R), \tag{10.4-55}$$

where $P_{\mathrm{avg},\,p_{n_a}}(\mathbf{r}_R)$ is the average power of the ambient noise given by (10.4-35). Substituting (10.4-54) and (10.4-55) into (10.4-48) yields

$$\boxed{\mathrm{SNR}_{\mathrm{a}} = \frac{E\left\{p^2_{\mathrm{rms},s}(\mathbf{r}_R)\right\}}{P_{\mathrm{avg},\,p_{n_a}}(\mathbf{r}_R)}} \tag{10.4-56}$$

And since the scattered acoustic pressure $p_s(t, \mathbf{r}_R)$ is time-harmonic in this analysis, the *acoustic signal-to-noise ratio at frequency f hertz* can be defined as

$$\boxed{\mathrm{SNR}_{\mathrm{a},f} \triangleq \frac{E\left\{p^2_{\mathrm{rms},s}(\mathbf{r}_R)\right\}}{P_{\mathrm{avg},\,p_{n_a},\,f}(\mathbf{r}_R)}} \tag{10.4-57}$$

where $P_{\mathrm{avg},\,p_{n_a},\,f}(\mathbf{r}_R)$ is the average power of the ambient noise $p_{n_a}(t, \mathbf{r}_R)$ at frequency f hertz given by (10.4-39). Therefore, substituting (10.4-39) into (10.4-57) yields

$$\boxed{\mathrm{SNR}_{\mathrm{a},f} = \frac{E\left\{p^2_{\mathrm{rms},s}(\mathbf{r}_R)\right\}}{2S_{p_{n_a}}(f, \mathbf{r}_R)\Delta f}} \tag{10.4-58}$$

where $S_{p_{n_a}}(f, \mathbf{r}_R)$ is the power spectrum (in pascals-squared per hertz) of $p_{n_a}(t, \mathbf{r}_R)$, and $\Delta f = 1\,\mathrm{Hz}$.

If (10.4-58) is rewritten as

$$\mathrm{SNR}_{\mathrm{a},f} = \frac{E\left\{p^2_{\mathrm{rms},s}(\mathbf{r}_R)\right\}\Big/P^2_{\mathrm{ref}}}{2S_{p_{n_a}}(f, \mathbf{r}_R)\Delta f\Big/P^2_{\mathrm{ref}}}, \tag{10.4-59}$$

then the acoustic signal-to-noise ratio (in decibels) at frequency f hertz is given by

$$\boxed{\mathrm{SNR}_{\mathrm{a},f}\,(\mathrm{dB}) = \overline{\mathrm{SPL}} - \mathrm{NL}} \tag{10.4-60}$$

where

$$\mathrm{SNR}_{\mathrm{a},f}\,(\mathrm{dB}) = 10\log_{10}\mathrm{SNR}_{\mathrm{a},f}, \tag{10.4-61}$$

$$\overline{\mathrm{SPL}} = 10\log_{10}\left[\frac{E\left\{p^2_{\mathrm{rms},s}(\mathbf{r}_R)\right\}}{P^2_{\mathrm{ref}}}\right]\ \mathrm{dB\ re}\ P^2_{\mathrm{ref}} \tag{10.4-62}$$

is the *average* sound-pressure level given by

$$\boxed{\overline{\mathrm{SPL}} = \mathrm{SL} - 20\log_{10}(r_{TS}r_{SR}) - \alpha'(f)(r_{TS} + r_{SR} - 1) + \overline{\mathrm{TS}}} \tag{10.4-63}$$

where $P_{\text{ref}} = 1\,\mu\text{Pa}$ (rms), SL is the source level given by (10.4-12), $\alpha'(f)$ is the attenuation coefficient in decibels per meter given by (10.4-15),

$$\overline{\text{TS}} = 10\log_{10}\left[\frac{\overline{\sigma_{\text{d}}}(f, \theta_{TS}, \psi_{TS}, \theta_{SR}, \psi_{SR})}{A_{\text{ref}}}\right] \text{dB re } A_{\text{ref}} \tag{10.4-64}$$

is the *average* target strength where $A_{\text{ref}} = 1\,\text{m}^2$ is the reference cross-sectional area,

$$\overline{\sigma_{\text{d}}}(f, \theta_{TS}, \psi_{TS}, \theta_{SR}, \psi_{SR}) = \frac{1}{(4\pi)^2} E\left\{\left|S_S(f, \theta_{TS}, \psi_{TS}, \theta_{SR}, \psi_{SR})\right|^2\right\} \tag{10.4-65}$$

is the *average* differential scattering cross-section in squared meters, and NL is the noise level given by (10.4-45).

Now that we have an equation for the average sound-pressure level $\overline{\text{SPL}}$, we can compute the *average* transmission loss ($\overline{\text{TL}}$) given by

$$\overline{\text{TL}} = \text{SL} - \overline{\text{SPL}}. \tag{10.4-66}$$

Substituting (10.4-63) into (10.4-66) yields

$$\boxed{\overline{\text{TL}} = 20\log_{10}(r_{TS} r_{SR}) + \alpha'(f)(r_{TS} + r_{SR} - 1) - \overline{\text{TS}}} \tag{10.4-67}$$

For the case of a monostatic (backscatter) geometry, where the transmitter (sound-source) and receiver are at the same location, $r_{SR} = r_{TS}$ [see (10.1-77)]. Therefore, (10.4-63) and (10.4-67) reduce as follows:

$$\boxed{\overline{\text{SPL}} = \text{SL} - 40\log_{10}(r_{TS}) - \alpha'(f)(2r_{TS} - 1) + \overline{\text{TS}}} \tag{10.4-68}$$

and

$$\boxed{\overline{\text{TL}} = 40\log_{10}(r_{TS}) + \alpha'(f)(2r_{TS} - 1) - \overline{\text{TS}}} \tag{10.4-69}$$

10.4.2 Direct Path

In this subsection we shall derive equations for the sound-pressure level (SPL), transmission loss (TL), and acoustic signal-to-noise ratio at the location of the receiver for the direct path shown in Fig. 10.3-1. The receiver is an omnidirectional point-element at $\mathbf{r}_R$. In order to derive equations for the SPL and TL, we first have to find the root-mean-square (rms) acoustic pressure for the direct path.

The time-harmonic acoustic pressure (in pascals) incident upon the receiver at $\mathbf{r}_R$ for the direct path is given by

$$p_d(t, \mathbf{r}_R) = p_{f,d}(\mathbf{r}_R)\exp(+j2\pi ft), \tag{10.4-70}$$

where, by substituting (10.3-3) into (10.3-2),

$$p_{f,d}(\mathbf{r}_R) = -jk\rho_0 c S_0 g_f(\mathbf{r}_R|\mathbf{r}_T). \tag{10.4-71}$$

Further substituting (10.3-5) into (10.4-71) yields

$$p_{f,d}(\mathbf{r}_R) = jk\rho_0 c\frac{S_0}{4\pi r_{TR}}\exp[-\alpha(f)r_{TR}]\exp(-jkr_{TR}). \tag{10.4-72}$$

Since the acoustic pressure $p_d(t, \mathbf{r}_R)$ is a time-harmonic acoustic field, its rms value is given by

$$p_{\text{rms},d}(\mathbf{r}_R) = \frac{\sqrt{2}}{2}|p_{f,d}(\mathbf{r}_R)|. \tag{10.4-73}$$

Computing the magnitude of (10.4-72) yields

$$|p_{f,d}(\mathbf{r}_R)| = k\rho_0 c\frac{|S_0|}{4\pi r_{TR}}\exp[-\alpha(f)r_{TR}], \tag{10.4-74}$$

and by substituting (10.1-56) into (10.4-74), we obtain

$$|p_{f,d}(\mathbf{r}_R)| = P_0\frac{R}{r_{TR}}\exp\left[-\alpha(f)(r_{TR}-R)\right]\Bigg|_{R=1\text{ m}}. \tag{10.4-75}$$

Therefore, substituting (10.4-75) into (10.4-73) yields the rms acoustic pressure for the direct path

$$p_{\text{rms},d}(\mathbf{r}_R) = \frac{\sqrt{2}}{2}P_0\frac{R}{r_{TR}}\exp\left[-\alpha(f)(r_{TR}-R)\right]\Bigg|_{R=1\text{ m}}. \tag{10.4-76}$$

Now that we have an equation for the rms acoustic pressure, we can begin the SPL calculation by dividing both sides of (10.4-76) by the rms reference pressure P_{ref}. Doing so yields

$$\frac{p_{\text{rms},d}(\mathbf{r}_R)}{P_{\text{ref}}} = \frac{\sqrt{2}P_0/2}{P_{\text{ref}}}\frac{R}{r_{TR}}\exp\left[-\alpha(f)(r_{TR}-R)\right]\Bigg|_{R=1\text{ m}} . \tag{10.4-77}$$

Note that both sides of (10.4-77) are *dimensionless*. Taking $20\log_{10}$ of (10.4-77) and setting $R = 1$ m yields

$$\text{SPL} = \text{SL} - 20\log_{10}(r_{TR}) + 20\log_{10}\left\{\exp\left[-\alpha(f)(r_{TR}-1)\right]\right\}, \tag{10.4-78}$$

where

$$\text{SPL} = 20\log_{10}\left[\frac{p_{\text{rms},d}(\mathbf{r}_R)}{P_{\text{ref}}}\right] \text{dB re } P_{\text{ref}} \tag{10.4-79}$$

is the sound-pressure level at the receiver, SL is the source level given by (10.4-12), $\alpha(f)$ is the real, nonnegative, frequency-dependent, attenuation coefficient of seawater in nepers per meter, and $P_{\text{ref}} = 1\,\mu\text{Pa (rms)}$ is the rms reference pressure used in underwater acoustics. By referring to (10.4-14) and (10.4-15),

$$20\log_{10}\left\{\exp\left[-\alpha(f)(r_{TR}-1)\right]\right\} = -\alpha'(f)(r_{TR}-1), \tag{10.4-80}$$

where $\alpha'(f)$ is the attenuation coefficient in decibels per meter. Substituting (10.4-80) into (10.4-78) yields the following equation for the sound-pressure level at the location of the receiver for the direct path shown in Fig. 10.3-1:

$$\boxed{\text{SPL} = \text{SL} - 20\log_{10}(r_{TR}) - \alpha'(f)(r_{TR}-1)} \tag{10.4-81}$$

Equation (10.4-81) is another example of an active sonar equation because of the intentional transmission of sound. However, if the sound-source is a submarine, for example, that is *unintentionally* radiating noise at a certain source level, then (10.4-81) would be referred to as a *passive sonar equation*. Substituting (10.4-81) into (10.4-17) yields the transmission loss

$$\boxed{\text{TL} = 20\log_{10}(r_{TR}) + \alpha'(f)(r_{TR}-1)} \tag{10.4-82}$$

Acoustic Signal-to-Noise Ratio

In the development of the sonar equations in this subsection, ambient noise in the ocean was not taken into account. In order to develop sonar equations that include ambient noise, we shall use the following definition for the *acoustic signal-to-noise ratio* (SNR_a) as a starting point:

$$\boxed{\text{SNR}_\text{a} \triangleq \frac{\left\langle \left| p_d(t, \mathbf{r}_R) \right|^2 \right\rangle}{E\left\{ \left| p_{n_a}(t, \mathbf{r}_R) \right|^2 \right\}}} \tag{10.4-83}$$

where $\left\langle \left| p_d(t, \mathbf{r}_R) \right|^2 \right\rangle$ is the time-average second moment (time-average power) of the *deterministic* (nonrandom), direct-path acoustic pressure $p_d(t, \mathbf{r}_R)$ incident upon the receiver at $\mathbf{r}_R$, and $E\left\{ \left| p_{n_a}(t, \mathbf{r}_R) \right|^2 \right\}$ is the statistical second moment (statistical average power) of the acoustic pressure due to ambient noise $p_{n_a}(t, \mathbf{r}_R)$ incident upon the receiver at $\mathbf{r}_R$, where $E\{\bullet\}$ is the ensemble-average (expectation) operator. We know from our discussion of the acoustic signal-to-noise ratio in Subsection 10.4.1 that the time-average of the magnitude-squared of a function of time is equal to its root-mean-square (rms) value squared. Therefore,

$$\left\langle \left| p_d(t, \mathbf{r}_R) \right|^2 \right\rangle = p^2_{\text{rms},\,d}(\mathbf{r}_R), \tag{10.4-84}$$

where $p_{\text{rms},\,d}(\mathbf{r}_R)$ is the rms value of the direct-path acoustic pressure given by (10.4-76).

Since the acoustic pressure due to ambient noise $p_{n_a}(t, \mathbf{r}_R)$ is assumed to be WSS in time, and is being treated as a real function of time and space, then from (10.4-55)

$$E\left\{ \left| p_{n_a}(t, \mathbf{r}_R) \right|^2 \right\} = P_{\text{avg},\,p_{n_a}}(\mathbf{r}_R), \tag{10.4-85}$$

where $P_{\text{avg},\,p_{n_a}}(\mathbf{r}_R)$ is the average power of the ambient noise given by (10.4-35). Substituting (10.4-84) and (10.4-85) into (10.4-83) yields

$$\boxed{\text{SNR}_\text{a} = \frac{p^2_{\text{rms},\,d}(\mathbf{r}_R)}{P_{\text{avg},\,p_{n_a}}(\mathbf{r}_R)}} \tag{10.4-86}$$

And since the direct-path acoustic pressure $p_d(t, \mathbf{r}_R)$ is time-harmonic in this analysis, the *acoustic signal-to-noise ratio at frequency* f *hertz* can be defined as

$$\boxed{\text{SNR}_{\text{a}, f} \triangleq \frac{p^2_{\text{rms}, d}(\mathbf{r}_R)}{P_{\text{avg}, p_{n_a}, f}(\mathbf{r}_R)}} \tag{10.4-87}$$

where $P_{\text{avg}, p_{n_a}, f}(\mathbf{r}_R)$ is the average power of the ambient noise $p_{n_a}(t, \mathbf{r}_R)$ at frequency f hertz given by (10.4-39). Therefore, substituting (10.4-39) into (10.4-87) yields

$$\boxed{\text{SNR}_{\text{a}, f} = \frac{p^2_{\text{rms}, d}(\mathbf{r}_R)}{2S_{p_{n_a}}(f, \mathbf{r}_R)\Delta f}} \tag{10.4-88}$$

where $S_{p_{n_a}}(f, \mathbf{r}_R)$ is the power spectrum (in pascals-squared per hertz) of $p_{n_a}(t, \mathbf{r}_R)$, and $\Delta f = 1\,\text{Hz}$.

If (10.4-88) is rewritten as

$$\text{SNR}_{\text{a}, f} = \frac{\left[p_{\text{rms}, d}(\mathbf{r}_R)/P_{\text{ref}}\right]^2}{2S_{p_{n_a}}(f, \mathbf{r}_R)\Delta f / P^2_{\text{ref}}}, \tag{10.4-89}$$

then the acoustic signal-to-noise ratio (in decibels) at frequency f hertz is given by

$$\boxed{\text{SNR}_{\text{a}, f}\,(\text{dB}) = \text{SPL} - \text{NL}} \tag{10.4-90}$$

where

$$\text{SNR}_{\text{a}, f}\,(\text{dB}) = 10\log_{10}\text{SNR}_{\text{a}, f}, \tag{10.4-91}$$

$$\text{SPL} = 20\log_{10}\left[\frac{p_{\text{rms}, d}(\mathbf{r}_R)}{P_{\text{ref}}}\right] \text{dB re } P_{\text{ref}} \tag{10.4-92}$$

is the sound-pressure level at the location of the receiver for the direct path and is given by (10.4-81), $P_{\text{ref}} = 1\,\mu\text{Pa (rms)}$ is the rms reference pressure used in underwater acoustics, and NL is the noise level given by (10.4-45).

Optimum Frequency

The solution for the direct path is not only part of the bistatic scattering problem, it can also be used to model underwater acoustic communication between a transmitter and a receiver. For a constant, fixed range between a transmitter and a receiver, the optimum value to use for the carrier frequency of a transmitted signal can be found as follows. Using (10.4-17) to solve for the SPL yields

$$\text{SPL} = \text{SL} - \text{TL}\,, \tag{10.4-93}$$

and by substituting (10.4-93) into (10.4-90), we obtain

$$\text{SNR}_{\text{a},f}\,(\text{dB}) = \text{SL} - (\,\text{TL} + \text{NL}\,)\,. \tag{10.4-94}$$

Next, plot $\text{TL} + \text{NL}$ in decibels versus frequency in hertz for a constant, fixed range in meters. The *optimum carrier frequency* is that frequency for which the $\text{TL} + \text{NL}$ curve is a *minimum*. Once the optimum carrier frequency has been determined, one can then solve for the SL required for a desired $\text{SNR}_{\text{a},f}\,(\text{dB})$:

$$\text{SL} = \text{SNR}_{\text{a},f}\,(\text{dB}) + \text{TL} + \text{NL}\,. \tag{10.4-95}$$

10.5 Broadband Solutions

Up to now in this chapter, we have dealt with time-harmonic acoustic fields whose values depend on a single frequency. The time-harmonic acoustic fields were produced by applying a time-harmonic, input electrical signal to an omnidirectional point-source. Since a time-harmonic electrical signal is assumed to exist for all time, its complex frequency spectrum contains a single frequency component [see (6C-16) and (6C-17)]. However, all real-world transmitted electrical signals have a finite duration or pulse length and are sometimes referred to as pulses. As a result, they have complex, bandpass frequency spectra that contain many significant frequency components. A bandpass frequency spectrum can be either narrowband or broadband (see Section 11.1). In this section we shall derive pulse solutions (commonly referred to as broadband solutions) for the velocity potentials and acoustic pressures of the scattered and direct-path acoustic fields by generalizing their time-harmonic solutions. The receiver is an omnidirectional point-element at $\mathbf{r}_R$. We begin with the scattered path.

10.5.1 Scattered Path

The time-harmonic, scalar velocity potential (in squared meters per second) of the scattered acoustic field incident upon the receiver at $\mathbf{r}_R$ is given by

$$\varphi_s(t, \mathbf{r}_R) = \varphi_{f,s}(\mathbf{r}_R)\exp(+j2\pi f t)\,, \tag{10.5-1}$$

where, by substituting (10.1-20) into (10.1-23),

$$\varphi_{f,s}(\mathbf{r}_R) = S_0\, g_f\left(\mathbf{r}_S \middle| \mathbf{r}_T\right) S_S(f, \theta_{TS}, \psi_{TS}, \theta_{SR}, \psi_{SR})\, g_f\left(\mathbf{r}_R \middle| \mathbf{r}_S\right). \tag{10.5-2}$$

Further substituting (10.1-6), (10.1-7), (10.1-24), and $k = 2\pi f/c$ into (10.5-2) yields

$$\varphi_{f,s}(\mathbf{r}_R) = \frac{A_x}{(4\pi)^2 r_{TS} r_{SR}} \mathcal{S}_T(f) S_S(f, \theta_{TS}, \psi_{TS}, \theta_{SR}, \psi_{SR}) \exp\left[-\alpha(f)(r_{TS} + r_{SR})\right] \times \exp\left(-j2\pi f \frac{r_{TS} + r_{SR}}{c}\right). \tag{10.5-3}$$

If we substitute (10.5-3) into (10.5-1), replace the complex amplitude A_x (in volts) of the time-harmonic, input electrical signal applied to the omnidirectional point-source with the complex frequency spectrum $X(f)$ (in volts per hertz) of a broadband input electrical signal, and integrate the right-hand side of the resulting equation over frequency f, then

$$\boxed{\varphi_s(t, \mathbf{r}_R) = \frac{1}{(4\pi)^2 r_{TS} r_{SR}} \int_{-\infty}^{\infty} X(f) \mathcal{S}_T(f) S_S(f, \theta_{TS}, \psi_{TS}, \theta_{SR}, \psi_{SR}) \times \exp\left[-\alpha(f)(r_{TS} + r_{SR})\right] \exp\left[+j2\pi f(t - \tau)\right] df} \tag{10.5-4}$$

is the broadband solution for the velocity potential of the scattered acoustic field for $t \geq \tau$, where

$$\tau = (r_{TS} + r_{SR})/c \tag{10.5-5}$$

is the *bistatic* time delay in seconds. Equation (10.5-4) indicates that the broadband solution $\varphi_s(t, \mathbf{r}_R)$ is equal to the "sum" of the contributions from each frequency component contained in the complex frequency spectrum $X(f)$. Note that (10.5-4) can be expressed as an inverse Fourier transform, that is,

$$\varphi_s(t, \mathbf{r}_R) = \frac{1}{(4\pi)^2 r_{TS} r_{SR}} F_f^{-1}\left\{X(f) \mathcal{S}_T(f) S_S(f, \theta_{TS}, \psi_{TS}, \theta_{SR}, \psi_{SR}) \times \exp\left[-\alpha(f)(r_{TS} + r_{SR})\right] \exp(-j2\pi f \tau)\right\}, \tag{10.5-6}$$

where τ is given by (10.5-5). Therefore, the complex frequency spectrum of the velocity potential of the scattered acoustic field at $\mathbf{r}_R$ is given by

$$\boxed{\begin{aligned}\Phi_s(f,\mathbf{r}_R) &= F_t\{\varphi_s(t,\mathbf{r}_R)\} \\ &= \frac{1}{(4\pi)^2 r_{TS} r_{SR}} X(f)\mathcal{S}_T(f)S_S(f,\theta_{TS},\psi_{TS},\theta_{SR},\psi_{SR})\times \\ &\quad \exp\left[-\alpha(f)(r_{TS}+r_{SR})\right]\exp(-j2\pi f\tau)\end{aligned}} \tag{10.5-7}$$

where $\Phi_s(f,\mathbf{r}_R)$ has units of $\left(\text{m}^2/\text{sec}\right)/\text{Hz}$.

The complex frequency spectrum of the output electrical signal from an omnidirectional point-element at $\mathbf{r}_R$ is given by [see (7.3-27)]

$$Y_s(f,\mathbf{r}_R) = \Phi_s(f,\mathbf{r}_R)\mathcal{S}_R(f), \tag{10.5-8}$$

where $Y_s(f,\mathbf{r}_R)$ has units of V/Hz, and $\mathcal{S}_R(f)$ is the complex, receiver sensitivity function with units of $\text{V}/\left(\text{m}^2/\text{sec}\right)$ (see Table 6.1-2 and Appendix 6B). The time-domain output electrical signal in volts is therefore

$$y_s(t,\mathbf{r}_R) = F_f^{-1}\{Y_s(f,\mathbf{r}_R)\} = F_f^{-1}\{\Phi_s(f,\mathbf{r}_R)\mathcal{S}_R(f)\}. \tag{10.5-9}$$

The time-harmonic, scattered acoustic pressure (in pascals) incident upon the receiver at $\mathbf{r}_R$ is given by

$$p_s(t,\mathbf{r}_R) = p_{f,s}(\mathbf{r}_R)\exp(+j2\pi ft), \tag{10.5-10}$$

where, by substituting (10.1-6) and $k = 2\pi f/c$ into (10.4-3),

$$p_{f,s}(\mathbf{r}_R) = -j\frac{f\rho_0 A_x}{8\pi r_{TS} r_{SR}}\mathcal{S}_T(f)S_S(f,\theta_{TS},\psi_{TS},\theta_{SR},\psi_{SR})\exp\left[-\alpha(f)(r_{TS}+r_{SR})\right]\times \\ \exp\left(-j2\pi f\frac{r_{TS}+r_{SR}}{c}\right). \tag{10.5-11}$$

If we substitute (10.5-11) into (10.5-10), replace the complex amplitude A_x (in volts) of the time-harmonic, input electrical signal applied to the omnidirectional point-source with the complex frequency spectrum $X(f)$ (in volts per hertz) of a broadband input electrical signal, and integrate the right-hand side of the resulting equation over frequency f, then

$$p_s(t,\mathbf{r}_R) = -j\frac{\rho_0}{8\pi r_{TS} r_{SR}} \int_{-\infty}^{\infty} f X(f) \mathcal{S}_T(f) S_S(f,\theta_{TS},\psi_{TS},\theta_{SR},\psi_{SR}) \times \exp\left[-\alpha(f)(r_{TS}+r_{SR})\right] \exp\left[+j2\pi f(t-\tau)\right] df \tag{10.5-12}$$

is the broadband solution for the scattered acoustic pressure for $t \geq \tau$, or

$$p_s(t,\mathbf{r}_R) = -j\frac{\rho_0}{8\pi r_{TS} r_{SR}} F_f^{-1}\left\{ f X(f) \mathcal{S}_T(f) S_S(f,\theta_{TS},\psi_{TS},\theta_{SR},\psi_{SR}) \times \exp\left[-\alpha(f)(r_{TS}+r_{SR})\right] \exp(-j2\pi f\tau) \right\}, \tag{10.5-13}$$

where τ is given by (10.5-5). Therefore, the complex frequency spectrum of the scattered acoustic pressure at $\mathbf{r}_R$ is given by

$$\begin{aligned} P_s(f,\mathbf{r}_R) &= F_t\{p_s(t,\mathbf{r}_R)\} \\ &= -j\frac{\rho_0}{8\pi r_{TS} r_{SR}} f X(f) \mathcal{S}_T(f) S_S(f,\theta_{TS},\psi_{TS},\theta_{SR},\psi_{SR}) \times \\ &\quad \exp\left[-\alpha(f)(r_{TS}+r_{SR})\right] \exp(-j2\pi f\tau) \end{aligned} \tag{10.5-14}$$

where $P_s(f,\mathbf{r}_R)$ has units of $\mathrm{Pa/Hz}$.

Since

$$P_s(f,\mathbf{r}_R) = -j2\pi f \rho_0 \Phi_s(f,\mathbf{r}_R), \tag{10.5-15}$$

the complex frequency spectrum of the output electrical signal from an omnidirectional point-element at $\mathbf{r}_R$ is given by

$$Y_s(f,\mathbf{r}_R) = \Phi_s(f,\mathbf{r}_R) \mathcal{S}_R(f) = j\frac{P_s(f,\mathbf{r}_R)}{2\pi f \rho_0} \mathcal{S}_R(f), \tag{10.5-16}$$

where $Y_s(f,\mathbf{r}_R)$ has units of $\mathrm{V/Hz}$, the ratio $P_s(f,\mathbf{r}_R)/(2\pi f \rho_0)$ has units of $\left(\mathrm{m^2/sec}\right)/\mathrm{Hz}$, and $\mathcal{S}_R(f)$ has units of $\mathrm{V}/\left(\mathrm{m^2/sec}\right)$. The time-domain output electrical signal in volts is therefore

$$y_s(t,\mathbf{r}_R) = F_f^{-1}\{Y_s(f,\mathbf{r}_R)\} = F_f^{-1}\left\{ j\frac{P_s(f,\mathbf{r}_R)}{2\pi f \rho_0} \mathcal{S}_R(f) \right\}. \tag{10.5-17}$$

10.5.2 Direct Path

The time-harmonic, scalar velocity potential (in squared meters per second) of the direct-path acoustic field incident upon the receiver at $\mathbf{r}_R$ is given by

$$\varphi_d(t,\mathbf{r}_R) = \varphi_{f,d}(\mathbf{r}_R)\exp(+j2\pi f t), \tag{10.5-18}$$

where, by substituting (10.3-4), (10.3-5), and $k = 2\pi f/c$ into (10.3-3),

$$\varphi_{f,d}(\mathbf{r}_R) = -\frac{A_x}{4\pi r_{TR}} \mathcal{S}_T(f)\exp[-\alpha(f) r_{TR}]\exp\left(-j2\pi f \frac{r_{TR}}{c}\right). \tag{10.5-19}$$

If we substitute (10.5-19) into (10.5-18), replace the complex amplitude A_x (in volts) of the time-harmonic, input electrical signal applied to the omnidirectional point-source with the complex frequency spectrum $X(f)$ (in volts per hertz) of a broadband input electrical signal, and integrate the right-hand side of the resulting equation over frequency f, then

$$\boxed{\varphi_d(t,\mathbf{r}_R) = -\frac{1}{4\pi r_{TR}} \int_{-\infty}^{\infty} X(f)\mathcal{S}_T(f)\exp[-\alpha(f) r_{TR}]\exp[+j2\pi f(t-\tau)]df} \tag{10.5-20}$$

is the broadband solution for the velocity potential of the direct-path acoustic field for $t \geq \tau$, or

$$\varphi_d(t,\mathbf{r}_R) = -\frac{1}{4\pi r_{TR}} F_f^{-1}\left\{X(f)\mathcal{S}_T(f)\exp[-\alpha(f) r_{TR}]\exp(-j2\pi f\tau)\right\}, \tag{10.5-21}$$

where

$$\tau = r_{TR}/c \tag{10.5-22}$$

is the direct-path time delay in seconds. Therefore, the complex frequency spectrum of the velocity potential of the direct-path acoustic field at $\mathbf{r}_R$ is given by

$$\boxed{\Phi_d(f,\mathbf{r}_R) = F_t\{\varphi_d(t,\mathbf{r}_R)\} = -\frac{1}{4\pi r_{TR}} X(f)\mathcal{S}_T(f)\exp[-\alpha(f)r_{TR}]\exp(-j2\pi f\tau)}$$

(10.5-23)

where $\Phi_d(f,\mathbf{r}_R)$ has units of $\left(\mathrm{m}^2/\mathrm{sec}\right)/\mathrm{Hz}$.

The complex frequency spectrum of the output electrical signal from an omnidirectional point-element at $\mathbf{r}_R$ is given by [see (7.3-27)]

$$Y_d(f,\mathbf{r}_R) = \Phi_d(f,\mathbf{r}_R)\mathcal{S}_R(f), \tag{10.5-24}$$

where $Y_d(f,\mathbf{r}_R)$ has units of $\mathrm{V/Hz}$, and $\mathcal{S}_R(f)$ is the complex, receiver sensitivity function with units of $\mathrm{V}/\left(\mathrm{m}^2/\mathrm{sec}\right)$ (see Table 6.1-2 and Appendix 6B). The time-domain output electrical signal in volts is therefore

$$y_d(t,\mathbf{r}_R) = F_f^{-1}\{Y_d(f,\mathbf{r}_R)\} = F_f^{-1}\{\Phi_d(f,\mathbf{r}_R)\mathcal{S}_R(f)\}. \tag{10.5-25}$$

The time-harmonic, direct-path acoustic pressure (in pascals) incident upon the receiver at $\mathbf{r}_R$ is given by

$$p_d(t,\mathbf{r}_R) = p_{f,d}(\mathbf{r}_R)\exp(+j2\pi ft), \tag{10.5-26}$$

where, by substituting (10.3-4) and $k = 2\pi f/c$ into (10.4-72),

$$p_{f,d}(\mathbf{r}_R) = j\frac{f\rho_0 A_x}{2r_{TR}}\mathcal{S}_T(f)\exp[-\alpha(f)r_{TR}]\exp\left(-j2\pi f\frac{r_{TR}}{c}\right). \tag{10.5-27}$$

If we substitute (10.5-27) into (10.5-26), replace the complex amplitude A_x (in volts) of the time-harmonic, input electrical signal applied to the omnidirectional point-source with the complex frequency spectrum $X(f)$ (in volts per hertz) of a broadband input electrical signal, and integrate the right-hand side of the resulting equation over frequency f, then

$$\boxed{p_d(t,\mathbf{r}_R) = j\frac{\rho_0}{2r_{TR}}\int_{-\infty}^{\infty} fX(f)\mathcal{S}_T(f)\exp[-\alpha(f)r_{TR}]\exp[+j2\pi f(t-\tau)]\,df}$$

(10.5-28)

is the broadband solution for the direct-path acoustic pressure for $t \geq \tau$, or

$$p_d(t, \mathbf{r}_R) = j\frac{\rho_0}{2r_{TR}} F_f^{-1}\left\{ f X(f) \mathcal{S}_T(f) \exp[-\alpha(f) r_{TR}] \exp(-j2\pi f \tau) \right\}, \tag{10.5-29}$$

where τ is given by (10.5-22). Therefore, the complex frequency spectrum of the direct-path acoustic pressure at $\mathbf{r}_R$ is given by

$$\boxed{P_d(f, \mathbf{r}_R) = F_t\{p_d(t, \mathbf{r}_R)\} = j\frac{\rho_0}{2r_{TR}} f X(f) \mathcal{S}_T(f) \exp[-\alpha(f) r_{TR}] \exp(-j2\pi f \tau)} \tag{10.5-30}$$

where $P_d(f, \mathbf{r}_R)$ has units of Pa/Hz.

Since

$$P_d(f, \mathbf{r}_R) = -j2\pi f \rho_0 \Phi_d(f, \mathbf{r}_R), \tag{10.5-31}$$

the complex frequency spectrum of the output electrical signal from an omnidirectional point-element at $\mathbf{r}_R$ is given by

$$Y_d(f, \mathbf{r}_R) = \Phi_d(f, \mathbf{r}_R) \mathcal{S}_R(f) = j\frac{P_d(f, \mathbf{r}_R)}{2\pi f \rho_0} \mathcal{S}_R(f), \tag{10.5-32}$$

where $Y_d(f, \mathbf{r}_R)$ has units of V/Hz, the ratio $P_d(f, \mathbf{r}_R)/(2\pi f \rho_0)$ has units of $\left(\text{m}^2/\text{sec}\right)/\text{Hz}$, and $\mathcal{S}_R(f)$ has units of $\text{V}/\left(\text{m}^2/\text{sec}\right)$. The time-domain output electrical signal in volts is therefore

$$y_d(t, \mathbf{r}_R) = F_f^{-1}\{Y_d(f, \mathbf{r}_R)\} = F_f^{-1}\left\{ j\frac{P_d(f, \mathbf{r}_R)}{2\pi f \rho_0} \mathcal{S}_R(f) \right\}. \tag{10.5-33}$$

10.6 A Statistical Model of the Scattering Function

In this section we shall develop a statistical model for the scattering function $S_S(f, \theta_{TS}, \psi_{TS}, \theta_{SR}, \psi_{SR})$ of an object by treating it as a *random function.* Therefore, for a given frequency and geometry, the value of the scattering function, which we shall call Z, will be a *complex, random variable* rather than a complex, nonrandom constant. As part of the model, we shall derive the probability density functions of the magnitude, magnitude-squared, and phase of Z.

We begin by equating the value of the scattering function Z to a sum of deterministic (nonrandom) and random components, that is,

$$Z = Z_D + Z_R, \tag{10.6-1}$$

where

$$Z = S_S(f, \theta_{TS}, \psi_{TS}, \theta_{SR}, \psi_{SR}) \tag{10.6-2}$$

is a complex, random variable (RV),

$$Z_D = X_D + jY_D \tag{10.6-3}$$

is a complex, deterministic (nonrandom) constant, where X_D and Y_D are real, nonrandom constants, and

$$Z_R = X_R + jY_R \tag{10.6-4}$$

is a complex RV, where X_R and Y_R are real RVs. The complex, nonrandom constant Z_D can be thought of as being the "signal" and the complex RV Z_R can be thought of as being "noise." Substituting (10.6-3) and (10.6-4) into (10.6-1) yields

$$Z = X + jY, \tag{10.6-5}$$

where

$$X = X_D + X_R \tag{10.6-6}$$

and

$$Y = Y_D + Y_R \tag{10.6-7}$$

are real RVs.

The next step is to decide how to model the complex RV Z_R. We shall model Z_R as a *complex, zero-mean, Gaussian RV*, where X_R and Y_R are *real, zero-mean, uncorrelated, Gaussian RVs*. Since [see (10.6-4)]

$$E\{Z_R\} = E\{X_R\} + jE\{Y_R\}, \tag{10.6-8}$$

if

$$E\{Y_R\} = E\{X_R\} = 0, \tag{10.6-9}$$

then

$$E\{Z_R\} = 0. \tag{10.6-10}$$

Also, since the variance of Z_R is given by

$$\begin{aligned}\sigma_{Z_R}^2 &= E\left\{|Z_R|^2\right\} - \left|E\{Z_R\}\right|^2 \\ &= E\left\{|Z_R|^2\right\} \\ &= E\left\{X_R^2\right\} + E\left\{Y_R^2\right\},\end{aligned} \tag{10.6-11}$$

if

$$E\left\{Y_R^2\right\} = E\left\{X_R^2\right\} = \sigma^2, \tag{10.6-12}$$

then

$$\sigma_{Z_R}^2 = 2\sigma^2. \tag{10.6-13}$$

Furthermore, since X_R and Y_R are zero-mean [see (10.6-9)], their variances are equal to their second moments given by (10.6-12), and because they are uncorrelated Gaussian RVs, they are also *statistically independent.*

If Z_R is a complex, zero-mean, Gaussian random variable (RV), then Z will be a *complex, non-zero mean, Gaussian RV* with mean value [see (10.6-1)]

$$m_Z = E\{Z\} = Z_D \tag{10.6-14}$$

and variance [see (10.6-1) and (10.6-13)]

$$\sigma_Z^2 = \sigma_{Z_R}^2 = 2\sigma^2, \tag{10.6-15}$$

where σ^2 is the variance of both X_R and Y_R [see (10.6-12)]. With the use of (10.6-2) and (10.6-5), the magnitude of Z can be expressed as

$$|Z| = \left|S_S(f, \theta_{TS}, \psi_{TS}, \theta_{SR}, \psi_{SR})\right| = \sqrt{X^2 + Y^2}, \tag{10.6-16}$$

where X and Y are Gaussian RVs because X_R and Y_R are Gaussian RVs [see (10.6-6) and (10.6-7)]. The statistics of X and Y are as follows:

$$m_X = E\{X\} = X_D, \tag{10.6-17}$$

$$\sigma_X^2 = \sigma_{X_R}^2 = \sigma^2, \tag{10.6-18}$$

$$m_Y = E\{Y\} = Y_D, \tag{10.6-19}$$

$$\sigma_Y^2 = \sigma_{Y_R}^2 = \sigma^2, \tag{10.6-20}$$

and

$$\begin{aligned}\text{Cov}(X,Y) &= E\{XY\} - m_X m_Y \\ &= E\{X_R Y_R\} \\ &= E\{X_R\}E\{Y_R\} \\ &= 0\end{aligned} \tag{10.6-21}$$

because X_R and Y_R are uncorrelated and zero-mean. Therefore, since X is $N(X_D, \sigma^2)$ and Y is $N(Y_D, \sigma^2)$, where X and Y have the *same* variance σ^2, and X and Y are *statistically independent* because they are uncorrelated Gaussian RVs [see (10.6-21)], $|Z|$ is *Rice* distributed with probability density function[6]

$$p_\zeta(\zeta) = \frac{\zeta}{\sigma^2}\exp\left[-\frac{1}{2\sigma^2}\left(\zeta^2 + \zeta_D^2\right)\right] I_0\left(\frac{\zeta_D}{\sigma^2}\zeta\right), \qquad \zeta \geq 0 \tag{10.6-22}$$

where the RV

$$\zeta = |Z| = \sqrt{X^2 + Y^2}\,, \tag{10.6-23}$$

the nonrandom constant

$$\zeta_D = |Z_D| = \sqrt{X_D^2 + Y_D^2}\,, \tag{10.6-24}$$

and

$$I_0(u) = \sum_{n=0}^{\infty} \frac{u^{2n}}{2^{2n}(n!)^2} \tag{10.6-25}$$

is the zeroth-order modified Bessel function of the first kind. Figure 10.6-1 is a plot of the scaled Rician probability density function (PDF) $\sigma p_\zeta(\zeta)$ versus ζ/σ for several different values of ζ_D/σ. The ratio $\zeta_D^2/(2\sigma^2) = |Z_D|^2/\sigma_{Z_R}^2$ can be thought of as being a signal-to-noise ratio (SNR).

If the nonrandom component of the scattering function $Z_D = 0$, then $Z = Z_R$, that is, the value of the scattering function is equal to a complex, zero-mean, Gaussian RV. Also, if $Z_D = 0$, then $\zeta_D = 0$, and since $I_0(0) = 1$, the Rician PDF given by (10.6-22) reduces to the *Rayleigh* PDF

$$p_\zeta(\zeta) = \frac{\zeta}{\sigma^2}\exp\left(-\frac{\zeta^2}{2\sigma^2}\right), \qquad \zeta \geq 0 \tag{10.6-26}$$

[6] T. A. Schonhoff and A. A. Giordano, *Detection and Estimation Theory*, Pearson Prentice Hall, Upper Saddle River, NJ, 2006, pg. 21.

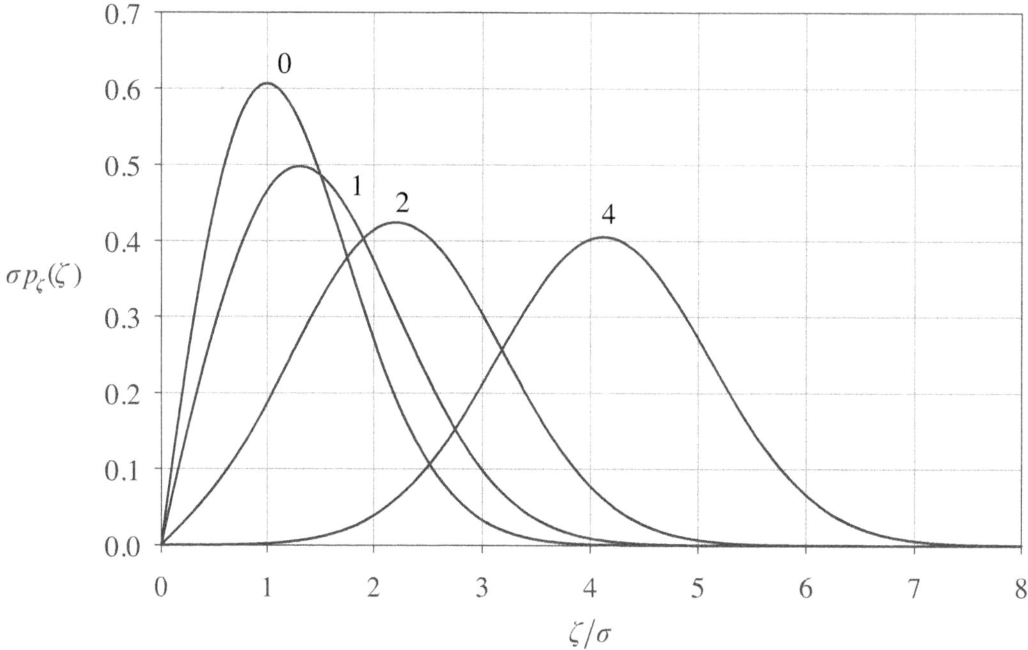

Figure 10.6-1 Plot of the scaled Rician probability density function $\sigma p_\zeta(\zeta)$ versus ζ/σ for $\zeta_D/\sigma = 0$, 1, 2, and 4.

Therefore, in the limit as $\zeta_D/\sigma \to 0$, or equivalently, as the SNR $\zeta_D^2/(2\sigma^2) \to 0$, $\zeta = |Z|$ approaches being Rayleigh distributed.

As ζ_D/σ becomes large, the Rician PDF given by (10.6-22) reduces to an approximate *Gaussian* PDF [see Fig. 10.6-1], as we shall show next. Since

$$I_0(u) \approx \frac{1}{\sqrt{2\pi u}} \exp(u), \qquad u >> 0.25\,, \tag{10.6-27}$$

or equivalently,

$$I_0(u) \approx \frac{1}{\sqrt{2\pi u}} \exp(u), \qquad u \geq 2.5\,, \tag{10.6-28}$$

$$I_0\left(\frac{\zeta_D}{\sigma^2}\zeta\right) \approx \frac{\sigma}{\sqrt{2\pi}\sqrt{\zeta_D \zeta}} \exp\left(\frac{\zeta_D}{\sigma^2}\zeta\right), \qquad \zeta \geq 2.5\frac{\sigma^2}{\zeta_D}. \tag{10.6-29}$$

Substituting (10.6-29) into (10.6-22) yields

$$\boxed{p_\zeta(\zeta) \approx \sqrt{\frac{\zeta}{\zeta_D}} \frac{1}{\sqrt{2\pi}\,\sigma} \exp\left[-\frac{1}{2\sigma^2}(\zeta - \zeta_D)^2\right], \qquad \zeta \geq 2.5\frac{\sigma^2}{\zeta_D}} \tag{10.6-30}$$

Therefore, as ζ_D/σ becomes large, or equivalently, as the SNR $\zeta_D^2/(2\sigma^2)$ becomes large, $\zeta = |Z|$ is approximately Gaussian distributed with mean value ζ_D and variance σ^2.

With the use of (10.6-2) and (10.6-5), the phase of the scattering function can be expressed as

$$\angle Z = \angle S_S(f, \theta_{TS}, \psi_{TS}, \theta_{SR}, \psi_{SR}) = \tan^{-1}(Y/X), \tag{10.6-31}$$

with probability density function[7]

$$p_\phi(\phi) = \frac{1}{2\pi}\exp\left(-\frac{\zeta_D^2}{2\sigma^2}\right) + \left\{\frac{\zeta_D}{\sqrt{2\pi}\,\sigma}\cos(\phi-\phi_D)\exp\left[-\frac{\zeta_D^2}{2\sigma^2}\sin^2(\phi-\phi_D)\right]\times \Phi\left[\frac{\zeta_D}{\sigma}\cos(\phi-\phi_D)\right]\right\}, \quad 0 \le \phi \le 2\pi \tag{10.6-32}$$

where the RV

$$\phi = \angle Z = \tan^{-1}(Y/X), \tag{10.6-33}$$

the nonrandom constant ζ_D is given by (10.6-24), the nonrandom constant

$$\phi_D = \angle Z_D = \tan^{-1}(Y_D/X_D), \tag{10.6-34}$$

and

$$\Phi(u) = \frac{1}{\sqrt{2\pi}}\int_{-\infty}^{u} \exp(-\xi^2/2)\,d\xi \tag{10.6-35}$$

is the cumulative distribution function (CDF) for a zero-mean, unit-variance, Gaussian RV. Figure 10.6-2 is a plot of (10.6-32) for $\phi_D = 210°$ and several different values of ζ_D/σ. If $Z_D = 0$, then $Z = Z_R$ and $\zeta_D = 0$. With $\zeta_D = 0$, the probability density function given by (10.6-32) reduces to the *uniform* probability density function

$$p_\phi(\phi) = \frac{1}{2\pi}, \quad 0 \le \phi \le 2\pi \tag{10.6-36}$$

---

[7] P. Z. Peebles, Jr., *Probability, Random Variables, and Random Signal Principles*, 4th ed., McGraw-Hill, New York, 2001, pp. 400-401.

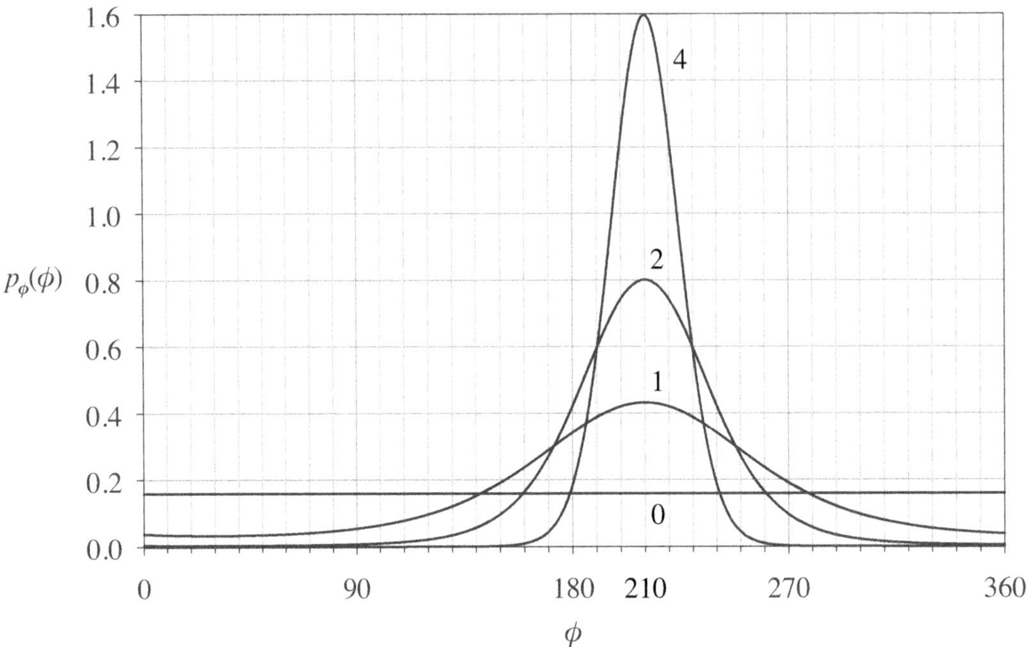

Figure 10.6-2 Plot of the probability density function $p_\phi(\phi)$ versus ϕ (in degrees) for $\phi_D = 210°$ and $\zeta_D/\sigma = 0$, 1, 2, and 4.

Therefore, in the limit as $\zeta_D/\sigma \to 0$, or equivalently, as the SNR $\zeta_D^2/(2\sigma^2) \to 0$, $\phi = \angle Z$ approaches being uniformly distributed in the interval $[0, 2\pi]$. However, as can be seen from Fig. 10.6-2, as ζ_D/σ increases in value, $p_\phi(\phi)$ becomes more concentrated about the deterministic phase angle $\phi_D = 210°$.

And finally, since X is $N(X_D, \sigma^2)$ and Y is $N(Y_D, \sigma^2)$, where X and Y have the *same* variance σ^2, and X and Y are *statistically independent* because they are uncorrelated Gaussian RVs, the magnitude-squared of the scattering function, given by

$$|Z|^2 = |S_S(f, \theta_{TS}, \psi_{TS}, \theta_{SR}, \psi_{SR})|^2 = X^2 + Y^2, \tag{10.6-37}$$

is a *noncentral, chi-squared RV* with 2 degrees of freedom and probability density function[8]

$$\boxed{p_\gamma(\gamma) = \frac{1}{2\sigma^2}\exp\left(-\frac{\gamma + \gamma_D}{2\sigma^2}\right) I_0\left(\frac{\sqrt{\gamma_D}}{\sigma^2}\sqrt{\gamma}\right), \qquad \gamma \geq 0} \tag{10.6-38}$$

[8] T. A. Schonhoff and A. A. Giordano, *Detection and Estimation Theory*, Pearson Prentice Hall, Upper Saddle River, NJ, 2006, pp. 602-604.

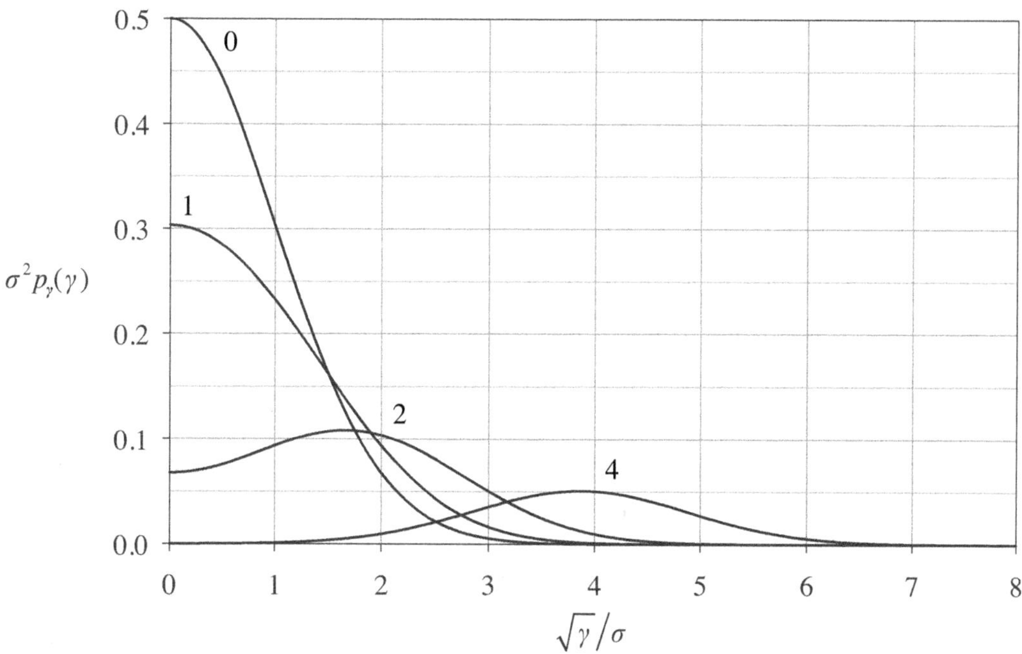

Figure 10.6-3 Plot of the scaled, noncentral, chi-squared probability density function $\sigma^2 p_\gamma(\gamma)$ versus $\sqrt{\gamma}/\sigma$ for $\sqrt{\gamma_D}/\sigma = 0$, 1, 2, and 4.

where the RV

$$\gamma = |Z|^2 = X^2 + Y^2, \tag{10.6-39}$$

and the nonrandom constant

$$\gamma_D = |Z_D|^2 = X_D^2 + Y_D^2 \tag{10.6-40}$$

is the *noncentrality parameter*. Figure 10.6-3 is a plot of the scaled, noncentral, chi-squared probability density function $\sigma^2 p_\gamma(\gamma)$ versus $\sqrt{\gamma}/\sigma$ for several different values of $\sqrt{\gamma_D}/\sigma$. The ratio $\gamma_D/(2\sigma^2) = |Z_D|^2/\sigma_{Z_R}^2$ can be thought of as being a SNR. If $Z_D = 0$, then $Z = Z_R$ and $\gamma_D = 0$. With $\gamma_D = 0$, and since $I_0(0) = 1$, the noncentral, chi-squared probability density function given by (10.6-38) reduces to the *exponential* probability density function

$$\boxed{p_\gamma(\gamma) = \frac{1}{2\sigma^2}\exp\left(-\frac{\gamma}{2\sigma^2}\right), \qquad \gamma \geq 0} \tag{10.6-41}$$

Therefore, in the limit as $\sqrt{\gamma_D}/\sigma \to 0$, or equivalently, as the SNR

$\gamma_D/(2\sigma^2)\to 0$, $\gamma=|Z|^2$ approaches being exponentially distributed.

In summary, if the nonrandom component of the scattering function $Z_D=0$, then $Z=Z_R$, that is, the value of the scattering function is equal to a complex, zero-mean, Gaussian RV, where the magnitude $|Z|=|Z_R|$ is Rayleigh distributed, the phase $\angle Z=\angle Z_R$ is uniformly distributed in the interval $[0,2\pi]$, and the magnitude-squared $|Z|^2=|Z_R|^2$ is exponentially distributed.

10.7 Moving Platforms

In this section we shall derive *exact* equations for the time delay, time-compression/time-expansion factor, and Doppler shift at the receiver for the bistatic scattering problem shown in Fig. 10.1-1 when the transmitter, scatterer (target), and receiver are in *three-dimensional*, *translational motion.* These equations shall be derived for both the scattered and direct paths. We shall also derive *exact* equations for the time-varying angles of incidence at the scatterer, and the time-varying angles of scatter at the receiver for the scattered path; and *exact* equations for the time-varying angles of incidence at the receiver for the direct path. For our purposes, the transmitter is an omnidirectional point-source, and the receiver is an omnidirectional point-element.

10.7.1 Scattered Path

The velocity vectors of the transmitter, $\mathbf{V}_T$, the scatterer, $\mathbf{V}_S$, and the receiver, $\mathbf{V}_R$, are given by

$$\mathbf{V}_T=V_T\hat{n}_{\mathbf{V}_T}, \tag{10.7-1}$$

$$\mathbf{V}_S=V_S\hat{n}_{\mathbf{V}_S}, \tag{10.7-2}$$

and

$$\mathbf{V}_R=V_R\hat{n}_{\mathbf{V}_R}, \tag{10.7-3}$$

where V_T, V_S, and V_R are the speeds in meters per second of the transmitter, scatterer, and receiver, respectively, and $\hat{n}_{\mathbf{V}_T}$, $\hat{n}_{\mathbf{V}_S}$, and $\hat{n}_{\mathbf{V}_R}$ are the dimensionless unit vectors in the directions of $\mathbf{V}_T$, $\mathbf{V}_S$, and $\mathbf{V}_R$, respectively (see Fig. 10.7-1). The velocity vectors given by (10.7-1) through (10.7-3) are *constants*, that is, the speeds and directions are *constants* – acceleration is *not* being modeled. Transmission begins at time $t=t_0$ seconds. The origin of the coordinate system is *not* in motion.

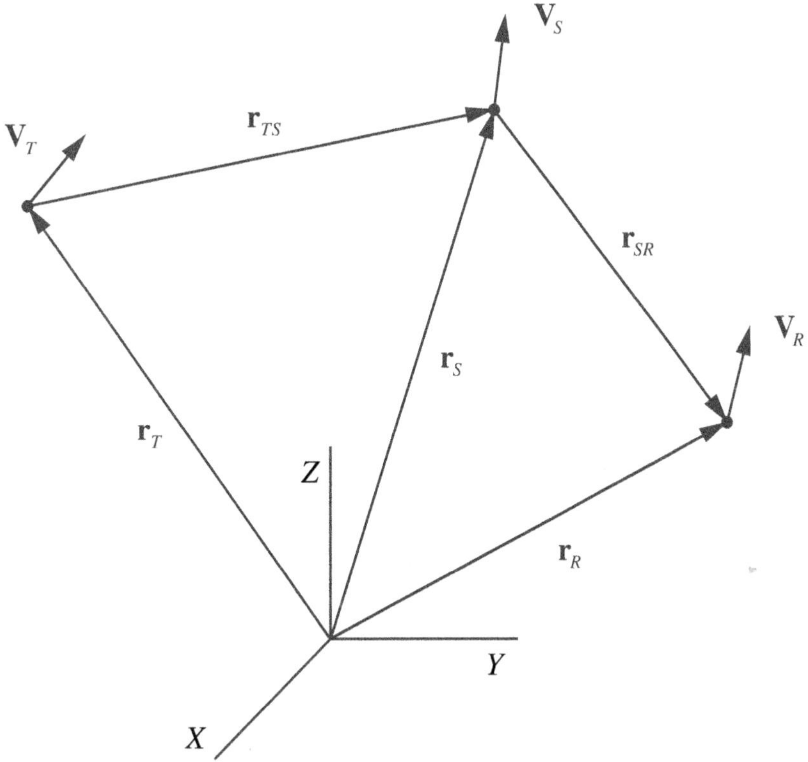

Figure 10.7-1 Bistatic scattering geometry when transmission begins at time $t = t_0$ seconds, where $\mathbf{V}_T$, $\mathbf{V}_S$, and $\mathbf{V}_R$ are the velocity vectors of the transmitter, scatterer, and receiver, respectively.

Since all three platforms are in motion, the position vectors from the origin to the transmitter, scatterer, and receiver – denoted by $\mathcal{R}_T(t)$, $\mathcal{R}_S(t)$, and $\mathcal{R}_R(t)$, respectively – are functions of time given by

$$\mathcal{R}_T(t) = \mathbf{r}_T + \mathbf{V}_T \Delta t, \qquad t \geq t_0, \tag{10.7-4}$$

$$\mathcal{R}_S(t) = \mathbf{r}_S + \mathbf{V}_S \Delta t, \qquad t \geq t_0, \tag{10.7-5}$$

and

$$\mathcal{R}_R(t) = \mathbf{r}_R + \mathbf{V}_R \Delta t, \qquad t \geq t_0, \tag{10.7-6}$$

where $\mathbf{r}_T = (x_T, y_T, z_T)$, $\mathbf{r}_S = (x_S, y_S, z_S)$, and $\mathbf{r}_R = (x_R, y_R, z_R)$ are the position vectors from the origin to the transmitter, scatterer, and receiver, respectively, when transmission begins (see Fig. 10.7-1), and

$$\Delta t = t - t_0, \qquad t \geq t_0. \tag{10.7-7}$$

Note that $\mathcal{R}_T(t_0) = \mathbf{r}_T$, $\mathcal{R}_S(t_0) = \mathbf{r}_S$, and $\mathcal{R}_R(t_0) = \mathbf{r}_R$.

The time-varying position vector from the transmitter to the scatterer is given by

$$\mathcal{R}_{TS}(t) = \mathcal{R}_S(t) - \mathcal{R}_T(t), \tag{10.7-8}$$

and by substituting (10.7-4) and (10.7-5) into (10.7-8), we obtain

$$\mathcal{R}_{TS}(t) = \mathbf{r}_{TS} + \mathbf{V}_{ST}\,\Delta t, \qquad t \geq t_0, \tag{10.7-9}$$

where

$$\mathbf{r}_{TS} = \mathbf{r}_S - \mathbf{r}_T \tag{10.7-10}$$

is the position vector from the transmitter to the scatterer when transmission begins (see Fig. 10.7-1), and

$$\mathbf{V}_{ST} = \mathbf{V}_S - \mathbf{V}_T = V_{ST}\,\hat{n}_{\mathbf{V}_{ST}} \tag{10.7-11}$$

is the velocity vector of the scatterer *relative to* the velocity vector of the transmitter, where $V_{ST} = |\mathbf{V}_{ST}|$ and $\hat{n}_{\mathbf{V}_{ST}}$ is the dimensionless unit vector in the direction of $\mathbf{V}_{ST}$. Note that $\mathcal{R}_{TS}(t_0) = \mathbf{r}_{TS}$.

Similarly, the time-varying position vector from the scatterer to the receiver is given by

$$\mathcal{R}_{SR}(t) = \mathcal{R}_R(t) - \mathcal{R}_S(t), \tag{10.7-12}$$

and by substituting (10.7-5) and (10.7-6) into (10.7-12), we obtain

$$\mathcal{R}_{SR}(t) = \mathbf{r}_{SR} + \mathbf{V}_{RS}\,\Delta t, \qquad t \geq t_0, \tag{10.7-13}$$

where

$$\mathbf{r}_{SR} = \mathbf{r}_R - \mathbf{r}_S \tag{10.7-14}$$

is the position vector from the scatterer to the receiver when transmission begins (see Fig. 10.7-1), and

$$\mathbf{V}_{RS} = \mathbf{V}_R - \mathbf{V}_S = V_{RS}\,\hat{n}_{\mathbf{V}_{RS}} \tag{10.7-15}$$

is the velocity vector of the receiver *relative to* the velocity vector of the scatterer, where $V_{RS} = |\mathbf{V}_{RS}|$ and $\hat{n}_{\mathbf{V}_{RS}}$ is the dimensionless unit vector in the direction of $\mathbf{V}_{RS}$. Note that $\mathcal{R}_{SR}(t_0) = \mathbf{r}_{SR}$.

Time Delay

When the transmitted acoustic field is *first* incident upon the scatterer at some time t' seconds, where $t' > t_0$, the position vector from the transmitter to the scatterer at time t' is given by [see (10.7-9)]

$$\boxed{\mathcal{R}_{TS}(t') = \mathbf{r}_{TS} + \mathbf{V}_{ST}\,\Delta t', \qquad t' > t_0} \tag{10.7-16}$$

where the position vector $\mathbf{r}_{TS}$ is given by (10.7-10), the relative velocity vector $\mathbf{V}_{ST}$ is given by (10.7-11), and

$$\Delta t' = t' - t_0, \qquad t' > t_0. \tag{10.7-17}$$

The position vector from the scatterer to the receiver at time t' is given by [see (10.7-13)]

$$\mathcal{R}_{SR}(t') = \mathbf{r}_{SR} + \mathbf{V}_{RS}\,\Delta t', \qquad t' > t_0, \tag{10.7-18}$$

where the position vector $\mathbf{r}_{SR}$ is given by (10.7-14), the relative velocity vector $\mathbf{V}_{RS}$ is given by (10.7-15), and $\Delta t'$ is given by (10.7-17).

When the scatterer is ensonified by the transmitted acoustic field, it begins to act like a sound source, scattering an acoustic field towards the receiver. In order to calculate when the scattered acoustic field arrives at the receiver, let

$$\mathcal{R}'_{SR}(t) = \mathbf{r}'_{SR} + \mathbf{V}_{RS}(t - t'), \qquad t \geq t', \tag{10.7-19}$$

where

$$\mathbf{r}'_{SR} = \mathcal{R}_{SR}(t'). \tag{10.7-20}$$

Note that $\mathcal{R}'_{SR}(t') = \mathbf{r}'_{SR} = \mathcal{R}_{SR}(t')$. When the scattered acoustic field is *first* incident upon the receiver at some time t'' seconds, where $t'' > t' > t_0$, the position vector from the scatterer to the receiver at time t'' is given by

$$\boxed{\mathcal{R}'_{SR}(t'') = \mathbf{r}'_{SR} + \mathbf{V}_{RS}(t'' - t'), \qquad t'' > t'} \tag{10.7-21}$$

Therefore, the time delay τ can be expressed as

$$\begin{aligned} \tau &= t'' - t_0 \\ &= t' - t_0 + t'' - t' \\ &= \Delta t' + (t'' - t'). \end{aligned} \tag{10.7-22}$$

Substituting

$$\Delta t' = t' - t_0 = \frac{\left|\mathcal{R}_{TS}(t')\right|}{c}, \qquad t' > t_0 \tag{10.7-23}$$

and

$$t'' - t' = \frac{\left|\boldsymbol{\mathcal{R}}'_{SR}(t'')\right|}{c}, \qquad t'' > t' \tag{10.7-24}$$

into (10.7-22) yields

$$\boxed{\tau = \frac{\left|\boldsymbol{\mathcal{R}}_{TS}(t')\right| + \left|\boldsymbol{\mathcal{R}}'_{SR}(t'')\right|}{c}} \tag{10.7-25}$$

where c is the constant speed of sound in meters per second. Equation (10.7-25) is the *time delay* (or travel time) in seconds associated with the scattered path when the scattered acoustic field is *first* incident upon the receiver. In order to evaluate (10.7-25), we need to find solutions for the ranges $\left|\boldsymbol{\mathcal{R}}_{TS}(t')\right|$ and $\left|\boldsymbol{\mathcal{R}}'_{SR}(t'')\right|$, which we shall do next.

Exact Solution for $\left|\boldsymbol{\mathcal{R}}_{TS}(t')\right|$

Since

$$\left|\boldsymbol{\mathcal{R}}_{TS}(t')\right|^2 = \boldsymbol{\mathcal{R}}_{TS}(t') \bullet \boldsymbol{\mathcal{R}}_{TS}(t'), \tag{10.7-26}$$

substituting (10.7-18) into (10.7-26) yields

$$\left|\boldsymbol{\mathcal{R}}_{TS}(t')\right|^2 = r_{TS}^2 + 2r_{TS}\left(\hat{r}_{TS} \bullet \mathbf{V}_{ST}\right)\Delta t' + V_{ST}^2\,(\Delta t')^2, \tag{10.7-27}$$

where $r_{TS} = \left|\mathbf{r}_{TS}\right|$ is given by (10.1-9), $\hat{r}_{TS}$ is the dimensionless unit vector in the direction of $\mathbf{r}_{TS}$, and $V_{ST} = \left|\mathbf{V}_{ST}\right|$, where $\mathbf{V}_{ST}$ is given by (10.7-11). Substituting (10.7-23) into (10.7-27) yields the second-order polynomial

$$A\left|\boldsymbol{\mathcal{R}}_{TS}(t')\right|^2 - B\left|\boldsymbol{\mathcal{R}}_{TS}(t')\right| - C = 0 \tag{10.7-28}$$

with *exact* solution

$$\boxed{\left|\boldsymbol{\mathcal{R}}_{TS}(t')\right| = \frac{B + \sqrt{B^2 + 4AC}}{2A}} \tag{10.7-29}$$

where

$$A = 1 - \left(V_{ST}/c\right)^2, \tag{10.7-30}$$

$$B = 2r_{TS}\frac{\hat{r}_{TS} \bullet \mathbf{V}_{ST}}{c}, \tag{10.7-31}$$

and

$$C = r_{TS}^2. \tag{10.7-32}$$

Equation (10.7-29) is the *exact* range between the transmitter and scatterer when the transmitted acoustic field is *first* incident upon the scatterer at time t', where $t' > t_0$. For the problems that we are interested in,

$$V_{ST}/c \ll 1, \tag{10.7-33}$$

and as a result, $A > 0$. The decision to use the plus sign in front of the square-root term in (10.7-29) instead of the minus sign was dictated by the fact that since $4AC > 0$, the square-root term is greater than B, and that the range $|\mathcal{R}_{TS}(t')|$ must be positive. Note that if the transmitter and scatterer are *not* in motion, then $\mathbf{V}_T = \mathbf{0}$, $\mathbf{V}_S = \mathbf{0}$, $\mathbf{V}_{ST} = \mathbf{0}$, $V_{ST} = 0$, and as expected, (10.7-29) reduces to $|\mathcal{R}_{TS}(t')| = r_{TS}$.

Exact Solution for $|\mathcal{R}'_{SR}(t'')|$

Since

$$|\mathcal{R}'_{SR}(t'')|^2 = \mathcal{R}'_{SR}(t'') \bullet \mathcal{R}'_{SR}(t''), \tag{10.7-34}$$

substituting (10.7-21) into (10.7-34) yields

$$|\mathcal{R}'_{SR}(t'')|^2 = (r'_{SR})^2 + 2r'_{SR}\left(\hat{r}'_{SR} \bullet \mathbf{V}_{RS}\right)(t'' - t') + V_{RS}^2\left(t'' - t'\right)^2, \tag{10.7-35}$$

where $r'_{SR} = |\mathbf{r}'_{SR}|$, $\mathbf{r}'_{SR}$ is given by (10.7-20), $\hat{r}'_{SR}$ is the dimensionless unit vector in the direction of $\mathbf{r}'_{SR}$, and $V_{RS} = |\mathbf{V}_{RS}|$, where $\mathbf{V}_{RS}$ is given by (10.7-15). Substituting (10.7-24) into (10.7-35) yields the second-order polynomial

$$\mathcal{A}|\mathcal{R}'_{SR}(t'')|^2 - \mathcal{B}|\mathcal{R}'_{SR}(t'')| - \mathcal{C} = 0 \tag{10.7-36}$$

with *exact* solution

$$\boxed{|\mathcal{R}'_{SR}(t'')| = \frac{\mathcal{B} + \sqrt{\mathcal{B}^2 + 4\mathcal{A}\mathcal{C}}}{2\mathcal{A}}} \tag{10.7-37}$$

where

$$\mathcal{A} = 1 - \left(V_{RS}/c\right)^2 , \tag{10.7-38}$$

$$\mathcal{B} = 2r'_{SR} \frac{\hat{r}'_{SR} \bullet \mathbf{V}_{RS}}{c} , \tag{10.7-39}$$

and

$$\mathcal{C} = (r'_{SR})^2 . \tag{10.7-40}$$

Equation (10.7-37) is the *exact* range between the scatterer and receiver when the scattered acoustic field is *first* incident upon the receiver at time t'', where $t'' > t' > t_0$. For the problems that we are interested in,

$$V_{RS}/c << 1 , \tag{10.7-41}$$

and as a result, $\mathcal{A} > 0$. The decision to use the plus sign in front of the square-root term in (10.7-37) instead of the minus sign was dictated by the fact that since $4\mathcal{A}\mathcal{C} > 0$, the square-root term is greater than $\mathcal{B}$, and that the range $\left|\mathcal{R}'_{SR}(t'')\right|$ must be positive. Note that if the scatterer and receiver are *not* in motion, then $\mathbf{V}_S = \mathbf{0}$, $\mathbf{V}_R = \mathbf{0}$, $\mathbf{V}_{RS} = \mathbf{0}$, $V_{RS} = 0$, and as expected, (10.7-37) reduces to $\left|\mathcal{R}'_{SR}(t'')\right| = r'_{SR} = r_{SR}$ [see (10.7-20) and (10.7-18)]. With the use of (10.7-29) and (10.7-37), the time delay τ given by (10.7-25) can now be evaluated.

Time-Compression/Time-Expansion Factor and Doppler Shift

The dimensionless *time-compression/time-expansion factor* s is defined as follows:

$$\boxed{s \triangleq \tau_0/\tau} \tag{10.7-42}$$

where for the scattered path

$$\tau_0 = (r_{TS} + r_{SR})/c \tag{10.7-43}$$

is the *bistatic* time delay in seconds when there is *no* motion, and τ is given by (10.7-25), where the ranges $\left|\mathcal{R}_{TS}(t')\right|$ and $\left|\mathcal{R}'_{SR}(t'')\right|$ are given by (10.7-29) and (10.7-37), respectively. Note that $s > 0$. The *Doppler shift* η_D in hertz is related to the dimensionless time-compression/time-expansion factor s and is defined as follows:

$$\boxed{\eta_D \triangleq (s-1) f_c} \tag{10.7-44}$$

where f_c is referred to as either the *center frequency* or *carrier frequency* in hertz of the transmitted electrical signal (see Chapter 11).

There are three ranges of values for the time-compression/time-expansion factor s; namely, $s=1$, $s>1$, and $0<s<1$. When there is *no* motion, $|\boldsymbol{\mathcal{R}}_{TS}(t')| = r_{TS}$ and $|\boldsymbol{\mathcal{R}}'_{SR}(t'')| = r_{SR}$ as we discussed, and as a result, $\tau = \tau_0$ [see (10.7-25) and (10.7-43)]. Substituting $\tau = \tau_0$ into (10.7-42) yields $s=1$. Substituting $s=1$ into (10.7-44) yields $\eta_D = 0$, that is, there is *zero* Doppler shift when there is no motion (and no relative motion). When $s=1$, there is *neither* time compression *nor* time expansion, that is, the duration of the received signal is *equal to* the duration of the transmitted signal resulting in *no* change in signal bandwidth at the receiver. If $\tau < \tau_0$, then $s>1$ (*time compression*) and $\eta_D > 0$ (*positive* Doppler shift). Time compression means that the duration of the received signal is *less than* the duration of the transmitted signal resulting in an *increase* in signal bandwidth. A positive Doppler shift means that the scatterer (target) and receiver are moving *towards* one another. If $\tau > \tau_0$, then $0<s<1$ (*time expansion*) and $\eta_D < 0$ (*negative* Doppler shift). Time expansion means that the duration of the received signal is *greater than* the duration of the transmitted signal resulting in a *decrease* in signal bandwidth. A negative Doppler shift means that the scatterer (target) and receiver are moving *away* from one another.

For a *monostatic* (*backscatter*) geometry, where both the transmitter and receiver are physically collocated,

$$\mathbf{V}_R = \mathbf{V}_T, \tag{10.7-45}$$

$$\mathbf{V}_{RS} = -\mathbf{V}_{ST}, \tag{10.7-46}$$

$$V_{RS} = V_{ST}, \tag{10.7-47}$$

$$\mathbf{r}_{SR} = -\mathbf{r}_{TS}, \tag{10.7-48}$$

$$r_{SR} = r_{TS}, \tag{10.7-49}$$

$$\hat{r}_{SR} = -\hat{r}_{TS}, \tag{10.7-50}$$

and

$$\tau_0 = 2 r_{TS} / c \tag{10.7-51}$$

is the *round-trip* time delay in seconds when there is *no* motion.

Example 10.7-1

In this example we shall demonstrate how to model a received signal using the time-compression/time-expansion factor s. For example purposes, let the transmitted signal $x(t)$ be a rectangular-envelope, CW (continuous-wave) pulse given by

$$x(t) = A\cos(2\pi f_c t + \theta_0)\operatorname{rect}\left[(t - 0.5T)/T\right], \tag{10.7-52}$$

where A is an amplitude factor in volts, $\cos(2\pi f_c t)$ is the carrier waveform, f_c is the carrier frequency in hertz, θ_0 is a phase term in radians,

$$\operatorname{rect}\left(\frac{t - 0.5T}{T}\right) = \begin{cases} 1, & 0 \leq t \leq T \\ 0, & \text{otherwise} \end{cases} \tag{10.7-53}$$

is the time-shifted rectangle function, and T is the pulse length in seconds.

A simple model for the received signal is given by

$$y(t) = Kx(s[t - \tau]), \tag{10.7-54}$$

or

$$y(t) = KA\cos\left(2\pi f_c s[t - \tau] + \theta_0\right)\operatorname{rect}\left[\left(s[t - \tau] - 0.5T\right)/T\right], \tag{10.7-55}$$

where $0 < K < 1$ is a dimensionless constant that is meant to model amplitude attenuation. Solving for s using (10.7-44) yields

$$s = 1 + \frac{\eta_D}{f_c} = \frac{f_c + \eta_D}{f_c}, \tag{10.7-56}$$

and by substituting (10.7-56) into the first term in the argument of the cosine function in (10.7-55), we obtain

$$y(t) = KA\cos\left(2\pi [f_c + \eta_D][t - \tau] + \theta_0\right)\operatorname{rect}\left[\left(s[t - \tau] - 0.5T\right)/T\right]. \tag{10.7-57}$$

Using the definition of the time-shifted rectangle function given by (10.7-53),

$$\operatorname{rect}\left(\frac{s[t - \tau] - 0.5T}{T}\right) = \begin{cases} 1, & 0 \leq s[t - \tau] \leq T \\ 0, & \text{otherwise,} \end{cases} \tag{10.7-58}$$

or

$$\text{rect}\left(\frac{s[t-\tau]-0.5T}{T}\right)=\begin{cases}1, & \tau\le t\le \tau+(T/s)\\ 0, & \text{otherwise.}\end{cases} \tag{10.7-59}$$

Therefore, (10.7-57) can be rewritten as

$$y(t)=KA\cos\left(2\pi[f_c+\eta_D][t-\tau]+\theta_0\right), \qquad \tau\le t\le \tau+T_y, \tag{10.7-60}$$

where the Doppler shift η_D is given by (10.7-44), the time delay τ is given by (10.7-25), and the pulse length of the received signal T_y is given by

$$\boxed{T_y=T/s} \tag{10.7-61}$$

where the time-compression/time-expansion factor s is given by (10.7-42). Equation (10.7-61) is a general result for any transmitted signal $x(t)$ with pulse length T seconds when the received signal $y(t)$ is modeled by (10.7-54). From (10.7-61), and as was stated before, if $s=1$ (neither time compression nor time expansion), $s>1$ (time compression), or $0<s<1$ (time expansion), then the pulse length of the received signal T_y is equal to, less than, or greater than the pulse length T of the transmitted signal, respectively. We shall compute the bandwidth of the received signal $y(t)$ next.

In order to compute the bandwidth of $y(t)$, we need the magnitude spectrum of $y(t)$ along the positive frequency axis. The Fourier transform of (10.7-54) is equal to

$$Y(\eta)=\frac{K}{s}X(\eta/s)\exp(-j2\pi\eta\tau), \tag{10.7-62}$$

with magnitude spectrum

$$|Y(\eta)|=\frac{K}{s}|X(\eta/s)|, \tag{10.7-63}$$

where η represents output (received) frequencies in hertz, $K>0$, and $s>0$. The magnitude spectrum along the positive frequency axis of the transmitted signal $x(t)$ given by (10.7-52) is equal to

$$|X(f)|=\frac{A}{2}T\left|\text{sinc}[(f-f_c)T]\right|, \qquad f>0, \tag{10.7-64}$$

where f represents input (transmitted) frequencies in hertz. Substituting (10.7-64) into (10.7-63) yields

$$|Y(\eta)| = \frac{K}{s}\frac{A}{2}T\left|\text{sinc}\left[\left(\frac{\eta - sf_c}{s}\right)T\right]\right|, \qquad \eta > 0, \tag{10.7-65}$$

or, with the use of (10.7-56),

$$|Y(\eta)| = \frac{K}{s}\frac{A}{2}T\left|\text{sinc}\left[\left(\frac{\eta - [f_c + \eta_D]}{s}\right)T\right]\right|, \qquad \eta > 0. \tag{10.7-66}$$

Equation (10.7-65), or equivalently (10.7-66), is the magnitude spectrum along the positive frequency axis of the received signal $y(t)$ given by (10.7-60).

The zero-crossings of the sinc function in (10.7-65) occur at

$$\frac{\eta}{s} - f_c = \pm\frac{i}{T}\ \text{Hz}, \qquad i = 1, 2, \ldots, \tag{10.7-67}$$

or

$$\eta = s\left(f_c \pm \frac{i}{T}\right)\text{Hz}, \qquad i = 1, 2, \ldots. \tag{10.7-68}$$

Since $|\text{sinc}(\eta T)| < 0.1$ for $\eta > 3/T$, estimates of the maximum and minimum frequency components are given by

$$\eta_{\max} = s\left(f_c + \frac{\text{NZC}}{T}\right) \tag{10.7-69}$$

and

$$\eta_{\min} = s\left(f_c - \frac{\text{NZC}}{T}\right), \tag{10.7-70}$$

respectively, where NZC is the integer number of zero-crossings of the sinc function that is used to estimate both the maximum and minimum frequency components. The parameter NZC should be at least 3, with 5 being a conservative choice. Therefore, the bandwidth BW_y (in hertz) of the received signal $y(t)$ given by (10.7-60) is

$$\text{BW}_y = \eta_{\max} - \eta_{\min} = s\,2\,\text{NZC}/T, \tag{10.7-71}$$

or

$$\boxed{\text{BW}_y = s\,\text{BW}_x} \tag{10.7-72}$$

where

$$\text{BW}_x = f_{\max} - f_{\min} = 2\,\text{NZC}/T \tag{10.7-73}$$

is the bandwidth (in hertz) of the transmitted signal $x(t)$ given by (10.7-52) [see (10.7-64)]. Equation (10.7-72) is a general result for any transmitted signal $x(t)$ with bandwidth BW_x hertz when the received signal $y(t)$ is modeled by (10.7-54). From (10.7-72), and as was stated before, if $s = 1$ (neither time compression nor time expansion), $s > 1$ (time compression), or $0 < s < 1$ (time expansion), then the bandwidth of the received signal BW_y is equal to, greater than, or less than the bandwidth BW_x of the transmitted signal, respectively.

Let us conclude this example by proving (10.7-61) and (10.7-72) for any transmitted signal $x(t)$ when the received signal $y(t)$ is modeled by (10.7-54). If $x(t)$ is defined in the time interval $0 \le t \le T$, then [see (10.7-54)]

$$y(\tau) = Kx(0) \tag{10.7-74}$$

and

$$y\left(\tau + [T/s]\right) = Kx(T). \tag{10.7-75}$$

Therefore, the pulse length of the received signal T_y is given by (10.7-61). If the maximum and minimum frequency components of $x(t)$ are $f_{\max}$ and $f_{\min}$, respectively, then [see (10.7-63)]

$$|Y(sf_{\max})| = \frac{K}{s}|X(f_{\max})| \tag{10.7-76}$$

and

$$|Y(sf_{\min})| = \frac{K}{s}|X(f_{\min})|. \tag{10.7-77}$$

Therefore, the bandwidth of the received signal BW_y is given by (10.7-72). ■

Time-Varying Ranges and Angles of Incidence and Scatter

The transmitted acoustic field is *first* incident upon the scatterer at time instant t'. The time-varying position vector from transmitter to scatterer for $t \ge t'$ is defined as follows:

$$\boxed{\boldsymbol{\mathcal{R}}'_{TS}(t) \triangleq \mathbf{r}'_{TS} + \mathbf{V}_{ST}(t - t'), \qquad t \ge t'} \tag{10.7-78}$$

where

$$\mathbf{r}'_{TS} = \boldsymbol{\mathcal{R}}_{TS}(t') \tag{10.7-79}$$

and $\mathcal{R}_{TS}(t')$ is given by (10.7-16). Note that $\mathcal{R}'_{TS}(t') = \mathbf{r}'_{TS} = \mathcal{R}_{TS}(t')$. The time instant t' is given by [see (10.7-23)]

$$t' = t_0 + \frac{\left|\mathcal{R}_{TS}(t')\right|}{c}, \tag{10.7-80}$$

where t_0 is the time instant when transmission begins, and $\left|\mathcal{R}_{TS}(t')\right|$ is given by (10.7-29). We shall derive equations for the time-varying angles of incidence of the transmitted acoustic field at the scatterer next.

The time-varying, dimensionless unit vector $\hat{r}'_{TS}(t)$ in the direction of $\mathcal{R}'_{TS}(t)$ is given by

$$\boxed{\hat{r}'_{TS}(t) = \frac{\mathcal{R}'_{TS}(t)}{\left|\mathcal{R}'_{TS}(t)\right|} = u'_{TS}(t)\hat{x} + v'_{TS}(t)\hat{y} + w'_{TS}(t)\hat{z}, \qquad t \geq t'} \tag{10.7-81}$$

where

$$u'_{TS}(t) = \sin\theta'_{TS}(t)\cos\psi'_{TS}(t), \tag{10.7-82}$$

$$v'_{TS}(t) = \sin\theta'_{TS}(t)\sin\psi'_{TS}(t), \tag{10.7-83}$$

and

$$w'_{TS}(t) = \cos\theta'_{TS}(t) \tag{10.7-84}$$

are time-varying, dimensionless direction cosines with respect to the X, Y, and Z axes, respectively. Therefore, the *time-varying angles of incidence* of the transmitted acoustic field at the scatterer for $t \geq t'$ are given by

$$\boxed{\theta'_{TS}(t) = \cos^{-1} w'_{TS}(t), \qquad t \geq t'} \tag{10.7-85}$$

and

$$\boxed{\psi'_{TS}(t) = \tan^{-1}\left[v'_{TS}(t)/u'_{TS}(t)\right], \qquad t \geq t'} \tag{10.7-86}$$

where t' is given by (10.7-80). Note that $\theta'_{TS}(t')$ and $\psi'_{TS}(t')$ are the angles of incidence at the scatterer when the transmitted acoustic field is first incident upon the scatterer at time t'. Also note that if the transmitter and scatterer are *not* in motion, then $\mathbf{V}_T = \mathbf{0}$, $\mathbf{V}_S = \mathbf{0}$, $\mathbf{V}_{ST} = \mathbf{0}$, and as a result, $\mathcal{R}'_{TS}(t) = \mathbf{r}_{TS}$ [see (10.7-78), (10.7-79), and (10.7-16)]. Therefore, (10.7-81) reduces to $\hat{r}'_{TS}(t) = \mathbf{r}_{TS}/r_{TS} = \hat{r}_{TS}$, where $\hat{r}_{TS}$ is the dimensionless unit vector in the direction of the position vector from the transmitter to the scatterer when the transmitter and scatterer are not in motion [see (10.1-8) through (10.1-10) and Fig. 10.1-1]. When the transmitter and scatterer are not in motion, the angles of incidence at the scatterer, θ_{TS} and ψ_{TS}, are given by (10.1-13) and (10.1-14), respectively.

The scattered acoustic field is *first* incident upon the receiver at time instant t''. The time-varying position vector from scatterer to receiver for $t \ge t''$ is defined as follows:

$$\boxed{\mathcal{R}''_{SR}(t) \triangleq \mathbf{r}''_{SR} + \mathbf{V}_{RS}(t - t''), \qquad t \ge t''} \tag{10.7-87}$$

where

$$\mathbf{r}''_{SR} = \mathcal{R}'_{SR}(t'') \tag{10.7-88}$$

and $\mathcal{R}'_{SR}(t'')$ is given by (10.7-21). Note that $\mathcal{R}''_{SR}(t'') = \mathbf{r}''_{SR} = \mathcal{R}'_{SR}(t'')$. The time instant t'' is given by [see (10.7-22)]

$$t'' = t_0 + \tau , \tag{10.7-89}$$

where t_0 is the time instant when transmission begins, and τ is the time delay given by (10.7-25). We shall derive equations for the time-varying angles of scatter at the receiver next.

The time-varying, dimensionless unit vector $\hat{r}''_{SR}(t)$ in the direction of $\mathcal{R}''_{SR}(t)$ is given by

$$\boxed{\hat{r}''_{SR}(t) = \frac{\mathcal{R}''_{SR}(t)}{\left|\mathcal{R}''_{SR}(t)\right|} = u''_{SR}(t)\hat{x} + v''_{SR}(t)\hat{y} + w''_{SR}(t)\hat{z}, \qquad t \ge t''} \tag{10.7-90}$$

where $\mathcal{R}''_{SR}(t)$ is given by (10.7-87), and

$$u''_{SR}(t) = \sin\theta''_{SR}(t)\cos\psi''_{SR}(t) , \tag{10.7-91}$$

$$v''_{SR}(t) = \sin\theta''_{SR}(t)\sin\psi''_{SR}(t) , \tag{10.7-92}$$

and

$$w''_{SR}(t) = \cos\theta''_{SR}(t) \tag{10.7-93}$$

are time-varying, dimensionless direction cosines with respect to the X, Y, and Z axes, respectively. Therefore, the *time-varying angles of scatter* at the receiver for $t \geq t''$ are given by

$$\boxed{\theta''_{SR}(t) = \cos^{-1} w''_{SR}(t), \qquad t \geq t''} \tag{10.7-94}$$

and

$$\boxed{\psi''_{SR}(t) = \tan^{-1}\left[v''_{SR}(t)/u''_{SR}(t)\right], \qquad t \geq t''} \tag{10.7-95}$$

where t'' is given by (10.7-89). Note that $\theta''_{SR}(t'')$ and $\psi''_{SR}(t'')$ are the angles of scatter at the receiver when the scattered acoustic field is first incident upon the receiver at time t''. Also note that if the scatterer and receiver are *not* in motion, then $\mathbf{V}_S = \mathbf{0}$, $\mathbf{V}_R = \mathbf{0}$, $\mathbf{V}_{RS} = \mathbf{0}$, and as a result, $\boldsymbol{\mathcal{R}}''_{SR}(t) = \mathbf{r}_{SR}$ [see (10.7-87), (10.7-88), (10.7-21), (10.7-20), and (10.7-18)]. Therefore, (10.7-90) reduces to $\hat{r}''_{SR}(t) = \mathbf{r}_{SR}/r_{SR} = \hat{r}_{SR}$, where $\hat{r}_{SR}$ is the dimensionless unit vector in the direction of the position vector from the scatterer to the receiver when the scatterer and receiver are not in motion [see (10.1-25) through (10.1-27) and Fig. 10.1-1]. When the scatterer and receiver are not in motion, the angles of scatter at the receiver, θ_{SR} and ψ_{SR}, are given by (10.1-31) and (10.1-32), respectively.

10.7.2 Direct Path

The time-varying position vector from the transmitter to the receiver is given by

$$\mathcal{R}_{TR}(t) = \mathcal{R}_R(t) - \mathcal{R}_T(t), \tag{10.7-96}$$

and by substituting (10.7-4) and (10.7-6) into (10.7-96), we obtain

$$\mathcal{R}_{TR}(t) = \mathbf{r}_{TR} + \mathbf{V}_{RT}\,\Delta t, \qquad t \geq t_0, \tag{10.7-97}$$

where

$$\mathbf{r}_{TR} = \mathbf{r}_R - \mathbf{r}_T \tag{10.7-98}$$

is the position vector from the transmitter to the receiver when transmission begins at time $t = t_0$ seconds (see Fig. 10.7-2),

$$\mathbf{V}_{RT} = \mathbf{V}_R - \mathbf{V}_T = V_{RT}\,\hat{n}_{\mathbf{V}_{RT}} \tag{10.7-99}$$

is the velocity vector of the receiver *relative to* the velocity vector of the

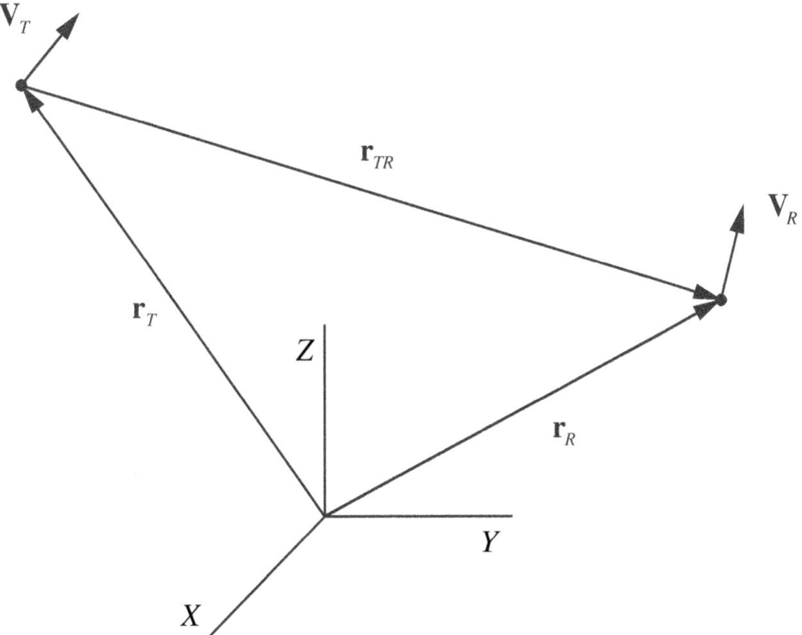

Figure 10.7-2 Direct path geometry when transmission begins at time $t = t_0$ seconds, where $\mathbf{V}_T$ and $\mathbf{V}_R$ are the velocity vectors of the transmitter and receiver, respectively.

transmitter, where $V_{RT} = |\mathbf{V}_{RT}|$ and $\hat{n}_{\mathbf{V}_{RT}}$ is the dimensionless unit vector in the direction of $\mathbf{V}_{RT}$, and Δt is given by (10.7-7). Note that $\mathcal{R}_{TR}(t_0) = \mathbf{r}_{TR}$.

Time Delay

When the transmitted acoustic field is *first* incident upon the receiver at some time t''' seconds, where $t''' > t_0$, the position vector from the transmitter to the receiver at time t''' is given by [see (10.7-97)]

$$\boxed{\mathcal{R}_{TR}(t''') = \mathbf{r}_{TR} + \mathbf{V}_{RT}\,\Delta t''', \qquad t''' > t_0} \tag{10.7-100}$$

where the position vector $\mathbf{r}_{TR}$ is given by (10.7-98), the relative velocity vector $\mathbf{V}_{RT}$ is given by (10.7-99), and

$$\Delta t''' = t''' - t_0, \qquad t''' > t_0. \tag{10.7-101}$$

Therefore, the time delay τ can be expressed as

$$\tau = \Delta t''' = t''' - t_0 . \qquad (10.7\text{-}102)$$

Substituting

$$\Delta t''' = t''' - t_0 = \frac{\left|\mathcal{R}_{TR}(t''')\right|}{c}, \qquad t''' > t_0 \qquad (10.7\text{-}103)$$

into (10.7-102) yields

$$\boxed{\tau = \frac{\left|\mathcal{R}_{TR}(t''')\right|}{c}} \qquad (10.7\text{-}104)$$

where c is the constant speed of sound in meters per second. Equation (10.7-104) is the *time delay* (or travel time) in seconds associated with the direct path when the transmitted acoustic field is *first* incident upon the receiver, In order to evaluate (10.7-104), we need to find the solution for the range $\left|\mathcal{R}_{TR}(t''')\right|$, which we shall do next.

Exact Solution for $\left|\mathcal{R}_{TR}(t''')\right|$

Since

$$\left|\mathcal{R}_{TR}(t''')\right|^2 = \mathcal{R}_{TR}(t''') \bullet \mathcal{R}_{TR}(t'''), \qquad (10.7\text{-}105)$$

substituting (10.7-100) into (10.7-105) yields

$$\left|\mathcal{R}_{TR}(t''')\right|^2 = r_{TR}^2 + 2r_{TR}\left(\hat{r}_{TR} \bullet \mathbf{V}_{RT}\right)\Delta t''' + V_{RT}^2\,(\Delta t''')^2, \qquad (10.7\text{-}106)$$

where $r_{TR} = \left|\mathbf{r}_{TR}\right|$ is given by (10.3-7), $\hat{r}_{TR}$ is the dimensionless unit vector in the direction of $\mathbf{r}_{TR}$, and $V_{RT} = \left|\mathbf{V}_{RT}\right|$, where $\mathbf{V}_{RT}$ is given by (10.7-99). Substituting (10.7-103) into (10.7-106) yields the second-order polynomial

$$A\left|\mathcal{R}_{TR}(t''')\right|^2 - B\left|\mathcal{R}_{TR}(t''')\right| - C = 0 \qquad (10.7\text{-}107)$$

with *exact* solution

$$\boxed{\left|\mathcal{R}_{TR}(t''')\right| = \frac{B + \sqrt{B^2 + 4AC}}{2A}} \qquad (10.7\text{-}108)$$

where

$$A = 1 - (V_{RT}/c)^2, \tag{10.7-109}$$

$$B = 2r_{TR}\frac{\hat{r}_{TR} \bullet \mathbf{V}_{RT}}{c}, \tag{10.7-110}$$

and

$$C = r_{TR}^2. \tag{10.7-111}$$

Equation (10.7-108) is the *exact* range between the transmitter and receiver when the transmitted acoustic field is *first* incident upon the receiver at time t''', where $t''' > t_0$. For the problems that we are interested in,

$$V_{RT}/c << 1, \tag{10.7-112}$$

and as a result, $A > 0$. The decision to use the plus sign in front of the square-root term in (10.7-108) instead of the minus sign was dictated by the fact that since $4AC > 0$, the square-root term is greater than B, and that the range $|\mathcal{R}_{TR}(t''')|$ must be positive. Note that if the transmitter and receiver are *not* in motion, then $\mathbf{V}_T = \mathbf{0}$, $\mathbf{V}_R = \mathbf{0}$, $\mathbf{V}_{RT} = \mathbf{0}$, $V_{RT} = 0$, and as expected, (10.7-108) reduces to $|\mathcal{R}_{TR}(t''')| = r_{TR}$.

Time-Compression/Time-Expansion Factor and Doppler Shift

The dimensionless *time-compression/time-expansion factor* s is defined as [see (10.7-42)]

$$\boxed{s \triangleq \tau_0/\tau} \tag{10.7-113}$$

where for the direct path

$$\tau_0 = r_{TR}/c \tag{10.7-114}$$

is the *one-way* time delay in seconds when there is *no* motion, and τ is given by (10.7-104), where the range $|\mathcal{R}_{TR}(t''')|$ is given by (10.7-108). Note that $s > 0$. The *Doppler shift* η_D in hertz is related to the dimensionless time-compression/time-expansion factor s and is defined as [see (10.7-44)]

$$\boxed{\eta_D \triangleq (s-1) f_c} \tag{10.7-115}$$

where f_c is referred to as either the *center frequency* or *carrier frequency* in hertz of the transmitted electrical signal (see Chapter 11).

Recall from the discussion of the scattered path that there are three ranges of values for s; namely, $s=1$, $s>1$, and $0<s<1$. When there is *no* motion, $|\mathcal{R}_{TR}(t''')|=r_{TR}$, $\tau=\tau_0$, $s=1$, and $\eta_D=0$, that is, there is *zero* Doppler shift when there is no motion (and no relative motion). When $s=1$, there is *neither* time compression *nor* time expansion, that is, the duration of the received signal is *equal to* the duration of the transmitted signal resulting in *no* change in signal bandwidth at the receiver. If $\tau<\tau_0$, then $s>1$ (*time compression*) and $\eta_D>0$ (*positive* Doppler shift). Time compression means that the duration of the received signal is *less than* the duration of the transmitted signal resulting in an *increase* in signal bandwidth. A positive Doppler shift means that the transmitter and receiver are moving *towards* one another. If $\tau>\tau_0$, then $0<s<1$ (*time expansion*) and $\eta_D<0$ (*negative* Doppler shift). Time expansion means that the duration of the received signal is *greater than* the duration of the transmitted signal resulting in a *decrease* in signal bandwidth. A negative Doppler shift means that the transmitter and receiver are moving *away* from one another.

Time-Varying Range and Angles of Incidence

The transmitted acoustic field is *first* incident upon the receiver at time instant t'''. The time-varying position vector from transmitter to receiver for $t\geq t'''$ is defined as follows:

$$\boxed{\mathcal{R}'''_{TR}(t)\triangleq\mathbf{r}'''_{TR}+\mathbf{V}_{RT}(t-t'''),\qquad t\geq t'''} \tag{10.7-116}$$

where

$$\mathbf{r}'''_{TR}=\mathcal{R}_{TR}(t''') \tag{10.7-117}$$

and $\mathcal{R}_{TR}(t''')$ is given by (10.7-100). Note that $\mathcal{R}'''_{TR}(t''')=\mathbf{r}'''_{TR}=\mathcal{R}_{TR}(t''')$. The time instant t''' is given by [see (10.7-102)]

$$t'''=t_0+\tau, \tag{10.7-118}$$

where t_0 is the time instant when transmission begins, and τ is the time delay given by (10.7-104). We shall derive equations for the time-varying angles of incidence of the transmitted acoustic field at the receiver next.

The time-varying, dimensionless unit vector $\hat{r}'''_{TR}(t)$ in the direction of $\mathcal{R}'''_{TR}(t)$ is given by

$$\boxed{\hat{r}'''_{TR}(t)=\frac{\mathcal{R}'''_{TR}(t)}{|\mathcal{R}'''_{TR}(t)|}=u'''_{TR}(t)\hat{x}+v'''_{TR}(t)\hat{y}+w'''_{TR}(t)\hat{z},\qquad t\geq t'''} \tag{10.7-119}$$

where $\mathcal{R}'''_{TR}(t)$ is given by (10.7-116), and

$$u'''_{TR}(t) = \sin\theta'''_{TR}(t)\cos\psi'''_{TR}(t), \tag{10.7-120}$$

$$v'''_{TR}(t) = \sin\theta'''_{TR}(t)\sin\psi'''_{TR}(t), \tag{10.7-121}$$

and

$$w'''_{TR}(t) = \cos\theta'''_{TR}(t) \tag{10.7-122}$$

are time-varying, dimensionless direction cosines with respect to the X, Y, and Z axes, respectively. Therefore, the *time-varying angles of incidence* of the transmitted acoustic field at the receiver for $t \geq t'''$ are given by

$$\boxed{\theta'''_{TR}(t) = \cos^{-1} w'''_{TR}(t), \qquad t \geq t'''} \tag{10.7-123}$$

and

$$\boxed{\psi'''_{TR}(t) = \tan^{-1}\left[v'''_{TR}(t) / u'''_{TR}(t) \right], \qquad t \geq t'''} \tag{10.7-124}$$

where t''' is given by (10.7-118). Note that $\theta'''_{TR}(t''')$ and $\psi'''_{TR}(t''')$ are the angles of incidence at the receiver when the transmitted acoustic field is first incident upon the receiver at time t'''. If the transmitter and receiver are *not* in motion, then $\mathbf{V}_T = \mathbf{0}$, $\mathbf{V}_R = \mathbf{0}$, $\mathbf{V}_{RT} = \mathbf{0}$, and as a result, $\mathcal{R}'''_{TR}(t) = \mathbf{r}_{TR}$ [see (10.7-116), (10.7-117), and (10.7-100)]. Therefore, (10.7-119) reduces to $\hat{r}'''_{TR}(t) = \mathbf{r}_{TR} / r_{TR} = \hat{r}_{TR}$, where $\hat{r}_{TR}$ is the dimensionless unit vector in the direction of the position vector from the transmitter to the receiver when the transmitter and receiver are not in motion [see (10.3-6) through (10.3-8) and Fig. 10.3-1]. When the transmitter and receiver are not in motion, the angles of incidence at the receiver, θ_{TR} and ψ_{TR}, are given by (10.3-11) and (10.3-12), respectively.

Problems

Section 10.4

10-1 A side-looking sonar (SLS) is being used to detect a mine on the ocean bottom. The frequency of operation and source level of the SLS are $30\ \text{kHz}$ and $215\ \text{dB re } 1\ \mu\text{Pa (rms)}$, respectively, and the range from the SLS to the mine is $218\ \text{m}$. If the sound-pressure level at the SLS after scatter from the mine is $98.79\ \text{dB re } 1\ \mu\text{Pa (rms)}$, the attenuation coefficient is $6.14\ \text{dB/km}$, and we treat this problem as a backscatter problem, then

(a) find the target strength of the mine.

(b) What is the differential scattering cross-section of the mine?

(c) What is the *magnitude* of the scattering function of the mine?

(d) What is the backscattering cross-section of the mine?

(e) What is the effective radius of the mine based on the backscattering cross-section?

(f) What is the transmission loss?

10-2 The range between a sound-source in the ocean and a receiver is 10 km. Find the transmission loss at the receiver if

(a) the frequency of operation of the sound-source is 200 Hz and the attenuation coefficient is 3.41×10^{-3} dB/km.

(b) the frequency of operation of the sound-source is 1 kHz and the attenuation coefficient is 5.7×10^{-2} dB/km.

(c) the frequency of operation of the sound-source is 10 kHz and the attenuation coefficient is 0.849 dB/km.

Section 10.7

10-3 Compute the Fourier transform of the received signal given by (10.7-54) using η as the output frequency in hertz. Compare your answer with (10.7-62).

Appendix 10A Radiation from a Time-Harmonic, Omnidirectional Point-Source

In order to derive an equation for the target strength of an object for the bistatic scattering problem shown in Fig. 10.1-1, we need some basic equations that describe the acoustic field radiated by a time-harmonic, omnidirectional point-source. An exact solution of the linear wave equation given by (10.1-2) for the source distribution given by (10.1-1) for free-space propagation in an ideal (nonviscous), homogeneous, fluid medium was derived in Appendix 6C and is summarized below:

$$\varphi(t,\mathbf{r}) = \varphi_f(\mathbf{r})\exp(+j2\pi ft), \tag{10A-1}$$

where $\varphi(t,\mathbf{r})$ is the time-harmonic, scalar velocity potential in squared meters per second of the radiated acoustic field,

$$\varphi_f(\mathbf{r}) = S_0\, g_f\left(\mathbf{r}\,|\,\mathbf{r}_0\right), \tag{10A-2}$$

$$S_0 = A_x \mathcal{S}_T(f) \tag{10A-3}$$

is the source strength of the omnidirectional point-source in cubic meters per second at frequency f hertz, A_x is the complex amplitude in volts of the time-harmonic, input electrical signal applied to the omnidirectional point-source, $\mathcal{S}_T(f)$ is the complex, transmitter sensitivity function of the omnidirectional point-source in $\left(\mathrm{m}^3/\mathrm{sec}\right)/\mathrm{V}$ (see Table 6.1-2 and Appendix 6B),

$$g_f\left(\mathbf{r}\,|\,\mathbf{r}_0\right) = -\frac{\exp\left(-jk\left|\mathbf{r}-\mathbf{r}_0\right|\right)}{4\pi\left|\mathbf{r}-\mathbf{r}_0\right|} = -\frac{\exp(-jkR)}{4\pi R} \tag{10A-4}$$

is the time-independent, free-space, Green's function of an unbounded, ideal (nonviscous), homogeneous, fluid medium with units of inverse meters,

$$k = 2\pi f/c = 2\pi/\lambda \tag{10A-5}$$

is the wavenumber in radians per meter, $c = f\lambda$ is the constant speed of sound in the fluid medium in meters per second, λ is the wavelength in meters,

$$\mathbf{r} = x\hat{x} + y\hat{y} + z\hat{z} \tag{10A-6}$$

is the position vector to a field point,

$$\mathbf{r}_0 = x_0\hat{x} + y_0\hat{y} + z_0\hat{z} \tag{10A-7}$$

is the position vector to the point-source, and

$$R = \left|\mathbf{r}-\mathbf{r}_0\right| = \sqrt{(x-x_0)^2 + (y-y_0)^2 + (z-z_0)^2} \tag{10A-8}$$

is the range in meters between the point-source and a field point. Frequency-dependent attenuation due to viscosity can be taken into account by replacing the real wavenumber k in (10A-4) with the complex wavenumber

$$K = k - j\alpha(f), \tag{10A-9}$$

where K has units of inverse meters and $\alpha(f)$ is the real, nonnegative, frequency-dependent, attenuation coefficient of seawater in nepers per meter (see Section 7.2 and Appendix 7A). Doing so yields

$$\boxed{g_f\left(\mathbf{r}\,\middle|\,\mathbf{r}_0\right)=-\frac{\exp\left(-jK\left|\mathbf{r}-\mathbf{r}_0\right|\right)}{4\pi\left|\mathbf{r}-\mathbf{r}_0\right|}=-\frac{\exp(-jKR)}{4\pi R}} \tag{10A-10}$$

or

$$\boxed{g_f\left(\mathbf{r}\,\middle|\,\mathbf{r}_0\right)=-\frac{\exp\left[-\alpha(f)\left|\mathbf{r}-\mathbf{r}_0\right|\right]}{4\pi\left|\mathbf{r}-\mathbf{r}_0\right|}\exp\left(-jk\left|\mathbf{r}-\mathbf{r}_0\right|\right)=-\frac{\exp[-\alpha(f)R]}{4\pi R}\exp(-jkR)} \tag{10A-11}$$

Note that if there is no viscosity, that is, if $\alpha(f)=0$, then (10A-9) reduces to (10A-5) and both (10A-10) and (10A-11) reduce to (10A-4).

The corresponding solution for the time-harmonic, radiated acoustic pressure $p(t,\mathbf{r})$ in pascals can be obtained from the scalar velocity potential $\varphi(t,\mathbf{r})$ in squared meters per second as follows:

$$p(t,\mathbf{r})=-\rho_0(\mathbf{r})\frac{\partial}{\partial t}\varphi(t,\mathbf{r})=-\rho_0\frac{\partial}{\partial t}\varphi(t,\mathbf{r}), \tag{10A-12}$$

and by substituting (10A-1) into (10A-12), we obtain

$$p(t,\mathbf{r})=p_f(\mathbf{r})\exp(+j2\pi ft), \tag{10A-13}$$

where

$$p_f(\mathbf{r})=-j2\pi f\rho_0\varphi_f(\mathbf{r})=-jk\rho_0 c\varphi_f(\mathbf{r}), \tag{10A-14}$$

k is the real wavenumber given by (10A-5), ρ_0 is the constant ambient (equilibrium) density of the fluid medium in kilograms per cubic meter, the factor $\rho_0 c$ is the characteristic impedance of the fluid medium in rayls ($1\text{ rayl}=1\text{ Pa-sec/m}$), $\varphi_f(\mathbf{r})$ is given by (10A-2), and $g_f\left(\mathbf{r}\,\middle|\,\mathbf{r}_0\right)$ is given by either (10A-10) or (10A-11). We also need an equation for the time-average intensity vector of the radiated acoustic field, which requires equations for the acoustic pressure – which we have [(10A-13) and (10A-14)] – and the acoustic fluid-velocity-vector (a.k.a. the acoustic particle-velocity-vector). Therefore, we shall compute the acoustic fluid-velocity-vector next.

The acoustic fluid-velocity-vector $\mathbf{u}(t,\mathbf{r})$ in meters per second can be obtained from the scalar velocity potential $\varphi(t,\mathbf{r})$ in squared meters per second as follows:

$$\mathbf{u}(t,\mathbf{r}) = \nabla\varphi(t,\mathbf{r}), \tag{10A-15}$$

where

$$\nabla = \frac{\partial}{\partial x}\hat{x} + \frac{\partial}{\partial y}\hat{y} + \frac{\partial}{\partial z}\hat{z} \tag{10A-16}$$

is the gradient expressed in the rectangular coordinates (x, y, z). Substituting (10A-1) and (10A-2) into (10A-15) yields

$$\mathbf{u}(t,\mathbf{r}) = S_0 \nabla\left[g_f\left(\mathbf{r}\,|\,\mathbf{r}_0\right)\right]\exp(+j2\pi ft). \tag{10A-17}$$

In Appendix 10B it is shown that the gradient of the Green's function given by (10A-10) is

$$\nabla g_f\left(\mathbf{r}\,|\,\mathbf{r}_0\right) = -\left(\frac{1}{R} + jK\right) g_f\left(\mathbf{r}\,|\,\mathbf{r}_0\right)\hat{R}, \tag{10A-18}$$

where

$$\hat{R} = \mathbf{R}/R \tag{10A-19}$$

is the dimensionless unit vector in the direction of the vector

$$\mathbf{R} = \mathbf{r} - \mathbf{r}_0 = (x - x_0)\hat{x} + (y - y_0)\hat{y} + (z - z_0)\hat{z}, \tag{10A-20}$$

and $R = |\mathbf{R}|$ is given by (10A-8). Substituting (10A-18) into (10A-17) yields the time-harmonic, radiated, acoustic fluid-velocity-vector in meters per second

$$\mathbf{u}(t,\mathbf{r}) = \mathbf{u}_f(\mathbf{r})\exp(+j2\pi ft), \tag{10A-21}$$

where

$$\mathbf{u}_f(\mathbf{r}) = -\left(\frac{1}{R} + jK\right)\varphi_f(\mathbf{r})\hat{R} \tag{10A-22}$$

and $\varphi_f(\mathbf{r})$ is given by (10A-2). We are now in a position to compute the time-average intensity vector.

Recall from Section 4.1 that the time-average intensity vector in watts per squared meter for time-harmonic acoustic fields is given by

$$\mathbf{I}_{\text{avg}}(\mathbf{r}) = \frac{1}{2}\text{Re}\left\{p_f(\mathbf{r})\mathbf{u}_f^*(\mathbf{r})\right\}, \tag{10A-23}$$

where $p_f(\mathbf{r})$ and $\mathbf{u}_f(\mathbf{r})$ are the spatial-dependent parts of the time-harmonic acoustic pressure and acoustic fluid-velocity-vector, respectively, and the asterisk denotes complex conjugate. Substituting (10A-14) and (10A-22) into (10A-23) yields

$$\mathbf{I}_{\text{avg}}(\mathbf{r}) = I_{\text{avg}}(\mathbf{r})\hat{R}, \tag{10A-24}$$

where

$$I_{\text{avg}}(\mathbf{r}) = \frac{1}{2}k^2 \rho_0 c \left|\varphi_f(\mathbf{r})\right|^2 \tag{10A-25}$$

is the magnitude of the time-average intensity vector of the radiated acoustic field, and $\varphi_f(\mathbf{r})$ is given by (10A-2). Since

$$\left|p_f(\mathbf{r})\right|^2 = k^2(\rho_0 c)^2\left|\varphi_f(\mathbf{r})\right|^2, \tag{10A-26}$$

where $p_f(\mathbf{r})$ is given by (10A-14), (10A-25) can be rewritten as

$$I_{\text{avg}}(\mathbf{r}) = \frac{\left|p_f(\mathbf{r})\right|^2}{2\rho_0 c}. \tag{10A-27}$$

And since

$$\left|\varphi_f(\mathbf{r})\right|^2 = \left(\frac{|S_0|}{4\pi R}\right)^2 \exp[-2\alpha(f)R], \tag{10A-28}$$

where $\varphi_f(\mathbf{r})$ is given by (10A-2) and $g_f\left(\mathbf{r}\middle|\mathbf{r}_0\right)$ given by (10A-11) was used, substituting (10A-28) into (10A-25) and (10A-26) yields

$$I_{\text{avg}}(\mathbf{r}) = \frac{1}{2}k^2 \rho_0 c \left(\frac{|S_0|}{4\pi R}\right)^2 \exp[-2\alpha(f)R] \tag{10A-29}$$

and

$$\left|p_f(\mathbf{r})\right|^2 = k^2(\rho_0 c)^2 \left(\frac{|S_0|}{4\pi R}\right)^2 \exp[-2\alpha(f)R], \tag{10A-30}$$

respectively.

The next set of equations that we need are equations for the spherical angles θ and ψ that describe the direction of the unit vector $\hat{R}$ as measured from the location of the omnidirectional point-source to a field point. Substituting (10A-20) into (10A-19) yields

$$\hat{R} = \frac{x - x_0}{R}\hat{x} + \frac{y - y_0}{R}\hat{y} + \frac{z - z_0}{R}\hat{z}. \tag{10A-31}$$

The unit vector $\hat{R}$ can also be expressed as

$$\hat{R} = u\hat{x} + v\hat{y} + w\hat{z}, \tag{10A-32}$$

where

$$u = \sin\theta\cos\psi, \tag{10A-33}$$

$$v = \sin\theta\sin\psi, \tag{10A-34}$$

and

$$w = \cos\theta \tag{10A-35}$$

are dimensionless direction cosines with respect to the X, Y, and Z axes, respectively. Equating the right-hand sides of (10A-31) and (10A-32) yields

$$u = \frac{x - x_0}{R}, \tag{10A-36}$$

$$v = \frac{y - y_0}{R}, \tag{10A-37}$$

and

$$w = \frac{z - z_0}{R}. \tag{10A-38}$$

Therefore, if the origin of a Cartesian coordinate system (with the same orientation as the one used to measure $\mathbf{r}_0$ and $\mathbf{r}$) is placed at $\mathbf{r}_0$, then the spherical angles θ and ψ, which are measured as shown in Fig. 1.2-2, can be computed as follows:

$$\theta = \cos^{-1} w = \cos^{-1}\left(\frac{z - z_0}{R}\right) \tag{10A-39}$$

and

$$\psi = \tan^{-1}\left(\frac{v}{u}\right) = \tan^{-1}\left(\frac{y - y_0}{x - x_0}\right). \tag{10A-40}$$

The last equation that we need is the equation for the time-average power radiated by a time-harmonic, omnidirectional point-source. If we enclose the time-harmonic, omnidirectional point-source at $\mathbf{r}_0$ with a sphere of radius R meters, where R is given by (10A-8), then the time-average radiated power in watts is given by

$$P_{\text{avg}} = \oint_S \mathbf{I}_{\text{avg}}(\mathbf{r}) \bullet \mathbf{dS}, \tag{10A-41}$$

where

$$\mathbf{dS} = R^2 \sin\theta \, d\theta \, d\psi \, \hat{R}. \tag{10A-42}$$

Substituting (10A-24), (10A-29), and (10A-42) into (10A-41) yields

$$P_{\text{avg}} = \frac{1}{2} k^2 \rho_0 c \left(\frac{|S_0|}{4\pi} \right)^2 \exp[-2\alpha(f)R] \int_0^{2\pi} \int_0^{\pi} \sin\theta \, d\theta \, d\psi , \tag{10A-43}$$

and since

$$\int_0^{2\pi} \int_0^{\pi} \sin\theta \, d\theta \, d\psi = 4\pi , \tag{10A-44}$$

(10A-43) reduces to

$$P_{\text{avg}} = \frac{1}{2} k^2 \rho_0 c \frac{|S_0|^2}{4\pi} \exp[-2\alpha(f)R] = 4\pi R^2 I_{\text{avg}}(\mathbf{r}), \tag{10A-45}$$

or

$$I_{\text{avg}}(\mathbf{r}) = \frac{P_{\text{avg}}}{4\pi R^2}, \tag{10A-46}$$

where $4\pi R^2$ is the surface area (in squared meters) of a sphere with radius R meters. Furthermore, by solving for the magnitude-squared of the source strength using (10A-30), we obtain

$$|S_0|^2 = \frac{(4\pi)^2 R^2}{k^2 (\rho_0 c)^2} \exp[+2\alpha(f)R] |p_f(\mathbf{r})|^2 , \tag{10A-47}$$

and if we let

$$P_0 = |p(t, \mathbf{r})|_{R=R_0=1\,\text{m}} = |p_f(\mathbf{r})|_{R=R_0=1\,\text{m}} , \tag{10A-48}$$

then

$$|S_0|^2 = \frac{(4\pi)^2 R_0^2}{k^2 (\rho_0 c)^2} \exp[+2\alpha(f)R_0] P_0^2 \Big|_{R_0=1\,\text{m}} . \tag{10A-49}$$

Substituting (10A-49) into (10A-45) yields

$$P_{\text{avg}} = 4\pi R_0^2 \frac{P_0^2}{2\rho_0 c} \exp\left[-2\alpha(f)(R - R_0)\right] \Big|_{R_0=1\,\text{m}} . \tag{10A-50}$$

Appendix 10B Gradient of the Time-Independent, Free-Space, Green's Function

In this Appendix we shall compute $\nabla g_f(\mathbf{r}|\mathbf{r}_0)$, where

$$\nabla = \frac{\partial}{\partial x}\hat{x} + \frac{\partial}{\partial y}\hat{y} + \frac{\partial}{\partial z}\hat{z} \tag{10B-1}$$

is the gradient expressed in the rectangular coordinates (x, y, z),

$$g_f(\mathbf{r}|\mathbf{r}_0) = -\frac{\exp(-jK|\mathbf{r}-\mathbf{r}_0|)}{4\pi|\mathbf{r}-\mathbf{r}_0|} = -\frac{\exp(-jKR)}{4\pi R} \tag{10B-2}$$

is the time-independent, free-space, Green's function with units of inverse meters,

$$K = k - j\alpha(f) \tag{10B-3}$$

is the complex wavenumber in inverse meters, and

$$R = |\mathbf{r}-\mathbf{r}_0| = \left[(x-x_0)^2 + (y-y_0)^2 + (z-z_0)^2\right]^{1/2} \tag{10B-4}$$

is the range in meters between source and field points. Expanding $\nabla g_f(\mathbf{r}|\mathbf{r}_0)$ using (10B-1) and (10B-2) yields

$$\begin{aligned} \nabla g_f(\mathbf{r}|\mathbf{r}_0) &= \nabla\left[-\frac{1}{4\pi R}\exp(-jKR)\right] \\ &= -\frac{1}{4\pi R}\nabla[\exp(-jKR)] + \exp(-jKR)\nabla\left(-\frac{1}{4\pi R}\right), \end{aligned} \tag{10B-5}$$

where

$$\nabla\exp(-jKR) = \frac{\partial}{\partial x}\exp(-jKR)\hat{x} + \frac{\partial}{\partial y}\exp(-jKR)\hat{y} + \frac{\partial}{\partial z}\exp(-jKR)\hat{z}, \tag{10B-6}$$

$$\nabla\left(-\frac{1}{4\pi R}\right) = -\frac{1}{4\pi}\nabla\frac{1}{R}, \tag{10B-7}$$

and

$$\nabla \frac{1}{R} = \frac{\partial}{\partial x}\frac{1}{R}\hat{x} + \frac{\partial}{\partial y}\frac{1}{R}\hat{y} + \frac{\partial}{\partial z}\frac{1}{R}\hat{z}. \tag{10B-8}$$

Next we shall compute $\nabla \exp(-jKR)$ as given by (10B-6).

Since

$$\frac{\partial}{\partial x}\exp(-jKR) = -jK\exp(-jKR)\frac{\partial R}{\partial x}, \tag{10B-9}$$

$$\frac{\partial}{\partial y}\exp(-jKR) = -jK\exp(-jKR)\frac{\partial R}{\partial y}, \tag{10B-10}$$

and

$$\frac{\partial}{\partial z}\exp(-jKR) = -jK\exp(-jKR)\frac{\partial R}{\partial z}, \tag{10B-11}$$

substituting (10B-9) through (10B-11) into (10B-6) yields

$$\nabla \exp(-jKR) = -jK\exp(-jKR)\nabla R, \tag{10B-12}$$

where

$$\nabla R = \frac{\partial R}{\partial x}\hat{x} + \frac{\partial R}{\partial y}\hat{y} + \frac{\partial R}{\partial z}\hat{z}. \tag{10B-13}$$

From (10B-4) it can be seen that

$$\frac{\partial R}{\partial x} = \frac{x - x_0}{R}, \tag{10B-14}$$

$$\frac{\partial R}{\partial y} = \frac{y - y_0}{R}, \tag{10B-15}$$

and

$$\frac{\partial R}{\partial z} = \frac{z - z_0}{R}. \tag{10B-16}$$

Substituting (10B-14) through (10B-16) into (10B-13) yields

$$\boxed{\nabla R = \hat{R} = \mathbf{R}/R} \tag{10B-17}$$

where $\hat{R}$ is the dimensionless unit vector in the direction of the vector

$$\mathbf{R} = \mathbf{r} - \mathbf{r}_0 = (x - x_0)\hat{x} + (y - y_0)\hat{y} + (z - z_0)\hat{z}, \tag{10B-18}$$

and $R = |\mathbf{R}|$ is given by (10B-4). And by substituting (10B-17) into (10B-12), we obtain

$$\boxed{\nabla \exp(-jKR) = -jK \exp(-jKR)\,\hat{R}} \tag{10B-19}$$

Next we shall compute $\nabla(1/R)$ as given by (10B-8).

Since

$$\frac{\partial}{\partial x}\frac{1}{R} = -\frac{1}{R^2}\frac{\partial R}{\partial x}, \tag{10B-20}$$

$$\frac{\partial}{\partial y}\frac{1}{R} = -\frac{1}{R^2}\frac{\partial R}{\partial y}, \tag{10B-21}$$

and

$$\frac{\partial}{\partial z}\frac{1}{R} = -\frac{1}{R^2}\frac{\partial R}{\partial z}, \tag{10B-22}$$

substituting (10B-20) through (10B-22), and (10B-17) into (10B-8) yields

$$\boxed{\nabla \frac{1}{R} = -\frac{1}{R^2}\nabla R = -\frac{1}{R^2}\hat{R}} \tag{10B-23}$$

and by substituting (10B-7), (10B-19), and (10B-23) into (10B-5), we finally obtain

$$\boxed{\nabla g_f(\mathbf{r}\,|\,\mathbf{r}_0) = -\left(\frac{1}{R} + jK\right) g_f(\mathbf{r}\,|\,\mathbf{r}_0)\,\hat{R}} \tag{10B-24}$$

Chapter 11

Real Bandpass Signals and Complex Envelopes

11.1 Definitions and Basic Relationships

Real-world transmitted electrical signals used in *active* sonar and radar systems, and communication systems, belong to a class of signals known as *amplitude-and-angle-modulated carriers*. Amplitude-and-angle-modulated carriers are *real bandpass signals*. In this section we shall discuss how to represent a real bandpass signal in terms of its *complex envelope*. As will be shown later, a complex envelope is, in general, a *complex signal* with a *lowpass (baseband)* frequency spectrum. One of the main advantages of expressing real bandpass signals in terms of their complex envelopes is that it provides for a simple representation of amplitude-and-angle-modulated carriers, which is very useful for doing analysis. For example, in Chapter 12, we will derive the *auto-ambiguity function*, which is used as a measure of the range and Doppler resolving capabilities of different transmitted electrical signals (amplitude-and-angle-modulated carriers) used in active sonar and radar systems. The complex envelope of a signal is required in order to compute its ambiguity function.

Let $x(t)$ be a *real bandpass signal* with magnitude spectrum $|X(f)|$ centered at $f = \pm f_c$ hertz (see Fig. 11.1-1), where

$$X(f) = F\{x(t)\} = \int_{-\infty}^{\infty} x(t)\exp(-j2\pi ft)\,dt \tag{11.1-1}$$

and f_c is referred to as either the *center frequency* or *carrier frequency*. If $x(t)$ has units of volts (or amperes), then $X(f)$ will have units of volts (or amperes) per hertz. As shown in Fig. 11.1-1, the bandwidth of $x(t)$ is $2W$ hertz. Bandwidth is always measured along the *positive* frequency axis. There are two kinds of bandpass signals, *narrowband* and *broadband* (a.k.a. wideband). A narrowband bandpass signal satisfies the inequality $2W/f_c << 1$, or equivalently, $2W/f_c \le 0.1$. Therefore, a broadband (wideband) bandpass signal satisfies the inequality $2W/f_c > 0.1$.

The *complex envelope* of $x(t)$, denoted by $\tilde{x}(t)$, is defined as follows:

$$\boxed{\tilde{x}(t) \triangleq x_p(t)\exp(-j2\pi f_c t)} \tag{11.1-2}$$

where

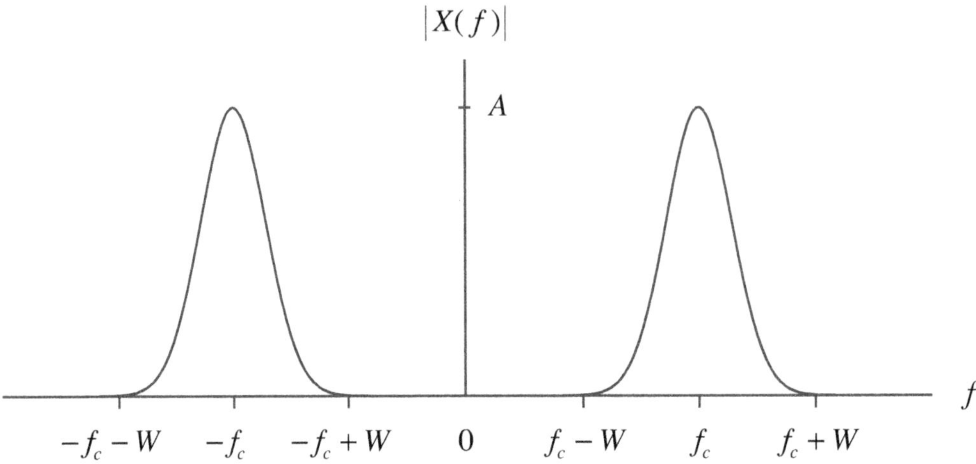

Figure 11.1-1 Magnitude spectrum of a real bandpass signal $x(t)$ with bandwidth $2W$ hertz.

$$\boxed{x_p(t) = x(t) + j\hat{x}(t)} \tag{11.1-3}$$

is the *pre-envelope* (a.k.a. the *analytic signal*) of $x(t)$, and $\hat{x}(t)$ is the *Hilbert transform* of $x(t)$. Solving for $x_p(t)$ using (11.1-2) yields

$$x_p(t) = \tilde{x}(t)\exp(+j2\pi f_c t), \tag{11.1-4}$$

and from (11.1-3),

$$x(t) = \text{Re}\{x_p(t)\}, \tag{11.1-5}$$

where Re means "take the real part." Substituting (11.1-4) into (11.1-5) yields

$$\boxed{x(t) = \text{Re}\{\tilde{x}(t)\exp(+j2\pi f_c t)\}} \tag{11.1-6}$$

Therefore, given a complex envelope $\tilde{x}(t)$, the corresponding real bandpass signal $x(t)$ can be obtained from (11.1-6).

The *envelope* of $x(t)$, denoted by $\mathcal{E}(t)$, is defined as follows:

$$\boxed{\mathcal{E}(t) \triangleq \text{Abs}\{|\tilde{x}(t)|\} \geq 0} \tag{11.1-7}$$

where Abs means "take the absolute value." At first glance it would seem that taking the absolute value of the magnitude of a complex signal is redundant.

However, since the magnitude of the complex envelope is, in general, a real-valued function of time that can take on both positive and negative values, in order to ensure that the envelope $\mathcal{E}(t)$ is nonnegative (analogous to the output from an envelope detector), we must also take the absolute value.

As can be seen from (11.1-2) and (11.1-3), in order to compute the complex envelope of $x(t)$, we first need to compute the Hilbert transform of $x(t)$. One way to compute the Hilbert transform is to perform the following inverse Fourier transform:

$$\boxed{\hat{x}(t) = F^{-1}\{-j\operatorname{sgn}(f)X(f)\}} \tag{11.1-8}$$

where

$$\operatorname{sgn}(f) = \begin{cases} 1, & f > 0 \\ 0, & f = 0 \\ -1, & f < 0 \end{cases} \tag{11.1-9}$$

is the *signum* or *sign function*, and $X(f)$ is the Fourier transform of $x(t)$ given by (11.1-1). Therefore, from (11.1-8), the Fourier transform of the Hilbert transform of $x(t)$ is given by

$$\boxed{\hat{X}(f) = F\{\hat{x}(t)\} = -j\operatorname{sgn}(f)X(f)} \tag{11.1-10}$$

The Hilbert transform is basically a 90° phase-shifter. This can easily be shown by using (11.1-8) and (11.1-9). For positive frequencies ($f > 0$),

$$\hat{x}(t) = F^{-1}\{-jX(f)\} = F^{-1}\{\exp(-j\pi/2)X(f)\}, \qquad f > 0, \tag{11.1-11}$$

which corresponds to a $-90°$ phase shift, and for negative frequencies ($f < 0$),

$$\hat{x}(t) = F^{-1}\{+jX(f)\} = F^{-1}\{\exp(+j\pi/2)X(f)\}, \qquad f < 0, \tag{11.1-12}$$

which corresponds to a $+90°$ phase shift. Therefore, the magnitude spectrum of a bandpass signal is not altered by a Hilbert transform – only the phase spectrum is. If we denote the Hilbert transform by $H\{\bullet\}$, then it can be shown that

$$H\{\cos(2\pi f_c t + \theta_0)\} = \sin(2\pi f_c t + \theta_0) = \cos\left(2\pi f_c t + \theta_0 - \frac{\pi}{2}\right) \tag{11.1-13}$$

and

$$H\{\sin(2\pi f_c t + \theta_0)\} = -\cos(2\pi f_c t + \theta_0) = \sin\left(2\pi f_c t + \theta_0 - \frac{\pi}{2}\right) \tag{11.1-14}$$

where θ_0 is a constant phase angle in radians. The right-hand sides of (11.1-13) and (11.1-14) are simply trigonometric identities.

Next we shall show that the complex envelope of a real bandpass signal has a lowpass (baseband) frequency spectrum. Taking the Fourier transform of (11.1-2) and (11.1-3) yields

$$\boxed{\tilde{X}(f) = X_p(f + f_c)} \tag{11.1-15}$$

and

$$X_p(f) = X(f) + j\hat{X}(f), \tag{11.1-16}$$

respectively, where

$$\tilde{X}(f) = F\{\tilde{x}(t)\} \tag{11.1-17}$$

and

$$X_p(f) = F\{x_p(t)\}. \tag{11.1-18}$$

Substituting (11.1-10) into (11.1-16) yields

$$X_p(f) = [1 + \text{sgn}(f)]X(f), \tag{11.1-19}$$

and with the use of (11.1-9),

$$\boxed{X_p(f) = \begin{cases} 2X(f), & f > 0 \\ 0, & f \le 0 \end{cases}} \tag{11.1-20}$$

The frequency spectrum of the pre-envelope given by (11.1-20) is known as a *one-sided spectrum* because it only contains positive frequency components (see Fig. 11.1-2). Note that

$$\begin{aligned} X_p(0) &= [1 + \text{sgn}(0)]X(0) \\ &= X(0) \\ &= 0 \end{aligned} \tag{11.1-21}$$

because $x(t)$ is a bandpass signal (see Fig. 11.1-1).

By referring to (11.1-15), it can be seen that the frequency spectrum of the complex envelope can be obtained by shifting the frequency spectrum of the pre-envelope to the left by an amount equal to the carrier frequency f_c, as shown in Fig. 11.1-3, thus creating a lowpass signal with bandwidth W hertz. This is another reason for representing a real bandpass signal in terms of its complex envelope because $\tilde{x}(t)$ can be sampled at the lower rate $f_S \ge 2W$ samples per second versus the higher rate $f_S \ge 2(f_c + W)$ samples per second for $x(t)$.

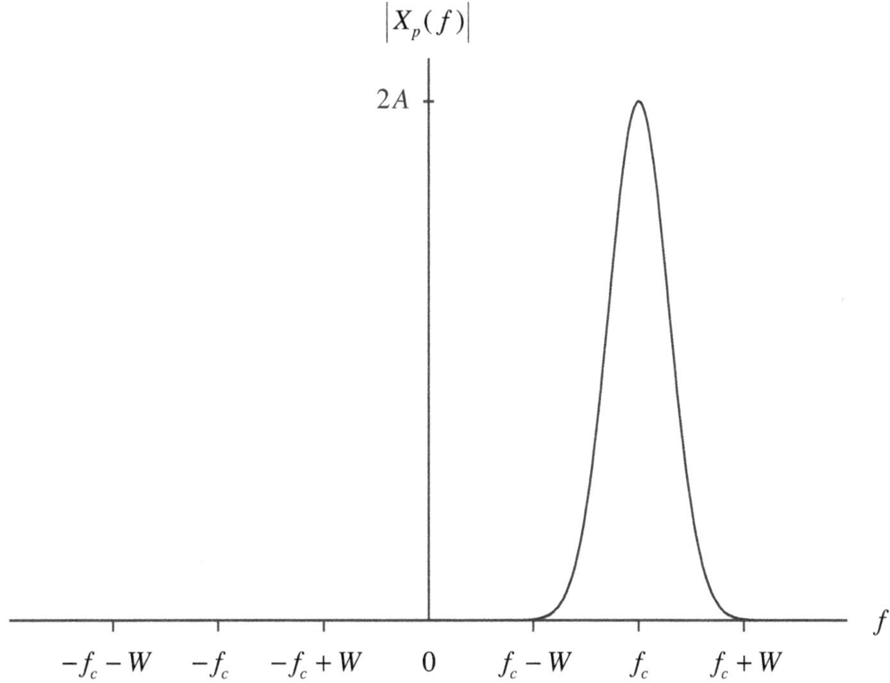

Figure 11.1-2 Magnitude spectrum of the pre-envelope $x_p(t)$ with bandwidth $2W$ hertz.

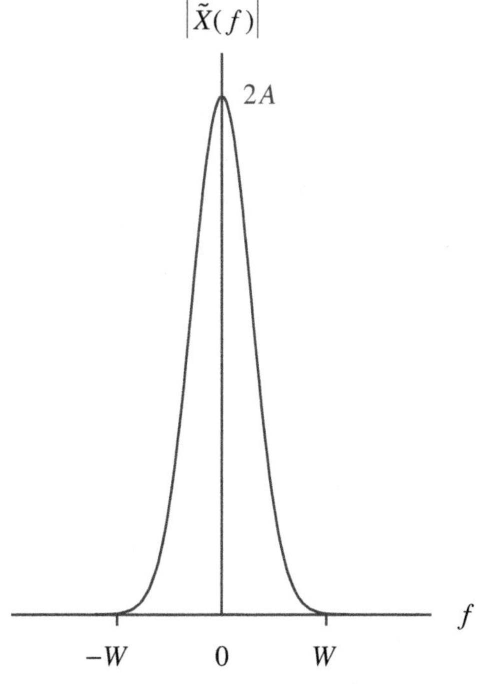

Figure 11.1-3 Magnitude spectrum of the lowpass complex envelope $\tilde{x}(t)$ with bandwidth W hertz.

We shall end our discussion in this section by deriving the relationship between the complex frequency spectrum $X(f)$ of the real bandpass signal $x(t)$ and the complex frequency spectrum $\tilde{X}(f)$ of its complex envelope $\tilde{x}(t)$. If Z is complex, then

$$\text{Re}\{Z\} = (Z + Z^*)/2. \tag{11.1-22}$$

With the use of (11.1-22), (11.1-6) can be rewritten as

$$x(t) = \frac{1}{2}\left[\tilde{x}(t)\exp(+j2\pi f_c t) + \tilde{x}^*(t)\exp(-j2\pi f_c t)\right]. \tag{11.1-23}$$

Taking the Fourier transform of (11.1-23) yields

$$\boxed{X(f) = \frac{1}{2}\left[\tilde{X}(f - f_c) + \tilde{X}^*(-[f + f_c])\right]} \tag{11.1-24}$$

where

$$F\{\tilde{x}(t)\exp(+j2\pi f_c t)\} = \tilde{X}(f - f_c), \tag{11.1-25}$$

$$F\{\tilde{x}^*(t)\} = \tilde{X}^*(-f), \tag{11.1-26}$$

and

$$F\{\tilde{x}^*(t)\exp(-j2\pi f_c t)\} = \tilde{X}^*(-[f + f_c]). \tag{11.1-27}$$

11.1.1 Signal Energy and Time-Average Power

In this subsection we shall relate the energy of a real bandpass signal (an amplitude-and-angle-modulated carrier) $x(t)$ to the energy of its complex envelope $\tilde{x}(t)$. We shall then use this energy relationship to compute the time-average power of $x(t)$.

From signal theory, the *energy* E_x of the real signal $x(t)$ is, by definition,

$$\boxed{E_x \triangleq \int_{-\infty}^{\infty} x^2(t)\,dt} \tag{11.1-28}$$

For example, if $x(t)$ is the instantaneous voltage (in volts) across a linear, time-invariant, resistive load, then E_x given by (11.1-28) has units of *joules-ohms*. However, if $x(t)$ is the instantaneous current (in amperes) flowing through a linear, time-invariant, resistive load, then E_x has units of *joules per ohm*. These units can easily be verified as follows: the instantaneous power $p(t)$ in watts is given by $p(t)=v(t)i(t)$, where $v(t)$ is the instantaneous voltage in volts, and $i(t)$ is the instantaneous current in amperes. For a linear, time-invariant, resistive load, Ohm's law states that $v(t)=Ri(t)$, where R is the constant resistance in ohms. Therefore, $p(t)=v^2(t)/R$ or $p(t)=Ri^2(t)$. Since energy E in joules is equal to the integral of instantaneous power $p(t)$ in watts, that is, since $E=\int p(t)dt$, then $E=R^{-1}\int v^2(t)dt$ has units of joules and $ER=\int v^2(t)dt$ has units of joules-ohms. In addition, $E=R\int i^2(t)dt$ has units of joules and $E/R=\int i^2(t)dt$ has units of joules per ohm. Therefore, if (11.1-28) is used to compute signal energy, divide or multiply your answer by the constant resistance R of the load to get energy in joules. For sonar applications, the load could be, for example, an electroacoustic transducer.

Similarly, the *energy* $E_{\tilde{x}}$ of the complex envelope $\tilde{x}(t)$ is, by definition,

$$\boxed{E_{\tilde{x}} \triangleq \int_{-\infty}^{\infty} |\tilde{x}(t)|^2 dt} \tag{11.1-29}$$

If $\tilde{x}(t)$ is the complex envelope of the instantaneous voltage (in volts) across a linear, time-invariant, resistive load, then $E_{\tilde{x}}$ given by (11.1-29) has units of *joules-ohms*. However, if $\tilde{x}(t)$ is the complex envelope of the instantaneous current (in amperes) flowing through a linear, time-invariant, resistive load, then $E_{\tilde{x}}$ has units of *joules per ohm*. By referring to (11.1-2) and (11.1-3), it can be seen that

$$|\tilde{x}(t)|^2 = x^2(t)+\hat{x}^2(t). \tag{11.1-30}$$

Substituting (11.1-30) into (11.1-29) yields

$$E_{\tilde{x}} = E_x + E_{\hat{x}}, \tag{11.1-31}$$

where

$$E_{\hat{x}} = \int_{-\infty}^{\infty} \hat{x}^2(t)dt \tag{11.1-32}$$

is the energy of the Hilbert transform of $x(t)$. With the use of *Parseval's Theorem*, (11.1-32) can be expressed as

$$E_{\hat{x}} = \int_{-\infty}^{\infty} |\hat{X}(f)|^2 df, \tag{11.1-33}$$

and by substituting (11.1-10) into (11.1-33), we obtain

$$E_{\hat{x}} = \int_{-\infty}^{\infty} |X(f)|^2 \, df = \int_{-\infty}^{\infty} x^2(t) \, dt \,, \tag{11.1-34}$$

or

$$\boxed{E_{\hat{x}} = E_x} \tag{11.1-35}$$

Therefore, the real bandpass signal $x(t)$ and its Hilbert transform $\hat{x}(t)$ have equal energy. This is not surprising because the magnitude spectrum of a bandpass signal is not altered by a Hilbert transform – only the phase spectrum is [see (11.1-11) and (11.1-12)]. Substituting (11.1-35) into (11.1-31) yields the following important relationship:

$$\boxed{E_x = E_{\tilde{x}}/2} \tag{11.1-36}$$

As we shall show in Section 11.2, it is much easier to compute the energy of an amplitude-and-angle-modulated carrier (a real bandpass signal) by first computing the energy of its complex envelope and then dividing that energy by a factor of two.

In real-world problems, $x(t)$ will have a finite duration. If $x(t)$ is defined in the closed time interval $[0, T]$, that is, for $0 \le t \le T$, where T is the *duration* or *pulse length* of $x(t)$ in seconds, then the energies given by (11.1-28) and (11.1-29) reduce as follows:

$$E_x = \int_0^T x^2(t) \, dt \tag{11.1-37}$$

and

$$E_{\tilde{x}} = \int_0^T |\tilde{x}(t)|^2 \, dt \,. \tag{11.1-38}$$

The *time-average power* $P_{\text{avg},x}$ of $x(t)$ in the time interval $[0, T]$ is given by

$$\boxed{P_{\text{avg},x} = \frac{E_x}{T} = \frac{1}{T}\int_0^T x^2(t) \, dt} \tag{11.1-39}$$

and the *time-average power* $P_{\text{avg},\tilde{x}}$ of $\tilde{x}(t)$ in the time interval $[0, T]$ is given by

$$\boxed{P_{\text{avg},\tilde{x}} = \frac{E_{\tilde{x}}}{T} = \frac{1}{T}\int_0^T |\tilde{x}(t)|^2 \, dt} \tag{11.1-40}$$

If the energies in (11.1-39) and (11.1-40) have units of joules-ohms or joules per ohm, then the time-average powers in (11.1-39) and (11.1-40) will have units of watts-ohms or watts per ohm, respectively. Finally, note that

$$\boxed{P_{\text{avg},\,x} = \frac{1}{2} P_{\text{avg},\,\tilde{x}} = \frac{1}{2}\frac{E_{\tilde{x}}}{T} = \frac{E_x}{T}} \tag{11.1-41}$$

Knowing the time-average power of a transmitted electrical signal is important because electroacoustic transducers have limits on the maximum time-average power that they can handle before being damaged, and it is one of the parameters that determines the signal-to-noise ratio at a receiver.

11.1.2 The Power Spectrum

In this subsection we shall relate the power spectrum of a real bandpass signal (an amplitude-and-angle-modulated carrier) $x(t)$ to the power spectrum of its complex envelope $\tilde{x}(t)$. We shall then use the power spectrums of $x(t)$ and $\tilde{x}(t)$ to compute the time-average powers of $x(t)$ and $\tilde{x}(t)$.

The definition of the power spectrum $S_x(f)$ of a finite duration, deterministic signal $x(t)$ is given by

$$\boxed{S_x(f) \triangleq \frac{1}{T}\left|X(f)\right|^2} \tag{11.1-42}$$

where T is the duration or pulse length of $x(t)$ in seconds. From (11.1-42) it can be seen that the power spectrum $S_x(f)$ is *real* and *nonnegative*, that is, $S_x(f) \geq 0$. If $x(t)$ has units of volts, then $S_x(f)$ has units of $(\text{W-}\Omega)/\text{Hz}$, and if $x(t)$ has units of amperes, then $S_x(f)$ has units of $(\text{W}/\Omega)/\text{Hz}$. Applying the identity

$$\left|A+B\right|^2 = \left|A\right|^2 + 2\operatorname{Re}\left\{AB^*\right\} + \left|B\right|^2 \tag{11.1-43}$$

to (11.1-24) yields

$$\begin{aligned} \left|X(f)\right|^2 &= \frac{1}{4}\left|\tilde{X}(f-f_c)\right|^2 + \frac{1}{4}\left|\tilde{X}^*(-[f+f_c])\right|^2 \\ &= \frac{1}{4}\left|\tilde{X}(f-f_c)\right|^2 + \frac{1}{4}\left|\tilde{X}(-[f+f_c])\right|^2 \end{aligned} \tag{11.1-44}$$

since

$$\tilde{X}(f-f_c)\tilde{X}(-[f+f_c]) = 0, \tag{11.1-45}$$

and by substituting (11.1-44) into (11.1-42), we obtain

$$S_x(f)=\frac{1}{4}S_{\tilde{x}}(f-f_c)+\frac{1}{4}S_{\tilde{x}}(-[f+f_c]) \quad (11.1\text{-}46)$$

where

$$S_{\tilde{x}}(f)=\frac{1}{T}\left|\tilde{X}(f)\right|^2 \quad (11.1\text{-}47)$$

is the power spectrum of the complex envelope $\tilde{x}(t)$. From (11.1-47) it can be seen that the power spectrum $S_{\tilde{x}}(f)$ of the complex envelope $\tilde{x}(t)$ is also *real* and *nonnegative* ($S_{\tilde{x}}(f)\geq 0$) even though $\tilde{x}(t)$ is *complex* in general.

The time-average power of $x(t)$ can be obtained by integrating its power spectrum $S_x(f)$, that is,

$$P_{\text{avg},x}=\int_{-\infty}^{\infty}S_x(f)df. \quad (11.1\text{-}48)$$

Since $x(t)$ is *real*, $S_x(f)$ is an *even* function of frequency f because the Fourier transform $X(f)$ satisfies the *conjugate symmetry property* – the magnitude spectrum $|X(f)|$ is an even function of frequency and the phase spectrum $\angle X(f)$ is an odd function of frequency. Therefore, (11.1-48) reduces to

$$P_{\text{avg},x}=2\int_{0}^{\infty}S_x(f)df \quad (11.1\text{-}49)$$

Equation (11.1-49) indicates that we only need to integrate the power spectrum along the positive frequency axis. With the use of (11.1-49), the time-average power of $x(t)$ in the frequency band $\Delta f=f_2-f_1$ hertz is given by

$$P_{\text{avg},x,\Delta f}=2\int_{f_1}^{f_2}S_x(f)df \quad (11.1\text{-}50)$$

Since the complex envelope $\tilde{x}(t)$ is, in general, a *complex* signal, its power spectrum $S_{\tilde{x}}(f)$ is *not*, in general, an even function of frequency. Therefore, in order to compute the time-average power of $\tilde{x}(t)$, we need to integrate its power spectrum $S_{\tilde{x}}(f)$ along both the negative and positive frequency axes, that is,

$$P_{\text{avg},\tilde{x}}=\int_{-\infty}^{\infty}S_{\tilde{x}}(f)df. \quad (11.1\text{-}51)$$

Therefore, substituting (11.1-51) into (11.1-41) yields

$$\boxed{P_{\text{avg},\,x} = \frac{1}{2} P_{\text{avg},\,\hat{x}} = \frac{1}{2}\int_{-\infty}^{\infty} S_{\hat{x}}(f)\,df} \tag{11.1-52}$$

where $P_{\text{avg},\,x}$ is the time-average power of $x(t)$.

11.1.3 Orthogonality Relationships

In this subsection we shall show that a real bandpass signal $x(t)$ and its Hilbert transform $\hat{x}(t)$ are *orthogonal*, that is, their *inner product* is equal to *zero*. As a result, their Fourier transforms are also orthogonal. The inner product of $x(t)$ and $\hat{x}(t)$ is, by definition,

$$\langle x(t), \hat{x}(t)\rangle \triangleq \int_{-\infty}^{\infty} x(t)\hat{x}^{*}(t)\,dt\,. \tag{11.1-53}$$

With the use of Parseval's Theorem, (11.1-53) can be rewritten as

$$\langle x(t), \hat{x}(t)\rangle = \langle X(f), \hat{X}(f)\rangle = \int_{-\infty}^{\infty} X(f)\hat{X}^{*}(f)\,df\,. \tag{11.1-54}$$

Substituting (11.1-10) into (11.1-54) yields

$$\langle x(t), \hat{x}(t)\rangle = \langle X(f), \hat{X}(f)\rangle = j\int_{-\infty}^{\infty} \operatorname{sgn}(f)\left|X(f)\right|^{2} df\,, \tag{11.1-55}$$

and since the sign function $\operatorname{sgn}(f)$ is an *odd* function of frequency f and the magnitude spectrum $|X(f)|$ is an *even* function of f because $x(t)$ is real (a consequence of the conjugate symmetry property for real signals), the integrand in (11.1-55) is an *odd* function of f, and as a result,

$$\boxed{\langle x(t), \hat{x}(t)\rangle = \langle X(f), \hat{X}(f)\rangle = 0} \tag{11.1-56}$$

Therefore, the real bandpass signal $x(t)$ and its Hilbert transform $\hat{x}(t)$ are orthogonal, as well as their Fourier transforms.

11.2 The Complex Envelope of an Amplitude-and-Angle-Modulated Carrier

In this section we shall use the definitions and basic relationships discussed in Section 11.1 and Subsection 11.1.1 to compute the complex envelope, envelope, energy, and time-average power of an amplitude-and-angle-modulated carrier (a real bandpass signal). An amplitude-and-angle-modulated carrier is given by

$$x(t) = a(t)\cos\left[2\pi f_c t + \theta(t)\right], \tag{11.2-1}$$

where $a(t)$ and $\theta(t)$ are real amplitude-modulating and angle-modulating functions, respectively, $\cos(2\pi f_c t)$ is the carrier waveform, and f_c is the carrier frequency in hertz. The amplitude-modulating function $a(t)$ has units of volts or amperes. The angle-modulating function $\theta(t)$ is also known as the *phase deviation* in radians.

The value of the carrier frequency f_c is determined by taking into account the following three main considerations: 1) the bandwidth W in hertz of the complex envelope of the amplitude-and-angle-modulated carrier (see Fig. 11.1-3), 2) the desired 3-dB beamwidth of the far-field beam pattern of the sonar system, and 3) frequency-dependent attenuation due to sound propagation in the ocean. First, in order to avoid signal distortion, $f_c > W$ so that the magnitude spectrum of the bandpass signal along the positive and negative frequency axes do not overlap (see Fig. 11.1-1). Second, when the size of an aperture (array) is fixed, the 3-dB beamwidth of the corresponding far-field beam pattern can be decreased if the operating frequency (carrier frequency f_c) is increased. Third, as the operating frequency increases, attenuation due to sound propagation in the ocean also increases. Therefore, after ensuring $f_c > W$ to avoid signal distortion, there is then a tradeoff between desired beamwidth and short-range versus long-range applications.

Before we begin the derivation of the complex envelope of (11.2-1), let us briefly review some of the basic terminology and equations associated with angle modulation. The *instantaneous phase* (in radians) of $x(t)$ is defined as the argument of the cosine function:

$$\boxed{\theta_i(t) \triangleq 2\pi f_c t + \theta(t)} \tag{11.2-2}$$

The *instantaneous radian frequency* (in radians per second) of $x(t)$ is defined as the time derivative of the instantaneous phase:

$$\boxed{\omega_i(t) \triangleq \frac{d}{dt}\theta_i(t) = 2\pi f_c + \frac{d}{dt}\theta(t)} \tag{11.2-3}$$

Therefore, the *instantaneous frequency* (in hertz) of $x(t)$ is defined as

$$\boxed{f_i(t) \triangleq \frac{1}{2\pi}\frac{d}{dt}\theta_i(t) = f_c + \frac{1}{2\pi}\frac{d}{dt}\theta(t)} \tag{11.2-4}$$

The time derivative $d\theta(t)/dt$ is known as the *frequency deviation* in radians per second.

There are two basic types of angle modulation – *phase modulation* (PM) and *frequency modulation* (FM). Phase modulation implies that the phase deviation $\theta(t)$ of the carrier is directly proportional to some *message* or *modulating signal* $m(t)$, that is,

$$\theta(t) = D_p\, m(t), \tag{11.2-5}$$

where D_p is the *phase-deviation constant* in radians per unit of $m(t)$. Frequency modulation implies that the frequency deviation $d\theta(t)/dt$ of the carrier is directly proportional to $m(t)$, that is,

$$\frac{d}{dt}\theta(t) = D_f\, m(t), \tag{11.2-6}$$

where D_f is the *frequency-deviation constant* in radians per second, per unit of $m(t)$. As a result, the phase deviation of a frequency-modulated carrier is given by

$$\theta(t) = \theta(t_0) + D_f \int_{t_0}^{t} m(\zeta)\,d\zeta, \tag{11.2-7}$$

where $\theta(t_0)$ is the phase deviation at $t = t_0$.

We begin the derivation of the complex envelope of (11.2-1) by using the trigonometric identity

$$\cos(\alpha+\beta) = \cos\alpha\cos\beta - \sin\alpha\sin\beta \tag{11.2-8}$$

to rewrite (11.2-1) as follows:

$$\boxed{x(t) = x_c(t)\cos(2\pi f_c t) - x_s(t)\sin(2\pi f_c t)} \tag{11.2-9}$$

where

$$\boxed{x_c(t) = a(t)\cos\theta(t)} \tag{11.2-10}$$

and

$$\boxed{x_s(t) = a(t)\sin\theta(t)} \tag{11.2-11}$$

are known as the *cosine* and *sine components* of $x(t)$, respectively. The functions $x_c(t)$ and $x_s(t)$ are also known as the *in-phase* and *quadrature-phase components*, respectively, because $x_c(t)$ is in-phase with the carrier waveform $\cos(2\pi f_c t)$, and $x_s(t)$ is 90° out-of-phase with the carrier waveform.

The next step is to compute the Hilbert transform of $x(t)$. First rewrite (11.2-9) as follows:

$$x(t) = x_c(t)x_{\text{BP1}}(t) - x_s(t)x_{\text{BP2}}(t), \tag{11.2-12}$$

where

$$x_{\text{BP1}}(t) = \cos(2\pi f_c t) \tag{11.2-13}$$

and

$$x_{\text{BP2}}(t) = \sin(2\pi f_c t) \tag{11.2-14}$$

are real bandpass functions with complex frequency spectra

$$X_{\text{BP1}}(f) = \frac{1}{2}\delta(f - f_c) + \frac{1}{2}\delta(f + f_c) \tag{11.2-15}$$

and

$$X_{\text{BP2}}(f) = \frac{1}{2}\exp(-j\pi/2)\delta(f - f_c) + \frac{1}{2}\exp(+j\pi/2)\delta(f + f_c). \tag{11.2-16}$$

Note that the bandpass spectra given by (11.2-15) and (11.2-16) are both centered at $f = \pm f_c$ hertz. Therefore, if both $x_c(t)$ and $x_s(t)$ are *lowpass* functions with bandwidth W hertz, that is, if

$$|X_c(f)| = 0, \qquad |f| \geq W, \tag{11.2-17}$$

and

$$|X_s(f)| = 0, \qquad |f| \geq W, \tag{11.2-18}$$

then the Hilbert transform of (11.2-12) is given by

$$\hat{x}(t) = x_c(t)\hat{x}_{\text{BP1}}(t) - x_s(t)\hat{x}_{\text{BP2}}(t) \tag{11.2-19}$$

provided that

$$\boxed{f_c > W} \tag{11.2-20}$$

Equation (11.2-20) guarantees that $|X_c(f)|$ and $|X_{\text{BP1}}(f)|$ do not overlap, and that $|X_s(f)|$ and $|X_{\text{BP2}}(f)|$ do not overlap. The following property of the Hilbert transform was used to obtain (11.2-19): if

$$x(t) = x_{\text{LP}}(t)x_{\text{BP}}(t), \tag{11.2-21}$$

where $x_{\text{LP}}(t)$ is a lowpass function and $x_{\text{BP}}(t)$ is a bandpass function with non-overlapping magnitude spectra, then

$$\boxed{\hat{x}(t) = x_{\text{LP}}(t)\hat{x}_{\text{BP}}(t)} \tag{11.2-22}$$

that is, only the bandpass function is Hilbert transformed. Since the Hilbert transforms of (11.2-13) and (11.2-14) are given by [see (11.1-13) and (11.1-14)]

$$\hat{x}_{\text{BP1}}(t) = \sin(2\pi f_c t) \tag{11.2-23}$$

and

$$\hat{x}_{\text{BP2}}(t) = -\cos(2\pi f_c t), \tag{11.2-24}$$

substituting (11.2-23) and (11.2-24) into (11.2-19) yields the Hilbert transform

$$\boxed{\hat{x}(t) = x_c(t)\sin(2\pi f_c t) + x_s(t)\cos(2\pi f_c t)} \tag{11.2-25}$$

The final two steps are to compute the pre-envelope and complex envelope of $x(t)$. Substituting (11.2-9) and (11.2-25) into (11.1-3) yields the pre-envelope

$$x_p(t) = \left[x_c(t) + jx_s(t)\right]\exp(+j2\pi f_c t), \tag{11.2-26}$$

and by substituting (11.2-26) into (11.1-2), we finally obtain the complex envelope

$$\boxed{\tilde{x}(t) = x_c(t) + jx_s(t)} \tag{11.2-27}$$

where the real, lowpass, cosine and sine components, $x_c(t)$ and $x_s(t)$, with bandwidths W hertz, are given by (11.2-10) and (11.2-11), respectively. Equation (11.2-27) is the complex envelope of the amplitude-and-angle-modulated carrier given by (11.2-1) and is, in general, a complex-valued function of time. Equation (11.2-27) expresses the complex envelope in rectangular form, that is, in terms of real and imaginary parts. The complex envelope can also be expressed in polar form as follows:

$$\tilde{x}(t) = \left|\tilde{x}(t)\right|\exp\left[+j\angle\tilde{x}(t)\right], \tag{11.2-28}$$

where

$$\begin{aligned} |\tilde{x}(t)| &= \sqrt{x_c^2(t)+x_s^2(t)} \\ &= \sqrt{a^2(t)\left[\cos^2\theta(t)+\sin^2\theta(t)\right]} \\ &= a(t) \end{aligned} \tag{11.2-29}$$

and

$$\begin{aligned} \angle\tilde{x}(t) &= \tan^{-1}\left[\frac{x_s(t)}{x_c(t)}\right] = \tan^{-1}\left[\frac{a(t)\sin\theta(t)}{a(t)\cos\theta(t)}\right] \\ &= \tan^{-1}[\tan\theta(t)] \\ &= \theta(t). \end{aligned} \tag{11.2-30}$$

Therefore, substituting (11.2-29) and (11.2-30) into (11.2-28) yields

$$\boxed{\tilde{x}(t) = a(t)\exp[+j\theta(t)]} \tag{11.2-31}$$

where $a(t)$ and $\theta(t)$ are real amplitude-modulating and angle-modulating functions. The polar form of the complex envelope given by (11.2-31) provides a simple way to represent the real amplitude-and-angle-modulated carrier given by (11.2-1). The polar form is the preferred form for doing analysis. Note that if (11.2-31) is substituted into (11.1-6), then we obtain (11.2-1).

If there is no angle modulation, that is, if $\theta(t)=0$, then $x(t)=a(t)\cos(2\pi f_c t)$,

$$x_c(t) = a(t), \tag{11.2-32}$$

$$x_s(t) = 0, \tag{11.2-33}$$

and

$$\tilde{x}(t) = a(t). \tag{11.2-34}$$

Therefore, the complex envelope is equal to the real amplitude-modulating function when there is no angle modulation.

The envelope of $x(t)$ can be obtained by taking the absolute value of (11.2-29) [see (11.1-7)]. Doing so yields

$$\boxed{\mathcal{E}(t) = |a(t)| \geq 0} \tag{11.2-35}$$

In communication theory, the envelope of a real amplitude-and-angle-modulated carrier is defined by (11.2-35) irrespective of complex envelopes. As was mentioned in Subsection 11.1.1, in real-world problems, $x(t)$ will have a finite duration. If $x(t)$ is defined for $0 \leq t \leq T$, where T is the duration or pulse length of $x(t)$ in seconds, then substituting (11.2-29) into (11.1-38) yields the energy of the complex envelope

$$\boxed{E_{\tilde{x}} = \int_0^T a^2(t)\,dt} \tag{11.2-36}$$

The energy and time-average power of an amplitude-and-angle-modulated carrier are given by [see (11.1-36) and (11.1-41)]

$$\boxed{E_x = \frac{1}{2}E_{\tilde{x}} = \frac{1}{2}\int_0^T a^2(t)\,dt} \tag{11.2-37}$$

and

$$\boxed{P_{\text{avg},\,x} = \frac{E_x}{T} = \frac{1}{2T}\int_0^T a^2(t)\,dt} \tag{11.2-38}$$

As can be seen from (11.2-37) and (11.2-38), the angle-modulating function $\theta(t)$ does *not* contribute to signal energy and, as a result, to time-average power.

11.2.1 The Bandpass Sampling Theorem

Since the complex envelope $\tilde{x}(t)$ is a lowpass (baseband) signal bandlimited to W hertz (see Fig. 11.1-3), the sampling theorem states that $\tilde{x}(t)$ can be reconstructed from its sampled values $\tilde{x}(nT_S)$ as follows:

$$\tilde{x}(t) = \sum_{n=-\infty}^{\infty} \tilde{x}(nT_S)\,\text{sinc}\left(\frac{t - nT_S}{T_S}\right), \tag{11.2-39}$$

where

$$f_S = \frac{1}{T_S} \geq 2W \tag{11.2-40}$$

is the sampling frequency in hertz (a.k.a. the sampling rate in samples per second), T_S is the sampling period in seconds, and

$$\text{sinc}(\alpha) \triangleq \frac{\sin(\pi\alpha)}{\pi\alpha}. \tag{11.2-41}$$

The *minimum* sampling rate is called the *Nyquist rate* and is equal to $2W$ samples per second.

Let $x(t)$ and, hence, $\tilde{x}(t)$ be defined for $0 \le t \le T$, where T is the duration or pulse length of $x(t)$ in seconds. With the use of (11.2-27), (11.2-39) can be rewritten as

$$\tilde{x}(t) = \sum_{n=0}^{N-1} \left[x_c(nT_S) + jx_s(nT_S) \right] \text{sinc}\left(\frac{t - nT_S}{T_S} \right), \tag{11.2-42}$$

where

$$T = (N-1)T_S \tag{11.2-43}$$

and $N = f_S T + 1$ is the number of samples taken of both $x_c(t)$ and $x_s(t)$ for a total of $2N$ samples. Substituting (11.2-42) into (11.1-6) yields

$$x(t) = \sum_{n=0}^{N-1} \text{Re}\left\{ \left[x_c(nT_S) + jx_s(nT_S) \right] \exp(+j2\pi f_c t) \right\} \text{sinc}\left(\frac{t - nT_S}{T_S} \right), \tag{11.2-44}$$

and by expanding the complex exponential using Euler's identity and taking the real part, we obtain

$$\boxed{x(t) = \sum_{n=0}^{N-1} \left[x_c(nT_S)\cos(2\pi f_c t) - x_s(nT_S)\sin(2\pi f_c t) \right] \text{sinc}\left(\frac{t - nT_S}{T_S} \right)} \tag{11.2-45}$$

Equation (11.2-45) is referred to as the *bandpass sampling theorem.* It states that an amplitude-and-angle-modulated carrier $x(t)$ (a real bandpass signal) can be reconstructed from sampled values of its lowpass (baseband) cosine and sine components, $x_c(t)$ and $x_s(t)$.

11.2.2 Orthogonality Relationships

In this subsection we shall show that the cosine and sine components, $x_c(t)$ and $x_s(t)$, of an amplitude-and-angle-modulated carrier $x(t)$ are *orthogonal*, that is, their *inner product* is equal to *zero*. As a result, their Fourier transforms are also orthogonal.

With the use of Parseval's Theorem and (11.2-27), the energy of the complex envelope given by (11.1-29) can also be expressed as

$$E_{\tilde{x}} \triangleq \int_{-\infty}^{\infty} |\tilde{x}(t)|^2 \, dt = \int_{-\infty}^{\infty} |\tilde{X}(f)|^2 \, df \tag{11.2-46}$$

or

$$E_{\tilde{x}} = \int_{-\infty}^{\infty} x_c^2(t)\,dt + \int_{-\infty}^{\infty} x_s^2(t)\,dt = \int_{-\infty}^{\infty} |X_c(f)|^2\,df + \int_{-\infty}^{\infty} |X_s(f)|^2\,df. \tag{11.2-47}$$

Equating the right-hand sides of (11.2-46) and (11.2-47) yields

$$\int_{-\infty}^{\infty} |\tilde{X}(f)|^2\,df = \int_{-\infty}^{\infty} |X_c(f)|^2\,df + \int_{-\infty}^{\infty} |X_s(f)|^2\,df. \tag{11.2-48}$$

Since [see (11.2-27)]

$$\tilde{X}(f) = X_c(f) + jX_s(f) \tag{11.2-49}$$

and

$$|\tilde{X}(f)|^2 = \tilde{X}(f)\tilde{X}^*(f), \tag{11.2-50}$$

substituting (11.2-49) into (11.2-50) yields

$$|\tilde{X}(f)|^2 = |X_c(f)|^2 + jX_c^*(f)X_s(f) - jX_c(f)X_s^*(f) + |X_s(f)|^2. \tag{11.2-51}$$

Therefore,

$$\begin{aligned} \int_{-\infty}^{\infty} |\tilde{X}(f)|^2\,df = \int_{-\infty}^{\infty} |X_c(f)|^2\,df + \int_{-\infty}^{\infty} |X_s(f)|^2\,df + \\ \left[-j\int_{-\infty}^{\infty} X_c(f)X_s^*(f)\,df\right] + \left[j\int_{-\infty}^{\infty} X_c^*(f)X_s(f)\,df\right] \end{aligned} \tag{11.2-52}$$

or [see (11.1-22)]

$$\begin{aligned} \int_{-\infty}^{\infty} |\tilde{X}(f)|^2\,df = \int_{-\infty}^{\infty} |X_c(f)|^2\,df + \int_{-\infty}^{\infty} |X_s(f)|^2\,df - \\ 2\,\mathrm{Re}\left\{j\int_{-\infty}^{\infty} X_c(f)X_s^*(f)\,df\right\}. \end{aligned} \tag{11.2-53}$$

In order for the right-hand sides of (11.2-48) and (11.2-53) to be equal,

$$\int_{-\infty}^{\infty} X_c(f)X_s^*(f)\,df = \int_{-\infty}^{\infty} x_c(t)x_s^*(t)\,dt = 0 \tag{11.2-54}$$

or

$$\boxed{\langle x_c(t), x_s(t)\rangle = \langle X_c(f), X_s(f)\rangle = 0} \tag{11.2-55}$$

Equation (11.2-55) indicates that the cosine and sine components and their Fourier transforms are orthogonal. In other words, their inner products are equal to zero.

11.3 The Quadrature Demodulator

The complex envelope given by (11.2-27) was obtained by evaluating the mathematical definition given by (11.1-2). However, in practical signal-processing applications, the complex envelope given by (11.2-27) can be obtained by passing a real amplitude-and-angle-modulated carrier through a *quadrature demodulator* (QD), if there are *no* frequency and phase offsets. The cosine and sine components can then be used to perform amplitude and angle demodulation. Furthermore, there is no need to compute a Hilbert transform and pre-envelope as we shall show next.

Let the amplitude-and-angle-modulated carrier

$$\begin{aligned} x(t) &= a(t)\cos\left[2\pi f_c t + \theta(t) + \varepsilon\right] \\ &= x_c(t)\cos(2\pi f_c t + \varepsilon) - x_s(t)\sin(2\pi f_c t + \varepsilon) \end{aligned} \tag{11.3-1}$$

be the input signal to the QD shown in Fig. 11.3-1, where $x_c(t)$ and $x_s(t)$ are the cosine and sine components of $x(t)$ given by (11.2-10) and (11.2-11), respectively, and ε is a phase term in radians. We begin by analyzing the *worst* case, that is, when there are both frequency and phase offsets. If there are *frequency offsets*, $f_1 \neq f_c$ and $f_2 \neq f_c$; and *phase offsets*, $\varepsilon_1 \neq \varepsilon$ and $\varepsilon_2 \neq \varepsilon$; then it can be shown that the input signal $x_1(t)$ to the ideal, lowpass filter (ILPF) in the upper channel of the QD is given by

$$x_1(t) = x_{1c}(t) + x_{1s}(t), \tag{11.3-2}$$

where

$$\begin{aligned} x_{1c}(t) = {} & x_c(t)\cos\left[2\pi(f_1 - f_c)t\right]\cos(\varepsilon_1 - \varepsilon) - x_c(t)\sin\left[2\pi(f_1 - f_c)t\right]\sin(\varepsilon_1 - \varepsilon) + \\ & x_c(t)\cos\left[2\pi(f_1 + f_c)t\right]\cos(\varepsilon_1 + \varepsilon) - x_c(t)\sin\left[2\pi(f_1 + f_c)t\right]\sin(\varepsilon_1 + \varepsilon) \end{aligned} \tag{11.3-3}$$

and

$$\begin{aligned} x_{1s}(t) = {} & x_s(t)\cos\left[2\pi(f_1 - f_c)t\right]\sin(\varepsilon_1 - \varepsilon) + x_s(t)\sin\left[2\pi(f_1 - f_c)t\right]\cos(\varepsilon_1 - \varepsilon) - \\ & x_s(t)\cos\left[2\pi(f_1 + f_c)t\right]\sin(\varepsilon_1 + \varepsilon) - x_s(t)\sin\left[2\pi(f_1 + f_c)t\right]\cos(\varepsilon_1 + \varepsilon), \end{aligned} \tag{11.3-4}$$

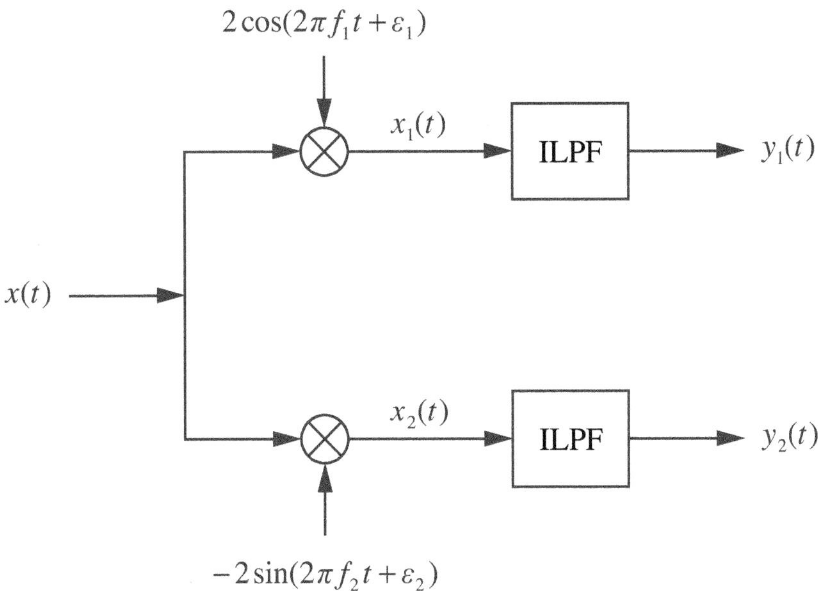

Figure 11.3-1 A quadrature demodulator. The upper channel is referred to as the *I* (in-phase) channel and the lower channel is referred to as the *Q* (quadrature-phase) channel. The acronym ILPF stands for ideal, lowpass filter.

and the input signal $x_2(t)$ to the ILPF in the lower channel of the QD is given by

$$x_2(t) = x_{2c}(t) + x_{2s}(t), \tag{11.3-5}$$

where

$$\begin{aligned} x_{2c}(t) = &-x_c(t)\cos\left[2\pi(f_2 - f_c)t\right]\sin(\varepsilon_2 - \varepsilon) - x_c(t)\sin\left[2\pi(f_2 - f_c)t\right]\cos(\varepsilon_2 - \varepsilon) - \\ &x_c(t)\cos\left[2\pi(f_2 + f_c)t\right]\sin(\varepsilon_2 + \varepsilon) - x_c(t)\sin\left[2\pi(f_2 + f_c)t\right]\cos(\varepsilon_2 + \varepsilon) \end{aligned} \tag{11.3-6}$$

and

$$\begin{aligned} x_{2s}(t) = &\, x_s(t)\cos\left[2\pi(f_2 - f_c)t\right]\cos(\varepsilon_2 - \varepsilon) - x_s(t)\sin\left[2\pi(f_2 - f_c)t\right]\sin(\varepsilon_2 - \varepsilon) - \\ &x_s(t)\cos\left[2\pi(f_2 + f_c)t\right]\cos(\varepsilon_2 + \varepsilon) + x_s(t)\sin\left[2\pi(f_2 + f_c)t\right]\sin(\varepsilon_2 + \varepsilon). \end{aligned} \tag{11.3-7}$$

The complex frequency response of an ILPF is given by

$$H(f) = G\exp(-j2\pi f t_0)\mathrm{rect}\left[f/(2B)\right], \tag{11.3-8}$$

where $G > 0$ is the dimensionless gain of the filter, t_0 is the constant time delay of the filter in seconds,

$$\text{rect}\left(\frac{f}{2B}\right) = \begin{cases} 1, & |f| \le B \\ 0, & |f| > B \end{cases} \tag{11.3-9}$$

and B is the bandwidth of the filter in hertz, also referred to as the passband. The magnitude and phase responses of the filter are given by

$$|H(f)| = G\,\text{rect}[f/(2B)] \tag{11.3-10}$$

and

$$\angle H(f) = -2\pi f t_0 \,\text{rect}[f/(2B)], \tag{11.3-11}$$

respectively (see Fig. 11.3-2). The slope of the straight line in the phase response is equal to $-2\pi t_0$.

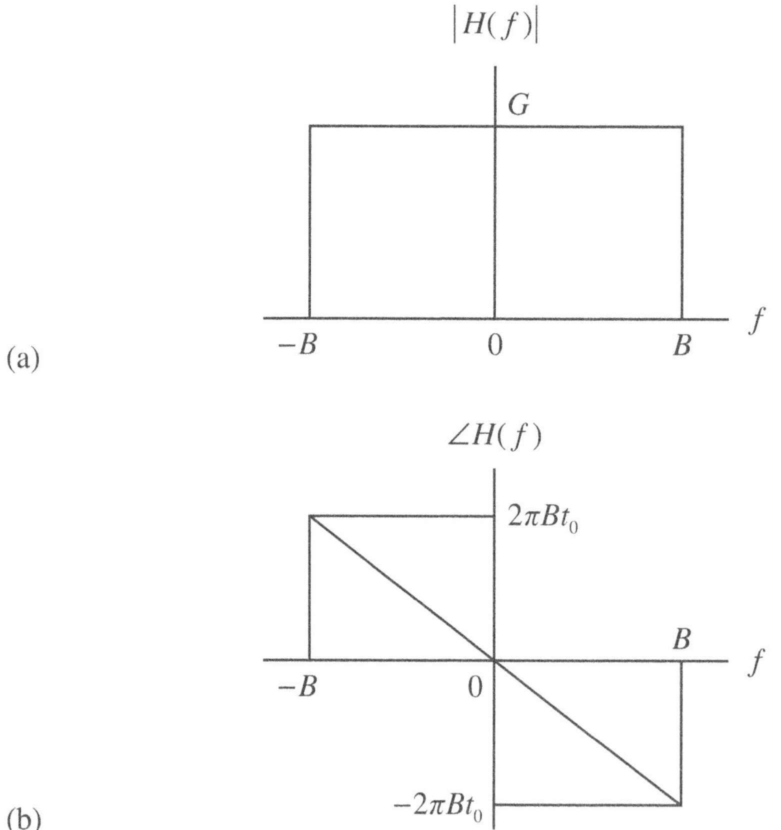

Figure 11.3-2 (a) Magnitude response and (b) phase response of an ideal, lowpass filter.

If $f_1 \approx f_c$ and $f_2 \approx f_c$, then $f_1 - f_c \approx 0$, $f_1 + f_c \approx 2f_c$, $f_2 - f_c \approx 0$, and $f_2 + f_c \approx 2f_c$. Therefore, if the bandwidth (passband) of the ILPFs $B = W$ Hz, which is the bandwidth of the lowpass cosine and sine components [see (11.2-17) and (11.2-18)], and $f_c > W$ [see (11.2-20)], then the output signal $y_1(t)$ from the ILPF in the upper channel (the I channel) of the QD is given by

$$y_1(t) = y_{1c}(t) + y_{1s}(t), \tag{11.3-12}$$

where

$$\begin{aligned} y_{1c}(t) \approx\ & Gx_c(t-t_0)\cos\left[2\pi(f_1 - f_c)(t-t_0)\right]\cos(\varepsilon_1 - \varepsilon) - \\ & Gx_c(t-t_0)\sin\left[2\pi(f_1 - f_c)(t-t_0)\right]\sin(\varepsilon_1 - \varepsilon) \end{aligned} \tag{11.3-13}$$

and

$$\begin{aligned} y_{1s}(t) \approx\ & Gx_s(t-t_0)\cos\left[2\pi(f_1 - f_c)(t-t_0)\right]\sin(\varepsilon_1 - \varepsilon) + \\ & Gx_s(t-t_0)\sin\left[2\pi(f_1 - f_c)(t-t_0)\right]\cos(\varepsilon_1 - \varepsilon), \end{aligned} \tag{11.3-14}$$

and the output signal $y_2(t)$ from the ILPF in the lower channel (the Q channel) of the QD is given by

$$y_2(t) = y_{2c}(t) + y_{2s}(t), \tag{11.3-15}$$

where

$$\begin{aligned} y_{2c}(t) \approx\ & -Gx_c(t-t_0)\cos\left[2\pi(f_2 - f_c)(t-t_0)\right]\sin(\varepsilon_2 - \varepsilon) - \\ & Gx_c(t-t_0)\sin\left[2\pi(f_2 - f_c)(t-t_0)\right]\cos(\varepsilon_2 - \varepsilon) \end{aligned} \tag{11.3-16}$$

and

$$\begin{aligned} y_{2s}(t) \approx\ & Gx_s(t-t_0)\cos\left[2\pi(f_2 - f_c)(t-t_0)\right]\cos(\varepsilon_2 - \varepsilon) - \\ & Gx_s(t-t_0)\sin\left[2\pi(f_2 - f_c)(t-t_0)\right]\sin(\varepsilon_2 - \varepsilon). \end{aligned} \tag{11.3-17}$$

The output signals given by (11.3-13) and (11.3-14), and (11.3-16) and (11.3-17), are shown as being approximate because $f_1 - f_c \neq 0$ and $f_2 - f_c \neq 0$. If $f_1 - f_c \neq 0$ and $f_2 - f_c \neq 0$, then the ILPFs will filter out some of the high frequency components contained in the input signals to the ILPFs involving $f_1 - f_c$ and $f_2 - f_c$ because the magnitude spectra of these input signals are *not* centered at $f = 0$ Hz. These output signals are unacceptable because of the frequency and phase offsets even though $f_1 - f_c \approx 0$ and $f_2 - f_c \approx 0$. Ideally, we want $y_1(t)$ to be equal to an amplitude-scaled and time-delayed version of $x_c(t)$ alone, and $y_2(t)$ to be equal to an amplitude-scaled and time-delayed version of $x_s(t)$ alone.

If there are *no* frequency offsets, that is, if $f_1 = f_c$ and $f_2 = f_c$, then

$$x_{1c}(t) = x_c(t)\cos(\varepsilon_1 - \varepsilon) + x_c(t)\cos\left[2\pi(2f_c)t\right]\cos(\varepsilon_1 + \varepsilon) - x_c(t)\sin\left[2\pi(2f_c)t\right]\sin(\varepsilon_1 + \varepsilon), \tag{11.3-18}$$

$$x_{1s}(t) = x_s(t)\sin(\varepsilon_1 - \varepsilon) - x_s(t)\cos\left[2\pi(2f_c)t\right]\sin(\varepsilon_1 + \varepsilon) - x_s(t)\sin\left[2\pi(2f_c)t\right]\cos(\varepsilon_1 + \varepsilon), \tag{11.3-19}$$

$$x_{2c}(t) = -x_c(t)\sin(\varepsilon_2 - \varepsilon) - x_c(t)\cos\left[2\pi(2f_c)t\right]\sin(\varepsilon_2 + \varepsilon) - x_c(t)\sin\left[2\pi(2f_c)t\right]\cos(\varepsilon_2 + \varepsilon), \tag{11.3-20}$$

and

$$x_{2s}(t) = x_s(t)\cos(\varepsilon_2 - \varepsilon) - x_s(t)\cos\left[2\pi(2f_c)t\right]\cos(\varepsilon_2 + \varepsilon) + x_s(t)\sin\left[2\pi(2f_c)t\right]\sin(\varepsilon_2 + \varepsilon). \tag{11.3-21}$$

Therefore,

$$y_{1c}(t) = Gx_c(t - t_0)\cos(\varepsilon_1 - \varepsilon), \tag{11.3-22}$$

$$y_{1s}(t) = Gx_s(t - t_0)\sin(\varepsilon_1 - \varepsilon), \tag{11.3-23}$$

$$y_{2c}(t) = -Gx_c(t - t_0)\sin(\varepsilon_2 - \varepsilon), \tag{11.3-24}$$

and

$$y_{2s}(t) = Gx_s(t - t_0)\cos(\varepsilon_2 - \varepsilon). \tag{11.3-25}$$

These output signals are still unacceptable because of the phase offsets.

Finally, if in addition to no frequency offsets there are *no* phase offsets, that is, if $\varepsilon_1 = \varepsilon$ and $\varepsilon_2 = \varepsilon$, then

$$x_{1c}(t) = x_c(t) + x_c(t)\cos\left[2\pi(2f_c)t\right]\cos(2\varepsilon) - x_c(t)\sin\left[2\pi(2f_c)t\right]\sin(2\varepsilon), \tag{11.3-26}$$

$$x_{1s}(t) = -x_s(t)\cos\left[2\pi(2f_c)t\right]\sin(2\varepsilon) - x_s(t)\sin\left[2\pi(2f_c)t\right]\cos(2\varepsilon), \tag{11.3-27}$$

$$x_{2c}(t) = -x_c(t)\cos\left[2\pi(2f_c)t\right]\sin(2\varepsilon) - x_c(t)\sin\left[2\pi(2f_c)t\right]\cos(2\varepsilon), \tag{11.3-28}$$

and

$$x_{2s}(t) = x_s(t) - x_s(t)\cos\left[2\pi(2f_c)t\right]\cos(2\varepsilon) + x_s(t)\sin\left[2\pi(2f_c)t\right]\sin(2\varepsilon). \tag{11.3-29}$$

Therefore,

$$\boxed{y_1(t) = Gx_c(t - t_0)} \tag{11.3-30}$$

and

$$\boxed{y_2(t) = Gx_s(t - t_0)} \tag{11.3-31}$$

where $G > 0$ is the dimensionless gain of the ILPFs in the QD, and t_0 is the constant time delay of the ILPFs.

Equations (11.3-30) and (11.3-31) indicate that when there are *no* frequency and phase offsets, the output signals $y_1(t)$ and $y_2(t)$ from the QD shown in Fig. 11.3-1 are amplitude-scaled and time-delayed versions of the cosine and sine components of the amplitude-and-angle-modulated carrier $x(t)$, given by (11.3-1), at the input to the QD. The cosine and sine components are the real and imaginary parts of the complex envelope $\tilde{x}(t)$ of $x(t)$ [see (11.2-27)]. Signal processing algorithms can be used to drive frequency and phase offsets to zero. By properly combining $y_1(t)$ and $y_2(t)$ given by (11.3-30) and (11.3-31), $x(t)$ can be *demodulated*. For example,

$$\boxed{\sqrt{y_1^2(t) + y_2^2(t)} = Ga(t - t_0)} \tag{11.3-32}$$

which is an amplitude-scaled and time-delayed version of the amplitude-modulating function $a(t)$, and

$$\boxed{\tan^{-1}\left[y_2(t) / y_1(t) \right] = \theta(t - t_0)} \tag{11.3-33}$$

which is a time-delayed version of the angle-modulating function $\theta(t)$. Since the values for $\theta(t - t_0)$ obtained from the arctangent function lie in the interval $[-\pi, \pi]$, that is, $-\pi \leq \theta(t - t_0) \leq \pi$, depending on the kind of angle modulation used, *phase unwrapping* may be necessary in order to obtain the full range of values for $\theta(t - t_0)$.

Problems

Section 11.1

11-1 Use (11.1-8) to verify

(a) (11.1-13)

(b) (11.1-14)

11-2 If the power spectrum $S_{\tilde{x}}(f)$ of the complex envelope $\tilde{x}(t)$ is given as shown in Fig. P11-2, then plot the power spectrum $S_x(f)$ of the real bandpass signal $x(t)$.

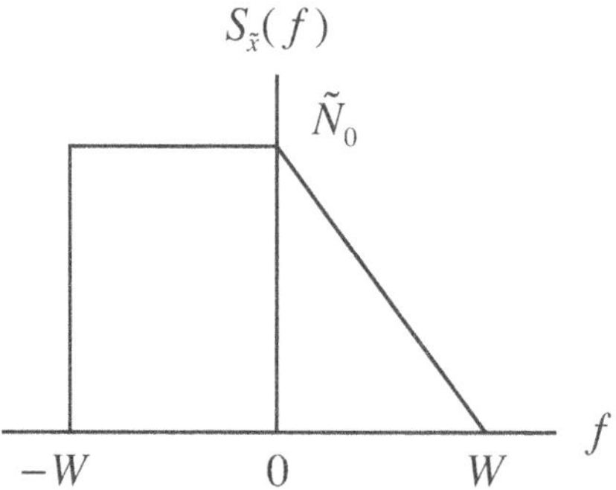

Figure P11-2

Section 11.2

11-3 A very common transmitted electrical signal used in active sonar (and radar) systems is the *rectangular-envelope*, *CW* (continuous-wave) *pulse* given by

$$x(t) = A\cos(2\pi f_c t), \qquad 0 \leq t \leq T,$$

where the amplitude factor A is a positive constant with units of volts, and T is the pulse length of $x(t)$ in seconds.

(a) Find the complex envelope of $x(t)$.

(b) Find the envelope of $x(t)$.

(c) Find the energy of $x(t)$ with the correct units.

(d) Find the time-average power of $x(t)$ with the correct units.

11-4 Another very common transmitted electrical signal used in active sonar (and radar) systems is the *rectangular-envelope*, *LFM* (linear-frequency-modulated) *pulse* given by

$$x(t) = A\cos[2\pi f_c t + \theta(t)], \qquad 0 \leq t \leq T,$$

where the amplitude factor A is a positive constant with units of volts,

$$\theta(t) = D_p\left(t - \frac{T}{2}\right)^2$$

is the angle-modulating function in radians, D_p is the phase-deviation constant in radians per second-squared, and T is the pulse length of $x(t)$ in seconds.

(a) Find the complex envelope of $x(t)$.

(b) Find the envelope of $x(t)$.

(c) Find the instantaneous frequency of $x(t)$ in hertz.

(d) Find the energy of $x(t)$ with the correct units.

(e) Find the time-average power of $x(t)$ with the correct units.

11-5 Use (11.1-24) to compute the complex frequency spectrum of the *double-sideband, suppressed-carrier* (DSBSC) waveform given by

$$x(t) = a(t)\cos(2\pi f_c t).$$

Section 11.3

11-6 Find the output signals from the I and Q channels of the quadrature demodulator shown in Fig. 11.3-1 if there are no frequency offsets and $\varepsilon_1 - \varepsilon = \pi/2$ and $\varepsilon_2 - \varepsilon = \pi/2$.

11-7 If the input signal to the quadrature demodulator shown in Fig. 11.3-1 is an amplitude-modulated carrier, that is, there is no angle-modulating function, then find the output signals from the I and Q channels

(a) if there are no frequency offsets.

(b) if there are no frequency offsets and $\varepsilon_1 - \varepsilon = \pi/2$ and $\varepsilon_2 - \varepsilon = \pi/2$.

Chapter 12

Target Detection in the Presence of Reverberation and Noise

When a sonar system goes active, some of the transmitted (radiated) acoustic power may be scattered by an object of interest. Scatter from an object of interest measured at a receiver is referred to as a *target return*. However, some of the transmitted acoustic power may also be scattered by unintentional objects (i.e., objects of no interest) such as a school of fish, marine mammals, air bubbles, or other particulate matter (examples of volume reverberation), the ocean surface (surface reverberation), and the ocean bottom (bottom reverberation). Scatter from objects of no interest measured at a receiver is referred to as a *reverberation return*. Since the transmitted electrical signals used in active sonar systems are amplitude-and-angle-modulated carriers (i.e., real *bandpass* signals), it will be more convenient to do the analysis in this chapter using complex envelopes (see Chapter 11).

12.1 A Binary Hypothesis-Testing Problem

In this section we shall consider the following *binary hypothesis-testing problem* of trying to detect a target return in the presence of *interference* (reverberation plus noise) for a monostatic (backscatter) geometry:

$$\mathrm{H}_0: \quad \tilde{r}(t) = \tilde{z}(t) \tag{12.1-1}$$

$$\mathrm{H}_1: \quad \tilde{r}(t) = \tilde{y}_{Trgt}(t) + \tilde{z}(t) \tag{12.1-2}$$

where

$$\tilde{z}(t) = \tilde{y}_{Rev}(t) + \tilde{y}_{n_a}(t) + \tilde{n}_r(t). \tag{12.1-3}$$

Hypothesis H_0, the null hypothesis, states that the complex envelope of the received signal, $\tilde{r}(t)$, is equal to the sum of the complex envelopes of the *reverberation return*, $\tilde{y}_{Rev}(t)$, ambient noise, $\tilde{y}_{n_a}(t)$, and receiver noise, $\tilde{n}_r(t)$. Hypothesis H_1 states that the complex envelope of the received signal is equal to the sum of the complex envelopes of the *target return*, $\tilde{y}_{Trgt}(t)$, reverberation return, ambient noise, and receiver noise. The functions $\tilde{y}_{Trgt}(t)$, $\tilde{y}_{Rev}(t)$, $\tilde{y}_{n_a}(t)$, and $\tilde{n}_r(t)$ are all random processes. Therefore, $\tilde{r}(t)$ is also a random process. The complex envelope of the received signal is due to processing the output electrical signals (in volts) from all the elements in a linear array. The linear array is composed of an odd number N of identical, equally-spaced, complex-weighted, omnidirectional point-elements lying along the X axis. The problem is to decide which hypothesis is correct.

The mathematical model that we shall use for the complex envelope of the target return is given by

$$\tilde{y}_{Trgt}(t) = \sum_{i=-N'}^{N'} \tilde{y}_{Trgt}(t, x_i), \tag{12.1-4}$$

where (see Appendix 12A)

$$\tilde{y}_{Trgt}(t, x_i) = a_i w_{Trgt} \tilde{x}(t - \tau_{Trgt}) \exp\left[+j2\pi\eta_{D,Trgt}(t - \tau_{Trgt})\right] \exp(-j2\pi f_c \tau_{Trgt}) \tag{12.1-5}$$

is the complex envelope of the output electrical signal from element i in the linear array due to the target after complex weighting for $t \geq \tau_{Trgt}$, a_i is the real, dimensionless, amplitude weight applied to element i,

$$w_{Trgt} = \frac{1}{(4\pi r_{Trgt})^2} \mathcal{S}_T(f_c) S_{Trgt}(f_c, \theta_{Trgt}, \psi_{Trgt}, \theta_{Trgt\,0}, \psi_{Trgt\,0}) \exp[-2\alpha(f_c) r_{Trgt}] \mathcal{S}_R(f_c) \tag{12.1-6}$$

is a dimensionless, complex, *nonzero-mean*, Gaussian, random variable (RV); $(r_{Trgt}, \theta_{Trgt}, \psi_{Trgt})$ are the spherical coordinates of the center of mass of the target, measured from the center of the array (see Fig. 12.1-1); $S_{Trgt}(\bullet)$ is the *target scattering function* in meters, and is treated as a complex, *nonzero-mean*, Gaussian RV (a statistical model of a scattering function is discussed in Section 10.6); $\theta_{Trgt\,0} = 180° - \theta_{Trgt}$, $\psi_{Trgt\,0} = 180° + \psi_{Trgt}$, $\mathcal{S}_T(f_c)$ and $\mathcal{S}_R(f_c)$ are the transmitter and receiver sensitivity functions of the identical, omnidirectional point-elements with units of $\left(\text{m}^3/\text{sec}\right)/\text{V}$ and $\text{V}/\left(\text{m}^2/\text{sec}\right)$, respectively; $\alpha(f_c)$ is the attenuation coefficient of seawater in nepers per meter, f_c is the carrier frequency in hertz, $\tilde{x}(t)$ is the complex envelope of the transmitted electrical signal in volts, τ_{Trgt} and $\eta_{D,Trgt}$ are the round-trip time delay in seconds and Doppler shift in hertz of the target measured from the center of the array, and

$$N' = (N-1)/2\,. \tag{12.1-7}$$

The parameters τ_{Trgt} and $\eta_{D,Trgt}$ shall be treated as *unknown, nonrandom constants* to be *estimated.* Equation (12.1-5) indicates that the complex envelopes of the output electrical signals from all the elements in the array due to the target

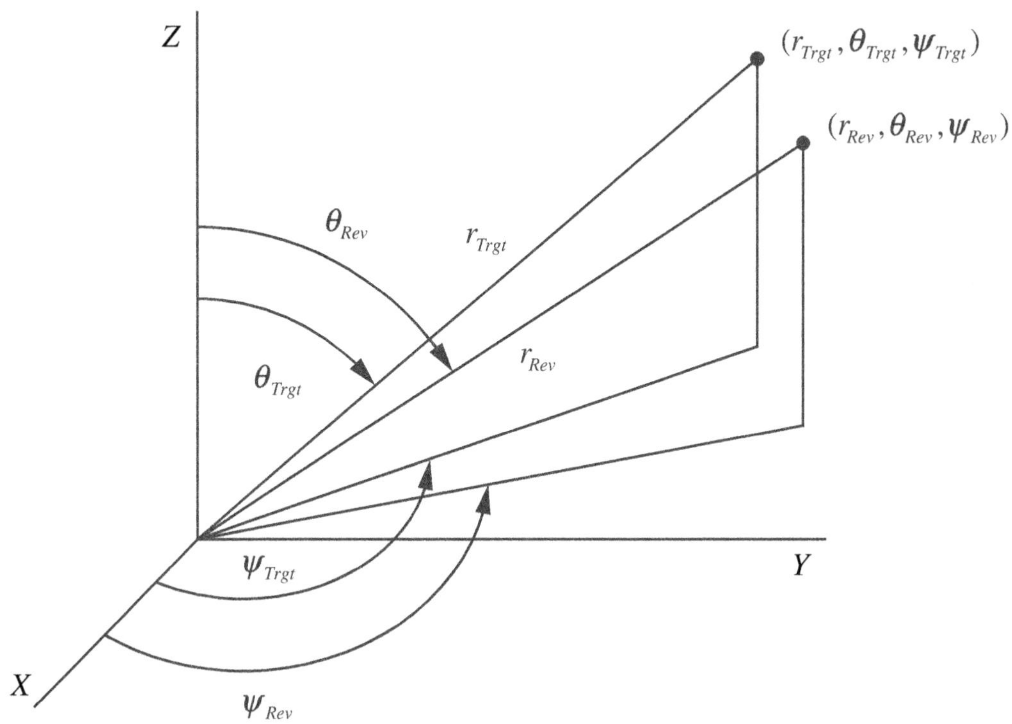

Figure 12.1-1 Spherical coordinates $(r_{Trgt}, \theta_{Trgt}, \psi_{Trgt})$ and $(r_{Rev}, \theta_{Rev}, \psi_{Rev})$ of the center of mass of the target and object responsible for the reverberation, respectively.

after complex weighting are *in-phase*. The target return $\tilde{y}_{Trgt}(t)$ is a complex, *nonzero-mean*, Gaussian, random process.

The mathematical model that we shall use for the complex envelope of the reverberation return is given by

$$\tilde{y}_{Rev}(t) = \sum_{i=-N'}^{N'} \tilde{y}_{Rev}(t, x_i), \tag{12.1-8}$$

where (see Appendix 12A)

$$\tilde{y}_{Rev}(t, x_i) = a_i w_{Rev} \tilde{x}\left(t - [\tau_{i,Rev} + \tau'_{i,Trgt}]\right) \exp\left\{+j2\pi\eta_{D,Rev}\left[t - (\tau_{i,Rev} + \tau'_{i,Trgt})\right]\right\} \times \exp\left[-j2\pi f_c(\tau_{i,Rev} + \tau'_{i,Trgt})\right] \tag{12.1-9}$$

is the complex envelope of the output electrical signal from element i in the linear array due to reverberation after complex weighting for $t \geq \tau_{i,Rev} + \tau'_{i,Trgt}$,

$$w_{Rev} = \frac{1}{(4\pi r_{Rev})^2} \mathcal{S}_T(f_c) S_{Rev}(f_c, \theta_{Rev}, \psi_{Rev}, \theta_{Rev0}, \psi_{Rev0}) \exp[-2\alpha(f_c) r_{Rev}] \mathcal{S}_R(f_c)$$

(12.1-10)

is a dimensionless, complex, *nonzero-mean*, Gaussian RV; $(r_{Rev}, \theta_{Rev}, \psi_{Rev})$ are the spherical coordinates of the center of mass of the object responsible for the reverberation, measured from the center of the array (see Fig. 12.1-1); $S_{Rev}(\bullet)$ is the *reverberation scattering function* in meters, and is treated as a complex, *nonzero-mean*, Gaussian RV (as mentioned earlier, a statistical model of a scattering function is discussed in Section 10.6); $\theta_{Rev0} = 180° - \theta_{Rev}$, $\psi_{Rev0} = 180° + \psi_{Rev}$, $\tau_{i,Rev}$ is the round-trip time delay in seconds from element i in the linear array to the object responsible for the reverberation, $\tau'_{i,Trgt}$ is the time-delay applied to element i via phase weighting in order to co-phase the output electrical signals from all the elements in the array due to the target (*not* the reverberation), and $\eta_{D,Rev}$ is the Doppler shift in hertz of the object responsible for the reverberation measured from the center of the array. Since $\tau_{i,Rev}$ and $\tau'_{i,Trgt}$ depend on index i, (12.1-9) indicates that the complex envelopes of the output electrical signals from all the elements in the array due to reverberation are *out-of-phase*. The reverberation return $\tilde{y}_{Rev}(t)$ is a complex, *nonzero-mean*, Gaussian, random process.

The mathematical model that we shall use for the complex envelope of ambient noise is given by

$$\tilde{y}_{n_a}(t) = \sum_{i=-N'}^{N'} \tilde{y}_{n_a}(t, x_i), \tag{12.1-11}$$

where (see Section 7.4)

$$\tilde{y}_{n_a}(t, x_i) = a_i \tilde{n}_a(t - \tau'_{i,Trgt}, x_i) \tag{12.1-12}$$

is the complex envelope of the output electrical signal from element i in the linear array due to ambient noise after complex weighting. Since $\tau'_{i,Trgt}$ depends on index i, (12.1-12) indicates that the complex envelopes of the output electrical signals from all the elements in the array due to ambient noise are *out-of-phase*. The complex envelope $\tilde{y}_{n_a}(t)$ is a complex, zero-mean, Gaussian, random process.

And finally, the mathematical model that we shall use for the complex envelope of receiver noise is given by

$$\tilde{n}_r(t) = \sum_{i=-N'}^{N'} \tilde{n}_r(t, x_i), \tag{12.1-13}$$

where (see Section 7.4)

$$\tilde{n}_r(t, x_i) = a_i \tilde{n}(t - \tau'_{i,Trgt}, x_i) \tag{12.1-14}$$

is the complex envelope of receiver noise at the output of element i in the linear array after complex weighting. Since $\tau'_{i,Trgt}$ depends on index i, (12.1-14) indicates that the complex envelopes of the output electrical signals from all the elements in the array due to receiver noise are *out-of-phase*. The complex envelope $\tilde{n}_r(t)$ is a complex, zero-mean, Gaussian, random process.

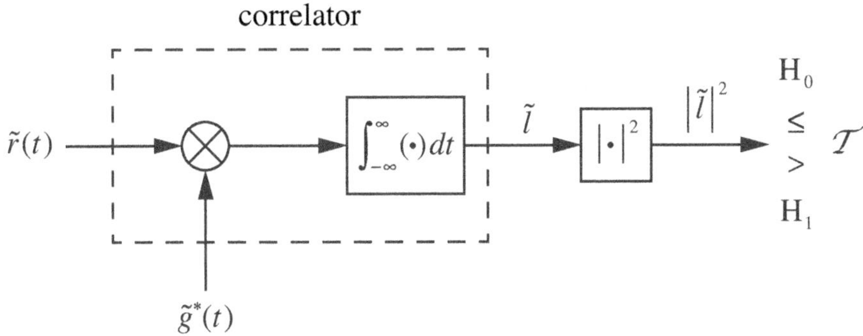

Figure 12.1-2 Correlator receiver followed by a magnitude-squared operation.

In order to decide whether or not a target return is present, we shall process $\tilde{r}(t)$ with the receiver shown in Fig. 12.1-2. The deterministic function $\tilde{g}(t)$ is referred to as the *processing waveform*, as of yet unspecified. The receiver performs the following test: choose hypothesis H_0 if

$$|\tilde{l}|^2 \leq \mathcal{T}, \tag{12.1-15}$$

and choose hypothesis H_1 if

$$|\tilde{l}|^2 > \mathcal{T}, \tag{12.1-16}$$

where

$$\tilde{l} = \langle \tilde{r}(t), \tilde{g}(t) \rangle \triangleq \int_{-\infty}^{\infty} \tilde{r}(t)\tilde{g}^*(t)dt \tag{12.1-17}$$

is the output from the correlator and is, in fact, the *inner product* of $\tilde{r}(t)$ and $\tilde{g}(t)$. It is also a complex, Gaussian RV. The inner product of two functions of time, denoted by $\langle \bullet, \bullet \rangle$, is analogous to the inner product of two vectors. The inner product is a linear operator, that is, it satisfies the principle of superposition. The real RV, $|\tilde{l}|^2$, is *non-Gaussian* and is referred to as the *sufficient statistic* or *test statistic* because it contains all the information necessary to make a decision. The *decision threshold* $\mathcal{T}$ is commonly chosen to satisfy the *Neyman-Pearson criterion*, which is to *maximize the probability of detection for a given probability of false alarm*. The Neyman-Pearson criterion is used in both sonar and radar signal processing. The decision threshold $\mathcal{T}$ required to satisfy the Neyman-Pearson criterion will be determined in Section 12.3.

12.2 The Signal-to-Interference Ratio

In this section we shall derive the output *signal-to-interference ratio* (SIR) of the correlator shown in Fig. 12.1-2, which is defined as follows:

$$\boxed{\text{SIR} \triangleq \frac{E\left\{\left|\tilde{l}_{Trgt}\right|^2\right\}}{E\left\{\left|\tilde{l}_Z\right|^2\right\}}} \tag{12.2-1}$$

where

$$\tilde{l}_{Trgt} = \left\langle \tilde{y}_{Trgt}(t), \tilde{g}(t) \right\rangle = \int_{-\infty}^{\infty} \tilde{y}_{Trgt}(t)\tilde{g}^*(t)dt \tag{12.2-2}$$

is the output from the correlator due to processing the complex envelope of the target return, and

$$\tilde{l}_Z = \left\langle \tilde{z}(t), \tilde{g}(t) \right\rangle = \int_{-\infty}^{\infty} \tilde{z}(t)\tilde{g}^*(t)dt \tag{12.2-3}$$

is the output from the correlator due to processing the sum of the complex envelopes of the reverberation return, ambient noise, and receiver noise [see (12.1-3)]. Both $\tilde{l}_{Trgt}$ and $\tilde{l}_Z$ are complex, *nonzero-mean*, Gaussian RVs. When modeling reverberation and noise, the term SIR is used rather than signal-to-noise ratio (SNR). Substituting (12.1-3) into (12.2-3) yields

$$\tilde{l}_Z = \tilde{l}_{Rev} + \tilde{l}_{n_a} + \tilde{l}_{n_r}, \tag{12.2-4}$$

where

$$\tilde{l}_{Rev} = \left\langle \tilde{y}_{Rev}(t), \tilde{g}(t) \right\rangle = \int_{-\infty}^{\infty} \tilde{y}_{Rev}(t)\tilde{g}^*(t)dt \tag{12.2-5}$$

is the output from the correlator due to processing the complex envelope of the reverberation return,

$$\tilde{l}_{n_a} = \left\langle \tilde{y}_{n_a}(t), \tilde{g}(t) \right\rangle = \int_{-\infty}^{\infty} \tilde{y}_{n_a}(t)\tilde{g}^*(t)dt \tag{12.2-6}$$

is the output from the correlator due to processing the complex envelope of the ambient noise, and

$$\tilde{l}_{n_r} = \langle \tilde{n}_r(t), \tilde{g}(t) \rangle = \int_{-\infty}^{\infty} \tilde{n}_r(t) \tilde{g}^*(t) dt \tag{12.2-7}$$

is the output from the correlator due to processing the complex envelope of the receiver noise, where $\tilde{l}_{Rev}$ is a complex, *nonzero-mean*, Gaussian RV, and both $\tilde{l}_{n_a}$ and $\tilde{l}_{n_r}$ are complex, zero-mean, Gaussian RVs.

In order to derive the SIR, we also need to specify the processing waveform $\tilde{g}(t)$. Substituting (12.1-5) into (12.1-4) yields

$$\tilde{y}_{Trgt}(t) = A_1 w_{Trgt} \tilde{x}(t - \tau_{Trgt}) \exp(+j2\pi \eta_{D,Trgt} t) \exp\left[-j2\pi(f_c + \eta_{D,Trgt})\tau_{Trgt}\right], \tag{12.2-8}$$

where

$$A_1 = \sum_{i=-N'}^{N'} a_i \,. \tag{12.2-9}$$

Therefore, let

$$\tilde{g}(t) = \tilde{x}(t - \hat{\tau}_{Trgt}) \exp(+j2\pi \hat{\eta}_{D,Trgt} t), \tag{12.2-10}$$

which is similar in form to the target return $\tilde{y}_{Trgt}(t)$ given by (12.2-8), where $\hat{\tau}_{Trgt}$ and $\hat{\eta}_{D,Trgt}$ are *estimates* of the round-trip time delay τ_{Trgt} and Doppler shift $\eta_{D,Trgt}$ of the target, respectively.

Now that we have an equation for the processing waveform $\tilde{g}(t)$, let us compute $\tilde{l}_{Trgt}$. Substituting (12.2-8) and (12.2-10) into (12.2-2) yields

$$\tilde{l}_{Trgt} = A_1 w_{Trgt} \int_{-\infty}^{\infty} \tilde{x}(t - \tau_{Trgt}) \tilde{x}^*(t - \hat{\tau}_{Trgt}) \exp\left[-j2\pi(\hat{\eta}_{D,Trgt} - \eta_{D,Trgt})t\right] dt \times \\ \exp\left[-j2\pi(f_c + \eta_{D,Trgt})\tau_{Trgt}\right]. \tag{12.2-11}$$

If we let $t' = t - \tau_{Trgt}$ in (12.2-11), then

$$\tilde{l}_{Trgt} = A_1 w_{Trgt} X(e_{\tau,Trgt}, e_{D,Trgt}) \exp\left[-j2\pi(f_c + \hat{\eta}_{D,Trgt})\tau_{Trgt}\right], \tag{12.2-12}$$

where

$$\begin{aligned} X(e_{\tau,Trgt}, e_{D,Trgt}) &= \int_{-\infty}^{\infty} \tilde{x}(t)\tilde{x}^*(t - e_{\tau,Trgt})\exp(-j2\pi e_{D,Trgt}t)dt \\ &= F\left\{\tilde{x}(t)\tilde{x}^*(t - e_{\tau,Trgt})\right\} \\ &= \left\langle \tilde{x}(t), \tilde{x}(t - e_{\tau,Trgt})\exp(+j2\pi e_{D,Trgt}t)\right\rangle \end{aligned} \tag{12.2-13}$$

is the *unnormalized, auto-ambiguity function* of the transmitted electrical signal $x(t)$, with complex envelope $\tilde{x}(t)$;

$$e_{\tau,Trgt} = \hat{\tau}_{Trgt} - \tau_{Trgt} \tag{12.2-14}$$

is the *round-trip time-delay estimation error of the target* in seconds, and

$$e_{D,Trgt} = \hat{\eta}_{D,Trgt} - \eta_{D,Trgt} \tag{12.2-15}$$

is the *Doppler-shift estimation error of the target* in hertz. The auto-ambiguity function is complex in general, and is a function of two estimation errors.

The auto-ambiguity function given by (12.2-13) has its *maximum* value at $e_{\tau,Trgt} = 0$ and $e_{D,Trgt} = 0$ where

$$\begin{aligned} X(0,0) &= \int_{-\infty}^{\infty} \left|\tilde{x}(t)\right|^2 dt \\ &= \left\langle \tilde{x}(t), \tilde{x}(t)\right\rangle \\ &= E_{\tilde{x}} \end{aligned} \tag{12.2-16}$$

is the energy of $\tilde{x}(t)$ in joules-ohms because $\tilde{x}(t)$ is a voltage signal. With the use of (12.2-16), the *normalized*, auto-ambiguity function is defined as follows:

$$X_N(e_{\tau,Trgt}, e_{D,Trgt}) \triangleq X(e_{\tau,Trgt}, e_{D,Trgt})/E_{\tilde{x}} \tag{12.2-17}$$

so that

$$X_N(0,0) = 1. \tag{12.2-18}$$

Substituting (12.2-17) into (12.2-12) and computing the second moment (mean-squared value) of $\tilde{l}_{Trgt}$ yields

$$E\left\{\left|\tilde{l}_{Trgt}\right|^2\right\} = A_1^2 E_{\tilde{x}}^2 E\left\{\left|w_{Trgt}\right|^2\right\}\left|X_N(e_{\tau,Trgt}, e_{D,Trgt})\right|^2 \tag{12.2-19}$$

where A_1 is given by (12.2-9), and from (12.1-6),

$$E\left\{\left|w_{Trgt}\right|^2\right\} = \frac{1}{(4\pi r_{Trgt})^4}\left|\mathcal{S}_T(f_c)\right|^2 E\left\{\left|S_{Trgt}(f_c, \theta_{Trgt}, \psi_{Trgt}, \theta_{Trgt\,0}, \psi_{Trgt\,0})\right|^2\right\}\times$$
$$\exp[-4\alpha(f_c)r_{Trgt}]\left|\mathcal{S}_R(f_c)\right|^2, \tag{12.2-20}$$

where $\theta_{Trgt\,0} = 180° - \theta_{Trgt}$ and $\psi_{Trgt\,0} = 180° + \psi_{Trgt}$. As can be seen from (12.2-19), the mean-squared value of $\tilde{l}_{Trgt}$, which is the numerator of the SIR, is directly proportional to the magnitude-squared of the normalized, auto-ambiguity function. Therefore, if the estimation errors $e_{\tau,Trgt} = 0$ and $e_{D,Trgt} = 0$, then $\left|\mathcal{X}_N(0,0)\right|^2 = 1$ [see (12.2-18)] and the numerator of the SIR is maximized. Conversely, as the magnitudes of the estimation errors $e_{\tau,Trgt}$ and $e_{D,Trgt}$ increase, $\left|\mathcal{X}_N(e_{\tau,Trgt}, e_{D,Trgt})\right|^2 \to 0$ and, as a result, the SIR $\to 0$, which means that the target won't be detected. Another way to explain this is as follows. Under hypothesis H_1, the output of the correlator $\tilde{l} = \tilde{l}_{Trgt} + \tilde{l}_Z$. However, as can be seen from (12.2-12), as $\mathcal{X}(e_{\tau,Trgt}, e_{D,Trgt}) \to 0$, $\tilde{l}_{Trgt} \to 0$. Therefore, $\tilde{l} \to \tilde{l}_Z$. As a result, the sufficient statistic $\left|\tilde{l}\right|^2$ won't exceed the decision threshold $\mathcal{T}$ and hypothesis H_0 will be chosen. Therefore, the target won't be detected.

The second moment (mean-squared value) of $\tilde{l}_Z$, which is the denominator of the SIR, is given by (see Appendix 12B)

$$E\left\{\left|\tilde{l}_Z\right|^2\right\} = E_{\tilde{x}}^2 E\left\{\left|w_{Rev}\right|^2\right\}\left|\sum_{i=-N'}^{N'} a_i \mathcal{X}_N(e_{\tau_i}, e_D)\exp\left[+j2\pi(f_c + \hat{\eta}_{D,Trgt})(\tau_{i,Trgt} - \tau_{i,Rev})\right]\right|^2 +$$
$$A_2 E_{\tilde{x}}\left[\tilde{N}_{a0} + \tilde{N}_{r0}\right] \tag{12.2-21}$$

where [see (12.1-10)]

$$E\left\{\left|w_{Rev}\right|^2\right\} = \frac{1}{(4\pi r_{Rev})^4}\left|\mathcal{S}_T(f_c)\right|^2 E\left\{\left|S_{Rev}(f_c, \theta_{Rev}, \psi_{Rev}, \theta_{Rev\,0}, \psi_{Rev\,0})\right|^2\right\}\times$$
$$\exp[-4\alpha(f_c)r_{Rev}]\left|\mathcal{S}_R(f_c)\right|^2, \tag{12.2-22}$$

$\theta_{Rev\,0} = 180° - \theta_{Rev}$, $\psi_{Rev\,0} = 180° + \psi_{Rev}$,

$$e_{\tau_i} = e_{\tau,\,Trgt} + (\tau_{i,\,Trgt} - \tau_{i,\,Rev}) \tag{12.2-23}$$

is a round-trip time-delay estimation error in seconds, where $e_{\tau,\,Trgt}$ is the round-trip time-delay estimation error of the target given by (12.2-14), and $\tau_{i,\,Trgt}$ and $\tau_{i,\,Rev}$ are the round-trip time delays in seconds from element i in the linear array to the target and the object responsible for the reverberation, respectively;

$$e_D = \hat{\eta}_{D,\,Trgt} - \eta_{D,\,Rev} \tag{12.2-24}$$

is a Doppler-shift estimation error in hertz,

$$A_2 = \sum_{i=-N'}^{N'} a_i^2 \,, \tag{12.2-25}$$

and $\tilde{N}_{a0}$ and $\tilde{N}_{r0}$ are the constant levels in $(\text{W-}\Omega)/\text{Hz}$ (or $\text{J-}\Omega$) of the lowpass (baseband), white-noise power spectra of $\tilde{n}_a(t, x_i)$ and $\tilde{n}(t, x_i)$ $\forall\ i$ before complex weighting. Equation (12.2-21) is based on the following set of statistical assumptions made in Appendix 12B: The random processes $\tilde{n}_a(t, x_i)$ and $\tilde{n}(t, x_i)$ are *zero-mean* and *wide-sense stationary* (WSS) *in time and space* $\forall\ i$, $\tilde{n}_a(t, x_i)$ and $\tilde{n}_a(t, x_j)$ are *uncorrelated* for $i \neq j$, $\tilde{n}(t, x_i)$ and $\tilde{n}(t, x_j)$ are *uncorrelated* for $i \neq j$, and $\tilde{n}_a(t, x_i)$ and $\tilde{n}(t, x_i)$ are *lowpass (baseband), white-noise* random processes in the time-domain $\forall\ i$. Since $\tilde{n}_a(t, x_i)$ and $\tilde{n}(t, x_i)$ are zero-mean and WSS in time $\forall\ i$, the random processes $\tilde{y}_{n_a}(t)$ and $\tilde{n}_r(t)$ are also *zero-mean* and *WSS*. It is also assumed that the random processes $\tilde{y}_{Rev}(t)$ and $\tilde{y}_{n_a}(t)$, $\tilde{y}_{Rev}(t)$ and $\tilde{n}_r(t)$, and $\tilde{y}_{n_a}(t)$ and $\tilde{n}_r(t)$ are *statistically independent*.

Substituting (12.2-19), (12.2-21), (12.2-9), and (12.2-25) into (12.2-1) yields the following expression for the signal-to-interference ratio (SIR):

$$\boxed{\text{SIR} = \frac{\text{AG}\, E_{\tilde{x}} E\left\{\left|w_{Trgt}\right|^2\right\}\left|X_N(e_{\tau,\,Trgt}, e_{D,\,Trgt})\right|^2}{\dfrac{E_{\tilde{x}}}{A_2} E\left\{\left|w_{Rev}\right|^2\right\}\left|\displaystyle\sum_{i=-N'}^{N'} a_i X_N(e_{\tau_i}, e_D)\exp\left[+j2\pi(f_c + \hat{\eta}_{D,\,Trgt})(\tau_{i,\,Trgt} - \tau_{i,\,Rev})\right]\right|^2 + \tilde{N}_{a0} + \tilde{N}_{r0}}}$$

(12.2-26)

where

$$\mathrm{AG} = \frac{\left[\sum_{i=-N'}^{N'} a_i\right]^2}{\sum_{i=-N'}^{N'} a_i^2} \tag{12.2-27}$$

is the *dimensionless* version of the *array gain* of the linear array (see Section 7.5). If rectangular amplitude weights are used, that is, if $a_i = 1 \;\forall\; i$, then $\mathrm{AG} = N$ and $A_2 = N$. As can be seen from (12.2-26), the transmitted signal's energy $E_{\tilde{x}}$ multiplies both the target factor $E\left\{\left|w_{Trgt}\right|^2\right\}$ in the numerator and the reverberation factor $E\left\{\left|w_{Rev}\right|^2\right\}$ in the denominator. Therefore, in a reverberant environment, in order to increase the SIR without increasing the strength of the reverberation return, the AG must be increased. If the estimation errors $e_{\tau,Trgt}$ and $e_{D,Trgt}$ are zero, that is, if $\hat{\tau}_{Trgt} = \tau_{Trgt}$ and $\hat{\eta}_{D,Trgt} = \eta_{D,Trgt}$, then (12.2-26) reduces to [see (12.2-18)]

$$\mathrm{SIR} = \frac{\mathrm{AG}\, E_{\tilde{x}}\, E\left\{\left|w_{Trgt}\right|^2\right\}}{\dfrac{E_{\tilde{x}}}{A_2} E\left\{\left|w_{Rev}\right|^2\right\} \left|\sum_{i=-N'}^{N'} a_i X_N(e_{\tau_i}, e_D) \exp\left[+j2\pi(f_c + \eta_{D,Trgt})(\tau_{i,Trgt} - \tau_{i,Rev})\right]\right|^2 + \tilde{N}_{a0} + \tilde{N}_{r0}}, \tag{12.2-28}$$

where

$$e_{\tau_i} = \tau_{i,Trgt} - \tau_{i,Rev} \tag{12.2-29}$$

because $e_{\tau,Trgt} = 0$ [see (12.2-23)], and

$$e_D = \eta_{D,Trgt} - \eta_{D,Rev} \tag{12.2-30}$$

because $\hat{\eta}_{D,Trgt} = \eta_{D,Trgt}$ [see (12.2-24)].

If *no* reverberation return is present at the receiver, then the SIR given by (12.2-26) reduces to the *signal-to-noise ratio* (SNR)

$$\boxed{\mathrm{SNR} = \left|X_N(e_{\tau,Trgt}, e_{D,Trgt})\right|^2 \mathrm{SNR}_{\max}} \tag{12.2-31}$$

where

$$\boxed{\mathrm{SNR}_{\max} = \mathrm{AG} \times \mathrm{SNR}_0} \tag{12.2-32}$$

is the maximum value of the SNR, and

$$\boxed{\mathrm{SNR}_0 = \frac{E_{\tilde{x}} E\left\{\left|w_{Trgt}\right|^2\right\}}{\tilde{N}_{a0} + \tilde{N}_{r0}}} \tag{12.2-33}$$

is the SNR due to processing the output electrical signal from a single element – element number 0 at the center of the linear array. Note that if the estimation errors $e_{\tau,Trgt}$ and $e_{D,Trgt}$ are zero, then $X_N(0,0)=1$ [see (12.2-18)] and $\mathrm{SNR}=\mathrm{SNR}_{\max}$. Therefore, in a reverberation free environment, the SNR can be increased by increasing the transmitted signal's energy $E_{\tilde{x}}$ and/or the AG.

We shall conclude our discussion in this section by examining the following theoretical limiting case of a target co-located with the object responsible for the reverberation and moving with the same velocity vector. If a target is co-located with the object responsible for the reverberation, then the spherical coordinates $(r_{Trgt}, \theta_{Trgt}, \psi_{Trgt})$ and $(r_{Rev}, \theta_{Rev}, \psi_{Rev})$ are equal, and the round-trip time delays from element i in the linear array to the target and object responsible for the reverberation are equal $\forall\ i$, that is,

$$\tau_{i,Rev} = \tau_{i,Trgt}, \qquad i = -N', \ldots, 0, \ldots, N'. \tag{12.2-34}$$

Substituting (12.2-34) into (12.2-23) yields

$$e_{\tau_i} = e_{\tau,Trgt}, \qquad i = -N', \ldots, 0, \ldots, N', \tag{12.2-35}$$

and by substituting (12.2-34), (12.2-35), and (12.2-25) into (12.2-26), the SIR reduces to

$$\mathrm{SIR} = \frac{\mathrm{AG}\, E_{\tilde{x}} E\left\{\left|w_{Trgt}\right|^2\right\}\left|X_N(e_{\tau,Trgt}, e_{D,Trgt})\right|^2}{\mathrm{AG}\, E_{\tilde{x}} E\left\{\left|w_{Rev}\right|^2\right\}\left|X_N(e_{\tau,Trgt}, e_D)\right|^2 + \tilde{N}_{a0} + \tilde{N}_{r0}}. \tag{12.2-36}$$

Note that in this case, the AG multiplies both the target factor $E\left\{\left|w_{Trgt}\right|^2\right\}$ in the numerator and the reverberation factor $E\left\{\left|w_{Rev}\right|^2\right\}$ in the denominator. Since the target is co-located with the object responsible for the reverberation, the directions of arrival of the acoustic fields incident upon the array due to the target and reverberation are identical. Therefore, the phase weights that will co-phase all the output electrical signals due to the target will also co-phase all the output electrical signals due to reverberation. Furthermore, if the Doppler-shifts of the target and object responsible for the reverberation are equal, that is, if

$$\eta_{D,\,Rev} = \eta_{D,\,Trgt}\,, \tag{12.2-37}$$

then substituting (12.2-37) into (12.2-24) yields

$$e_D = e_{D,\,Trgt}\,, \tag{12.2-38}$$

and by substituting (12.2-38) into (12.2-36),

$$\text{SIR} = \frac{\text{AG}\,E_{\tilde{x}}\,E\left\{\left|w_{Trgt}\right|^2\right\}\left|\chi_N(e_{\tau,\,Trgt}, e_{D,\,Trgt})\right|^2}{\text{AG}\,E_{\tilde{x}}\,E\left\{\left|w_{Rev}\right|^2\right\}\left|\chi_N(e_{\tau,\,Trgt}, e_{D,\,Trgt})\right|^2 + \tilde{N}_{a0} + \tilde{N}_{r0}}\,. \tag{12.2-39}$$

If the estimation errors $e_{\tau,\,Trgt}$ and $e_{D,\,Trgt}$ are zero, then (12.2-39) reduces to [see (12.2-18)]

$$\text{SIR} = \frac{\text{AG}\,E_{\tilde{x}}\,E\left\{\left|w_{Trgt}\right|^2\right\}}{\text{AG}\,E_{\tilde{x}}\,E\left\{\left|w_{Rev}\right|^2\right\} + \tilde{N}_{a0} + \tilde{N}_{r0}}\,, \tag{12.2-40}$$

and if

$$\text{AG}\,E_{\tilde{x}}\,E\left\{\left|w_{Rev}\right|^2\right\} >> \tilde{N}_{a0} + \tilde{N}_{r0}\,, \tag{12.2-41}$$

then (12.2-40) reduces to

$$\text{SIR} = \frac{E\left\{\left|w_{Trgt}\right|^2\right\}}{E\left\{\left|w_{Rev}\right|^2\right\}} = \frac{E\left\{\left|S_{Trgt}(f_c, \theta_{Trgt}, \psi_{Trgt}, \theta_{Trgt\,0}, \psi_{Trgt\,0})\right|^2\right\}}{E\left\{\left|S_{Rev}(f_c, \theta_{Trgt}, \psi_{Trgt}, \theta_{Trgt\,0}, \psi_{Trgt\,0})\right|^2\right\}} \tag{12.2-42}$$

since the spherical coordinates $(r_{Trgt}, \theta_{Trgt}, \psi_{Trgt})$ and $(r_{Rev}, \theta_{Rev}, \psi_{Rev})$ are equal. Equation (12.2-42) is the ratio between the mean-squared values of the target and reverberation scattering functions. Approximate examples of this theoretical limiting case are a motionless target lying on or near the ocean bottom (bottom reverberation), or a very slow moving target near the ocean surface (surface reverberation).

12.3 Probability of False Alarm and Decision Threshold

In this section we shall determine the decision threshold $\mathcal{T}$ shown in Fig. 12.1-2 for a Neyman-Pearson decision criterion which is to maximize the probability of detection for a given probability of false alarm. The decision threshold can be obtained by satisfying a probability of false alarm constraint.

We begin by rewriting the binary hypothesis-testing problem given by (12.1-1) and (12.1-2) in terms of the output of the correlator $\tilde{l}$ as follows:

$$\mathrm{H}_0: \ \tilde{l} = \tilde{l}_Z \tag{12.3-1}$$

$$\mathrm{H}_1: \ \tilde{l} = \tilde{l}_{Trgt} + \tilde{l}_Z \tag{12.3-2}$$

where $\tilde{l}_Z$ and $\tilde{l}_{Trgt}$ are given by (12.2-4) and (12.2-12), respectively. For hypothesis H_0, $\tilde{l}$ shall be expressed in rectangular form as follows:

$$\mathrm{H}_0: \ \tilde{l} = \tilde{l}_Z = X_0 + jY_0 . \tag{12.3-3}$$

Since $\tilde{l}$ is a complex, Gaussian, random variable (RV), X_0 and Y_0 are real, Gaussian RVs. In order to compute the probability of false alarm, we first need to determine some of the statistics of $\tilde{l}$ given by (12.3-3).

The mean value (average value) of $\tilde{l}$ given hypothesis H_0 is true is given by

$$E\left\{\tilde{l}\,\middle|\,\mathrm{H}_0\right\} = m_{X_0} + jm_{Y_0} , \tag{12.3-4}$$

where

$$m_{X_0} = E\left\{X_0\right\} \tag{12.3-5}$$

and

$$m_{Y_0} = E\left\{Y_0\right\}. \tag{12.3-6}$$

However, from (12.3-1) and (12.2-4),

$$E\left\{\tilde{l}\,\middle|\,\mathrm{H}_0\right\} = E\left\{\tilde{l}_Z\right\} = E\left\{\tilde{l}_{Rev}\right\} \tag{12.3-7}$$

since [see (12.2-6)]

$$E\left\{\tilde{l}_{n_a}\right\} = \int_{-\infty}^{\infty} E\left\{\tilde{y}_{n_a}(t)\right\} \tilde{g}^*(t)\,dt = 0 \tag{12.3-8}$$

and [see (12.2-7)]

$$E\left\{\tilde{l}_{n_r}\right\} = \int_{-\infty}^{\infty} E\left\{\tilde{n}_r(t)\right\} \tilde{g}^*(t)\,dt = 0 \tag{12.3-9}$$

because $\tilde{y}_{n_a}(t)$ and $\tilde{n}_r(t)$ are zero-mean. Substituting (12B-26) into (12.3-7) yields

$$E\left\{\tilde{l}\middle|\mathrm{H}_0\right\}=E\left\{\tilde{l}_Z\right\}$$
$$=E\left\{\tilde{l}_{Rev}\right\}$$
$$=E_{\tilde{x}}E\left\{w_{Rev}\right\}\exp\left[-j2\pi(f_c+\hat{\eta}_{D,Trgt})\tau_{Trgt}\right]\times$$
$$\sum_{i=-N'}^{N'}a_i X_N(e_{\tau_i},e_D)\exp\left[+j2\pi(f_c+\hat{\eta}_{D,Trgt})(\tau_{i,Trgt}-\tau_{i,Rev})\right],$$

(12.3-10)

where [see (12.1-10)]

$$E\left\{w_{Rev}\right\}=\frac{1}{(4\pi r_{Rev})^2}\mathcal{S}_T(f_c)E\left\{S_{Rev}(f_c,\theta_{Rev},\psi_{Rev},\theta_{Rev\,0},\psi_{Rev\,0})\right\}\times$$
$$\exp[-2\alpha(f_c)r_{Rev}]\mathcal{S}_R(f_c).$$

(12.3-11)

Therefore, m_{X_0} and m_{Y_0} are the real and imaginary parts of (12.3-10), respectively [see (12.3-4)].

The variance of $\tilde{l}$ given hypothesis H_0 is true is given by

$$\mathrm{Var}\left(\tilde{l}\middle|\mathrm{H}_0\right)=E\left\{\left|\tilde{l}\right|^2\middle|\mathrm{H}_0\right\}-\left|E\left\{\tilde{l}\middle|\mathrm{H}_0\right\}\right|^2, \quad (12.3\text{-}12)$$

where [see (12.3-3)]

$$E\left\{\left|\tilde{l}\right|^2\middle|\mathrm{H}_0\right\}=E\left\{X_0^2\right\}+E\left\{Y_0^2\right\} \quad (12.3\text{-}13)$$

and [see (12.3-4)]

$$\left|E\left\{\tilde{l}\middle|\mathrm{H}_0\right\}\right|^2=m_{X_0}^2+m_{Y_0}^2. \quad (12.3\text{-}14)$$

Substituting (12.3-13) and (12.3-14) into (12.3-12) yields

$$\mathrm{Var}\left(\tilde{l}\middle|\mathrm{H}_0\right)=\sigma_{X_0}^2+\sigma_{Y_0}^2, \quad (12.3\text{-}15)$$

where

$$\sigma_{X_0}^2=E\left\{X_0^2\right\}-m_{X_0}^2 \quad (12.3\text{-}16)$$

is the variance of X_0, and

$$\sigma_{Y_0}^2=E\left\{Y_0^2\right\}-m_{Y_0}^2 \quad (12.3\text{-}17)$$

is the variance of Y_0. However, from (12.3-1),

$$\mathrm{Var}\left(\tilde{l}\,\middle|\,\mathrm{H}_0\right)=\mathrm{Var}\left(\tilde{l}_Z\right)=E\left\{\left|\tilde{l}_Z\right|^2\right\}-\left|E\left\{\tilde{l}_Z\right\}\right|^2, \tag{12.3-18}$$

and by substituting (12.3-7) into (12.3-18),

$$\mathrm{Var}\left(\tilde{l}\,\middle|\,\mathrm{H}_0\right)=\mathrm{Var}\left(\tilde{l}_Z\right)=E\left\{\left|\tilde{l}_Z\right|^2\right\}-\left|E\left\{\tilde{l}_{Rev}\right\}\right|^2. \tag{12.3-19}$$

With the use of (12.3-18), (12.3-19), (12.2-21), and (12.3-10), $\mathrm{Var}\left(\tilde{l}\,\middle|\,\mathrm{H}_0\right)$ can also be expressed as

$$\begin{aligned}\mathrm{Var}\left(\tilde{l}\,\middle|\,\mathrm{H}_0\right)&=\mathrm{Var}\left(\tilde{l}_Z\right)\\&=E\left\{\left|\tilde{l}_Z\right|^2\right\}-\left|E\left\{\tilde{l}_Z\right\}\right|^2\\&=E\left\{\left|\tilde{l}_Z\right|^2\right\}-\left|E\left\{\tilde{l}_{Rev}\right\}\right|^2\\&=E_{\tilde{x}}^2\,\sigma_{w_{Rev}}^2\left|\sum_{i=-N'}^{N'}a_i X_N(e_{\tau_i},e_D)\exp\left[+j2\pi(f_c+\hat{\eta}_{D,Trgt})(\tau_{i,Trgt}-\tau_{i,Rev})\right]\right|^2+\\&\quad A_2E_{\tilde{x}}\left[\tilde{N}_{a0}+\tilde{N}_{r0}\right],\end{aligned} \tag{12.3-20}$$

where

$$\sigma_{w_{Rev}}^2=E\left\{\left|w_{Rev}\right|^2\right\}-\left|E\left\{w_{Rev}\right\}\right|^2 \tag{12.3-21}$$

is the variance of w_{Rev}, $E\left\{\left|w_{Rev}\right|^2\right\}$ is given by (12.2-22), $E\left\{w_{Rev}\right\}$ is given by (12.3-11), and A_2 is given by (12.2-25). Also, since [see (12.2-4) and (12.3-3)]

$$\mathrm{H}_0:\ \tilde{l}=\tilde{l}_Z=\tilde{l}_{Rev}+\tilde{l}_{n_a}+\tilde{l}_{n_r}=X_0+jY_0, \tag{12.3-22}$$

X_0 and Y_0 are *uncorrelated* and *statistically independent* because $\tilde{l}_{Rev}$ and $\tilde{l}_{n_a}$, $\tilde{l}_{Rev}$ and $\tilde{l}_{n_r}$, and $\tilde{l}_{n_a}$ and $\tilde{l}_{n_r}$ are *uncorrelated* and *statistically independent* [see (12B-9), (12B-15), and (12B-20), respectively].

Our next problem is to determine the probability density function (PDF) of the test statistic $\left|\tilde{l}\right|^2$ given hypothesis H_0 is true. Let us begin with the magnitude of $\tilde{l}$, where from (12.3-3),

$$\mathrm{H}_0\text{:}\quad \left|\tilde{l}\right| = \left|\tilde{l}_Z\right| = \sqrt{X_0^2 + Y_0^2}\,. \tag{12.3-23}$$

Since X_0 is $N(m_{X_0}, \sigma_{X_0}^2)$ and Y_0 is $N(m_{Y_0}, \sigma_{Y_0}^2)$, $\left|\tilde{l}\right|$ is a *Beckmann* RV.[1] In our case, X_0 and Y_0 are uncorrelated Gaussian RVs with correlation coefficient $\rho_{X_0Y_0} = 0$. However, even if X_0 and Y_0 are correlated (i.e., $-1 \le \rho_{X_0Y_0} \le 1$), $\left|\tilde{l}\right|$ is still a Beckmann RV. The set of equations for the PDF and cumulative distribution function (CDF) for a Beckmann RV are lengthy and are not closed-form expressions.[2]

In order to obtain more tractable results that will give us a good insight into our problem, we shall make the following simplifying assumption: let

$$\boxed{\sigma_{Y_0}^2 = \sigma_{X_0}^2 = \sigma_0^2} \tag{12.3-24}$$

Therefore, substituting (12.3-24) into (12.3-15) yields

$$\mathrm{Var}\left(\tilde{l}\,\middle|\,\mathrm{H}_0\right) = 2\sigma_0^2, \tag{12.3-25}$$

where $\mathrm{Var}\left(\tilde{l}\,\middle|\,\mathrm{H}_0\right)$ is given by (12.3-20). Since [see (12.3-23)]

$$\mathrm{H}_0\text{:}\quad \left|\tilde{l}\right|^2 = \left|\tilde{l}_Z\right|^2 = X_0^2 + Y_0^2, \tag{12.3-26}$$

where X_0 is $N(m_{X_0}, \sigma_0^2)$, Y_0 is $N(m_{Y_0}, \sigma_0^2)$, X_0 and Y_0 have the *same* variance σ_0^2, and X_0 and Y_0 are *statistically independent* because they are uncorrelated Gaussian RVs, $\left|\tilde{l}\right|^2$ is a *noncentral, chi-squared* RV with 2 degrees of freedom and PDF[3]

$$\boxed{p_0(\gamma) = \frac{1}{2\sigma_0^2}\exp\left(-\frac{\gamma + \gamma_0}{2\sigma_0^2}\right) I_0\left(\frac{\sqrt{\gamma_0}}{\sigma_0^2}\sqrt{\gamma}\right), \qquad \gamma \ge 0} \tag{12.3-27}$$

---

[1] *Mathematica*, Version 12.2.0, Wolfram Research, Inc., 2020.

[2] P. Dharmawansa, N. Rajatheva, and C. Tellambura, "Envelope and Phase Distribution of Two Correlated Gaussian Variables," *IEEE Trans. Commun.*, vol. 57, pp. 915-921, April 2009.

[3] T. A. Schonhoff and A. A. Giordano, *Detection and Estimation Theory*, Pearson Prentice Hall, Upper Saddle River, NJ, 2006, pp. 602-604.

where the RV

$$\gamma = \left|\tilde{l}\right|^2 = X_0^2 + Y_0^2 , \tag{12.3-28}$$

the nonrandom constant [see (12.3-14)]

$$\gamma_0 = \left|E\left\{\tilde{l}\,\middle|\,\mathrm{H}_0\right\}\right|^2 = m_{X_0}^2 + m_{Y_0}^2 \tag{12.3-29}$$

is the *noncentrality parameter*, where m_{X_0} and m_{Y_0} are the real and imaginary parts of (12.3-10), respectively [see (12.3-4)],

$$2\sigma_0^2 = \mathrm{Var}\left(\tilde{l}\,\middle|\,\mathrm{H}_0\right), \tag{12.3-30}$$

where $\mathrm{Var}\left(\tilde{l}\,\middle|\,\mathrm{H}_0\right)$ is given by (12.3-20) [see (12.3-25)],

$$I_0(u) = \sum_{n=0}^{\infty} \frac{u^{2n}}{2^{2n}(n!)^2} \tag{12.3-31}$$

is the zeroth-order modified Bessel function of the first kind, and

$$p_0(\gamma) = p_{\gamma|\mathrm{H}_0}\left(\gamma\,\middle|\,\mathrm{H}_0\right). \tag{12.3-32}$$

See Fig. 10.6-3 for plots of a scaled, noncentral, chi-squared PDF with 2 degrees of freedom.

The probability of false alarm P_{FA} is given by

$$P_{\mathrm{FA}} = P\left(\left|\tilde{l}\right|^2 > \mathcal{T}\,\middle|\,\mathrm{H}_0\right), \tag{12.3-33}$$

or

$$P_{\mathrm{FA}} = P\left(\gamma > \mathcal{T}\,\middle|\,\mathrm{H}_0\right) = \int_{\mathcal{T}}^{\infty} p_0(\gamma)\,d\gamma , \tag{12.3-34}$$

where $\mathcal{T}$ is the decision threshold. Substituting (12.3-27) into (12.3-34) yields

$$P_{\mathrm{FA}} = \frac{1}{2\sigma_0^2} \int_{\mathcal{T}}^{\infty} \exp\left(-\frac{\gamma + \gamma_0}{2\sigma_0^2}\right) I_0\left(\frac{\sqrt{\gamma_0}}{\sigma_0^2}\sqrt{\gamma}\right) d\gamma . \tag{12.3-35}$$

If we let

$$x^2 = \gamma / \sigma_0^2 \tag{12.3-36}$$

and

$$\alpha_0^2 = \gamma_0 / \sigma_0^2 , \tag{12.3-37}$$

then (12.3-35) can be rewritten as

$$\boxed{P_{\text{FA}} = Q\left(\alpha_0, \sqrt{\mathcal{T}} / \sigma_0\right) = \int_{\sqrt{\mathcal{T}}/\sigma_0}^{\infty} \exp\left[-(x^2 + \alpha_0^2)/2\right] I_0(\alpha_0 x)\, x\, dx} \tag{12.3-38}$$

where

$$\boxed{Q(a, b) = \int_b^{\infty} \exp\left[-(x^2 + a^2)/2\right] I_0(ax)\, x\, dx} \tag{12.3-39}$$

is the *Marcum Q-function*. See Example 12.3-1 for a solution for the decision threshold $\mathcal{T}$ that satisfies a P_{FA} constraint using (12.3-38).

If the reverberation scattering function is *zero-mean*, that is, if

$$E\left\{S_{Rev}(f_c, \theta_{Rev}, \psi_{Rev}, \theta_{Rev\,0}, \psi_{Rev\,0})\right\} = 0 , \tag{12.3-40}$$

then [see (12.3-11)]

$$E\{w_{Rev}\} = 0 \tag{12.3-41}$$

and [see (12.3-10)]

$$E\left\{\tilde{l} \,\middle|\, \text{H}_0\right\} = E\left\{\tilde{l}_Z\right\} = E\left\{\tilde{l}_{Rev}\right\} = 0 . \tag{12.3-42}$$

As a result, $m_{X_0} = 0$, $m_{Y_0} = 0$, and $\gamma_0 = 0$ [see (12.3-4) and (12.3-29)]. Therefore, the statistically independent Gaussian RVs X_0 and Y_0 are now $N(0, \sigma_0^2)$, and $\gamma = |\tilde{l}|^2$ given by (12.3-28) is an *exponential* RV. If $\gamma_0 = 0$, and since $I_0(0) = 1$, the noncentral, chi-squared PDF given by (12.3-27) reduces to the *exponential* PDF

$$\boxed{p_0(\gamma) = \frac{1}{2\sigma_0^2} \exp\left(-\frac{\gamma}{2\sigma_0^2}\right), \qquad \gamma \geq 0} \tag{12.3-43}$$

where $2\sigma_0^2 = \text{Var}\left(\tilde{l} \,\middle|\, \text{H}_0\right)$ [see (12.3-30)], but now, because of (12.3-41) and

(12.3-42), $\operatorname{Var}\left(\tilde{l}\,\middle|\,\mathrm{H}_0\right)$ given by (12.3-20) reduces to

$$\begin{aligned}
\operatorname{Var}\left(\tilde{l}\,\middle|\,\mathrm{H}_0\right) &= \operatorname{Var}\left(\tilde{l}_{\mathrm{Z}}\right) \\
&= E\left\{\left|\tilde{l}_{\mathrm{Z}}\right|^2\right\} \\
&= E_{\tilde{x}}^2 E\left\{\left|w_{Rev}\right|^2\right\}\left|\sum_{i=-N'}^{N'} a_i X_N(e_{\tau_i}, e_D)\exp\left[+j2\pi(f_c+\hat{\eta}_{D,Trgt})(\tau_{i,Trgt}-\tau_{i,Rev})\right]\right|^2 + \\
&\quad A_2 E_{\tilde{x}}\left[\tilde{N}_{a0}+\tilde{N}_{r0}\right].
\end{aligned} \tag{12.3-44}$$

Therefore, in the limit as $\sqrt{\gamma_0}/\sigma_0 \to 0$, or equivalently, as

$$\frac{\gamma_0}{2\sigma_0^2} = \frac{\left|E\left\{\tilde{l}\,\middle|\,\mathrm{H}_0\right\}\right|^2}{\operatorname{Var}\left(\tilde{l}\,\middle|\,\mathrm{H}_0\right)} \to 0, \tag{12.3-45}$$

$\gamma = \left|\tilde{l}\right|^2$ approaches being an exponential RV.

If the exponential PDF given by (12.3-43) is substituted into (12.3-34), then

$$P_{\mathrm{FA}} = \frac{1}{2\sigma_0^2}\int_{\mathcal{T}}^{\infty}\exp\left(-\frac{\gamma}{2\sigma_0^2}\right)d\gamma, \tag{12.3-46}$$

or

$$\boxed{P_{\mathrm{FA}} = \exp\left(-\frac{\mathcal{T}}{2\sigma_0^2}\right)} \tag{12.3-47}$$

Solving for the decision threshold $\mathcal{T}$ yields

$$\boxed{\mathcal{T} = 2\sigma_0^2\ln\left(1/P_{\mathrm{FA}}\right)} \tag{12.3-48}$$

where $2\sigma_0^2 = \operatorname{Var}\left(\tilde{l}\,\middle|\,\mathrm{H}_0\right)$ and $\operatorname{Var}\left(\tilde{l}\,\middle|\,\mathrm{H}_0\right)$ is given by (12.3-44). From (12.3-48) it can be seen that $\mathcal{T}$ is inversely proportional to P_{FA}. Therefore, for a given value of σ_0^2, as the $P_{\mathrm{FA}} \to 0$, $\mathcal{T} \to \infty$; and as the $P_{\mathrm{FA}} \to 1$, $\mathcal{T} \to 0$.

Example 12.3-1 Nonzero-Mean Reverberation Scattering Function

In this example we shall determine the decision threshold $\mathcal{T}$ that satisfies a P_{FA} constraint using (12.3-38). We first need to generate a table of values for the Marcum Q-function $Q(a,b)$ given by (12.3-39) by deciding on several different values for the parameter a. Toward this end, if we divide (12.3-37) by a factor of 2, then

$$\frac{\alpha_0^2}{2} = \frac{\gamma_0}{2\sigma_0^2} = \frac{\left|E\left\{\tilde{l}\,\middle|\,\text{H}_0\right\}\right|^2}{\text{Var}\left(\tilde{l}\,\middle|\,\text{H}_0\right)}, \tag{12.3-49}$$

and by setting $\alpha_0^2/2$ equal to several different representative values as shown in Table 12.3-1, we can solve for α_0, and hence, several different representative values for parameter a. Values of $Q(a,b)$ for $a = 1/2$, $\sqrt{2}/2$, 1, $\sqrt{2}$, 2, $2\sqrt{2}$, and 4 for $b = 0$ to 5.5 are given in Table 12C-1 in Appendix 12C. The case $\alpha_0 = 0$ ($a = 0$) has already been solved since $\alpha_0 = 0$ when $\gamma_0 = 0$, and when $\gamma_0 = 0$, the decision threshold is given by (12.3-48). Equation (12.3-49) can be thought of as a measure of the randomness of the output of the correlator $\tilde{l}$ given hypothesis H_0 is true. For example, as $\alpha_0^2/2 \to 0$, $\tilde{l}$ becomes more random. Conversely, as $\alpha_0^2/2 \to \infty$, $\tilde{l}$ becomes more deterministic.

Table 12.3-1 Values of α_0 Used to Determine Values of Parameter a in the Marcum Q-Function $Q(a,b)$

$\alpha_0^2/2$	$10\log_{10}(\alpha_0^2/2)$	α_0
0.125	-9 dB	$1/2$
0.25	-6 dB	$\sqrt{2}/2$
0.5	-3 dB	1
1	0 dB	$\sqrt{2}$
2	3 dB	2
4	6 dB	$2\sqrt{2}$
8	9 dB	4

Now that we have a table of values for the Marcum Q-function, let us determine the decision threshold $\mathcal{T}$ for a $P_{\text{FA}} = 0.001$ when $\alpha_0 = \sqrt{2}/2$. Since [see (12.3-38)]

$$P_{\text{FA}} = Q\left(\sqrt{2}/2, \sqrt{\mathcal{T}}/\sigma_0\right) = 0.001, \tag{12.3-50}$$

set $a = \alpha_0 = \sqrt{2}/2$, and find the value of b that yields $Q\left(\sqrt{2}/2, b\right) = 0.001$ in Table 12C-1 in Appendix 12C. From Table 12C-1:

	$a = \sqrt{2}/2$
b	$Q(a,b)$
4	0.0012837
4.1	0.0009038

Next, use linear interpolation to find b.

If we let

$$Q(a,b) = mb + c, \tag{12.3-51}$$

then the slope m is

$$\begin{aligned} m &= \frac{Q(a,b_2) - Q(a,b_1)}{b_2 - b_1} \\ &= \frac{Q\left(\sqrt{2}/2, 4.1\right) - Q\left(\sqrt{2}/2, 4\right)}{4.1 - 4} \\ &= \frac{0.0009038 - 0.0012837}{0.1} \\ &= -0.003799 \end{aligned} \tag{12.3-52}$$

and the y-intercept is

$$\begin{aligned} c &= Q(a,b_1) - mb_1 \\ &= Q\left(\sqrt{2}/2, 4\right) - (-0.003799)4 \\ &= 0.0012837 + 0.015196 \\ &= 0.0164797. \end{aligned} \tag{12.3-53}$$

Solving for b from (12.3-51) yields

$$b = \frac{Q(a,b) - c}{m}, \tag{12.3-54}$$

and by substituting (12.3-52), (12.3-53), and $Q(a,b) = 0.001$ into (12.3-54), we obtain

$$b \approx 4.0747 . \tag{12.3-55}$$

By referring to (12.3-38) and (12.3-39),

$$b = \sqrt{\mathcal{T}} / \sigma_0 , \tag{12.3-56}$$

or

$$\mathcal{T} = b^2 \sigma_0^2 , \tag{12.3-57}$$

where $2\sigma_0^2 = \text{Var}\left(\tilde{l} \middle| \text{H}_0\right)$ [see (12.3-30)] and $\text{Var}\left(\tilde{l} \middle| \text{H}_0\right)$ is given by (12.3-20). Substituting (12.3-55) into (12.3-57) yields

$$\mathcal{T} = 16.603\sigma_0^2 . \tag{12.3-58}$$

The numerical factor 16.603 in (12.3-58) is based on the specific values of $\alpha_0 = \sqrt{2}/2$ and $P_{\text{FA}} = 0.001$. The equation for the decision threshold $\mathcal{T}$ can be generalized for any value of α_0 and P_{FA} as follows. By referring to (12.3-54) and noting that $P_{\text{FA}} = Q(a,b)$, (12.3-54) can be rewritten as

$$\boxed{b_{\text{FA}} = \frac{P_{\text{FA}} - c_{\alpha_0}}{m_{\alpha_0}}} \tag{12.3-59}$$

where [see (12.3-52) and (12.3-53)]

$$m_{\alpha_0} = \frac{Q(\alpha_0, b_2) - Q(\alpha_0, b_1)}{b_2 - b_1} , \tag{12.3-60}$$

$$c_{\alpha_0} = Q(\alpha_0, b_1) - m_{\alpha_0} b_1 , \tag{12.3-61}$$

and

$$Q(\alpha_0, b_2) < P_{\text{FA}} < Q(\alpha_0, b_1) . \tag{12.3-62}$$

Therefore, (12.3-57) can be rewritten as

$$\boxed{\mathcal{T} = b_{\text{FA}}^2 \sigma_0^2} \tag{12.3-63}$$

where $2\sigma_0^2 = \text{Var}\left(\tilde{l} \middle| \text{H}_0\right)$ [see (12.3-30)] and $\text{Var}\left(\tilde{l} \middle| \text{H}_0\right)$ is given by (12.3-20). As can be seen from (12.3-59), for a given value of α_0, b_{FA} will change value as the P_{FA} changes value. As the P_{FA} decreases, b_{FA} increases, and for a given value of σ_0^2, $\mathcal{T}$ increases. ■

12.4 Probability of Detection and Receiver Operating Characteristic Curves

In this section we shall derive an equation for the probability of detection P_{D}. We begin by expressing the complex, Gaussian, random variable (RV) $\tilde{l}_{Trgt}$ in rectangular form as follows:

$$\tilde{l}_{Trgt} = X + jY \,, \tag{12.4-1}$$

where X and Y are real, Gaussian RVs. Since [see (12.3-3)]

$$\tilde{l}_Z = X_0 + jY_0 \,, \tag{12.4-2}$$

where X_0 and Y_0 are real, Gaussian RVs, substituting (12.4-1) and (12.4-2) into (12.3-2) yields

$$\mathrm{H}_1\text{:} \quad \tilde{l} = \tilde{l}_{Trgt} + \tilde{l}_Z = X_1 + jY_1 \,, \tag{12.4-3}$$

where

$$X_1 = X + X_0 \tag{12.4-4}$$

and

$$Y_1 = Y + Y_0 \tag{12.4-5}$$

are real, Gaussian RVs. In order to compute the P_{D}, we first need to determine some of the statistics of $\tilde{l}$ given by (12.4-3).

The mean value (average value) of $\tilde{l}_{Trgt}$ is given by

$$E\left\{\tilde{l}_{Trgt}\right\} = m_X + jm_Y \,, \tag{12.4-6}$$

where

$$m_X = E\{X\} \tag{12.4-7}$$

and

$$m_Y = E\{Y\} \,. \tag{12.4-8}$$

The mean value of $\tilde{l}$ given hypothesis H_1 is true is given by

$$E\left\{\tilde{l} \,\middle|\, \mathrm{H}_1\right\} = m_{X_1} + jm_{Y_1} \,, \tag{12.4-9}$$

where

$$m_{X_1} = E\{X_1\} = m_X + m_{X_0} \tag{12.4-10}$$

and

$$m_{Y_1} = E\{Y_1\} = m_Y + m_{Y_0}. \tag{12.4-11}$$

However, from (12.4-3) and (12.3-7),

$$E\{\tilde{l}\,|\,\mathrm{H}_1\} = E\{\tilde{l}_{Trgt}\} + E\{\tilde{l}_Z\} = E\{\tilde{l}_{Trgt}\} + E\{\tilde{l}_{Rev}\}. \tag{12.4-12}$$

Computing the expected value of $\tilde{l}_{Trgt}$ given by (12.2-12) yields

$$E\{\tilde{l}_{Trgt}\} = A_1 E_{\tilde{x}} E\{w_{Trgt}\} X_N(e_{\tau,Trgt}, e_{D,Trgt}) \exp\left[-j2\pi(f_c + \hat{\eta}_{D,Trgt})\tau_{Trgt}\right], \tag{12.4-13}$$

where A_1 is given by (12.2-9) and [see (12.1-6)]

$$E\{w_{Trgt}\} = \frac{1}{(4\pi r_{Trgt})^2} \mathcal{S}_T(f_c) E\{\mathcal{S}_{Trgt}(f_c, \theta_{Trgt}, \psi_{Trgt}, \theta_{Trgt\,0}, \psi_{Trgt\,0})\} \times \exp[-2\alpha(f_c) r_{Trgt}] \mathcal{S}_R(f_c). \tag{12.4-14}$$

Substituting (12.4-13) and (12.3-10) into (12.4-12) yields

$$\begin{aligned} E\{\tilde{l}\,|\,\mathrm{H}_1\} &= E\{\tilde{l}_{Trgt}\} + E\{\tilde{l}_Z\} \\ &= E\{\tilde{l}_{Trgt}\} + E\{\tilde{l}_{Rev}\} \\ &= A_1 E_{\tilde{x}} E\{w_{Trgt}\} X_N(e_{\tau,Trgt}, e_{D,Trgt}) \exp\left[-j2\pi(f_c + \hat{\eta}_{D,Trgt})\tau_{Trgt}\right] + \\ &\quad E_{\tilde{x}} E\{w_{Rev}\} \exp\left[-j2\pi(f_c + \hat{\eta}_{D,Trgt})\tau_{Trgt}\right] \times \\ &\quad \sum_{i=-N'}^{N'} a_i X_N(e_{\tau_i}, e_D) \exp\left[+j2\pi(f_c + \hat{\eta}_{D,Trgt})(\tau_{i,Trgt} - \tau_{i,Rev})\right]. \end{aligned} \tag{12.4-15}$$

Therefore, m_{X_1} and m_{Y_1} are the real and imaginary parts of (12.4-15), respectively [see (12.4-9)].

The variance of $\tilde{l}$ given hypothesis H_1 is true is given by

$$\operatorname{Var}\left(\tilde{l}\middle|\mathrm{H}_1\right)=E\left\{\left|\tilde{l}\right|^2\middle|\mathrm{H}_1\right\}-\left|E\left\{\tilde{l}\middle|\mathrm{H}_1\right\}\right|^2, \tag{12.4-16}$$

where [see (12.4-3)]

$$E\left\{\left|\tilde{l}\right|^2\middle|\mathrm{H}_1\right\}=E\left\{X_1^2\right\}+E\left\{Y_1^2\right\} \tag{12.4-17}$$

and [see (12.4-9)]

$$\left|E\left\{\tilde{l}\middle|\mathrm{H}_1\right\}\right|^2=m_{X_1}^2+m_{Y_1}^2. \tag{12.4-18}$$

Substituting (12.4-17) and (12.4-18) into (12.4-16) yields

$$\operatorname{Var}\left(\tilde{l}\middle|\mathrm{H}_1\right)=\sigma_{X_1}^2+\sigma_{Y_1}^2, \tag{12.4-19}$$

where

$$\sigma_{X_1}^2=E\left\{X_1^2\right\}-m_{X_1}^2 \tag{12.4-20}$$

is the variance of X_1, and

$$\sigma_{Y_1}^2=E\left\{Y_1^2\right\}-m_{Y_1}^2 \tag{12.4-21}$$

is the variance of Y_1. However, from (12.4-3),

$$\operatorname{Var}\left(\tilde{l}\middle|\mathrm{H}_1\right)=\operatorname{Var}\left(\tilde{l}_{Trgt}\right)+\operatorname{Var}\left(\tilde{l}_Z\right)+2\operatorname{Cov}\left(\tilde{l}_{Trgt},\tilde{l}_Z\right). \tag{12.4-22}$$

Assuming that $\tilde{l}_{Trgt}$ and $\tilde{l}_Z=\tilde{l}_{Rev}+\tilde{l}_{n_a}+\tilde{l}_{n_r}$ are *uncorrelated*, (12.4-22) reduces to

$$\begin{aligned}\operatorname{Var}\left(\tilde{l}\middle|\mathrm{H}_1\right)&=\operatorname{Var}\left(\tilde{l}_{Trgt}\right)+\operatorname{Var}\left(\tilde{l}_Z\right)\\&=\operatorname{Var}\left(\tilde{l}_{Trgt}\right)+\operatorname{Var}\left(\tilde{l}\middle|\mathrm{H}_0\right),\end{aligned} \tag{12.4-23}$$

where

$$\operatorname{Var}\left(\tilde{l}_{Trgt}\right)=E\left\{\left|\tilde{l}_{Trgt}\right|^2\right\}-\left|E\left\{\tilde{l}_{Trgt}\right\}\right|^2 \tag{12.4-24}$$

and $\operatorname{Var}\left(\tilde{l}\middle|\mathrm{H}_0\right)$ is given by (12.3-20). Substituting (12.2-19) and (12.4-13) into (12.4-24) yields

$$\text{Var}\left(\tilde{l}_{Trgt}\right) = A_1^2 E_{\tilde{x}}^2 \sigma_{w_{Trgt}}^2 \left| X_N(e_{\tau, Trgt}, e_{D, Trgt}) \right|^2, \tag{12.4-25}$$

where A_1 is given by (12.2-9),

$$\sigma_{w_{Trgt}}^2 = E\left\{ \left| w_{Trgt} \right|^2 \right\} - \left| E\left\{ w_{Trgt} \right\} \right|^2 \tag{12.4-26}$$

is the variance of w_{Trgt}, $E\left\{ \left| w_{Trgt} \right|^2 \right\}$ is given by (12.2-20), and $E\left\{ w_{Trgt} \right\}$ is given by (12.4-14). Substituting (12.4-25) and (12.3-20) into (12.4-23) yields

$$\begin{aligned} \text{Var}\left(\tilde{l} \,\middle|\, \text{H}_1\right) &= \text{Var}\left(\tilde{l}_{Trgt}\right) + \text{Var}\left(\tilde{l}_Z\right) \\ &= \text{Var}\left(\tilde{l}_{Trgt}\right) + \text{Var}\left(\tilde{l} \,\middle|\, \text{H}_0\right) \\ &= A_1^2 E_{\tilde{x}}^2 \sigma_{w_{Trgt}}^2 \left| X_N(e_{\tau, Trgt}, e_{D, Trgt}) \right|^2 + \\ &\quad E_{\tilde{x}}^2 \sigma_{w_{Rev}}^2 \left| \sum_{i=-N'}^{N'} a_i X_N(e_{\tau_i}, e_D) \exp\left[+j2\pi (f_c + \hat{\eta}_{D, Trgt})(\tau_{i, Trgt} - \tau_{i, Rev}) \right] \right|^2 + \\ &\quad A_2 E_{\tilde{x}} \left[\tilde{N}_{a0} + \tilde{N}_{r0} \right]. \end{aligned} \tag{12.4-27}$$

Also, since it was assumed that $\tilde{l}_{Trgt}$ and $\tilde{l}_Z$ are uncorrelated, X_1 and Y_1 are *uncorrelated* and *statistically independent* because they are Gaussian RVs [see (12.4-3)].

Our next problem is to determine the probability density function (PDF) of the test statistic $\left| \tilde{l} \right|^2$ given hypothesis H_1 is true. Using the same argument that was made in Section 12.3, in order to obtain more tractable results that will give us a good insight into our problem, we shall make the following simplifying assumption: let

$$\boxed{\sigma_{Y_1}^2 = \sigma_{X_1}^2 = \sigma_1^2} \tag{12.4-28}$$

Therefore, substituting (12.4-28) into (12.4-19) yields

$$\text{Var}\left(\tilde{l} \,\middle|\, \text{H}_1\right) = 2\sigma_1^2, \tag{12.4-29}$$

where $\text{Var}\left(\tilde{l} \,\middle|\, \text{H}_1\right)$ is given by (12.4-27). Since [see (12.4-3)]

$$\text{H}_1: \quad \left|\tilde{l}\right|^2 = \left|\tilde{l}_{Trgt} + \tilde{l}_Z\right|^2 = X_1^2 + Y_1^2, \tag{12.4-30}$$

where X_1 is $N(m_{X_1}, \sigma_1^2)$, Y_1 is $N(m_{Y_1}, \sigma_1^2)$, X_1 and Y_1 have the *same* variance σ_1^2, and X_1 and Y_1 are *statistically independent* because they are uncorrelated Gaussian RVs, $\left|\tilde{l}\right|^2$ is a *noncentral, chi-squared* RV with 2 degrees of freedom and PDF[3]

$$\boxed{p_1(\gamma) = \frac{1}{2\sigma_1^2}\exp\left(-\frac{\gamma+\gamma_1}{2\sigma_1^2}\right) I_0\left(\frac{\sqrt{\gamma_1}}{\sigma_1^2}\sqrt{\gamma}\right), \qquad \gamma \geq 0} \tag{12.4-31}$$

where the RV

$$\gamma = \left|\tilde{l}\right|^2 = X_1^2 + Y_1^2, \tag{12.4-32}$$

the nonrandom constant [see (12.4-9)]

$$\gamma_1 = \left|E\left\{\tilde{l}\,\middle|\,\text{H}_1\right\}\right|^2 = m_{X_1}^2 + m_{Y_1}^2 \tag{12.4-33}$$

is the *noncentrality parameter*, where m_{X_1} and m_{Y_1} are the real and imaginary parts of (12.4-15), respectively [see (12.4-9)],

$$2\sigma_1^2 = \text{Var}\left(\tilde{l}\,\middle|\,\text{H}_1\right), \tag{12.4-34}$$

where $\text{Var}\left(\tilde{l}\,\middle|\,\text{H}_1\right)$ is given by (12.4-27) [see (12.4-29)], $I_0(\bullet)$ is the zeroth-order modified Bessel function of the first kind given by (12.3-31), and

$$p_1(\gamma) = p_{\gamma|\text{H}_1}\left(\gamma\,\middle|\,\text{H}_1\right). \tag{12.4-35}$$

See Fig. 10.6-3 for plots of a scaled, noncentral, chi-squared PDF with 2 degrees of freedom.

The probability of detection P_D is given by

$$P_\text{D} = P\left(\left|\tilde{l}\right|^2 > \mathcal{T}\,\middle|\,\text{H}_1\right), \tag{12.4-36}$$

or

$$P_\text{D} = P\left(\gamma > \mathcal{T}\,\middle|\,\text{H}_1\right) = \int_{\mathcal{T}}^{\infty} p_1(\gamma)\,d\gamma\,, \tag{12.4-37}$$

where $\mathcal{T}$ is the decision threshold. Substituting (12.4-31) into (12.4-37) yields

$$P_{\mathrm{D}} = \frac{1}{2\sigma_1^2} \int_{\mathcal{T}}^{\infty} \exp\left(-\frac{\gamma + \gamma_1}{2\sigma_1^2} \right) I_0\left(\frac{\sqrt{\gamma_1}}{\sigma_1^2} \sqrt{\gamma} \right) d\gamma . \tag{12.4-38}$$

If we let

$$x^2 = \gamma / \sigma_1^2 \tag{12.4-39}$$

and

$$\alpha_1^2 = \gamma_1 / \sigma_1^2 , \tag{12.4-40}$$

then (12.4-38) can be rewritten as

$$\boxed{P_{\mathrm{D}} = Q\left(\alpha_1, \sqrt{\mathcal{T}} / \sigma_1\right) = \int_{\sqrt{\mathcal{T}}/\sigma_1}^{\infty} \exp\left[-(x^2 + \alpha_1^2)/2 \right] I_0(\alpha_1 x) x \, dx} \tag{12.4-41}$$

where $Q(a, b)$ is the Marcum Q-function given by (12.3-39). See Example 12.4-1 for a solution for the P_{D} using (12.4-41) and the decision threshold $\mathcal{T}$ obtained in Example 12.3-1.

If the target scattering function is *zero-mean*, that is, if

$$E\left\{ S_{Trgt}(f_c, \theta_{Trgt}, \psi_{Trgt}, \theta_{Trgt\,0}, \psi_{Trgt\,0}) \right\} = 0 , \tag{12.4-42}$$

then [see (12.4-14)]

$$E\left\{ w_{Trgt} \right\} = 0 \tag{12.4-43}$$

and [see (12.4-13)]

$$E\left\{ \tilde{l}_{Trgt} \right\} = 0 . \tag{12.4-44}$$

As a result, $m_X = 0$ and $m_Y = 0$ [see (12.4-6)]. If the reverberation scattering function is also *zero-mean*, then $m_{X_0} = 0$ and $m_{Y_0} = 0$ (see Section 12.3). Therefore, substituting $m_X = 0$ and $m_{X_0} = 0$ into (12.4-10) yields $m_{X_1} = 0$, and substituting $m_Y = 0$ and $m_{Y_0} = 0$ into (12.4-11) yields $m_{Y_1} = 0$. And by substituting $m_{X_1} = 0$ and $m_{Y_1} = 0$ into (12.4-9), we obtain [see (12.4-12)]

$$\begin{aligned} E\left\{ \tilde{l} \,|\, \mathrm{H}_1 \right\} &= E\left\{ \tilde{l}_{Trgt} \right\} + E\left\{ \tilde{l}_Z \right\} \\ &= E\left\{ \tilde{l}_{Trgt} \right\} + E\left\{ \tilde{l}_{Rev} \right\} \\ &= 0, \end{aligned} \tag{12.4-45}$$

and by substituting $m_{X_1}=0$ and $m_{Y_1}=0$ into (12.4-33), we obtain $\gamma_1=0$. Therefore, the statistically independent Gaussian RVs X_1 and Y_1 are now $N(0,\sigma_1^2)$, and $\boldsymbol{\gamma}=|\tilde{l}|^2$ given by (12.4-32) is an *exponential* RV. If $\gamma_1=0$, and since $I_0(0)=1$, the noncentral, chi-squared PDF given by (12.4-31) reduces to the *exponential* PDF

$$\boxed{p_1(\gamma)=\frac{1}{2\sigma_1^2}\exp\left(-\frac{\gamma}{2\sigma_1^2}\right), \qquad \gamma\geq 0} \tag{12.4-46}$$

where $2\sigma_1^2=\mathrm{Var}\left(\tilde{l}\,\middle|\,\mathrm{H}_1\right)$ [see (12.4-34)], but now, because of (12.3-41) and (12.4-43), $\mathrm{Var}\left(\tilde{l}\,\middle|\,\mathrm{H}_1\right)$ given by (12.4-27) reduces to

$$\begin{aligned}\mathrm{Var}\left(\tilde{l}\,\middle|\,\mathrm{H}_1\right)&=\mathrm{Var}\left(\tilde{l}_{Trgt}\right)+\mathrm{Var}\left(\tilde{l}_Z\right)\\&=E\left\{\left|\tilde{l}_{Trgt}\right|^2\right\}+E\left\{\left|\tilde{l}_Z\right|^2\right\}\\&=A_1^2E_{\tilde{x}}^2E\left\{\left|w_{Trgt}\right|^2\right\}\left|\chi_N(e_{\tau,Trgt},e_{D,Trgt})\right|^2+\\&\quad E_{\tilde{x}}^2E\left\{\left|w_{Rev}\right|^2\right\}\left|\sum_{i=-N'}^{N'}a_i\chi_N(e_{\tau_i},e_D)\exp\left[+j2\pi(f_c+\hat{\eta}_{D,Trgt})(\tau_{i,Trgt}-\tau_{i,Rev})\right]\right|^2+\\&\quad A_2E_{\tilde{x}}\left[\tilde{N}_{a0}+\tilde{N}_{r0}\right].\end{aligned} \tag{12.4-47}$$

Therefore, in the limit as $\sqrt{\gamma_1}/\sigma_1\to 0$, or equivalently, as

$$\frac{\gamma_1}{2\sigma_1^2}=\frac{\left|E\left\{\tilde{l}\,\middle|\,\mathrm{H}_1\right\}\right|^2}{\mathrm{Var}\left(\tilde{l}\,\middle|\,\mathrm{H}_1\right)}\to 0, \tag{12.4-48}$$

$\boldsymbol{\gamma}=|\tilde{l}|^2$ approaches being an exponential RV.

If the exponential PDF given by (12.4-46) is substituted into (12.4-37), then

$$P_\mathrm{D}=\frac{1}{2\sigma_1^2}\int_{\mathcal{T}}^{\infty}\exp\left(-\frac{\gamma}{2\sigma_1^2}\right)d\gamma, \tag{12.4-49}$$

or

$$P_\mathrm{D}=\exp\left(-\frac{\mathcal{T}}{2\sigma_1^2}\right). \tag{12.4-50}$$

Since [see (12.3-30) and (12.3-44)]

$$2\sigma_0^2 = \mathrm{Var}\left(\tilde{l}\,\middle|\,\mathrm{H}_0\right) = E\left\{\left|\tilde{l}_Z\right|^2\right\} \tag{12.4-51}$$

and [see (12.4-34) and (12.4-47)]

$$2\sigma_1^2 = \mathrm{Var}\left(\tilde{l}\,\middle|\,\mathrm{H}_1\right) = E\left\{\left|\tilde{l}_{Trgt}\right|^2\right\} + E\left\{\left|\tilde{l}_Z\right|^2\right\}, \tag{12.4-52}$$

(12.4-52) can be rewritten as

$$2\sigma_1^2 = 2\sigma_0^2\,(\mathrm{SIR}+1), \tag{12.4-53}$$

where the signal-to-interference ratio (SIR) is defined by (12.2-1). Substituting (12.4-53) into (12.4-50) yields

$$P_{\mathrm{D}} = \left[\exp\left(-\frac{\mathcal{T}}{2\sigma_0^2}\right)\right]^{\frac{1}{\mathrm{SIR}+1}}, \tag{12.4-54}$$

and by substituting (12.3-47) into (12.4-54), we obtain

$$\boxed{P_{\mathrm{D}} = P_{\mathrm{FA}}^{1/(\mathrm{SIR}+1)}} \tag{12.4-55}$$

where the SIR is given by (12.2-26). From (12.4-55) it can be seen that as the $\mathrm{SIR} \rightarrow \infty$, the $P_{\mathrm{D}} \rightarrow P_{\mathrm{FA}}^0 = 1$; and as the $\mathrm{SIR} \rightarrow 0$, the $P_{\mathrm{D}} \rightarrow P_{\mathrm{FA}}$.

Taking the logarithm (base 10) of (12.4-55) yields

$$\log_{10} P_{\mathrm{D}} = \frac{1}{\mathrm{SIR}+1} \log_{10} P_{\mathrm{FA}}, \tag{12.4-56}$$

and by solving for the SIR, we obtain

$$\boxed{\mathrm{SIR} = \frac{\log_{10} P_{\mathrm{FA}}}{\log_{10} P_{\mathrm{D}}} - 1} \tag{12.4-57}$$

Equation (12.4-57) can be used to compute the SIR that is required in order to obtain a desired P_{D} for a given P_{FA}. Note that the SIR computed according to (12.4-57) is *dimensionless*.

Example 12.4-1 Nonzero-Mean Target and Reverberation Scattering Functions

In this example we shall determine the P_D using (12.4-41) and the decision threshold $\mathcal{T}$ obtained in Example 12.3-1 for $\alpha_0 = \sqrt{2}/2$ and a $P_{FA} = 0.001$. If we divide (12.4-40) by a factor of 2, then

$$\frac{\alpha_1^2}{2} = \frac{\gamma_1}{2\sigma_1^2} = \frac{\left|E\{\tilde{l}|\mathrm{H}_1\}\right|^2}{\mathrm{Var}\left(\tilde{l}|\mathrm{H}_1\right)}. \tag{12.4-58}$$

Several different representative values for α_1, and hence, parameter a in the Marcum Q-function $Q(a,b)$, are given in Table 12.4-1. The case $\alpha_1 = 0$ ($a = 0$) has already been solved since $\alpha_1 = 0$ when $\gamma_1 = 0$, and when $\gamma_1 = 0$, the P_D is given by (12.4-55). Equation (12.4-58) can be thought of as a measure of the randomness of the output of the correlator $\tilde{l}$ given hypothesis H_1 is true. For example, as $\alpha_1^2/2 \to 0$, $\tilde{l}$ becomes more random. Conversely, as $\alpha_1^2/2 \to \infty$, $\tilde{l}$ becomes more deterministic.

Table 12.4-1 Values of α_1 Used to Determine Values of Parameter a in the Marcum Q-Function $Q(a,b)$

$\alpha_1^2/2$	$10\log_{10}\left(\alpha_1^2/2\right)$	α_1
0.125	−9 dB	$1/2$
0.25	−6 dB	$\sqrt{2}/2$
0.5	−3 dB	1
1	0 dB	$\sqrt{2}$
2	3 dB	2
4	6 dB	$2\sqrt{2}$
8	9 dB	4

By referring to (12.4-41), parameter b in $Q(a,b)$ is given by

$$b = \sqrt{\mathcal{T}}/\sigma_1. \tag{12.4-59}$$

Substituting (12.3-55) into (12.3-56) yields

$$\sqrt{\mathcal{T}} = 4.0747\,\sigma_0\,, \tag{12.4-60}$$

and by substituting (12.4-60) into (12.4-59), we obtain

$$b = 4.0747\frac{\sigma_0}{\sigma_1}\,, \tag{12.4-61}$$

where (see Appendix 12D)

$$\boxed{\frac{\sigma_0}{\sigma_1} = \sqrt{\frac{\gamma_0}{\gamma_1}\frac{\alpha_1}{\alpha_0}} < 1} \tag{12.4-62}$$

$$\boxed{\sqrt{\gamma_0/\gamma_1} < \alpha_0/\alpha_1} \tag{12.4-63}$$

and

$$\boxed{\sqrt{\gamma_0/\gamma_1} < 1} \tag{12.4-64}$$

In order to determine the P_D, we need to choose values for α_1 and $\sqrt{\gamma_0/\gamma_1}$. Since $\alpha_0 = \sqrt{2}/2$ in Example 12.3-1, let $\alpha_1 = 2\sqrt{2}$ so that $\alpha_1 > \alpha_0$ (see Table 12.4-1). Therefore, since $\alpha_0/\alpha_1 = 0.25$, by referring to (12.4-63) and (12.4-64), let $\sqrt{\gamma_0/\gamma_1} = 0.125$. Substituting $\sqrt{\gamma_0/\gamma_1} = 0.125$ and $\alpha_1/\alpha_0 = 4$ into (12.4-62) yields $\sigma_0/\sigma_1 = 0.5$, and by substituting $\sigma_0/\sigma_1 = 0.5$ into (12.4-61), we obtain $b = 2.03735$. Since $a = \alpha_1 = 2\sqrt{2}$, from Table 12C-1 in Appendix 12C:

	$a = 2\sqrt{2}$
b	$Q(a,b)$
2	0.8519364
2.1	0.8262441

Next, use linear interpolation to find the value of $Q\left(2\sqrt{2}, 2.03735\right)$ and, thus, the P_D.

If we let

$$Q(a,b) = mb + c\,, \tag{12.4-65}$$

then the slope m is

$$
\begin{aligned}
m &= \frac{Q(a,b_2)-Q(a,b_1)}{b_2-b_1} \\
&= \frac{Q\left(2\sqrt{2},2.1\right)-Q\left(2\sqrt{2},2\right)}{2.1-2} \\
&= \frac{0.8262441-0.8519364}{0.1} \\
&= -0.256923
\end{aligned}
\tag{12.4-66}
$$

and the y-intercept is

$$
\begin{aligned}
c &= Q(a,b_1)-mb_1 \\
&= Q\left(2\sqrt{2},2\right)-(-0.256923)2 \\
&= 0.8519364+0.513846 \\
&= 1.3657824.
\end{aligned}
\tag{12.4-67}
$$

Substituting (12.4-66), (12.4-67), and $a=2\sqrt{2}$ into (12.4-65) yields

$$
Q\left(2\sqrt{2},b\right)=-0.256923b+1.3657824\,, \tag{12.4-68}
$$

and by evaluating (12.4-68) at $b=2.03735$, we obtain

$$
P_{\mathrm{D}}=Q\left(2\sqrt{2},2.03735\right)=0.842\,. \tag{12.4-69}
$$

The formula for computing the P_{D} can be generalized as follows. By referring to (12.4-59), let

$$
b_{\mathrm{D}}=\sqrt{\mathcal{T}}\big/\sigma_1\,. \tag{12.4-70}
$$

Substituting (12.3-63) into (12.4-70) yields

$$
\boxed{b_{\mathrm{D}}=b_{\mathrm{FA}}\frac{\sigma_0}{\sigma_1}} \tag{12.4-71}
$$

where b_{FA} depends on the values of α_0 and P_{FA} and is given by (12.3-59), and σ_0/σ_1 depends on the values of α_0, α_1, and $\sqrt{\gamma_0/\gamma_1}$, and is given by (12.4-62). By referring to (12.4-65) and noting that $P_{\mathrm{D}}=Q(\alpha_1,b_{\mathrm{D}})$, we can write that

$$\boxed{P_{\mathrm{D}} = m_{\alpha_1} b_{\mathrm{D}} + c_{\alpha_1}} \tag{12.4-72}$$

where

$$m_{\alpha_1} = \frac{Q(\alpha_1, b_2) - Q(\alpha_1, b_1)}{b_2 - b_1}, \tag{12.4-73}$$

$$c_{\alpha_1} = Q(\alpha_1, b_1) - m_{\alpha_1} b_1, \tag{12.4-74}$$

and

$$b_1 < b_{\mathrm{D}} < b_2. \tag{12.4-75}$$

Therefore, the P_{D} can be plotted versus the P_{FA} for a fixed value of σ_0/σ_1 (i.e., fixed values of α_0, α_1, and $\sqrt{\gamma_0/\gamma_1}$) because for a fixed value of α_0, b_{FA} will change value as the P_{FA} changes value [see (12.3-59)], and as b_{FA} changes value, b_{D} changes value [see (12.4-71)] and, thus, the P_{D} changes value [see (12.4-72)]. As a result, *receiver operating characteristic* (ROC) *curves* can be generated by plotting the P_{D} versus the P_{FA} for different values of σ_0/σ_1. ■

Example 12.4-2 Receiver Operating Characteristic Curves – Zero-Mean Target Scattering Function and No Reverberation Return

Recall that if the reverberation scattering function is *zero-mean*, then the decision threshold $\mathcal{T}$ is given by (12.3-48). However, if *no* reverberation return is present at the receiver, then (12.3-48) reduces to

$$\boxed{\mathcal{T} = A_2 E_{\tilde{x}} \left[\tilde{N}_{a0} + \tilde{N}_{r0} \right] \ln\left(1/P_{\mathrm{FA}}\right)} \tag{12.4-76}$$

where A_2 is given by (12.2-25), $E_{\tilde{x}}$ is the energy of the complex envelope of the transmitted electrical signal in J-Ω, and $\tilde{N}_{a0}$ and $\tilde{N}_{r0}$ are the constant levels in $(\text{W-}\Omega)/\text{Hz}$ (or J-Ω) of the lowpass (baseband), white-noise power spectra of the complex envelopes of the ambient noise and receiver noise $\tilde{n}_a(t, x_i)$ and $\tilde{n}(t, x_i)$, respectively, $\forall \; i$, before complex weighting (see Appendix 12B). If rectangular amplitude weights are used, then $A_2 = N$ where N is the total odd number of elements in the linear array.

Also recall that if both the target and reverberation scattering functions are *zero-mean*, then the relationship between the P_{D}, P_{FA}, and SIR is given by (12.4-55), where the SIR is given by (12.2-26). However, if *no* reverberation return is present at the receiver, then (12.4-55) and (12.2-26) reduce to

$$\boxed{P_{\mathrm{D}} = P_{\mathrm{FA}}^{1/(\mathrm{SNR}+1)}} \tag{12.4-77}$$

and

$$\boxed{\mathrm{SNR} = \left|\mathcal{X}_N(e_{\tau,Trgt}, e_{D,Trgt})\right|^2 \mathrm{SNR}_{\max}} \tag{12.4-78}$$

respectively, where $\mathcal{X}_N(e_{\tau,Trgt}, e_{D,Trgt})$ is the normalized, auto-ambiguity function of the transmitted electrical signal [see (12.2-17) and (12.2-13)],

$$\boxed{\mathrm{SNR}_{\max} = \mathrm{AG} \times \mathrm{SNR}_0} \tag{12.4-79}$$

is the maximum value of the SNR, AG is the *dimensionless* version of the array gain of the linear array given by (12.2-27),

$$\boxed{\mathrm{SNR}_0 = \frac{E_{\tilde{x}} E\left\{\left|w_{Trgt}\right|^2\right\}}{\tilde{N}_{a0} + \tilde{N}_{r0}}} \tag{12.4-80}$$

is the SNR due to processing the output electrical signal from a single element – element number 0 at the center of the linear array – and $E\left\{\left|w_{Trgt}\right|^2\right\}$ is given by (12.2-20). If the estimation errors $e_{\tau,Trgt}$ and $e_{D,Trgt}$ are zero, then $\mathcal{X}_N(0,0) = 1$ [see (12.2-18)] and $\mathrm{SNR} = \mathrm{SNR}_{\max}$. If rectangular amplitude weights are used, then $\mathrm{AG} = N$. Equations (12.4-78) through (12.4-80) are identical to (12.2-31) through (12.2-33), respectively.

Taking the logarithm (base 10) of (12.4-77) and solving for the SNR yields

$$\boxed{\mathrm{SNR} = \frac{\log_{10} P_{\mathrm{FA}}}{\log_{10} P_{\mathrm{D}}} - 1} \tag{12.4-81}$$

Equation (12.4-81) can be used to compute the SNR that is required in order to obtain a desired P_{D} for a given P_{FA}. Note that the SNR computed according to (12.4-81) is *dimensionless*.

As was previously mentioned, if the estimation errors $e_{\tau,Trgt}$ and $e_{D,Trgt}$ are zero, then $\mathrm{SNR} = \mathrm{SNR}_{\max}$ where $\mathrm{SNR}_{\max}$ is given by (12.4-79). Substituting $\mathrm{SNR} = \mathrm{SNR}_{\max}$ into (12.4-77) yields the corresponding maximum value for the probability of detection, which is desirable, since we want the probability of declaring that we detected a target with the *correct* round-trip time delay (range) and Doppler-shift values to be maximized. Conversely, we would like the SNR to *decrease* as the magnitude of the estimation errors *increase*. A small SNR results in a small P_{D} [see (12.4-77)], which is desirable, since whenever we make large estimation errors, we want the probability of declaring that we detected a target with the *wrong* round-trip time delay (range) and Doppler-shift values to be *small*. To ensure this kind of performance and, hence, to avoid any *ambiguity* concerning the estimates of τ_{Trgt} and $\eta_{D,Trgt}$, the normalized, auto-ambiguity function should have as narrow a mainlobe about the origin $(e_{\tau,Trgt}=0, e_{D,Trgt}=0)$ as possible so that as the estimation errors increase, the value of $\left|\chi_N(e_{\tau,Trgt}, e_{D,Trgt})\right|$ decreases rapidly and, as a result, both the SNR and the P_{D} decrease rapidly. In addition, it is also desirable that $\left|\chi_N(e_{\tau,Trgt}, e_{D,Trgt})\right|$ have sidelobe levels as low as possible. Auto-ambiguity functions are discussed in detail in Chapter 13.

Let us conclude this example by plotting several *receiver operating characteristic* (ROC) curves. For a Neyman-Pearson test, a ROC curve is a plot of the probability of detection P_{D} versus the probability of false alarm P_{FA} for a given value of some parameter. In this example, the parameter of interest is the signal-to-noise ratio (SNR).

In order to simplify the results, assume that the estimation errors $e_{\tau,Trgt}$ and $e_{D,Trgt}$ are *zero* so that $\chi_N(0,0)=1$, and that *rectangular amplitude weights* are used so that $\mathrm{AG}=N$. Therefore, (12.4-78) reduces to

$$\boxed{\mathrm{SNR}=N\times\mathrm{SNR}_0} \tag{12.4-82}$$

First consider the case where $N=1$ (a single element). Figure 12.4-1 (a) is a plot of (12.4-77) for $\mathrm{SNR}=\mathrm{SNR}_0$ where $\mathrm{SNR}_0=0.5\,(-3.01\text{ dB})$, $\mathrm{SNR}_0=1\,(0\text{ dB})$, $\mathrm{SNR}_0=2\,(3.01\text{ dB})$, and $\mathrm{SNR}_0=10\,(10\text{ dB})$. Now consider the case of a linear array with $N=101$ elements. As a result, the array gain $\mathrm{AG}=101\,(20.04\text{ dB})$. Figure 12.4-1 (b) is a plot of (12.4-77) for $\mathrm{SNR}=101\times\mathrm{SNR}_0$ where SNR_0 is equal to the same values used for Fig. 12.4-1 (a). Therefore, $\mathrm{SNR}=50.5\,(17.03\text{ dB})$, $\mathrm{SNR}=101\,(20.04\text{ dB})$, $\mathrm{SNR}=202\,(23.05\text{ dB})$, and $\mathrm{SNR}=1010\,(30.04\text{ dB})$. By comparing Figs. 12.4-1 (a) and 12.4-1 (b), it can clearly be seen how AG greatly increases the P_{D} for a given P_{FA}. ■

Several additional comments regarding the practical importance of (12.4-82) are in order. Solving for N in (12.4-82) yields

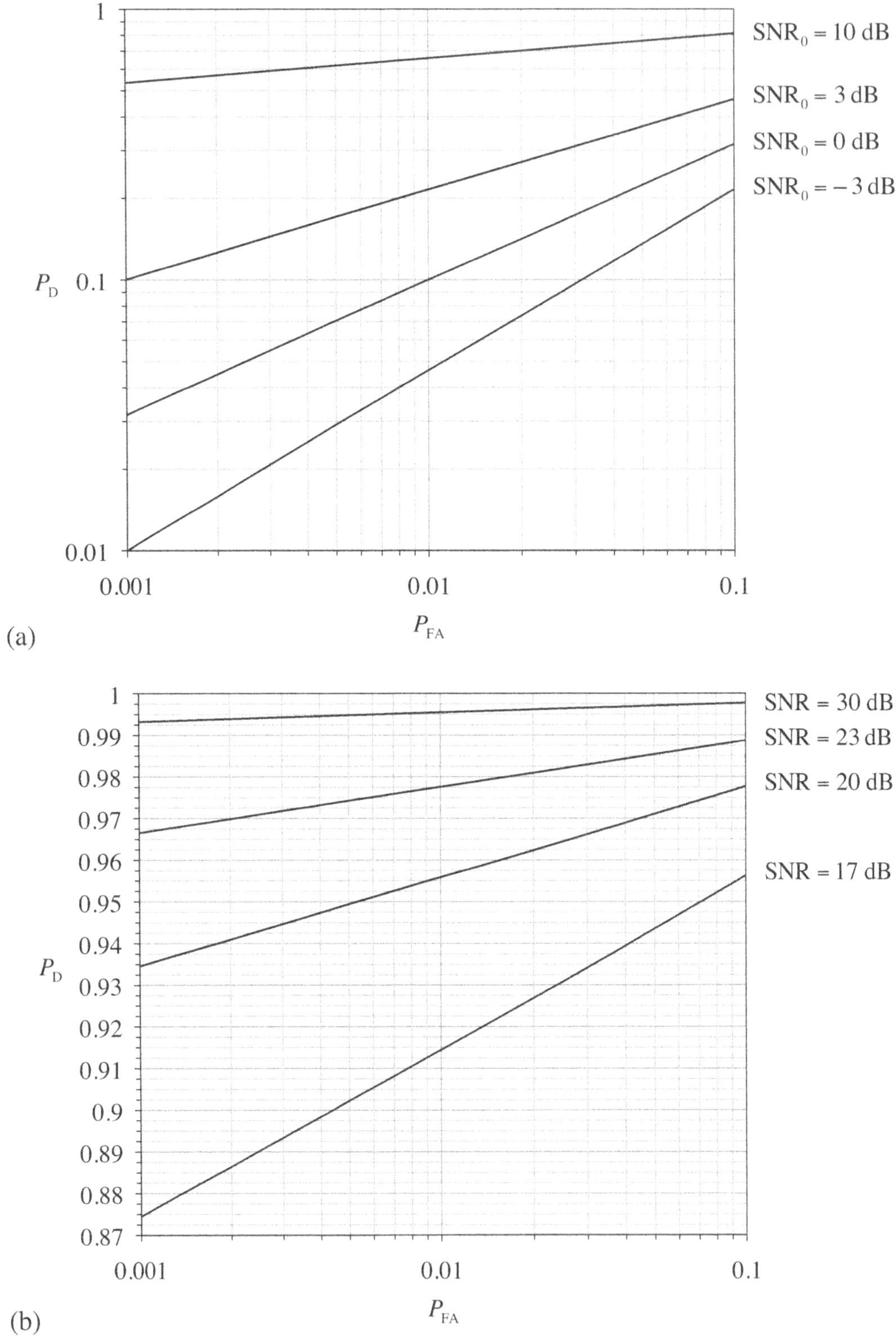

Figure 12.4-1 Receiver operating characteristic (ROC) curves for (a) a single element and (b) a linear array of 101 elements.

$$\boxed{N = \text{SNR}/\text{SNR}_0} \tag{12.4-83}$$

which is the total number of elements required in order to obtain a desired P_{D} for a given P_{FA}, where SNR is given by (12.4-81) and SNR_0 is given by (12.4-80). Note that the value of N obtained from (12.4-83) may have to be *rounded up* to the nearest *odd* number for our linear array. Once N has been determined, and the interelement spacing d is known (e.g., in order to avoid grating lobes for all possible directions of beam steering, $d < \lambda_{\text{min}}/2$), then the overall length of the linear array L_A that is required in order to satisfy the P_{D} and P_{FA} constraints is given by $L_A = (N-1)d$. Once L_A has been determined, the range to the near-field/far-field boundary can be computed for a given operating frequency (wavelength) since $r_{\text{NF/FF}} = \pi R_A^2/\lambda$, where $R_A = L_A/2$ is the maximum radial extent of a linear array.

Problems

Section 12.2

12-1 Use the Schwarz inequality

$$\left|\int_{-\infty}^{\infty} A(\zeta)B(\zeta)d\zeta\right|^2 \le \int_{-\infty}^{\infty} |A(\zeta)|^2 d\zeta \int_{-\infty}^{\infty} |B(\zeta)|^2 d\zeta ,$$

where $A(\zeta)$ and $B(\zeta)$ are finite energy, complex functions in general, to prove that $\left|X(e_{\tau,\,Trgt}, e_{D,\,Trgt})\right| \le E_{\tilde{x}}$.

Section 12.3

12-2 Show that the variance of w_{Rev} depends on the variance of the reverberation scattering function.

Section 12.4

12-3 Show that the variance of w_{Trgt} depends on the variance of the target scattering function.

12-4 Consider the case where the target scattering function is zero-mean and there is no reverberation. If the $P_{\text{FA}} = 0.01$, then find the output SNR in *decibels* that is needed for a

(a) $P_{\text{D}} = 0.9$

(b) $P_{\text{D}} = 0.95$

(c) $P_{\text{D}} = 0.99$

(d) Repeat Parts (a) through (c) for a $P_{\text{FA}} = 0.001$.

12-5 Consider the case where the target scattering function is zero-mean and there is no reverberation. If the estimation errors $e_{\tau,Trgt}$ and $e_{D,Trgt}$ are *zero* and *rectangular amplitude weights* are used, then

(a) how many array elements (an odd number) are required in order to achieve a $P_{\text{D}} = 0.9$ for a $P_{\text{FA}} = 0.01$ if $\text{SNR}_0 = 0.25\ (-6\text{ dB})$?

(b) Find the corresponding value of AG in decibels.

(c) If the frequency of operation is 1 kHz and half-wavelength interelement spacing is used, then what is the overall length of the linear array? Use $c = 1500\text{ m/sec}$.

(d) What is the range to the near-field/far-field boundary for this linear array?

Appendix 12A Mathematical Models of the Target Return and Reverberation Return

In order to derive mathematical models of the target return and reverberation return, we first need to derive an equation for the complex envelope, $\tilde{y}_s(t, \mathbf{r}_R)$, of the output electrical signal, $y_s(t, \mathbf{r}_R)$, from an omnidirectional point-element at $\mathbf{r}_R$ for the bistatic scattering problem discussed in Chapter 10 (see Fig. 10.1-1 and Section 10.5). We begin by expressing $y_s(t, \mathbf{r}_R)$ as an inverse Fourier transform [see (10.5-9)]:

$$\begin{aligned} y_s(t, \mathbf{r}_R) &= F_f^{-1}\{Y_s(f, \mathbf{r}_R)\} \\ &= \int_{-\infty}^{\infty} Y_s(f, \mathbf{r}_R)\exp(+j2\pi ft)\,df \\ &= \int_{-\infty}^{0} Y_s(f, \mathbf{r}_R)\exp(+j2\pi ft)\,df + \int_{0}^{\infty} Y_s(f, \mathbf{r}_R)\exp(+j2\pi ft)\,df. \end{aligned} \tag{12A-1}$$

If we let $f' = -f$ in the first integral in (12A-1), then $f = -f'$, $df = -df'$, and when $f = -\infty$, $f' = \infty$; and when $f = 0$, $f' = 0$. Therefore,

$$\begin{aligned}\int_{-\infty}^{0} Y_s(f, \mathbf{r}_R)\exp(+j2\pi ft)df &= -\int_{\infty}^{0} Y_s(-f', \mathbf{r}_R)\exp(-j2\pi f't)df' \\ &= \int_{0}^{\infty} Y_s(-f', \mathbf{r}_R)\exp(-j2\pi f't)df'.\end{aligned} \tag{12A-2}$$

Replacing f' with f yields

$$\int_{-\infty}^{0} Y_s(f, \mathbf{r}_R)\exp(+j2\pi ft)df = \int_{0}^{\infty} Y_s(-f, \mathbf{r}_R)\exp(-j2\pi ft)df\,, \tag{12A-3}$$

and by substituting (12A-3) into (12A-1), we obtain

$$y_s(t, \mathbf{r}_R) = \int_{0}^{\infty} Y_s(-f, \mathbf{r}_R)\exp(-j2\pi ft)df + \int_{0}^{\infty} Y_s(f, \mathbf{r}_R)\exp(+j2\pi ft)df\,. \tag{12A-4}$$

Since $y_s(t, \mathbf{r}_R)$ is real, its time-domain Fourier transform satisfies the conjugate symmetry property, that is,

$$Y_s(-f, \mathbf{r}_R) = Y_s^*(f, \mathbf{r}_R)\,. \tag{12A-5}$$

Therefore, substituting (12A-5) into (12A-4) yields

$$y_s(t, \mathbf{r}_R) = \int_{0}^{\infty} \left[Y_s(f, \mathbf{r}_R)\exp(+j2\pi ft)\right]^* df + \int_{0}^{\infty} Y_s(f, \mathbf{r}_R)\exp(+j2\pi ft)df\,. \tag{12A-6}$$

If we let

$$Z = \int_{0}^{\infty} Y_s(f, \mathbf{r}_R)\exp(+j2\pi ft)df\,, \tag{12A-7}$$

and since

$$2\,\mathrm{Re}\{Z\} = Z + Z^*, \tag{12A-8}$$

(12A-6) reduces to

$$y_s(t, \mathbf{r}_R) = 2\,\mathrm{Re}\left\{\int_{0}^{\infty} Y_s(f, \mathbf{r}_R)\exp(+j2\pi ft)df\right\}. \tag{12A-9}$$

The next step is to express (12A-9) in terms of the complex frequency spectrum $X(f)$ of the real transmitted electrical signal $x(t)$. Substituting (10.5-7) into (10.5-8) yields

$$Y_s(f,\mathbf{r}_R) = X(f)H\left(f,\mathbf{r}_R\middle|\mathbf{r}_T\right), \tag{12A-10}$$

where

$$H\left(f,\mathbf{r}_R\middle|\mathbf{r}_T\right) = \frac{1}{(4\pi)^2 r_{TS} r_{SR}} \mathcal{S}_T(f) S_S(f,\theta_{TS},\psi_{TS},\theta_{SR},\psi_{SR}) \exp\left[-\alpha(f)(r_{TS}+r_{SR})\right] \times \\ \exp(-j2\pi f\tau)\mathcal{S}_R(f) \tag{12A-11}$$

is the overall complex frequency response for the bistatic scattering problem, and

$$\tau = (r_{TS}+r_{SR})/c \tag{12A-12}$$

is the bistatic time delay. Therefore, substituting (12A-10) into (12A-9) yields

$$y_s(t,\mathbf{r}_R) = 2\,\mathrm{Re}\left\{\int_0^\infty X(f)H\left(f,\mathbf{r}_R\middle|\mathbf{r}_T\right)\exp(+j2\pi ft)df\right\}. \tag{12A-13}$$

We are now in a position to derive the complex envelope of $y_s(t,\mathbf{r}_R)$.

Substituting (11.1-24) into (12A-13) yields

$$y_s(t,\mathbf{r}_R) = \mathrm{Re}\left\{\int_0^\infty \tilde{X}(f-f_c)H\left(f,\mathbf{r}_R\middle|\mathbf{r}_T\right)\exp(+j2\pi ft)df\right\}, \tag{12A-14}$$

where $\tilde{X}(f)$ is the complex frequency spectrum of the lowpass (baseband) complex envelope $\tilde{x}(t)$ of the real bandpass signal $x(t)$. If we let $f' = f - f_c$ in (12A-14), then $f = f' + f_c$, $df = df'$, and when $f = 0$, $f' = -f_c$; and when $f = \infty$, $f' = \infty$. Therefore,

$$y_s(t,\mathbf{r}_R) = \mathrm{Re}\left\{\int_{-f_c}^\infty \tilde{X}(f')H\left(f'+f_c,\mathbf{r}_R\middle|\mathbf{r}_T\right)\exp(+j2\pi f't)df' \exp(+j2\pi f_c t)\right\}. \tag{12A-15}$$

Since $\tilde{X}(f)$ is lowpass with bandwidth W hertz,

$$\left|\tilde{X}(f)\right| = 0, \qquad |f| \geq W, \tag{12A-16}$$

and since the carrier frequency $f_c > W$ [see (11.2-20)], $-f_c < -W$. Therefore, by replacing f' with f, (12A-15) can be rewritten as

$$y_s(t, \mathbf{r}_R) = \text{Re}\left\{\int_{-W}^{W} \tilde{X}(f) H\left(f + f_c, \mathbf{r}_R \middle| \mathbf{r}_T\right) \exp(+j2\pi f t) df \exp(+j2\pi f_c t)\right\},$$

(12A-17)

or

$$y_s(t, \mathbf{r}_R) = \text{Re}\left\{\int_{-\infty}^{\infty} \tilde{X}(f) H\left(f + f_c, \mathbf{r}_R \middle| \mathbf{r}_T\right) \exp(+j2\pi f t) df \exp(+j2\pi f_c t)\right\}.$$

(12A-18)

By comparing (12A-18) with (11.1-6), (12A-18) can be expressed as

$$\boxed{y_s(t, \mathbf{r}_R) = \text{Re}\left\{\tilde{y}_s(t, \mathbf{r}_R) \exp(+j2\pi f_c t)\right\}} \tag{12A-19}$$

where

$$\boxed{\begin{aligned} \tilde{y}_s(t, \mathbf{r}_R) &= F_f^{-1}\left\{\tilde{X}(f) H\left(f + f_c, \mathbf{r}_R \middle| \mathbf{r}_T\right)\right\} \\ &= \int_{-\infty}^{\infty} \tilde{X}(f) H\left(f + f_c, \mathbf{r}_R \middle| \mathbf{r}_T\right) \exp(+j2\pi f t) df \\ &= \int_{-W}^{W} \tilde{X}(f) H\left(f + f_c, \mathbf{r}_R \middle| \mathbf{r}_T\right) \exp(+j2\pi f t) df \end{aligned}} \tag{12A-20}$$

is the complex envelope of $y_s(t, \mathbf{r}_R)$, and [see (12A-11)]

$$\begin{aligned} H\left(f + f_c, \mathbf{r}_R \middle| \mathbf{r}_T\right) = \frac{1}{(4\pi)^2 r_{TS} r_{SR}} &\mathcal{S}_T(f + f_c) S_S(f + f_c, \theta_{TS}, \psi_{TS}, \theta_{SR}, \psi_{SR}) \times \\ &\exp\left[-\alpha(f + f_c)(r_{TS} + r_{SR})\right] \exp\left[-j2\pi(f + f_c)\tau\right] \times \\ &\mathcal{S}_R(f + f_c). \end{aligned}$$

(12A-21)

In order to obtain a closed-form, analytical solution for the complex envelope $\tilde{y}_s(t, \mathbf{r}_R)$, we shall make the following approximation in the frequency interval $[f_c - W, f_c + W]$:

$$\begin{aligned} H\left(f + f_c, \mathbf{r}_R \middle| \mathbf{r}_T\right) \approx \frac{1}{(4\pi)^2 r_{TS} r_{SR}} &\mathcal{S}_T(f_c) S_S(f_c, \theta_{TS}, \psi_{TS}, \theta_{SR}, \psi_{SR}) \times \\ &\exp\left[-\alpha(f_c)(r_{TS} + r_{SR})\right] \exp\left[-j2\pi(f + f_c)\tau\right] \mathcal{S}_R(f_c), \\ &\qquad f_c - W \le f \le f_c + W. \end{aligned}$$

(12A-22)

Therefore, substituting (12A-22) into (12A-20) yields

$$\boxed{\tilde{y}_s(t,\mathbf{r}_R) \approx w_S\,\tilde{x}(t-\tau)\exp(-j2\pi f_c\tau)} \tag{12A-23}$$

for $t \geq \tau$, where

$$\tilde{x}(t-\tau) = F_f^{-1}\left\{\tilde{X}(f)\exp(-j2\pi f\tau)\right\} = \int_{-\infty}^{\infty} \tilde{X}(f)\exp(-j2\pi f\tau)\exp(+j2\pi ft)\,df \tag{12A-24}$$

and

$$\boxed{w_S = \frac{1}{(4\pi)^2 r_{TS} r_{SR}} \mathcal{S}_T(f_c) S_S(f_c, \theta_{TS}, \psi_{TS}, \theta_{SR}, \psi_{SR}) \exp\left[-\alpha(f_c)(r_{TS}+r_{SR})\right] \mathcal{S}_R(f_c)} \tag{12A-25}$$

In the case of a monostatic (backscatter) geometry, (12A-23) is still applicable, however (see Section 10.1),

$$r_{SR} = r_{TS}. \tag{12A-26}$$

Therefore, w_S given by (12A-25) reduces to

$$\boxed{w_S = \frac{1}{(4\pi r_{TS})^2} \mathcal{S}_T(f_c) S_S(f_c, \theta_{TS}, \psi_{TS}, \theta_{SR}, \psi_{SR}) \exp[-2\alpha(f_c) r_{TS}] \mathcal{S}_R(f_c)} \tag{12A-27}$$

where (see Section 10.1 and Figs. 10.1-2 and 10.1-3)

$$\theta_{SR} = 180° - \theta_{TS} \tag{12A-28}$$

and

$$\psi_{SR} = 180° + \psi_{TS}, \tag{12A-29}$$

and the bistatic time delay given by (12A-12) reduces to the round-trip time delay

$$\tau = 2r_{TS}/c. \tag{12A-30}$$

The complex envelope given by (12A-23) is applicable to both bistatic and monostatic scattering problems when none of the platforms are in motion. However, we need to include the effects of platform motion in our target return and reverberation return models. In other words, we need to include a Doppler shift in the model given by (12A-23). If time-compression/time-expansion effects are ignored, then

$$\boxed{\tilde{y}_s(t,\mathbf{r}_R) \approx w_S\,\tilde{x}(t-\tau)\exp\left[+j2\pi\eta_D(t-\tau)\right]\exp(-j2\pi f_c\tau)} \tag{12A-31}$$

for $t \geq \tau$, where η_D is the *Doppler shift* in hertz. Equation (12A-31) is the simplest mathematical model for the complex envelope of the output electrical signal from an omnidirectional point-element at $\mathbf{r}_R$ for the bistatic scattering problem discussed in Chapter 10 when one, two, or all three of the platforms shown in Fig. 10.1-1 are in motion. Since the complex envelope of the transmitted amplitude-and-angle-modulated carrier is given by [see (11.2-31)]

$$\tilde{x}(t) = a(t)\exp[+j\theta(t)], \tag{12A-32}$$

where $a(t)$ and $\theta(t)$ are real amplitude-modulating and angle-modulating functions, respectively, if we represent w_S in polar form as

$$w_S = |w_S|\exp(+j\angle w_S), \tag{12A-33}$$

then substituting (12A-32) and (12A-33) into (12A-31) yields

$$\begin{aligned}\tilde{y}_s(t,\mathbf{r}_R) \approx |w_S|\,a(t-\tau)\exp\left[+j\theta(t-\tau)\right]\exp\left[+j2\pi\eta_D(t-\tau)\right]\exp(+j\angle w_S)\times \\ \exp(-j2\pi f_c\tau).\end{aligned} \tag{12A-34}$$

And by substituting (12A-34) into (12A-19), we obtain the real bandpass signal

$$\boxed{y_s(t,\mathbf{r}_R) \approx |w_S|\,a(t-\tau)\cos\left[2\pi(f_c+\eta_D)(t-\tau)+\theta(t-\tau)+\angle w_S\right]} \tag{12A-35}$$

for $t \geq \tau$. We shall now use (12A-31) to model the complex envelopes of the output electrical signals at each element in a linear array for a *backscatter* geometry.

Consider a linear array composed of an odd number N of identical, equally-spaced, complex-weighted, omnidirectional point-elements lying along the X axis. The position vector to element i in the array is given by

$$\mathbf{r}_i = x_i\hat{x} = id\hat{x}, \qquad i = -N',\ldots,0,\ldots,N', \tag{12A-36}$$

where

$$x_i = id \tag{12A-37}$$

is the x coordinate of the center of element i, d is the interelement spacing in meters, and

$$N' = (N-1)/2 . \tag{12A-38}$$

For a *backscatter* geometry,

$$\mathbf{r}_R = \mathbf{r}_T = \mathbf{r}_i , \tag{12A-39}$$

where $\mathbf{r}_i$ is given by (12A-36).

The position vector to a scatterer (target or object) is given by either

$$\mathbf{r}_S = x_S \hat{x} + y_S \hat{y} + z_S \hat{z} \tag{12A-40}$$

or

$$\mathbf{r}_S = r_S \hat{r}_S , \tag{12A-41}$$

where

$$r_S = |\mathbf{r}_S| = \sqrt{x_S^2 + y_S^2 + z_S^2} \tag{12A-42}$$

is the range to the scatterer in meters measured from the origin of the coordinate system as shown in Fig. 10.1-1 (the center of the linear array in our problem),

$$\hat{r}_S = u_S \hat{x} + v_S \hat{y} + w_S \hat{z} \tag{12A-43}$$

is the unit vector in the direction of $\mathbf{r}_S$, and

$$u_S = \sin\theta_S \cos\psi_S , \tag{12A-44}$$

$$v_S = \sin\theta_S \sin\psi_S , \tag{12A-45}$$

and

$$w_S = \cos\theta_S \tag{12A-46}$$

are dimensionless direction cosines with respect to the X, Y, and Z axes, respectively. The spherical coordinates of the scatterer's location (r_S, θ_S, ψ_S) are *unknown* a priori. Therefore, the range between the transmitter at $\mathbf{r}_T$ and scatterer at $\mathbf{r}_S$ given in Section 10.1 as

$$r_{TS} = |\mathbf{r}_S - \mathbf{r}_T| = \sqrt{(x_S - x_T)^2 + (y_S - y_T)^2 + (z_S - z_T)^2} \tag{12A-47}$$

can be rewritten for our problem as

$$r_{TS} = r_{iS} , \tag{12A-48}$$

where

$$\begin{aligned} r_{iS} &= \left| \mathbf{r}_S - \mathbf{r}_i \right| \\ &= \sqrt{(x_S - x_i)^2 + (y_S - y_i)^2 + (z_S - z_i)^2} \\ &= \sqrt{(x_S - x_i)^2 + y_S^2 + z_S^2} \end{aligned} \tag{12A-49}$$

since $\mathbf{r}_T = \mathbf{r}_i = (x_i, 0, 0)$ [see (12A-39) and (12A-36)]. By expanding the radicand in (12A-49), and since

$$x_S = r_S u_S, \tag{12A-50}$$

(12A-49) can also be expressed as

$$r_{iS} = \sqrt{r_S^2 - 2 r_S u_S x_i + x_i^2}, \tag{12A-51}$$

where r_S is given by (12A-42), u_S is given by (12A-44), and x_i is given by (12A-37).

Equation (12A-48) can also be used to rewrite the equations for w_S and τ given by (12A-27) and (12A-30), respectively. Substituting (12A-48) into (12A-27) yields

$$w_S = \frac{1}{(4\pi r_{iS})^2} \mathcal{S}_T(f_c) S_S(f_c, \theta_{iS}, \psi_{iS}, \theta_{Si}, \psi_{Si}) \exp[-2\alpha(f_c) r_{iS}] \mathcal{S}_R(f_c), \tag{12A-52}$$

where [see (12A-28) and (12A-29)]

$$\theta_{Si} = 180^\circ - \theta_{iS} \tag{12A-53}$$

and

$$\psi_{Si} = 180^\circ + \psi_{iS}. \tag{12A-54}$$

The factor w_S given by (12A-52) for a backscatter geometry can be approximated by removing its dependency on element number i as follows:

$$\boxed{w_S \approx \frac{1}{(4\pi r_S)^2} \mathcal{S}_T(f_c) S_S(f_c, \theta_S, \psi_S, \theta_{S0}, \psi_{S0}) \exp[-2\alpha(f_c) r_S] \mathcal{S}_R(f_c)} \tag{12A-55}$$

where $r_S = r_{0S}$ is the range to the scatterer measured from the origin of the coordinate system (i.e., from element $i = 0$ at the center of the linear array), $\theta_S = \theta_{0S}$ and $\psi_S = \psi_{0S}$ are the spherical angles measured from the origin of the coordinate system (the center of the linear array) to the scatterer, and $\theta_{S0} = 180° - \theta_S$ and $\psi_{S0} = 180° + \psi_S$ [see (12A-53) and (12A-54), respectively]. And, by substituting (12A-48) into (12A-30), we obtain

$$\boxed{\tau_i = 2r_{iS}/c} \tag{12A-56}$$

which is the round-trip time delay from element i in the array to the scatterer, where the range r_{iS} is given by either (12A-49) or (12A-51). Therefore, the complex envelope of the output electrical signal from element i in the linear array for a backscatter geometry *before* complex weighting is given by [see (12A-31)]

$$\tilde{y}_s(t, x_i) \approx w_S\, \tilde{x}(t - \tau_i)\exp\left[+j2\pi\eta_D(t - \tau_i)\right]\exp(-j2\pi f_c \tau_i), \tag{12A-57}$$

for $t \geq \tau_i$, where w_S is given by (12A-55) and τ_i is given by (12A-56).

In order to obtain a model for the complex envelope after complex weighting, we first have to compute the Fourier transform of (12A-57). Doing so yields

$$\tilde{Y}_s(\eta, x_i) \approx w_S\, \tilde{X}(\eta - \eta_D)\exp\left[-j2\pi(\eta + f_c)\tau_i\right], \tag{12A-58}$$

since

$$F_t\{\tilde{x}(t - \tau_i)\exp(+j2\pi\eta_D t)\} = \tilde{X}(\eta - \eta_D)\exp(-j2\pi\eta\tau_i)\exp(+j2\pi\eta_D\tau_i), \tag{12A-59}$$

where η represents *output (received)* frequencies in hertz, and because of platform motion, $\eta \neq f$, where f represents *input (transmitted)* frequencies in hertz. If there is no platform motion, then $\eta = f$. Multiplying the right-hand side of (12A-58) by the complex weight

$$c_i(\eta) = a_i \exp[+j\theta_i(\eta)] = a_i \exp\left[-j2\pi(\eta + f_c)\tau_i'\right] \tag{12A-60}$$

yields

$$\tilde{Y}_s(\eta, x_i) \approx a_i w_S\, \tilde{X}(\eta - \eta_D)\exp\left[-j2\pi\eta(\tau_i + \tau_i')\right]\exp\left[-j2\pi f_c(\tau_i + \tau_i')\right], \tag{12A-61}$$

and by inverse Fourier transforming (12A-61), we obtain

$$\boxed{\tilde{y}_s(t, x_i) \approx a_i w_S\, \tilde{x}(t - [\tau_i + \tau_i'])\exp\left\{+j2\pi\eta_D\left[t - (\tau_i + \tau_i')\right]\right\}\exp\left[-j2\pi f_c(\tau_i + \tau_i')\right]} \tag{12A-62}$$

for $t \ge \tau_i + \tau_i'$. Equation (12A-62) is the complex envelope of the output electrical signal from element i in the linear array for a backscatter geometry *after* complex weighting, where a_i is the amplitude weight applied to element i, w_S is given by (12A-55), τ_i is given by (12A-56), and τ_i' is the time-delay applied to element i via phase weighting in order to co-phase the output electrical signals from all the elements in the array due to the scatterer. Substituting (12A-32) and (12A-33) into (12A-62) yields

$$\tilde{y}_s(t, x_i) \approx a_i |w_S| a(t - [\tau_i + \tau_i']) \exp[+j\theta(t - [\tau_i + \tau_i'])] \exp\{+j2\pi\eta_D [t - (\tau_i + \tau_i')]\} \times \exp(+j\angle w_S) \exp[-j2\pi f_c(\tau_i + \tau_i')], \tag{12A-63}$$

and by substituting (12A-63) into (12A-19), we obtain the real bandpass signal

$$\boxed{y_s(t, x_i) \approx a_i |w_S| a(t - [\tau_i + \tau_i']) \cos[2\pi(f_c + \eta_D)(t - [\tau_i + \tau_i']) + \theta(t - [\tau_i + \tau_i']) + \angle w_S]} \tag{12A-64}$$

for $t \ge \tau_i + \tau_i'$.

If the scatterer is in the Fresnel or far-field region of the linear array, and if the far-field beam pattern of the array is focused at the range to the scatterer and steered in the direction of the acoustic field incident upon the array (Fresnel region case), or if the far-field beam pattern is simply steered in the direction of the incident acoustic field (far-field region case), then (see Section 7.3)

$$\tau_i + \tau_i' = \tau_S, \qquad i = -N', \ldots, 0, \ldots, N', \tag{12A-65}$$

where

$$\tau_S = 2r_S / c \tag{12A-66}$$

is the round-trip time delay associated with the path length between the scatterer and the center of the array, and $r_S = r_{0S}$ is the range to the scatterer measured from the center of the array (i.e., from element $i = 0$). Substituting (12A-65) into (12A-62) and (12A-64) yields

$$\boxed{\tilde{y}_s(t, x_i) \approx a_i w_S \tilde{x}(t - \tau_S) \exp[+j2\pi\eta_D(t - \tau_S)] \exp(-j2\pi f_c \tau_S)} \tag{12A-67}$$

and

$$\boxed{y_s(t, x_i) \approx a_i |w_S| a(t - \tau_S) \cos[2\pi(f_c + \eta_D)(t - \tau_S) + \theta(t - \tau_S) + \angle w_S]} \tag{12A-68}$$

respectively, for $t \geq \tau_S$, where w_S is given by (12A-55). Therefore, if the scatterer is an object of interest (i.e., a target), and if correct focusing and/or beam steering is done so that [see (12A-65)]

$$\tau_{i,Trgt} + \tau'_{i,Trgt} = \tau_{Trgt}, \qquad i = -N', \ldots, 0, \ldots, N', \tag{12A-69}$$

then

$$\tilde{y}_{Trgt}(t, x_i) \approx a_i w_{Trgt} \tilde{x}(t - \tau_{Trgt}) \exp\left[+j2\pi\eta_{D,Trgt}(t - \tau_{Trgt})\right] \exp(-j2\pi f_c \tau_{Trgt}) \tag{12A-70}$$

and

$$y_{Trgt}(t, x_i) \approx a_i \left| w_{Trgt} \right| a(t - \tau_{Trgt}) \cos\left[2\pi(f_c + \eta_{D,Trgt})(t - \tau_{Trgt}) + \theta(t - \tau_{Trgt}) + \angle w_{Trgt}\right] \tag{12A-71}$$

are the lowpass (baseband) complex envelope and real bandpass output electrical signals from element i in the linear array due to a target for $t \geq \tau_{Trgt}$, where τ_{Trgt} and $\eta_{D,Trgt}$ are the round-trip time delay and Doppler shift of the target measured from the center of the array,

$$w_{Trgt} \approx \frac{1}{(4\pi r_{Trgt})^2} \mathcal{S}_T(f_c) S_{Trgt}(f_c, \theta_{Trgt}, \psi_{Trgt}, \theta_{Trgt\,0}, \psi_{Trgt\,0}) \exp[-2\alpha(f_c) r_{Trgt}] \mathcal{S}_R(f_c) \tag{12A-72}$$

is a complex, nonzero-mean, Gaussian, random variable (RV); $(r_{Trgt}, \theta_{Trgt}, \psi_{Trgt})$ are the spherical coordinates of the center of mass of the target, measured from the center of the array; $S_{Trgt}(\bullet)$ is the scattering function of the target, and is treated as a complex, nonzero-mean, Gaussian RV; $\theta_{Trgt\,0} = 180° - \theta_{Trgt}$, and $\psi_{Trgt\,0} = 180° + \psi_{Trgt}$. However, if there is also an object of no interest nearby the target, then

$$\begin{aligned}\tilde{y}_{Rev}(t, x_i) \approx a_i w_{Rev} \tilde{x}\left(t - [\tau_{i,Rev} + \tau'_{i,Trgt}]\right) \exp\left\{+j2\pi\eta_{D,Rev}\left[t - (\tau_{i,Rev} + \tau'_{i,Trgt})\right]\right\} \times \\ \exp\left[-j2\pi f_c(\tau_{i,Rev} + \tau'_{i,Trgt})\right]\end{aligned} \tag{12A-73}$$

and

$$y_{Rev}(t, x_i) \approx a_i |w_{Rev}| a\left(t - [\tau_{i,Rev} + \tau'_{i,Trgt}]\right) \times \cos\left[2\pi(f_c + \eta_{D,Rev})\left(t - [\tau_{i,Rev} + \tau'_{i,Trgt}]\right) + \theta\left(t - [\tau_{i,Rev} + \tau'_{i,Trgt}]\right) + \angle w_{Rev}\right]$$

(12A-74)

are the lowpass (baseband) complex envelope and real bandpass output electrical signals from element i in the linear array due to reverberation for $t \geq \tau_{i,Rev} + \tau'_{i,Trgt}$, where $\tau_{i,Rev}$ is the round-trip time delay from element i in the linear array to the object responsible for the reverberation [see (12A-56)], $\eta_{D,Rev}$ is the Doppler shift of the object responsible for the reverberation measured from the center of the array,

$$w_{Rev} \approx \frac{1}{(4\pi r_{Rev})^2} \boldsymbol{\mathcal{S}}_T(f_c) S_{Rev}(f_c, \theta_{Rev}, \psi_{Rev}, \theta_{Rev0}, \psi_{Rev0}) \exp[-2\alpha(f_c) r_{Rev}] \boldsymbol{\mathcal{S}}_R(f_c)$$

(12A-75)

is a complex, nonzero-mean, Gaussian RV, $(r_{Rev}, \theta_{Rev}, \psi_{Rev})$ are the spherical coordinates of the center of mass of the object responsible for the reverberation, measured from the center of the array; $S_{Rev}(\bullet)$ is the scattering function of the object responsible for the reverberation, and is treated as a complex, nonzero-mean, Gaussian RV; $\theta_{Rev0} = 180° - \theta_{Rev}$, and $\psi_{Rev0} = 180° + \psi_{Rev}$.

Appendix 12B Derivation of the Denominator of the Signal-to-Interference Ratio

In this appendix we shall derive the second moment (mean-squared value) of $\tilde{l}_Z$, which is the denominator of the signal-to-interference ratio (SIR) given by (12.2-1). Computing the magnitude-squared of $\tilde{l}_Z$ given by (12.2-4) yields

$$|\tilde{l}_Z|^2 = |\tilde{l}_{Rev}|^2 + |\tilde{l}_{n_a}|^2 + |\tilde{l}_{n_r}|^2 + 2\,\mathrm{Re}\{\tilde{l}_{Rev}\tilde{l}^*_{n_a}\} + 2\,\mathrm{Re}\{\tilde{l}_{Rev}\tilde{l}^*_{n_r}\} + 2\,\mathrm{Re}\{\tilde{l}_{n_a}\tilde{l}^*_{n_r}\},$$

(12B-1)

where use was made of the identity

$$Z + Z^* = 2\,\mathrm{Re}\{Z\}. \tag{12B-2}$$

Therefore,

$$E\left\{\left|\tilde{l}_Z\right|^2\right\} = E\left\{\left|\tilde{l}_{Rev}\right|^2\right\} + E\left\{\left|\tilde{l}_{n_a}\right|^2\right\} + E\left\{\left|\tilde{l}_{n_r}\right|^2\right\} + \\ 2\,\mathrm{Re}\left\{E\left\{\tilde{l}_{Rev}\tilde{l}^*_{n_a}\right\}\right\} + 2\,\mathrm{Re}\left\{E\left\{\tilde{l}_{Rev}\tilde{l}^*_{n_r}\right\}\right\} + 2\,\mathrm{Re}\left\{E\left\{\tilde{l}_{n_a}\tilde{l}^*_{n_r}\right\}\right\}. \tag{12B-3}$$

Consider the fourth term in (12B-3). With the use of (12.2-5) and (12.2-6),

$$E\left\{\tilde{l}_{Rev}\tilde{l}^*_{n_a}\right\} = \int_{-\infty}^{\infty}\int_{-\infty}^{\infty} \tilde{g}^*(t) R_{\tilde{y}_{Rev}\tilde{y}_{n_a}}(t,t')\tilde{g}(t')\,dt\,dt', \tag{12B-4}$$

where

$$R_{\tilde{y}_{Rev}\tilde{y}_{n_a}}(t,t') = E\left\{\tilde{y}_{Rev}(t)\tilde{y}^*_{n_a}(t')\right\} \tag{12B-5}$$

is the cross-correlation function of $\tilde{y}_{Rev}(t)$ and $\tilde{y}_{n_a}(t)$. If $\tilde{y}_{Rev}(t)$ and $\tilde{y}_{n_a}(t)$ are *statistically independent* random processes and $\tilde{y}_{n_a}(t)$ is *zero-mean*, then

$$R_{\tilde{y}_{Rev}\tilde{y}_{n_a}}(t,t') = E\left\{\tilde{y}_{Rev}(t)\right\}E\left\{\tilde{y}^*_{n_a}(t')\right\} = 0, \tag{12B-6}$$

and, as a result,

$$E\left\{\tilde{l}_{Rev}\tilde{l}^*_{n_a}\right\} = 0. \tag{12B-7}$$

Furthermore, since [see (12.2-6)]

$$E\left\{\tilde{l}_{n_a}\right\} = \int_{-\infty}^{\infty} E\left\{\tilde{y}_{n_a}(t)\right\}\tilde{g}^*(t)\,dt = 0, \tag{12B-8}$$

$$\mathrm{Cov}\left(\tilde{l}_{Rev}, \tilde{l}_{n_a}\right) = E\left\{\tilde{l}_{Rev}\tilde{l}^*_{n_a}\right\} - E\left\{\tilde{l}_{Rev}\right\}\left[E\left\{\tilde{l}_{n_a}\right\}\right]^* = 0, \tag{12B-9}$$

that is, $\tilde{l}_{Rev}$ and $\tilde{l}_{n_a}$ are *uncorrelated*, and since they are Gaussian RVs, they are also *statistically independent*.

Consider the fifth term in (12B-3). With the use of (12.2-5) and (12.2-7),

$$E\left\{\tilde{l}_{Rev}\tilde{l}^*_{n_r}\right\} = \int_{-\infty}^{\infty}\int_{-\infty}^{\infty} \tilde{g}^*(t) R_{\tilde{y}_{Rev}\tilde{n}_r}(t,t')\tilde{g}(t')\,dt\,dt', \tag{12B-10}$$

where

$$R_{\tilde{y}_{Rev}\tilde{n}_r}(t,t') = E\left\{\tilde{y}_{Rev}(t)\tilde{n}^*_r(t')\right\} \tag{12B-11}$$

is the cross-correlation function of $\tilde{y}_{Rev}(t)$ and $\tilde{n}_r(t)$. If $\tilde{y}_{Rev}(t)$ and $\tilde{n}_r(t)$ are *statistically independent* random processes and $\tilde{n}_r(t)$ is *zero-mean*, then

$$R_{\tilde{y}_{Rev}\tilde{n}_r}(t,t') = E\{\tilde{y}_{Rev}(t)\}E\{\tilde{n}_r^*(t')\} = 0\,, \tag{12B-12}$$

and, as a result,

$$E\{\tilde{l}_{Rev}\tilde{l}_{n_r}^*\} = 0\,. \tag{12B-13}$$

Furthermore, since [see (12.2-7)]

$$E\{\tilde{l}_{n_r}\} = \int_{-\infty}^{\infty} E\{\tilde{n}_r(t)\}\tilde{g}^*(t)\,dt = 0\,, \tag{12B-14}$$

$$\text{Cov}(\tilde{l}_{Rev}, \tilde{l}_{n_r}) = E\{\tilde{l}_{Rev}\tilde{l}_{n_r}^*\} - E\{\tilde{l}_{Rev}\}\left[E\{\tilde{l}_{n_r}\}\right]^* = 0\,, \tag{12B-15}$$

that is, $\tilde{l}_{Rev}$ and $\tilde{l}_{n_r}$ are *uncorrelated*, and since they are Gaussian RVs, they are also *statistically independent.*

Consider the sixth term in (12B-3). With the use of (12.2-6) and (12.2-7),

$$E\{\tilde{l}_{n_a}\tilde{l}_{n_r}^*\} = \int_{-\infty}^{\infty}\int_{-\infty}^{\infty} \tilde{g}^*(t)R_{\tilde{y}_{n_a}\tilde{n}_r}(t,t')\tilde{g}(t')\,dt\,dt'\,, \tag{12B-16}$$

where

$$R_{\tilde{y}_{n_a}\tilde{n}_r}(t,t') = E\{\tilde{y}_{n_a}(t)\tilde{n}_r^*(t')\} \tag{12B-17}$$

is the cross-correlation function of $\tilde{y}_{n_a}(t)$ and $\tilde{n}_r(t)$. If $\tilde{y}_{n_a}(t)$ and $\tilde{n}_r(t)$ are *statistically independent* random processes and $\tilde{n}_r(t)$ is *zero-mean*, then

$$R_{\tilde{y}_{n_a}\tilde{n}_r}(t,t') = E\{\tilde{y}_{n_a}(t)\}E\{\tilde{n}_r^*(t')\} = 0\,, \tag{12B-18}$$

and, as a result,

$$E\{\tilde{l}_{n_a}\tilde{l}_{n_r}^*\} = 0\,. \tag{12B-19}$$

Furthermore, since $\tilde{l}_{n_a}$ and $\tilde{l}_{n_r}$ are zero-mean [see (12B-8) and (12B-14)],

$$\text{Cov}(\tilde{l}_{n_a}, \tilde{l}_{n_r}) = E\{\tilde{l}_{n_a}\tilde{l}_{n_r}^*\} - E\{\tilde{l}_{n_a}\}\left[E\{\tilde{l}_{n_r}\}\right]^* = 0\,, \tag{12B-20}$$

that is, $\tilde{l}_{n_a}$ and $\tilde{l}_{n_r}$ are *uncorrelated*, and since they are Gaussian RVs, they are also *statistically independent.* Therefore, substituting (12B-7), (12B-13), and (12B-19) into (12B-3) yields

$$E\left\{\left|\tilde{l}_Z\right|^2\right\} = E\left\{\left|\tilde{l}_{Rev}\right|^2\right\} + E\left\{\left|\tilde{l}_{n_a}\right|^2\right\} + E\left\{\left|\tilde{l}_{n_r}\right|^2\right\}. \tag{12B-21}$$

Consider the first term in (12B-21). Substituting (12.1-9) into (12.1-8) yields

$$\tilde{y}_{Rev}(t) = w_{Rev} \sum_{i=-N'}^{N'} a_i \tilde{x}(t-\Delta_i)\exp(+j2\pi\eta_{D,Rev}t)\exp\left[-j2\pi(f_c+\eta_{D,Rev})\Delta_i\right], \tag{12B-22}$$

where

$$\Delta_i = \tau_{i,Rev} + \tau'_{i,Trgt}, \tag{12B-23}$$

and by substituting (12A-69) into (12B-23), we obtain

$$\Delta_i = \tau_{Trgt} + (\tau_{i,Rev} - \tau_{i,Trgt}). \tag{12B-24}$$

Substituting (12B-22) and (12.2-10) into (12.2-5) yields

$$\tilde{l}_{Rev} = w_{Rev} \sum_{i=-N'}^{N'} a_i \int_{-\infty}^{\infty} \tilde{x}(t-\Delta_i)\tilde{x}^*(t-\hat{\tau}_{Trgt})\exp\left[-j2\pi(\hat{\eta}_{D,Trgt}-\eta_{D,Rev})t\right]dt \times \exp\left[-j2\pi(f_c+\eta_{D,Rev})\Delta_i\right]. \tag{12B-25}$$

If we let $t' = t - \Delta_i$ in (12B-25), and make use of (12B-24), then

$$\tilde{l}_{Rev} = w_{Rev}\exp\left[-j2\pi(f_c+\hat{\eta}_{D,Trgt})\tau_{Trgt}\right]\times \sum_{i=-N'}^{N'} a_i X(e_{\tau_i}, e_D)\exp\left[+j2\pi(f_c+\hat{\eta}_{D,Trgt})(\tau_{i,Trgt}-\tau_{i,Rev})\right], \tag{12B-26}$$

where

$$X(e_{\tau_i}, e_D) = \int_{-\infty}^{\infty} \tilde{x}(t)\tilde{x}^*(t-e_{\tau_i})\exp(-j2\pi e_D t)dt \tag{12B-27}$$

is the unnormalized, auto-ambiguity function of the transmitted electrical signal $x(t)$, with complex envelope $\tilde{x}(t)$;

$$e_{\tau_i} = e_{\tau,Trgt} + (\tau_{i,Trgt} - \tau_{i,Rev}) \tag{12B-28}$$

is a round-trip time-delay estimation error in seconds, where $e_{\tau,Trgt}$ is the round-trip time-delay estimation error of the target given by (12.2-14), and

$$e_D = \hat{\eta}_{D,\,Trgt} - \eta_{D,\,Rev} \tag{12B-29}$$

is a Doppler-shift estimation error in hertz. Therefore,

$$\boxed{E\left\{\left|\tilde{l}_{Rev}\right|^2\right\} = E_{\tilde{x}}^2\, E\left\{\left|w_{Rev}\right|^2\right\}\left|\sum_{i=-N'}^{N'} a_i X_N(e_{\tau_i}, e_D)\exp\left[+j2\pi(f_c + \hat{\eta}_{D,\,Trgt})(\tau_{i,\,Trgt} - \tau_{i,\,Rev})\right]\right|^2} \tag{12B-30}$$

where the normalized, auto-ambiguity function $X_N(e_{\tau_i}, e_D)$ was obtained by dividing $X(e_{\tau_i}, e_D)$ by the energy $E_{\tilde{x}}$ of $\tilde{x}(t)$, where $E_{\tilde{x}}$ has units of joules-ohms [see (12.2-17)].

Consider the second term in (12B-21). With the use of (12.2-6),

$$E\left\{\left|\tilde{l}_{n_a}\right|^2\right\} = \int_{-\infty}^{\infty}\int_{-\infty}^{\infty} \tilde{g}^*(t) R_{\tilde{y}_{n_a}}(t, t')\tilde{g}(t')\,dt\,dt', \tag{12B-31}$$

where

$$R_{\tilde{y}_{n_a}}(t, t') = E\left\{\tilde{y}_{n_a}(t)\tilde{y}_{n_a}^*(t')\right\} \tag{12B-32}$$

is the autocorrelation function of $\tilde{y}_{n_a}(t)$. Substituting (12.1-12) into (12.1-11) yields

$$\tilde{y}_{n_a}(t) = \sum_{i=-N'}^{N'} a_i \tilde{n}_a(t - \tau'_{i,\,Trgt}, x_i), \tag{12B-33}$$

and by substituting (12B-33) into (12B-32), we obtain

$$R_{\tilde{y}_{n_a}}(t, t') = \sum_{i=-N'}^{N'}\sum_{j=-N'}^{N'} a_i a_j R_{\tilde{n}_a}(t - \tau'_{i,\,Trgt}, t' - \tau'_{j,\,Trgt}, x_i, x_j), \tag{12B-34}$$

or

$$\begin{aligned} R_{\tilde{y}_{n_a}}(t, t') = {} & \sum_{i=-N'}^{N'} a_i^2 R_{\tilde{n}_a}(t - \tau'_{i,\,Trgt}, t' - \tau'_{i,\,Trgt}, x_i, x_i) + \\ & \sum_{\substack{i=-N' \\ i\neq j}}^{N'}\sum_{j=-N'}^{N'} a_i a_j R_{\tilde{n}_a}(t - \tau'_{i,\,Trgt}, t' - \tau'_{j,\,Trgt}, x_i, x_j), \end{aligned} \tag{12B-35}$$

where

$$R_{\tilde{n}_a}(t - \tau'_{i,\,Trgt}, t' - \tau'_{j,\,Trgt}, x_i, x_j) = E\left\{\tilde{n}_a(t - \tau'_{i,\,Trgt}, x_i)\tilde{n}_a^*(t' - \tau'_{j,\,Trgt}, x_j)\right\}. \tag{12B-36}$$

If the random process $\tilde{n}_a(t, x_i)$ is *wide-sense stationary* (WSS) *in time and space* $\forall\ i$, then (12B-35) reduces to

$$R_{\tilde{y}_{n_a}}(\Delta t) = R_{\tilde{n}_a}(\Delta t, 0) \sum_{i=-N'}^{N'} a_i^2 + \sum_{\substack{i=-N' \\ i \neq j}}^{N'} \sum_{j=-N'}^{N'} a_i a_j R_{\tilde{n}_a}\left(\Delta t - [\tau'_{i,Trgt} - \tau'_{j,Trgt}], x_i - x_j\right), \tag{12B-37}$$

where $\Delta t = t - t'$. Furthermore, if $\tilde{n}_a(t, x_i)$ is *zero-mean* $\forall\ i$, and $\tilde{n}_a(t, x_i)$ and $\tilde{n}_a(t, x_j)$ are *uncorrelated* for $i \neq j$, then

$$R_{\tilde{n}_a}\left(\Delta t - [\tau'_{i,Trgt} - \tau'_{j,Trgt}], x_i - x_j\right) = 0, \qquad i \neq j, \tag{12B-38}$$

and as a result, (12B-37) reduces to

$$R_{\tilde{y}_{n_a}}(\Delta t) = R_{\tilde{n}_a}(\Delta t, 0) \sum_{i=-N'}^{N'} a_i^2, \tag{12B-39}$$

where

$$R_{\tilde{n}_a}(\Delta t, 0) = E\left\{\tilde{n}_a(t, x_i)\tilde{n}_a^*(t', x_i)\right\} \tag{12B-40}$$

is the autocorrelation function of $\tilde{n}_a(t, x_i)$ at the output of any element i in the linear array. Substituting (12B-39) into (12B-31) yields

$$E\left\{\left|\tilde{l}_{n_a}\right|^2\right\} = \sum_{i=-N'}^{N'} a_i^2 \int_{-\infty}^{\infty} \int_{-\infty}^{\infty} \tilde{g}^*(t) R_{\tilde{n}_a}(\Delta t, 0) \tilde{g}(t') dt\, dt'. \tag{12B-41}$$

Since $\tilde{n}_a(t, x_i)$ is WSS,

$$R_{\tilde{n}_a}(\Delta t, 0) = F_\eta^{-1}\left\{S_{\tilde{n}_a}(\eta)\right\} = \int_{-\infty}^{\infty} S_{\tilde{n}_a}(\eta) \exp(+j2\pi\eta\Delta t) d\eta, \tag{12B-42}$$

where $S_{\tilde{n}_a}(\eta)$ is the power spectrum of $\tilde{n}_a(t, x_i)$ $\forall\ i$. If $\tilde{n}_a(t, x_i)$ is a *lowpass (baseband), white-noise* random process in the time-domain, then

$$S_{\tilde{n}_a}(\eta) = \tilde{N}_{a0} \operatorname{rect}\left[\eta/(2B)\right], \tag{12B-43}$$

where the constant $\tilde{N}_{a0}$ has units of $(\text{W-}\Omega)/\text{Hz}$ (or J-Ω),

$$\operatorname{rect}\left(\frac{\eta}{2B}\right) = \begin{cases} 1, & |\eta| \leq B \\ 0, & |\eta| > B \end{cases} \tag{12B-44}$$

and B is the bandwidth in hertz of the ideal, lowpass filters in a quadrature demodulator used to compute complex envelopes (see Section 11.3). Therefore,

$$R_{\tilde{n}_a}(\Delta t, 0) = 2\tilde{N}_{a0} B \operatorname{sinc}(2B\Delta t), \tag{12B-45}$$

where

$$\operatorname{sinc}(x) \triangleq \frac{\sin(\pi x)}{\pi x}. \tag{12B-46}$$

The constant variance of the zero-mean ambient noise $\tilde{n}_a(t, x_i)$ is given by

$$\sigma^2_{\tilde{n}_a} = R_{\tilde{n}_a}(0, 0) = E\left\{\left|\tilde{n}_a(t, x_i)\right|^2\right\} = 2\tilde{N}_{a0} B \tag{12B-47}$$

since $\operatorname{sinc}(0) = 1$. The variance $\sigma^2_{\tilde{n}_a}$ is the average power of $\tilde{n}_a(t, x_i)$ in watts-ohms.

Substituting (12B-45) into (12B-41) and changing the order of integration and rearranging factors yields

$$E\left\{\left|\tilde{I}_{n_a}\right|^2\right\} = 2\tilde{N}_{a0} B \sum_{i=-N'}^{N'} a_i^2 \int_{-\infty}^{\infty} \int_{-\infty}^{\infty} \tilde{g}(t') \operatorname{sinc}[2B(t - t')] dt' \, \tilde{g}^*(t) dt, \tag{12B-48}$$

or

$$E\left\{\left|\tilde{I}_{n_a}\right|^2\right\} = 2\tilde{N}_{a0} B \sum_{i=-N'}^{N'} a_i^2 \int_{-\infty}^{\infty} \tilde{f}(t) \tilde{g}^*(t) dt, \tag{12B-49}$$

where

$$\tilde{f}(t) = \tilde{g}(t) * \operatorname{sinc}(2Bt) = \int_{-\infty}^{\infty} \tilde{g}(t') \operatorname{sinc}[2B(t - t')] dt'. \tag{12B-50}$$

By using Parseval's Theorem, the integral on the right-hand side of (12B-49) can be rewritten as

$$\int_{-\infty}^{\infty} \tilde{f}(t) \tilde{g}^*(t) dt = \int_{-\infty}^{\infty} \tilde{F}(\eta) \tilde{G}^*(\eta) d\eta, \tag{12B-51}$$

where, from (12B-50),

$$\tilde{F}(\eta) = \frac{1}{2B} \tilde{G}(\eta) \operatorname{rect}\left(\frac{\eta}{2B}\right). \tag{12B-52}$$

Substituting (12B-52) and (12B-44) into (12B-51) yields

$$\int_{-\infty}^{\infty} \tilde{f}(t)\tilde{g}^*(t)dt = \frac{1}{2B}\int_{-B}^{B} \left|\tilde{G}(\eta)\right|^2 d\eta . \qquad (12B\text{-}53)$$

From (12.2-10),

$$\tilde{G}(\eta) = \tilde{X}(\eta - \hat{\eta}_{D,Trgt})\exp\left[-j2\pi(\eta - \hat{\eta}_{D,Trgt})\hat{\tau}_{Trgt}\right], \qquad (12B\text{-}54)$$

and by substituting (12B-54) into (12B-53), we obtain

$$\int_{-\infty}^{\infty} \tilde{f}(t)\tilde{g}^*(t)dt = \frac{1}{2B}\int_{-B}^{B} \left|\tilde{X}(\eta - \hat{\eta}_{D,Trgt})\right|^2 d\eta . \qquad (12B\text{-}55)$$

Recall that the frequency spectrum $\tilde{X}(\eta)$ of the complex envelope of the transmitted electrical signal $\tilde{x}(t)$ is lowpass (baseband) with bandwidth W hertz. If the bandwidth B of the ideal, lowpass filters in a quadrature demodulator is computed from the formula

$$\boxed{B = W + \max\left|\hat{\eta}_{D,Trgt}\right|} \qquad (12B\text{-}56)$$

then (12B-55) can be rewritten as

$$\int_{-\infty}^{\infty} \tilde{f}(t)\tilde{g}^*(t)dt = \frac{1}{2B}\int_{-\infty}^{\infty} \left|\tilde{X}(\eta - \hat{\eta}_{D,Trgt})\right|^2 d\eta = \frac{1}{2B}\int_{-\infty}^{\infty} \left|\tilde{X}(\eta)\right|^2 d\eta , \qquad (12B\text{-}57)$$

and with the use of Parseval's Theorem,

$$\int_{-\infty}^{\infty} \tilde{f}(t)\tilde{g}^*(t)dt = \frac{1}{2B}\int_{-\infty}^{\infty} \left|\tilde{x}(t)\right|^2 dt = \frac{E_{\tilde{x}}}{2B}, \qquad (12B\text{-}58)$$

where $E_{\tilde{x}}$ is the energy of $\tilde{x}(t)$ in joules-ohms. Therefore, substituting (12B-58) into (12B-49) finally yields

$$\boxed{E\left\{\left|\tilde{l}_{n_a}\right|^2\right\} = E_{\tilde{x}}\tilde{N}_{a0}\sum_{i=-N'}^{N'} a_i^2} \qquad (12B\text{-}59)$$

Consider the third term in (12B-21). With the use of (12.2-7),

$$E\left\{\left|\tilde{l}_{n_r}\right|^2\right\} = \int_{-\infty}^{\infty}\int_{-\infty}^{\infty} \tilde{g}^*(t)R_{\tilde{n}_r}(t,t')\tilde{g}(t')dt\,dt' , \qquad (12B\text{-}60)$$

where

$$R_{\tilde{n}_r}(t,t') = E\{\tilde{n}_r(t)\tilde{n}_r^*(t')\} \tag{12B-61}$$

is the autocorrelation function of $\tilde{n}_r(t)$. Substituting (12.1-14) into (12.1-13) yields

$$\tilde{n}_r(t) = \sum_{i=-N'}^{N'} a_i \tilde{n}(t - \tau'_{i,Trgt}, x_i), \tag{12B-62}$$

and by substituting (12B-62) into (12B-61), we obtain

$$R_{\tilde{n}_r}(t,t') = \sum_{i=-N'}^{N'} \sum_{j=-N'}^{N'} a_i a_j R_{\tilde{n}}(t - \tau'_{i,Trgt}, t' - \tau'_{j,Trgt}, x_i, x_j), \tag{12B-63}$$

or

$$\begin{aligned} R_{\tilde{n}_r}(t,t') = & \sum_{i=-N'}^{N'} a_i^2 R_{\tilde{n}}(t - \tau'_{i,Trgt}, t' - \tau'_{i,Trgt}, x_i, x_i) + \\ & \sum_{\substack{i=-N' \\ i \neq j}}^{N'} \sum_{j=-N'}^{N'} a_i a_j R_{\tilde{n}}(t - \tau'_{i,Trgt}, t' - \tau'_{j,Trgt}, x_i, x_j), \end{aligned} \tag{12B-64}$$

where

$$R_{\tilde{n}}(t - \tau'_{i,Trgt}, t' - \tau'_{j,Trgt}, x_i, x_j) = E\{\tilde{n}(t - \tau'_{i,Trgt}, x_i)\tilde{n}^*(t' - \tau'_{j,Trgt}, x_j)\}. \tag{12B-65}$$

If the same assumptions made about the ambient noise are made about the receiver noise, that is, if the random process $\tilde{n}(t, x_i)$ is *zero-mean* and *wide-sense stationary* (WSS) *in time and space* $\forall\ i$, $\tilde{n}(t, x_i)$ and $\tilde{n}(t, x_j)$ are *uncorrelated* for $i \neq j$, and $\tilde{n}(t, x_i)$ is a *lowpass (baseband), white-noise* random process in the time-domain with power spectrum

$$S_{\tilde{n}}(\eta) = \tilde{N}_{r0} \operatorname{rect}[\eta/(2B)] \tag{12B-66}$$

at the output of any element i in the linear array, where the constant $\tilde{N}_{r0}$ has units of $(\text{W-}\Omega)/\text{Hz}$ (or $\text{J-}\Omega$), then by following the same procedure used to derive (12B-37) through (12B-59), the autocorrelation function of the receiver noise $\tilde{n}(t, x_i)$ is given by

$$R_{\tilde{n}}(\Delta t, 0) = E\{\tilde{n}(t, x_i)\tilde{n}^*(t', x_i)\} = 2\tilde{N}_{r0} B \operatorname{sinc}(2B\Delta t) \tag{12B-67}$$

$\forall\ i$, where $\Delta t = t - t'$, the constant variance of the zero-mean receiver noise

$\tilde{n}(t, x_i)$ is given by

$$\sigma_{\tilde{n}}^2 = R_{\tilde{n}}(0,0) = E\left\{\left|\tilde{n}(t, x_i)\right|^2\right\} = 2\tilde{N}_{r0}B\,, \tag{12B-68}$$

where $\sigma_{\tilde{n}}^2$ is the average power of $\tilde{n}(t, x_i)$ in watts-ohms, and

$$\boxed{E\left\{\left|\tilde{l}_{n_r}\right|^2\right\} = E_{\tilde{x}}\tilde{N}_{r0}\sum_{i=-N'}^{N'} a_i^2} \tag{12B-69}$$

where $E_{\tilde{x}}$ is the energy of $\tilde{x}(t)$ in joules-ohms.

Substituting (12B-30), (12B-59), and (12B-69) into (12B-21) yields

$$\boxed{\begin{aligned} E\left\{\left|\tilde{l}_Z\right|^2\right\} = E_{\tilde{x}}^2 E\left\{\left|w_{Rev}\right|^2\right\}\left|\sum_{i=-N'}^{N'} a_i X_N(e_{\tau_i}, e_D)\exp\left[+j2\pi(f_c + \hat{\eta}_{D,Trgt})(\tau_{i,Trgt} - \tau_{i,Rev})\right]\right|^2 + \\ A_2 E_{\tilde{x}}\left[\tilde{N}_{a0} + \tilde{N}_{r0}\right] \end{aligned}} \tag{12B-70}$$

where

$$A_2 = \sum_{i=-N'}^{N'} a_i^2\,. \tag{12B-71}$$

Equation (12B-70) is the denominator of the signal-to-interference ratio (SIR).

Appendix 12C Table 12C-1 Marcum Q-Function $Q(a, b)$

Table 12C-1 Marcum Q-Function $Q(a, b)$

	$a=1/2$	$a=\sqrt{2}/2$	$a=1$	$a=\sqrt{2}$	$a=2$	$a=2\sqrt{2}$	$a=4$
b	$Q(a,b)$	$Q(a,b)$	$Q(a,b)$	$Q(a,b)$	$Q(a,b)$	$Q(a,b)$	$Q(a,b)$
0	1.0000000	1.0000000	1.0000000	1.0000000	1.0000000	1.0000000	1.0000000
0.1	0.9955972	0.9961133	0.9969711	0.9981606	0.9993216	0.9999077	0.9999983
0.2	0.9825036	0.9845403	0.9879299	0.9926427	0.9972664	0.9996227	0.9999928
0.3	0.9610594	0.9655391	0.9730120	0.9834482	0.9937749	0.9991199	0.9999824
0.4	0.9318151	0.9395302	0.9524414	0.9705849	0.9887516	0.9983574	0.9999652
0.5	0.8955086	0.9070812	0.9265274	0.9540725	0.9820694	0.9972757	0.9999378
0.6	0.8530342	0.8688875	0.8956601	0.9339501	0.9735767	0.9957971	0.9998959
0.7	0.8054067	0.8257489	0.8603040	0.9102852	0.9631057	0.9938250	0.9998328
0.8	0.7537215	0.7785424	0.8209899	0.8831822	0.9504818	0.9912438	0.9997396
0.9	0.6991130	0.7281940	0.7783047	0.8527903	0.9355343	0.9879192	0.9996041
1	0.6427142	0.6756493	0.7328798	0.8193100	0.9181077	0.9836985	0.9994101
1.1	0.5856192	0.6218454	0.6853771	0.7829968	0.8980728	0.9784124	0.9991361
1.2	0.5288489	0.5676836	0.6364742	0.7441625	0.8753379	0.9718777	0.9987544
1.3	0.4733238	0.5140058	0.5868485	0.7031730	0.8498581	0.9639007	0.9982296
1.4	0.4198429	0.4615742	0.5371608	0.6604429	0.8216440	0.9542822	0.9975171
1.5	0.3690690	0.4110555	0.4880400	0.6164276	0.7907678	0.9428235	0.9965616
1.6	0.3215222	0.3630093	0.4400679	0.5716114	0.7573667	0.9293330	0.9952957
1.7	0.2775793	0.3178823	0.3937666	0.5264953	0.7216440	0.9136346	0.9936381
1.8	0.2374792	0.2760064	0.3495881	0.4815821	0.6838668	0.8955758	0.9914930
1.9	0.2013335	0.2376015	0.3079061	0.4373617	0.6443605	0.8750364	0.9887485
2	0.1691406	0.2027822	0.2690121	0.3942969	0.6035010	0.8519364	0.9852765
2.1	0.1408026	0.1715675	0.2331134	0.3528097	0.5617037	0.8262441	0.9809334
2.2	0.1161430	0.1438927	0.2003353	0.3132710	0.5194115	0.7979823	0.9755602
2.3	0.0949259	0.1196235	0.1707252	0.2759910	0.4770802	0.7672322	0.9689856
2.4	0.0768731	0.0985698	0.1442599	0.2412136	0.4351646	0.7341366	0.9610280
2.5	0.0616811	0.0805003	0.1208541	0.2091139	0.3941039	0.6988994	0.9515004
2.6	0.0490350	0.0651560	0.1003710	0.1797978	0.3543082	0.6617828	0.9402153
2.7	0.0386213	0.0522629	0.0826325	0.1533048	0.3161467	0.6231024	0.9269909
2.8	0.0301372	0.0415425	0.0674306	0.1296139	0.2799382	0.5832189	0.9116584
2.9	0.0232984	0.0327216	0.0545376	0.1086497	0.2459437	0.5425284	0.8940693
3	0.0178437	0.0255386	0.0437160	0.0902915	0.2143621	0.5014507	0.8741039
3.1	0.0135384	0.0197498	0.0347266	0.0743825	0.1853284	0.4604161	0.8516786
3.2	0.0101758	0.0151327	0.0273362	0.0607388	0.1589150	0.4198524	0.8267536
3.3	0.0075766	0.0114879	0.0213228	0.0491588	0.1351350	0.3801714	0.7993388
3.4	0.0055883	0.0086401	0.0164801	0.0394319	0.1139476	0.3417567	0.7694987
3.5	0.0040830	0.0064378	0.0126201	0.0313457	0.0952653	0.3049531	0.7373555
3.6	0.0029549	0.0047521	0.0095750	0.0246927	0.0789620	0.2700573	0.7030894
3.7	0.0021183	0.0034749	0.0071973	0.0192750	0.0648811	0.2373114	0.6669370
3.8	0.0015042	0.0025171	0.0053597	0.0149086	0.0528448	0.2068995	0.6291874
3.9	0.0010580	0.0018061	0.0039539	0.0114254	0.0426617	0.1789451	0.5901752
4	0.0007370	0.0012837	0.0028895	0.0086753	0.0341348	0.1535135	0.5502721

	$a=1/2$	$a=\sqrt{2}/2$	$a=1$	$a=\sqrt{2}$	$a=2$	$a=2\sqrt{2}$	$a=4$
b	$Q(a,b)$	$Q(a,b)$	$Q(a,b)$	$Q(a,b)$	$Q(a,b)$	$Q(a,b)$	$Q(a,b)$
4	0.0007370	0.0012837	0.0028895	0.0086753	0.0341348	0.1535135	0.5502721
4.1	0.0005086	0.0009038	0.0020918	0.0065262	0.0270679	0.1306137	0.5098758
4.2	0.0003476	0.0006302	0.0015000	0.0048638	0.0212707	0.1102048	0.4693988
4.3	0.0002353	0.0004353	0.0010655	0.0035910	0.0165638	0.0922015	0.4292554
4.4	0.0001577	0.0002978	0.0007497	0.0026265	0.0127809	0.0764824	0.3898493
4.5	0.0001047	0.0002018	0.0005224	0.0019030	0.0097718	0.0628974	0.3515611
4.6	0.0000689	0.0001354	0.0003606	0.0013658	0.0074024	0.0512764	0.3147382
4.7	0.0000449	0.0000900	0.0002465	0.0009710	0.0055557	0.0414366	0.2796846
4.8	0.0000289	0.0000593	0.0001669	0.0006838	0.0041311	0.0331895	0.2466545
4.9	0.0000185	0.0000386	0.0001120	0.0004770	0.0030432	0.0263476	0.2158473
5	0.0000117	0.0000249	0.0000744	0.0003295	0.0022208	0.0207290	0.1874047
5.1	0.0000073	0.0000159	0.0000489	0.0002255	0.0016055	0.0161618	0.1614116
5.2	0.0000045	0.0000101	0.0000319	0.0001528	0.0011498	0.0124868	0.1378976
5.3	0.0000028	0.0000063	0.0000206	0.0001026	0.0008156	0.0095596	0.1168417
5.4	0.0000017	0.0000039	0.0000131	0.0000682	0.0005731	0.0072517	0.0981776
5.5	0.0000010	0.0000024	0.0000083	0.0000449	0.0003989	0.0054504	0.0818009

Appendix 12D How to Compute Values for σ_0/σ_1

In this appendix we shall derive an equation for the ratio σ_0/σ_1. Substituting (12.3-25) and (12.4-29) into (12.4-23) yields

$$\sigma_1^2 = \frac{\operatorname{Var}\left(\tilde{l}_{Trgt}\right)}{2} + \sigma_0^2, \tag{12D-1}$$

from which we can conclude that

$$\boxed{\sigma_1^2 > \sigma_0^2} \tag{12D-2}$$

Since [see (12.3-49)]

$$\alpha_0^2 = \gamma_0/\sigma_0^2 \tag{12D-3}$$

and [see (12.4-58)]

$$\alpha_1^2 = \gamma_1/\sigma_1^2, \tag{12D-4}$$

$$\frac{\alpha_1^2}{\alpha_0^2} = \frac{\gamma_1}{\gamma_0}\frac{\sigma_0^2}{\sigma_1^2}, \tag{12D-5}$$

or

$$\boxed{\frac{\sigma_0^2}{\sigma_1^2} = \frac{\gamma_0}{\gamma_1}\frac{\alpha_1^2}{\alpha_0^2} < 1} \tag{12D-6}$$

From (12D-6),

$$\frac{\gamma_0}{\gamma_1}\frac{\alpha_1^2}{\alpha_0^2} < 1, \tag{12D-7}$$

or

$$\boxed{\gamma_0/\gamma_1 < \alpha_0^2/\alpha_1^2} \tag{12D-8}$$

We shall establish the relationship between γ_0 and γ_1 next.

Substituting (12.3-7) into (12.3-29) yields

$$\gamma_0 = \left|E\left\{\tilde{l}\middle|\mathrm{H}_0\right\}\right|^2 = \left|E\left\{\tilde{l}_{Rev}\right\}\right|^2, \tag{12D-9}$$

and substituting (12.4-12) into (12.4-33) yields

$$\gamma_1 = \left|E\{\tilde{l}\,|\,\mathrm{H}_1\}\right|^2 = \left|E\{\tilde{l}_{Trgt}\}\right|^2 + \left|E\{\tilde{l}_{Rev}\}\right|^2 + 2\,\mathrm{Re}\left\{E\{\tilde{l}_{Trgt}\}E\{\tilde{l}^*_{Rev}\}\right\}, \tag{12D-10}$$

where use was made of the identity

$$|A+B|^2 = |A|^2 + 2\,\mathrm{Re}\{AB^*\} + |B|^2. \tag{12D-11}$$

Note that $\gamma_0 > 0$ and $\gamma_1 > 0$ because the target and reverberation scattering functions are *nonzero-mean* in this analysis. Substituting (12D-9) into (12D-10) yields

$$\gamma_1 = \gamma_0 + \left|E\{\tilde{l}_{Trgt}\}\right|^2 + 2\,\mathrm{Re}\left\{E\{\tilde{l}_{Trgt}\}E\{\tilde{l}^*_{Rev}\}\right\}, \tag{12D-12}$$

and since it can be shown that (see Appendix 12E)

$$\left|E\{\tilde{l}_{Trgt}\}\right|^2 + 2\,\mathrm{Re}\left\{E\{\tilde{l}_{Trgt}\}E\{\tilde{l}^*_{Rev}\}\right\} > 0 \tag{12D-13}$$

whether $\alpha_1^2/2 > \alpha_0^2/2$ or $\alpha_0^2/2 > \alpha_1^2/2$,

$$\boxed{\gamma_1 > \gamma_0} \tag{12D-14}$$

Therefore, from (12D-6),

$$\boxed{\frac{\sigma_0}{\sigma_1} = \sqrt{\frac{\gamma_0}{\gamma_1}}\,\frac{\alpha_1}{\alpha_0} < 1} \tag{12D-15}$$

where [see (12D-8)],

$$\boxed{\sqrt{\gamma_0/\gamma_1} < \alpha_0/\alpha_1} \tag{12D-16}$$

and [see (12D-14)],

$$\boxed{\sqrt{\gamma_0/\gamma_1} < 1} \tag{12D-17}$$

Appendix 12E

In Appendix 12D it was shown that [see (12D-12)]

$$\gamma_1 = \gamma_0 + \left|E\{\tilde{I}_{Trgt}\}\right|^2 + 2\,\mathrm{Re}\left\{E\{\tilde{I}_{Trgt}\}E\{\tilde{I}^*_{Rev}\}\right\}, \tag{12E-1}$$

where $\gamma_0 > 0$ and $\gamma_1 > 0$ because the target and reverberation scattering functions are *nonzero-mean* in Appendix 12D. In this appendix we shall show that

$$\left|E\{\tilde{I}_{Trgt}\}\right|^2 + 2\,\mathrm{Re}\left\{E\{\tilde{I}_{Trgt}\}E\{\tilde{I}^*_{Rev}\}\right\} > 0 \tag{12E-2}$$

whether $\alpha_1^2/2 > \alpha_0^2/2$ or $\alpha_0^2/2 > \alpha_1^2/2$, and as a result [see (12D-14)],

$$\boxed{\gamma_1 > \gamma_0} \tag{12E-3}$$

Substituting (12D-9) into (12.3-49) yields

$$\frac{\alpha_0^2}{2} = \frac{\left|E\{\tilde{I}_{Rev}\}\right|^2}{2\sigma_0^2}, \tag{12E-4}$$

and by substituting (12D-10) into (12.4-58), we obtain

$$\frac{\alpha_1^2}{2} = \frac{\left|E\{\tilde{I}_{Trgt}\}\right|^2 + \left|E\{\tilde{I}_{Rev}\}\right|^2 + 2\,\mathrm{Re}\left\{E\{\tilde{I}_{Trgt}\}E\{\tilde{I}^*_{Rev}\}\right\}}{2\sigma_1^2}. \tag{12E-5}$$

If $\alpha_1^2/2 > \alpha_0^2/2$, then

$$\frac{\left|E\{\tilde{I}_{Trgt}\}\right|^2 + \left|E\{\tilde{I}_{Rev}\}\right|^2 + 2\,\mathrm{Re}\left\{E\{\tilde{I}_{Trgt}\}E\{\tilde{I}^*_{Rev}\}\right\}}{2\sigma_1^2} > \frac{\left|E\{\tilde{I}_{Rev}\}\right|^2}{2\sigma_0^2}, \tag{12E-6}$$

or

$$\left|E\{\tilde{I}_{Trgt}\}\right|^2 + 2\,\mathrm{Re}\left\{E\{\tilde{I}_{Trgt}\}E\{\tilde{I}^*_{Rev}\}\right\} > \left(\frac{\sigma_1^2}{\sigma_0^2} - 1\right)\left|E\{\tilde{I}_{Rev}\}\right|^2. \tag{12E-7}$$

Since $\sigma_1^2/\sigma_0^2 > 1$ [see (12D-2)],

$$\left|E\{\tilde{I}_{Trgt}\}\right|^2 + 2\,\mathrm{Re}\left\{E\{\tilde{I}_{Trgt}\}E\{\tilde{I}^*_{Rev}\}\right\} > \left(\frac{\sigma_1^2}{\sigma_0^2} - 1\right)\left|E\{\tilde{I}_{Rev}\}\right|^2 > 0, \tag{12E-8}$$

or

$$\left|E\left\{\tilde{l}_{Trgt}\right\}\right|^2 + 2\,\mathrm{Re}\left\{E\left\{\tilde{l}_{Trgt}\right\}E\left\{\tilde{l}^*_{Rev}\right\}\right\} > 0\,, \tag{12E-9}$$

which is the desired result [see (12E-2)].

Similarly, if $\alpha_0^2/2 > \alpha_1^2/2$, then

$$\frac{\left|E\left\{\tilde{l}_{Rev}\right\}\right|^2}{2\sigma_0^2} > \frac{\left|E\left\{\tilde{l}_{Trgt}\right\}\right|^2 + \left|E\left\{\tilde{l}_{Rev}\right\}\right|^2 + 2\,\mathrm{Re}\left\{E\left\{\tilde{l}_{Trgt}\right\}E\left\{\tilde{l}^*_{Rev}\right\}\right\}}{2\sigma_1^2}\,, \tag{12E-10}$$

or

$$\frac{\sigma_1^2}{\sigma_0^2} > 1 + \frac{\left|E\left\{\tilde{l}_{Trgt}\right\}\right|^2 + 2\,\mathrm{Re}\left\{E\left\{\tilde{l}_{Trgt}\right\}E\left\{\tilde{l}^*_{Rev}\right\}\right\}}{\left|E\left\{\tilde{l}_{Rev}\right\}\right|^2}\,. \tag{12E-11}$$

Since $\sigma_1^2/\sigma_0^2 > 1$ [see (12D-2)],

$$\left|E\left\{\tilde{l}_{Trgt}\right\}\right|^2 + 2\,\mathrm{Re}\left\{E\left\{\tilde{l}_{Trgt}\right\}E\left\{\tilde{l}^*_{Rev}\right\}\right\} > 0\,, \tag{12E-12}$$

which is the desired result [see (12E-2)].

Chapter 13

The Auto-Ambiguity Function and Signal Design

Transmitted electrical signals used in *active* sonar (and radar) systems are amplitude-and-angle-modulated carriers. The *auto-ambiguity function* of a transmitted electrical signal contains information about the range and Doppler resolving capabilities of the signal. As such, it is a very important function that is used to design transmit signals to satisfy range and Doppler resolution requirements and to analyze the performance of active sonar (and radar) systems.

In Sections 13.1 and 13.2, we shall derive and discuss the normalized, auto-ambiguity functions of two very common transmitted electrical signals used in active sonar systems, namely, the *rectangular-envelope*, *CW* (continuous-wave) *pulse* and the *rectangular-envelope*, *LFM* (linear-frequency-modulated) *pulse*, respectively. We shall then use the information obtained from the ambiguity functions to design rectangular-envelope, CW and LFM pulses to satisfy range and Doppler resolution requirements. Since the auto-ambiguity function depends on the complex envelope of a transmitted signal (see Section 12.2), different signals have different ambiguity functions and, in general, different range and Doppler resolving capabilities.

In Section 13.3, we shall discuss the fundamentals of a *hyperbolic-frequency-modulated* (HFM) pulse. A HFM pulse is a very important signal because it is *Doppler-invariant.* As a result, it will be shown in Example 13.3-3 that a rectangular-envelope, HFM pulse is superior to a rectangular-envelope, LFM pulse in its ability to detect a target and estimate time delay in the presence of time-compression/time-expansion effects (see Section 10.7). We shall also discuss how to design a HFM pulse to minimize time-delay estimation error.

13.1 The Rectangular-Envelope CW Pulse

A CW (continuous-wave) pulse is nothing more than a finite duration, *double-sideband, suppressed-carrier* (DSBSC) waveform given by

$$\boxed{x(t) = a(t)\cos(2\pi f_c t)\text{rect}\left[(t-0.5T)/T\right]} \tag{13.1-1}$$

where $a(t)$ is a real amplitude-modulating function, $\cos(2\pi f_c t)$ is the carrier waveform, f_c is the carrier frequency in hertz, and

$$\text{rect}\left(\frac{t-0.5T}{T}\right) = \begin{cases} 1, & 0 \le t \le T \\ 0, & \text{otherwise} \end{cases} \tag{13.1-2}$$

is the time-shifted rectangle function (see Fig. 13.1-1), where T is the pulse length in seconds. An example of a CW pulse is shown in Fig. 13.1-2.

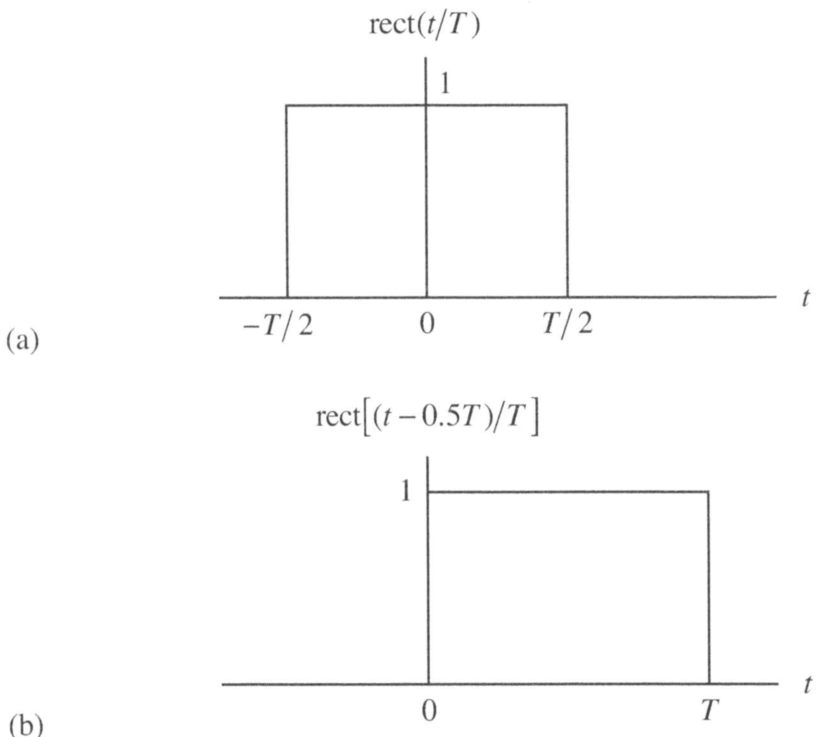

Figure 13.1-1 (a) Rectangle function and (b) time-shifted rectangle function.

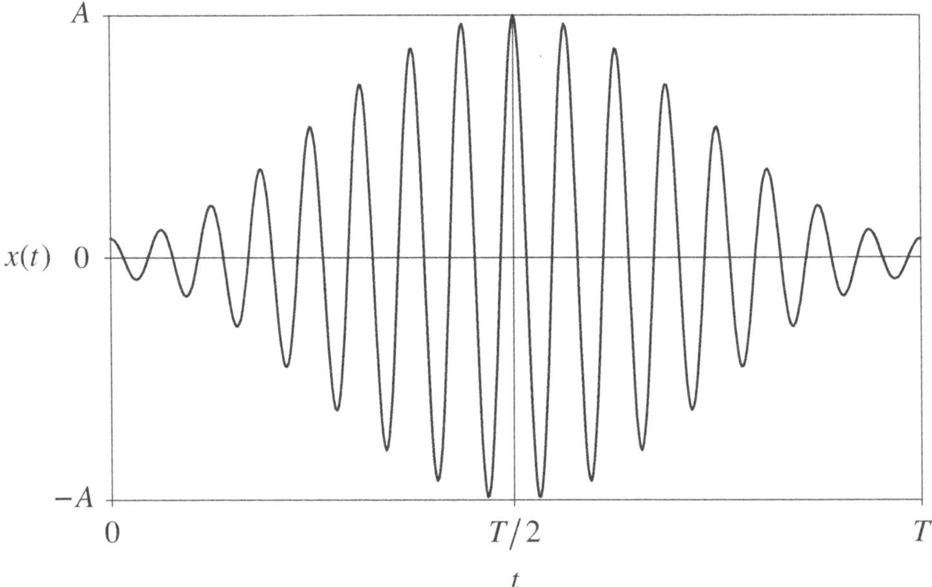

Figure 13.1-2 A CW pulse with pulse length T seconds.

The complex envelope and envelope of (13.1-1) are given by

$$\boxed{\tilde{x}(t) = a(t)\operatorname{rect}[(t - 0.5T)/T]} \tag{13.1-3}$$

and

$$\boxed{\mathcal{E}(t) = |a(t)|\operatorname{rect}[(t - 0.5T)/T]} \tag{13.1-4}$$

respectively (see Section 11.2). In addition, the energies of $\tilde{x}(t)$ and $x(t)$, and the time-average power of $x(t)$ are given by

$$\boxed{E_{\tilde{x}} = \int_0^T a^2(t)\,dt} \tag{13.1-5}$$

$$\boxed{E_x = \frac{1}{2}E_{\tilde{x}} = \frac{1}{2}\int_0^T a^2(t)\,dt} \tag{13.1-6}$$

and

$$\boxed{P_{\text{avg},x} = \frac{E_x}{T} = \frac{1}{2T}\int_0^T a^2(t)\,dt} \tag{13.1-7}$$

respectively (see Section 11.2). If $\tilde{x}(t)$ and $x(t)$ are voltage signals, then $E_{\tilde{x}}$ and E_x have units of *joules-ohms*. However, if $\tilde{x}(t)$ and $x(t)$ are current signals, then $E_{\tilde{x}}$ and E_x have units of *joules per ohm* (see Subsection 11.1.1). If E_x has units of joules-ohms or joules per ohm, then $P_{\text{avg},x}$ will have units of watts-ohms or watts per ohm, respectively.

If the real amplitude-modulating function $a(t)$ is set equal to a *positive constant* A, that is, if $a(t) = A > 0$, then the CW pulse given by (13.1-1) reduces to the *rectangular-envelope, CW pulse*

$$\boxed{x(t) = A\cos(2\pi f_c t)\operatorname{rect}[(t - 0.5T)/T]} \tag{13.1-8}$$

An example of a rectangular-envelope, CW pulse is shown in Fig. 13.1-3. The complex envelope and envelope of (13.1-8) are given by

$$\boxed{\tilde{x}(t) = A\operatorname{rect}[(t - 0.5T)/T]} \tag{13.1-9}$$

and

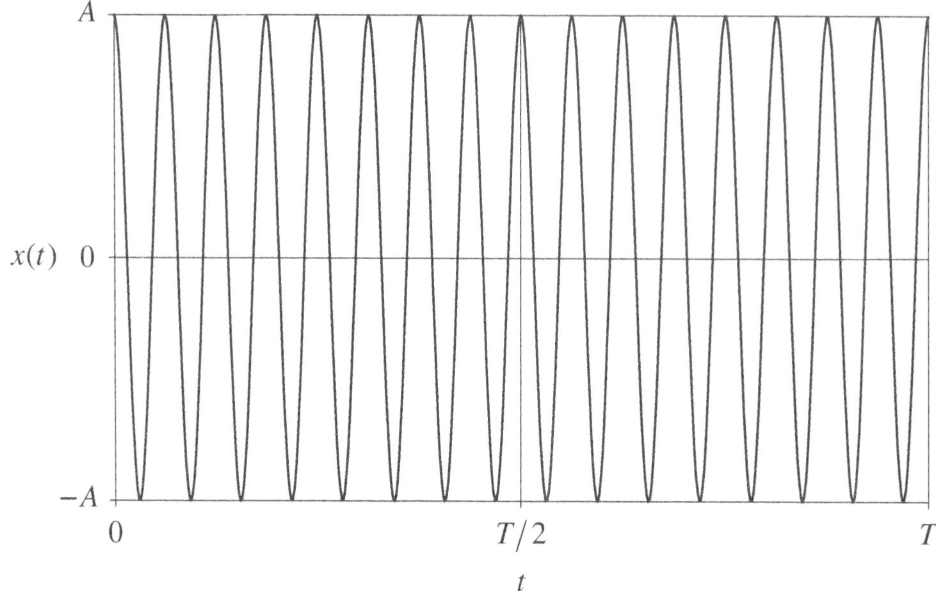

Figure 13.1-3 A rectangular-envelope, CW pulse with pulse length T seconds.

$$\boxed{\mathcal{E}(t) = A\,\mathrm{rect}\left[(t - 0.5T)/T\right]} \tag{13.1-10}$$

respectively [see (13.1-3) and (13.1-4)]. Since the envelope given by (13.1-10) is a rectangle, hence the terminology *rectangular-envelope*, CW pulse. The energy of $\tilde{x}(t)$ and $x(t)$, and the time-average power of $x(t)$ for a rectangular-envelope, CW pulse are given by

$$\boxed{E_{\tilde{x}} = A^2 T} \tag{13.1-11}$$

$$\boxed{E_x = \frac{1}{2} E_{\tilde{x}} = \frac{A^2}{2} T} \tag{13.1-12}$$

and

$$\boxed{P_{\text{avg},\,x} = \frac{E_x}{T} = \frac{A^2}{2}} \tag{13.1-13}$$

respectively [see (13.1-5) through (13.1-7)]. The unnormalized and normalized, auto-ambiguity functions of a rectangular-envelope, CW pulse shall be derived next.

The *unnormalized*, auto-ambiguity function of a transmitted electrical signal $x(t)$, with complex envelope $\tilde{x}(t)$, is given by (see Section 12.2)

$$\boxed{\begin{aligned} X(e_{\tau,Trgt}, e_{D,Trgt}) &= \int_{-\infty}^{\infty} \tilde{x}(t)\tilde{x}^*(t-e_{\tau,Trgt})\exp(-j2\pi e_{D,Trgt}t)dt \\ &= F\left\{\tilde{x}(t)\tilde{x}^*(t-e_{\tau,Trgt})\right\} \\ &= \left\langle \tilde{x}(t), \tilde{x}(t-e_{\tau,Trgt})\exp(+j2\pi e_{D,Trgt}t)\right\rangle \end{aligned}} \tag{13.1-14}$$

where

$$\boxed{e_{\tau,Trgt} = \hat{\tau}_{Trgt} - \tau_{Trgt}} \tag{13.1-15}$$

is the *round-trip time-delay estimation error of the target* in seconds, where $\hat{\tau}_{Trgt}$ is the estimate of τ_{Trgt}, and

$$\boxed{e_{D,Trgt} = \hat{\eta}_{D,Trgt} - \eta_{D,Trgt}} \tag{13.1-16}$$

is the *Doppler-shift estimation error of the target* in hertz, where $\hat{\eta}_{D,Trgt}$ is the estimate of $\eta_{D,Trgt}$. Two other very important functions related to the auto-ambiguity function are the range and Doppler profiles. By setting $e_{D,Trgt}=0$ in (13.1-14), we obtain the *unnormalized, round-trip time-delay (range) profile*:

$$\boxed{\begin{aligned} X(e_{\tau,Trgt}, 0) &= \int_{-\infty}^{\infty} \tilde{x}(t)\tilde{x}^*(t-e_{\tau,Trgt})dt \\ &= \tilde{x}(e_{\tau,Trgt}) * \tilde{x}^*(-e_{\tau,Trgt}) \\ &= F^{-1}\left\{\tilde{X}(f)\tilde{X}^*(f)\right\} \end{aligned}} \tag{13.1-17}$$

where

$$\tilde{x}(e_{\tau,Trgt}) \leftrightarrow \tilde{X}(f) \tag{13.1-18}$$

and

$$\tilde{x}^*(-e_{\tau,Trgt}) \leftrightarrow \tilde{X}^*(f). \tag{13.1-19}$$

The range profile $X(e_{\tau,Trgt}, 0)$ given by (13.1-17) is the *time-domain autocorrelation function* of $\tilde{x}(t)$. Furthermore, $X(e_{\tau,Trgt}, 0) = F^{-1}\left\{\left|\tilde{X}(f)\right|^2\right\}$. Similarly, by setting $e_{\tau,Trgt}=0$ in (13.1-14), we obtain the *unnormalized, Doppler profile*:

$$\boxed{\begin{aligned} \chi(0, e_{D,Trgt}) &= \int_{-\infty}^{\infty} \tilde{x}(t)\tilde{x}^*(t)\exp(-j2\pi e_{D,Trgt}t)dt \\ &= F\{\tilde{x}(t)\tilde{x}^*(t)\} \\ &= \tilde{X}(e_{D,Trgt}) * \tilde{X}^*(-e_{D,Trgt}) \\ &= \int_{-\infty}^{\infty} \tilde{X}(f)\tilde{X}^*(f - e_{D,Trgt})df \end{aligned}} \tag{13.1-20}$$

where

$$\tilde{x}(t) \leftrightarrow \tilde{X}(e_{D,Trgt}) \tag{13.1-21}$$

and

$$\tilde{x}^*(t) \leftrightarrow \tilde{X}^*(-e_{D,Trgt}). \tag{13.1-22}$$

The Doppler profile $\chi(0, e_{D,Trgt})$ given by (13.1-20) is the *frequency-domain autocorrelation function* of $\tilde{X}(f)$. Furthermore, $\chi(0, e_{D,Trgt}) = F\left\{|\tilde{x}(t)|^2\right\} = F\left\{a^2(t)\right\}$. As we shall show later, both the range and Doppler profiles contain very important information that is pertinent to signal design.

In order to derive the unnormalized, auto-ambiguity function of a rectangular-envelope, CW pulse, we start by substituting its complex envelope given by (13.1-9) into (13.1-14). Doing so yields

$$\chi(e_{\tau,Trgt}, e_{D,Trgt}) = A^2 \int_{-\infty}^{\infty} \text{rect}\left(\frac{t - 0.5T}{T}\right)\text{rect}\left[\frac{t - (0.5T + e_{\tau,Trgt})}{T}\right] \times \exp(-j2\pi e_{D,Trgt}t)dt, \tag{13.1-23}$$

or

$$\chi(e_{\tau,Trgt}, e_{D,Trgt}) = A^2 F\left\{\text{rect}\left(\frac{t - 0.5T}{T}\right)\text{rect}\left[\frac{t - (0.5T + e_{\tau,Trgt})}{T}\right]\right\}. \tag{13.1-24}$$

By referring to Fig. 13.1-4, it can be seen that for $0 \le e_{\tau,Trgt} \le T$, (13.1-23) reduces to

$$\chi(e_{\tau,Trgt}, e_{D,Trgt}) = A^2 \int_{e_{\tau,Trgt}}^{T} \exp(-j2\pi e_{D,Trgt}t)dt, \qquad 0 \le e_{\tau,Trgt} \le T, \tag{13.1-25}$$

which simplifies to

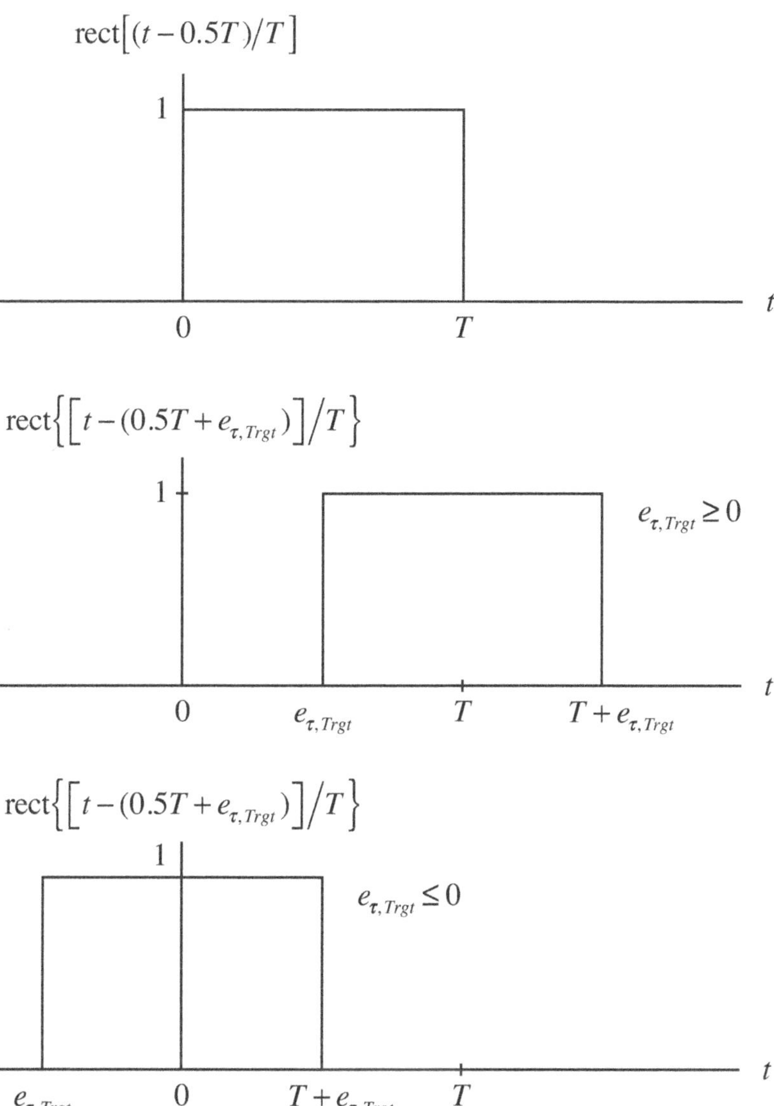

Figure 13.1-4 Time-shifted rectangle functions.

$$X(e_{\tau,Trgt}, e_{D,Trgt}) = A^2(T - e_{\tau,Trgt})\text{sinc}\left[e_{D,Trgt}(T - e_{\tau,Trgt})\right]\times \exp\left[-j\pi e_{D,Trgt}(T + e_{\tau,Trgt})\right], \qquad 0 \le e_{\tau,Trgt} \le T. \tag{13.1-26}$$

Similarly, for $-T \le e_{\tau,Trgt} \le 0$, (13.1-23) reduces to

$$X(e_{\tau,Trgt}, e_{D,Trgt}) = A^2 \int_0^{T+e_{\tau,Trgt}} \exp(-j2\pi e_{D,Trgt}t)dt, \qquad -T \le e_{\tau,Trgt} \le 0, \tag{13.1-27}$$

which simplifies to

$$\chi(e_{\tau,Trgt}, e_{D,Trgt}) = A^2(T + e_{\tau,Trgt})\operatorname{sinc}\left[e_{D,Trgt}(T + e_{\tau,Trgt})\right]\times$$
$$\exp\left[-j\pi e_{D,Trgt}(T + e_{\tau,Trgt})\right], \qquad -T \le e_{\tau,Trgt} \le 0.$$

(13.1-28)

Equations (13.1-26) and (13.1-28) can be combined into the following single equation:

$$\chi(e_{\tau,Trgt}, e_{D,Trgt}) = A^2 T\left[1 - \frac{\left|e_{\tau,Trgt}\right|}{T}\right]\operatorname{sinc}\left\{e_{D,Trgt}\left[T - \left|e_{\tau,Trgt}\right|\right]\right\}\times$$
$$\exp\left[-j\pi e_{D,Trgt}(T + e_{\tau,Trgt})\right], \qquad -T \le e_{\tau,Trgt} \le T.$$

(13.1-29)

Equation (13.1-29) is the *unnormalized*, auto-ambiguity function of a rectangular-envelope, CW pulse.

In order to normalize an unnormalized, auto-ambiguity function, we need to divide it by the energy of the complex envelope of the transmitted electrical signal (see Section 12.2). Therefore, dividing (13.1-29) by (13.1-11) yields

$$\chi_N(e_{\tau,Trgt}, e_{D,Trgt}) = \left[1 - \frac{\left|e_{\tau,Trgt}\right|}{T}\right]\operatorname{sinc}\left\{e_{D,Trgt}\left[T - \left|e_{\tau,Trgt}\right|\right]\right\}\times$$
$$\exp\left[-j\pi e_{D,Trgt}(T + e_{\tau,Trgt})\right], \qquad -T \le e_{\tau,Trgt} \le T$$

(13.1-30)

which is the *normalized, auto-ambiguity function of a rectangular-envelope, CW pulse*. By setting $e_{D,Trgt} = 0$ in (13.1-30) and taking the magnitude of the resulting expression, we obtain the magnitude of the *normalized, round-trip time-delay (range) profile*:

$$\left|\chi_N(e_{\tau,Trgt}, 0)\right| = 1 - \frac{\left|e_{\tau,Trgt}\right|}{T}, \qquad -T \le e_{\tau,Trgt} \le T \qquad (13.1\text{-}31)$$

since $\operatorname{sinc}(0) = 1$. Equation (13.1-31) is the equation of a triangle (see Fig. 13.1-5). Similarly, by setting $e_{\tau,Trgt} = 0$ in (13.1-30) and taking the magnitude of the resulting expression, we obtain the magnitude of the *normalized, Doppler profile* (see Fig. 13.1-6):

$$\left|X_N(0, e_{D,Trgt})\right| = \left|\text{sinc}(e_{D,Trgt}\, T)\right| \tag{13.1-32}$$

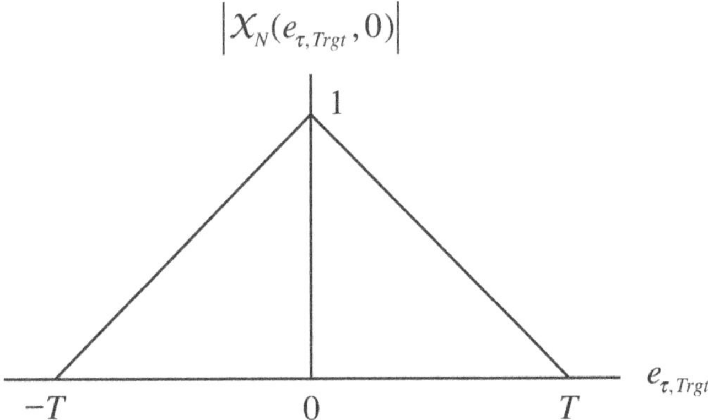

Figure 13.1-5 Magnitude of the normalized, round-trip time-delay (range) profile of a rectangular-envelope, CW pulse with pulse length T seconds.

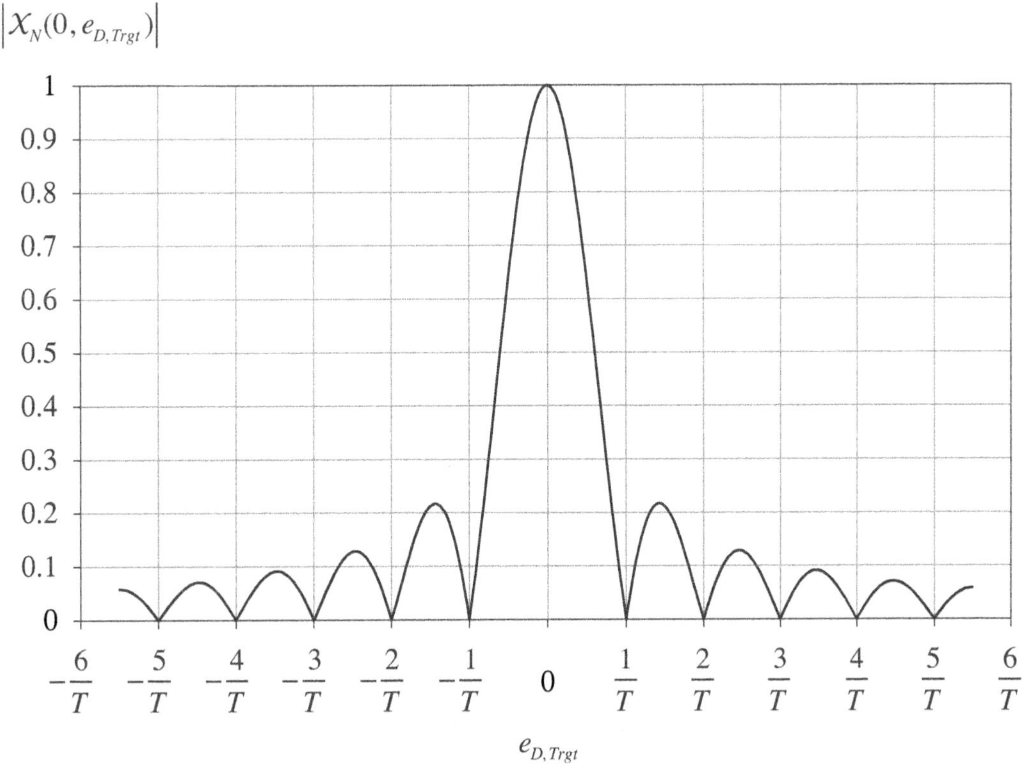

Figure 13.1-6 Magnitude of the normalized, Doppler profile of a rectangular-envelope, CW pulse with pulse length T seconds.

By inspecting the range profile given by (13.1-31) and referring to Fig. 13.1-5, it can be seen that the width of the mainlobe of the ambiguity function along the $e_{\tau,Trgt}$ axis is *directly proportional to the pulse length* T since the range profile decreases to zero at $e_{\tau,Trgt} = \pm T$ seconds. However, by inspecting the Doppler profile given by (13.1-32) and referring to Fig. 13.1-6, it can be seen that the width of the mainlobe of the ambiguity function along the $e_{D,Trgt}$ axis is *inversely proportional to the pulse length* T since the locations of the *first* zero crossings of the Doppler profile are at $e_{D,Trgt} = \pm 1/T$ hertz. A single parameter, the pulse length T, controls both the round-trip time-delay (range) and Doppler resolving capabilities of a rectangular-envelope, CW pulse. One can make the width of the mainlobe of the ambiguity function arbitrarily small in either direction by varying T, but *not* in both directions simultaneously. As a result, *a short duration rectangular-envelope, CW pulse has good range resolution but poor Doppler resolution, and a long duration rectangular-envelope, CW pulse has poor range resolution but good Doppler resolution.*

Besides wanting, in general, an ambiguity function to have as narrow a mainlobe about the origin as possible, it is also desirable that an ambiguity function have sidelobe levels as low as possible. The sidelobe levels of an ambiguity function can be reduced by using an amplitude-modulating function $a(t)$ other than rectangular (i.e., a constant), analogous to using different amplitude weights to reduce the sidelobe levels of the far-field beam pattern of an array. In fact, one can use time-domain versions of the same continuous amplitude windows discussed in Section 2.2 for $a(t)$. However, as with far-field beam patterns, there is a trade-off, since the width of the mainlobe of an ambiguity function generally increases as the sidelobe levels decrease.

Example 13.1-1 Design of a Rectangular-Envelope, CW Pulse

A rectangular-envelope, CW pulse cannot provide good range and Doppler resolution simultaneously. Therefore, we shall first design a rectangular-envelope, CW pulse to satisfy a range-resolution requirement and then compute the resulting Doppler resolution. In general, in signal-design problems, satisfying a range-resolution requirement is of primary concern. We will then reverse the problem, that is, we will design a rectangular-envelope, CW pulse to satisfy a Doppler-resolution requirement and then compute the resulting range resolution. We will also estimate the bandwidth of the resulting pulses in both cases and solve for the amplitude of a rectangular-envelope, CW pulse that is needed to satisfy a time-average-power requirement or constraint. Being able to estimate the bandwidth of a transmitted electrical signal is important because one needs to know if the sensitivity functions of the electroacoustic transducers in a sonar array have sufficient bandwidth so as not to filter out any important frequency components in the input electrical signals to the transducers.

We begin the design procedure by introducing the *range estimation error of the target* in meters given by

$$\boxed{e_{r,Trgt} = \hat{r}_{Trgt} - r_{Trgt}} \tag{13.1-33}$$

where $\hat{r}_{Trgt}$ is the estimate of the range to the target r_{Trgt}. The round-trip time-delay estimation error $e_{\tau,Trgt}$ is related to the range estimation error $e_{r,Trgt}$ by

$$e_{\tau,Trgt} = 2\,e_{r,Trgt}/c\,, \tag{13.1-34}$$

where c is the constant speed of sound in meters per second. As can be seen from (13.1-34), an error in estimating the range to a target will result in an error in estimating the round-trip time delay, and vice versa. Since estimation errors can be either positive or negative in value, we take the magnitude of (13.1-34) to obtain

$$\left|e_{\tau,Trgt}\right| = 2\left|e_{r,Trgt}\right|/c\,. \tag{13.1-35}$$

Solving for $\left|e_{r,Trgt}\right|$ yields

$$\left|e_{r,Trgt}\right| = c\left|e_{\tau,Trgt}\right|/2\,. \tag{13.1-36}$$

In signal design, the widths of the mainlobe of an auto-ambiguity function along the $e_{\tau,Trgt}$ and $e_{D,Trgt}$ axes are used as measures of the range and Doppler resolutions, respectively, of a transmitted electrical signal. For a rectangular-envelope, CW pulse, the round-trip time-delay (range) profile decreases to zero at [see (13.1-31)]

$$e_{\tau,Trgt} = \pm T\,, \tag{13.1-37}$$

and the locations of the first zero crossings of the Doppler profile are at [see (13.1-32)]

$$e_{D,Trgt} = \pm 1/T\,, \tag{13.1-38}$$

where T is the pulse length in seconds. Therefore, from (13.1-37),

$$\left|e_{\tau,Trgt}\right| = T\,, \tag{13.1-39}$$

and from (13.1-38),

$$\left|e_{D,Trgt}\right| = 1/T\,. \tag{13.1-40}$$

Let us now design a rectangular-envelope, CW pulse to satisfy a range-resolution requirement.

Specifying a value for range resolution is equivalent to specifying a desired value for the magnitude of the range estimation error of the target $|e_{r,Trgt}|$ that corresponds to the location of the first zero crossing along the positive $e_{\tau,Trgt}$ axis. If (13.1-39) is substituted into (13.1-35), then

$$\boxed{T = 2|e_{r,Trgt}|/c} \tag{13.1-41}$$

Equation (13.1-41) is used to compute the pulse length T in seconds that is required for a rectangular-envelope, CW pulse to provide a range resolution equal to $|e_{r,Trgt}|$ meters. The resulting Doppler resolution is obtained by substituting the value of T from (13.1-41) into (13.1-40). For example, if $|e_{r,Trgt}| = 1\text{ m}$ ($e_{r,Trgt} = \pm 1\text{ m}$) and $c = 1500\text{ m/sec}$, then $T = 1.333\text{ msec}$ and $|e_{D,Trgt}| = 750\text{ Hz}$ ($e_{D,Trgt} = \pm 750\text{ Hz}$). Also, if $|e_{r,Trgt}| = 1\text{ m}$, then $|e_{\tau,Trgt}| = 1.333\text{ msec}$ [see (13.1-39)]. The magnitude of the normalized, auto-ambiguity function $|\chi_N(e_{\tau,Trgt}, e_{D,Trgt})|$ of this signal is shown in Fig. 13.1-7. As can be seen from the figure, if the round-trip time-delay estimation error $e_{\tau,Trgt}$ is zero or very small, $|\chi_N(e_{\tau,Trgt}, e_{D,Trgt})|$ remains "big" even if the Doppler estimation error $e_{D,Trgt}$ is "big". For example, from Fig. 13.1-7, $|\chi_N(0, \pm 400\text{ Hz})| \approx 0.7$. This waveform is an example of a *Doppler-tolerant* waveform, that is, a waveform that is tolerant of Doppler estimation errors. In other words, if a very good estimate of the range to a target is made so that $e_{\tau,Trgt}$ is zero or very small, then $|\chi_N(e_{\tau,Trgt}, e_{D,Trgt})|$ remains "big" even if $e_{D,Trgt}$ is "big". As a result, the target can still be detected. Recall that the numerator of the signal-to-interference ratio (SIR) is directly proportional to the magnitude-squared of the normalized, auto-ambiguity function (see Section 12.2). Therefore, as $|\chi_N(e_{\tau,Trgt}, e_{D,Trgt})| \to 0$, the $\text{SIR} \to 0$ and, as a result, the target won't be detected.

Before we estimate the bandwidth of a rectangular-envelope, CW pulse, let us state the following two relationships that pertain to all amplitude-and-angle-modulated carriers whose magnitude spectra are symmetrical about f_c:

$$\boxed{f_c > \text{BW}_{\tilde{x}}} \tag{13.1-42}$$

and

$$\boxed{\text{BW}_x = 2\,\text{BW}_{\tilde{x}}} \tag{13.1-43}$$

where f_c is the carrier frequency in hertz, $\text{BW}_{\tilde{x}}$ is the bandwidth in hertz of the

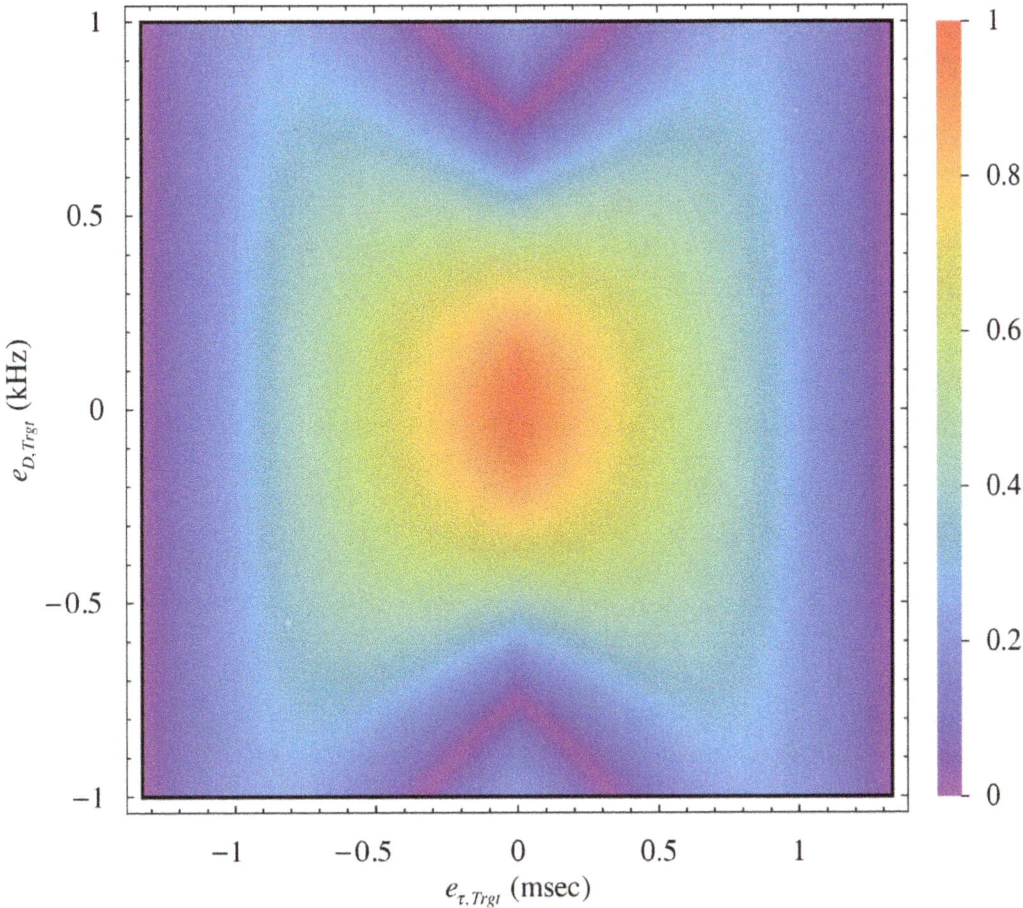

Figure 13.1-7 Magnitude of the normalized, auto-ambiguity function of a rectangular-envelope, CW pulse with pulse length $T = 1.333$ msec.

lowpass (baseband) complex envelope $\tilde{x}(t)$, and BW_x is the bandwidth in hertz of the bandpass, amplitude-and-angle-modulated carrier $x(t)$. As can be seen from (13.1-42), $\text{BW}_{\tilde{x}}$ is the lower bound for determining a value for the carrier frequency f_c. The inequality in (13.1-42) must be satisfied in order to avoid signal distortion.

The complex envelope of a rectangular-envelope, CW pulse is given by (13.1-9) with unnormalized and normalized Fourier transforms

$$\tilde{X}(f) = AT\,\text{sinc}(fT)\exp(-j\pi fT) \tag{13.1-44}$$

and

$$\tilde{X}_N(f) = \text{sinc}(fT)\exp(-j\pi fT), \tag{13.1-45}$$

respectively. Since

$$\max\left|\tilde{X}_N(f)\right| = 1, \qquad f = 0, \tag{13.1-46}$$

and

$$\left|\tilde{X}_N(f)\right| < 0.1, \qquad f > 3/T, \tag{13.1-47}$$

a conservative estimate of the bandwidth in hertz of the complex envelope of a rectangular-envelope, CW pulse is given by (see Fig. 13.1-8)

$$\boxed{\mathrm{BW}_{\tilde{x}} \approx 5/T} \tag{13.1-48}$$

Substituting (13.1-48) into (13.1-42) and (13.1-43) yields

$$\boxed{f_c > 5/T} \tag{13.1-49}$$

and

$$\boxed{\mathrm{BW}_x \approx 10/T} \tag{13.1-50}$$

for a rectangular-envelope, CW pulse. For example, for the previously designed rectangular-envelope, CW pulse with pulse length $T = 1.333\,\text{msec}$; $\mathrm{BW}_{\tilde{x}} \approx 3750\,\text{Hz}$, $f_c > 3750\,\text{Hz}$, and $\mathrm{BW}_x \approx 7500\,\text{Hz}$.

Next we shall design a rectangular-envelope, CW pulse to satisfy a Doppler-resolution requirement. Specifying a value for Doppler resolution is equivalent to specifying a desired value for the magnitude of the Doppler estimation error of the target $\left|e_{D,Trgt}\right|$ that corresponds to the location of the first zero crossing along the positive $e_{D,Trgt}$ axis. Solving for T from (13.1-40) yields

$$\boxed{T = 1/\left|e_{D,Trgt}\right|} \tag{13.1-51}$$

and by substituting (13.1-39) into (13.1-36), we obtain

$$\left|e_{r,Trgt}\right| = cT/2. \tag{13.1-52}$$

Equation (13.1-51) is used to compute the pulse length T in seconds that is required for a rectangular-envelope, CW pulse to provide a Doppler resolution equal to $\left|e_{D,Trgt}\right|$ hertz. The resulting range resolution is obtained by substituting the value of T from (13.1-51) into (13.1-52). For example, if $\left|e_{D,Trgt}\right| = 1\,\text{Hz}$ ($e_{D,Trgt} = \pm 1\,\text{Hz}$) and $c = 1500\,\text{m/sec}$, then $T = 1\,\text{sec}$, $\left|e_{r,Trgt}\right| = 750\,\text{m}$

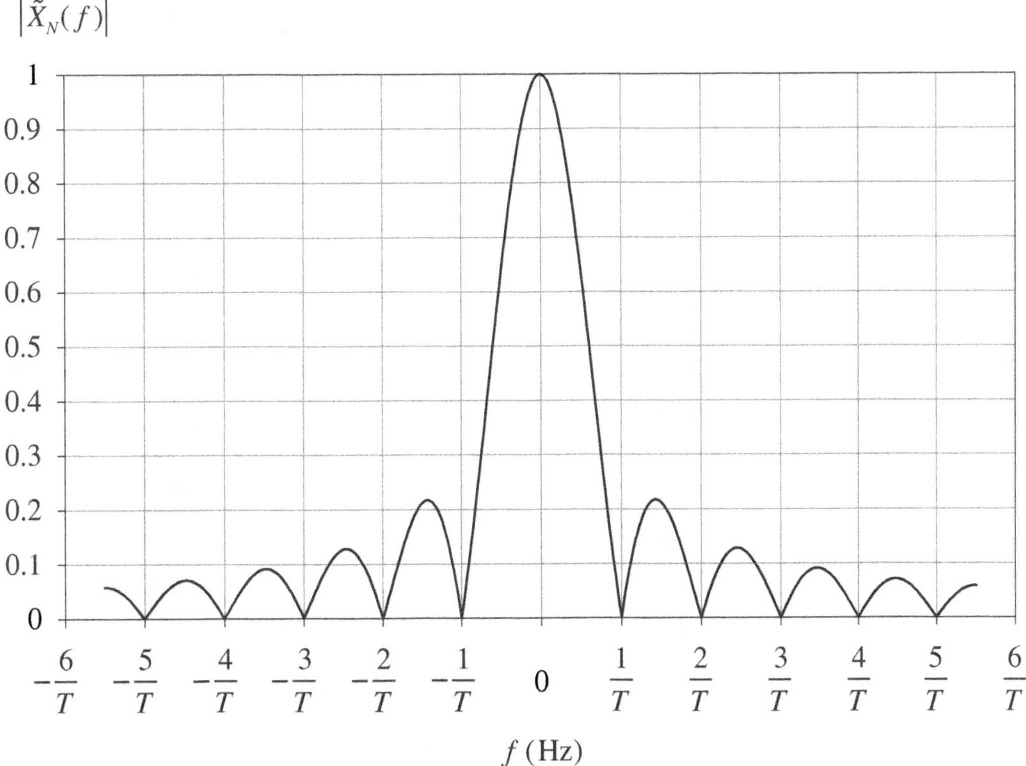

Figure 13.1-8 Normalized magnitude spectrum of the complex envelope of a rectangular-envelope, CW pulse with pulse length T seconds.

($e_{r,Trgt} = \pm 750$ m), $\text{BW}_{\tilde{x}} \approx 5$ Hz, $f_c > 5$ Hz, and $\text{BW}_x \approx 10$ Hz. Also, if $|e_{r,Trgt}| = 750$ m, then $|e_{\tau,Trgt}| = 1$ sec [see (13.1-39)]. The magnitude of the normalized, auto-ambiguity function $|X_N(e_{\tau,Trgt}, e_{D,Trgt})|$ of this signal is shown in Fig. 13.1-9. As was previously mentioned, since the numerator of the SIR is directly proportional to the magnitude-squared of the normalized, auto-ambiguity function (see Section 12.2), as $|X_N(e_{\tau,Trgt}, e_{D,Trgt})| \to 0$, the $\text{SIR} \to 0$ and, as a result, the target won't be detected.

Our last calculation is to solve for the amplitude A of a rectangular-envelope, CW pulse that will satisfy a time-average-power requirement or constraint. Since the time-average power $P_{\text{avg},x}$ of a rectangular-envelope, CW pulse is given by [see (13.1-13)]

$$P_{\text{avg},x} = A^2/2, \tag{13.1-53}$$

solving for the amplitude A yields

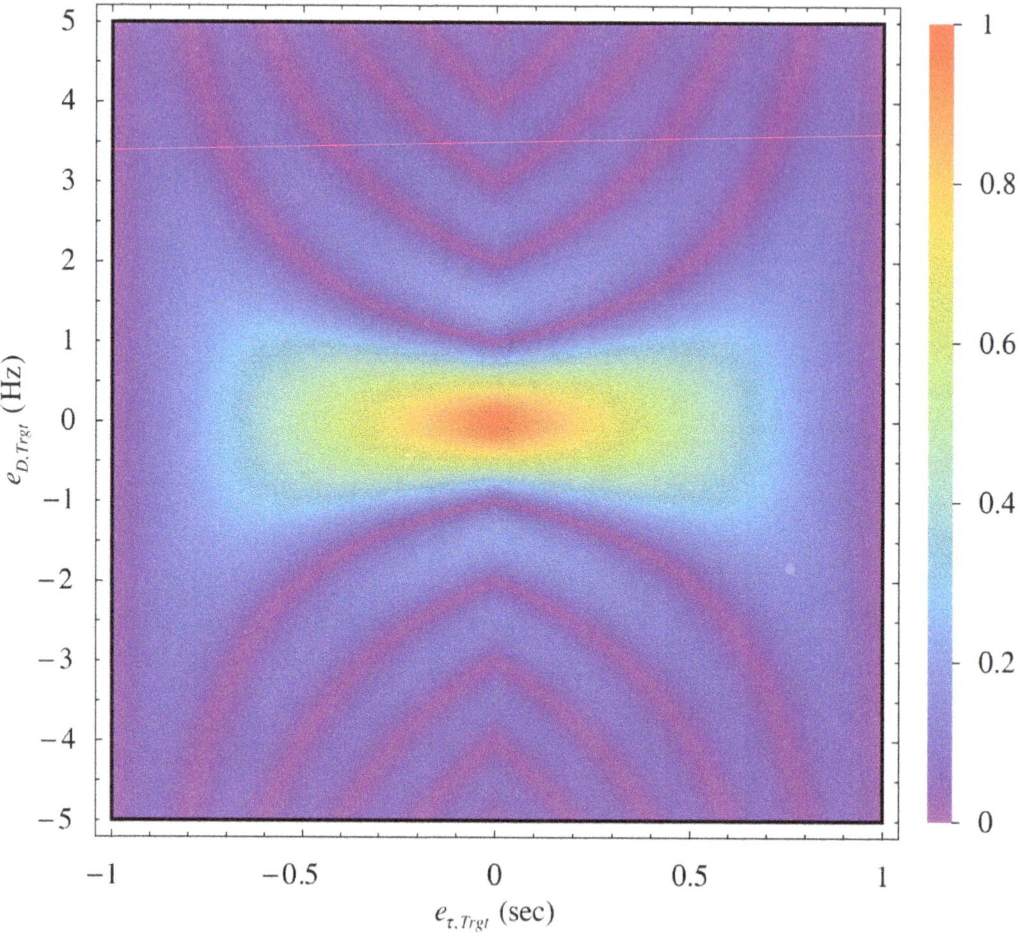

Figure 13.1-9 Magnitude of the normalized, auto-ambiguity function of a rectangular-envelope, CW pulse with pulse length $T = 1$ sec .

$$\boxed{A = \sqrt{2P_{\text{avg},x}}} \tag{13.1-54}$$

If $x(t)$ is a voltage signal, then A has units of volts and $P_{\text{avg},x}$ has units of watts-ohms. Once the pulse length T has been calculated using either (13.1-41) or (13.1-51), and the amplitude A has been calculated using (13.1-54), the energies $E_{\tilde{x}}$ and E_x of the rectangular-envelope, CW pulse can be calculated using (13.1-11) and (13.1-12), respectively. If $x(t)$ is a voltage signal, then $E_{\tilde{x}}$ and E_x have units of joules-ohms.

Finally, we know that the range resolution of a rectangular-envelope, CW pulse improves as pulse length T decreases. However, as T decreases, signal bandwidth BW_x increases [see (13.1-50)] and signal energy E_x decreases [see

(13.1-12)]. In order to increase E_x, signal amplitude A must be increased which increases the time-average power $P_{\text{avg},x}$ of the signal [see (13.1-53)]. ■

13.2 The Rectangular-Envelope LFM Pulse

It was shown in Section 13.1 that a rectangular-envelope, CW pulse cannot provide good range and Doppler resolution simultaneously. In order to obtain good range and Doppler resolution *simultaneously*, a more complicated signal must be used that contains several parameters that can be varied. A rectangular-envelope, LFM (linear-frequency-modulated) pulse is such a signal and shall be discussed next.

If we begin with a finite duration, amplitude-and-angle-modulated carrier

$$x(t) = a(t)\cos\left[2\pi f_c t + \theta(t)\right]\text{rect}\left[(t - 0.5T)/T\right], \tag{13.2-1}$$

and if we let the real angle-modulating function (phase deviation) be given by

$$\theta(t) = D_p\left[t - (T/2)\right]^2 \tag{13.2-2}$$

so that the frequency deviation

$$\frac{d}{dt}\theta(t) = 2D_p\left[t - (T/2)\right] \tag{13.2-3}$$

is equal to a *linear* function of time, then the signal

$$\boxed{x(t) = a(t)\cos\left\{2\pi f_c t + D_p\left[t - (T/2)\right]^2\right\}\text{rect}\left[(t - 0.5T)/T\right]} \tag{13.2-4}$$

is known as a *linear-frequency-modulated* (LFM) *pulse*, where $a(t)$ is a real amplitude-modulating function, $\cos(2\pi f_c t)$ is the carrier waveform, f_c is the carrier frequency in hertz, D_p is the phase-deviation constant with units of radians per second-squared, T is the pulse length in seconds, and the time-shifted rectangle function is given by (13.1-2). An example of a LFM pulse is shown in Fig. 13.2-1. The complex envelope and envelope of (13.2-4) are given by

$$\boxed{\tilde{x}(t) = a(t)\exp\left\{+jD_p\left[t - (T/2)\right]^2\right\}\text{rect}\left[(t - 0.5T)/T\right]} \tag{13.2-5}$$

and

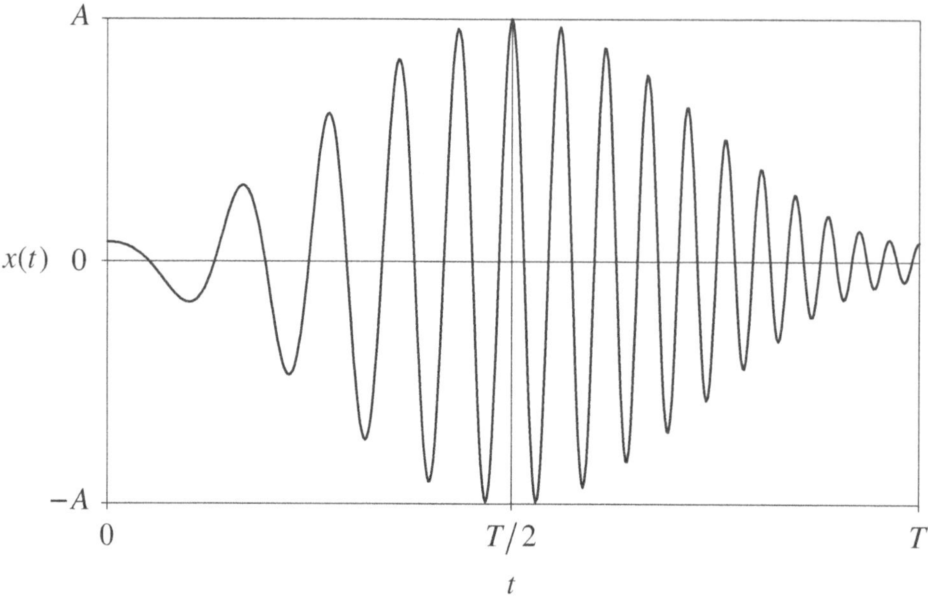

Figure 13.2-1 A LFM pulse with pulse length T seconds.

$$\boxed{\mathcal{E}(t) = |a(t)| \operatorname{rect}[(t - 0.5T)/T]} \tag{13.2-6}$$

respectively (see Section 11.2). The energies of $\tilde{x}(t)$ and $x(t)$, and the time-average power of $x(t)$ for a LFM pulse are identical to those for a CW pulse and are given by (13.1-5) through (13.1-7), respectively.

If the real amplitude-modulating function $a(t)$ is set equal to a *positive constant* A, that is, if $a(t) = A > 0$, then the LFM pulse given by (13.2-4) reduces to the *rectangular-envelope, LFM pulse*

$$\boxed{x(t) = A\cos\left\{2\pi f_c t + D_p\left[t - (T/2)\right]^2\right\} \operatorname{rect}[(t - 0.5T)/T]} \tag{13.2-7}$$

An example of a rectangular-envelope, LFM pulse is shown in Fig. 13.2-2. The complex envelope and envelope of (13.2-7) are given by

$$\boxed{\tilde{x}(t) = A\exp\left\{+jD_p\left[t - (T/2)\right]^2\right\} \operatorname{rect}[(t - 0.5T)/T]} \tag{13.2-8}$$

and

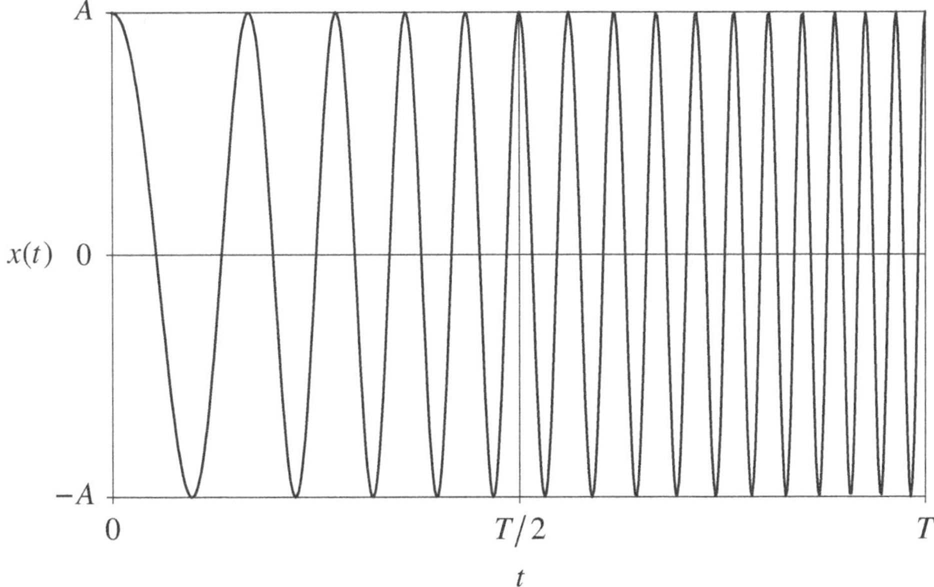

Figure 13.2-2 A rectangular-envelope, LFM pulse with pulse length T seconds.

$$\boxed{\mathcal{E}(t) = A\,\text{rect}\left[(t - 0.5T)/T\right]} \tag{13.2-9}$$

respectively [see (13.2-5) and (13.2-6)]. Since the envelope given by (13.2-9) is a rectangle, hence the terminology *rectangular-envelope*, LFM pulse. The energies of $\tilde{x}(t)$ and $x(t)$, and the time-average power of $x(t)$ for a rectangular-envelope, LFM pulse are identical to those for a rectangular-envelope, CW pulse and are given by (13.1-11) through (13.1-13), respectively.

The instantaneous phase in radians and instantaneous frequency in hertz of both the LFM pulse given by (13.2-4) and the rectangular-envelope, LFM pulse given by (13.2-7) are given by

$$\begin{aligned} \theta_i(t) &\triangleq 2\pi f_c t + \theta(t) \\ &= 2\pi f_c t + D_p\left[t - (T/2)\right]^2, \qquad 0 \le t \le T \end{aligned} \tag{13.2-10}$$

and

$$\begin{aligned} f_i(t) &\triangleq \frac{1}{2\pi}\frac{d}{dt}\theta_i(t) \\ &= f_c + \frac{1}{\pi}D_p\left[t - (T/2)\right], \qquad 0 \le t \le T, \end{aligned} \tag{13.2-11}$$

respectively (see Fig. 13.2-3). From (13.2-11) or Fig. 13.2-3,

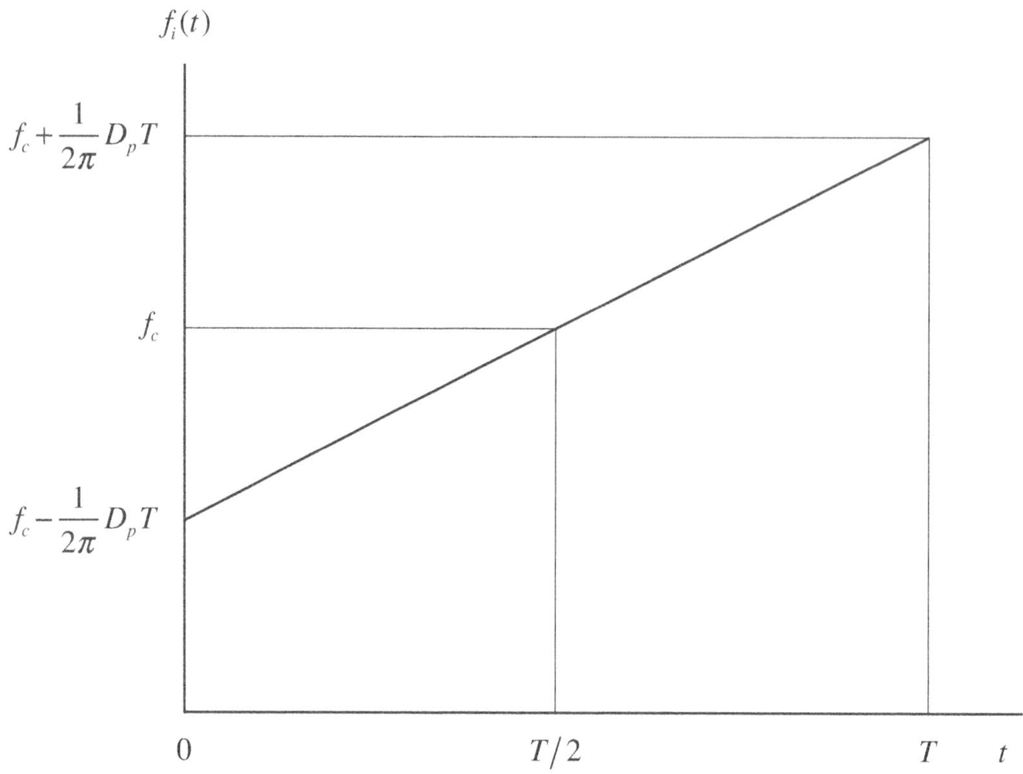

Figure 13.2-3 Instantaneous frequency $f_i(t)$ in hertz versus time t in seconds of a LFM pulse with pulse length T seconds and phase-deviation constant $D_p > 0$.

$$f_i(T) - f_i(0) = f_c + \frac{1}{2\pi} D_p T - \left[f_c - \frac{1}{2\pi} D_p T \right] = \frac{1}{\pi} D_p T . \qquad (13.2\text{-}12)$$

Therefore, the *swept-bandwidth* BW_{swept} in hertz is defined as

$$\boxed{\text{BW}_{\text{swept}} \triangleq \frac{1}{\pi} |D_p| T} \qquad (13.2\text{-}13)$$

where the absolute value of the phase-deviation constant D_p is used in (13.2-13) in order to keep the swept-bandwidth positive since D_p can be negative. A LFM pulse is also known as a *chirp pulse*. When the phase-deviation constant $D_p > 0$, a LFM pulse is called an "up-chirp" because the instantaneous frequency $f_i(t)$ *increases* with time; and when $D_p < 0$, it is called a "down-chirp" because $f_i(t)$ *decreases* with time.

In order to derive the unnormalized, auto-ambiguity function of a rectangular-envelope, LFM pulse, we start by substituting its complex envelope given by (13.2-8) into (13.1-14). Doing so yields

$$\chi(e_{\tau,Trgt}, e_{D,Trgt}) = A^2 \exp\left(-jD_p e^2_{\tau,Trgt}\right)\exp\left(-jD_p T e_{\tau,Trgt}\right)\times \int_{-\infty}^{\infty} \text{rect}\left(\frac{t-0.5T}{T}\right)\text{rect}\left[\frac{t-(0.5T+e_{\tau,Trgt})}{T}\right]\times \exp\left[-j2\pi\left(e_{D,Trgt}-\frac{1}{\pi}D_p e_{\tau,Trgt}\right)t\right]dt, \tag{13.2-14}$$

or

$$\chi(e_{\tau,Trgt}, e_{D,Trgt}) = A^2 \exp\left(-jD_p e^2_{\tau,Trgt}\right)\exp\left(-jD_p T e_{\tau,Trgt}\right)\times F\left\{\text{rect}\left(\frac{t-0.5T}{T}\right)\text{rect}\left[\frac{t-(0.5T+e_{\tau,Trgt})}{T}\right]\exp\left(+j2D_p e_{\tau,Trgt} t\right)\right\}. \tag{13.2-15}$$

If we designate (13.1-23) for the rectangular-envelope, CW pulse as $\chi_{\text{RECW}}(e_{\tau,Trgt}, e_{D,Trgt})$, then (13.2-14) can be expressed as

$$\chi(e_{\tau,Trgt}, e_{D,Trgt}) = \chi_{\text{RECW}}\left(e_{\tau,Trgt}, e_{D,Trgt}-\frac{1}{\pi}D_p e_{\tau,Trgt}\right)\exp\left(-jD_p e^2_{\tau,Trgt}\right)\times \exp\left(-jD_p T e_{\tau,Trgt}\right). \tag{13.2-16}$$

Therefore, by replacing $e_{D,Trgt}$ with $e_{D,Trgt}-\left(D_p e_{\tau,Trgt}/\pi\right)$ in (13.1-29) for the unnormalized, auto-ambiguity function of a rectangular-envelope, CW pulse, and then substituting the resulting expression into (13.2-16) yields

$$\chi(e_{\tau,Trgt}, e_{D,Trgt}) = A^2 T\left[1-\frac{|e_{\tau,Trgt}|}{T}\right]\text{sinc}\left\{\left[e_{D,Trgt}-\frac{1}{\pi}D_p e_{\tau,Trgt}\right]\left[T-|e_{\tau,Trgt}|\right]\right\}\times \exp\left[-j\pi e_{D,Trgt}(T+e_{\tau,Trgt})\right], \quad -T \le e_{\tau,Trgt} \le T. \tag{13.2-17}$$

Equation (13.2-17) is the *unnormalized*, auto-ambiguity function of a rectangular-envelope, LFM pulse.

Since the energy of the complex envelope of a rectangular-envelope, LFM pulse is A^2T, dividing (13.2-17) by A^2T yields

$$\boxed{\begin{aligned} X_N(e_{\tau,Trgt}, e_{D,Trgt}) = \left[1 - \frac{\left|e_{\tau,Trgt}\right|}{T}\right] \mathrm{sinc}\left\{\left[e_{D,Trgt} - \frac{1}{\pi} D_p e_{\tau,Trgt}\right]\left[T - \left|e_{\tau,Trgt}\right|\right]\right\} \times \\ \exp\left[-j\pi e_{D,Trgt}(T + e_{\tau,Trgt})\right], \qquad -T \le e_{\tau,Trgt} \le T \end{aligned}}$$

(13.2-18)

which is the *normalized, auto-ambiguity function of a rectangular-envelope, LFM pulse*. The magnitude of (13.2-18) has $|\mathrm{sinc}(x)|$ type profiles along every line with a constant $e_{\tau,Trgt}$ value, that is, along every line parallel to and including the $e_{D,Trgt}$ axis. For a given value of $e_{\tau,Trgt}$, the maximum value of the corresponding profile is $1-\left(\left|e_{\tau,Trgt}\right|/T\right)$, and this maximum value is located at $e_{D,Trgt} = \left(D_p e_{\tau,Trgt}/\pi\right)$, which is the equation of a straight line in the $e_{\tau,Trgt}\, e_{D,Trgt}$ plane that passes through the origin. Therefore, when $D_p > 0$ ("up-chirp"), $\left|X_N(e_{\tau,Trgt}, e_{D,Trgt})\right|$ will be concentrated mainly in the first and third quadrants of the $e_{\tau,Trgt}\, e_{D,Trgt}$ plane; and when $D_p < 0$ ("down-chirp"), $\left|X_N(e_{\tau,Trgt}, e_{D,Trgt})\right|$ will be concentrated mainly in the second and fourth quadrants. In order to cover all four quadrants, one needs to transmit an up-chirp pulse followed by a down-chirp pulse or vice versa.

By setting $e_{D,Trgt} = 0$ in (13.2-18) and taking the magnitude of the resulting expression, we obtain the magnitude of the *normalized, round-trip time-delay (range) profile*:

$$\boxed{\left|X_N(e_{\tau,Trgt}, 0)\right| = \left[1 - \frac{\left|e_{\tau,Trgt}\right|}{T}\right] \left|\mathrm{sinc}\left\{\frac{1}{\pi} D_p e_{\tau,Trgt}\left[T - \left|e_{\tau,Trgt}\right|\right]\right\}\right|, \qquad -T \le e_{\tau,Trgt} \le T}$$

(13.2-19)

since $\mathrm{sinc}(-x) = \mathrm{sinc}(x)$. Similarly, by setting $e_{\tau,Trgt} = 0$ in (13.2-18) and taking the magnitude of the resulting expression, we obtain the magnitude of the *normalized, Doppler profile*:

$$\boxed{\left|X_N(0, e_{D,Trgt})\right| = \left|\mathrm{sinc}(e_{D,Trgt}\, T)\right|} \qquad (13.2\text{-}20)$$

which is identical to the magnitude of the normalized, Doppler profile of a rectangular-envelope, CW pulse given by (13.1-32) (see Fig. 13.1-6).

Since $\text{sinc}(\pm 1) = 0$, the range profile given by (13.2-19) has its *first* zero crossings along the $e_{\tau,\,Trgt}$ axis when the magnitude of the argument of the sinc function is equal to 1, that is,

$$\left| \frac{1}{\pi} D_p e_{\tau,\,Trgt} \left[T - \left| e_{\tau,\,Trgt} \right| \right] \right| = 1, \qquad \left| e_{\tau,\,Trgt} \right| \le T\,, \tag{13.2-21}$$

or

$$\left| e_{\tau,\,Trgt} \right|^2 - T \left| e_{\tau,\,Trgt} \right| + \frac{\pi}{\left| D_p \right|} = 0\,. \tag{13.2-22}$$

Using the quadratic formula to solve (13.2-22) yields

$$\left| e_{\tau,\,Trgt} \right| = \frac{T}{2} \left[1 \pm \sqrt{1 - \frac{4\pi}{\left| D_p \right| T^2}} \right], \tag{13.2-23}$$

and by choosing the minus sign (since we want the locations of the *first* zero crossings),

$$\boxed{\left| e_{\tau,\,Trgt} \right| = \frac{T}{2} \left[1 - \sqrt{1 - \frac{4\pi}{\left| D_p \right| T^2}} \right]} \tag{13.2-24}$$

where the condition

$$\frac{4\pi}{\left| D_p \right| T^2} \le 1\,, \tag{13.2-25}$$

or

$$\boxed{T \times \text{BW}_{\text{swept}} \ge 4} \tag{13.2-26}$$

must be satisfied at all times in order to ensure that $\left| e_{\tau,\,Trgt} \right|$ is real and not complex, where the swept-bandwidth BW_{swept} is given by (13.2-13). However, if we impose the more stringent condition

$$\frac{4\pi}{\left| D_p \right| T^2} \le 0.1\,, \tag{13.2-27}$$

or

$$\boxed{T \times \mathrm{BW}_{\text{swept}} \geq 40} \tag{13.2-28}$$

then we can use the first two terms in a *binomial expansion* of the square root in (13.2-24). Doing so yields

$$\left|e_{\tau,\,Trgt}\right| \approx \frac{T}{2}\left[1-\left(1-\frac{1}{2}\frac{4\pi}{\left|D_p\right|T^2}\right)\right], \tag{13.2-29}$$

$$\left|e_{\tau,\,Trgt}\right| \approx \frac{\pi}{\left|D_p\right|T}, \tag{13.2-30}$$

or

$$\boxed{e_{\tau,\,Trgt} \approx \pm 1/\mathrm{BW}_{\text{swept}}} \tag{13.2-31}$$

Equation (13.2-31) gives the locations of the *first* zero crossings of the range profile when (13.2-28) is satisfied. As a result, the width of the mainlobe of the ambiguity function along the $e_{\tau,\,Trgt}$ axis is *inversely proportional to the swept-bandwidth* $\mathrm{BW}_{\text{swept}}$ when (13.2-28) is satisfied.

By inspecting the Doppler profile given by (13.2-20) and referring to Fig. 13.1-6, it can be seen that the width of the mainlobe of the ambiguity function along the $e_{D,\,Trgt}$ axis is *inversely proportional to the pulse length* T since the locations of the *first* zero crossings of the Doppler profile are at

$$\boxed{e_{D,\,Trgt} = \pm 1/T} \tag{13.2-32}$$

By inspecting (13.2-30) and (13.2-32), it can also be seen that the width of the mainlobe of the ambiguity function along *both* the $e_{\tau,\,Trgt}$ and $e_{D,\,Trgt}$ axes is *inversely proportional to the pulse length* T.

Now that we have two parameters that we can vary, namely, the phase-deviation constant D_p and the pulse length T, we can *simultaneously* control both the range and Doppler resolving capabilities of a rectangular-envelope, LFM pulse subject to the constraint given by (13.2-28).

Example 13.2-1 Design of a Rectangular-Envelope, LFM Pulse

As was mentioned in Example 13.1-1, the widths of the mainlobe of the ambiguity function along the $e_{\tau,\,Trgt}$ and $e_{D,\,Trgt}$ axes are used as measures of the range and Doppler resolutions, respectively, of a transmitted electrical signal. The locations of the first zero crossings of the round-trip time-delay (range) profile of a rectangular-envelope, LFM pulse are given by (13.2-31) provided that (13.2-28) is satisfied, and the locations of the first zero crossings of the Doppler profile are given by (13.2-32). Therefore, from (13.2-31),

$$\left|e_{\tau,Trgt}\right| \approx 1/\mathrm{BW}_{\mathrm{swept}}, \tag{13.2-33}$$

and from (13.2-32),

$$\left|e_{D,Trgt}\right| = 1/T. \tag{13.2-34}$$

Let us now design a rectangular-envelope, LFM pulse to satisfy a range-resolution requirement. In general, in signal-design problems, satisfying a range-resolution requirement is of primary concern.

Specifying a value for range resolution is equivalent to specifying a desired value for the magnitude of the range estimation error of the target $\left|e_{r,Trgt}\right|$ that corresponds to the location of the first zero crossing along the positive $e_{\tau,Trgt}$ axis. If (13.2-33) is substituted into (13.1-35), then

$$\boxed{\mathrm{BW}_{\mathrm{swept}} \approx \frac{c}{2\left|e_{r,Trgt}\right|}} \tag{13.2-35}$$

Equation (13.2-35) is used to compute the swept-bandwidth $\mathrm{BW}_{\mathrm{swept}}$ in hertz that is required for a rectangular-envelope, LFM pulse to provide a range resolution equal to $\left|e_{r,Trgt}\right|$ meters. The pulse length T in seconds must then satisfy the condition [see (13.2-28)]

$$\boxed{T \geq 40/\mathrm{BW}_{\mathrm{swept}}} \tag{13.2-36}$$

The resulting Doppler resolution is obtained by substituting the value of T from (13.2-36) into (13.2-34). Since we have a choice as to how big to make T, as T increases, Doppler resolution gets better, that is, $\left|e_{D,Trgt}\right|$ gets smaller, without changing the range resolution $\left|e_{r,Trgt}\right|$. Furthermore, as pulse length T increases, signal bandwidth BW_x decreases and signal energy E_x increases [see (13.1-12)], but the waveform becomes less Doppler-tolerant because the Doppler resolution gets better. Therefore, if the minimum value of pulse length is used in order to maintain some Doppler tolerance, signal energy can be increased by increasing the signal amplitude A, which increases the time-average power $P_{\mathrm{avg},x}$ [see (13.1-13)]. And by solving for $\left|D_p\right|$ from (13.2-13), we obtain

$$\left|D_p\right| = \pi\,\mathrm{BW}_{\text{swept}}/T\,, \tag{13.2-37}$$

or

$$\boxed{D_p = \pm\pi\,\mathrm{BW}_{\text{swept}}/T} \tag{13.2-38}$$

Equation (13.2-38) is used to compute the phase-deviation constant D_p in radians per second-squared that is required for a rectangular-envelope, LFM pulse to provide a range resolution equal to $\left|e_{r,Trgt}\right|$ meters, and a Doppler resolution equal to $\left|e_{D,Trgt}\right|$ hertz. For example, if $\left|e_{r,Trgt}\right| = 1\text{ m}$ ($e_{r,Trgt} = \pm 1\text{ m}$) and $c = 1500\text{ m/sec}$, then $\mathrm{BW}_{\text{swept}} = 750\text{ Hz}$ and $T \geq 53.333\text{ msec}$. Also, if $\left|e_{r,Trgt}\right| = 1\text{ m}$, then $\left|e_{\tau,Trgt}\right| = 1.333\text{ msec}$ [see (13.1-35)]. If we let $T = 54\text{ msec}$, then $\left|e_{D,Trgt}\right| = 18.519\text{ Hz}$ ($e_{D,Trgt} = \pm 18.519\text{ Hz}$) and $D_p = \pm 43{,}633.231\text{ rad/sec}^2$. The magnitude of the normalized, auto-ambiguity function $\left|X_N(e_{\tau,Trgt}, e_{D,Trgt})\right|$ of this signal is shown in Fig. 13.2-4. Figure 13.2-5 is a plot of the magnitude of the normalized, round-trip time-delay (range) profile of the auto-ambiguity function shown in Fig. 13.2-4. Since the numerator of the signal-to-interference ratio (SIR) is directly proportional to the magnitude-squared of the normalized, auto-ambiguity function (see Section 12.2), as $\left|X_N(e_{\tau,Trgt}, e_{D,Trgt})\right| \to 0$, the SIR $\to 0$ and, as a result, the target won't be detected.

Unlike the complex envelope of a rectangular-envelope, CW pulse; an exact, closed-form, analytical expression for the Fourier transform of the complex envelope of a rectangular-envelope, LFM pulse does not exist – the Fourier transform must be computed numerically. However, a conservative estimate of the bandwidth in hertz of the complex envelope of a rectangular-envelope, LFM pulse when (13.2-28) is satisfied is given by

$$\boxed{\mathrm{BW}_{\tilde{x}} \approx \mathrm{BW}_{\text{swept}} + \frac{5}{T}} \tag{13.2-39}$$

Substituting (13.2-39) into (13.1-42) and (13.1-43) yields

$$\boxed{f_c > \mathrm{BW}_{\text{swept}} + \frac{5}{T}} \tag{13.2-40}$$

and

$$\boxed{\mathrm{BW}_x \approx 2\,\mathrm{BW}_{\text{swept}} + \frac{10}{T}} \tag{13.2-41}$$

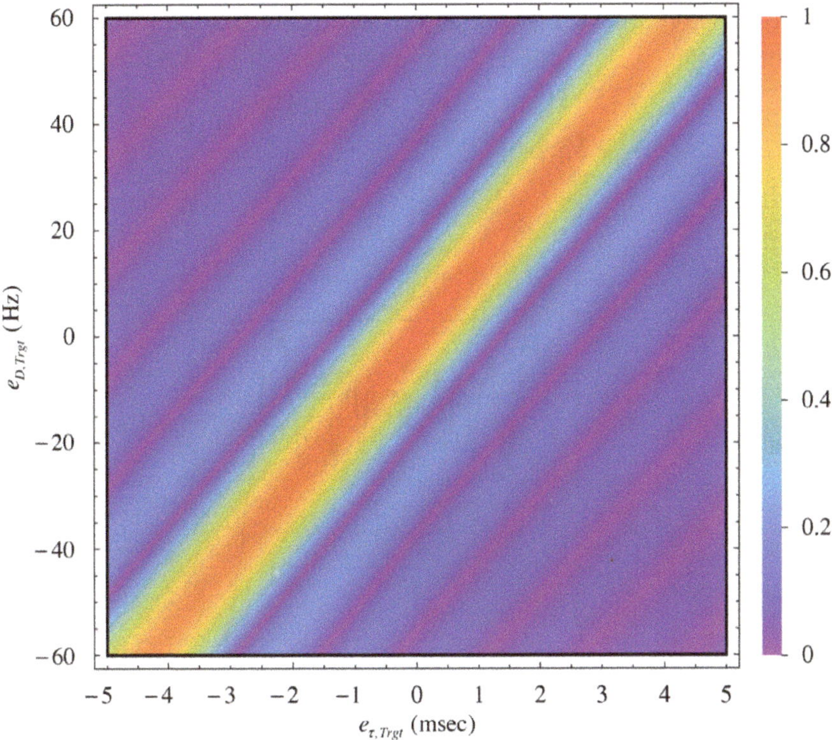

Figure 13.2-4 Magnitude of the normalized, auto-ambiguity function of a rectangular-envelope, up-chirp, LFM pulse with pulse length $T = 54$ msec and swept-bandwidth $\text{BW}_{\text{swept}} = 750$ Hz.

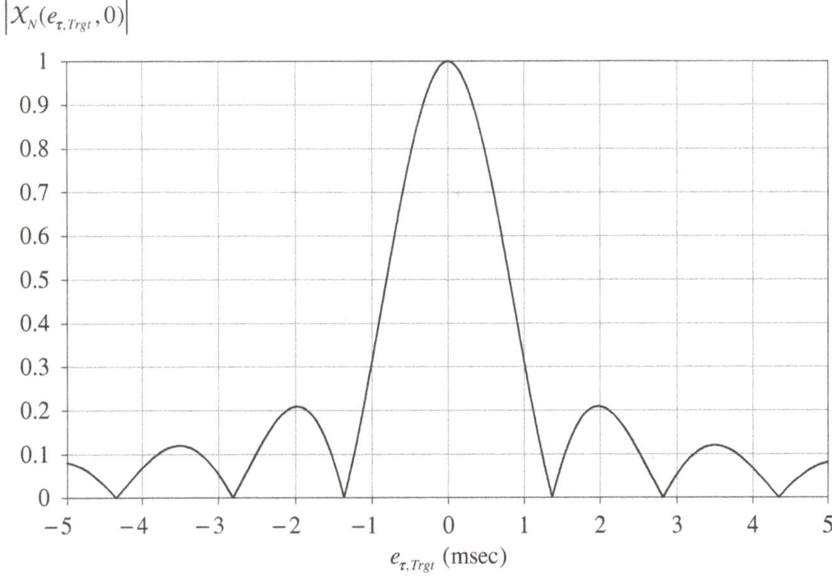

Figure 13.2-5 Magnitude of the normalized, round-trip time-delay (range) profile of the auto-ambiguity function shown in Fig. 13.2-4.

for a rectangular-envelope, LFM pulse. For example, for the previously designed rectangular-envelope, LFM pulse with pulse length $T = 54$ msec and swept-bandwidth $\text{BW}_{\text{swept}} = 750$ Hz; $\text{BW}_{\tilde{x}} \approx 842.6$ Hz, $f_c > 842.6$ Hz, and $\text{BW}_x \approx 1685.2$ Hz (see Fig. 13.2-6).

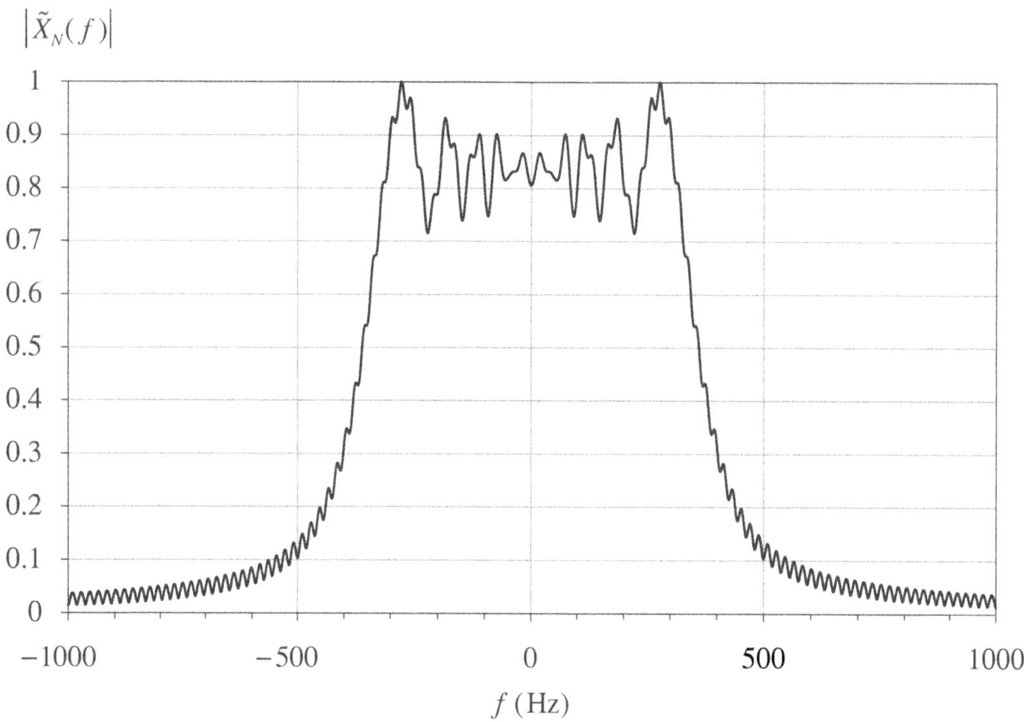

Figure 13.2-6 Normalized magnitude spectrum of the complex envelope of a rectangular-envelope, LFM pulse with pulse length $T = 54$ msec and swept-bandwidth $\text{BW}_{\text{swept}} = 750$ Hz.

Next we shall design a rectangular-envelope, LFM pulse to satisfy a Doppler-resolution requirement. Specifying a value for Doppler resolution is equivalent to specifying a desired value for the magnitude of the Doppler estimation error of the target $|e_{D,Trgt}|$ that corresponds to the location of the first zero crossing along the positive $e_{D,Trgt}$ axis. Solving for T from (13.2-34) yields

$$\boxed{T = 1/|e_{D,Trgt}|} \tag{13.2-42}$$

Equation (13.2-42) is used to compute the pulse length T in seconds that is required for a rectangular-envelope, LFM pulse to provide a Doppler resolution equal to $\left|e_{D,Trgt}\right|$ hertz. The swept-bandwidth BW_{swept} must then satisfy the condition [see (13.2-28)]

$$\boxed{\text{BW}_{\text{swept}} \geq 40/T} \tag{13.2-43}$$

And by solving for $\left|e_{r,Trgt}\right|$ from (13.2-35), we obtain

$$\left|e_{r,Trgt}\right| \approx \frac{c}{2\,\text{BW}_{\text{swept}}}. \tag{13.2-44}$$

The resulting range resolution is obtained by substituting the value of BW_{swept} from (13.2-43) into (13.2-44). Since we have a choice as to how big to make BW_{swept}, as BW_{swept} increases, range resolution gets better, that is, $\left|e_{r,Trgt}\right|$ gets smaller, without changing the Doppler resolution $\left|e_{D,Trgt}\right|$. However, as swept-bandwidth BW_{swept} increases, signal bandwidth BW_x increases [see (13.2-41)]. After T and BW_{swept} have been determined, the required value for the phase-deviation constant D_p in radians per second-squared is computed from (13.2-38). For example, if $\left|e_{D,Trgt}\right| = 1\text{ Hz}$ ($e_{D,Trgt} = \pm 1\text{ Hz}$) and $c = 1500\text{ m/sec}$, then $T = 1\text{ sec}$ and $\text{BW}_{\text{swept}} \geq 40\text{ Hz}$. If we let $\text{BW}_{\text{swept}} = 40\text{ Hz}$, then $\left|e_{r,Trgt}\right| \approx 18.75\text{ m}$ ($e_{r,Trgt} \approx \pm 18.75\text{ m}$), $D_p = \pm 125.664\text{ rad/sec}^2$, $\text{BW}_{\tilde{x}} \approx 45\text{ Hz}$, $f_c > 45\text{ Hz}$, and $\text{BW}_x \approx 90\text{ Hz}$ (see Fig. 13.2-7). Also, if $\left|e_{r,Trgt}\right| \approx 18.75\text{ m}$, then $\left|e_{\tau,Trgt}\right| \approx 25\text{ msec}$ [see (13.1-35)]. The magnitude of the normalized, auto-ambiguity function $\left|\chi_N(e_{\tau,Trgt}, e_{D,Trgt})\right|$ of this signal is shown in Fig. 13.2-8. Figure 13.2-9 is a plot of the magnitude of the normalized, round-trip time-delay (range) profile of the auto-ambiguity function shown in Fig. 13.2-8. As was previously mentioned, since the numerator of the SIR is directly proportional to the magnitude-squared of the normalized, auto-ambiguity function (see Section 12.2), as $\left|\chi_N(e_{\tau,Trgt}, e_{D,Trgt})\right| \to 0$, the SIR $\to 0$ and, as a result, the target won't be detected.

Finally, since the energy and time-average power of a rectangular-envelope, LFM and CW pulse are equal, (13.1-54) can be used to solve for the amplitude A of a rectangular-envelope, LFM pulse that will satisfy a time-average-power requirement. Tables 13.2-1 and 13.2-2 summarize the four different signal designs discussed in Examples 13.1-1 and 13.2-1. From Table 13.2-1 it can be seen that the rectangular-envelope, LFM pulse has more energy than the rectangular-envelope, CW pulse because for the same value of signal amplitude A, the *minimum* value of pulse length used for the rectangular-envelope, LFM pulse is $T = 54\text{ msec}$, which is much greater than the *required* value of $T = 1.333\text{ msec}$ for the rectangular-envelope, CW pulse. ■

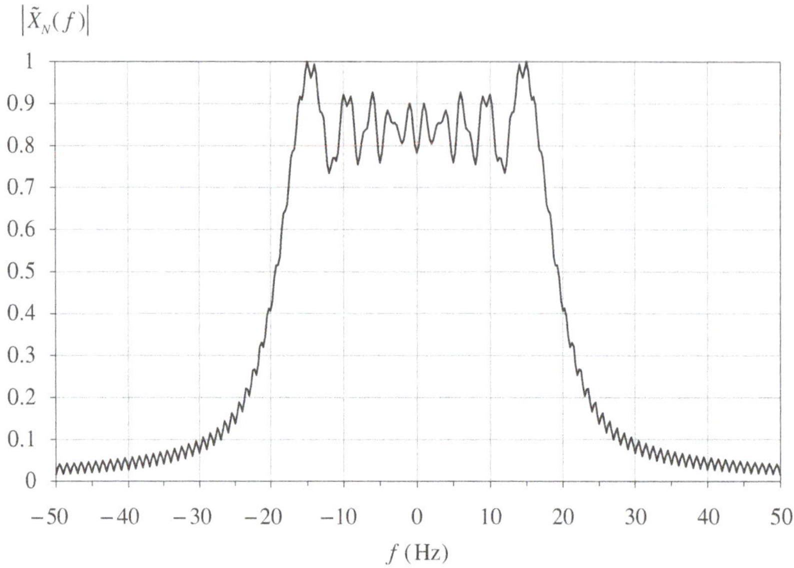

Figure 13.2-7 Normalized magnitude spectrum of the complex envelope of a rectangular-envelope, LFM pulse with pulse length $T = 1\,\text{sec}$ and swept-bandwidth $\text{BW}_{\text{swept}} = 40\text{ Hz}$.

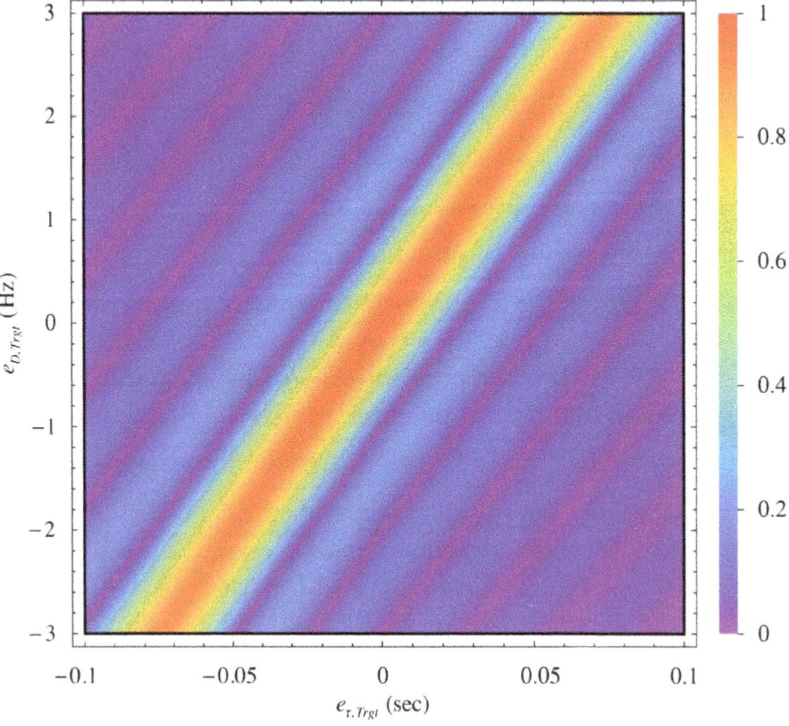

Figure 13.2-8 Magnitude of the normalized, auto-ambiguity function of a rectangular-envelope, up-chirp, LFM pulse with pulse length $T = 1\,\text{sec}$ and swept-bandwidth $\text{BW}_{\text{swept}} = 40\text{ Hz}$.

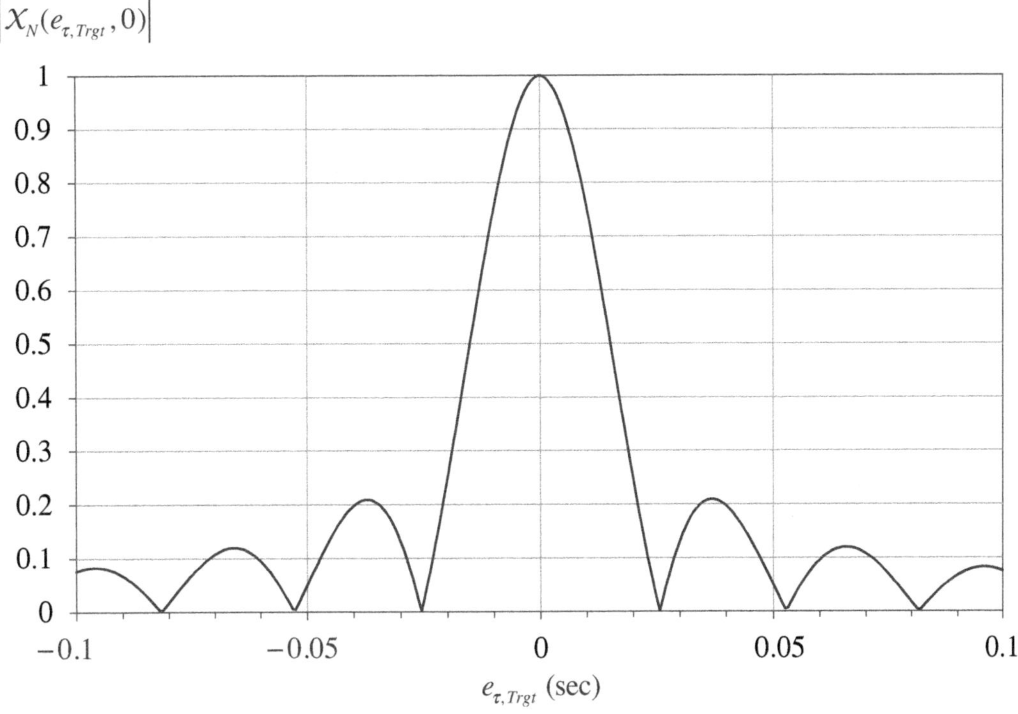

Figure 13.2-9 Magnitude of the normalized, round-trip time-delay (range) profile of the auto-ambiguity function shown in Fig. 13.2-8.

Table 13.2-1 Parameter Values for Rectangular-Envelope, CW and LFM Pulses Designed for a Range Resolution of 1 m

	T	BW_{swept}	BW_x	$\lvert e_{r,Trgt}\rvert$	$\lvert e_{D,Trgt}\rvert$
RE CW Pulse	1.333 msec	NA	7500 Hz	1 m	750 Hz
RE LFM Pulse	54 msec	750 Hz	1685.2 Hz	1 m	18.519 Hz

Table 13.2-2 Parameter Values for Rectangular-Envelope, CW and LFM Pulses Designed for a Doppler Resolution of 1 Hz

	T	BW_{swept}	BW_x	$\lvert e_{r,Trgt}\rvert$	$\lvert e_{D,Trgt}\rvert$
RE CW Pulse	1 sec	NA	10 Hz	750 m	1 Hz
RE LFM Pulse	1 sec	40 Hz	90 Hz	18.75 m	1 Hz

13.3 The Rectangular-Envelope HFM Pulse

In this section we shall discuss two different equations that are used to describe a *hyperbolic-frequency-modulated* (HFM) pulse. We shall show how both of these equations can be transformed into the form of an amplitude-and-angle-modulated carrier which makes it easy to identify the angle-modulating function and to compute the complex envelope. The second equation, which is more common in the recent literature, allows one to pick values for the beginning and ending frequencies of the HFM swept-bandwidth.

A HFM pulse is a very important signal because it is *Doppler-invariant* (see Subsection 13.3.3). As a result, it will be shown in Example 13.3-3 that a rectangular-envelope, HFM pulse is superior to a rectangular-envelope, linear-frequency-modulated (LFM) pulse in its ability to detect a target and estimate time delay in the presence of time-compression/time-expansion effects (see Section 10.7). In Subsection 13.3.4, we shall discuss how to design a HFM pulse to minimize time-delay estimation error.

13.3.1 First Equation Description

The first equation that we shall discuss to describe a HFM pulse takes on several different forms. We begin by using a modified version of the equation for this waveform given by Thor[1] in order to compare a HFM pulse with a LFM pulse as discussed in Section 13.2:

$$x(t)=a(t)\cos\left[\frac{(2\pi f_c)^2}{2D_p}\ln\left[1-\frac{2D_p}{2\pi f_c}(t-0.5T)\right]\right]\text{rect}\left[(t-0.5T)/T\right] \tag{13.3-1}$$

or, equivalently,

$$\boxed{x(t)=a(t)\cos\left[-\frac{(2\pi f_c)^2}{2D_p}\ln\left[1-\frac{2D_p}{2\pi f_c}(t-0.5T)\right]\right]\text{rect}\left[(t-0.5T)/T\right]} \tag{13.3-2}$$

where $a(t)$ is a real amplitude-modulating function (with units of volts or amperes), f_c is the carrier frequency in hertz, D_p is the phase-deviation constant with units of radians per second-squared, T is the pulse length in seconds, and

[1] R. C. Thor, "A Large Time-Bandwidth Product Pulse-Compression Technique," *IRE Transactions on Military Electronics*, vol. MIL-6, Issue 2, pp. 169-173, April 1962.

the time-shifted rectangle function is given by (13.1-2). If $a(t)$ is set equal to a *positive constant* A, that is, if $a(t)=A>0$, then a HFM pulse reduces to a *rectangular-envelope, HFM pulse.* The reason for introducing the minus sign in the argument of the cosine function (which is an even function) shall become apparent next.

The instantaneous phase (in radians) of $x(t)$ given by (13.3-2) is

$$\theta_i(t)=-\frac{(2\pi f_c)^2}{2D_p}\ln\left[1-\frac{2D_p}{2\pi f_c}(t-0.5T)\right], \tag{13.3-3}$$

and since

$$\ln(1-x)=-\left[x+\frac{x^2}{2}+\frac{x^3}{3}+\frac{x^4}{4}+\cdots\right], \qquad |x|<1, \tag{13.3-4}$$

$$\theta_i(t)=2\pi f_c(t-0.5T)+D_p(t-0.5T)^2+\frac{1}{3}\frac{(2D_p)^2}{2\pi f_c}(t-0.5T)^3+$$
$$\frac{1}{4}\frac{(2D_p)^3}{(2\pi f_c)^2}(t-0.5T)^4+\ldots, \qquad \frac{|D_p|/\pi}{f_c}|t-0.5T|<1. \tag{13.3-5}$$

A valid series expansion of $\theta_i(t)$ for any time t in the closed time interval $[0,T]$ can be obtained by setting $t=T$ in the inequality on the right-hand side of (13.3-5). Doing so yields

$$\frac{|D_p|T/\pi}{2f_c}<1, \tag{13.3-6}$$

or

$$\frac{|\mathrm{SBW}_{\mathrm{LFM}}|}{2f_c}<1 \tag{13.3-7}$$

where

$$\boxed{\mathrm{SBW}_{\mathrm{LFM}}=D_pT/\pi} \tag{13.3-8}$$

is the swept-bandwidth in hertz of a LFM pulse [compare with (13.2-13)]. Recall that since the phase-deviation constant D_p can be either positive or negative in value, $\mathrm{SBW}_{\mathrm{LFM}}$ given by (13.3-8) can be either positive or negative in value. Therefore,

$$\theta_i(t) = -\frac{(2\pi f_c)^2}{2D_p}\ln\left[1 - \frac{2D_p}{2\pi f_c}(t-0.5T)\right]$$
$$= -\pi f_c T + 2\pi f_c t + D_p(t-0.5T)^2 + \frac{1}{3}\frac{(2D_p)^2}{2\pi f_c}(t-0.5T)^3 +$$
$$\frac{1}{4}\frac{(2D_p)^3}{(2\pi f_c)^2}(t-0.5T)^4 + \ldots, \qquad \frac{|\text{SBW}_{\text{LFM}}|}{2f_c} < 1.$$
(13.3-9)

Note that the second and third terms on the right-hand side of (13.3-9) correspond to the instantaneous phase of a LFM pulse [see (13.2-10)].

In order to easily compute the complex envelope of a HFM pulse, we first need to express it in the standard form of an amplitude-and-angle-modulated carrier. Therefore,

$$\boxed{x(t) = a(t)\cos\left[2\pi f_c t + \theta(t)\right]\text{rect}\left[(t-0.5T)/T\right]} \qquad (13.3\text{-}10)$$

where, by referring to (13.3-9), the real angle-modulating function (a.k.a. the phase deviation) with units of radians is given by

$$\theta(t) = -\pi f_c T + 2\pi f_c \sum_{k=1}^{k_{\max}} \frac{1}{k+1}\left[\frac{D_p/\pi}{f_c}\right]^k (t-0.5T)^{k+1}, \qquad \frac{|\text{SBW}_{\text{LFM}}|}{2f_c} < 1,$$
(13.3-11)

or, with the use of (13.3-8),

$$\boxed{\theta(t) = -\pi f_c T + 2\pi f_c (t-0.5T)\sum_{k=1}^{k_{\max}} \frac{1}{k+1}\left[\frac{\text{SBW}_{\text{LFM}}}{f_c T}(t-0.5T)\right]^k, \qquad \frac{|\text{SBW}_{\text{LFM}}|}{2f_c} < 1}$$

(13.3-12)

The argument of the cosine function in (13.3-10) is an *estimate* of the instantaneous phase of a HFM pulse, that is,

$$\hat{\theta}_i(t) = 2\pi f_c t + \theta(t), \qquad (13.3\text{-}13)$$

where $\theta(t)$ is given by (13.3-12), because the upper limit of summation in (13.3-12) is $k_{\max}$, not ∞. If $a(t)$ is set equal to a positive constant A, then the HFM pulse given by (13.3-10) reduces to the rectangular-envelope, HFM pulse

$$x(t) = A\cos\left[2\pi f_c t + \theta(t)\right]\text{rect}\left[(t-0.5T)/T\right], \tag{13.3-14}$$

where $\theta(t)$ is given by (13.3-12). Note that if $k_{\max} = 1$, then (13.3-12) reduces to

$$\theta(t) = -\pi f_c T + D_p (t-0.5T)^2, \tag{13.3-15}$$

where the first term on the right-hand side of (13.3-15) is a constant phase and the second term is the angle-modulating function for a LFM pulse [see (13.2-2)]. We shall discuss how to determine a value for $k_{\max}$ later.

The complex envelopes of a HFM pulse and a rectangular-envelope, HFM pulse are

$$\tilde{x}(t) = a(t)\exp[+j\theta(t)]\text{rect}\left[(t-0.5T)/T\right] \tag{13.3-16}$$

and

$$\tilde{x}(t) = A\exp[+j\theta(t)]\text{rect}\left[(t-0.5T)/T\right], \tag{13.3-17}$$

respectively, where $\theta(t)$ is given by (13.3-12). The envelope, the energies of $\tilde{x}(t)$ and $x(t)$, and the time-average power of $x(t)$ for HFM and LFM pulses are identical to those for a continuous-wave (CW) pulse and are given by (13.1-4) through (13.1-7), respectively. The envelope, the energies of $\tilde{x}(t)$ and $x(t)$, and the time-average power of $x(t)$ for rectangular-envelope, HFM and LFM pulses are identical to those for a rectangular-envelope, CW pulse and are given by (13.1-10) through (13.1-13), respectively.

The instantaneous frequency (in hertz) of $x(t)$ given by (13.3-2) can be obtained by using (see Section 11.2)

$$f_i(t) = \frac{1}{2\pi}\frac{d}{dt}\theta_i(t). \tag{13.3-18}$$

Substituting (13.3-3) into (13.3-18) yields

$$f_i(t) = -\frac{1}{2\pi}\frac{(2\pi f_c)^2}{2D_p}\frac{d}{dt}\ln\left[1 - \frac{2D_p}{2\pi f_c}(t-0.5T)\right], \tag{13.3-19}$$

and since

$$\frac{d}{dt}\ln u = \frac{1}{u}\frac{du}{dt}, \tag{13.3-20}$$

$$f_i(t) = \frac{f_c}{1 - \dfrac{D_p/\pi}{f_c}(t - 0.5T)} = \frac{f_c}{1 - \dfrac{\text{SBW}_{\text{LFM}}}{f_c T}(t - 0.5T)} \tag{13.3-21}$$

Equation (13.3-21) is one form of an equation for a *hyperbola* – hence the terminology hyperbolic-frequency-modulation. Hyperbolic-frequency-modulation is an example of nonlinear frequency modulation. As can be seen from (13.3-21), if $\text{SBW}_{\text{LFM}} > 0$, then $f_i(t)$ increases with time; and if $\text{SBW}_{\text{LFM}} < 0$, then $f_i(t)$ decreases with time. Note that $f_i(0.5T) = f_c$, analogous to a LFM pulse (see Fig. 13.2-3).

The *instantaneous period* (in seconds) is defined as the reciprocal of the instantaneous frequency:

$$p_i(t) \triangleq \frac{1}{f_i(t)} = \frac{1}{f_c}\left[1 - \frac{\text{SBW}_{\text{LFM}}}{f_c T}(t - 0.5T)\right] \tag{13.3-22}$$

Since the instantaneous period is a linear function of time, HFM is also referred to as *linear-period-modulation* (LPM). If $\text{SBW}_{\text{LFM}} > 0$, then $p_i(t)$ decreases with time; and if $\text{SBW}_{\text{LFM}} < 0$, then $p_i(t)$ increases with time.

The *swept-bandwidth* in hertz of the HFM pulse given by (13.3-2) can be computed using (13.3-21) as follows:

$$\text{SBW}_{\text{HFM}} = f_i(T) - f_i(0) = \frac{f_c}{1 - \dfrac{\text{SBW}_{\text{LFM}}}{2f_c}} - \frac{f_c}{1 + \dfrac{\text{SBW}_{\text{LFM}}}{2f_c}}, \tag{13.3-23}$$

which simplifies to

$$\text{SBW}_{\text{HFM}} = \frac{\text{SBW}_{\text{LFM}}}{1 - [\text{SBW}_{\text{LFM}}/(2f_c)]^2}, \qquad \frac{|\text{SBW}_{\text{LFM}}|}{2f_c} < 1 \tag{13.3-24}$$

or

$$\frac{\text{SBW}_{\text{HFM}}}{f_c} = \frac{\text{SBW}_{\text{LFM}}/f_c}{1 - [\text{SBW}_{\text{LFM}}/(2f_c)]^2}, \qquad \frac{|\text{SBW}_{\text{LFM}}|}{2f_c} < 1 \tag{13.3-25}$$

If $\text{SBW}_{\text{LFM}} > 0$ (up-sweep), then $\text{SBW}_{\text{HFM}} > 0$ (up-sweep); and if $\text{SBW}_{\text{LFM}} < 0$ (down-sweep), then $\text{SBW}_{\text{HFM}} < 0$ (down-sweep).

If we let

$$z = \mathrm{SBW}_{\mathrm{LFM}}/(2f_c), \tag{13.3-26}$$

then (13.3-24) can be rewritten as

$$\mathrm{SBW}_{\mathrm{HFM}}\, z^2 + 2f_c z - \mathrm{SBW}_{\mathrm{HFM}} = 0. \tag{13.3-27}$$

Using the quadratic formula to solve (13.3-27) yields

$$z = \frac{-1 \pm \sqrt{1 + (\mathrm{SBW}_{\mathrm{HFM}}/f_c)^2}}{\mathrm{SBW}_{\mathrm{HFM}}/f_c}, \tag{13.3-28}$$

and since it is required that [see (13.3-26) and (13.3-7)]

$$|z| = |\mathrm{SBW}_{\mathrm{LFM}}|/(2f_c) < 1, \tag{13.3-29}$$

choosing the plus sign in (13.3-28) and substituting (13.3-26) yields

$$\boxed{\mathrm{SBW}_{\mathrm{LFM}} = 2f_c \frac{\left[-1 + \sqrt{1 + (\mathrm{SBW}_{\mathrm{HFM}}/f_c)^2}\right]}{\mathrm{SBW}_{\mathrm{HFM}}/f_c}} \tag{13.3-30}$$

For completeness, note that by using (13.3-8), (13.3-1) can be rewritten as (see Skolnik[2])

$$x(t) = a(t)\cos\left[\frac{2\pi f_c}{\mathrm{SBW}_{\mathrm{LFM}}/(f_c T)} \ln\left[1 - \frac{\mathrm{SBW}_{\mathrm{LFM}}}{f_c T}(t - 0.5T)\right]\right] \mathrm{rect}[(t - 0.5T)/T] \tag{13.3-31}$$

or, equivalently, as our *first equation* description of a HFM pulse given by

$$\boxed{x(t) = a(t)\cos\left[-\frac{2\pi f_c}{\mathrm{SBW}_{\mathrm{LFM}}/(f_c T)} \ln\left[1 - \frac{\mathrm{SBW}_{\mathrm{LFM}}}{f_c T}(t - 0.5T)\right]\right] \mathrm{rect}[(t - 0.5T)/T]} \tag{13.3-32}$$

---

[2] M. I. Skolnik, *Introduction to Radar Systems*, 3rd ed., McGraw-Hill, New York, 2001, pp. 359-360.

Equation (13.3-32) can also be rewritten in the standard form of an amplitude-and-angle-modulated carrier as given by (13.3-10) where the angle-modulating function $\theta(t)$ is given by (13.3-12).

The instantaneous frequency (in hertz) of $x(t)$ given by (13.3-10) and (13.3-12) can be obtained by using (see Section 11.2)

$$\hat{f}_i(t) = f_c + \frac{1}{2\pi}\frac{d}{dt}\theta(t). \tag{13.3-33}$$

The instantaneous frequency $\hat{f}_i(t)$ is designated as being an *estimate* because $\theta(t)$ given by (13.3-12) depends on the value used for $k_{\max}$. Substituting (13.3-11) instead of (13.3-12) into (13.3-33), and using (13.3-8) yields

$$\boxed{\hat{f}_i(t) = f_c\left[1 + \sum_{k=1}^{k_{\max}}\left[\frac{\text{SBW}_{\text{LFM}}}{f_c T}(t - 0.5T)\right]^k\right]} \tag{13.3-34}$$

Estimates of the swept-bandwidth in hertz and the instantaneous period in seconds of a HFM pulse are given by

$$\text{SBW}_{\text{HFM}} = \hat{f}_i(T) - \hat{f}_i(0) \tag{13.3-35}$$

and

$$\hat{p}_i(t) = 1/\hat{f}_i(t), \tag{13.3-36}$$

respectively. The value of $k_{\max}$ is determined when the estimate of the instantaneous frequency $\hat{f}_i(t)$ given by (13.3-34) matches the exact value of the instantaneous frequency $f_i(t)$ given by (13.3-21) to a desired accuracy.

Example 13.3-1 Design of a Rectangular-Envelope, HFM Pulse

As was mentioned in Examples 13.1-1 and 13.2-1, satisfying a range-resolution requirement is, in general, of primary concern in signal-design problems. Exact, closed-form, analytical solutions exist for the magnitudes of the normalized, round-trip time-delay (range) profiles for rectangular-envelope, CW and LFM pulses [see (13.1-31) and (13.2-19), respectively]. Formulas derived from these two profiles are used to design these two waveforms to satisfy range-resolution requirements as shown in Examples 13.1-1 and 13.2-1. Such is not the case for a rectangular-envelope, HFM pulse.

One can design the rectangular-envelope, HFM pulse given by (13.3-14) and (13.3-12) by designing a rectangular-envelope, LFM pulse to satisfy a range-resolution requirement because $\theta(t)$ given by (13.3-12) depends on SBW_{LFM}. For example, using the same range-resolution requirement as in Example 13.2-1, if $|e_{r,Trgt}| = 1\text{ m}$, and $c = 1500\text{ m/sec}$, then $\text{SBW}_{\text{LFM}} = 750\text{ Hz}$, $T \geq 53.333\text{ msec}$, and $|e_{\tau,Trgt}| = 1.333\text{ msec}$. For example purposes, let the pulse length $T = 54\text{ msec}$ as in Example 13.2-1.

We next have to choose a value for the carrier frequency f_c. If we let

$$x = \text{SBW}_{\text{LFM}}/(2f_c) \tag{13.3-37}$$

and

$$y = \text{SBW}_{\text{HFM}}/f_c, \tag{13.3-38}$$

then (13.3-25) can be rewritten as

$$y = \frac{2x}{1-x^2}, \qquad |x| < 1, \tag{13.3-39}$$

where

$$|x| = \frac{|\text{SBW}_{\text{LFM}}|}{2f_c} < 1 \tag{13.3-40}$$

must be satisfied in order for a valid series expansion of $\theta(t)$ given by (13.3-12). From (13.3-39), if $0 < x < 1$ (LFM up-sweep), then $y > 0$ (HFM up-sweep); and if $-1 < x < 0$ (LFM down-sweep), then $y < 0$ (HFM down-sweep).

For the case $y = 1$,

$$\frac{2x}{1-x^2} = 1, \qquad 0 < x < 1. \tag{13.3-41}$$

Solving for x yields

$$x = \sqrt{2} - 1 \approx 0.414. \tag{13.3-42}$$

Therefore, in order for

$$|y| = \frac{|\text{SBW}_{\text{HFM}}|}{f_c} < 1, \tag{13.3-43}$$

it is required that

$$|x| = \frac{|\text{SBW}_{\text{LFM}}|}{2f_c} < 0.414 \tag{13.3-44}$$

or

$$\boxed{f_c > 1.208|\text{SBW}_{\text{LFM}}|} \tag{13.3-45}$$

Satisfying (13.3-45), which is equivalent to satisfying (13.3-44), and hence, (13.3-40), guarantees that the angle-modulating function $\theta(t)$ given by (13.3-12) is valid.

Substituting $\text{SBW}_{\text{LFM}} = 750 \text{ Hz}$ into (13.3-45) yields $f_c > 906 \text{ Hz}$. For example purposes, let $f_c = 1000 \text{ Hz}$. The swept-bandwidth for the rectangular-envelope, HFM pulse can now be obtained by substituting $\text{SBW}_{\text{LFM}} = 750 \text{ Hz}$ and $f_c = 1000 \text{ Hz}$ into (13.3-24). Doing so yields $\text{SBW}_{\text{HFM}} \approx 872.7 \text{ Hz}$. As a result, $\text{SBW}_{\text{HFM}} / f_c \approx 0.873$, and with pulse length $T = 54 \text{ msec}$, the time-bandwidth product is $T \times \text{SBW}_{\text{HFM}} \approx 47.1$.

Figures 13.3-1 and 13.3-2 are plots of the exact instantaneous frequency and period of the rectangular-envelope, HFM pulse obtained by substituting $\text{SBW}_{\text{LFM}} = 750 \text{ Hz}$, $T = 54 \text{ msec}$, and $f_c = 1000 \text{ Hz}$ into (13.3-21) and (13.3-22), respectively. Using $k_{\max} = 18$, estimates of the instantaneous frequency and period from (13.3-34) and (13.3-36), respectively, match the exact values for the instantaneous frequency and period from (13.3-21) and (13.3-22), respectively, to at least three and six decimal places, respectively. The HFM swept-bandwidth values computed from (13.3-24) and (13.3-35) are both approximately equal to 872.7 Hz. Figure 13.3-3 is a plot of $x(t)$ using (13.3-14) and (13.3-12) with $A = 10 \text{ V}$ and the aforementioned parameter values.

Similar to a rectangular-envelope, LFM pulse; exact, closed-form, analytical expressions for the Fourier transforms of a rectangular-envelope, HFM pulse and its complex envelope do not exist – the Fourier transforms must be computed numerically. Figure 13.3-4 is a plot of the normalized magnitude spectrum of $x(t)$ using (13.3-14) and (13.3-12), which (by looking at the data) matches exactly the normalized magnitude spectrum of $x(t)$ using (13.3-32) with $a(t) = 10 \text{ V}$. Figure 13.3-5 is a plot of the normalized magnitude spectrum of $\tilde{x}(t)$ using (13.3-17) and (13.3-12). Note that the magnitude spectra are *asymmetric* about $f_c = 1000 \text{ Hz}$ and 0 Hz, respectively.

Figure 13.3-6 is a plot of the magnitudes of the normalized, round-trip time-delay (range) profiles of a rectangular-envelope, LFM pulse with swept-bandwidth $\text{SBW}_{\text{LFM}} = 750 \text{ Hz}$ and pulse length $T = 54 \text{ msec}$ (dashed curve) and a rectangular-envelope, HFM pulse with swept-bandwidth $\text{SBW}_{\text{HFM}} \approx 872.7 \text{ Hz}$ and pulse length $T = 54 \text{ msec}$ (solid curve), using (13.2-19) and (13.1-17), respectively. As can be seen from Fig. 13.3-6, the rectangular-envelope, HFM pulse has a slightly better range resolution as measured at the -3-dB down point (magnitude of $\sqrt{2}/2 \approx 0.707$). Otherwise, its round-trip time-delay (range) profile is not very impressive compared to that of the rectangular-envelope, LFM pulse.

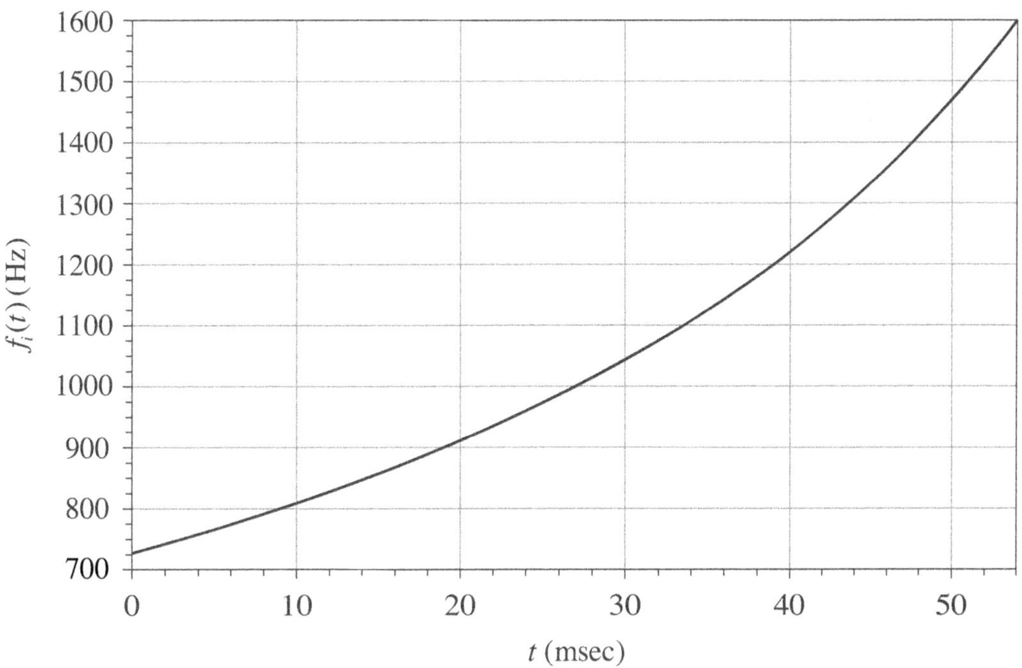

Figure 13.3-1 Instantaneous frequency $f_i(t)$ of a HFM pulse with $T = 54$ msec, $f_c = 1000$ Hz, and $\text{SBW}_{\text{HFM}} \approx 872.7$ Hz.

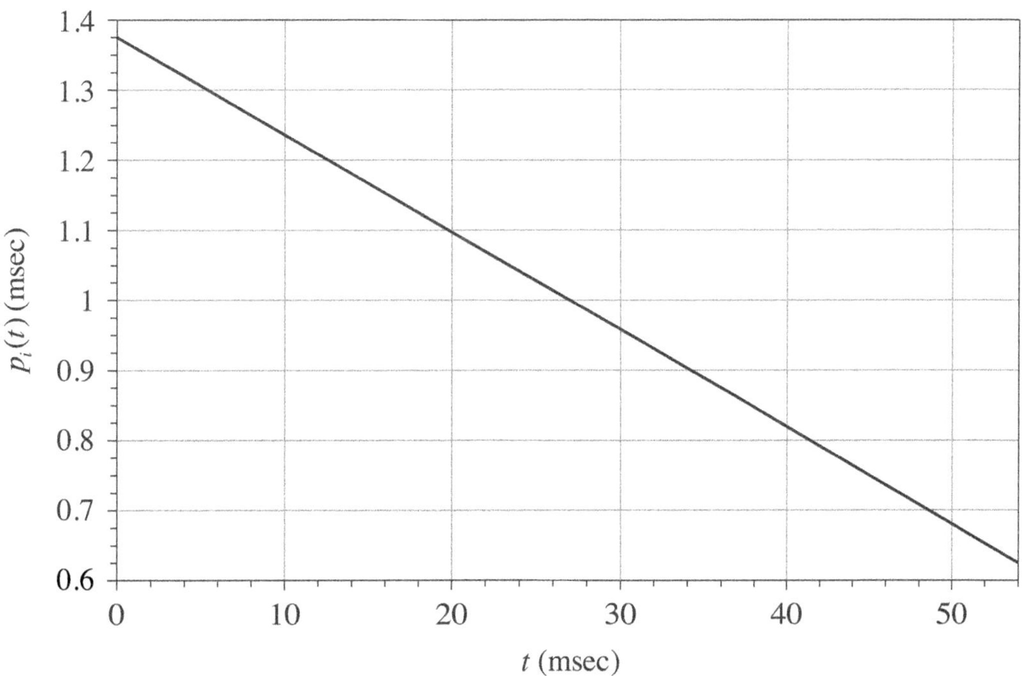

Figure 13.3-2 Instantaneous period $p_i(t)$ of a HFM pulse with $T = 54$ msec, $f_c = 1000$ Hz, and $\text{SBW}_{\text{HFM}} \approx 872.7$ Hz.

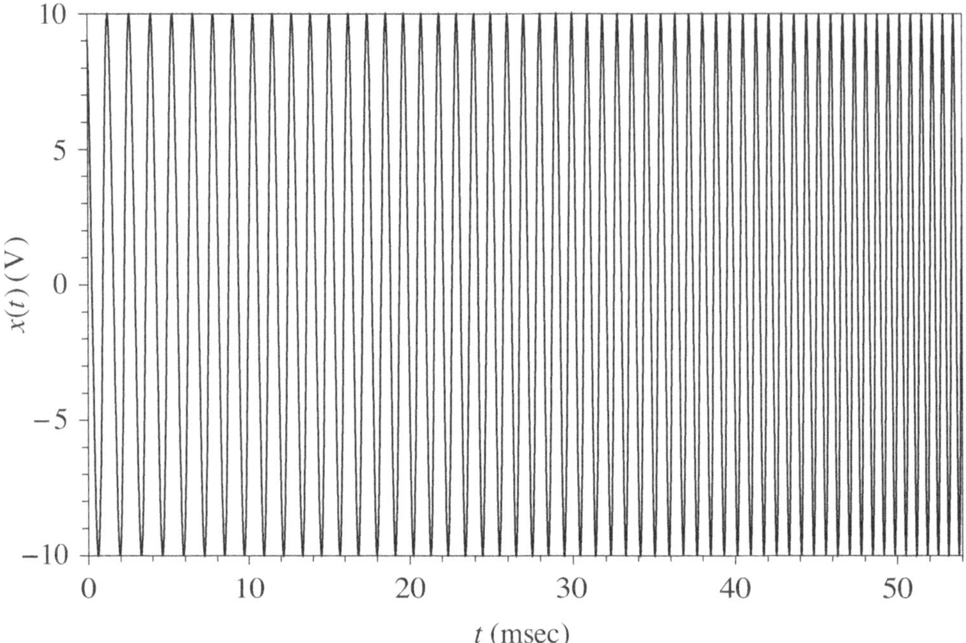

Figure 13.3-3 A rectangular-envelope, HFM pulse with $T = 54$ msec, $f_c = 1000$ Hz, and $\text{SBW}_{\text{HFM}} \approx 872.7$ Hz.

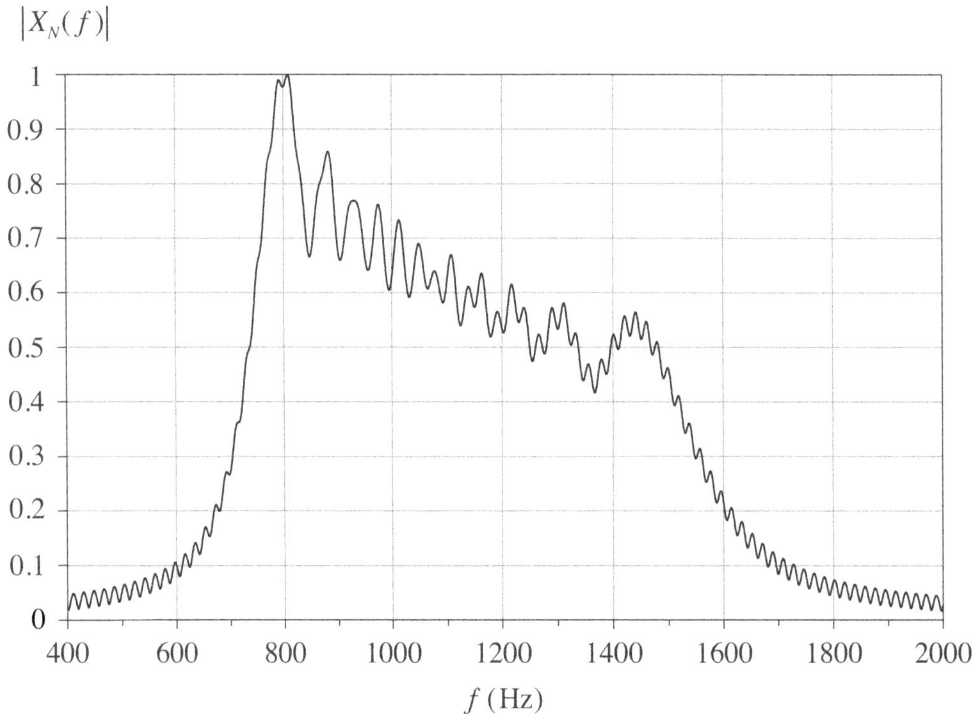

Figure 13.3-4 Normalized magnitude spectrum of a rectangular-envelope, HFM pulse with $T = 54$ msec, $f_c = 1000$ Hz, and $\text{SBW}_{\text{HFM}} \approx 872.7$ Hz.

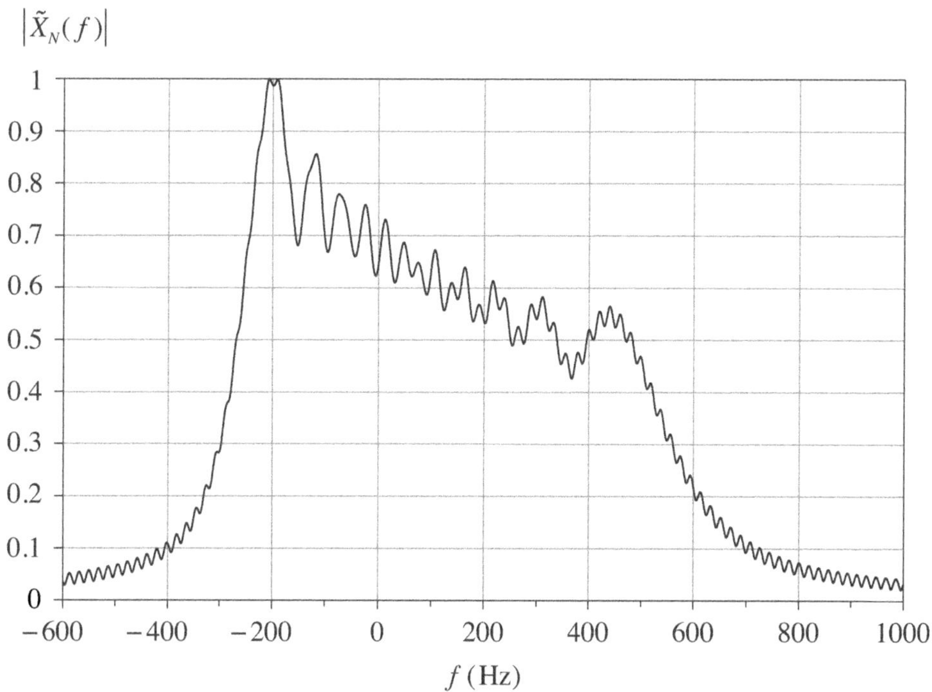

Figure 13.3-5 Normalized magnitude spectrum of the complex envelope of a rectangular-envelope, HFM pulse with $T = 54$ msec and $\text{SBW}_{\text{HFM}} \approx 872.7$ Hz.

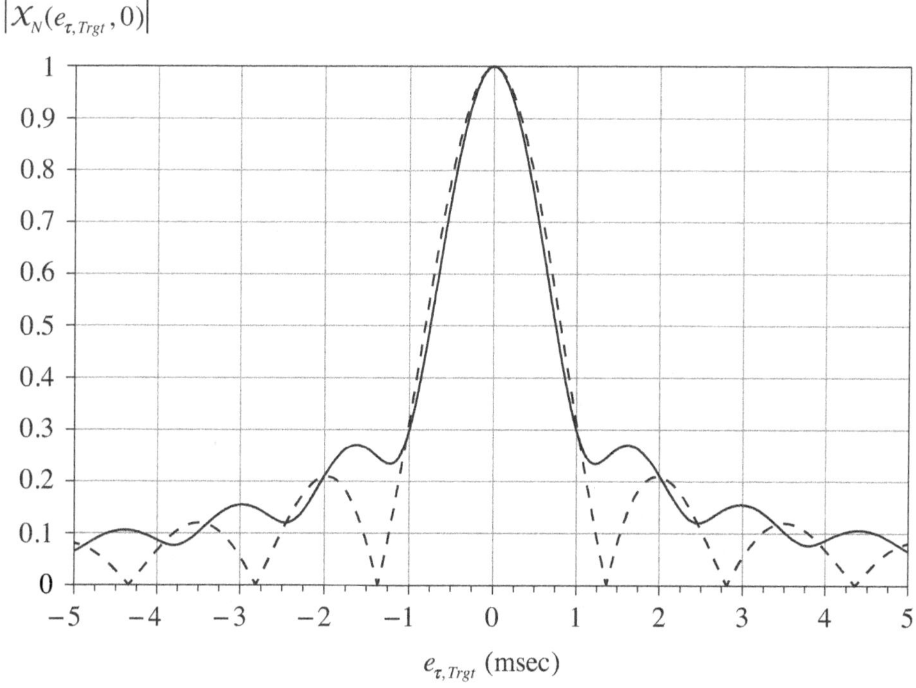

Figure 13.3-6 Magnitudes of the normalized, round-trip time-delay (range) profiles of rectangular-envelope, LFM and HFM pulses (dashed and solid curves, respectively) with $\text{SBW}_{\text{LFM}} = 750$ Hz, $\text{SBW}_{\text{HFM}} \approx 872.7$ Hz, and $T = 54$ msec.

However, as will be shown in Example 13.3-3, a rectangular-envelope, HFM pulse is superior to a rectangular-envelope, LFM pulse in its ability to detect a target and estimate time delay in the presence of time-compression/time-expansion effects.

The Doppler profile of a rectangular-envelope, HFM pulse is the same as that for rectangular-envelope, CW and LFM pulses and is given by either (13.1-32) or (13.2-20). Therefore, the Doppler resolution of the rectangular-envelope, HFM pulse with pulse length $T = 54$ msec is $|e_{D,Trgt}| \approx 18.5$ Hz [see (13.2-34)], which is the approximate location of the first zero crossing along the positive $e_{D,Trgt}$ axis. ■

13.3.2 Second Equation Description

The instantaneous frequency (in hertz) of a HFM pulse was shown by Williams and Battestin[3] to be

$$f_i(t) = \frac{2f_1 f_2}{f_1 + f_2 - \frac{2\,\text{SBW}_{\text{HFM}}}{T}(t - 0.5T)}, \qquad 0 \le t \le T, \tag{13.3-46}$$

or, upon simplifying,

$$\boxed{f_i(t) = \frac{f_1}{1 - \frac{\text{SBW}_{\text{HFM}}}{f_2 T}t}, \qquad 0 \le t \le T} \tag{13.3-47}$$

where f_1 and f_2 are the beginning and ending frequencies of the frequency-sweep, that is, $f_i(0) = f_1$ and $f_i(T) = f_2$,

$$\boxed{\text{SBW}_{\text{HFM}} = f_2 - f_1} \tag{13.3-48}$$

is the swept-bandwidth, and T is the pulse length in seconds. Note that if $f_2 > f_1$, then $\text{SBW}_{\text{HFM}} > 0$ (up-sweep); and if $f_2 < f_1$, then $\text{SBW}_{\text{HFM}} < 0$ (down-sweep).

Solving for the instantaneous phase $\theta_i(t)$ (in radians) from (13.3-18) yields

---

[3] R. E. Williams and H. F. Battestin, "Time coherence of acoustic signals transmitted over resolved paths in the deep ocean," *J. Acoust. Soc. Am.*, vol. 59, no. 2, pp. 312-328, February 1976.

$$\theta_i(t) = 2\pi \int f_i(t)\,dt\,, \tag{13.3-49}$$

and by substituting (13.3-47) into (13.3-49) and ignoring the constant of integration, we obtain

$$\theta_i(t) = -2\pi \frac{f_1 f_2}{\text{SBW}_{\text{HFM}}} T \ln\left(f_2 - \frac{\text{SBW}_{\text{HFM}}}{T} t \right). \tag{13.3-50}$$

Rewriting (13.3-50) as

$$\theta_i(t) = -2\pi \frac{f_1 f_2}{\text{SBW}_{\text{HFM}}} T \ln f_2 - 2\pi \frac{f_1 f_2}{\text{SBW}_{\text{HFM}}} T \ln\left(1 - \frac{\text{SBW}_{\text{HFM}}}{f_2 T} t \right) \tag{13.3-51}$$

and ignoring the constant phase term finally yields

$$\boxed{\theta_i(t) = -2\pi \frac{f_1 f_2}{\text{SBW}_{\text{HFM}}} T \ln\left(1 - \frac{\text{SBW}_{\text{HFM}}}{f_2 T} t \right)} \tag{13.3-52}$$

where SBW_{HFM} is given by (13.3-48). Substituting either (13.3-50) or (13.3-52) into (13.3-18) yields (13.3-47). Therefore, with the use of (13.3-52), our *second equation* description of a HFM pulse can be expressed as

$$\boxed{x(t) = a(t)\cos\left[-2\pi \frac{f_1 f_2}{\text{SBW}_{\text{HFM}}} T \ln\left(1 - \frac{\text{SBW}_{\text{HFM}}}{f_2 T} t \right) \right] \text{rect}\left[(t - 0.5T)/T \right]} \tag{13.3-53}$$

Compare (13.3-53) with (13.3-2) and (13.3-32). Note that a carrier frequency f_c does not appear in the argument of the cosine function in (13.3-53). When using (13.3-53), one can place the frequency spectrum of $x(t)$ along the frequency axis by specifying values for f_1 and f_2, which results in the swept-bandwidth given by (13.3-48).

Although a carrier frequency f_c does not appear in (13.3-53), it can be computed. Recall that from (13.3-21), $f_i(0.5T) = f_c$, analogous to a LFM pulse (see Fig. 13.2-3). Therefore, evaluating either (13.3-46) or (13.3-47) at $t = 0.5T$ yields

$$\boxed{f_c = 2\frac{f_1 f_2}{f_1 + f_2}} \tag{13.3-54}$$

as the carrier frequency associated with the HFM pulse $x(t)$ given by (13.3-53). We shall discuss how to transform (13.3-53) into the form of an amplitude-and-angle-modulated carrier in Example 13.3-2.

Example 13.3-2 Alternate Design of a Rectangular-Envelope, HFM Pulse

Specifying values for f_1 and f_2 determines the swept-bandwidth of a HFM pulse SBW_{HFM} given by (13.3-48) and the carrier frequency f_c given by (13.3-54). Substituting SBW_{HFM} and f_c into (13.3-30) yields the swept-bandwidth of a LFM pulse SBW_{LFM}. The angle-modulating function $\theta(t)$ given by (13.3-12) can then be evaluated using f_c and SBW_{LFM}, and by specifying a value for the pulse length T and k_{max}. Therefore, the HFM pulse given by (13.3-53) can be expressed as an amplitude-and-angle-modulated carrier as given by (13.3-10), with angle-modulating function given by (13.3-12), which makes it easy to compute the complex envelope given by (13.3-16). For a rectangular-envelope, HFM pulse, (13.3-14), (13.3-12), and (13.3-17) are used. We shall demonstrate this procedure next for a rectangular-envelope, HFM pulse.

For example, let $A = 10\text{ V}$, $f_1 = 9\text{ kHz}$, $f_2 = 14\text{ kHz}$, and $T = 50\text{ msec}$. Therefore, $\text{SBW}_{\text{HFM}} = 5\text{ kHz}$, $f_c \approx 10{,}956.5\text{ Hz}$, and $\text{SBW}_{\text{LFM}} \approx 4763.7\text{ Hz}$. As a result, $\text{SBW}_{\text{HFM}}/f_c \approx 0.456$ and the time-bandwidth product $T \times \text{SBW}_{\text{HFM}} = 250$. The instantaneous frequency values from (13.3-47) match the instantaneous frequency values from (13.3-21), and (13.3-34) using $k_{\text{max}} = 18$, to at least three decimal places, where SBW_{HFM}, f_c, and SBW_{LFM} are computed using (13.3-48), (13.3-54), and (13.3-30), respectively. The corresponding instantaneous period values match to at least six decimal places.

Figure 13.3-7 is a plot of the normalized magnitude spectrum of $x(t)$ using (13.3-53) with $a(t) = 10\text{ V}$. Figure 13.3-8 is a plot of the normalized magnitude spectrum of $x(t)$ using (13.3-14) and (13.3-12), which (by looking at the data) closely matches the normalized magnitude spectrum shown in Fig. 13.3-7. Note that both magnitude spectra are *asymmetric* about f_c. Figure 13.3-9 is a plot of the normalized magnitude spectrum of $\tilde{x}(t)$ using (13.3-17) and (13.3-12). Note that the magnitude spectrum is *asymmetric* about 0 Hz.

Figure 13.3-10 is a plot of the magnitudes of the normalized, round-trip time-delay (range) profiles of a rectangular-envelope, LFM pulse with swept-bandwidth $\text{SBW}_{\text{LFM}} \approx 4763.7\text{ Hz}$ and pulse length $T = 50\text{ msec}$ (dashed curve),

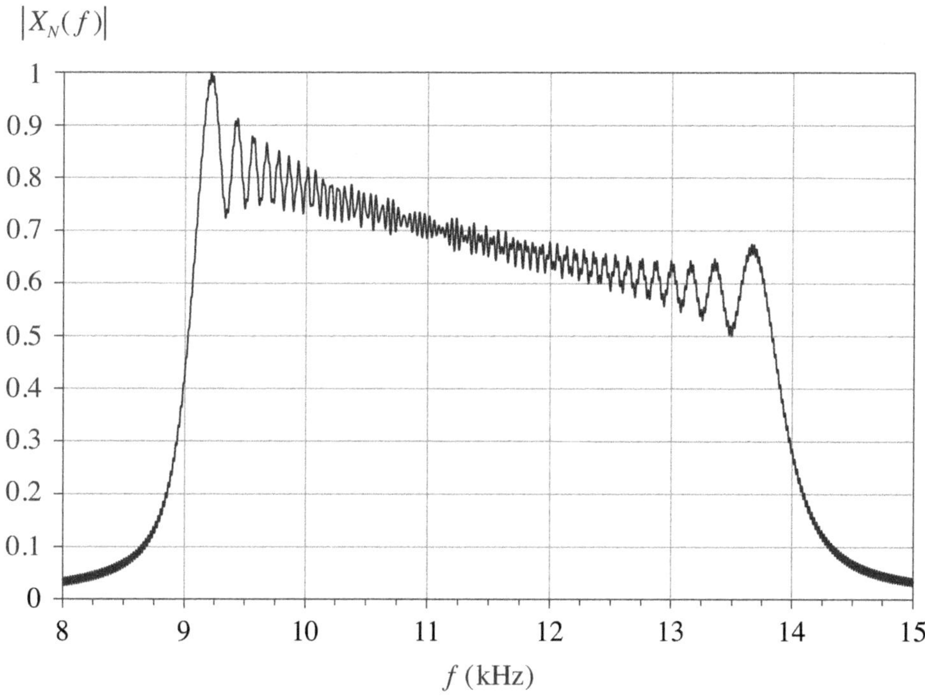

Figure 13.3-7 Normalized magnitude spectrum of a rectangular-envelope, HFM pulse with $T = 50$ msec and $\text{SBW}_{\text{HFM}} = 5$ kHz using (13.3-53).

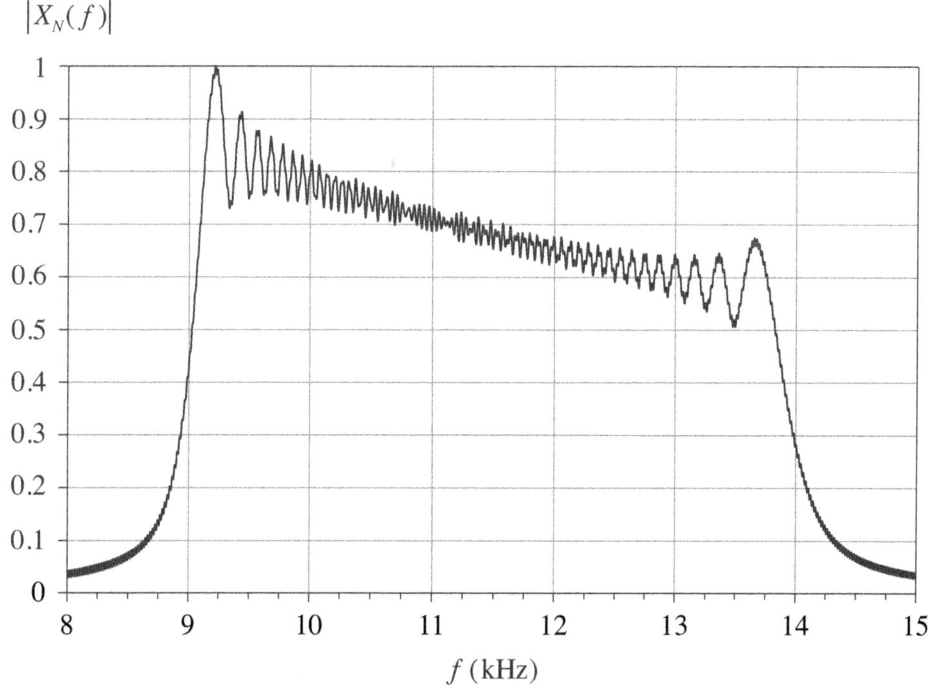

Figure 13.3-8 Normalized magnitude spectrum of a rectangular-envelope, HFM pulse with $T = 50$ msec and $\text{SBW}_{\text{HFM}} = 5$ kHz using (13.3-14) and (13.3-12).

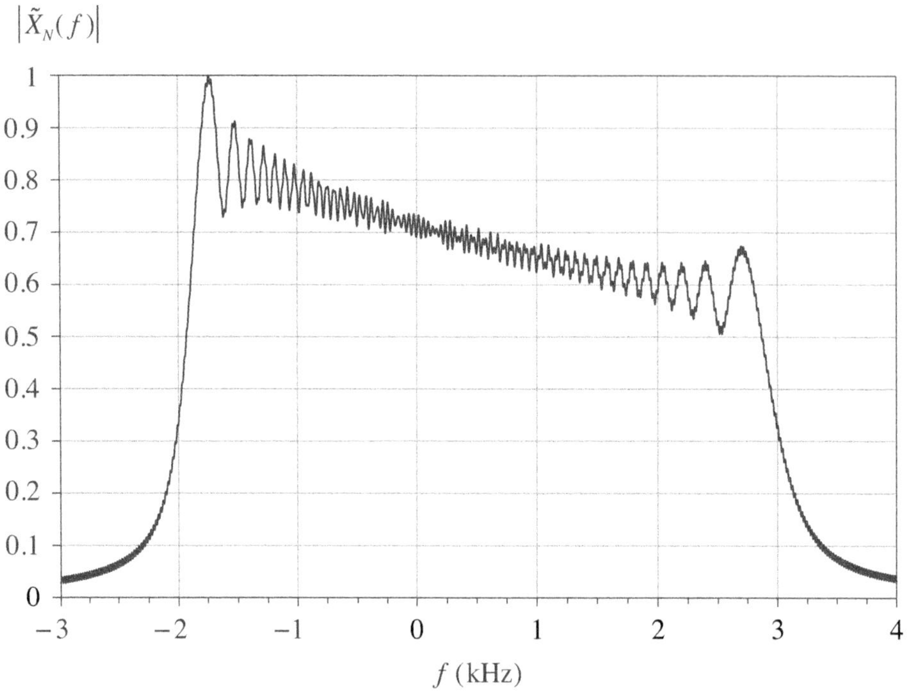

Figure 13.3-9 Normalized magnitude spectrum of the complex envelope of a rectangular-envelope, HFM pulse with $T = 50$ msec and $\text{SBW}_{\text{HFM}} = 5$ kHz.

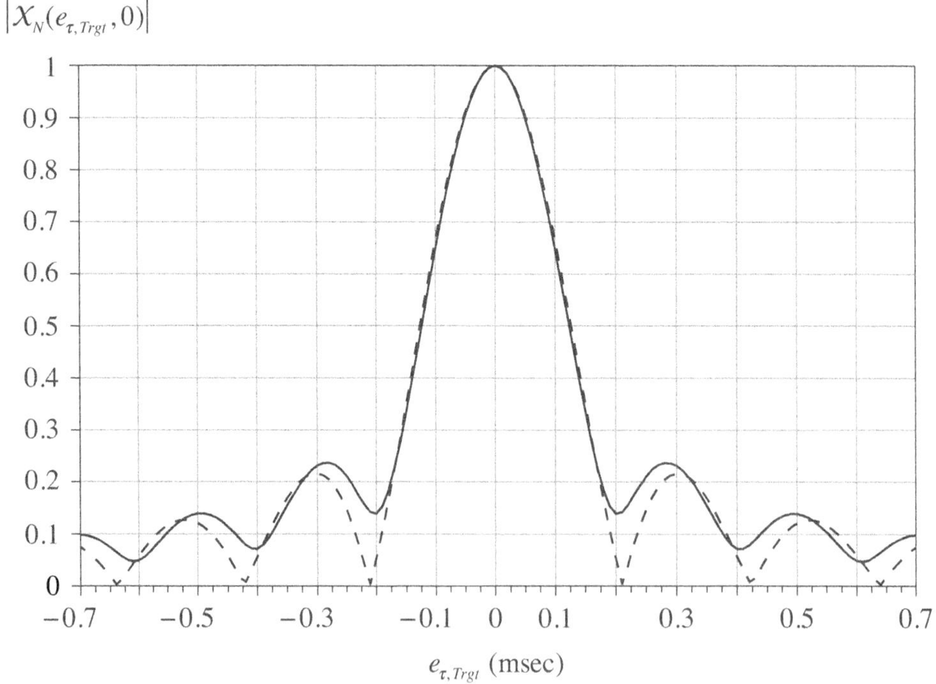

Figure 13.3-10 Magnitudes of the normalized, round-trip time-delay (range) profiles of rectangular-envelope, LFM and HFM pulses (dashed and solid curves, respectively) with $\text{SBW}_{\text{LFM}} \approx 4763.7$ Hz, $\text{SBW}_{\text{HFM}} = 5$ kHz, and $T = 50$ msec.

and a rectangular-envelope, HFM pulse with swept-bandwidth $\text{SBW}_{\text{HFM}} = 5$ kHz and pulse length $T = 50$ msec (solid curve), using (13.2-19) and (13.1-17), respectively. The range resolution of a rectangular-envelope, LFM pulse with $\text{SBW}_{\text{LFM}} \approx 4763.7$ Hz and $c = 1500$ m/sec is $|e_{r,Trgt}| \approx 0.157$ m [see (13.2-44)], and the round-trip time-delay resolution is $|e_{\tau,Trgt}| \approx 0.21$ msec [see (13.2-33)], which is the approximate location of the first zero crossing along the positive $e_{\tau,Trgt}$ axis.

As can be seen from Fig. 13.3-10, the rectangular-envelope, HFM pulse has a slightly better range resolution as measured at the -3-dB down point (magnitude of $\sqrt{2}/2 \approx 0.707$). Otherwise, its round-trip time-delay (range) profile is similar to that of the rectangular-envelope, LFM pulse in this example. Compare Figs. 13.3-6 and 13.3-10. However, as was mentioned in Example 13.3-1, it will be shown in Example 13.3-3 that a rectangular-envelope, HFM pulse is superior to a rectangular-envelope, LFM pulse in its ability to detect a target and estimate time delay in the presence of time-compression/time-expansion effects.

The Doppler profile of a rectangular-envelope, HFM pulse is the same as that for rectangular-envelope, CW and LFM pulses and is given by either (13.1-32) or (13.2-20). Therefore, the Doppler resolution of the rectangular-envelope, HFM pulse with pulse length $T = 50$ msec is $|e_{D,Trgt}| = 20$ Hz [see (13.2-34)], which is the location of the first zero crossing along the positive $e_{D,Trgt}$ axis. ■

13.3.3 Doppler-Invariant Property of a HFM Pulse

Consider a bistatic scattering problem where the transmitter, scatterer (target), and receiver are in three-dimensional, translational motion as discussed in Subsection 10.7.1. A simple model for the received signal is given by

$$y(t) = Kx(s[t-\tau]) = Kx(st-\tau_0), \qquad \tau \leq t \leq \tau + T_y, \tag{13.3-55}$$

where $0 < K < 1$ is a dimensionless constant that is meant to model amplitude attenuation, $x(t)$ is the transmitted signal,

$$s = \tau_0/\tau \tag{13.3-56}$$

is the dimensionless *time-compression/time-expansion factor*, τ_0 is the time delay in seconds when there is *no* motion, τ is the time delay (or travel time) in seconds associated with the scattered path when the scattered acoustic field is *first* incident upon the receiver,

$$T_y = T/s \tag{13.3-57}$$

is the pulse length of the received signal $y(t)$, and T is the pulse length of the transmitted signal $x(t)$ (see Example 10.7-1). Note that a monostatic (backscatter) geometry can be obtained from a bistatic scattering geometry.

A waveform is said to be *Doppler-invariant* if a *constant* time delay Δ exists such that it satisfies the following equation[3]:

$$\boxed{f_{i,y}(t) = f_i(t-\Delta)} \tag{13.3-58}$$

where $f_{i,y}(t)$ and $f_i(t)$ are the instantaneous frequencies in hertz of the received signal $y(t)$ given by (13.3-55), and the transmitted signal $x(t)$, respectively.

First let us show that a LFM pulse does *not* satisfy (13.3-58). Substituting (13.2-4) into (13.3-55) yields

$$y(t) = Ka(s[t-\tau])\cos\left\{2\pi f_c s(t-\tau) + D_p\left[s(t-\tau)-0.5T\right]^2\right\}, \qquad \tau \le t \le \tau + T_y. \tag{13.3-59}$$

The instantaneous phase of $y(t)$ is

$$\theta_{i,y}(t) = 2\pi f_c s(t-\tau) + D_p\left[s(t-\tau)-0.5T\right]^2, \tag{13.3-60}$$

and since

$$f_{i,y}(t) = \frac{1}{2\pi}\frac{d}{dt}\theta_{i,y}(t), \tag{13.3-61}$$

the instantaneous frequency of $y(t)$ is

$$f_{i,y}(t) = sf_c + \frac{D_p}{\pi}\left[s^2(t-\tau) - 0.5sT\right]. \tag{13.3-62}$$

Substituting (13.2-11) and (13.3-62) into (13.3-58) and solving for Δ yields

$$\Delta = (1-s^2)t + s^2\tau + (1-s)\left[\left(\pi f_c / D_p\right) - 0.5T\right]. \tag{13.3-63}$$

As can be seen from (13.3-63), Δ is not a constant, it is a function of time. Therefore, a LFM pulse is *not* a Doppler-invariant waveform.

Next let us show that a HFM pulse *does* satisfy (13.3-58). Substituting (13.3-53) into (13.3-55) yields

$$y(t) = Ka(s[t-\tau])\cos\left\{-2\pi\frac{f_1 f_2}{\text{SBW}_{\text{HFM}}}T\ln\left[1-\frac{\text{SBW}_{\text{HFM}}}{f_2 T}s(t-\tau)\right]\right\}, \qquad \tau \le t \le \tau + T_y. \tag{13.3-64}$$

The instantaneous phase and instantaneous frequency of the received signal $y(t)$ given by (13.3-64) are

$$\theta_{i,y}(t) = -2\pi\frac{f_1 f_2}{\text{SBW}_{\text{HFM}}}T\ln\left[1-\frac{\text{SBW}_{\text{HFM}}}{f_2 T}s(t-\tau)\right] \tag{13.3-65}$$

and

$$f_{i,y}(t) = \frac{sf_1}{1-\dfrac{\text{SBW}_{\text{HFM}}}{f_2 T}s(t-\tau)}, \tag{13.3-66}$$

respectively. Substituting (13.3-47) and (13.3-66) into (13.3-58) and solving for Δ yields

$$\boxed{\Delta = \tau + \frac{1-s}{s\left[\text{SBW}_{\text{HFM}}/(f_2 T)\right]}} \tag{13.3-67}$$

Since Δ is a constant, a HFM pulse *is* Doppler-invariant.

However, as shall be shown in Example 13.3-3, when computing the cross-correlation between received and transmitted HFM signals in order to estimate the time delay τ, the magnitude of the cross-correlation function will have its maximum value at Δ, *not* at τ. Since the instantaneous frequency of the received signal is delayed by Δ seconds relative to the instantaneous frequency of the transmitted signal [see (13.3-58)], the frequency spectra of the transmitted and received signals will overlap at a delay of Δ seconds, producing a maximum value for the magnitude of the cross-correlation function. The *error* in the time-delay estimate is given by

$$\boxed{\Delta_s = \frac{1-s}{s\left[\text{SBW}_{\text{HFM}}/(f_2 T)\right]}} \tag{13.3-68}$$

where the time-compression/time-expansion factor s is unknown a priori. As can be seen from (13.3-67) and (13.3-68), when there is no motion, that is, when $s=1$, $\Delta=\tau$ and $\Delta_s=0$, which means that the magnitude of the cross-correlation function will have its maximum value at the actual time delay τ. For $s>1$ (time compression), if $\text{SBW}_{\text{HFM}}>0$, then $\Delta_s<0$; and if $\text{SBW}_{\text{HFM}}<0$, then $\Delta_s>0$. For $0<s<1$ (time expansion), if $\text{SBW}_{\text{HFM}}>0$, then $\Delta_s>0$; and if $\text{SBW}_{\text{HFM}}<0$, then $\Delta_s<0$.

13.3.4 Designing a HFM Pulse to Minimize Time-Delay Estimation Error

The $|\Delta_s|$ can be made small by making $|\text{SBW}_{\text{HFM}}|/(f_2T)$ big by increasing the $|\text{SBW}_{\text{HFM}}|$ and decreasing the pulse length T of the transmitted signal. Furthermore, $|\text{SBW}_{\text{HFM}}|/(f_2T)$ can be made even bigger for the same values of $|\text{SBW}_{\text{HFM}}|$ and pulse length T by simply interchanging the values of f_1 and f_2 to create a negative swept-bandwidth. For example, if $s = 0.975$, $f_1 = 9\text{ kHz}$ and $f_2 = 14\text{ kHz}$ so that $\text{SBW}_{\text{HFM}} = 5\text{ kHz}$, and $T = 50\text{ msec}$, then $\Delta_s \approx 3.59\text{ msec}$. However, if $f_1 = 14\text{ kHz}$ and $f_2 = 9\text{ kHz}$ so that $\text{SBW}_{\text{HFM}} = -5\text{ kHz}$, then $\Delta_s \approx -2.31\text{ msec}$, or $|\Delta_s| \approx 2.31\text{ msec}$, since with $f_2 = 9\text{ kHz}$ instead of 14 kHz, $|\text{SBW}_{\text{HFM}}|/(f_2T)$ increases, and as a result, $|\Delta_s|$ decreases. Therefore, a *negative* swept-bandwidth HFM pulse would be best to use. Next we shall discuss a design procedure for a HFM pulse that will yield a desired value for $|\Delta_s|$ or less.

Substituting (13.3-48) into (13.3-68) yields

$$\boxed{f_2 = \frac{f_1}{1 - [(1-s)/(s\Delta_s)]T}} \tag{13.3-69}$$

and in order to ensure that $f_2 > 0$, the pulse length T must satisfy the following inequality:

$$\boxed{T < \frac{s\Delta_s}{1-s}} \tag{13.3-70}$$

For the case $s > 1$, substitute a worst case value for s and a desired *negative* value for Δ_s into (13.3-70) and then choose a value for *T*. Next, substitute a desired value for f_1 and your choice for T into (13.3-69) and then compute f_2. For $s > 1$ and $\Delta_s < 0$, $\text{SBW}_{\text{HFM}} > 0$, that is, $f_2 > f_1$. As was previously discussed, the $|\Delta_s|$ can be further reduced by simply interchanging the values of f_1 and f_2 to create a negative swept-bandwidth.

For the case $0 < s < 1$, substitute a worst case value for s and a desired *positive* value for Δ_s into (13.3-70) and then choose a value for *T*. Next, substitute a desired value for f_1 and your choice for T into (13.3-69) and then compute f_2. For $0 < s < 1$ and $\Delta_s > 0$, $\text{SBW}_{\text{HFM}} > 0$. The $|\Delta_s|$ can be further reduced by simply interchanging the values of f_1 and f_2 to create a negative swept-bandwidth.

In both cases, the $|\Delta_s|$ will be smaller if the actual value of s is closer to one than the worst case, design value [see (13.3-68)].

Example 13.3-3 Time-Delay Estimation via Cross-Correlation

In this example we shall compute the cross-correlation between received and transmitted HFM signals, and received and transmitted LFM signals in order to compare the performance of both waveforms in estimating the actual time delay τ in the presence of time-compression/time-expansion effects. In order to compute the cross-correlation function, we shall use the following mathematical model for the complex envelope of the received signal:

$$\tilde{y}(t) = K\tilde{x}(s[t-\tau])\exp(+j2\pi\eta_D[t-\tau]), \qquad \tau \le t \le \tau + T_y, \tag{13.3-71}$$

or, equivalently,

$$\tilde{y}(t) = K\tilde{x}(st-\tau_0)\exp(+j2\pi\eta_D[t-\tau]), \qquad \tau \le t \le \tau + T_y, \tag{13.3-72}$$

where

$$y(t) = \text{Re}\{\tilde{y}(t)\exp(+j2\pi f_c[t-\tau])\}, \qquad \tau \le t \le \tau + T_y. \tag{13.3-73}$$

The cross-correlation function of two complex signals $g_1(t)$ and $g_2(t)$ is given by

$$\mathsf{R}_{12}(\delta) = \int_{-\infty}^{\infty} g_1(t)g_2^*(t-\delta)\,dt. \tag{13.3-74}$$

In this example, $g_1(t)$ is the complex envelope of the received signal $\tilde{y}(t)$ given by (13.3-72), and $g_2(t)$ is the complex envelope of the transmitted signal $\tilde{x}(t)$. The two transmitted signals are the rectangular-envelope, HFM pulse and its equivalent rectangular-envelope, LFM pulse as described in Example 13.3-2. The following parameter values were also used: $K=1$, $s=0.975$, $\tau_0 = 58.5$ msec, and $\tau = 60$ msec.

The value of 0.975 for s was chosen by evaluating the equations in Subsection 10.7.1 for the following two different *backscatter* problems: 1) the transmitter/receiver (T/R) was not in motion, but the scatterer (target) was moving away from the T/R, in the line-of-sight (LOS) direction, at the relatively high speed of 20 m/sec resulting in $s \approx 0.980$, and 2) both the T/R and target were moving away from each other in opposite, collinear directions, each with speeds of 20 m/sec, for a speed of the target relative to the T/R of 40 m/sec resulting in $s \approx 0.960$. Therefore, 0.975 is just one *worst case* value for backscatter in the range $0.960 < s < 0.980$.

Since the Doppler shift η_D is given by (see Subsection 10.7.1),

$$\eta_D = (s-1)f_c, \tag{13.3-75}$$

and the pulse length of the received signal T_y is given by (13.3-57), with $s = 0.975$, and $f_c \approx 10{,}956.5$ Hz and $T = 50$ msec from Example 13.3-2, $\eta_D \approx -273.9$ Hz and $T_y \approx 51.3$ msec. Recall our discussion in Subsection 10.7.1 that if $\tau > \tau_0$, then $0 < s < 1$ (*time expansion*) and $\eta_D < 0$ (*negative* Doppler shift). Time expansion means that the duration of the received signal is *greater than* the duration of the transmitted signal resulting in a *decrease* in signal bandwidth. A negative Doppler shift means that the scatterer (target) and receiver are moving *away* from one another.

Figure 13.3-11 is a plot of the normalized magnitudes of the cross-correlation functions of a rectangular-envelope, LFM pulse with swept-bandwidth $\text{SBW}_{\text{LFM}} \approx 4763.7$ Hz and pulse length $T = 50$ msec (dashed curve) and a rectangular-envelope, HFM pulse with swept-bandwidth $\text{SBW}_{\text{HFM}} = 5$ kHz ($f_1 = 9$ kHz and $f_2 = 14$ kHz) and pulse length $T = 50$ msec (solid curve) for $s = 0.975$. The normalization factor that was used was the energy contained in the complex envelopes of both transmitted signals which is the same, namely, $E_{\tilde{x}} = A^2 T$. With $A = 10$ V and $T = 50$ msec, $E_{\tilde{x}} = 5$ J-Ω.

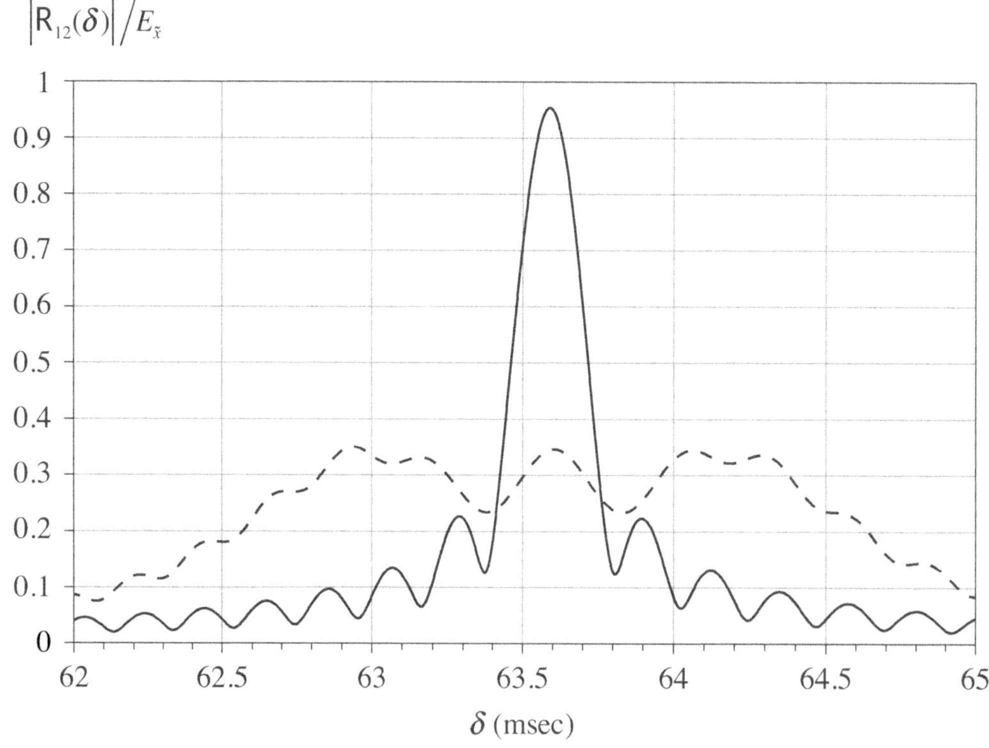

Figure 13.3-11 Normalized magnitudes of the cross-correlation functions of rectangular-envelope, LFM and HFM pulses (dashed and solid curves, respectively) with $\text{SBW}_{\text{LFM}} \approx 4763.7$ Hz, $\text{SBW}_{\text{HFM}} = 5$ kHz, and $T = 50$ msec for $s = 0.975$.

The maximum normalized magnitude of the cross-correlation function of the rectangular-envelope, HFM pulse is approximately 0.954, which occurs at $\delta = \Delta \approx 63.59$ msec, resulting in an error of $\Delta_s \approx 3.59$ msec. The values for Δ and Δ_s obtained by evaluating the cross-correlation function agree with the values obtained for Δ and Δ_s using (13.3-67) and (13.3-68), respectively. The reason that the max. normalized mag. of the cross-correlation function does not equal one is because both waveforms do *not* overlap completely at $\delta = \Delta$. The time interval of overlap at $\delta = \Delta$ is $\Delta \leq t \leq \tau + T_y$. Using $A = 10$ V, $\tau = 0.06$ sec, $T_y = 0.0512821$ sec, and $\Delta = 0.0635897$ sec, the max. unnormalized magnitude of the cross-correlation function is equal to $A^2(\tau + T_y - \Delta) = 4.76924$ J-Ω. Dividing this number by $E_{\tilde{x}} = 5$ J-Ω yields approximately 0.954.

As can be seen from Fig. 13.3-11, although the estimate of the time delay τ is in error, the rectangular-envelope, HFM pulse can still detect the target compared to the rectangular-envelope, LFM pulse. For the rectangular-envelope, HFM pulse, the $|\Delta_s|$ can be further reduced by simply interchanging the values of f_1 and f_2 to create a negative swept-bandwidth, while keeping all other parameter values the same. Therefore, if $f_1 = 14$ kHz and $f_2 = 9$ kHz, then $\text{SBW}_{\text{HFM}} = -5$ kHz and $\Delta_s \approx -2.31$ msec (an underestimate), or $|\Delta_s| \approx 2.31$ msec (see Subsection 13.3.4).

Figure 13.3-12 is a plot of the normalized magnitudes of the cross-correlation functions of a rectangular-envelope, LFM pulse with swept-bandwidth $\text{SBW}_{\text{LFM}} \approx -4763.7$ Hz and pulse length $T = 50$ msec (dashed curve) and a rectangular-envelope, HFM pulse with swept-bandwidth $\text{SBW}_{\text{HFM}} = -5$ kHz and pulse length $T = 50$ msec (solid curve) for $s = 0.975$. The maximum normalized magnitude of the cross-correlation function of the rectangular-envelope, HFM pulse is approximately 0.954, which occurs at $\delta = \Delta \approx 57.69$ msec, resulting in an error of $\Delta_s \approx -2.31$ msec (an underestimate). The values for Δ and Δ_s obtained by evaluating the cross-correlation function agree with the values obtained for Δ and Δ_s using (13.3-67) and (13.3-68), respectively.

As discussed before, the reason that the maximum normalized magnitude of the cross-correlation function does not equal one is because both waveforms do *not* overlap completely at $\delta = \Delta$. Since $\Delta < \tau$, the time interval of overlap at $\delta = \Delta$ is now $\tau \leq t \leq \Delta + T$. Using $A = 10$ V, $\Delta = 0.0576923$ sec, $T = 0.05$ sec, and $\tau = 0.06$ sec, the maximum unnormalized magnitude of the cross-correlation function is equal to $A^2(\Delta + T - \tau) = 4.76923$ J-Ω. Dividing this number by $E_{\tilde{x}} = 5$ J-Ω yields approximately 0.954. The $|\Delta_s|$ can also be reduced to a desired level or less by following the additional procedure discussed in Subsection 13.3.4. If the value of the time-compression/time-expansion factor s is known, then the time-delay error Δ_s can be removed from the estimate. ■

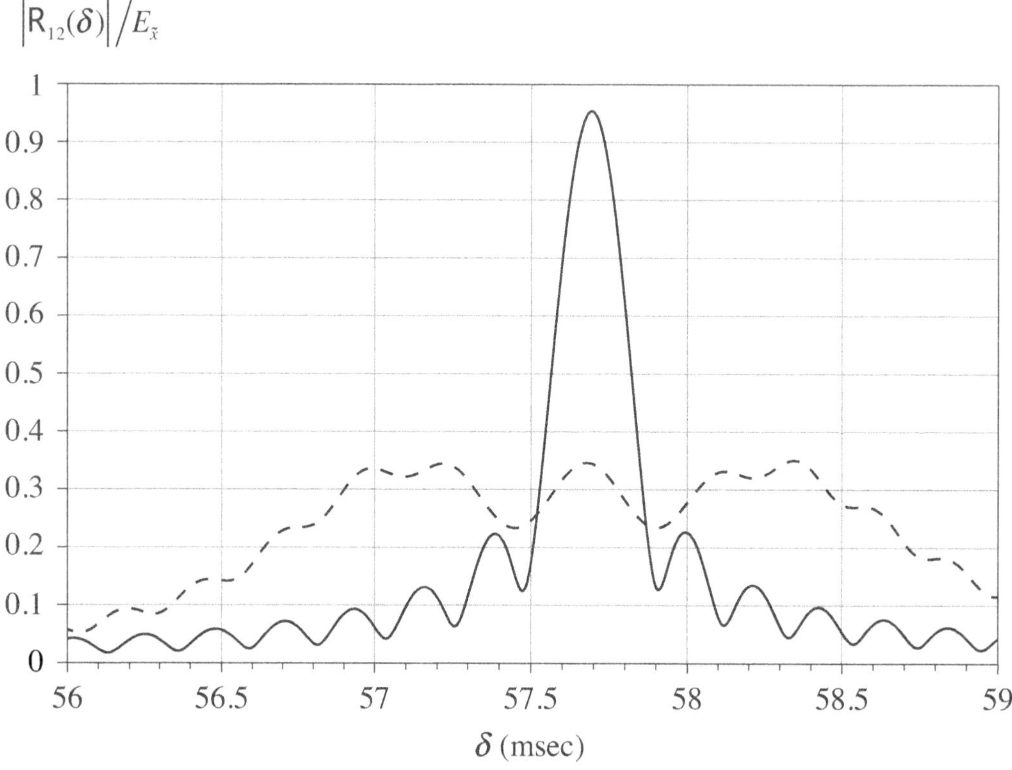

Figure 13.3-12 Normalized magnitudes of the cross-correlation functions of rectangular-envelope, LFM and HFM pulses (dashed and solid curves, respectively) with $\mathrm{SBW}_{\mathrm{LFM}} \approx -4763.7$ Hz, $\mathrm{SBW}_{\mathrm{HFM}} = -5$ kHz, and $T = 50$ msec for $s = 0.975$.

Problems

Section 13.1

13-1 Design a rectangular-envelope, CW pulse $x(t)$ with a range resolution of ± 0.15 m. In addition, $x(t)$ must have a time-average power of 200 W-Ω. Use $c = 1500$ m/sec.

(a) Find the required pulse length in seconds.

(b) Find the required amplitude in volts.

(c) What is the energy of $x(t)$ in joules-ohms?

(d) What is the Doppler resolution of $x(t)$ in hertz?

(e) What is the bandwidth of $x(t)$ in hertz?

(f) What is the minimum allowed value for the carrier frequency in hertz?

13-2 A rectangular-envelope, CW pulse $x(t)$ has an amplitude of 15 V and a pulse length of 100 msec. If $c = 1500$ m/sec, then

(a) what is the range resolution of $x(t)$ in meters?

(b) what is the Doppler resolution of $x(t)$ in hertz?

(c) what is the energy of $x(t)$ in joules-ohms?

(d) what is the time-average power of $x(t)$ in watts-ohms?

(e) what is the bandwidth of $x(t)$ in hertz?

(f) what is the minimum allowed value for the carrier frequency in hertz?

Section 13.2

13-3 Design a rectangular-envelope, LFM pulse $x(t)$ with a range resolution of ± 0.15 m. In addition, $x(t)$ must have a time-average power of 450 W-Ω. Use $c = 1500$ m/sec.

(a) Find the required swept-bandwidth in hertz.

(b) Find the required amplitude in volts.

(c) Find the minimum allowed pulse length in seconds.

If the minimum allowed pulse length is used, find

(d) the energy of $x(t)$ in joules-ohms.

(e) the phase-deviation constant in radians per second-squared for a down-chirp.

(f) the Doppler resolution of $x(t)$ in hertz.

(g) the bandwidth of $x(t)$ in hertz.

(h) the minimum allowed value for the carrier frequency in hertz.

13-4 A rectangular-envelope, up-chirp, LFM pulse $x(t)$ has an amplitude of 50 V, a pulse length of 2 sec, and a swept-bandwidth of 1.5 kHz. If $c = 1500$ m/sec, then

(a) what is the range resolution of $x(t)$ in meters?

(b) what is the Doppler resolution of $x(t)$ in hertz?

(c) what is the phase-deviation constant in radians per second-squared?

(d) what is the energy of $x(t)$ in joules-ohms?

(e) what is the time-average power of $x(t)$ in watts-ohms?

(f) what is the bandwidth of $x(t)$ in hertz?

(g) what is the minimum allowed value for the carrier frequency in hertz?

Section 13.3

13-5 The HFM pulse discussed in Subsection 13.3.4 and Example 13.3-3, with $T = 50$ msec, and $f_1 = 9$ kHz and $f_2 = 14$ kHz so that $\text{SBW}_{\text{HFM}} = 5$ kHz, produced a time-delay estimation error $\Delta_s \approx 3.59$ msec for $s = 0.975$. If you want $\Delta_s = 1$ msec,

(a) find the requirement for the pulse length.

(b) Using $f_1 = 9$ kHz and $T = 30$ msec, find f_2 and the resulting swept-bandwidth.

(c) Find the $|\Delta_s|$ for $s = 0.975$ if you use $T = 30$ msec and simply interchange the values of f_1 and f_2 from Part (b) in order to create a HFM pulse with a negative swept-bandwidth.

(d) Using $f_1 = 9$ kHz and $T = 20$ msec, find f_2 and the resulting swept-bandwidth.

(e) Find the $|\Delta_s|$ for $s = 0.975$ if you use $T = 20$ msec and simply interchange the values of f_1 and f_2 from Part (d) in order to create a HFM pulse with a negative swept-bandwidth.

Chapter 14

Underwater Acoustic Communication Signals

The transmission of digital information corresponds to the transmission of *binary words* which are encoded by the *alphabet* of a *symbol*. Binary words are created by processing sampled values of an analog (continuous-time) message signal by an analog-to-digital (A/D) converter. Binary words encode the output voltages from a quantizer in an A/D converter.

In this chapter we shall discuss the basic principles of three important digital modulation techniques that are not only used to transmit digital information in traditional communication systems, but are also used for underwater acoustic communication. The three digital modulation techniques are 1) *M*-ary Frequency-Shift Keying (MFSK), 2) *M*-ary Quadrature Amplitude Modulation (MQAM), and 3) Orthogonal Frequency-Division Multiplexing (OFDM). For each waveform we shall provide a time-domain description, derive its frequency spectrum, derive bandwidth and time-average power formulas, and discuss how to demodulate the waveform. In this way we can compare the advantages and disadvantages of each digital modulation technique.

14.1 *M*-ary Frequency-Shift Keying

14.1.1 Time-Domain Description

If digital information is being transmitted using *M*-ary frequency-shift keying (MFSK), then a time-domain description of the transmitted signal $x(t)$ is given as follows:

$$x(t) = \sum_{n=1}^{N} x_n(t - t_n), \qquad 0 \le t \le T_d, \tag{14.1-1}$$

where N is the total number of transmitted pulses (symbols),

$$x_n(t) = A\cos\left(2\pi[f_c + \Delta f_n]t + \varepsilon_n\right)\text{rect}\left[(t - 0.5T)/T\right] \tag{14.1-2}$$

is the nth pulse where A is the amplitude factor in volts, f_c is the carrier frequency in hertz,

$$\Delta f_n = k_n / T, \qquad k_n \in \{\pm 1, \pm 2, \ldots, \pm M/2\} \tag{14.1-3}$$

is the *frequency offset* in hertz where the *symbol* k_n is an *integer* (positive or negative),

$$T = T_{\text{sym}} = n_b T_b \tag{14.1-4}$$

is the pulse length of an individual pulse, which is equal to the *symbol duration* T_{sym} in seconds – also known as (a.k.a.) the duration of a binary word; n_b is the number of bits per symbol – a.k.a. the number of bits per binary word; T_b is the *bit duration* in seconds,

$$M = 2^{n_b} \tag{14.1-5}$$

is the total *even* number of *unique* symbol values – the different unique symbol values are known as the *alphabet*; ε_n is a possible, unwanted phase shift in radians at the transmitter,

$$t_n = (n-1)T \tag{14.1-6}$$

is the time instant in seconds when the nth pulse (symbol) begins, and

$$T_d = NT \tag{14.1-7}$$

is the total duration in seconds of the transmitted signal $x(t)$. Note that the total number of transmitted pulses (symbols) N is greater than or equal to the total even number of unique symbol values M, that is,

$$N \geq M. \tag{14.1-8}$$

In Subsection 14.1.4 it is shown that if the frequency offset Δf_n is given by (14.1-3), and the product $f_c T$ is equal to an *integer*, or $f_c T >> 1$ if $f_c T$ does not equal an integer, then the set of functions $x_n(t)$, $n = 1, 2, \ldots, N$, given by (14.1-2) is an *orthogonal* set of functions in the time interval $[0, T]$. The frequency offset Δf_n given by (14.1-3) is required for orthogonality when performing *noncoherent* demodulation and detection of MFSK signals.[1] In noncoherent demodulation and detection, no attempt is made to estimate the unknown phase of a received signal.

For the special case known as *binary frequency-shift keying* (BFSK), $n_b = 1$. Therefore, $M = 2$ and (14.1-3) reduces to

$$\Delta f_n = \frac{k_n}{T}, \qquad k_n = \begin{cases} +1, & \text{binary 1} \\ -1, & \text{binary 0.} \end{cases} \tag{14.1-9}$$

As can be seen from (14.1-1), the MFSK signal $x(t)$ is a *pulse train* where

$$\text{PRI} = T \tag{14.1-10}$$

[1] J. G. Proakis and M. Salehi, *Digital Communications*, 5th ed., McGraw-Hill, New York, 2008, pg. 215.

is the pulse-repetition interval (PRI) in seconds. Since the $\text{PRI} = T$, the time-shifted rectangle function in (14.1-2) is nonzero in the interval $[0, T)$ for the first $N-1$ pulses. Therefore, for $n = 1, 2, \ldots, N-1$,

$$\text{rect}\left(\frac{t-0.5T}{T}\right) = \begin{cases} 1, & 0 \le t < T \\ 0, & \text{otherwise.} \end{cases} \tag{14.1-11}$$

For the last pulse $n = N$, the time-shifted rectangle function is nonzero in the interval $[0, T]$, that is,

$$\text{rect}\left(\frac{t-0.5T}{T}\right) = \begin{cases} 1, & 0 \le t \le T \\ 0, & \text{otherwise.} \end{cases} \tag{14.1-12}$$

14.1.2 Frequency Spectrum and Bandwidth

In order to derive an equation for the bandwidth of the MFSK signal $x(t)$ given by (14.1-1), we first have to find its frequency spectrum. Taking the Fourier transform of (14.1-1) yields

$$X(f) = \sum_{n=1}^{N} X_n(f)\exp(-j2\pi f t_n), \tag{14.1-13}$$

and by substituting (14.1-6) into (14.1-13) we obtain

$$X(f) = \exp(+j2\pi f T)\sum_{n=1}^{N} X_n(f)\exp(-j2\pi f nT), \tag{14.1-14}$$

where

$$X(f) = F\{x(t)\} \tag{14.1-15}$$

and

$$X_n(f) = F\{x_n(t)\} \tag{14.1-16}$$

is the frequency spectrum of the nth pulse $x_n(t)$ given by (14.1-2). The Fourier transform of $x_n(t)$ can be obtained by using the Fourier-transform pair

$$\begin{aligned} A\cos(2\pi f_c t + \theta_0)\,\text{rect}\left(\frac{t-0.5T}{T}\right) \leftrightarrow\ & \frac{A}{2}T\,\text{sinc}\big[(f-f_c)T\big]\exp\big[-j\pi(f-f_c)T\big]\exp(+j\theta_0) + \\ & \frac{A}{2}T\,\text{sinc}\big[(f+f_c)T\big]\exp\big[-j\pi(f+f_c)T\big]\exp(-j\theta_0) \end{aligned} \tag{14.1-17}$$

and replacing f_c with $f_c + \Delta f_n$, and θ_0 with ε_n. Doing so and substituting the resulting expression into (14.1-14) yields the frequency spectrum of the MFSK signal $x(t)$ given by (14.1-1):

$$\boxed{\begin{aligned} X(f) = {} & \frac{A}{2} T c_1(f) \sum_{n=1}^{N} \operatorname{sinc}\{[f-(f_c+\Delta f_n)]T\} \exp(+j\pi \Delta f_n T) \exp(-j2\pi f n T) \exp(+j\varepsilon_n) + \\ & \frac{A}{2} T c_2(f) \sum_{n=1}^{N} \operatorname{sinc}\{[f+(f_c+\Delta f_n)]T\} \exp(-j\pi \Delta f_n T) \exp(-j2\pi f n T) \exp(-j\varepsilon_n) \end{aligned}}$$

(14.1-18)

where

$$c_1(f) = \exp[+j\pi(f+f_c)T] \tag{14.1-19}$$

and

$$c_2(f) = \exp[+j\pi(f-f_c)T]. \tag{14.1-20}$$

The units of $X(f)$ are volts per hertz.

Since bandwidth is always measured along the positive frequency axis, we shall work with the first term in (14.1-18), that is,

$$\begin{aligned} X(f) = \frac{A}{2} T c_1(f) \sum_{n=1}^{N} \operatorname{sinc}\{[f-(f_c+\Delta f_n)]T\} \exp(+j\pi \Delta f_n T) \times \\ \exp(-j2\pi f n T) \exp(+j\varepsilon_n), \qquad f \geq 0. \end{aligned}$$

(14.1-21)

By examining the argument of the sinc function in (14.1-21) and referring to (14.1-3), estimates of the maximum and minimum frequency components are given by

$$f_{\max} = f_c + \max \Delta f_n + \frac{\text{NZC}}{T} = f_c + \frac{M}{2}\frac{1}{T} + \frac{\text{NZC}}{T} \tag{14.1-22}$$

and

$$f_{\min} = f_c + \min \Delta f_n - \frac{\text{NZC}}{T} = f_c - \frac{M}{2}\frac{1}{T} - \frac{\text{NZC}}{T}, \tag{14.1-23}$$

where NZC is the integer number of zero-crossings of the sinc function that is used to estimate both the maximum and minimum frequency components $f_{\max}$ and $f_{\min}$, respectively. Therefore, the bandwidth BW_x (in hertz) of the MFSK signal is given by

$$\text{BW}_x = f_{\max} - f_{\min}, \tag{14.1-24}$$

and by substituting (14.1-22), (14.1-23), and (14.1-4) into (14.1-24), we obtain

$$\boxed{\mathrm{BW}_x = (M + 2\,\mathrm{NZC})D} \tag{14.1-25}$$

where

$$\boxed{D = \frac{1}{T_{\mathrm{sym}}} = \frac{1}{n_b T_b} = \frac{R_b}{n_b}} \tag{14.1-26}$$

is the *symbol rate (baud)* with units of symbols per second, and

$$\boxed{R_b = 1/T_b} \tag{14.1-27}$$

is the *bit rate* in bits per second. Since $\max|\mathrm{sinc}(fT)| = 1$ for $f = 0$; $\mathrm{sinc}(fT) = 0$ for $f = i/T$, where $i = \pm 1, \pm 2, \ldots$; and $|\mathrm{sinc}(fT)| < 0.1$ for $f > 3/T$; NZC should be at least 3, with 5 being a conservative choice.

14.1.3 Signal Energy and Time-Average Power

In this subsection we shall compute the energy and time-average power of the MFSK signal $x(t)$ given by (14.1-1). We begin by computing the energy of the nth pulse $x_n(t)$ given by (14.1-2). Note that $x_n(t)$ is an amplitude-and-angle-modulated carrier with amplitude and angle-modulating functions

$$a_n(t) = A\,\mathrm{rect}\big[(t - 0.5T)/T\big] \tag{14.1-28}$$

and

$$\theta_n(t) = 2\pi\Delta f_n t + \varepsilon_n, \tag{14.1-29}$$

respectively (see Section 11.2), where the time-shifted rectangle function is given by (14.1-11) and (14.1-12). In order to compute the energy of $x_n(t)$, we shall first compute the energy of its complex envelope $\tilde{x}_n(t)$.

Since the general form of the complex envelope of $x_n(t)$ is given by (see Section 11.2)

$$\tilde{x}_n(t) = a_n(t)\exp[+j\theta_n(t)], \tag{14.1-30}$$

substituting (14.1-28) and (14.1-29) into (14.1-30) yields

$$\tilde{x}_n(t) = A\exp\big[+j(2\pi\Delta f_n t + \varepsilon_n)\big]\mathrm{rect}\big[(t - 0.5T)/T\big]. \tag{14.1-31}$$

The energy $E_{\tilde{x}_n}$ of $\tilde{x}_n(t)$ is given by

$$E_{\tilde{x}_n} = \int_{-\infty}^{\infty} a_n^2(t)\,dt\,, \tag{14.1-32}$$

and the energy E_{x_n} of $x_n(t)$ is given by (see Section 11.2)

$$E_{x_n} = E_{\tilde{x}_n}/2\,. \tag{14.1-33}$$

Therefore, substituting (14.1-28) into (14.1-32) yields

$$E_{\tilde{x}_n} = A^2 T\,, \tag{14.1-34}$$

which is the energy (in joules-ohms) of the complex envelope $\tilde{x}_n(t)$ given by (14.1-31) $\forall\ n$, and substituting (14.1-34) into (14.1-33) yields

$$\boxed{E_{x_n} = \frac{A^2}{2}T} \tag{14.1-35}$$

which is the energy (in joules-ohms) of $x_n(t)$ given by (14.1-2) $\forall\ n$. The energy E_{x_n} is also referred to as the energy per symbol because the pulse length T is equal to the symbol duration T_{sym} [see (14.1-4)]. The time-average power of $x_n(t)$ (in watts-ohms) $\forall\ n$ is given by

$$\boxed{P_{\text{avg},\,x_n,\,T} = \frac{E_{x_n}}{T} = \frac{A^2}{2}} \tag{14.1-36}$$

where $P_{\text{avg},\,x_n,\,T}$ is also referred to as the time-average power per symbol.

The energy per bit E_b can also be derived using the energy E_{x_n} given by (14.1-35). Substituting (14.1-4) into (14.1-35) yields

$$E_{x_n} = \frac{A^2}{2}T_b n_b\,, \tag{14.1-37}$$

which can be rewritten as

$$\boxed{E_{x_n} = E_b n_b} \tag{14.1-38}$$

where

$$\boxed{E_b = \frac{A^2}{2}T_b} \tag{14.1-39}$$

is the *energy per bit* in joules-ohms.

From signal theory, the energy E_x of a signal $x(t)$ is defined as

$$E_x \triangleq \int_{-\infty}^{\infty} |x(t)|^2 dt . \tag{14.1-40}$$

Using (14.1-1),

$$|x(t)|^2 = x(t)x^*(t) = \sum_{m=1}^{N}\sum_{n=1}^{N} x_m(t-t_m)x_n^*(t-t_n), \tag{14.1-41}$$

and since

$$x_m(t-t_m)x_n^*(t-t_n) = \begin{cases} |x_n(t-t_n)|^2, & m=n \\ 0, & m \neq n, \end{cases} \tag{14.1-42}$$

(14.1-41) reduces to

$$|x(t)|^2 = \sum_{n=1}^{N} |x_n(t-t_n)|^2 . \tag{14.1-43}$$

Substituting (14.1-43) into (14.1-40) yields

$$E_x = \sum_{n=1}^{N} \int_{-\infty}^{\infty} |x_n(t-t_n)|^2 dt , \tag{14.1-44}$$

and by letting $\alpha = t - t_n$ in (14.1-44),

$$E_x = \sum_{n=1}^{N} \int_{-\infty}^{\infty} |x_n(\alpha)|^2 d\alpha = \sum_{n=1}^{N} E_{x_n} . \tag{14.1-45}$$

Substituting (14.1-35) into (14.1-45) yields

$$E_x = N\frac{A^2}{2}T , \tag{14.1-46}$$

or [see (14.1-7)]

$$\boxed{E_x = \frac{A^2}{2}T_d} \tag{14.1-47}$$

which is the energy (in joules-ohms) of the MFSK signal $x(t)$ given by (14.1-1). Therefore, the time-average power of $x(t)$ in watts-ohms is given by

$$\boxed{P_{\text{avg},x,T_d} = \frac{E_x}{T_d} = \frac{A^2}{2}} \tag{14.1-48}$$

14.1.4 Orthogonality Conditions

In this subsection we shall show that if the frequency offset Δf_n is given by (14.1-3), and the product $f_c T$ is equal to an *integer*, then the set of functions $x_n(t)$, $n = 1, 2, \ldots, N$, given by (14.1-2) is an *orthogonal* set of functions in the time interval $[0, T]$. We begin by computing the following *inner product*:

$$\langle x_m(t), x_n(t) \rangle = \int_0^T x_m(t) x_n^*(t)\, dt\,, \tag{14.1-49}$$

where

$$x_m(t) = A\cos(2\pi[f_c + \Delta f_m]t + \varepsilon_m)\operatorname{rect}[(t - 0.5T)/T] \tag{14.1-50}$$

and

$$x_n(t) = A\cos(2\pi[f_c + \Delta f_n]t + \varepsilon_n)\operatorname{rect}[(t - 0.5T)/T]. \tag{14.1-51}$$

Substituting (14.1-50) and (14.1-51) into (14.1-49) yields

$$\begin{aligned}\int_0^T x_m(t) x_n^*(t)\, dt = {} & \frac{A^2}{2} T \operatorname{sinc}[2(\Delta f_m - \Delta f_n)T]\cos(\varepsilon_m - \varepsilon_n) - \\ & A^2 \frac{\pi(\Delta f_m - \Delta f_n)T^2}{2} \operatorname{sinc}^2[(\Delta f_m - \Delta f_n)T]\sin(\varepsilon_m - \varepsilon_n) + \\ & \frac{A^2}{2} T \operatorname{sinc}[4 f_c T + 2(\Delta f_m + \Delta f_n)T]\cos(\varepsilon_m + \varepsilon_n) - \\ & A^2 \frac{[2\pi f_c + \pi(\Delta f_m + \Delta f_n)]T^2}{2} \operatorname{sinc}^2[2 f_c T + (\Delta f_m + \Delta f_n)T]\sin(\varepsilon_m + \varepsilon_n).\end{aligned} \tag{14.1-52}$$

If the inner product given by (14.1-52) is equal to *zero* when $m \neq n$, then $x_m(t)$ and $x_n(t)$ are said to be *orthogonal*.

By referring to (14.1-3), it can be seen that $\Delta f_m \pm \Delta f_n = i/T$, where i is either a positive or negative *integer*, including zero ($\Delta f_m - \Delta f_n = 0$ when $m = n$). If the product $f_c T$ is also equal to an *integer*, then (14.1-52) reduces to

$$\langle x_m(t), x_n(t) \rangle = \int_0^T x_m(t) x_n^*(t)\, dt = \begin{cases} \dfrac{A^2}{2} T, & m = n \\ 0, & m \neq n, \end{cases} \tag{14.1-53}$$

or

$$\boxed{\langle x_m(t), x_n(t)\rangle = \int_0^T x_m(t)x_n^*(t)\,dt = E_{x_n}\delta_{mn}, \qquad m, n = 1, 2, \ldots, N}$$

(14.1-54)

where

$$E_{x_n} = \langle x_n(t), x_n(t)\rangle = \int_0^T |x_n(t)|^2\,dt = \frac{A^2}{2}T, \qquad n = 1, 2, \ldots, N, \quad (14.1\text{-}55)$$

is the energy (in joules-ohms) of $x_n(t)$ given by (14.1-2) $\forall\ n$ [see (14.1-35)], and

$$\delta_{mn} = \begin{cases} 1, & m = n \\ 0, & m \neq n \end{cases} \quad (14.1\text{-}56)$$

is the Kronecker delta. Therefore, the set of functions $x_n(t)$, $n = 1, 2, \ldots, N$, given by (14.1-2) is an *orthogonal* set of functions in the time interval $[0, T]$. If the product $f_c T$ does *not* equal an integer, but $f_c T >> 1$, then

$$\langle x_m(t), x_n(t)\rangle \approx E_{x_n}\delta_{mn}, \qquad m, n = 1, 2, \ldots, N. \quad (14.1\text{-}57)$$

14.1.5 Demodulation

Demodulation is a procedure meant to retrieve a transmitted message at a receiver. In this subsection we shall briefly discuss two different methods to demodulate a MFSK signal at a receiver. The first method takes advantage of the orthogonality of the pulses $x_n(t)$, $n = 1, 2, \ldots, N$, given by (14.1-2). The second method is based on computing the Fourier transform of each pulse in the received signal.

In order to discuss demodulation, we first need to model a MFSK signal at a receiver, which we shall designate as $y(t)$. We shall use the following simple model for $y(t)$:

$$y(t) = Kx(t - \tau), \qquad \tau \leq t \leq \tau + T_d, \quad (14.1\text{-}58)$$

where the transmitted signal $x(t)$ is given by (14.1-1). As can be seen from (14.1-58), $y(t)$ is modeled as an amplitude-scaled, time-delayed version of $x(t)$ (distortionless transmission) where $K > 0$ is a dimensionless constant. Substituting (14.1-1) into (14.1-58) yields

$$y(t) = \sum_{n=1}^{N} y_n(t - t_n), \qquad \tau \le t \le \tau + T_d, \tag{14.1-59}$$

where

$$y_n(t) = KA\cos(2\pi[f_c + \Delta f_n]t + \delta_n)\text{rect}\left[\left(t - [0.5T + \tau]\right)/T\right] \tag{14.1-60}$$

is the nth pulse of the received signal,

$$\Delta f_n = k_n / T, \qquad k_n \in \{\pm 1, \pm 2, \ldots, \pm M/2\} \tag{14.1-61}$$

is the frequency offset in hertz,

$$\delta_n = \varepsilon_n - 2\pi(f_c + \Delta f_n)\tau \tag{14.1-62}$$

is a phase shift in radians that takes into account the time delay τ in seconds, the time-shifted rectangle function is given by (14.1-11) and (14.1-12), and [see (14.1-6)]

$$t_n = (n-1)T\ . \tag{14.1-63}$$

In order to demodulate $y(t)$, compute the following inner product for *each* received pulse $y_n(t - t_n)$, $n = 1, 2, \ldots, N$: For *each* value of n, compute

$$\langle y_n(t - t_n), g_m(t - t_n) \rangle = \int_{\tau + t_n}^{\tau + t_n + T} y_n(t - t_n) g_m^*(t - t_n)\, dt \tag{14.1-64}$$

$\forall\, m$, where

$$g_m(t) = A\cos(2\pi[f_c + \Delta f_m]t + \phi_m)\text{rect}\left[\left(t - [0.5T + \tau]\right)/T\right], \tag{14.1-65}$$

$$\Delta f_m = m/T, \qquad m = \pm 1, \pm 2, \ldots, \pm M/2, \tag{14.1-66}$$

and ϕ_m is a possible, unwanted phase shift in radians. Evaluating the integral on the right-hand side of (14.1-64) is equivalent to evaluating the integral I where

$$\begin{aligned} I &= KA^2 \int_0^T \cos(2\pi[f_c + \Delta f_n]t + \delta_n)\cos(2\pi[f_c + \Delta f_m]t + \phi_m)\, dt \\ &= KA^2 \int_0^T \cos(2\pi[f_c + \Delta f_m]t + \phi_m)\cos(2\pi[f_c + \Delta f_n]t + \delta_n)\, dt. \end{aligned} \tag{14.1-67}$$

By referring to (14.1-52) and replacing ε_m and ε_n with ϕ_m and δ_n, respectively, it can be seen that

$$I = \begin{cases} K\dfrac{A^2}{2}T\cos(\phi_m - \delta_n), & \Delta f_m = \Delta f_n \\ 0, & \Delta f_m \neq \Delta f_n. \end{cases} \tag{14.1-68}$$

Therefore,

$$\boxed{\langle y_n(t-t_n), g_m(t-t_n)\rangle = \begin{cases} K\dfrac{A^2}{2}T\cos(\phi_m - \delta_n), & \Delta f_m = \Delta f_n \\ 0, & \Delta f_m \neq \Delta f_n \end{cases}} \tag{14.1-69}$$

For example, if $n=4$ and $m=-2$ so that $\Delta f_{-2} = \Delta f_4$, then the symbol value that was transmitted by the 4th pulse is $k_4 = -2$ because

$$\left|\langle y_4(t-t_4), g_{-2}(t-t_4)\rangle\right| = K\frac{A^2}{2}T\left|\cos(\phi_{-2} - \delta_4)\right| \tag{14.1-70}$$

and

$$\left|\langle y_4(t-t_4), g_m(t-t_4)\rangle\right| = 0, \qquad \forall\, m \neq -2. \tag{14.1-71}$$

However, if $\phi_{-2} - \delta_4 = \pm\pi/2$, then $\left|\langle y_4(t-t_4), g_{-2}(t-t_4)\rangle\right| = 0$, and as a result, the symbol value that was transmitted by the 4th pulse will not be detected. Once all the transmitted symbol values have been determined, decode the symbol values back into binary words, and then decode the binary words back into quantized voltages. One can then use the quantized voltages to approximate the original analog (continuous-time) message using the reconstruction formula from the Sampling Theorem.

Another easier way to demodulate $y(t)$ is to compute the Fourier transform of each received pulse $y_n(t-t_n)$, $n = 1, 2, \ldots, N$. In other words, compute the Fourier transform of each T seconds worth of data. For example,

$$F\{y_n(t-t_n)\} = Y_n(f)\exp(-j2\pi f t_n), \tag{14.1-72}$$

where $Y_n(f)$ is the complex frequency spectrum of the nth received pulse $y_n(t)$ given by (14.1-60). The magnitude spectrum of $y_n(t)$ along the positive frequency axis is given by

$$|Y_n(f)| = K\frac{A}{2}T\left|\text{sinc}\{[f - (f_c + \Delta f_n)]T\}\right|, \qquad f \geq 0, \tag{14.1-73}$$

which is equal to the magnitude spectrum of the nth transmitted pulse $x_n(t)$ along the positive frequency axis multiplied by the factor K. As can be seen from (14.1-73), the maximum value of $|Y_n(f)|$ is at $f = f_c + \Delta f_n$. The value of this frequency determines the binary word that was transmitted by the nth pulse (see Table 14.1-3 in Example 14.1-1). For all other frequencies such that the argument of the sinc function is equal to a *nonzero integer*, the sinc function is equal to zero. A numerical estimate of the Fourier transform can be obtained by using the algorithm discussed in Appendix 7B. This method of demodulation is demonstrated in Example 14.1-1.

Example 14.1-1 Gray-Encoded Quaternary FSK

In this example we shall consider the problem of transmitting the following binary words

00 11 01 10 00 11 10 00 01 11

using MFSK where a binary word is equal to $n_b = 2$ bits. Since there are 10 binary words (20 bits), the total number of pulses (symbols) to be transmitted is $N = 10$. Also, since $n_b = 2$, the total even number of *unique* symbol values $M = 2^2 = 4$. Therefore, this example corresponds to 4-ary FSK or *quaternary FSK*. Substituting $M = 4$ into (14.1-3) yields

$$\Delta f_n = k_n / T, \qquad k_n \in \{\pm 1, \pm 2\}. \tag{14.1-74}$$

As can be seen from (14.1-74), there are 4 unique symbol values, where the set of numbers $\{\pm 1, \pm 2\}$ is the alphabet.

The next step is to assign a unique symbol value to each of the 4 *unique* binary words (4 possible combinations of 2 bits). For example, see Table 14.1-1 where the 4 unique binary words are arranged in a *Gray code*. In a Gray code, the leftmost or most-significant-bit (MSB) is the *sign bit*. If the MSB is 0, then a negative number is being represented. If the MSB is 1, then a positive number is being represented. Also, in a Gray code, adjacent binary words differ by only *one* bit.

In order to simulate a time-domain MFSK signal using the information given up to this point, we need values for the amplitude factor A, the carrier frequency f_c, and the pulse length T of an individual pulse. For example, if $A = 100\ \text{V}$, $f_c = 1\ \text{kHz}$, and $T = T_{\text{sym}} = n_b T_b = 20\ \text{msec}$, then we obtain the parameter values shown in Table 14.1-2 for the MFSK signal. Note that the values for A and T were chosen so that the factor $AT/2 = 1$ in (14.1-18). Also note that $f_c T = 20$ is an integer (see Subsection 14.1.4). Using the symbol

Table 14.1-1 Four Unique Symbol Values Assigned to Four Unique Binary Words

Gray Code	Symbol k_n	Frequency Offset Δf_n (Hz)
00	-2	$-2/T$
01	-1	$-1/T$
11	1	$1/T$
10	2	$2/T$

Table 14.1-2 Parameter Values of the Transmitted MFSK Signal

Parameter	Value
Amplitude factor A	100 V
Carrier frequency f_c	1 kHz
Pulse length of an individual pulse T	20 msec
Frequency offset $\Delta f_n = k_n/T$	$50\,k_n$ Hz
Total number of transmitted pulses (symbols) N	10
Total duration of the transmitted signal $T_d = NT$	200 msec
Symbol duration T_{sym}	20 msec
Baud (symbol rate) $D = 1/T_{\text{sym}}$	50 symbols/sec
Number of bits per symbol (binary word) n_b	2
Number of *unique* symbol values $M = 2^{n_b}$	4
Bit duration $T_b = T_{\text{sym}}/n_b$	10 msec
Bit rate $R_b = 1/T_b$	100 bps
Time-average power $P_{\text{avg},x,T_d} = A^2/2$	5000 W-Ω

assignments for the 4 unique binary words in Table 14.1-1, and $\Delta f_n = 50\,k_n$ Hz from Table 14.1-2, Table 14.1-3 shows the frequencies that are assigned to the 4 unique binary words, and Table 14.1-4 shows the frequencies of the 10 transmitted pulses associated with the 10 binary words, where the values for the phase term ε_n were made up for example purposes.

Table 14.1-3 Frequencies Assigned to the Four Unique Binary Words

Binary Word	k_n	$\Delta f_n = 50k_n$ (Hz)	$f_c + \Delta f_n$ (Hz)
00	-2	-100	900
01	-1	-50	950
11	1	50	1050
10	2	100	1100

Table 14.1-4 Frequencies of the Ten Transmitted Pulses

n	Binary Word	k_n	$\Delta f_n = 50k_n$ (Hz)	$f_c + \Delta f_n$ (Hz)	ε_n (deg)
1	00	-2	-100	900	5
2	11	1	50	1050	10
3	01	-1	-50	950	15
4	10	2	100	1100	20
5	00	-2	-100	900	25
6	11	1	50	1050	30
7	10	2	100	1100	35
8	00	-2	-100	900	40
9	01	-1	-50	950	45
10	11	1	50	1050	50

Figure 14.1-1 is a plot of the MFSK signal $x(t)$ given by (14.1-1), and Fig. 14.1-2 is a plot of the magnitude of the MFSK frequency spectrum $X(f)$ given by (14.1-21), using the parameter values shown in Tables 14.1-2 and 14.1-4. The bandwidth of $x(t)$ can be computed using (14.1-25). Since $M=4$ and $D=50$ symbols/sec, if $\text{NZC}=1$, 3, and 5, then $\text{BW}_x = 300$ Hz, 500 Hz, and 700 Hz, respectively. As can be seen from Fig. 14.1-2, a bandwidth of 500 Hz is a good estimate, whereas a bandwidth of 700 Hz is more conservative.

As was previously mentioned, an easy way to demodulate the received MFSK signal $y(t)$ is to compute the Fourier transform of each received pulse. The magnitude spectrum of each received pulse was shown to be directly proportional to the magnitude spectrum of each transmitted pulse [see (14.1-73)]. Therefore, for example purposes, we shall compute the Fourier transforms of the first four transmitted pulses shown in Fig. 14.1-1 using the algorithm discussed in Appendix 7B.

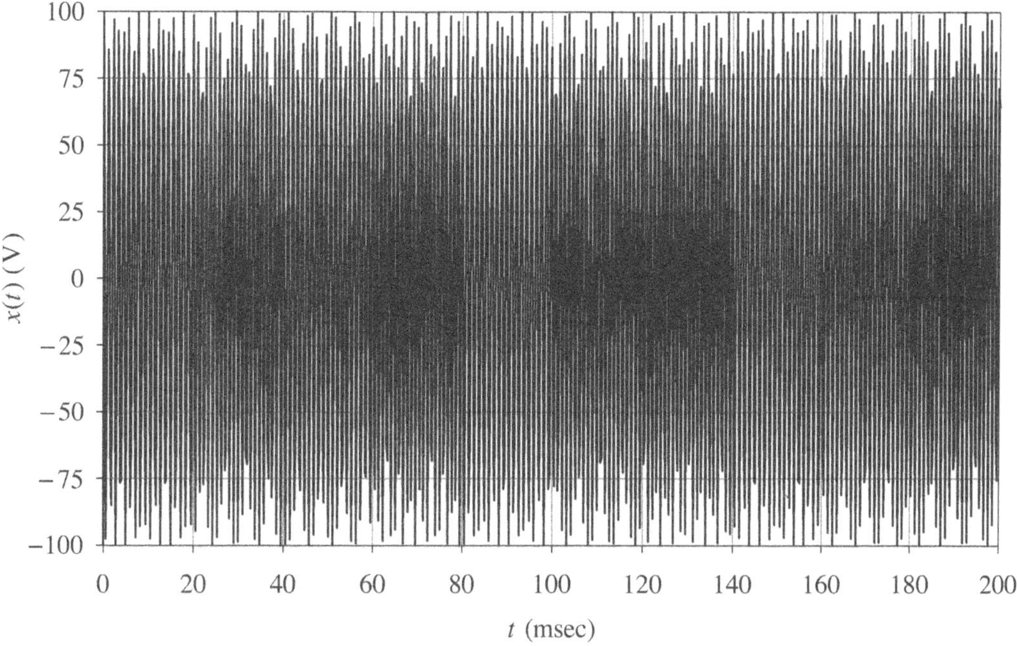

Figure 14.1-1 MFSK signal $x(t)$ given by (14.1-1) using the parameter values shown in Tables 14.1-2 and 14.1-4.

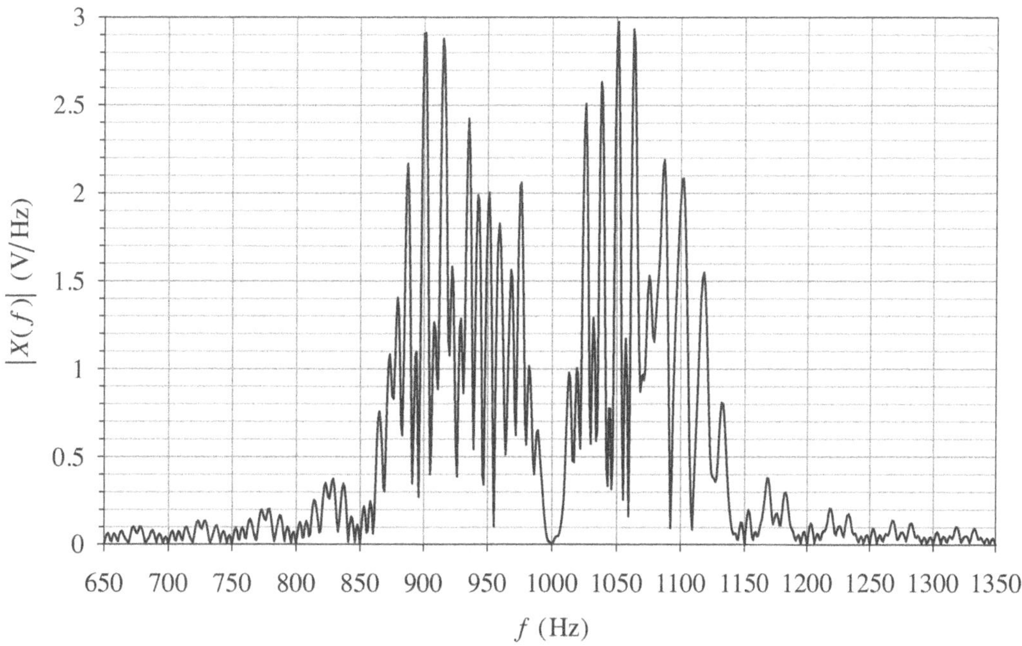

Figure 14.1-2 Magnitude of the MFSK frequency spectrum $X(f)$ given by (14.1-21) using the parameter values shown in Tables 14.1-2 and 14.1-4.

In order to compute the Fourier transforms, a sampling frequency of 4 kHz was used. Since each pulse has a pulse length of 20 msec, the total number of samples taken in a time interval of 20 msec was

$$N = f_S T = 4000\text{ Hz} \times 0.02\text{ sec} = 80. \tag{14.1-75}$$

If padding-with-zeros is not done, then $Z = 0$, the fundamental period $T_0 = T$, and the DFT bin spacing is

$$f_0 = \frac{1}{T_0} = \frac{1}{T} = \frac{1}{0.02\text{ sec}} = 50\text{ Hz}. \tag{14.1-76}$$

Since the four frequencies of interest, 900 Hz, 950 Hz, 1050 Hz, and 1100 Hz, are located at DFT bins when $Z = 0$, padding-with-zeros was not required. From (14.1-21), it can be seen that at $f = f_c + \Delta f_n$,

$$|X_n(f)| = \frac{A}{2}T = \frac{100\text{ V}}{2}(0.02\text{ sec}) = 1\text{ V/Hz}. \tag{14.1-77}$$

Table 14.1-5 Estimates of the Magnitude Spectra of the First Four Transmitted Pulses

		$n=1$	$n=2$	$n=3$	$n=4$
q	f (Hz)	$\lvert\hat{X}_1(f)\rvert$ (V/Hz)	$\lvert\hat{X}_2(f)\rvert$ (V/Hz)	$\lvert\hat{X}_3(f)\rvert$ (V/Hz)	$\lvert\hat{X}_4(f)\rvert$ (V/Hz)
18	900	1	0	0	0
19	950	0	0	1	0
20	1000	0	0	0	0
21	1050	0	1	0	0
22	1100	0	0	0	1

Estimates of the magnitude spectra of the first four transmitted pulses are shown in Table 14.1-5. In Table 14.1-5, integer q is the DFT bin number, frequency $f = qf_0$ Hz where the DFT bin spacing f_0 is given by (14.1-76), and $\hat{X}_n(f)$ is the estimate of the Fourier transform of $x_n(t)$ using the algorithm discussed in Appendix 7B. As can be seen in Table 14.1-5, $|\hat{X}_n(f)| = 1\text{ V/Hz}$, $n = 1, 2, 3, 4$, which agrees with the theoretical value given by (14.1-77). Since frequency 900 Hz is present in pulse 1, binary word 00 was transmitted (see

Table 14.1-3). Since frequency 1050 Hz is present in pulse 2, binary word 11 was transmitted. Since frequency 950 Hz is present in pulse 3, binary word 01 was transmitted. And since frequency 1100 Hz is present in pulse 4, binary word 10 was transmitted.

Once all the transmitted binary words have been determined, decode the binary words back into quantized voltages. One can then use the quantized voltages to approximate the original analog (continuous-time) message using the reconstruction formula from the Sampling Theorem. ■

14.2 *M*-ary Quadrature Amplitude Modulation

14.2.1 Time-Domain Description

If digital information is being transmitted using *M*-ary quadrature amplitude-modulation (MQAM), then a time-domain description of the transmitted signal $x(t)$ is given as follows:

$$x(t)=\sum_{n=1}^{N} x_n(t), \qquad 0 \le t \le T_d, \tag{14.2-1}$$

where N is the total number of transmitted pulses (symbols),

$$x_n(t)=A|w_n|\cos(2\pi f_c t+\angle w_n)\,p(t-t_n) \tag{14.2-2}$$

is the *n*th pulse where A is the amplitude factor in volts, $|w_n|$ and $\angle w_n$ are the magnitude and phase of the *complex symbol* (see Fig. 14.2-1)

$$w_n=|w_n|\exp(+j\angle w_n), \tag{14.2-3}$$

f_c is the carrier frequency in hertz, $p(t)$ is the pulse-shape function where

$$T=T_{\text{sym}}=n_b T_b \tag{14.2-4}$$

is the pulse length of an individual pulse, which is equal to the *symbol duration* T_{sym} in seconds – also known as (a.k.a.) the duration of a binary word; n_b is the number of bits per symbol – a.k.a. the number of bits per binary word; T_b is the *bit duration* in seconds,

$$M=2^{n_b} \tag{14.2-5}$$

is the total *even* number of *unique* symbol values – the different unique symbol

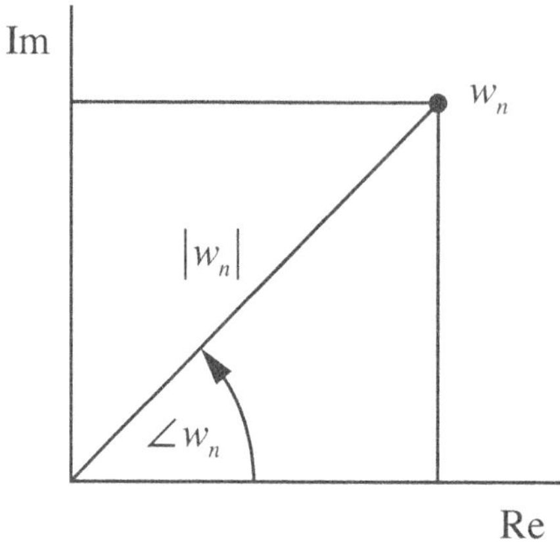

Figure 14.2-1 Complex symbol w_n in the complex plane.

values are known as the *alphabet*;

$$t_n = (n-1)T \tag{14.2-6}$$

is the time instant in seconds when the *n*th pulse (symbol) begins, and

$$T_d = NT \tag{14.2-7}$$

is the total duration in seconds of the transmitted signal $x(t)$. Note that the total number of transmitted pulses (symbols) N is greater than or equal to the total even number of unique symbol values M, that is,

$$N \geq M. \tag{14.2-8}$$

Because a MQAM signal uses a complex symbol in general, MQAM is a combination of amplitude-shift keying (ASK) via $|w_n|$, and *M*-ary phase-shift keying (MPSK) via $\angle w_n$.

By referring to (14.2-2), it can be seen that there is *no* frequency offset associated with the *n*th pulse. Also, the MQAM signal $x(t)$ is a *pulse train* where

$$\text{PRI} = T \tag{14.2-9}$$

is the pulse-repetition interval (PRI) in seconds. In this section, the pulse-shape function $p(t)$ is equal to the time-shifted rectangle function. Since the $\text{PRI} = T$, $p(t)$ is nonzero in the interval $[0, T)$ for the first $N-1$ pulses. Therefore, for $n = 1, 2, \ldots, N-1$,

$$p(t) = \text{rect}\left(\frac{t-0.5T}{T}\right) = \begin{cases} 1, & 0 \leq t < T \\ 0, & \text{otherwise.} \end{cases} \tag{14.2-10}$$

For the last pulse $n=N$, $p(t)$ is nonzero in the interval $[0,T]$, that is,

$$p(t)=\text{rect}\left(\frac{t-0.5T}{T}\right)=\begin{cases}1, & 0\leq t\leq T\\ 0, & \text{otherwise.}\end{cases} \tag{14.2-11}$$

By properly choosing values for $|w_n|$ and $\angle w_n$, several well-known special cases can be obtained as shall be discussed next.

Amplitude-Shift Keying (ASK)

If we set $\angle w_n = 0 \ \forall \ n$, and replace $|w_n|$ with w_n, then (14.2-2) reduces to

$$x_n(t)=Aw_n\cos(2\pi f_c t)\,p(t-t_n), \tag{14.2-12}$$

where

$$w_n\in\{2i-1-M\}_{i=1}^{M}. \tag{14.2-13}$$

For example, if $n_b=2$, then $M=2^{n_b}=4$ and $w_n\in\{\pm 1,\pm 3\}$. For the special case known as *binary amplitude-shift keying* (BASK), $n_b=1$. Therefore, $M=2$ and

$$w_n=\begin{cases}1, & \text{binary 1}\\ -1, & \text{binary 0.}\end{cases} \tag{14.2-14}$$

For *on-off keying* (OOK), use

$$w_n=\begin{cases}1, & \text{binary 1}\\ 0, & \text{binary 0.}\end{cases} \tag{14.2-15}$$

M-ary Phase-Shift Keying (MPSK)

If we set $|w_n|=1 \ \forall \ n$ so that $w_n=\exp(+j\angle w_n)$, then (14.2-2) reduces to

$$x_n(t)=A\cos\left(2\pi f_c t+\angle w_n\right)p(t-t_n), \tag{14.2-16}$$

where

$$\angle w_n\in\left\{\frac{2\pi}{M}(i-1)\right\}_{i=1}^{M} \tag{14.2-17}$$

or

$$\angle w_n\in\left\{\frac{2\pi}{M}(i-1)+\frac{\pi}{M}\right\}_{i=1}^{M}. \tag{14.2-18}$$

For example, for the special case known as 4-ary PSK, also known as *quadrature phase-shift keying* (QPSK), $n_b = 2$. Therefore, $M = 4$ and

$$\angle w_n \in \left\{0, \frac{\pi}{2}, \pi, \frac{3\pi}{2}\right\}, \tag{14.2-19}$$

or

$$\angle w_n \in \left\{\frac{\pi}{4}, \frac{3\pi}{4}, \frac{5\pi}{4}, \frac{7\pi}{4}\right\}. \tag{14.2-20}$$

For the special case known as *binary phase-shift keying* (BPSK), $n_b = 1$. Therefore, $M = 2$ and

$$\angle w_n = \begin{cases} 0, & \text{binary 1} \\ \pi, & \text{binary 0} \end{cases} \tag{14.2-21}$$

or

$$\angle w_n = \begin{cases} \pi/2, & \text{binary 1} \\ 3\pi/2, & \text{binary 0.} \end{cases} \tag{14.2-22}$$

Equations (14.2-1) and (14.2-2) represent one way to describe a MQAM signal in the time domain. An alternate time-domain description can be obtained by rewriting the equation for the *n*th transmitted pulse $x_n(t)$ given by (14.2-2) in terms of its cosine and sine components. Since $x_n(t)$ is an amplitude-and-angle-modulated carrier with amplitude-modulating function

$$a_n(t) = A|w_n|\, p(t - t_n), \tag{14.2-23}$$

and angle-modulating function

$$\theta_n(t) = \angle w_n, \tag{14.2-24}$$

it can be rewritten as

$$x_n(t) = x_{cn}(t)\cos(2\pi f_c t) - x_{sn}(t)\sin(2\pi f_c t), \tag{14.2-25}$$

where

$$x_{cn}(t) = a_n(t)\cos\theta_n(t) = A|w_n|\cos(\angle w_n)\, p(t - t_n) \tag{14.2-26}$$

is the cosine component of $x_n(t)$, and

$$x_{sn}(t) = a_n(t)\sin\theta_n(t) = A|w_n|\sin(\angle w_n)\, p(t - t_n) \tag{14.2-27}$$

is the sine component of $x_n(t)$ (see Section 11.2). By referring to (14.2-3), it can be seen that

$$\text{Re}\{w_n\} = |w_n| \cos(\angle w_n) \tag{14.2-28}$$

and

$$\text{Im}\{w_n\} = |w_n| \sin(\angle w_n). \tag{14.2-29}$$

Substituting (14.2-28) into (14.2-26), and (14.2-29) into (14.2-27) yields

$$x_{cn}(t) = A\,\text{Re}\{w_n\}\,p(t - t_n) \tag{14.2-30}$$

and

$$x_{sn}(t) = A\,\text{Im}\{w_n\}\,p(t - t_n), \tag{14.2-31}$$

respectively, and by substituting (14.2-30) and (14.2-31) into (14.2-25), we obtain

$$x_n(t) = A\,\text{Re}\{w_n\}\,p(t - t_n)\cos(2\pi f_c t) - A\,\text{Im}\{w_n\}\,p(t - t_n)\sin(2\pi f_c t). \tag{14.2-32}$$

Therefore, substituting (14.2-32) into (14.2-1) yields

$$x(t) = x_c(t)\cos(2\pi f_c t) - x_s(t)\sin(2\pi f_c t), \qquad 0 \le t \le T_d, \tag{14.2-33}$$

where

$$x_c(t) = A \sum_{n=1}^{N} \text{Re}\{w_n\}\,p(t - t_n) \tag{14.2-34}$$

is the cosine component of $x(t)$, and

$$x_s(t) = A \sum_{n=1}^{N} \text{Im}\{w_n\}\,p(t - t_n) \tag{14.2-35}$$

is the sine component of $x(t)$. Equations (14.2-33) through (14.2-35) represent an alternative way of describing a MQAM signal in the time domain.

Example 14.2-1 Gray-Encoded 8-QAM

In this example we shall consider the problem of transmitting binary words using MQAM where a binary word is equal to $n_b = 3$ bits. Since $n_b = 3$, there are $M = 2^3 = 8$ unique *pairs* of symbol values $(\text{Re}\{w_n\}, \text{Im}\{w_n\})$, and 8 unique binary words (8 possible combinations of 3 bits).

The next step is to assign a unique pair of symbol values to each of the 8 unique binary words. For example, see Table 14.2-1 where the 8 unique binary

Table 14.2-1 Eight Unique Pairs of Symbol Values Assigned to Eight Unique Binary Words for a Rectangular Signal-Space Constellation

Gray Code	$\text{Re}\{w_n\}$	$\text{Im}\{w_n\}$
000	−3	−1
001	−1	−1
011	1	−1
010	3	−1
100	−3	1
101	−1	1
111	1	1
110	3	1

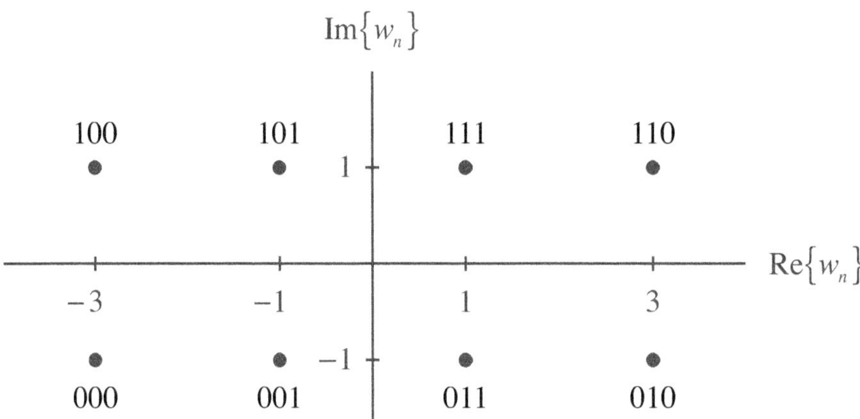

Figure 14.2-2 Rectangular signal-space constellation for Gray-encoded 8-QAM.

words are arranged in a Gray code to form a rectangular signal-space constellation. Figure 14.2-2 is a pictorial representation of Table 14.2-1. As can be seen from Fig. 14.2-2, only one bit is changed going from one signal-point to an adjacent signal-point in the constellation, either in the horizontal or vertical direction. Also, the distance between adjacent signal-points, either in the horizontal or vertical direction, is 2.

The 8 unique binary words can also be encoded as shown in Table 14.2-2. Figure 14.2-3 is a pictorial representation of Table 14.2-2. As can be seen from Fig. 14.2-3, only one bit is changed going from one signal-point to an adjacent signal-point on the inner circle in the constellation, and from one signal-point to an adjacent signal-point on the outer circle in the constellation. The distance between adjacent signal-points on the inner circle, either in the horizontal or

Table 14.2-2 Eight Unique Pairs of Symbol Values Assigned to Eight Unique Binary Words for a Concentrically-Circular Signal-Space Constellation

Inner Circle			Outer Circle		
Gray Code	$\mathrm{Re}\{w_n\}$	$\mathrm{Im}\{w_n\}$	Gray Code	$\mathrm{Re}\{w_n\}$	$\mathrm{Im}\{w_n\}$
001	-1	-1	000	$-(1+\sqrt{3})$	0
011	1	-1	010	0	$-(1+\sqrt{3})$
111	1	1	110	$1+\sqrt{3}$	0
101	-1	1	100	0	$1+\sqrt{3}$

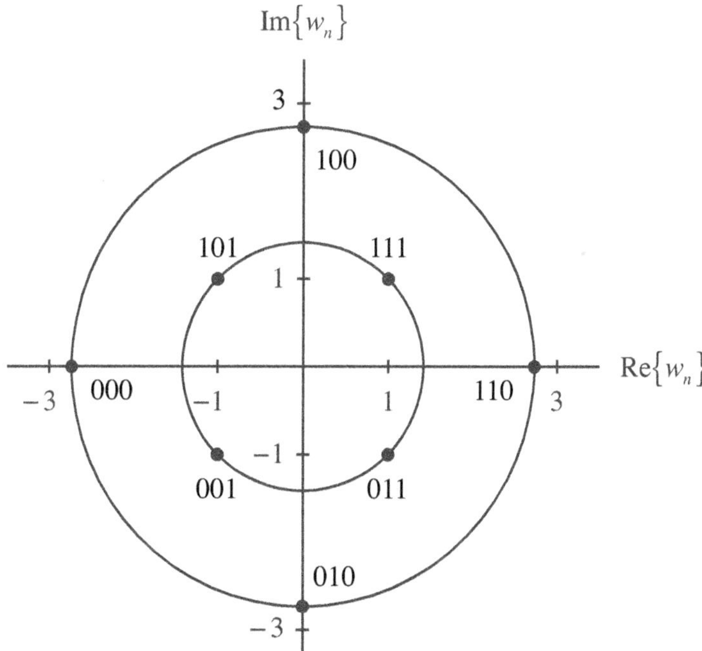

Figure 14.2-3 Concentrically-circular signal-space constellation for Gray-encoded 8-QAM.

vertical direction, is 2. The distance between adjacent signal-points on the inner and outer circles, measured diagonally, is also 2 (see Fig. 14.2-4). Figure 14.2-3 is the *optimum* 8-QAM signal-space constellation because it requires the smallest time-average power for a given distance between adjacent signal points[2] (see Example 14.2-2).

[2] J. G. Proakis and M. Salehi, *Digital Communications*, 5th ed., McGraw-Hill, New York, 2008, pg. 197.

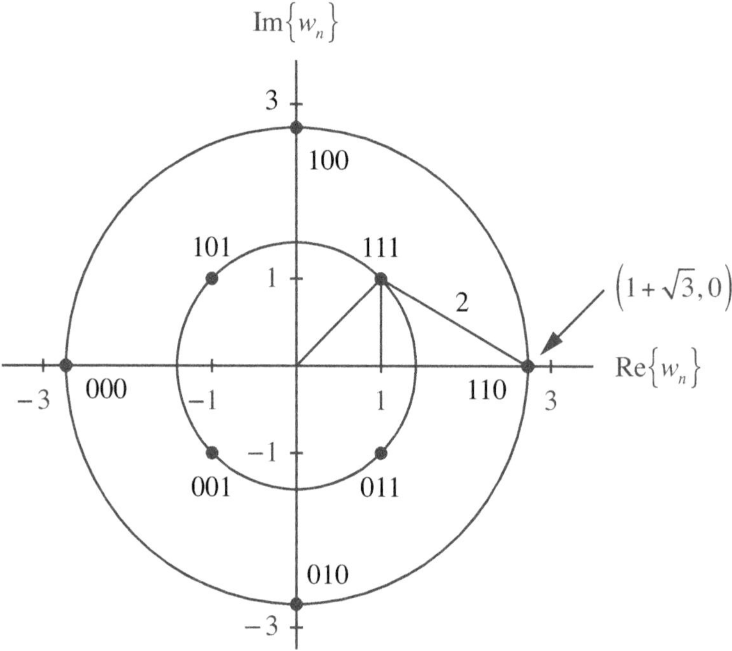

Figure 14.2-4 Right triangles used to compute the radii of the inner and outer circles.

Figure 14.2-4 shows the right triangles used to compute the radii of the inner and outer circles. By referring to Fig. 14.2-4, it can be seen that the radius of the inner circle is

$$r = \sqrt{1^2 + 1^2} = \sqrt{2} . \tag{14.2-36}$$

Since the base of the right triangle with hypotenuse 2 is

$$b = \sqrt{2^2 - 1^2} = \sqrt{3} , \tag{14.2-37}$$

the radius of the outer circle is

$$r = 1 + \sqrt{3} . \tag{14.2-38}$$

And finally, by referring to Fig. 14.2-1:

$$|w_n| = \sqrt{\left(\text{Re}\{w_n\}\right)^2 + \left(\text{Im}\{w_n\}\right)^2} \tag{14.2-39}$$

and

$$\angle w_n = \tan^{-1}\left[\text{Im}\{w_n\}/\text{Re}\{w_n\}\right]. \tag{14.2-40}$$

So although Tables 14.2-1 and 14.2-2 specify values for the real and imaginary parts of the complex symbol w_n, with the use of (14.2-39) and (14.2-40), the magnitude and phase of w_n can also be computed. ■

14.2.2 Frequency Spectrum and Bandwidth

In order to derive an equation for the bandwidth of the MQAM signal $x(t)$ given by (14.2-1), we first have to find its frequency spectrum. Taking the Fourier transform of (14.2-1) yields

$$X(f) = \sum_{n=1}^{N} X_n(f), \tag{14.2-41}$$

where

$$X(f) = F\{x(t)\} \tag{14.2-42}$$

and

$$X_n(f) = F\{x_n(t)\} \tag{14.2-43}$$

is the frequency spectrum of the nth pulse $x_n(t)$ given by (14.2-2). The Fourier transform of $x_n(t)$ can be obtained by using the Fourier-transform pair

$$a(t)\cos(2\pi f_c t + \theta_0) \leftrightarrow \frac{1}{2}A(f - f_c)\exp(+j\theta_0) + \frac{1}{2}A(f + f_c)\exp(-j\theta_0), \tag{14.2-44}$$

where

$$a(t) = A|w_n|\, p(t - t_n) \tag{14.2-45}$$

and $\theta_0 = \angle w_n$. Taking the Fourier transform of (14.2-45) yields

$$A(f) = A|w_n| P(f)\exp(-j2\pi f t_n), \tag{14.2-46}$$

and by substituting (14.2-6) into (14.2-46), we obtain

$$A(f) = A|w_n| P(f)\exp(-j2\pi f nT)\exp(+j2\pi f T). \tag{14.2-47}$$

Since the Fourier transform of the time-shifted rectangle function $p(t)$ is given by

$$P(f) = T\,\text{sinc}(fT)\exp(-j\pi f T), \tag{14.2-48}$$

substituting (14.2-48) into (14.2-47) yields

$$A(f) = A|w_n| T \operatorname{sinc}(fT) \exp(-j2\pi f n T) \exp(+j\pi f T). \tag{14.2-49}$$

Therefore, substituting (14.2-49), $\theta_0 = \angle w_n$, and (14.2-3) into the right-hand side of (14.2-44) yields

$$X_n(f) = \frac{A}{2} T w_n \operatorname{sinc}[(f - f_c)T] \exp[-j2\pi(f - f_c)nT] \exp[+j\pi(f - f_c)T] + \\ \frac{A}{2} T w_n^* \operatorname{sinc}[(f + f_c)T] \exp[-j2\pi(f + f_c)nT] \exp[+j\pi(f + f_c)T], \tag{14.2-50}$$

and by substituting (14.2-50) into (14.2-41), we finally obtain the frequency spectrum of the MQAM signal $x(t)$ given by (14.2-1):

$$X(f) = \frac{A}{2} T \left[\sum_{n=1}^{N} w_n \exp[-j2\pi(f - f_c)nT] \right] \operatorname{sinc}[(f - f_c)T] \exp[+j\pi(f - f_c)T] + \\ \frac{A}{2} T \left[\sum_{n=1}^{N} w_n^* \exp[-j2\pi(f + f_c)nT] \right] \operatorname{sinc}[(f + f_c)T] \exp[+j\pi(f + f_c)T] \tag{14.2-51}$$

The units of $X(f)$ are volts per hertz.

Since bandwidth is always measured along the positive frequency axis, we shall work with the first term in (14.2-51), that is,

$$X(f) = \frac{A}{2} T \left[\sum_{n=1}^{N} w_n \exp[-j2\pi(f - f_c)nT] \right] \operatorname{sinc}[(f - f_c)T] \times \\ \exp[+j\pi(f - f_c)T], \qquad f \geq 0. \tag{14.2-52}$$

By examining the argument of the sinc function in (14.2-52), estimates of the maximum and minimum frequency components are given by

$$f_{\max} = f_c + \frac{\text{NZC}}{T} \tag{14.2-53}$$

and

$$f_{\min} = f_c - \frac{\text{NZC}}{T}, \tag{14.2-54}$$

where NZC is the integer number of zero-crossings of the sinc function that is used to estimate both the maximum and minimum frequency components $f_{\max}$ and $f_{\min}$, respectively. Therefore, the bandwidth BW_x (in hertz) of the MQAM signal is given by

$$\mathrm{BW}_x = f_{\max} - f_{\min}, \tag{14.2-55}$$

and by substituting (14.2-53), (14.2-54), and (14.2-4) into (14.2-55), we obtain

$$\boxed{\mathrm{BW}_x = 2\,\mathrm{NZC} \times D} \tag{14.2-56}$$

where

$$\boxed{D = \frac{1}{T_{\text{sym}}} = \frac{1}{n_b T_b} = \frac{R_b}{n_b}} \tag{14.2-57}$$

is the *symbol rate (baud)* with units of symbols per second, and

$$\boxed{R_b = 1/T_b} \tag{14.2-58}$$

is the *bit rate* in bits per second. Since $\max|\mathrm{sinc}(fT)| = 1$ for $f = 0$; $\mathrm{sinc}(fT) = 0$ for $f = i/T$, where $i = \pm 1, \pm 2, \ldots$; and $|\mathrm{sinc}(fT)| < 0.1$ for $f > 3/T$; NZC should be at least 3, with 5 being a conservative choice.

14.2.3 Signal Energy and Time-Average Power

In this subsection we shall compute the energy and time-average power of the MQAM signal $x(t)$ given by (14.2-1). We begin by computing the energy of the nth pulse $x_n(t)$ given by (14.2-2). Note that $x_n(t)$ is an amplitude-and-angle-modulated carrier with amplitude and angle-modulating functions

$$a_n(t) = A|w_n|\, p(t - t_n) \tag{14.2-59}$$

and

$$\theta_n(t) = \angle w_n, \tag{14.2-60}$$

respectively (see Section 11.2), where the pulse-shape function $p(t)$ is given by (14.2-10) and (14.2-11). In order to compute the energy of $x_n(t)$, we shall first compute the energy of its complex envelope $\tilde{x}_n(t)$.

Since the general form of the complex envelope of $x_n(t)$ is given by (see Section 11.2)

$$\tilde{x}_n(t) = a_n(t)\exp[+j\theta_n(t)], \tag{14.2-61}$$

substituting (14.2-59) and (14.2-60) into (14.2-61) yields

$$\tilde{x}_n(t) = A|w_n|\exp(+j\angle w_n)\,p(t-t_n) = Aw_n\,p(t-t_n). \tag{14.2-62}$$

The energy $E_{\tilde{x}_n}$ of $\tilde{x}_n(t)$ is given by

$$E_{\tilde{x}_n} = \int_{-\infty}^{\infty} a_n^2(t)\,dt, \tag{14.2-63}$$

and the energy E_{x_n} of $x_n(t)$ is given by (see Section 11.2)

$$E_{x_n} = E_{\tilde{x}_n}/2. \tag{14.2-64}$$

Therefore, substituting (14.2-59) into (14.2-63) yields

$$E_{\tilde{x}_n} = A^2|w_n|^2 T, \tag{14.2-65}$$

which is the energy (in joules-ohms) of the complex envelope $\tilde{x}_n(t)$ given by (14.2-62), and substituting (14.2-65) into (14.2-64) yields

$$\boxed{E_{x_n} = \frac{A^2}{2}|w_n|^2 T} \tag{14.2-66}$$

which is the energy (in joules-ohms) of $x_n(t)$ given by (14.2-2), where $|w_n|$ is given by (14.2-39). The time-average power of $x_n(t)$ in watts-ohms is given by

$$\boxed{P_{\text{avg},\,x_n,\,T} = \frac{E_{x_n}}{T} = \frac{A^2}{2}|w_n|^2} \tag{14.2-67}$$

From signal theory, the energy E_x of a signal $x(t)$ is defined as

$$E_x \triangleq \int_{-\infty}^{\infty} |x(t)|^2\,dt. \tag{14.2-68}$$

Using (14.2-1),

$$|x(t)|^2 = x(t)x^*(t) = \sum_{m=1}^{N}\sum_{n=1}^{N} x_m(t)x_n^*(t), \tag{14.2-69}$$

and since

$$x_m(t)x_n^*(t) = \begin{cases} |x_n(t)|^2, & m = n \\ 0, & m \neq n \end{cases} \tag{14.2-70}$$

because $p(t-t_m)p(t-t_n)=0$, $m \neq n$; (14.2-69) reduces to

$$|x(t)|^2 = \sum_{n=1}^{N} |x_n(t)|^2 . \tag{14.2-71}$$

Substituting (14.2-71) into (14.2-68) yields

$$E_x = \sum_{n=1}^{N} \int_{-\infty}^{\infty} |x_n(t)|^2 dt = \sum_{n=1}^{N} E_{x_n} . \tag{14.2-72}$$

Substituting (14.2-66) into (14.2-72) yields

$$\boxed{E_x = \frac{A^2}{2} T \sum_{n=1}^{N} |w_n|^2} \tag{14.2-73}$$

which is the energy (in joules-ohms) of the MQAM signal $x(t)$ given by (14.2-1), where $|w_n|$ is given by (14.2-39). Therefore, the time-average power of $x(t)$ (in watts-ohms) in the time interval $[0, T_d]$ is given by

$$\boxed{P_{\text{avg}, x, T_d} = \frac{E_x}{T_d} = \frac{A^2}{2} \frac{1}{N} \sum_{n=1}^{N} |w_n|^2} \tag{14.2-74}$$

where T_d is given by (14.2-7). Table 14.2-3 compares the signal bandwidths and time-average powers of MFSK and MQAM signals. As can be seen from Table 14.2-3, MQAM has a smaller bandwidth but a larger time-average power, in general, compared to MFSK. For the special case of MPSK where $|w_n| = 1 \ \forall \ n$, the time-average powers of MFSK and MQAM signals are equal. In addition, both signals have the same total duration T_d. Recall that the pulse length T is equal to the symbol duration T_{sym} in both MFSK and MQAM signals [see (14.1-4) and (14.2-4)].

Table 14.2-3 Signal Bandwidth and Time-Average Power of MFSK and MQAM Signals

	Signal Bandwidth BW_x (Hz)	Time-Average Power (W-Ω)
MFSK	$(M+2\text{NZC})D$	$P_{\text{avg},x,T_d} = \frac{A^2}{2}$ $T_d = NT_{\text{sym}}$
MQAM	$2\text{NZC}\times D$	$P_{\text{avg},x,T_d} = \frac{A^2}{2}\frac{1}{N}\sum_{n=1}^{N}\lvert w_n\rvert^2$ $T_d = NT_{\text{sym}}$

Example 14.2-2

In this example we shall compute the time-average power of a MQAM signal using the 8-QAM rectangular signal-space constellation shown in Table 14.2-1 and Fig. 14.2-2, and the optimum 8-QAM concentrically-circular signal-space constellation shown in Table 14.2-2 and Fig. 14.2-3. By substituting (14.2-39) into (14.2-74), the time-average power of a MQAM signal can be expressed as

$$P_{\text{avg},x,T_d} = \frac{A^2}{2}\frac{1}{N}\sum_{n=1}^{N}\left[\left(\text{Re}\{w_n\}\right)^2 + \left(\text{Im}\{w_n\}\right)^2\right]. \tag{14.2-75}$$

For example purposes, let the total number of transmitted pulses (symbols) $N = M = 8$. If the 8 transmitted symbols are those shown in Table 14.2-1, then

$$P_{\text{avg},x,T_d} = 3A^2 \text{ W-}\Omega, \tag{14.2-76}$$

and if the 8 transmitted symbols are those shown in Table 14.2-2, then

$$P_{\text{avg},x,T_d} = 2.37A^2 \text{ W-}\Omega. \tag{14.2-77}$$

Therefore, the 8-QAM concentrically-circular signal-space constellation requires less time-average power than the 8-QAM rectangular signal-space constellation.

■

Example 14.2-3 Gray-Encoded 4-QAM

In this example we shall consider the problem of transmitting the following binary words

00 11 01 10 00 11 10 00 01 11

using MQAM where a binary word is equal to $n_b = 2$ bits. This is the same problem that was considered in Example 14.1-1 using MFSK. Since there are 10 binary words (20 bits), the total number of pulses (symbols) to be transmitted is $N = 10$. Also, since $n_b = 2$, there are $M = 2^2 = 4$ unique *pairs* of symbol values $(\text{Re}\{w_n\}, \text{Im}\{w_n\})$, and 4 unique binary words (4 possible combinations of 2 bits).

The next step is to assign a unique pair of symbol values to each of the 4 unique binary words. For example, see Table 14.2-4 where the 4 unique binary words are arranged in a Gray code to form a rectangular signal-space constellation. The set of pairs of numbers $\{(-1,-1), (1,-1), (1,1), (-1,1)\}$ is the alphabet. Figure 14.2-5 is a pictorial representation of Table 14.2-4. As can be seen from Fig. 14.2-5, only one bit is changed going from one signal-point to an adjacent signal-point in the constellation, either in the horizontal or vertical direction. Also, the distance between adjacent signal-points, either in the horizontal or vertical direction, is 2.

Table 14.2-4 Four Unique Pairs of Symbol Values Assigned to Four Unique Binary Words for a Rectangular Signal-Space Constellation

Gray Code	$\text{Re}\{w_n\}$	$\text{Im}\{w_n\}$
00	−1	−1
01	1	−1
11	1	1
10	−1	1

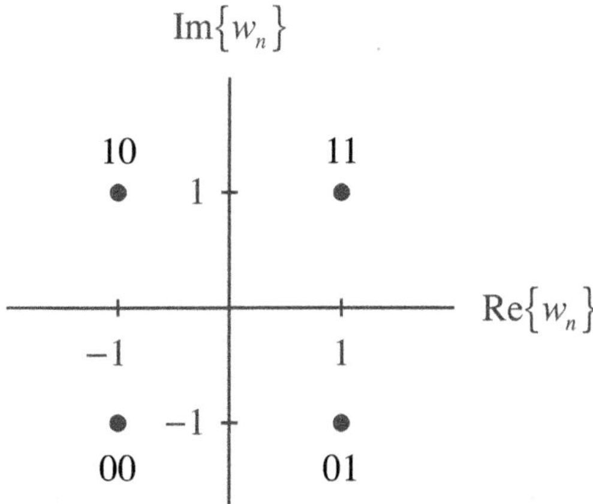

Figure 14.2-5 Rectangular signal-space constellation for Gray-encoded 4-QAM.

In order to simulate a time-domain MQAM signal using the information given up to this point, we need values for the amplitude factor A, the carrier frequency f_c, and the pulse length T of an individual pulse. For example, if $A = 100$ V, $f_c = 1$ kHz, and $T = T_{\text{sym}} = n_b T_b = 20$ msec, then we obtain the parameter values shown in Table 14.2-5 for the MQAM signal. Note that the values for A and T were chosen so that the factor $AT/2 = 1$ in (14.2-51). Using the pairs of symbol values assigned to the 4 unique binary words in Table 14.2-4, Table 14.2-6 shows the pairs of symbol values for the 10 transmitted pulses associated with the 10 binary words.

Table 14.2-5 Parameter Values of the Transmitted MQAM Signal

Amplitude factor A	100 V
Carrier frequency f_c	1 kHz
Pulse length of an individual pulse T	20 msec
Total number of transmitted pulses (symbols) N	10
Total duration of the transmitted signal $T_d = NT$	200 msec
Symbol duration T_{sym}	20 msec
Baud (symbol rate) $D = 1/T_{\text{sym}}$	50 symbols/sec
Number of bits per symbol (binary word) n_b	2
Number of *unique* symbol values $M = 2^{n_b}$	4
Bit duration $T_b = T_{\text{sym}}/n_b$	10 msec
Bit rate $R_b = 1/T_b$	100 bps
Time-average power P_{avg, x, T_d}	10,000 W-Ω

Table 14.2-6 Pairs of Symbol Values for the Ten Transmitted Pulses

n	Binary Word	$\text{Re}\{w_n\}$	$\text{Im}\{w_n\}$
1	00	−1	−1
2	11	1	1
3	01	1	−1
4	10	−1	1
5	00	−1	−1
6	11	1	1
7	10	−1	1
8	00	−1	−1
9	01	1	−1
10	11	1	1

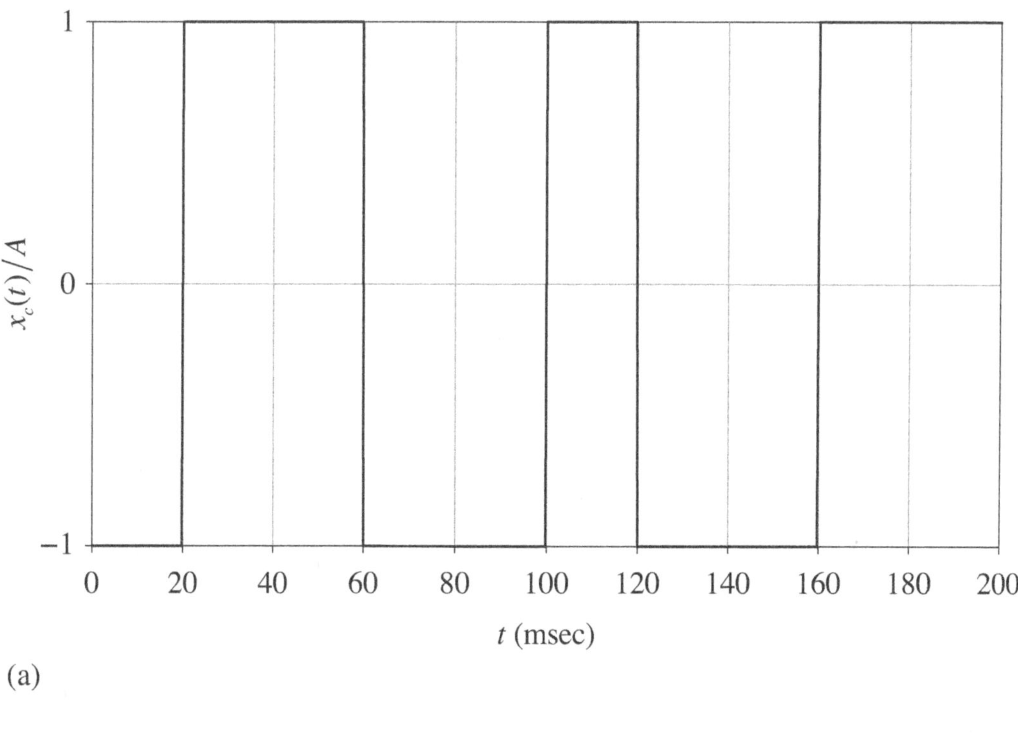

(a)

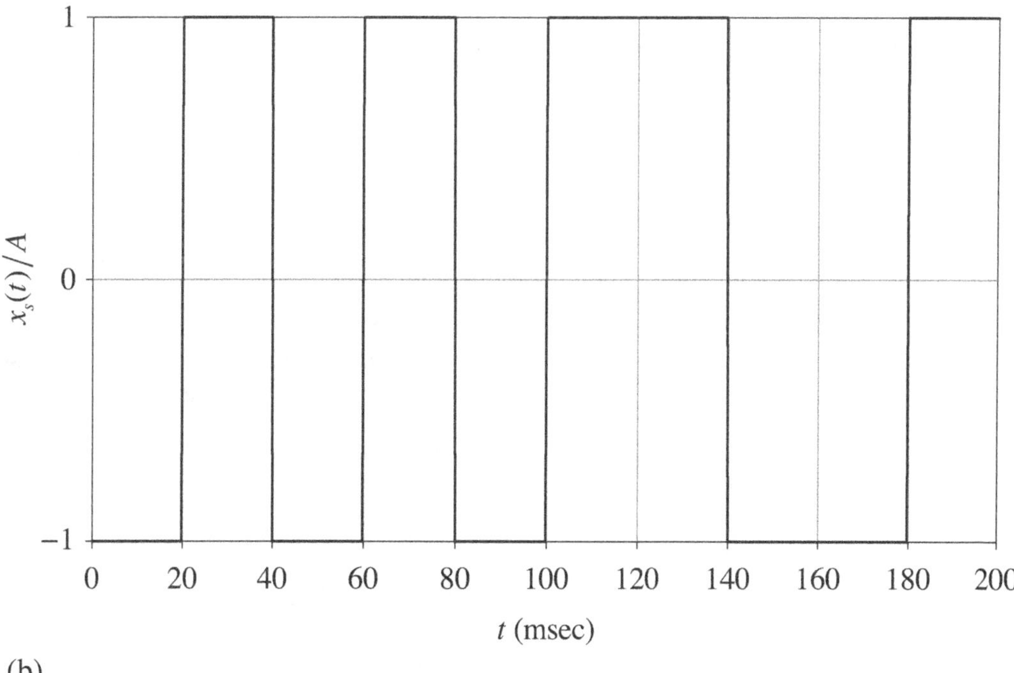

(b)

Figure 14.2-6 (a) Normalized MQAM cosine component, and (b) normalized MQAM sine component using the parameter values shown in Tables 14.2-5 and 14.2-6.

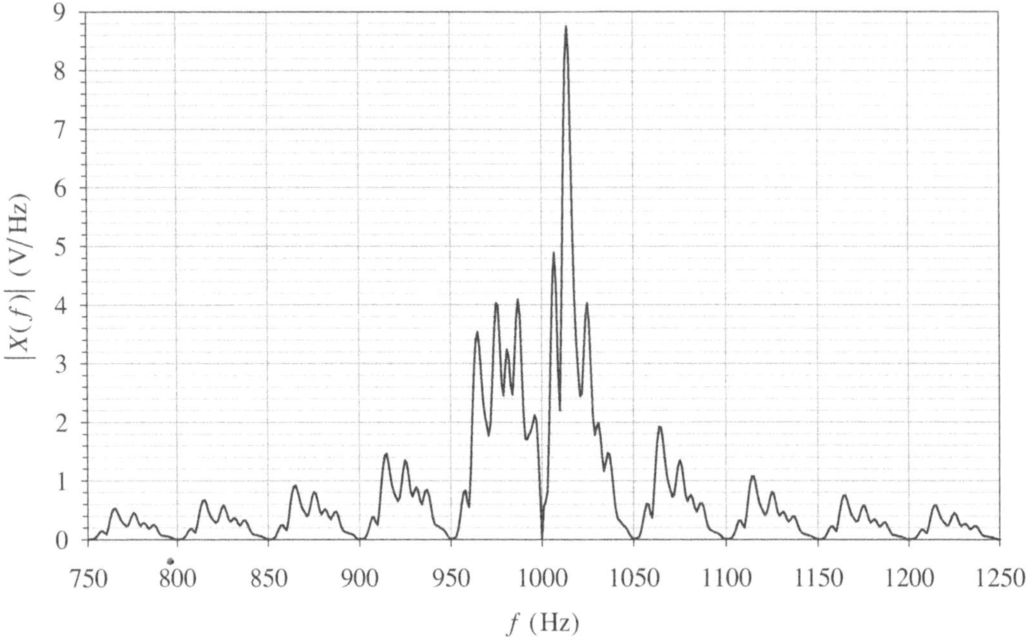

Figure 14.2-7 Magnitude of the MQAM frequency spectrum $X(f)$ given by (14.2-52) using the parameter values shown in Tables 14.2-5 and 14.2-6.

Figure 14.2-6 (a) is a plot of the normalized MQAM cosine component $x_c(t)/A$, Fig. 14.2-6 (b) is a plot of the normalized MQAM sine component $x_s(t)/A$, and Fig. 14.2-7 is a plot of the magnitude of the MQAM frequency spectrum $X(f)$ given by (14.2-52), using the parameter values shown in Tables 14.2-5 and 14.2-6. The bandwidth of $x(t)$ can be computed using (14.2-56). Since $D = 50$ symbols/sec, if $\text{NZC} = 1$, 3, and 5, then $\text{BW}_x = 100$ Hz, 300 Hz, and 500 Hz, respectively. As can be seen from Fig. 14.2-7, a bandwidth of 300 Hz is a good estimate, whereas a bandwidth of 500 Hz is more conservative. ■

14.2.4 Demodulation

As was mentioned in Subsection 14.1.5, demodulation is a procedure meant to retrieve a transmitted message at a receiver. In order to discuss demodulation, we first need to model a MQAM signal at a receiver, which we shall designate as $y(t)$. We shall use the following simple model for $y(t)$:

$$y(t) = Kx(t-\tau), \qquad \tau \le t \le \tau + T_d, \tag{14.2-78}$$

where the transmitted signal $x(t)$ is given by (14.2-33). As can be seen from

(14.2-78), $y(t)$ is modeled as an amplitude-scaled, time-delayed version of $x(t)$ (distortionless transmission) as was done for a MFSK signal at a receiver (see Subsection 14.1.5) where $K > 0$ is a dimensionless constant. Substituting (14.2-33) into (14.2-78) yields

$$y(t) = Kx_c(t-\tau)\cos(2\pi f_c t + \delta) - Kx_s(t-\tau)\sin(2\pi f_c t + \delta), \qquad \tau \le t \le \tau + T_d, \tag{14.2-79}$$

where [see (14.2-34) and (14.2-35)]

$$x_c(t-\tau) = A\sum_{n=1}^{N} \mathrm{Re}\{w_n\}\, p(t-[\tau+t_n]), \tag{14.2-80}$$

$$x_s(t-\tau) = A\sum_{n=1}^{N} \mathrm{Im}\{w_n\}\, p(t-[\tau+t_n]), \tag{14.2-81}$$

$p(t)$ is given by (14.2-10) and (14.2-11),

$$t_n = (n-1)T\,, \tag{14.2-82}$$

and

$$\delta = -2\pi f_c \tau \tag{14.2-83}$$

is a phase shift in radians that takes into account the time delay τ in seconds.

The received signal $y(t)$ can be demodulated by passing it through a quadrature-demodulator (QD) (see Section 11.3). As shown in Section 11.3, if there are *no* frequency and phase offsets, the outputs from the QD are

$$y_1(t) = GKx_c(t-[\tau+t_0]) \tag{14.2-84}$$

and

$$y_2(t) = GKx_s(t-[\tau+t_0]), \tag{14.2-85}$$

where $G > 0$ is the dimensionless gain and t_0 is the constant time delay of the ideal, lowpass filters in the QD. Substituting (14.2-80) into (14.2-84), and (14.2-81) into (14.2-85) yields

$$y_1(t) = GKA\sum_{n=1}^{N} \mathrm{Re}\{w_n\}\, p(t-[\tau+t_0+t_n]) \tag{14.2-86}$$

and

$$y_2(t) = GKA\sum_{n=1}^{N} \mathrm{Im}\{w_n\}\, p(t-[\tau+t_0+t_n]), \tag{14.2-87}$$

respectively. Each T seconds worth of data from $y_1(t)$ and $y_2(t)$ contains the values of $\mathrm{Re}\{w_n\}$ and $\mathrm{Im}\{w_n\}$ in the nth transmitted pulse, respectively. Once all the transmitted pairs of symbol values $(\mathrm{Re}\{w_n\}, \mathrm{Im}\{w_n\})$ have been determined, decode the pairs of symbol values back into binary words, and then decode the binary words back into quantized voltages. One can then use the quantized voltages to approximate the original analog (continuous-time) message using the reconstruction formula from the Sampling Theorem.

14.3 Orthogonal Frequency-Division Multiplexing

14.3.1 Time-Domain Description

If digital information is being transmitted using orthogonal frequency-division multiplexing (OFDM), then a time-domain description of the transmitted signal $x(t)$ is given as follows:

$$x(t) = \sum_{n=0}^{N-1} x_n(t), \qquad 0 \le t \le T, \tag{14.3-1}$$

where N is the total number of pulses (symbols) that are transmitted *simultaneously*,

$$x_n(t) = A|w_n|\cos\left(2\pi[f_c + \Delta f_n]t + \angle w_n + \varepsilon_n\right)\mathrm{rect}\left[(t - 0.5T)/T\right] \tag{14.3-2}$$

is the nth pulse, also known as (a.k.a.) the nth *subcarrier*, where A is the amplitude factor in volts; $|w_n|$ and $\angle w_n$ are the magnitude and phase of the *complex symbol* (see Fig. 14.2-1)

$$w_n = |w_n|\exp(+j\angle w_n), \tag{14.3-3}$$

f_c is the carrier frequency in hertz, the *frequency offset* in hertz is given by either

$$\Delta f_n = \frac{n - N'}{T}, \qquad N' = \begin{cases} N/2, & N \text{ even} \\ (N-1)/2, & N \text{ odd} \end{cases} \tag{14.3-4}$$

or

$$\Delta f_n = n/T, \tag{14.3-5}$$

$$T = NT_{\mathrm{sym}} \tag{14.3-6}$$

is the pulse length of an individual pulse, which is equal to the total duration of the transmitted signal in seconds;

$$T_{\text{sym}} = n_b T_b \tag{14.3-7}$$

is the *input symbol duration* in seconds – a.k.a. the duration of a binary word; n_b is the number of bits per symbol – a.k.a. the number of bits per binary word; T_b is the *bit duration* in seconds,

$$M = 2^{n_b} \tag{14.3-8}$$

is the total *even* number of *unique* symbol values – the different unique symbol values are known as the *alphabet*; ε_n is a possible, unwanted phase shift in radians at the transmitter, and

$$\text{rect}\left(\frac{t-0.5T}{T}\right) = \begin{cases} 1, & 0 \leq t \leq T \\ 0, & \text{otherwise} \end{cases} \tag{14.3-9}$$

is the time-shifted rectangle function. Note that the total number of transmitted pulses (symbols) N is greater than or equal to the total even number of unique symbol values M, that is,

$$N \geq M. \tag{14.3-10}$$

Because the numerator of the frequency offset Δf_n given by either (14.3-4) or (14.3-5) is an *integer*, if the product $f_c T$ is also equal to an *integer*, or $f_c T >> 1$ if $f_c T$ does not equal an integer, then the set of functions $x_n(t)$, $n = 0, 1, \ldots, N-1$, given by (14.3-2) is an *orthogonal* set of functions in the time interval $[0, T]$ (see Subsection 14.1.4). Whereas MFSK and MQAM signals are pulse trains, OFDM signals are not. Because an OFDM signal uses a complex symbol in general, OFDM is a combination of amplitude-shift keying (ASK) via $|w_n|$, and *M*-ary phase-shift keying (MPSK) via $\angle w_n$. See Subsection 14.2.1 for how to choose w_n for the special cases of amplitude-shift keying (ASK) and *M*-ary Phase-Shift Keying (MPSK).

14.3.2 Frequency Spectrum and Bandwidth

In order to derive an equation for the bandwidth of the OFDM signal $x(t)$ given by (14.3-1), we first have to find its frequency spectrum. Taking the Fourier transform of (14.3-1) yields

$$X(f) = \sum_{n=0}^{N-1} X_n(f), \tag{14.3-11}$$

where

$$X(f) = F\{x(t)\} \tag{14.3-12}$$

and

$$X_n(f) = F\{x_n(t)\} \tag{14.3-13}$$

is the frequency spectrum of the nth pulse $x_n(t)$ given by (14.3-2). The Fourier transform of $x_n(t)$ can be obtained by using the Fourier-transform pair

$$\begin{aligned}A\cos(2\pi f_c t+\theta_0)\operatorname{rect}\left(\frac{t-0.5T}{T}\right) \leftrightarrow &\frac{A}{2}T\operatorname{sinc}\left[(f-f_c)T\right]\exp\left[-j\pi(f-f_c)T\right]\exp(+j\theta_0)+\\ &\frac{A}{2}T\operatorname{sinc}\left[(f+f_c)T\right]\exp\left[-j\pi(f+f_c)T\right]\exp(-j\theta_0)\end{aligned} \tag{14.3-14}$$

and replacing A with $A|w_n|$, f_c with $f_c+\Delta f_n$, and θ_0 with $\angle w_n+\varepsilon_n$. Doing so and substituting the resulting expression into (14.3-11) yields the frequency spectrum of the OFDM signal $x(t)$ given by (14.3-1):

$$\boxed{\begin{aligned}X(f)=&\frac{A}{2}Tc_1(f)\sum_{n=0}^{N-1}|w_n|\operatorname{sinc}\left\{\left[f-(f_c+\Delta f_n)\right]T\right\}\exp(+j\pi\Delta f_n T)\exp(+j\angle w_n)\exp(+j\varepsilon_n)+\\ &\frac{A}{2}Tc_2(f)\sum_{n=0}^{N-1}|w_n|\operatorname{sinc}\left\{\left[f+(f_c+\Delta f_n)\right]T\right\}\exp(-j\pi\Delta f_n T)\exp(-j\angle w_n)\exp(-j\varepsilon_n)\end{aligned}} \tag{14.3-15}$$

where

$$c_1(f) = \exp\left[-j\pi(f-f_c)T\right] \tag{14.3-16}$$

and

$$c_2(f) = \exp\left[-j\pi(f+f_c)T\right]. \tag{14.3-17}$$

The units of $X(f)$ are volts per hertz.

Since bandwidth is always measured along the positive frequency axis, we shall work with the first term in (14.3-15), that is,

$$\begin{aligned}X(f)=\frac{A}{2}Tc_1(f)\sum_{n=0}^{N-1}|w_n|\operatorname{sinc}\left\{\left[f-(f_c+\Delta f_n)\right]T\right\}\exp(+j\pi\Delta f_n T)\exp(+j\angle w_n)\times\\ \exp(+j\varepsilon_n), \qquad f\geq 0.\end{aligned} \tag{14.3-18}$$

By examining the argument of the sinc function in (14.3-18), estimates of the maximum and minimum frequency components are given by

$$f_{\max} = f_c + \max \Delta f_n + \frac{\text{NZC}}{T} \tag{14.3-19}$$

and

$$f_{\min} = f_c + \min \Delta f_n - \frac{\text{NZC}}{T}, \tag{14.3-20}$$

where NZC is the integer number of zero-crossings of the sinc function that is used to estimate both the maximum and minimum frequency components $f_{\max}$ and $f_{\min}$, respectively. Since the frequency offset Δf_n is given by either (14.3-4) or (14.3-5), we shall use (14.3-4) first in order to estimate $f_{\max}$ and $f_{\min}$.

By referring to (14.3-4), it can be seen that for N even,

$$\max \Delta f_n = \Delta f_{N-1} = \left(\frac{N}{2} - 1\right)\frac{1}{T} \tag{14.3-21}$$

and

$$\min \Delta f_n = \Delta f_0 = -\frac{N}{2}\frac{1}{T}, \tag{14.3-22}$$

and for N odd,

$$\max \Delta f_n = \Delta f_{N-1} = \frac{N-1}{2}\frac{1}{T} \tag{14.3-23}$$

and

$$\min \Delta f_n = \Delta f_0 = -\frac{N-1}{2}\frac{1}{T}. \tag{14.3-24}$$

Therefore, substituting (14.3-21) into (14.3-19), and (14.3-22) into (14.3-20) yields for N even,

$$f_{\max} = f_c + \left(\frac{N}{2} - 1 + \text{NZC}\right)\frac{1}{T} \tag{14.3-25}$$

and

$$f_{\min} = f_c - \left(\frac{N}{2} + \text{NZC}\right)\frac{1}{T}, \tag{14.3-26}$$

and substituting (14.3-23) into (14.3-19), and (14.3-24) into (14.3-20) yields for N odd,

$$f_{\max} = f_c + \left(\frac{N-1}{2} + \text{NZC}\right)\frac{1}{T} \tag{14.3-27}$$

and

$$f_{\min} = f_c - \left(\frac{N-1}{2} + \text{NZC} \right) \frac{1}{T} . \tag{14.3-28}$$

Next we shall use (14.3-5) to estimate $f_{\max}$ and $f_{\min}$.

By referring to (14.3-5), it can be seen that

$$\max \Delta f_n = \Delta f_{N-1} = \frac{N-1}{T} \tag{14.3-29}$$

and

$$\min \Delta f_n = \Delta f_0 = 0 . \tag{14.3-30}$$

Therefore, substituting (14.3-29) into (14.3-19), and (14.3-30) into (14.3-20) yields

$$f_{\max} = f_c + (N + \text{NZC} - 1) \frac{1}{T} \tag{14.3-31}$$

and

$$f_{\min} = f_c - \frac{\text{NZC}}{T} . \tag{14.3-32}$$

The bandwidth BW_x (in hertz) of the OFDM signal is given by

$$\text{BW}_x = f_{\max} - f_{\min} . \tag{14.3-33}$$

Substituting (14.3-25) and (14.3-26), or (14.3-27) and (14.3-28), or (14.3-31) and (14.3-32), and (14.3-6) into (14.3-33) yields the same result

$$\boxed{\text{BW}_x = \left(1 + \frac{2\,\text{NZC} - 1}{N} \right) D} \tag{14.3-34}$$

where

$$\boxed{D = \frac{1}{T_{\text{sym}}} = \frac{1}{n_b T_b} = \frac{R_b}{n_b}} \tag{14.3-35}$$

is the *input symbol rate (input baud)* with units of symbols per second, and

$$\boxed{R_b = 1/T_b} \tag{14.3-36}$$

is the *bit rate* in bits per second. Since $\max|\text{sinc}(fT)| = 1$ for $f = 0$; $\text{sinc}(fT) = 0$ for $f = i/T$, where $i = \pm 1, \pm 2, \ldots$; and $|\text{sinc}(fT)| < 0.1$ for

$f > 3/T$; NZC should be at least 3, with 5 being a conservative choice. Substituting $\text{NZC} = 5$ into (14.3-34) yields

$$\text{BW}_x = \left(1 + \frac{9}{N}\right) D \tag{14.3-37}$$

or

$$\text{BW}_x \approx D, \quad N > 90. \tag{14.3-38}$$

Therefore, if the total number of transmitted pulses (symbols) $N > 90$, then the bandwidth of an OFDM signal is approximately equal to the input baud D.

14.3.3 Signal Energy and Time-Average Power

In this subsection we shall compute the energy and time-average power of the OFDM signal $x(t)$ given by (14.3-1). We begin by computing the energy of the nth pulse $x_n(t)$ given by (14.3-2). Note that $x_n(t)$ is an amplitude-and-angle-modulated carrier with amplitude and angle-modulating functions

$$a_n(t) = A|w_n| \operatorname{rect}\left[(t - 0.5T)/T\right] \tag{14.3-39}$$

and

$$\theta_n(t) = 2\pi \Delta f_n t + \angle w_n + \varepsilon_n, \tag{14.3-40}$$

respectively (see Section 11.2), where the time-shifted rectangle function is given by (14.3-9). In order to compute the energy of $x_n(t)$, we shall first compute the energy of its complex envelope $\tilde{x}_n(t)$.

Since the general form of the complex envelope of $x_n(t)$ is given by (see Section 11.2)

$$\tilde{x}_n(t) = a_n(t) \exp[+j\theta_n(t)], \tag{14.3-41}$$

substituting (14.3-39) and (14.3-40) into (14.3-41) yields

$$\begin{aligned} \tilde{x}_n(t) &= A|w_n| \exp\left[+j(2\pi \Delta f_n t + \angle w_n + \varepsilon_n)\right] \operatorname{rect}\left[(t - 0.5T)/T\right] \\ &= A w_n \exp\left[+j(2\pi \Delta f_n t + \varepsilon_n)\right] \operatorname{rect}\left[(t - 0.5T)/T\right]. \end{aligned} \tag{14.3-42}$$

The energy $E_{\tilde{x}_n}$ of $\tilde{x}_n(t)$ is given by

$$E_{\tilde{x}_n} = \int_{-\infty}^{\infty} a_n^2(t)\, dt, \tag{14.3-43}$$

and the energy E_{x_n} of $x_n(t)$ is given by (see Section 11.2)

$$E_{x_n} = E_{\tilde{x}_n}/2 . \tag{14.3-44}$$

Therefore, substituting (14.3-39) into (14.3-43) yields

$$E_{\tilde{x}_n} = A^2 |w_n|^2 T , \tag{14.3-45}$$

which is the energy (in joules-ohms) of the complex envelope $\tilde{x}_n(t)$ given by (14.3-42), and substituting (14.3-45) into (14.3-44) yields

$$\boxed{E_{x_n} = \frac{A^2}{2} |w_n|^2 T} \tag{14.3-46}$$

which is the energy (in joules-ohms) of $x_n(t)$ given by (14.3-2). The time-average power of $x_n(t)$ in watts-ohms is given by

$$\boxed{P_{\text{avg}, x_n, T} = \frac{E_{x_n}}{T} = \frac{A^2}{2} |w_n|^2} \tag{14.3-47}$$

From signal theory, the energy E_x of a signal $x(t)$ is defined as

$$E_x \triangleq \int_{-\infty}^{\infty} |x(t)|^2 dt . \tag{14.3-48}$$

Using (14.3-1),

$$|x(t)|^2 = x(t)x^*(t) = \sum_{m=0}^{N-1} \sum_{n=0}^{N-1} x_m(t) x_n^*(t), \qquad 0 \le t \le T , \tag{14.3-49}$$

and by substituting (14.3-49) into (14.3-48), we obtain

$$E_x = \sum_{m=0}^{N-1} \sum_{n=0}^{N-1} \int_0^T x_m(t) x_n^*(t) dt . \tag{14.3-50}$$

As was mentioned in Subsection 14.3.1, because the numerator of the frequency offset Δf_n given by either (14.3-4) or (14.3-5) is an *integer*, if the product $f_c T$ is also equal to an *integer*, or $f_c T >> 1$ if $f_c T$ does not equal an integer, then the set of functions $x_n(t)$, $n = 0, 1, \ldots, N-1$, given by (14.3-2) is an *orthogonal* set of functions in the time interval $[0, T]$. Therefore,

$$\int_0^T x_m(t)x_n^*(t)dt = E_{x_n}\delta_{mn}, \tag{14.3-51}$$

where the energy of the nth pulse E_{x_n} is given by (14.3-46), and δ_{mn} is the Kronecker delta given by (14.1-56). Substituting (14.3-51) into (14.3-50) yields

$$E_x = \sum_{m=0}^{N-1}\sum_{n=0}^{N-1} E_{x_n}\delta_{mn} = \sum_{n=0}^{N-1} E_{x_n}\sum_{m=0}^{N-1}\delta_{mn} = \sum_{n=0}^{N-1} E_{x_n}, \tag{14.3-52}$$

or

$$\boxed{E_x = \frac{A^2}{2}T\sum_{n=0}^{N-1}|w_n|^2} \tag{14.3-53}$$

which is the energy (in joules-ohms) of the OFDM signal $x(t)$ given by (14.3-1). Therefore, the time-average power of $x(t)$ in watts-ohms is given by

$$\boxed{P_{\text{avg},x,T} = \frac{E_x}{T} = \frac{A^2}{2}\sum_{n=0}^{N-1}|w_n|^2} \tag{14.3-54}$$

Table 14.3-1 Signal Bandwidth and Time-Average Power of MFSK, MQAM, and OFDM Signals

	Signal Bandwidth BW_x (Hz)	Time-Average Power (W-Ω)
MFSK	$(M + 2\text{NZC})D$	$P_{\text{avg},x,T_d} = \frac{A^2}{2}$ $T_d = NT_{\text{sym}}$
MQAM	$2\text{NZC} \times D$	$P_{\text{avg},x,T_d} = \frac{A^2}{2}\frac{1}{N}\sum_{n=1}^{N}|w_n|^2$ $T_d = NT_{\text{sym}}$
OFDM	$\left(1 + \frac{2\text{NZC} - 1}{N}\right)D$	$P_{\text{avg},x,T} = \frac{A^2}{2}\sum_{n=0}^{N-1}|w_n|^2$ $T = NT_{\text{sym}}$

Table 14.3-1 compares the signal bandwidths and time-average powers of MFSK, MQAM, and OFDM signals. As can be seen from Table 14.3-1, MFSK has the largest bandwidth and smallest time-average power, whereas OFDM has the smallest bandwidth and largest time-average power. All three signals have the same total duration of NT_{sym} sec. Recall that the pulse length T is equal to the symbol duration T_{sym} in both MFSK and MQAM signals [see (14.1-4) and (14.2-4)], whereas the pulse length T is equal to the total duration of an OFDM signal, that is, $T = NT_{\text{sym}}$.

14.3.4 Demodulation

As was mentioned in Subsections 14.1.5 and 14.2.4, demodulation is a procedure meant to retrieve a transmitted message at a receiver. In order to discuss demodulation, we first need to model an OFDM signal at a receiver, which we shall designate as $y(t)$. We shall use the following simple model for $y(t)$:

$$y(t) = Kx(t-\tau), \qquad \tau \le t \le \tau + T, \tag{14.3-55}$$

where the transmitted signal $x(t)$ is given by (14.3-1). As can be seen from (14.3-55), $y(t)$ is modeled as an amplitude-scaled, time-delayed version of $x(t)$ (distortionless transmission) as was done for MFSK and MQAM signals at a receiver (see Subsections 14.1.5 and 14.2.4, respectively) where $K > 0$ is a dimensionless constant. Substituting (14.3-1) into (14.3-55) yields

$$y(t) = \sum_{n=0}^{N-1} y_n(t), \qquad \tau \le t \le \tau + T, \tag{14.3-56}$$

where

$$y_n(t) = KA|w_n|\cos\left(2\pi[f_c + \Delta f_n](t-\tau) + \angle w_n + \varepsilon_n\right)\text{rect}\left[\left(t - [0.5T+\tau]\right)/T\right] \tag{14.3-57}$$

is the nth pulse of the received signal, and

$$\text{rect}\left(\frac{t-[0.5T+\tau]}{T}\right) = \begin{cases} 1, & \tau \le t \le \tau + T \\ 0, & \text{otherwise.} \end{cases} \tag{14.3-58}$$

The received OFDM signal $y(t)$ is demodulated by computing its Fourier transform. Taking the Fourier transform of (14.3-55) yields

$$Y(f) = KX(f)\exp(-j2\pi f\tau), \tag{14.3-59}$$

where $X(f)$ is the Fourier transform of the transmitted signal $x(t)$ given by (14.3-15). Substituting (14.3-18) for $X(f)$ along the positive frequency axis into (14.3-59) yields

$$Y(f) = K\frac{A}{2}T c_1(f)\exp(-j2\pi f\tau)\times \sum_{n=0}^{N-1}|w_n|\text{sinc}\{[f-(f_c+\Delta f_n)]T\}\exp[+j(\pi\Delta f_n T+\angle w_n+\varepsilon_n)], \qquad f \ge 0. \tag{14.3-60}$$

The demodulation of $y(t)$ is demonstrated in Example 14.3-1.

Example 14.3-1 Gray-Encoded QPSK

In this example we shall consider the problem of transmitting the following binary words

00 11 01 10 00 11 10 00 01 11

using OFDM where a binary word is equal to $n_b = 2$ bits. This is the same problem that was considered in Example 14.1-1 using MFSK, and Example 14.2-3 using MQAM. Since there are 10 binary words (20 bits), the total number of pulses (symbols) to be transmitted is $N = 10$. Also, since $n_b = 2$, the total even number of *unique* symbol values $M = 2^2 = 4$.

Table 14.3-2 Four Unique Symbol Values Assigned to Four Unique Binary Words

Gray Code	Symbol w_n
00	$1\exp(+j225°)$
01	$1\exp(+j315°)$
11	$1\exp(+j45°)$
10	$1\exp(+j135°)$

The next step is to assign a unique symbol value to each of the 4 *unique* binary words (4 possible combinations of 2 bits). For example, see Table 14.3-2 where the 4 unique binary words are arranged in a Gray code using 4-ary PSK, also known as quadrature PSK (QPSK). The set of complex numbers $\{1\exp(+j45°), 1\exp(+j135°), 1\exp(+j225°), 1\exp(+j315°)\}$ is the alphabet. Note that $\pi/4 = 45°$, $3\pi/4 = 135°$, $5\pi/4 = 225°$, and $7\pi/4 = 315°$ [see (14.2-20)]. Since QPSK is being used, $|w_n| = 1$ is not important, only the values of $\angle w_n$ encode the 4 unique binary words. Figure 14.3-1 is a pictorial representation of Table 14.3-2. As can be seen from Fig. 14.3-1, only one bit is changed going from one signal-point to an adjacent signal-point in the constellation.

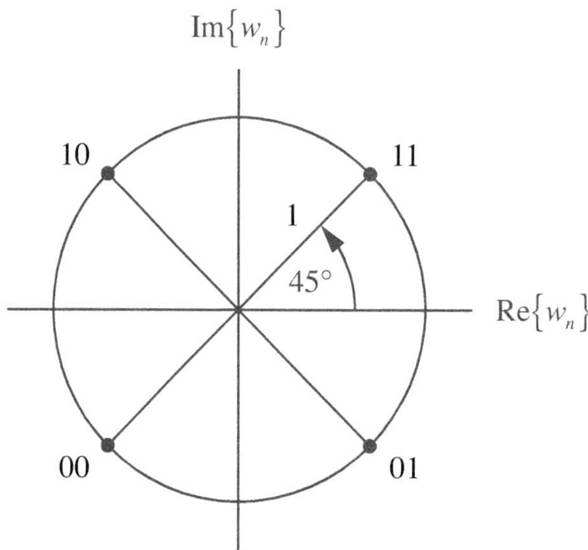

Figure 14.3-1 Signal-space constellation for Gray-encoded QPSK.

In order to simulate a time-domain OFDM signal using the information given up to this point, we need values for the amplitude factor A, the carrier frequency f_c, the input symbol duration T_{sym}, and to decide which equation to use for the frequency offset Δf_n. For example, if $A = 10\text{ V}$, $f_c = 1\text{ kHz}$, and $T_{\text{sym}} = n_b T_b = 20\text{ msec}$, then $T = NT_{\text{sym}} = 200\text{ msec}$, the factor $AT/2 = 1$ in (14.3-15), and $f_c T = 200$ is an integer (see Subsection 14.1.4). In this example we shall use the frequency offset given by (14.3-4). Substituting $N = 10$ and $T = 200\text{ msec}$ into (14.3-4) yields

$$\Delta f_n = 5(n-5)\text{ Hz}. \tag{14.3-61}$$

Table 14.3-3 summarizes the parameter values of the OFDM signal. Using the symbol values for the 4 unique binary words in Table 14.3-2, and Δf_n given by

(14.3-61), Table 14.3-4 shows the symbol values and frequencies of the 10 transmitted pulses associated with the 10 binary words, where the phase term $\varepsilon_n = 0 \;\; \forall \;\; n$ for example purposes.

Table 14.3-3 Parameter Values of the Transmitted OFDM Signal

Amplitude factor A	10 V
Carrier frequency f_c	1 kHz
Input symbol duration T_{sym}	20 msec
Total number of transmitted pulses (symbols) N	10
Pulse length of an individual pulse $T = NT_{\text{sym}}$	200 msec
Total duration of the transmitted signal $T = NT_{\text{sym}}$	200 msec
Frequency offset Δf_n	$5(n-5)$ Hz
Input baud (input symbol rate) $D = 1/T_{\text{sym}}$	50 symbols/sec
Number of bits per symbol (binary word) n_b	2
Number of *unique* symbol values $M = 2^{n_b}$	4
Bit duration $T_b = T_{\text{sym}}/n_b$	10 msec
Bit rate $R_b = 1/T_b$	100 bps
Time-average power $P_{\text{avg},x,T}$	500 W-Ω

Table 14.3-4 Symbol Values and Frequencies of the Ten Transmitted Pulses

n	Binary Word	w_n	$\Delta f_n = 5(n-5)$ (Hz)	$f_c + \Delta f_n$ (Hz)	ε_n (deg)
0	00	$1\exp(+j225°)$	−25	975	0
1	11	$1\exp(+j45°)$	−20	980	0
2	01	$1\exp(+j315°)$	−15	985	0
3	10	$1\exp(+j135°)$	−10	990	0
4	00	$1\exp(+j225°)$	−5	995	0
5	11	$1\exp(+j45°)$	0	1000	0
6	10	$1\exp(+j135°)$	5	1005	0
7	00	$1\exp(+j225°)$	10	1010	0
8	01	$1\exp(+j315°)$	15	1015	0
9	11	$1\exp(+j45°)$	20	1020	0

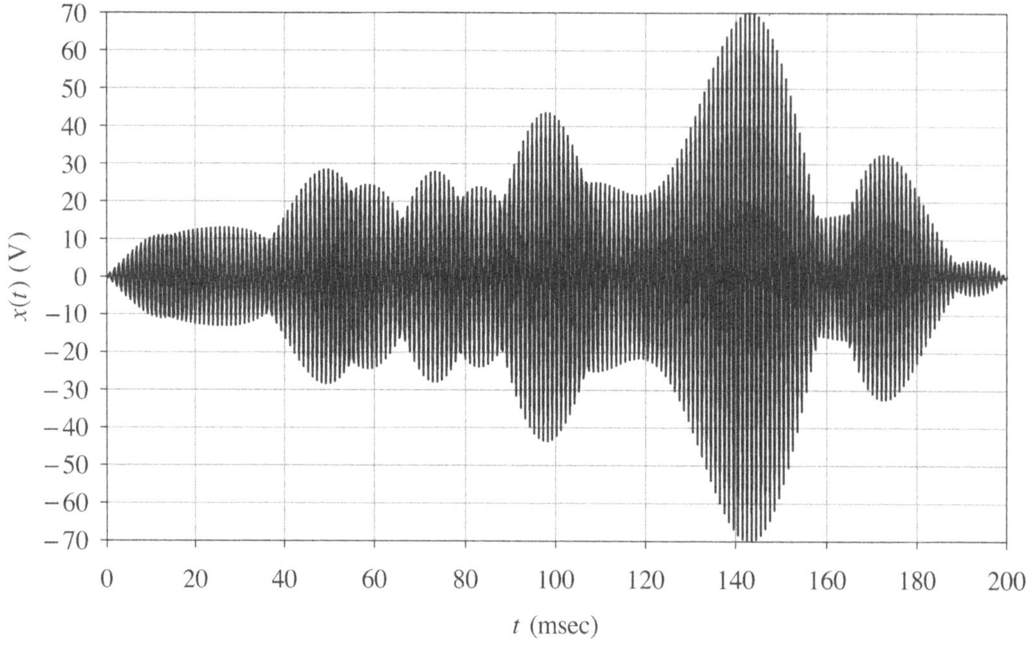

Figure 14.3-2 OFDM signal $x(t)$ given by (14.3-1) using the parameter values shown in Tables 14.3-3 and 14.3-4.

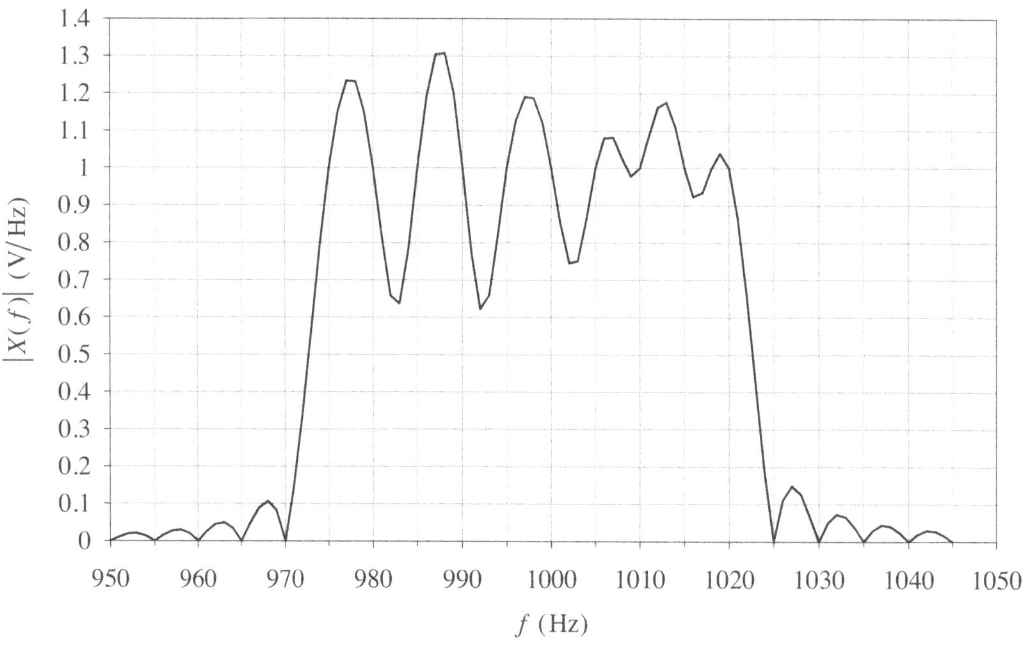

Figure 14.3-3 Magnitude of the OFDM frequency spectrum $X(f)$ given by (14.3-18) using the parameter values shown in Tables 14.3-3 and 14.3-4.

Figure 14.3-2 is a plot of the OFDM signal $x(t)$ given by (14.3-1), and Fig. 14.3-3 is a plot of the magnitude of the OFDM frequency spectrum $X(f)$ given by (14.3-18), using the parameter values shown in Tables 14.3-3 and 14.3-4. Although the amplitude factor $A = 10\text{ V}$, by inspecting Fig. 14.3-2, it can be seen that the amplitude of $x(t)$ grows to 70 V as a result of all the pulses being transmitted simultaneously and constructive interference. This is a disadvantage of OFDM. The bandwidth of $x(t)$ can be computed using (14.3-34). Since $N = 10$ and $D = 50\text{ symbols/sec}$, if $\text{NZC} = 1$, 3, and 5, then $\text{BW}_x = 55\text{ Hz}$, 75 Hz, and 95 Hz, respectively. As can be seen from Fig. 14.3-3, a bandwidth of 75 Hz is a good estimate, whereas a bandwidth of 95 Hz is more conservative.

As was previously mentioned, the received OFDM signal $y(t)$ is demodulated by computing its Fourier transform. Since $Y(f)$ was shown to be directly proportional to $X(f)$ [see (14.3-59)], for example purposes, we shall compute the Fourier transform of the transmitted signal $x(t)$ shown in Fig. 14.3-2 using the algorithm discussed in Appendix 7B.

In order to compute the Fourier transform of $x(t)$, a sampling frequency of 4 kHz was used. Since the total duration of $x(t)$ is $T = 200\text{ msec}$, the total number of samples taken was

$$N = f_S T = 4000\text{ Hz} \times 0.2\text{ sec} = 800. \tag{14.3-62}$$

If padding-with-zeros is not done, then $Z = 0$, the fundamental period $T_0 = T$, and the DFT bin spacing is

$$f_0 = \frac{1}{T_0} = \frac{1}{T} = \frac{1}{0.2\text{ sec}} = 5\text{ Hz}. \tag{14.3-63}$$

Since the frequencies of interest $\{975\text{ Hz}, 980\text{ Hz}, \ldots, 1015\text{ Hz}, 1020\text{ Hz}\}$ are located at DFT bins when $Z = 0$, padding-with-zeros was not required. Note that the frequency offset Δf_n, given by either (14.3-4) or (14.3-5), is an integer multiple of the DFT bin spacing f_0 given by (14.3-63).

Estimates of the magnitude and phase spectra of $x(t)$ are shown in Table 14.3-5 where integer n is the pulse (subcarrier) number, integer q is the DFT bin number, frequency $f = qf_0\text{ Hz}$ where the DFT bin spacing f_0 is given by (14.3-63), and $\hat{X}(f)$ is the estimate of the Fourier transform of $x(t)$ using the algorithm discussed in Appendix 7B. Each frequency component in Table 14.3-5 corresponds to an individual transmitted pulse (subcarrier) where $|\hat{X}(f)| = |w_n|$ because $AT/2 = 1$ and $\angle\hat{X}(f) = \angle w_n$ because $\varepsilon_n = 0\ \forall\ n$. Note that the phase angles $225°$ and $315°$ are equivalent to $-135°$ and $-45°$, respectively. The magnitude and phase values in Table 14.3-5 match exactly the symbol values in Table 14.3-4. As was mentioned earlier, since QPSK is used in this example, the value of $|w_n|$ for each pulse is not important, only the value of $\angle w_n$ encodes a binary word.

Table 14.3-5 Estimates of the Magnitude and Phase Spectra of the Transmitted OFDM Signal

n	q	f (Hz)	$\lvert\hat{X}(f)\rvert$ (V/Hz)	$\angle\hat{X}(f)$ (deg)
0	195	975	1	225
1	196	980	1	45
2	197	985	1	315
3	198	990	1	135
4	199	995	1	225
5	200	1000	1	45
6	201	1005	1	135
7	202	1010	1	225
8	203	1015	1	315
9	204	1020	1	45

By computing the Fourier transform of $x(t)$, we were able to demonstrate that values of the symbol w_n, that is, $|w_n|$ and $\angle w_n$ where $w_n = |w_n| \exp(+j\angle w_n)$, are located at DFT bins corresponding to frequencies $f = f_c + \Delta f_n$, $n = 0, 1, \ldots, N-1$. That is why the received OFDM signal $y(t)$ is demodulated by computing its Fourier transform using the algorithm discussed in Appendix 7B. Once all the transmitted symbol values have been determined, decode the symbol values back into binary words, and then decode the binary words back into quantized voltages. One can then use the quantized voltages to approximate the original analog (continuous-time) message using the reconstruction formula from the Sampling Theorem. ■

Problems

Section 14.1

14-1 If a total of 50 symbols are transmitted using MFSK, where each symbol represents a 4-bit binary word, and if the total duration of the transmitted signal is 250 msec, then

(a) how many unique symbol values are there?

(b) what is the baud?

(c) what is the bit rate?

(d) what is the bit duration?

(e) what is the frequency offset and alphabet of the symbol?

(f) what is the bandwidth of the signal if a factor of 3 is used for the number of zero-crossings parameter?

14-2 If a total of 100 symbols are transmitted using MFSK, and if the baud is 1000 symbols/sec and there are 16 unique symbol values, then find the bandwidth of the transmitted signal using a factor of 3 for the number of zero-crossings parameter.

Section 14.2

14-3 Repeat 14-2 using MQAM.

14-4 Digital information is being transmitted using MQAM, where symbol values represent 3-bit binary words. If ASK is being used to determine symbol values, then find the alphabet of the symbol.

Section 14.3

14-5 Repeat 14-2 using OFDM.

Chapter 15

Synthetic-Aperture Sonar

In this chapter we shall discuss some of the fundamentals concerning the design and analysis of a *synthetic-aperture sonar* (SAS). We shall also discuss the various assumptions and approximations that are made in order to derive some of the classic results for a SAS. A SAS is an active sonar system that shares some of the same principles of operation as a *side-looking sonar* (SLS) because a SLS is used to create a *synthetic aperture* (SA), as discussed in Section 15.1. As a result, our discussion in this chapter will rely heavily on the material covered in Chapter 5.

One of the biggest differences between a SAS and a SLS is the *along-track (azimuthal) resolution* capability of each sonar system. In Section 5.3, it was shown that the along-track resolution of a SLS depends on the 3-dB beamwidth of the horizontal, far-field beam pattern of the transmit aperture of the SLS, and the cross-range coordinate of a point on the ocean bottom. As the slant-range to a point on the ocean bottom increases, the cross-range increases, and the ability of a SLS to resolve closely-spaced objects on the ocean bottom at the same cross-range *decreases* (see Fig. 5.3-3). However, as will be shown in Section 15.2, the along-track resolution of a SAS is *independent* of the 3-dB beamwidth of the horizontal, far-field beam pattern of the transmit aperture of the SLS, and slant-range. As a result, a SAS is capable of producing very high resolution underwater images of large areas on the ocean bottom, much better than a SLS.

15.1 Creating a Synthetic Aperture

In order to create a synthetic aperture (SA), a side-looking sonar (SLS) mounted on an underwater vehicle is placed in motion with a constant velocity vector (constant speed and direction), at a constant height (altitude) above the ocean bottom. We shall use the same geometry and model for a SLS as shown in Fig. 5.1-1, which is reproduced as Fig. 15.1-1 for convenience, where the X axis is cross-range (the *cross-track direction*), the Y axis is depth, and the Z axis is down-range (the *along-track direction*). The SLS is modeled as a *single* planar aperture, rectangular in shape, lying in the YZ plane with dimensions L_Y and L_Z meters in the Y and Z directions. It is used alternately in the active (transmit) mode and passive (receive) mode. A SLS composed of a single planar aperture used alternately in the transmit and receive modes of operation to create a SA is known as a *single-element*, synthetic-aperture sonar (SESAS). It should be noted that a SLS may be composed of two separate planar apertures – a transmit aperture and a receive aperture. This kind of SLS used to create a SA is known as a *multi-element*, synthetic-aperture sonar (MESAS). A MESAS will be discussed in Subsection 15.4.1. Side-looking sonars are usually mounted on both sides of an underwater vehicle. Scatterers lie in the XZ plane on the ocean bottom with rectangular coordinates (x, y, z) or spherical coordinates (r, θ, ψ) (see Fig. 15.1-1).

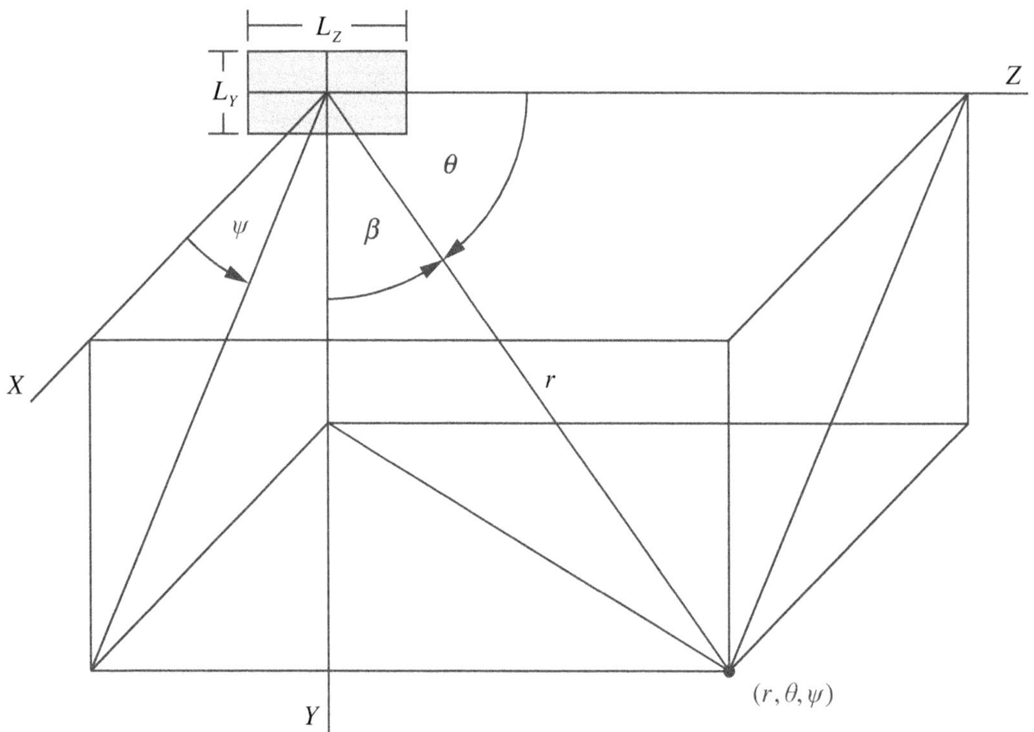

Figure 15.1-1 A SLS (planar aperture, rectangular in shape) lying in the YZ plane. Also shown is a field point in three-dimensional space with spherical coordinates (r, θ, ψ) as measured from the center of the aperture, and the angle β which is measured from the positive Y axis.

A SLS is commonly referred to as the *physical aperture* or the *real aperture* in the SAS literature. We shall call it the physical aperture (PA). As shown in Fig. 15.1-2, the PA is traveling in the positive Z direction (the along-track direction). When the PA transmits a pulse for the first time at time t_1 seconds, the scatterer on the ocean bottom at slant-range $r_\perp$ meters and z coordinate z_{CPA} meters begins to appear at the edge of the 3-dB beamwidth $\Delta\theta_{\text{PA}}$ of the horizontal, far-field beam pattern of the PA, where $r_\perp$ is measured *perpendicular* to the Z axis, and CPA is the acronym for *closest point of approach*. The z coordinate z_{CPA} is determined when the PA transmits a pulse at

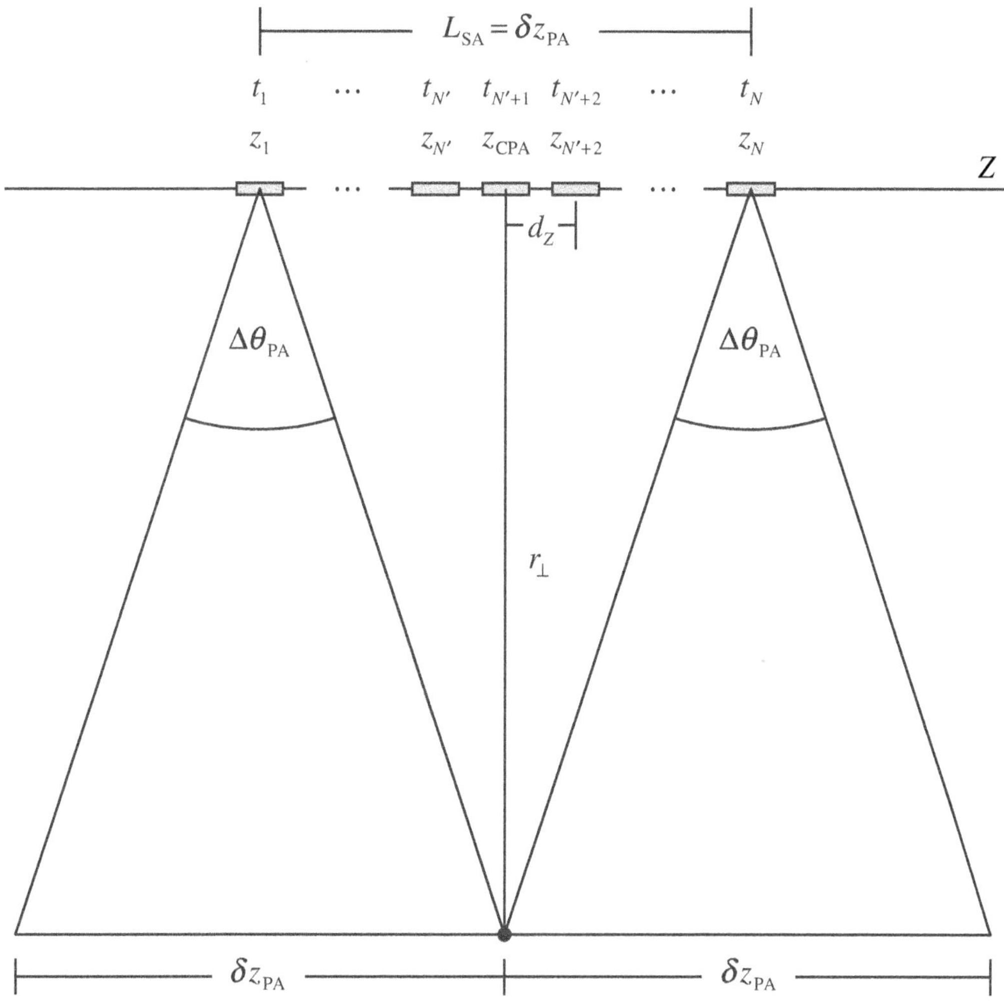

Figure 15.1-2 Creating a synthetic aperture.

time $t_{N'+1}$, where

$$N' = (N-1)/2 \tag{15.1-1}$$

and N is an *odd* integer equal to the total number of transmitted pulses, which is equal to the total number of elements in the *synthetic array* (SA). As successive pulses are transmitted every d_Z meters (d_Z will be determined later in Section 15.3), the scatterer remains within the 3-dB beamwidth $\Delta\theta_{PA}$ of the PA as it moves along the Z axis, until the last transmission at time t_N. At time t_N, the scatterer is at the edge of the 3-dB beamwidth $\Delta\theta_{PA}$ of the PA. Each time the PA transmits a pulse, its location along the Z axis can be considered the location of an element in a linear array. As shown in Fig. 15.1-2, the length of the SA, L_{SA}, is equal to the along-track (azimuthal) resolution δz_{PA} of the PA. Equations for δz_{PA} and L_{SA} shall be derived next.

A realistic SLS composed of a single planar aperture is a *rectangular array* of identical, equally-spaced, omnidirectional, electroacoustic transducers with *rectangular amplitude weights*. A very good approximation of this kind of array is a *rectangular piston* (see Sections 5.6 and 3.2). If a SLS (the PA) is modeled as a single rectangular piston lying in the YZ plane as shown in Fig. 15.1-1, then the 3-dB beamwidth of the horizontal, far-field beam pattern in the XZ plane is given by [see (5.7-6)]

$$\Delta\theta = 2\sin^{-1}\left(0.443\frac{\lambda}{L_Z}\right), \qquad 0.443\frac{\lambda}{L_Z} \le 1, \tag{15.1-2}$$

where $\lambda = c/f$ is the wavelength in meters, c is the speed of sound in meters per second, and f is frequency in hertz. It is common in the SLS and SAS literature to approximate (15.1-2). Since

$$\sin^{-1}(x) = x + \frac{1}{2}\frac{x^3}{3} + \frac{3}{8}\frac{x^5}{5} + \ldots, \qquad |x| \le 1, \tag{15.1-3}$$

if $0.443\lambda/L_Z < 1$, then by using only the first term in the series expansion of $\sin^{-1}(x)$, (15.1-2) can be approximated as

$$\Delta\theta \approx 0.886\lambda/L_Z, \qquad 0.443\lambda/L_Z < 1, \tag{15.1-4}$$

or, as is done in the SLS and SAS literature,

$$\Delta\theta \approx \lambda/L_Z, \qquad 0.443\lambda/L_Z < 1, \tag{15.1-5}$$

where $\Delta\theta$ is in radians (dimensionless). The values for $\Delta\theta$ obtained from (15.1-4) are in very good agreement with those from (15.1-2) – the values from (15.1-5) are not (see Table 15.1-1).

Note that the *Rayleigh beamwidth* of the horizontal, far-field beam pattern of a rectangular piston, which is the beamwidth to the *first null* – one-half the *null-to-null beamwidth* – is approximately equal to λ/L_Z if $\lambda/L_Z < 1$ (see Appendix 15A). The values for $\Delta\theta$ obtained from (15.1-5) are actually in better agreement with those from (15A-8) – the Rayleigh beamwidth of the horizontal, far-field beam pattern of a rectangular piston, $\Delta\theta_{\text{R}}$ (see Table 15.1-2). Figure 15.1-3 shows a comparison of the 3-dB beamwidth $\Delta\theta$, the Rayleigh beamwidth $\Delta\theta_{\text{R}}$, and the null-to-null beamwidth $\Delta\theta_{\text{NN}}$.

Table 15.1-1 Three-dB Horizontal Beamwidth Calculations Using (15.1-2), (15.1-4), and (15.1-5), Respectively

λ/L_Z	$\Delta\theta = 2\sin^{-1}(0.443\lambda/L_Z)$	$\Delta\theta \approx 0.886\lambda/L_Z$	$\Delta\theta \approx \lambda/L_Z$
0.02	1.015°	1.015°	1.146°
0.03	1.523°	1.523°	1.719°
0.04	2.031°	2.031°	2.292°
0.1	5.078°	5.076°	5.730°
0.2	10.166°	10.153°	11.459°
0.3	15.274°	15.229°	17.189°
0.35	17.839°	17.767°	20.054°

Table 15.1-2 Rayleigh Horizontal Beamwidth Calculations Using (15A-8) and (15.1-5), Respectively

λ/L_Z	$\Delta\theta_R = \sin^{-1}(\lambda/L_Z)$	$\Delta\theta \approx \lambda/L_Z$
0.02	1.146°	1.146°
0.03	1.719°	1.719°
0.04	2.292°	2.292°
0.1	5.739°	5.730°
0.2	11.537°	11.459°
0.3	17.458°	17.189°
0.35	20.487°	20.054°

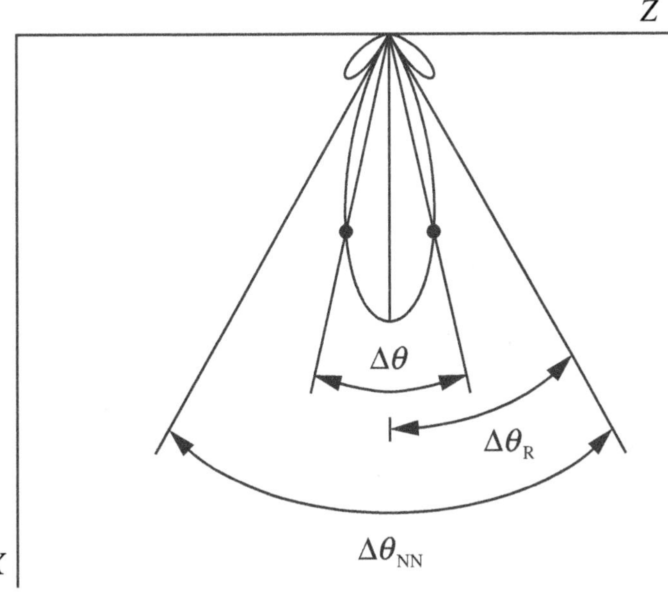

Figure 15.1-3 Comparison of the 3-dB beamwidth $\Delta\theta$, the Rayleigh beamwidth $\Delta\theta_R$, and the null-to-null beamwidth $\Delta\theta_{NN}$.

The along-track (azimuthal) resolution Δz in meters at slant-range r and cross-range x as shown in Fig. 5.3-1 for a SLS is given by (5.3-13). However, when analyzing a SLS or SAS, it is also customary to compute Δz as an *arc length*. Therefore, by changing notation from Δz to δz in order to avoid confusion,

$$\delta z = r\Delta\theta , \tag{15.1-6}$$

where the slant-range r (in meters) is measured from the center of the SLS and is *perpendicular* to the Z axis as shown in Fig. 5.3-1, and $\Delta\theta$ is the 3-dB beamwidth of the horizontal, far-field beam pattern of the SLS. The following additional changes in notation are made in order to make a clear distinction between the PA (the SLS) and the SA:

$$\delta z_{\text{PA}} = r_{\perp}\Delta\theta_{\text{PA}} , \tag{15.1-7}$$

where $\delta z_{\text{PA}} = \delta z$, $r_{\perp} = r$, and $\Delta\theta_{\text{PA}} = \Delta\theta$. The 3-dB beamwidth of the horizontal, far-field beam pattern of the PA [see (15.1-2)]

$$\Delta\theta_{\text{PA}} = 2\sin^{-1}\left(0.443\frac{\lambda}{L_{\text{PA}}}\right), \qquad 0.443\frac{\lambda}{L_{\text{PA}}} \le 1 , \tag{15.1-8}$$

can be approximated by using either [see (15.1-4)]

$$\Delta\theta_{\text{PA}} \approx 0.886\lambda/L_{\text{PA}} , \qquad 0.443\lambda/L_{\text{PA}} < 1 , \tag{15.1-9}$$

or [see (15.1-5)]

$$\Delta\theta_{\text{PA}} \approx \lambda/L_{\text{PA}} , \qquad 0.443\lambda/L_{\text{PA}} < 1 , \tag{15.1-10}$$

where $L_{\text{PA}} = L_Z$. Equation (15.1-9) is the better approximation for the 3-dB beamwidth, whereas (15.1-10) is a better approximation for the Rayleigh beamwidth (not the 3-dB beamwidth). Although (15.1-9) is the better approximation for the 3-dB beamwidth, (15.1-10) is commonly used in the SLS and SAS literature.

By referring to Fig. 15.1-2, the length of the SA in meters is given by

$$\boxed{L_{\text{SA}} = \delta z_{\text{PA}} = r_{\perp}\Delta\theta_{\text{PA}}} \tag{15.1-11}$$

where the slant-range $r_{\perp}$ is *perpendicular* to the Z axis. As can be seen from (15.1-11), *the length of a SA is a function of slant-range* $r_{\perp}$ *and the 3-dB beamwidth of the horizontal, far-field beam pattern of the PA,* $\Delta\theta_{PA}$, which is a function of the ratio λ/L_{PA} [see (15.1-8) through (15.1-10)]. Substituting (15.1-9) into (15.1-11) yields

$$L_{SA} \approx 0.886 r_{\perp} \lambda / L_{PA}, \tag{15.1-12}$$

or, by substituting (15.1-10) into (15.1-11), we obtain the following equation for the length of a SA that appears often in the SAS literature:

$$\boxed{L_{SA} \approx r_{\perp} \lambda / L_{PA}} \tag{15.1-13}$$

15.2 Along-Track (Azimuthal) Resolution

In this section we shall derive an equation for the along-track (azimuthal) resolution of a synthetic-aperture sonar (SAS), δz_{SA}, by computing the *Doppler bandwidth* of a SAS for a stationary (motionless) scatterer. However, in order to compute the Doppler bandwidth, we first need to compute the *Doppler history*, which we shall do next.

The location of the center of the physical aperture (PA) along the Z axis at time t is given by

$$z(t) = z_1 + V(t - t_1), \qquad t_1 \le t \le t_N, \tag{15.2-1}$$

where V is the constant speed of the PA in meters per second, and the first transmission of the PA occurs at time t_1 seconds (see Fig. 15.1-2). By referring to Fig. 15.2-1, it can be seen that the slant-range at time t between the center of the PA and a scatterer on the ocean bottom with cross-range coordinate x is

$$r(t) = \sqrt{r_{\perp}^2 + [z(t) - z_{CPA}]^2}, \tag{15.2-2}$$

where CPA is the acronym for *closest point of approach*. Equation (15.2-2) can be approximated by first rewriting it as

$$r(t) = r_{\perp}\sqrt{1 + b(t)}, \tag{15.2-3}$$

where

$$b(t) = [z(t) - z_{CPA}]^2 / r_{\perp}^2 . \tag{15.2-4}$$

Next, by using a binomial expansion of the square-root factor in (15.2-3), we obtain

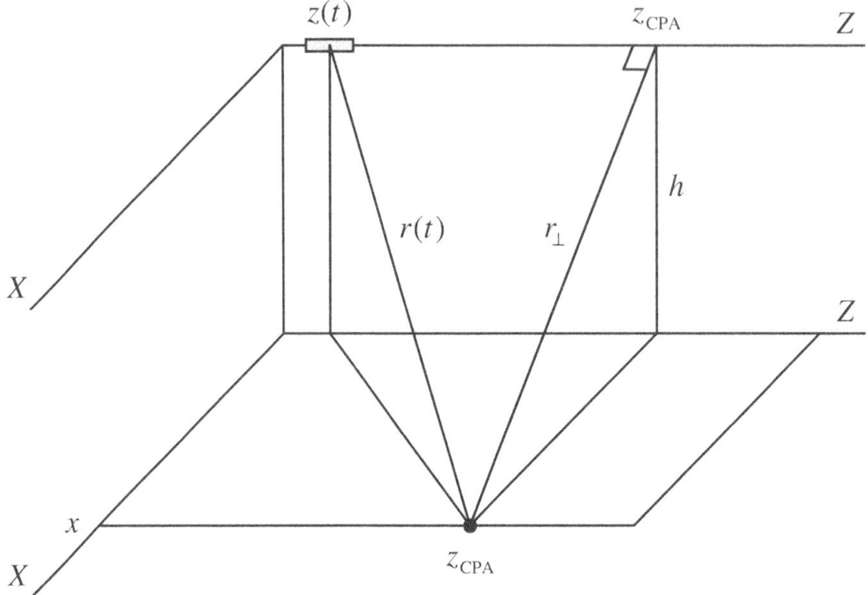

Figure 15.2-1 Slant-range $r(t)$ at time t from the center of the physical aperture to a stationary (motionless) scatterer on the ocean bottom with cross-range coordinate x.

$$r(t) = r_\perp \sqrt{1 + b(t)} = r_\perp \left[1 + \frac{b(t)}{2} - \frac{b^2(t)}{8} + \cdots \right], \qquad |b(t)| < 1. \tag{15.2-5}$$

Finally, substituting (15.2-4) into (15.2-5) and using only the first two terms in the binomial expansion yields the following approximation of $r(t)$:

$$\boxed{r(t) \approx r_\perp + \frac{[z(t) - z_{\text{CPA}}]^2}{2r_\perp}, \qquad \left[\frac{z(t) - z_{\text{CPA}}}{r_\perp} \right]^2 < 0.1} \tag{15.2-6}$$

The factor of 0.1 was used in the inequality on the right-hand side of (15.2-6) instead of 1 because only the first two terms in the binomial expansion were used. The factor of 0.1 was used instead of $<< 1$.

In order to proceed further, we shall make the *stop-and-hop assumption*,[1] that is, if it is assumed that the PA is stationary (motionless) during the time of transmission of a pulse and reception of a scattered return from a scatterer, then

[1] M. A. Richards, *Fundamentals of Radar Signal Processing*, McGraw-Hill, New York, 2005, pg. 99.

the phase factor $\exp[-jk2r(t)]$ can be used to model the *two-way* propagation path between the PA and a scatterer, where

$$k = 2\pi f / c = 2\pi / \lambda \tag{15.2-7}$$

is the wavenumber in radians per meter. After the PA stops to transmit and receive at one location along the Z axis, it then hops to the next location where it stops to transmit and receive again, etc. Therefore, the *two-way*, time-varying phase (in radians) is given by

$$\phi(t) = -k2r(t). \tag{15.2-8}$$

Substituting (15.2-6) and (15.2-7) into (15.2-8) yields

$$\boxed{\phi(t) = -\frac{4\pi}{\lambda} r_{\perp} - 2\pi \frac{[z(t) - z_{\text{CPA}}]^2}{r_{\perp}\lambda}, \qquad t_1 \le t \le t_N} \tag{15.2-9}$$

The time-varying *Doppler shift* $\eta_D(t)$ in hertz can be obtained from (15.2-9) as follows:

$$\eta_D(t) = \frac{1}{2\pi}\frac{d}{dt}\phi(t). \tag{15.2-10}$$

Substituting (15.2-9) into (15.2-10), and using $z(t)$ given by (15.2-1) yields

$$\boxed{\eta_D(t) = -\frac{2V}{r_{\perp}\lambda}[z(t) - z_{\text{CPA}}], \qquad t_1 \le t \le t_N} \tag{15.2-11}$$

Equation (15.2-11) is the *Doppler history* of a SAS for a stationary (motionless) scatterer based on the stop-and-hop assumption. Note that the instantaneous phase $\phi(t)$ is a *quadratic* function of time and the instantaneous Doppler shift $\eta_D(t)$ is a *linear* function of time analogous to the instantaneous phase in radians and instantaneous frequency in hertz of a linear-frequency-modulated (LFM) pulse, respectively (see Section 13.2). From (15.2-11) it can be seen that $\eta_D(t) > 0$ (positive Doppler shift) when $z(t) < z_{\text{CPA}}$, $\eta_D(t) = 0$ (zero Doppler shift) when $z(t) = z_{\text{CPA}}$, and $\eta_D(t) < 0$ (negative Doppler shift) when $z(t) > z_{\text{CPA}}$ (see Fig. 15.2-1). Also, with the use of (15.2-1),

$$\frac{d}{dt}\eta_D(t) = -\frac{2V^2}{r_{\perp}\lambda}, \qquad t_1 \le t \le t_N. \tag{15.2-12}$$

Equation (15.2-12) is the rate of change of Doppler shift across the synthetic aperture (SA) for a stationary (motionless) scatterer based on the stop-and-hop assumption.

Referring to Fig. 15.1-2, the beginning of the SA coincides with the transmission of a pulse at time t_1 where

$$z_1 = z(t_1) = z_{\text{CPA}} - \left(L_{\text{SA}}/2\right) \tag{15.2-13}$$

and

$$z_{N'+1} = z(t_{N'+1}) = z_{\text{CPA}} \,. \tag{15.2-14}$$

The end of the SA coincides with the transmission of a pulse at time t_N where

$$z_N = z(t_N) = z_{\text{CPA}} + \left(L_{\text{SA}}/2\right). \tag{15.2-15}$$

Evaluating (15.2-11) at $t = t_1$, $t_{N'+1}$, and t_N; and using (15.2-13) through (15.2-15), yields

$$\eta_D(t_1) = \frac{VL_{\text{SA}}}{r_\perp \lambda}, \tag{15.2-16}$$

$$\eta_D(t_{N'+1}) = 0\,, \tag{15.2-17}$$

and

$$\eta_D(t_N) = -\frac{VL_{\text{SA}}}{r_\perp \lambda}. \tag{15.2-18}$$

Therefore, the magnitude of the change in Doppler shift across the SA [a.k.a. the *Doppler bandwidth* (in hertz) of a SAS] for a stationary (motionless) scatterer based on the stop-and-hop assumption is given by

$$\Delta\eta_D = \left|\eta_D(t_N) - \eta_D(t_1)\right| = 2\frac{VL_{\text{SA}}}{r_\perp \lambda}. \tag{15.2-19}$$

Substituting (15.1-12) – which is based on an approximation of the horizontal, 3-dB beamwidth of the PA – into (15.2-19) yields

$$\Delta\eta_D = \left|\eta_D(t_N) - \eta_D(t_1)\right| = 1.772\,V/L_{\text{PA}}\,, \tag{15.2-20}$$

or, by substituting (15.1-13) – which is based on an approximation of the horizontal, Rayleigh beamwidth of the PA – into (15.2-19),

$$\boxed{\Delta\eta_D = |\eta_D(t_N) - \eta_D(t_1)| = 2V/L_{\text{PA}}} \tag{15.2-21}$$

Figure 15.2-2 is a plot of the Doppler history using (15.2-16) through (15.2-18) and (15.1-13) for L_{SA}. As the PA approaches a scatterer, the Doppler shift is positive but decreases in value. At the CPA, the Doppler shift is zero, and as the PA recedes from a scatterer, the Doppler shift becomes more negative.

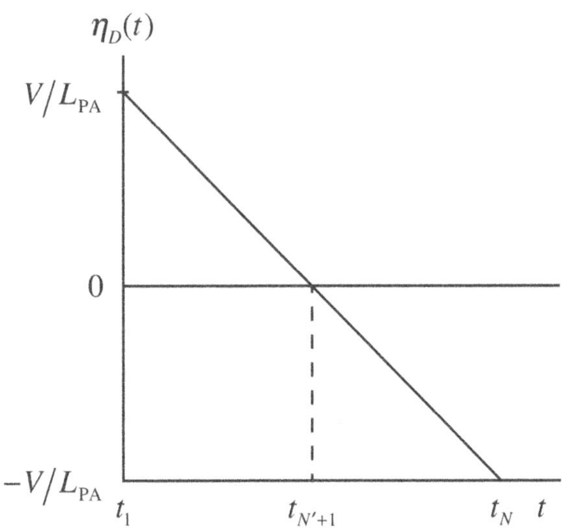

Figure 15.2-2 Doppler history (Doppler shift $\eta_D(t)$ in hertz versus time t in seconds) of a synthetic-aperture sonar for a stationary (motionless) scatterer based on the stop-and-hop assumption.

Equations (15.2-20) and (15.2-21) can also be derived by considering the amount of time it takes to create a SA:

$$\boxed{T_{\text{SA}} = L_{\text{SA}}/V} \tag{15.2-22}$$

Substituting (15.1-12) into (15.2-22) yields

$$T_{\text{SA}} = 0.886\frac{r_\perp \lambda}{VL_{\text{PA}}}, \tag{15.2-23}$$

or, by substituting (15.1-13) into (15.2-22),

$$\boxed{T_{\text{SA}} = \frac{r_\perp \lambda}{VL_{\text{PA}}}} \tag{15.2-24}$$

The parameter T_{SA} is referred to as the *time-on-target*, *dwell time*, *coherent integration time*, or *aperture time* in seconds. Note that multiplying (15.2-12) by (15.2-23) and taking the magnitude of the resulting product yields (15.2-20), and multiplying (15.2-12) by (15.2-24) and taking the magnitude of the resulting product yields (15.2-21).

As was mentioned earlier, the instantaneous phase $\phi(t)$ given by (15.2-9) is a quadratic function of time and the instantaneous Doppler shift $\eta_D(t)$ given by (15.2-11) is a linear function of time analogous to the instantaneous phase in radians and instantaneous frequency in hertz of a linear-frequency-modulated (LFM) pulse, respectively. The range resolution (in meters) of a rectangular-envelope, LFM pulse is given by (see Example 13.2-1)

$$\left|e_{r,Trgt}\right| \approx \frac{c}{2\,\text{BW}_{\text{swept}}}, \tag{15.2-25}$$

where c is the speed of sound in meters per second, and BW_{swept} is the swept-bandwidth in hertz, which is a good estimate of the bandwidth of the *lowpass (baseband), complex envelope* of a rectangular-envelope, LFM pulse. Therefore, $2\,\text{BW}_{\text{swept}}$ is a good estimate of the bandwidth of a *real*, *bandpass*, rectangular-envelope, LFM pulse. Based on (15.2-25), the along-track (azimuthal) resolution in meters of a SAS is defined as[2, 3]

$$\boxed{\delta z_{\text{SA}} \triangleq V/\Delta\eta_D} \tag{15.2-26}$$

where V is the constant speed of the PA in meters per second, and $\Delta\eta_D$ is the Doppler bandwidth in hertz. Therefore, substituting (15.2-19) into (15.2-26) yields

$$\delta z_{\text{SA}} = \frac{r_\perp \lambda}{2L_{\text{SA}}}. \tag{15.2-27}$$

Substituting (15.1-12) – which is based on an approximation of the horizontal, 3-dB beamwidth of the PA – into (15.2-27) yields

$$\delta z_{\text{SA}} = L_{\text{PA}}/1.772\,, \tag{15.2-28}$$

---

[2] J. J. Kovaly, "Radar Techniques for Planetary Mapping with Orbiting Vehicle," *Annals of the New York Academy of Sciences*, vol. 187, pp. 154-176, 1972.

[3] K. Tomiyasu, "Tutorial Review of Synthetic-Aperture Radar (SAR) with Applications to Imaging of the Ocean Surface," *Proc. IEEE*, vol. 66, no. 5, pp. 563-583, 1978.

or, by substituting (15.1-13) – which is based on an approximation of the horizontal, Rayleigh beamwidth of the PA – into (15.2-27), we obtain the classic result

$$\boxed{\delta z_{\mathrm{SA}} = L_{\mathrm{PA}}/2} \tag{15.2-29}$$

where $L_{\mathrm{PA}} = L_Z$ is the length of the PA [the side-looking sonar (SLS)] in the Z direction (the along-track direction). As can be seen from both (15.2-28) and (15.2-29), *the along-track (azimuthal) resolution of a SAS is independent of slant-range and frequency*, and as L_{PA} gets *smaller*, the resolution δz_{SA} gets *better*. Equations (15.2-28) and (15.2-29) apply to a stationary (motionless) scatterer and are based on the stop-and-hop assumption. Furthermore, although not obvious here, in order to obtain the resolution given by either (15.2-28) or (15.2-29), *the far-field array factor of the SA has to be focused at the slant-range* $r_\perp$ *of interest* as will be shown in Section 15.3.

And finally, from (15.2-27), we obtain the following alternate equation for computing the length of a SA, L_{SA}, as a function of slant-range $r_\perp$, wavelength λ, and desired along-track (azimuthal) resolution of the SA, δz_{SA}:

$$\boxed{L_{\mathrm{SA}} = \frac{r_\perp \lambda}{2\delta z_{\mathrm{SA}}}} \tag{15.2-30}$$

15.3 Far-Field Beam Pattern of a Linear Synthetic Array

Figure 15.3-1 shows the linear synthetic array (SA) created by the moving physical aperture (PA) as discussed in Section 15.1. The linear array is composed of an *odd* number N of identical, equally-spaced elements lying along the Z axis. The elements (the PA) are identical rectangular pistons lying in the YZ plane (see Fig. 15.1-1). The z coordinate of the center of element n is given by

$$z_n = nd_Z, \qquad n = -N', \ldots, 0, \ldots, N', \tag{15.3-1}$$

where d_Z is the interelement spacing in meters, and

$$N' = (N-1)/2\,. \tag{15.3-2}$$

Note that for purposes of deriving the far-field beam pattern of the SA, the z coordinates of the centers of the elements in the array given by (15.3-1) are different from the z coordinates obtained from (15.2-1) [e.g., see (15.2-13) through (15.2-15)].

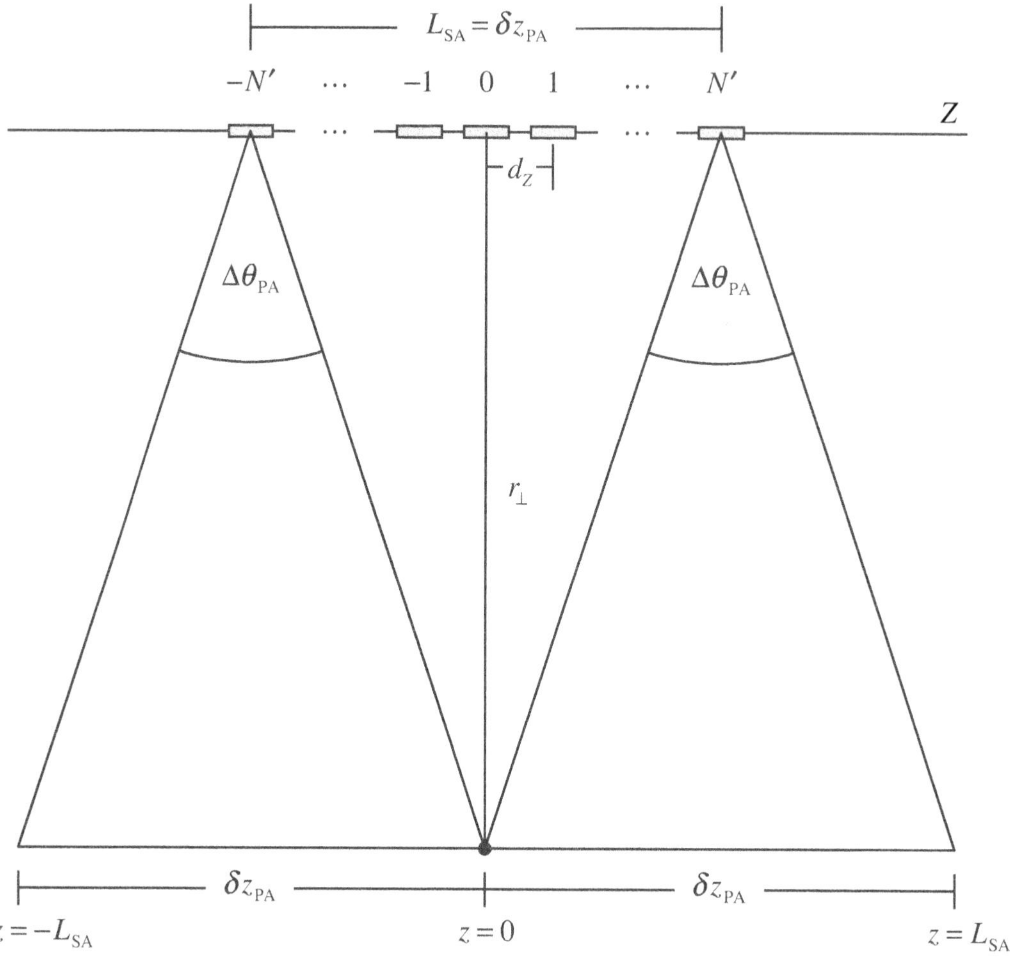

Figure 15.3-1 Linear synthetic array.

Recall from our discussion of swath width (SW) in Section 5.1, which is directly applicable to a synthetic-aperture sonar (SAS), that it is the *vertical*, far-field beam pattern of a side-looking sonar (SLS) that is steered toward the ocean bottom in order to create a SW. Since the vertical, far-field beam pattern lies in the XY plane where $\theta = 90°$, beam steering is only done in the Y (depth) direction. Also recall that the PA (SLS) is used alternately in the active (transmit) mode and passive (receive) mode. As mentioned in Section 15.1, this kind of arrangement used to create a synthetic aperture (SA) is known as a single-element, synthetic-aperture sonar (SESAS). Therefore, the *unnormalized*, transmit and receive, far-field beam patterns of an element (a rectangular piston) lying in the YZ plane are given by (see Sections 3.2 and 3.4)

$$E_T(f, f_Y, f_Z) = a_T(f) L_Y L_Z \operatorname{sinc}[(f_Y - f_Y')L_Y] \operatorname{sinc}(f_Z L_Z) \qquad (15.3\text{-}3)$$

and

$$E_R(f, f_Y, f_Z) = a_R(f) L_Y L_Z \operatorname{sinc}[(f_Y - f_Y')L_Y] \operatorname{sinc}(f_Z L_Z), \tag{15.3-4}$$

respectively, where $a_T(f)$, with units of $\left(\left(\text{m}^3/\text{sec}\right)/\text{V}\right)/\text{m}^2$, and $a_R(f)$, with units of $\left(\text{V}/\left(\text{m}^2/\text{sec}\right)\right)/\text{m}^2$, are real, nonnegative *amplitude* responses of the rectangular piston in the transmit and receive modes of operation (see Table 3.1-1),

$$f_Y = v/\lambda = \sin\theta \sin\psi/\lambda \tag{15.3-5}$$

and

$$f_Z = w/\lambda = \cos\theta/\lambda \tag{15.3-6}$$

are spatial frequencies in the Y and Z directions with units of cycles per meter, and setting $\theta' = 90°$ (XY plane),

$$f_Y' = v'/\lambda = \sin\psi'/\lambda, \tag{15.3-7}$$

where ψ' is the beam-steer angle in the XY plane (see Fig. 5.1-2). The units of the far-field beam patterns $E_T(f, f_Y, f_Z)$ and $E_R(f, f_Y, f_Z)$ are $\left(\text{m}^3/\text{sec}\right)/\text{V}$ and $\text{V}/\left(\text{m}^2/\text{sec}\right)$, respectively (see Table 3.1-1). The vertical, far-field beam patterns of $E_T(f, f_Y, f_Z)$ and $E_R(f, f_Y, f_Z)$ lie in the XY plane where $\theta = 90°$.

The far-field beam pattern of the SA can be obtained by using the Product Theorem. When using the Product Theorem to compute the far-field beam pattern of the SA, it is important to use the two-way, far-field beam pattern of the PA, and the two-way, dimensionless, far-field array factor *focused* at a slant-range $r_\perp$ of interest to account for wave propagation from the transmitter to the scatterer, and from the scatterer back to the receiver.[3] Therefore, the *unnormalized, two-way*, far-field beam pattern of the linear SA is given by

$$D_2(f, f_Y, f_Z) = E_2(f, f_Y, f_Z) S_{Z2}(f, f_Z), \tag{15.3-8}$$

where

$$E_2(f, f_Y, f_Z) = E_T(f, f_Y, f_Z) E_R(f, f_Y, f_Z) \tag{15.3-9}$$

is the *two-way*, far-field beam pattern of the PA, and $S_{Z2}(f, f_Z)$ is the *two-way*, dimensionless, far-field array factor focused at a slant-range $r_\perp$ of interest. The units of $E_2(f, f_Y, f_Z)$ and, hence, $D_2(f, f_Y, f_Z)$ are meters. We need to find the array factor $S_{Z2}(f, f_Z)$ next.

In order to derive an equation for the two-way, far-field array factor $S_{Z2}(f, f_Z)$, we first need to derive an equation for the output electrical signal from element n in the SA using the geometry shown in Fig. 15.3-2, which is based on Fig. 15.1-1. The spherical coordinates of a scatterer on the ocean bottom as measured from the center of the array (element $n=0$, $z_0=0$) are (r,θ,ψ), and the rectangular coordinates are (x,h,z). We shall follow, in general, the procedure used by McCord.[4]

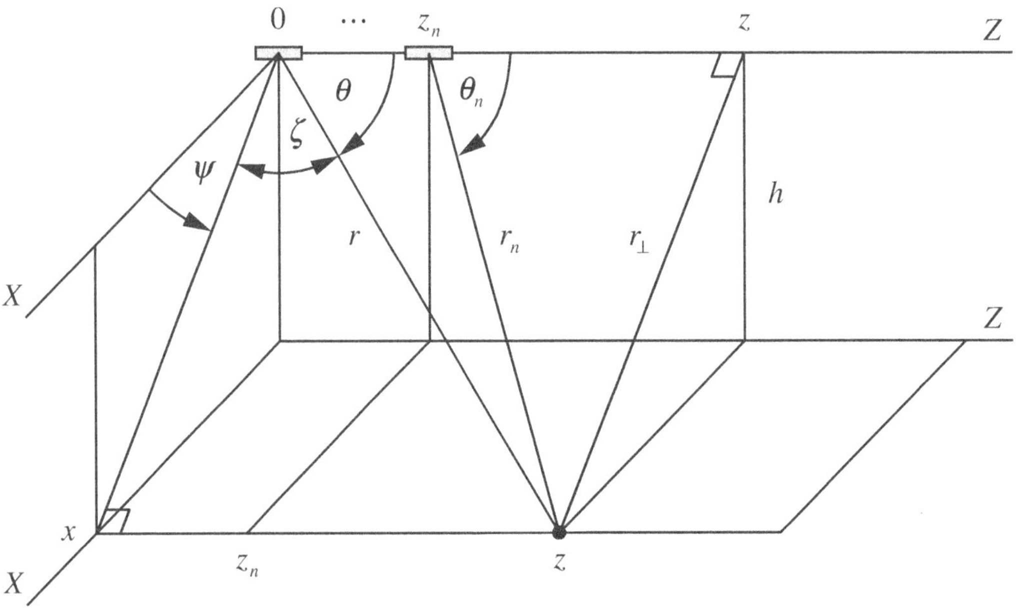

Figure 15.3-2 Geometry used for deriving the two-way array factor for a linear synthetic array.

As shown in Fig. 15.3-2, when element n begins to transmit a pulse, it is at a range r_n meters from the scatterer. Although element n remains in motion while it is transmitting and receiving, if we make the stop-and-hop assumption as discussed in Section 15.2, then the two-way (round-trip) time delay in seconds is given by

$$\tau_n = 2r_n / c\,, \tag{15.3-10}$$

where

$$r_n = \sqrt{r_\perp^2 + (z - z_n)^2}\,, \tag{15.3-11}$$

[4] H. L. McCord, "The Equivalence Among Three Approaches to Deriving Synthetic Array Patterns and Analyzing Processing Techniques," *IRE Trans. on Military Electronics*, vol. MIL-6, pp. 116-119, April 1962.

$$r_{\perp}=\sqrt{x^2+h^2}\,, \tag{15.3-12}$$

and h is the height (altitude) in meters that the center of the PA is above the ocean bottom.

Equation (15.3-11) can be approximated by first rewriting it as

$$r_n=r_{\perp}\sqrt{1+b}\,, \tag{15.3-13}$$

where

$$b=(z-z_n)^2/r_{\perp}^2\,. \tag{15.3-14}$$

Next, by using a binomial expansion of the square-root factor in (15.3-13), we obtain

$$r_n=r_{\perp}\sqrt{1+b}=r_{\perp}\left[1+\frac{b}{2}-\frac{b^2}{8}+\cdots\right], \qquad |b|<1\,. \tag{15.3-15}$$

Finally, substituting (15.3-14) into (15.3-15) and using only the first two terms in the binomial expansion yields the following approximation of r_n:

$$\boxed{r_n\approx r_{\perp}+\frac{(z-z_n)^2}{2r_{\perp}}, \qquad \left[(z-z_n)/r_{\perp}\right]^2<0.1} \tag{15.3-16}$$

The factor of 0.1 was used on the right-hand side of (15.3-16) because only the first two terms in the binomial expansion were used. Furthermore, by substituting (15.3-16) into (15.3-10) and expanding $(z-z_n)^2$, we obtain the following approximation of τ_n:

$$\boxed{\tau_n\approx\frac{2r_{\perp}}{c}+\frac{z^2}{r_{\perp}c}-2\frac{zz_n}{r_{\perp}c}+\frac{z_n^2}{r_{\perp}c}} \tag{15.3-17}$$

Note that (15.3-16) is similar to (15.2-6) because r_n, z_n, and z in (15.3-16) (see Fig. 15.3-2) are similar to $r(t)$, $z(t)$, and z_{CPA}, respectively, in (15.2-6) (see Fig. 15.2-1). The inequality on the right-hand side of (15.3-16) – and (15.2-6) – is equivalent to requiring that $\Delta\theta_{\text{PA}}<36.2°$ (see Appendix 15B). It is also shown in Appendix 15C that $r_n\approx r$ and $\theta_n\approx\theta$ if $r>2.739\,L_{\text{SA}}$ (which is equivalent to $\Delta\theta_{\text{PA}}<21°$), and $\psi_n=\psi$. However, $\Delta\theta_{\text{PA}}$ *must satisfy an even more stringent constraint*, namely, $\Delta\theta_{\text{PA}}<18.1°$ ($\lambda/L_{\text{PA}}<0.355$) (see Appendix 15D).

The model that we shall use for the output electrical signal (in volts) from element n in the SA *before* complex weighting, for a monostatic (backscatter) scattering geometry between the nth element and a scatterer on the ocean bottom based on the stop-and-hop assumption, is given by (see Appendix 15C)

$$y_s'(t, z_n) = s(t - \tau_n), \tag{15.3-18}$$

where

$$s(t) = F^{-1}\{S(f)\} \tag{15.3-19}$$

has units of volts, the complex frequency spectrum

$$S(f) \approx \frac{1}{(4\pi r)^2} X(f) E_T(f, \theta, \psi) S_S(f, \theta, \psi, \theta_{SR}, \psi_{SR}) E_R(f, \theta, \psi) \exp[-2\alpha(f) r] \tag{15.3-20}$$

has units of volts per hertz,

$$X(f) = F\{x(t)\} \tag{15.3-21}$$

is the complex frequency spectrum (in volts per hertz) of the transmitted electrical signal $x(t)$ in volts, $S_S(f, \theta, \psi, \theta_{SR}, \psi_{SR})$ is the scattering function of the scatterer in meters, where

$$\theta_{SR} = 180° - \theta \tag{15.3-22}$$

and

$$\psi_{SR} = 180° + \psi, \tag{15.3-23}$$

and $\alpha(f)$ is the real, nonnegative, frequency-dependent, attenuation coefficient of seawater in nepers per meter. For a scatterer located at $(r_\perp, z)$, the corresponding spherical coordinates (r, θ, ψ) needed to evaluate $S(f)$ are given by (see Fig. 15.3-2)

$$r = r(r_\perp, z) = \sqrt{r_\perp^2 + z^2}, \tag{15.3-24}$$

$$\theta = \theta(r_\perp, z) = \tan^{-1}(r_\perp / z), \tag{15.3-25}$$

and

$$\psi = \psi(r_\perp) = \sin^{-1}(h / r_\perp). \tag{15.3-26}$$

Therefore, the frequency spectrum $S(f)$ is not only a function of frequency, but it is also a function of $(r_\perp, z)$.

Now that we have a time-domain model for the output electrical signal $y_s'(t,z_n)$ from element n in the SA before complex weighting, the second step is to apply a complex weight. In order to apply a complex weight, we need to get into the frequency domain (e.g., see the beamforming procedure illustrated in Fig. 7.1-1 or Fig. 7.3-1). Therefore, taking the time-domain Fourier transform of (15.3-18) yields

$$Y_s'(f,z_n) = S(f)\exp(-j2\pi f\tau_n), \tag{15.3-27}$$

where $S(f)$ is given by (15.3-20) and τ_n is given by (15.3-17). Multiplying (15.3-27) by the complex weight

$$c_n(f) = a_n \exp[+j\theta_n(f)] = a_n \exp(-j2\pi f\tau_n') \tag{15.3-28}$$

yields

$$Y_s(f,z_n) = c_n(f)Y_s'(f,z_n), \tag{15.3-29}$$

and by substituting (15.3-27) into (15.3-29), we obtain

$$Y_s(f,z_n) = c_n(f)S(f)\exp(-j2\pi f\tau_n), \tag{15.3-30}$$

where a_n is a dimensionless amplitude weight, and either the phase weight $\theta_n(f)$ in radians or the equivalent time delay τ_n' in seconds is used for array focusing and beam steering. Equation (15.3-30) is the complex frequency spectrum (in volts per hertz) of the output electrical signal from element n in the SA *after* complex weighting.

In practical applications, the Fourier transforms of the output electrical signals from the elements in a SA are computed by taking properly scaled, time-domain, discrete Fourier transforms (DFTs) of time samples of the output electrical signals (see Appendix 7B). A DFT can be evaluated very quickly by using a fast-Fourier-transform (FFT) algorithm. The sampling frequency used is typically greater than the Nyquist rate. The time samples are referred to as *fast-time* data. The sampling frequency and sampling period are fast-time parameters. *Slow-time* parameters are discussed in Section 15.4.

The third step is to compute the complex frequency spectrum of the *total* output electrical signal from the SA after complex weighting. Therefore, substituting (15.3-17) into (15.3-30) and summing $Y_s(f,z_n)$ over all n yields

$$\boxed{Y_s(f,r_\perp,z) = S(f)\exp\left[-jk\left(2r_\perp + \frac{z^2}{r_\perp}\right)\right]S_{Z2}(f,r_\perp,z)} \tag{15.3-31}$$

where $Y_s(f, r_\perp, z)$ has units of volts per hertz, $S(f)$ is given by (15.3-20),

$$\boxed{\mathcal{S}_{Z2}(f, r_\perp, z) = \sum_{n=-N'}^{N'} c_n(f) \exp\left[+j\frac{2k}{r_\perp} z z_n\right] \exp\left(-j\frac{k}{r_\perp} z_n^2\right)} \tag{15.3-32}$$

is the *two-way*, *dimensionless*, *near-field* array factor of the linear SA, $c_n(f)$ is given by (15.3-28),

$$k = 2\pi f / c = 2\pi / \lambda \tag{15.3-33}$$

is the wavenumber in radians per meter, $r_\perp$ is given by (15.3-12), and z_n is given by (15.3-1). Scatterers with the same cross-range (cross-track) coordinate x but different down-range (along-track) coordinate z will have the same slant-range $r_\perp$ perpendicular to the Z axis [see Fig. 15.3-2 and (15.3-12)]. With the exception of differences in notation, the frequency spectrum $Y_s(f, r_\perp, z)$ given by (15.3-31) and the array factor $\mathcal{S}_{Z2}(f, r_\perp, z)$ given by (15.3-32) are analogous to the frequency spectrum and array factor derived by McCord.[4]

The fourth step is to rewrite the complex exponential

$$\exp\left[+j\frac{2k}{r_\perp} z z_n\right]$$

that appears inside the summation in (15.3-32). From Fig. 15.3-2,

$$z = r\cos\theta, \tag{15.3-34}$$

where r is given by (15.3-24). Therefore, with the use of (15.3-33) and (15.3-34),

$$\exp\left[+j\frac{2k}{r_\perp} z z_n\right] = \exp\left[+j2\pi\frac{r}{r_\perp} f_Z(2z_n)\right], \tag{15.3-35}$$

where [see (15.3-6)]

$$f_Z = w/\lambda = \cos\theta/\lambda \tag{15.3-36}$$

is the spatial frequency in the Z direction with units of cycles per meter. Since (see Appendix 15D)

$$\boxed{r \approx r_\perp, \quad \Delta\theta_{\text{PA}} < 18.1^\circ} \tag{15.3-37}$$

(15.3-35) reduces to

$$\exp\left[+j\frac{2k}{r_\perp}zz_n\right]=\exp\left[+j2\pi f_Z(2z_n)\right], \tag{15.3-38}$$

and by substituting (15.3-38) into (15.3-32), we obtain

$$\boxed{S_{Z2}(f,r_\perp,f_Z)=\sum_{n=-N'}^{N'}c_n(f)\exp\left(-j\frac{k}{r_\perp}z_n^2\right)\exp\left[+j2\pi f_Z(2z_n)\right]} \tag{15.3-39}$$

Note that if $\Delta\theta_{\text{PA}}<18.1°$ ($\lambda/L_{\text{PA}}<0.355$) (see Appendix 15D), then because (15.3-16) and (15.2-6) are similar, as was discussed earlier, (15.3-16) and (15.2-6) are very good approximations because they only require that $\Delta\theta_{\text{PA}}<36.2°$. In addition, (15C-23) and (15C-26) are also very good approximations because they only require that $\Delta\theta_{\text{PA}}<21°$.

Finally, the far-field array factor of the linear SA can be *focused* at slant-range $r_\perp$ by using the following set of quadratic phase weights:

$$\theta_n(f)=\frac{k}{r_\perp}z_n^2, \qquad n=-N',\ldots,0,\ldots,N', \tag{15.3-40}$$

where $r_\perp$ is given by (15.3-12). Substituting (15.3-28), (15.3-40), and (15.3-1) into (15.3-39) yields

$$\boxed{S_{Z2}(f,f_Z)=\sum_{n=-N'}^{N'}a_n\exp\left[+j2\pi f_Z(2nd_Z)\right]} \tag{15.3-41}$$

Equation (15.3-41) is the *unnormalized, two-way, dimensionless, far-field* array factor of the linear SA. Substituting (15.3-41) into (15.3-8) yields the *unnormalized, two-way*, far-field beam pattern of the linear SA. If $a_n>0\ \forall\ n$, then the normalization factor for the far-field array factor is given by (see Appendix 6A)

$$S_{Z2\max}=\max\left|S_{Z2}(f,f_Z)\right|=\sum_{n=-N'}^{N'}a_n. \tag{15.3-42}$$

The *normalized* far-field array factor can be obtained by dividing (15.3-41) by (15.3-42).

If rectangular amplitude weights are used, that is, if

$$a_n = 1, \qquad n = -N', \ldots, 0, \ldots, N', \tag{15.3-43}$$

then the far-field array factor given by (15.3-41) reduces to

$$S_{Z2}(f, f_Z) = \sum_{n=-N'}^{N'} \exp(+jn4\pi f_Z d_Z), \tag{15.3-44}$$

and by using the identity

$$\sum_{k=-K}^{K} \exp(+jk\alpha) = \frac{\sin[(2K+1)\alpha/2]}{\sin(\alpha/2)}, \tag{15.3-45}$$

$$\boxed{S_{Z2}(f, f_Z) = \frac{\sin(2\pi f_Z N d_Z)}{\sin(2\pi f_Z d_Z)}} \tag{15.3-46}$$

Equation (15.3-46) is the *unnormalized*, two-way, dimensionless, far-field array factor of the linear SA when rectangular amplitude weights are used. By comparing the two-way array factor given by (15.3-46) with the one-way array factor given by (6.2-15), one can see that the only difference is that the two-way array factor has a factor of two in the arguments of both sine functions.[3] Substituting (15.3-46) into (15.3-8) yields the unnormalized, two-way, far-field beam pattern of the linear SA when rectangular amplitude weights are used.

With the use of (15.3-36), (15.3-46) can also be expressed as

$$S_{Z2}(f, w) = \frac{\sin\left(2\pi \frac{N d_Z}{\lambda} w\right)}{\sin\left(2\pi \frac{d_Z}{\lambda} w\right)} \tag{15.3-47}$$

or

$$S_{Z2}(f, \theta) = \frac{\sin\left(2\pi \frac{N d_Z}{\lambda} \cos\theta\right)}{\sin\left(2\pi \frac{d_Z}{\lambda} \cos\theta\right)}. \tag{15.3-48}$$

Furthermore, since (see Fig. 15.3-2)

$$\cos\theta = \sin\zeta, \tag{15.3-49}$$

substituting (15.3-49) into (15.3-48) yields

$$S_{Z2}(f,\zeta)=\frac{\sin\left(2\pi\frac{Nd_Z}{\lambda}\sin\zeta\right)}{\sin\left(2\pi\frac{d_Z}{\lambda}\sin\zeta\right)}, \tag{15.3-50}$$

where ζ is the angle measured from broadside.[5] Since the normalization factor for rectangular amplitude weights is [see (15.3-42) and (15.3-2)]

$$S_{Z2\max}=N, \tag{15.3-51}$$

dividing (15.3-50) by N yields

$$\boxed{S_{NZ2}(f,\zeta)=\frac{\sin\left(2\pi\frac{Nd_Z}{\lambda}\sin\zeta\right)}{N\sin\left(2\pi\frac{d_Z}{\lambda}\sin\zeta\right)}} \tag{15.3-52}$$

which is the *normalized*, two-way, dimensionless, far-field array factor of the linear SA when rectangular amplitude weights are used.[3] Therefore, after the far-field array factor of the SA is focused at a slant-range $r_\perp$ of interest, the complex frequency spectrum of the *total* output electrical signal from the SA given by (15.3-31) reduces to

$$\boxed{Y_s(f,r_\perp,z)=S(f)\exp\left[-jk\left(2r_\perp+\frac{z^2}{r_\perp}\right)\right]S_{Z2}(f,\zeta)} \tag{15.3-53}$$

where $S(f)$ is given by (15.3-20), $S_{Z2}(f,\zeta)$ is given by (15.3-50), and (see Fig. 15.3-2)

$$\zeta=\zeta(r_\perp,z)=\tan^{-1}(z/r_\perp). \tag{15.3-54}$$

Recall that $S(f)$ is not only a function of frequency, but it is also a function of $(r_\perp,z)$ [see (15.3-24) through (15.3-26)]. We shall derive an approximate formula for computing the Rayleigh beamwidth of (15.3-52) next.

The *first null* of $S_{Z2}(f,\zeta)$ and $S_{NZ2}(f,\zeta)$ occurs at $\zeta=\Delta\theta_{\mathrm{R,AF}}$ where

$$2\pi\frac{Nd_Z}{\lambda}\sin(\Delta\theta_{\mathrm{R,AF}})=\pi, \tag{15.3-55}$$

[5] M. A. Richards, *Fundamentals of Radar Signal Processing*, McGraw-Hill, New York, 2005, pp. 390-404.

or

$$\Delta\theta_{\text{R,AF}} = \sin^{-1}\left(\frac{\lambda}{2Nd_Z}\right), \qquad \frac{\lambda}{2Nd_Z} \le 1, \tag{15.3-56}$$

where $\Delta\theta_{\text{R,AF}}$ is the *Rayleigh beamwidth*, which is the beamwidth to the first null (see Fig. 15.1-3). By using only the first term in the series expansion of $\sin^{-1}(x)$ given by (15.1-3), (15.3-56) can be approximated as

$$\Delta\theta_{\text{R,AF}} \approx \frac{\lambda}{2Nd_Z}, \qquad \frac{\lambda}{2Nd_Z} < 1. \tag{15.3-57}$$

If the length of the SA along the Z axis is measured from the centers of elements $-N'$ and N' [see Fig. 15.3-1 and (15C-17)], then

$$L_{\text{SA}} = 2N'd_Z. \tag{15.3-58}$$

Substituting (15.3-2) into (15.3-58) yields

$$L_{\text{SA}} = (N-1)d_Z = N\left(1 - \frac{1}{N}\right)d_Z, \tag{15.3-59}$$

or

$$L_{\text{SA}} \approx Nd_Z, \qquad N > 10. \tag{15.3-60}$$

Therefore, by substituting (15.3-60) into (15.3-57), we obtain the following approximate formula for computing the *Rayleigh beamwidth* of the *two-way*, far-field array factor of the linear SA when rectangular amplitude weights are used[5]:

$$\boxed{\Delta\theta_{\text{R,AF}} \approx \frac{\lambda}{2L_{\text{SA}}}, \qquad \frac{\lambda}{2L_{\text{SA}}} < 1} \tag{15.3-61}$$

Equation (15.3-61) is the Rayleigh beamwidth of both the unnormalized and normalized, far-field array factors $360°$ around the Z axis. Therefore, it is the horizontal, Rayleigh beamwidth of the far-field array factor in the XZ plane and along the ocean bottom. For comparison purposes, we shall next compute the horizontal, Rayleigh beamwidth in the XZ plane of the two-way, far-field beam pattern of the PA, $E_2(f, f_Y, f_Z)$, given by (15.3-9).

In the XZ plane, $\psi = 0°$ and $\psi' = 0°$ (see Fig. 15.3-2). Therefore, $f_Y = 0$ and $f_Y' = 0$ [see (15.3-5) and (15.3-7), respectively]. Substituting (15.3-3), (15.3-

4), $f_Y = 0$, and $f'_Y = 0$ into (15.3-9) yields

$$E_2(f, 0, f_Z) = a_T(f) a_R(f) (L_Y L_Z)^2 \operatorname{sinc}^2(f_Z L_Z). \tag{15.3-62}$$

As can be seen from (15.3-62), the Rayleigh beamwidth of $E_2(f, 0, f_Z)$ depends on $\operatorname{sinc}^2(f_Z L_Z)$. Using (15.3-6) and (15.3-49), $\operatorname{sinc}^2(f_Z L_Z)$ can be rewritten as

$$\operatorname{sinc}^2\left(\frac{L_Z}{\lambda}\sin\zeta\right) = \left[\sin\left(\pi\frac{L_Z}{\lambda}\sin\zeta\right) \Big/ \left(\pi\frac{L_Z}{\lambda}\sin\zeta\right)\right]^2. \tag{15.3-63}$$

The *first null* of (15.3-63) occurs at $\zeta = \Delta\theta_{\text{R}}$ where

$$\pi\frac{L_Z}{\lambda}\sin(\Delta\theta_{\text{R}}) = \pi, \tag{15.3-64}$$

or

$$\Delta\theta_{\text{R,PA}} = \sin^{-1}\left(\lambda/L_{\text{PA}}\right), \qquad \lambda/L_{\text{PA}} \le 1, \tag{15.3-65}$$

since $L_{\text{PA}} = L_Z$. By using only the first term in the series expansion of $\sin^{-1}(x)$ given by (15.1-3), (15.3-65) can be approximated as

$$\boxed{\Delta\theta_{\text{R,PA}} \approx \lambda/L_{\text{PA}}, \quad \lambda/L_{\text{PA}} < 1} \tag{15.3-66}$$

Equation (15.3-66) is the approximate *horizontal, Rayleigh beamwidth* in the *XZ* plane (and along the ocean bottom) of the *two-way*, far-field beam pattern of the PA, $E_2(f, 0, f_Z)$, given by (15.3-62). Note that the two-way, horizontal, Rayleigh beamwidths of the PA given by either (15.3-65) or (15.3-66) are identical to the one-way, horizontal, Rayleigh beamwidths of the PA given by either (15A-8) or (15A-9), respectively, since $L_{\text{PA}} = L_Z$. However, the one-way and two-way *3-dB beamwidths* of the PA are *not* equal, that is, the 3-dB beamwidths of sinc(•) and sinc^2(•) are different. Also, since $\Delta\theta_{\text{R,AF}}$ given by (15.3-61) is less than $\Delta\theta_{\text{R,PA}}$ given by (15.3-66), the approximate *horizontal, Rayleigh beamwidth* of the *two-way*, far-field beam pattern of the linear SA is given by

$$\boxed{\Delta\theta_{\text{R,SA}} \approx \frac{\lambda}{2L_{\text{SA}}}, \qquad \frac{\lambda}{2L_{\text{SA}}} < 1} \tag{15.3-67}$$

By referring to (15.3-47), it can be seen that the two-way, far-field array factor of the linear SA when rectangular amplitude weights are used has grating lobes in direction cosine space at $w = w_g$, where

$$w_g = i\frac{\lambda}{2d_Z}, \qquad i = \pm 1, \pm 2, \pm 3, \ldots, \tag{15.3-68}$$

if $|w_g| \leq 1$, that is, if w_g is in the *visible region. Azimuthal ambiguity* is caused by grating lobes.

The nulls of the two-way, horizontal, far-field beam pattern of the PA in the XZ plane, $E_2(f, 0, f_Z)$, given by (15.3-62), depend on

$$\mathrm{sinc}^2(f_Z L_Z) = \mathrm{sinc}^2\left(wL_{\mathrm{PA}}/\lambda\right) \tag{15.3-69}$$

since $L_{\mathrm{PA}} = L_Z$. The nulls are located in direction cosine space at $w = w_{\mathrm{N}}$, where

$$w_{\mathrm{N}} = i\lambda/L_{\mathrm{PA}}, \qquad i = \pm 1, \pm 2, \pm 3, \ldots \tag{15.3-70}$$

If the *maximum* value of interelement spacing is

$$\boxed{\max d_Z = L_{\mathrm{PA}}/2} \tag{15.3-71}$$

then substituting (15.3-71) into (15.3-68) yields

$$w_g = i\lambda/L_{\mathrm{PA}}, \qquad i = \pm 1, \pm 2, \pm 3, \ldots \tag{15.3-72}$$

By comparing (15.3-70) and (15.3-72), the nulls of the two-way, horizontal, far-field beam pattern of the PA will cancel the grating lobes of the two-way, far-field array factor, eliminating the azimuthal ambiguity problem. Therefore, in order to avoid azimuthal ambiguities for a SESAS, the PA must transmit a pulse every $\max d_Z$ meters or *less*.

Equation (15.3-71) can be generalized as follows:

$$\boxed{d_Z = L_{\mathrm{PA}}/(2k), \qquad k = 1, 2, 3, \ldots} \tag{15.3-73}$$

Note, if $k = 1$, then (15.3-73) reduces to (15.3-71), and as k *increases*, d_Z *decreases*. As d_Z decreases, the scattered returns from the ocean bottom are sampled more frequently increasing the amount of signal processing that needs to be done. Typical values for k are $k = 1$ and $k = 2$. Choosing a value for d_Z

according to (15.3-73) will still guarantee that the nulls of the two-way, horizontal, far-field beam pattern of the PA will cancel the grating lobes of the two-way, far-field array factor. This can be seen by substituting (15.3-73) into (15.3-68). Doing so yields

$$\boxed{w_g = ik\lambda/L_{\text{PA}}, \quad i = \pm1, \pm2, \pm3, \ldots, \quad k = 1, 2, 3, \ldots} \tag{15.3-74}$$

For a fixed value of λ/L_{PA}, as k *increases*, the number of grating lobes in the visible region *decreases* and adjacent grating lobes are spaced farther apart. However, as mentioned earlier, as k increases, d_Z decreases, and the amount of signal processing that needs to be done increases.

The *normalized*, *two-way*, *horizontal*, far-field beam pattern of the linear SA in the XZ plane (where $f_Y = 0$ and $f'_Y = 0$) is given by [see (15.3-8)]

$$D_{N2}(f, 0, w) = E_{N2}(f, 0, w) S_{NZ2}(f, w), \tag{15.3-75}$$

where [see (15.3-62)]

$$E_{N2}(f, 0, w) = \text{sinc}^2\left(wL_{\text{PA}}/\lambda\right) \tag{15.3-76}$$

since $L_{\text{PA}} = L_Z$, and [see (15.3-47), (15.3-51), and (15.3-71)]

$$S_{NZ2}(f, w) = \frac{\sin\left(N\pi \dfrac{L_{\text{PA}}}{\lambda} w\right)}{N \sin\left(\pi \dfrac{L_{\text{PA}}}{\lambda} w\right)}. \tag{15.3-77}$$

Substituting (15.3-76) and (15.3-77) into (15.3-75) yields

$$\boxed{D_{N2}(f, 0, w) = \text{sinc}\left(\frac{L_{\text{PA}}}{\lambda} w\right) \text{sinc}\left(N \frac{L_{\text{PA}}}{\lambda} w\right)} \tag{15.3-78}$$

Figure 15.3-3 is a plot of the normalized, two-way, horizontal, far-field beam pattern of the physical aperture given by (15.3-76) for $\lambda/L_{\text{PA}} = 0.2$. Figure 15.3-4 is a plot of the magnitude of the normalized, two-way, horizontal, far-field array factor of the linear synthetic array when rectangular amplitude weights are used given by (15.3-77) for $\lambda/L_{\text{PA}} = 0.2$ and $N = 25$. And Fig. 15.3-5 is a plot of the magnitude of the normalized, two-way, horizontal, far-field beam pattern of the linear synthetic array when rectangular amplitude weights are used given by

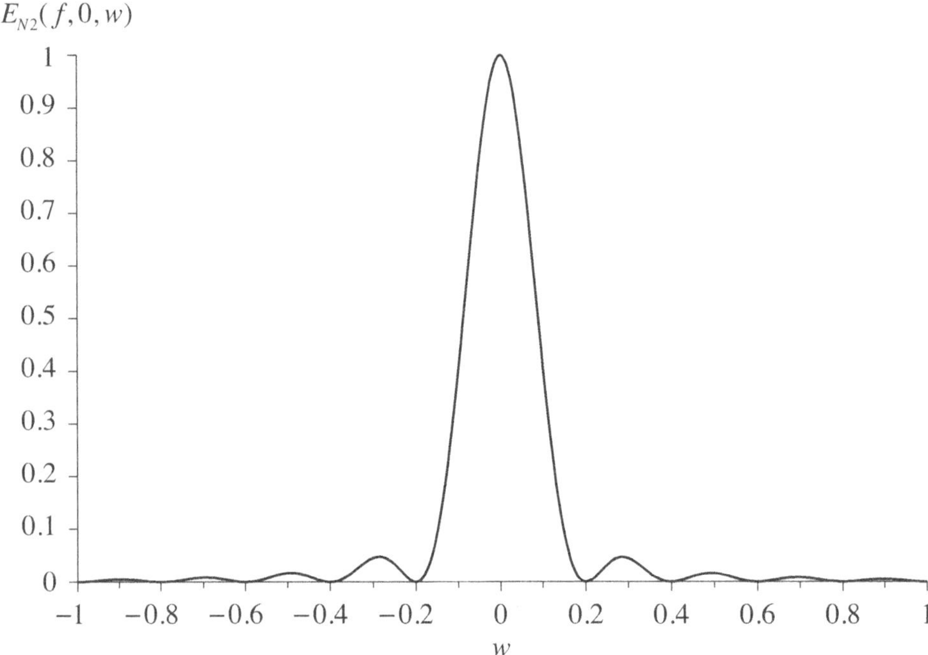

Figure 15.3-3 Normalized, two-way, horizontal, far-field beam pattern of the physical aperture for $\lambda/L_{\text{PA}} = 0.2$.

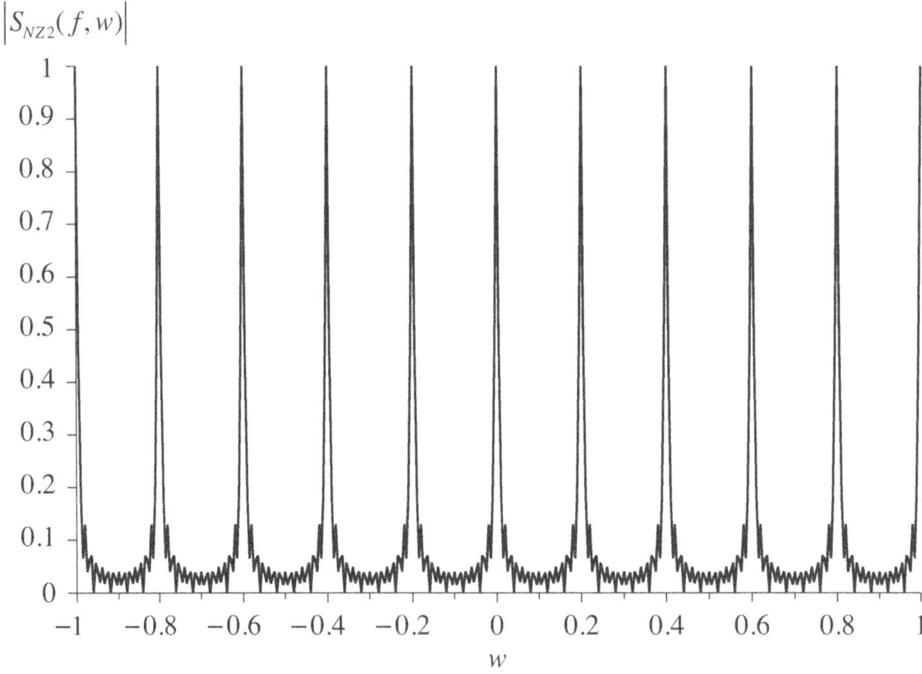

Figure 15.3-4 Magnitude of the normalized, two-way, horizontal, far-field array factor of the linear synthetic array when rectangular amplitude weights are used for $\lambda/L_{\text{PA}} = 0.2$ and $N = 25$.

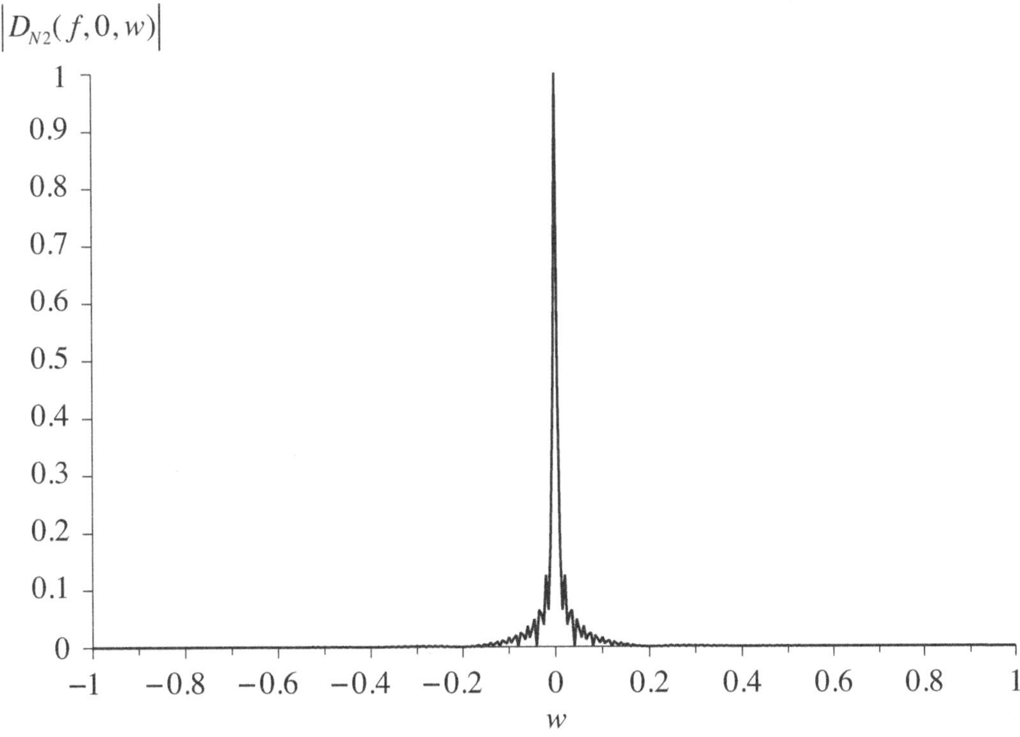

Figure 15.3-5 Magnitude of the normalized, two-way, horizontal, far-field beam pattern of the linear synthetic array when rectangular amplitude weights are used for $\lambda/L_{\text{PA}}=0.2$ and $N=25$.

(15.3-78) for $\lambda/L_{\text{PA}}=0.2$ and $N=25$. Figure 15.3-5 is equal to the product of Figs. 15.3-3 and 15.3-4. The parameter values $\lambda/L_{\text{PA}}=0.2$ and $N=25$ were chosen for example purposes.

We conclude our discussion in this section by computing the along-track (azimuthal) resolution of a SAS, δz_{SA}, by using the approximate horizontal, Rayleigh beamwidth of the two-way, far-field beam pattern of the linear SA, $\Delta\theta_{\text{R,SA}}$, given by (15.3-67). If we let

$$\delta z_{\text{SA}} = r_{\perp}\Delta\theta_{\text{R,SA}}, \tag{15.3-79}$$

then substituting (15.3-67) into (15.3-79) yields

$$\delta z_{\text{SA}} = r_{\perp}\frac{\lambda}{2L_{\text{SA}}}. \tag{15.3-80}$$

By substituting (15.1-13) – which is based on an approximation of the horizontal,

Rayleigh beamwidth of the PA – into (15.3-80), we obtain

$$\delta z_{\text{SA}} = L_{\text{PA}}/2\,, \tag{15.3-81}$$

which is identical to (15.2-29).

15.4 Slant-Range and Azimuthal Ambiguity

The material discussed in Section 5.4 regarding slant-range ambiguity for a side-looking sonar (SLS) is directly applicable to a synthetic-aperture sonar (SAS). In Section 5.4, a lower bound was derived for the pulse repetition interval (PRI) in seconds, also known as the pulse repetition period (PRP), and equivalently, the upper bound for the pulse repetition frequency (PRF) in pulses per second (pps), also known as the pulse repetition rate (PRR), for a SLS in order to avoid slant-range ambiguities. The equations for PRI_{min} and PRF_{max} given by (5.4-5) and (5.4-11), respectively, and $\text{PRI}_{\text{min,op}}$ and $\text{PRF}_{\text{max,op}}$ given by (5.4-7) and (5.4-12), respectively, for a SLS are directly applicable to a SAS. In this section we shall derive the upper bounds for the PRI, and equivalently, the lower bounds for the PRF for a single-element SAS (SESAS) and a multi-element SAS (MESAS) in order to avoid azimuthal ambiguities. In the SAS literature, PRI and PRF are referred to as *slow-time* parameters because their values are small compared to the *fast-time* parameters sampling period and sampling frequency, respectively, used for time-domain sampling of the output electrical signals from the elements in a SA. Fast-time data and parameters are discussed in Section 15.3.

In Section 15.3, it was shown that in order to avoid azimuthal ambiguities for a SESAS, the physical aperture (PA) must transmit a pulse every $\max d_Z$ meters or *less* [see (15.3-71)]. Therefore,

$$\text{PRI} \le \text{PRI}_{\text{max}}\,, \tag{15.4-1}$$

where

$$\boxed{\text{PRI}_{\text{max}} = \max d_Z / V = L_{\text{PA}}/(2V)} \tag{15.4-2}$$

$L_{\text{PA}} = L_Z$ is the length of the PA in the along-track direction, and V is the constant speed of the PA (the SLS) in meters per second. As a result,

$$\text{PRF} \ge \text{PRF}_{\text{min}}\,, \tag{15.4-3}$$

where, for a SESAS,

$$\boxed{\text{PRF}_{\text{min}} = \frac{1}{\text{PRI}_{\text{max}}} = \frac{2V}{L_{\text{PA}}}} \tag{15.4-4}$$

In summary, in order for a SESAS to avoid both *slant-range* and *azimuthal ambiguities*,

$$\boxed{\text{PRI}_{\text{min, op}} \le \text{PRI} \le \text{PRI}_{\text{max}}} \tag{15.4-5}$$

or, substituting (5.4-7) and (15.4-2) into (15.4-5),

$$\boxed{\left(2r_{\text{max}}/c\right) + T \le \text{PRI} \le L_{\text{PA}}/(2V)} \tag{15.4-6}$$

where r_{max} is the slant-range (in meters) to the end or far-edge of the swath width (SW), c is the speed of sound in meters per second, and T is the pulse length in seconds of the transmitted signal. Equivalently,

$$\boxed{\text{PRF}_{\text{min}} \le \text{PRF} \le \text{PRF}_{\text{max, op}}} \tag{15.4-7}$$

or, substituting (5.4-12) and (15.4-4) into (15.4-7),

$$\boxed{\frac{2V}{L_{\text{PA}}} \le \text{PRF} \le \frac{1}{\left(2r_{\text{max}}/c\right) + T}} \tag{15.4-8}$$

As discussed in Section 5.5, in order to fully ensonify the ocean bottom within the SW (without leaving *gaps* in the coverage) as the PA (the SLS) moves in the along-track direction (the positive Z direction), the PA must transmit (ping) every Δz_{min} meters or *less*, where [see (5.3-14)]

$$\Delta z_{\text{min}} = 2x_{\text{min}} \tan\left(\Delta\theta_{\text{PA}}/2\right) \tag{15.4-9}$$

is the minimum value of along-track (azimuthal) resolution due to the PA at cross-range x_{min} [the beginning or near-edge of the SW given by (5.1-22)], and $\Delta\theta_{\text{PA}}$ is the 3-dB beamwidth of the horizontal, far-field beam pattern of the PA given by (15.1-8). Therefore, in order for a SESAS to avoid both azimuthal ambiguities and gaps in the coverage on the ocean bottom,

$$\max d_Z \le \Delta z_{\text{min}}\,, \tag{15.4-10}$$

where $\max d_Z$ is given by (15.3-71).

15.4.1 Multi-Element Synthetic-Aperture Sonar

A major disadvantage of a SESAS is its inability to simultaneously provide both high resolution and a reasonable value for $r_{\max}$. In order to show this, note that $\text{PRI}_{\max}$ must satisfy the following inequality [see Fig. 5.4-1]:

$$\text{PRI}_{\max} > \frac{2r_{\max}}{c} + T \,. \tag{15.4-11}$$

Solving for $r_{\max}$ yields

$$r_{\max} < c(\text{PRI}_{\max} - T)/2 \,. \tag{15.4-12}$$

If $L_{\text{PA}} = 5$ cm so that the along-track resolution of the focused SAS $\delta z_{\text{SA}} = 2.5$ cm (≈ 1 in) [see (15.2-29)], and $V = 2$ m/sec (≈ 4 knots), then using (15.4-2) yields $\text{PRI}_{\max} = 12.5$ msec. If $\text{PRI}_{\max} = 12.5$ msec, $c = 1500$ m/sec, and $T = 5$ msec are substituted into (15.4-12), then $r_{\max} < 5.625$ m, which is too small. A small value for $r_{\max}$ implies a small value for the SW, and thus, the area coverage rate (ACR) (see Section 5.1). From (15.4-12), it can be seen that $r_{\max}$ can be increased by increasing $\text{PRI}_{\max}$ which, according to (15.4-2), can be increased by either increasing L_{PA} (which increases δz_{SA} resulting in poorer resolution), and/or decreasing V (which decreases the ACR). Both options are undesirable.

A common solution to this problem is to use a SLS with two separate apertures – a transmit aperture and a receive aperture.[6] The dimensions of the transmit and receive apertures are different, and the receive aperture is composed of multiple elements (see Fig. 15.4-1). This kind of SLS is known as a *multi-element* SAS (MESAS). The physical transmit aperture (PTA) and the individual physical receive elements (PREs) in the physical receive aperture (PRA) are modeled as rectangular pistons. The length of the PTA in the along-track direction, L_{PTA}, determines the along-track (azimuthal) resolution of the focused MESAS[7], that is [see (15.2-29)],

$$\boxed{\delta z_{\text{SA}} = L_{\text{PTA}}/2} \tag{15.4-13}$$

where $L_{\text{PTA}} = L_{\text{PTA}_Z}$ (see Fig. 15.4-1). In order to determine the length of the PRA

[6] M. P. Hayes and P. T. Gough, "Synthetic Aperture Sonar: A Review of Current Status," *IEEE J. Oceanic Eng*., vol. 34, no. 3, pp. 207-224, 2009.

[7] M. Pinto, "Design of Synthetic Aperture Sonar Systems for High-Resolution Seabed Imaging," tutorial slides presented at the *MTS/IEEE OCEANS Conf.*, Boston, MA, September 2006.

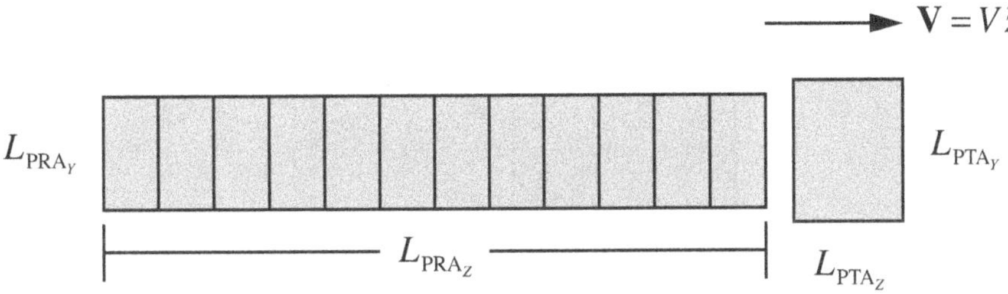

Figure 15.4-1 A side-looking sonar with separate physical transmit and receive apertures, traveling in the positive Z direction (the along-track direction). The physical receive aperture is composed of multiple physical receive elements.

in the along-track direction, L_{PRA}, that is required for a desired $r_{\max}$, (15.4-2) is substituted into (15.4-11) yielding

$$\boxed{L_{\text{PRA}} > \frac{4Vr_{\max}}{c} + 2VT} \tag{15.4-14}$$

or, as it often appears in the SAS literature,

$$\boxed{L_{\text{PRA}} > 4Vr_{\max}/c} \tag{15.4-15}$$

where L_{PA} was replaced by L_{PRA}, and $L_{\text{PRA}} = L_{\text{PRA}_Z}$ (see Fig. 15.4-1).

In a SESAS, the single PA transmits and receives every d_Z meters under the stop-and-hop assumption, creating a SA. However, in a MESAS, only the PREs in the PRA measure the scattered returns from the ocean bottom every d_Z meters between transmissions of the PTA. The PREs create a linear synthetic array (SA) with an interelement spacing of d_Z meters.

In a MESAS, the length of a PRE in the along-track direction in the PRA is less than the length of the PTA in the along-track direction.[7] If the length of a PRE in the along-track direction is

$$L_{\text{PRE}} = L_{\text{PTA}}/2, \tag{15.4-16}$$

then the centers of adjacent PREs will be separated by

$$\boxed{d_Z = L_{\text{PRE}} = L_{\text{PTA}}/2} \tag{15.4-17}$$

analogous to (15.3-71), where $L_{\text{PRE}} = L_{\text{PRE}_Z}$ (see Fig. 15.4-2). The length of the

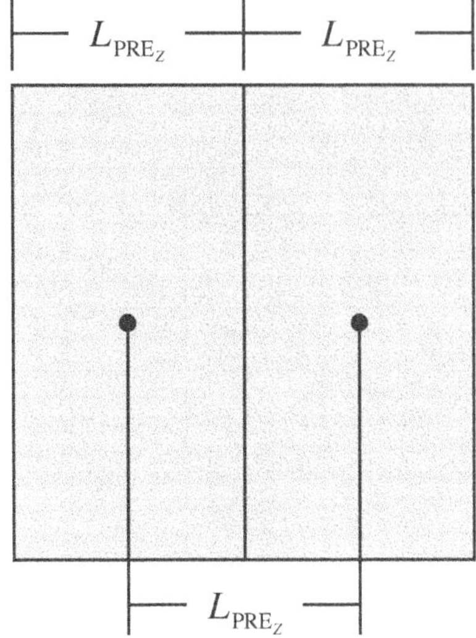

Figure 15.4-2 The centers of adjacent physical receive elements are separated by L_{PRE_Z} meters.

PRA in the along-track direction is

$$L_{\text{PRA}} = N_{\text{PRE}} L_{\text{PRE}}, \tag{15.4-18}$$

where N_{PRE} is the total *integer* number of PREs. Solving for N_{PRE} yields

$$N_{\text{PRE}} = L_{\text{PRA}} / L_{\text{PRE}}, \tag{15.4-19}$$

and by substituting (15.4-16) into (15.4-19), we obtain

$$\boxed{N_{\text{PRE}} = 2L_{\text{PRA}} / L_{\text{PTA}}} \tag{15.4-20}$$

or, by substituting (15.4-13) into (15.4-20),

$$\boxed{N_{\text{PRE}} = L_{\text{PRA}} / \delta z_{\text{SA}}} \tag{15.4-21}$$

Note, when using either (15.4-20) or (15.4-21), it may be necessary to *round-up* to the nearest integer.

The distance traveled during one PRI in the along-track direction is

$$\Delta z_{\text{PRI}} = V \times \text{PRI}. \tag{15.4-22}$$

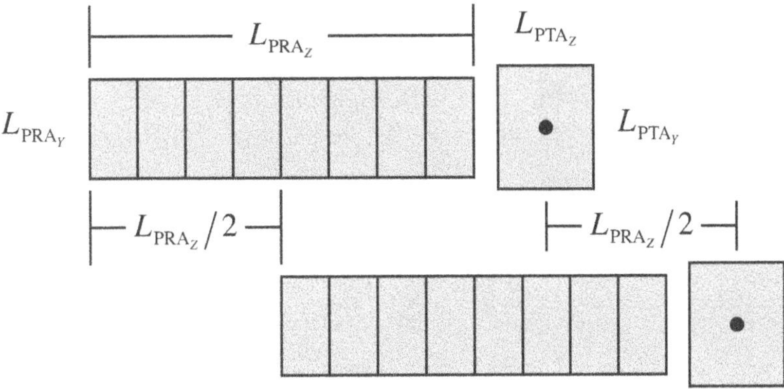

Figure 15.4-3 The physical transmit aperture transmits a pulse every $L_{\text{PRA}}/2$ meters where $L_{\text{PRA}} = L_{\text{PRA}_Z}$.

In the normal mode of operation of a MESAS, the PTA transmits a pulse every $L_{\text{PRA}}/2$ meters where $L_{\text{PRA}} = L_{\text{PRA}_Z}$ (see Fig. 15.4-3), that is,

$$\boxed{\Delta z_{\text{PRI}} = L_{\text{PRA}}/2} \tag{15.4-23}$$

Therefore, substituting (15.4-23) into (15.4-22) yields

$$\boxed{\text{PRI} = L_{\text{PRA}}/(2V)} \tag{15.4-24}$$

Furthermore, in order to fully ensonify the ocean bottom within the SW (without leaving *gaps* in the coverage) as the PTA (the SLS) moves in the along-track direction (the positive Z direction),

$$\Delta z_{\text{PRI}} \le \Delta z_{\min}, \tag{15.4-25}$$

where

$$\Delta z_{\min} = 2x_{\min}\tan\left(\Delta\theta_{\text{PTA}}/2\right) \tag{15.4-26}$$

is the minimum value of along-track (azimuthal) resolution due to the PTA at cross-range $x_{\min}$ [the beginning or near-edge of the SW given by (5.1-22)], and $\Delta\theta_{\text{PTA}}$ is the 3-dB beamwidth of the horizontal, far-field beam pattern of the PTA given by

$$\Delta\theta_{\text{PTA}} = 2\sin^{-1}\left(0.443\frac{\lambda}{L_{\text{PTA}}}\right), \qquad 0.443\frac{\lambda}{L_{\text{PTA}}} \le 1. \tag{15.4-27}$$

Example 15.4-1 Multi-Element Synthetic-Aperture Sonar

The specifications for a short-range, multi-element, synthetic-aperture sonar (MESAS) are given as follows[7]: $r_{\max} = 150\text{ m}$, $V = 2\text{ m/sec}$, $\delta z_{\text{SA}} = 5\text{ cm}$, and $c = 1500\text{ m/sec}$. Find L_{PTA}, L_{PRA}, L_{PRE}, N_{PRE}, PRI, and PRF.

Solving for L_{PTA} from (15.4-13) yields

$$L_{\text{PTA}} = 2\delta z_{\text{SA}} = 2 \times 5\text{ cm} = 10\text{ cm}. \tag{15.4-28}$$

Using (15.4-15) yields

$$L_{\text{PRA}} > \frac{4Vr_{\max}}{c} = \frac{4 \times (2\text{ m/sec}) \times 150\text{ m}}{1500\text{ m/sec}} = 0.8\text{ m}. \tag{15.4-29}$$

Using (15.4-16) yields

$$L_{\text{PRE}} = L_{\text{PTA}}/2 = 10\text{ cm}/2 = 5\text{ cm}. \tag{15.4-30}$$

And finally, using the minimum value of L_{PRA} in (15.4-20) and (15.4-24) yields

$$N_{\text{PRE}} = 2\frac{L_{\text{PRA}}}{L_{\text{PTA}}} = 2\frac{0.8\text{ m}}{0.1\text{ m}} = 16 \tag{15.4-31}$$

and

$$\text{PRI} = \frac{L_{\text{PRA}}}{2V} = \frac{0.8\text{ m}}{2 \times (2\text{ m/sec})} = 0.2\text{ sec}, \tag{15.4-32}$$

respectively. Therefore,

$$\text{PRF} = \frac{1}{\text{PRI}} = \frac{1}{0.2\text{ sec}} = 5\text{ pps}. \tag{15.4-33}$$

■

15.5 Stripmap Synthetic-Aperture Sonar

We know from Section 15.1 that the length of a synthetic aperture (SA) is a function of slant-range $r_{\perp}$ and the 3-dB beamwidth of the horizontal, far-field beam pattern of a physical transmit aperture (PTA) $\Delta\theta_{\text{PTA}}$ [replace $\Delta\theta_{\text{PA}}$ with $\Delta\theta_{\text{PTA}}$ in (15.1-11)]:

$$L_{\text{SA}} = r_{\perp}\Delta\theta_{\text{PTA}}. \tag{15.5-1}$$

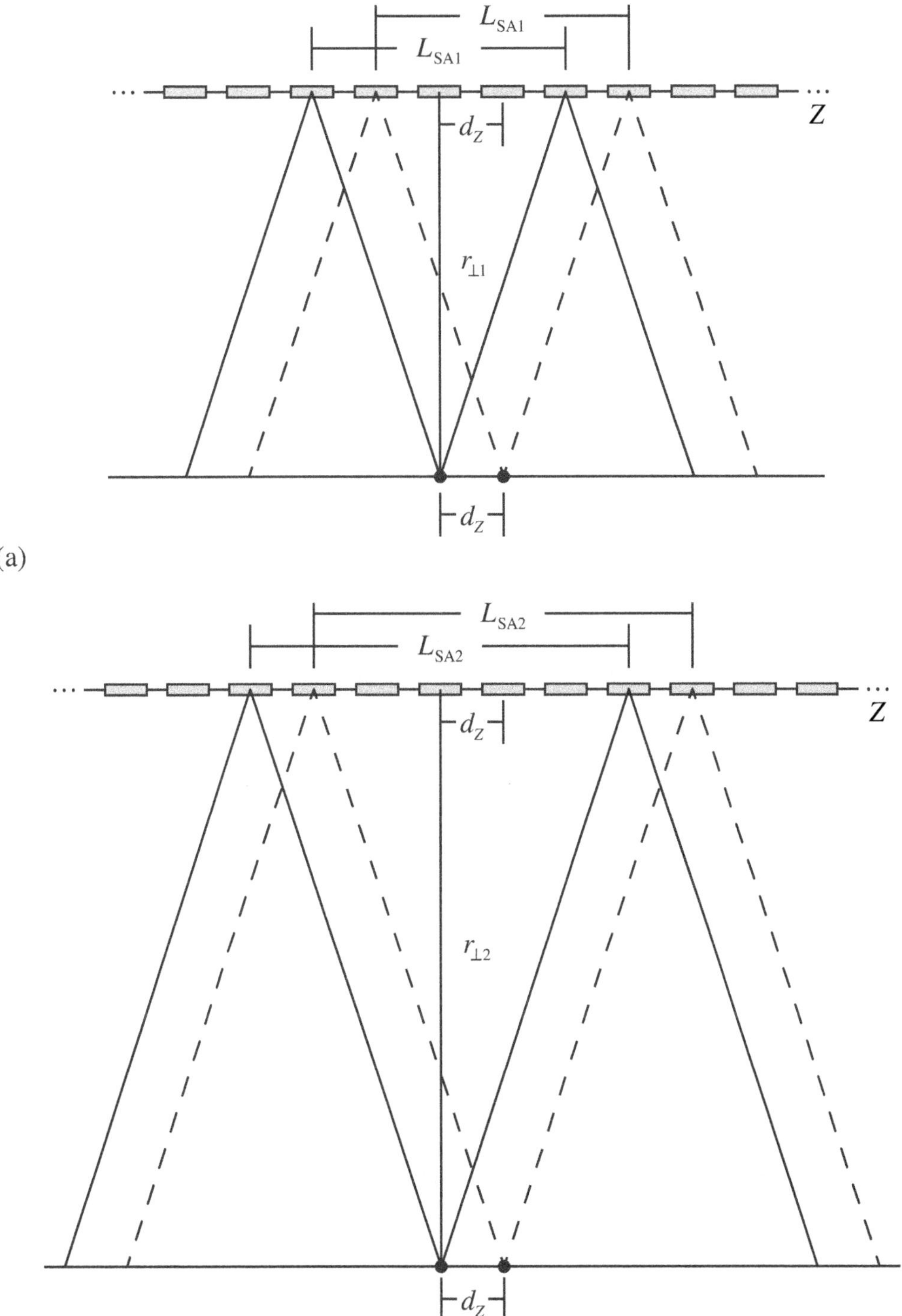

Figure 15.5-1 Illustration of the operation of a stripmap, synthetic-aperture sonar. Synthetic subarrays of lengths (a) L_{SA1} needed to image scatterers at slant-range $r_{\perp 1}$, and (b) L_{SA2} needed to image scatterers at slant-range $r_{\perp 2}$.

The slant-range $r_\perp$ takes on values in the interval

$$r_{\min} \le r_\perp \le r_{\max}, \tag{15.5-2}$$

where $r_{\min}$ and $r_{\max}$ are the slant-ranges to the beginning (near-edge) and end (far-edge) of the swath width (SW), respectively (see Fig. 5.1-2). The horizontal, 3-dB beamwidth $\Delta\theta_{\text{PTA}}$ is fixed. Therefore,

$$\min L_{\text{SA}} = r_{\min}\,\Delta\theta_{\text{PTA}} \tag{15.5-3}$$

and

$$\max L_{\text{SA}} = r_{\max}\,\Delta\theta_{\text{PTA}}\,. \tag{15.5-4}$$

A *stripmap*, synthetic-aperture sonar (SAS) first creates a synthetic array (SA) equal to the maximum length $\max L_{\text{SA}}$ given by (15.5-4), recording the output electrical signals from all the physical receive elements (PREs) in the SA. Synthetic subarrays of different lengths are then created from the overall SA by using an appropriate subset of elements. Each synthetic subarray length corresponds to a particular value of slant-range $r_\perp$ [see (15.5-1)]. As shown in Fig. 15.5-1, where $L_{\text{SA2}} > L_{\text{SA1}}$ and $r_{\perp 2} > r_{\perp 1}$, by shifting one element to the right while keeping the total number of elements in a subarray constant, which keeps the length of the subarray constant, different scatterers at the same slant-range $r_\perp$ but different along-track coordinates z can be imaged.

Problems

Section 15.1

15-1 Using (15.1-12) and (15.1-13), which are based on approximations of the horizontal, 3-dB and Rayleigh beamwidths of the PA, respectively; and slant-range $r_\perp = 200$ m, find the required lengths of a SA for the following values of λ/L_{PA} (see Table 15.1-1 where $L_Z = L_{\text{PA}}$):

(a) 0.1

(b) 0.2

(c) 0.3

Section 15.2

15-2 Using the stop-and-hop assumption, derive an equation for the time-varying Doppler shift $\eta_D(t)$ in hertz (a.k.a. the Doppler history of a SAS) using the exact equation for the time-varying slant-range $r(t)$.

15-3 Using your answers for Problem 15-1, compute the "time on target" if the speed of a SAS $V = 2$ m/sec (≈ 4 knots).

Section 15.3

15-4 Using your answers for Problem 15-1, verify that (15C-22) is satisfied if r is set equal to its minimum value $r_\perp = 200$ m because $r_\perp = 200$ m was used in Problem 15-1.

15-5 Verify that the right-hand side of (15.3-20) has units of volts per hertz.

15-6 Find the locations of the nulls of the two-way, horizontal, far-field beam pattern of the PA, and the grating lobes of the two-way, far-field array factor of the linear SA when rectangular amplitude weights are used, for

(a) $i = 1$ to 6 in (15.3-70) and (15.3-74), and $k = 1$ in (15.3-74).

(b) $i = 1$ to 6 in (15.3-70), and $i = 1$ to 3 and $k = 2$ in (15.3-74).

Section 15.4

15-7 The specifications for a MESAS are given as follows: $r_{\max} = 220$ m, $V = 2$ m/sec, $\delta z_{\text{SA}} = 2.5$ cm, and $c = 1500$ m/sec. Find L_{PTA}, L_{PRA}, L_{PRE}, N_{PRE}, PRI, and PRF.

Appendix 15A Rayleigh Beamwidth of the Horizontal, Far-Field Beam Pattern of a Rectangular Piston

Since the SLS (the PA) is modeled as a single rectangular piston lying in the YZ plane, the normalized, horizontal, far-field beam pattern in the XZ plane is given by [see (5.6-5)]

$$D_N(f,0,w) = \text{sinc}\left(\frac{L_Z}{\lambda}w\right), \tag{15A-1}$$

where

$$w = \cos\theta. \tag{15A-2}$$

The location of the *first null* along the positive direction cosine w axis is

$$w = w_+ = \lambda / L_Z \tag{15A-3}$$

since

$$\operatorname{sinc}\left(\frac{L_Z}{\lambda} w_+ \right) = \operatorname{sinc}(1) = 0 . \tag{15A-4}$$

Using (5.3-6),

$$w_+ = \sin\left(\Delta\theta_{\text{NN}} / 2 \right) = \lambda / L_Z , \tag{15A-5}$$

or

$$\boxed{\Delta\theta_{\text{NN}} = 2\sin^{-1}\left(\lambda / L_Z \right), \qquad \lambda / L_Z \le 1} \tag{15A-6}$$

where $\Delta\theta_{\text{NN}}$ is the *null-to-null beamwidth*. By using only the first term in the series expansion of $\sin^{-1}(x)$ given by (15.1-3), (15A-6) can be approximated as follows:

$$\boxed{\Delta\theta_{\text{NN}} \approx 2\lambda / L_Z , \qquad \lambda / L_Z < 1} \tag{15A-7}$$

Therefore, the *Rayleigh beamwidth* $\Delta\theta_{\text{R}}$, which is the beamwidth to the *first null* – one-half the null-to-null beamwidth – is given by

$$\boxed{\Delta\theta_{\text{R}} = \sin^{-1}\left(\lambda / L_Z \right), \qquad \lambda / L_Z \le 1} \tag{15A-8}$$

or

$$\boxed{\Delta\theta_{\text{R}} \approx \lambda / L_Z , \qquad \lambda / L_Z < 1} \tag{15A-9}$$

Appendix 15B

In this appendix we shall analyze the inequality

$$\left[(z - z_n) / r_{\perp} \right]^2 < 0.1 \tag{15B-1}$$

that is associated with (15.3-16), which is repeated below for convenience:

$$\boxed{r_n \approx r_{\perp} + \frac{(z - z_n)^2}{2 r_{\perp}}, \qquad \left[(z - z_n) / r_{\perp} \right]^2 < 0.1} \tag{15.3-16}$$

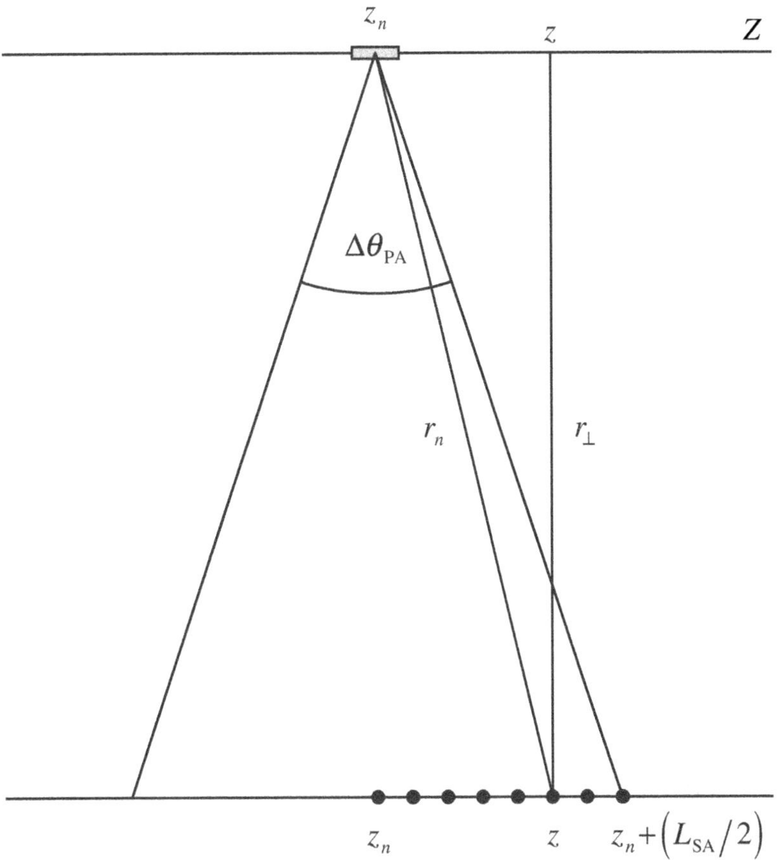

Figure 15B-1

From Fig. 15B-1, it can be seen that when element n at $z=z_n$ transmits a pulse, the maximum value of the z coordinate of interest for a scatterer on the ocean bottom, measured from the center of element n, is

$$\max z = z_n + (L_{SA}/2). \tag{15B-2}$$

Therefore, for a fixed value of $r_\perp$, the worst case for the inequality given by (15B-1) is

$$\left[\frac{\max z - z_n}{r_\perp}\right]^2 = \left(\frac{L_{SA}}{2r_\perp}\right)^2 < 0.1. \tag{15B-3}$$

Substituting (15.1-11) into (15B-3) and solving for $\Delta\theta_{PA}$ yields

$$\Delta\theta_{PA} < 2\sqrt{0.1}\ \text{rad}, \tag{15B-4}$$

or

$$\Delta\theta_{\mathrm{PA}} < 36.2° . \tag{15B-5}$$

Appendix 15C

In this appendix we shall derive an equation for the output electrical signal from element n in a linear SA for a monostatic (backscatter) scattering geometry between the nth element and a scatterer on the ocean bottom based on the stop-and-hop assumption.

From Example 1.3-1, the time-harmonic velocity potential (in squared meters per second) incident upon a scatterer at (r_n, θ_n, ψ_n), where $\psi_n = \psi$ (see Fig. 15.3-2), is given by

$$\varphi_i(t, r_n, \theta_n, \psi_n) = \varphi_{f,i}(r_n, \theta_n, \psi_n)\exp(+j2\pi ft), \tag{15C-1}$$

where

$$\varphi_{f,i}(r_n, \theta_n, \psi_n) = -\frac{A_x}{4\pi r_n} E_T(f, \theta_n, \psi_n)\exp[-\alpha(f)r_n]\exp(-jkr_n), \tag{15C-2}$$

A_x is the complex amplitude in volts of a time-harmonic, input electrical signal applied to element n with transmit, far-field beam pattern $E_T(f, \theta_n, \psi_n)$, and $\alpha(f)$ is the real, nonnegative, frequency-dependent, attenuation coefficient of seawater in nepers per meter.

Since the time-varying source strength $s_0(t)$ of the ensonified scatterer acting like a sound-source is time-harmonic in our problem, that is, since

$$s_0(t) = S_0' \exp(+j2\pi ft), \tag{15C-3}$$

the frequency spectrum of $s_0(t)$ is given by

$$S_0(\eta) = S_0' \,\delta(\eta - f), \tag{15C-4}$$

and the output electrical signal from element n before complex weighting is (see Example 1.3-2)

$$y_s'(t, z_n) = -\frac{S_0'}{4\pi r_n} E_R(f, \theta_n, \psi_n)\exp[-\alpha(f)r_n]\exp(-jkr_n)\exp(+j2\pi ft), \tag{15C-5}$$

where S_0' is the source strength in cubic meters per second at frequency f hertz, and $E_R(f,\theta_n,\psi_n)$ is the receive, far-field beam pattern of element n. From Section 10.1,

$$S_0' = \varphi_{f,i}(r_n,\theta_n,\psi_n)S_S(f,\theta_n,\psi_n,\theta_{SR},\psi_{SR}), \tag{15C-6}$$

where $S_S(f,\theta_n,\psi_n,\theta_{SR},\psi_{SR})$ is the scattering function of the scatterer in meters,

$$\theta_{SR} = 180° - \theta_n, \tag{15C-7}$$

and

$$\psi_{SR} = 180° + \psi_n, \tag{15C-8}$$

where $\psi_n = \psi$. Substituting (15C-6) and (15C-2) into (15C-5) yields

$$y_s'(t,z_n) = \frac{A_x}{(4\pi r_n)^2} E_T(f,\theta_n,\psi_n)S_S(f,\theta_n,\psi_n,\theta_{SR},\psi_{SR})E_R(f,\theta_n,\psi_n)\times \\ \exp[-2\alpha(f)r_n]\exp(-j2\pi f\tau_n)\exp(+j2\pi ft), \tag{15C-9}$$

where

$$\tau_n = 2r_n/c \tag{15C-10}$$

is the two-way (round-trip) time delay in seconds.

The next step is to simplify (15C-9) by determining what condition must be satisfied in order to replace the spherical coordinates (r_n,θ_n,ψ_n) with (r,θ,ψ). By referring to Fig. 15.3-2, it can be seen that

$$r^2 = r_\perp^2 + z^2 \tag{15C-11}$$

and

$$r_n^2 = r_\perp^2 + (z-z_n)^2 = r_\perp^2 + z^2 - 2zz_n + z_n^2. \tag{15C-12}$$

Substituting (15C-11) into (15C-12) and simplifying yields

$$r_n^2 = r^2 - (2z - z_n)z_n. \tag{15C-13}$$

If

$$r^2 >> |(2z-z_n)z_n|, \tag{15C-14}$$

or

$$r^2 > 10|(2z-z_n)z_n|, \tag{15C-15}$$

then

$$r_n \approx r. \tag{15C-16}$$

From Fig. 15B-1, it can be seen that when element n at $z = z_n$ transmits a pulse, the maximum value of the z coordinate of interest for a scatterer on the ocean bottom, measured from the center of element n, is

$$\max z = z_n + \left(L_{SA}/2 \right). \tag{15C-17}$$

If we let $n = N'$, then

$$z_{N'} = N' d_Z = L_{SA}/2. \tag{15C-18}$$

Therefore, evaluating (15C-17) at $n = N'$ yields

$$\max z = L_{SA}. \tag{15C-19}$$

The worst case for the inequality given by (15C-15) is

$$r^2 > 10 \left| (2 \max z - z_{N'}) z_{N'} \right|. \tag{15C-20}$$

Substituting (15C-18) and (15C-19) into (15C-20) yields

$$r > \sqrt{7.5}\, L_{SA}, \tag{15C-21}$$

or

$$r > 2.739\, L_{SA}. \tag{15C-22}$$

Therefore,

$$\boxed{r_n \approx r, \quad r > 2.739\, L_{SA}} \tag{15C-23}$$

From Fig. 15.3-2,

$$\sin\theta_n = r_\perp / r_n \tag{15C-24}$$

and

$$\sin\theta = r_\perp / r. \tag{15C-25}$$

Using (15C-23) yields

$$\boxed{\theta_n \approx \theta, \quad r > 2.739\, L_{SA}} \tag{15C-26}$$

As already noted, from Fig. 15.3-2,

$$\boxed{\psi_n = \psi} \tag{15C-27}$$

Using (15C-23), (15C-26), and (15C-27), (15C-9) can be simplified by replacing (r_n, θ_n, ψ_n) with (r, θ, ψ), which are the spherical coordinates of a scatterer on the ocean bottom as measured from the center of the array (element $n = 0$) (see Fig. 15.3-2). Doing so yields

$$y_s'(t, z_n) \approx \frac{A_x}{(4\pi r)^2} E_T(f, \theta, \psi) S_S(f, \theta, \psi, \theta_{SR}, \psi_{SR}) E_R(f, \theta, \psi) \times \exp[-2\alpha(f) r] \exp(-j2\pi f \tau_n) \exp(+j2\pi f t), \tag{15C-28}$$

where

$$\theta_{SR} = 180° - \theta \tag{15C-29}$$

and

$$\psi_{SR} = 180° + \psi . \tag{15C-30}$$

Note that r_n is *not* replaced by r in τ_n given by (15C-10). If we also replace the complex amplitude A_x (in volts) of the time-harmonic, input electrical signal with the complex frequency spectrum $X(f)$ (in volts per hertz) of a bandpass input electrical signal, and integrate the right-hand side of the resulting equation over frequency f, then

$$y_s'(t, z_n) = F^{-1}\{S(f) \exp(-j2\pi f \tau_n)\} = \int_{-\infty}^{\infty} S(f) \exp(-j2\pi f \tau_n) \exp(+j2\pi f t) df , \tag{15C-31}$$

or

$$y_s'(t, z_n) = s(t - \tau_n), \tag{15C-32}$$

where

$$S(f) \approx \frac{1}{(4\pi r)^2} X(f) E_T(f, \theta, \psi) S_S(f, \theta, \psi, \theta_{SR}, \psi_{SR}) E_R(f, \theta, \psi) \exp[-2\alpha(f) r] \tag{15C-33}$$

and θ_{SR} and ψ_{SR} are given by (15C-29) and (15C-30), respectively.

We conclude our discussion in this appendix by further analyzing the inequality given by (15C-21). Substituting (15.1-12) – which is based on an approximation of the horizontal, 3-dB beamwidth of the PA – into (15C-21) yields

$$r > 2.426 r_{\perp} \left(\lambda / L_{\text{PA}}\right). \tag{15C-34}$$

If r is set equal to its minimum value $r_{\perp}$ in (15C-34), then (15C-34) reduces to

$$\lambda / L_{\text{PA}} < 0.412 . \tag{15C-35}$$

Substituting (15C-35) into (15.1-8) yields the following constraint on the 3-dB beamwidth of the horizontal, far-field beam pattern of the PA:

$$\Delta\theta_{\text{PA}} < 21^\circ . \tag{15C-36}$$

Therefore, (15C-21), (15C-22), (15C-35), and (15C-36) are equivalent.

Appendix 15D

From Fig. 15.3-2,

$$r = \sqrt{r_\perp^2 + z^2} . \tag{15D-1}$$

Equation (15D-1) can be approximated by first rewriting it as

$$r = r_\perp \sqrt{1+b} , \tag{15D-2}$$

where

$$b = z^2 / r_\perp^2 . \tag{15D-3}$$

Next, by using a binomial expansion of the square-root factor in (15D-2), we obtain

$$r = r_\perp \sqrt{1+b} = r_\perp \left[1 + \frac{b}{2} - \frac{b^2}{8} + \cdots \right], \qquad |b| < 1 . \tag{15D-4}$$

Finally, substituting (15D-3) into (15D-4) and using only the first two terms in the binomial expansion yields the following approximation of r:

$$\boxed{r \approx r_\perp + \frac{z^2}{2r_\perp}, \qquad (z/r_\perp)^2 < 0.1} \tag{15D-5}$$

The factor of 0.1 was used on the right-hand side of (15D-5) because only the first two terms in the binomial expansion were used.

Next we shall analyze the inequality

$$(z/r_\perp)^2 < 0.1 \tag{15D-6}$$

that is associated with (15D-5). From Fig. 15B-1, it can be seen that when element n at $z = z_n$ transmits a pulse, the maximum value of the z coordinate of interest for a scatterer on the ocean bottom, measured from the center of element n, is

$$\max z = z_n + \left(L_{\text{SA}}/2\right). \tag{15D-7}$$

If we let $n = N'$, then

$$z_{N'} = N'd_Z = L_{\text{SA}}/2. \tag{15D-8}$$

Therefore, evaluating (15D-7) at $n = N'$ yields

$$\max z = L_{\text{SA}}. \tag{15D-9}$$

The worst case for the inequality given by (15D-6) for a fixed value of $r_\perp$ is

$$\left(\max z/r_\perp\right)^2 = \left(L_{\text{SA}}/r_\perp\right)^2 < 0.1, \tag{15D-10}$$

or, by substituting (15.1-11) into (15D-10),

$$\left(\max z/r_\perp\right)^2 = (\Delta\theta_{\text{PA}})^2 < 0.1 \text{ rad}, \tag{15D-11}$$

from which we obtain

$$\boxed{\Delta\theta_{\text{PA}} < 18.1^\circ} \tag{15D-12}$$

In addition, using (15.1-8) to solve for the ratio λ/L_{PA} yields

$$\frac{\lambda}{L_{\text{PA}}} = \frac{\sin\left(\Delta\theta_{\text{PA}}/2\right)}{0.443}, \qquad 0.443\frac{\lambda}{L_{\text{PA}}} \le 1. \tag{15D-13}$$

Therefore, in order to satisfy (15D-12),

$$\boxed{\lambda/L_{\text{PA}} < 0.355} \tag{15D-14}$$

If the 3-dB beamwidth of the horizontal, far-field beam pattern of the PA satisfies (15D-12), then (15D-5) is a valid approximation of range r even if the absolute value of z is equal to its maximum value. The smaller the value for $\Delta\theta_{\text{PA}}$, the better the approximation for r. We shall use (15D-11) to simply (15D-5).

We begin by rewriting (15D-5) as follows:

$$r \approx r_{\perp}\left\{1+\left[(z/r_{\perp})^2/2\right]\right\}, \qquad (z/r_{\perp})^2 < 0.1. \tag{15D-15}$$

The worst case for the approximation for a fixed value of $r_{\perp}$ is given by

$$r \approx r_{\perp}\left\{1+\left[(\max z/r_{\perp})^2/2\right]\right\}, \qquad (\max z/r_{\perp})^2 < 0.1, \tag{15D-16}$$

or, by substituting (15D-11) into (15D16),

$$r \approx r_{\perp}\left\{1+\left[(\Delta\theta_{\text{PA}})^2/2\right]\right\}, \qquad (\Delta\theta_{\text{PA}})^2 < 0.1 \text{ rad}. \tag{15D-17}$$

Substituting the worst case value $(\Delta\theta_{\text{PA}})^2 = 0.1$ rad (which corresponds to $\Delta\theta_{\text{PA}} = 18.1°$) into (15D-17) yields

$$r \approx r_{\perp}(1+0.05). \tag{15D-18}$$

However, for more typical horizontal, 3-dB beamwidths of the PA used in SASs; for example, $\Delta\theta_{\text{PA}} = 2°$, $5°$, $10°$, and $15°$,

$$r \approx r_{\perp}(1+0.000609), \tag{15D-19}$$

$$r \approx r_{\perp}(1+0.003808), \tag{15D-20}$$

$$r \approx r_{\perp}(1+0.015231), \tag{15D-21}$$

and

$$r \approx r_{\perp}(1+0.034269), \tag{15D-22}$$

respectively. Therefore,

$$\boxed{r \approx r_{\perp}, \quad \Delta\theta_{\text{PA}} < 18.1°} \tag{15D-23}$$

The smaller the value for $\Delta\theta_{\text{PA}}$, the better the approximation for r.

Bibliography

M. A. Ainslie, *Principles of Sonar Performance Modeling*, Springer-Verlag, Berlin, 2010.

A. Barbagelata et al., "Thirty Years of Towed Arrays at NURC," *Oceanography*, vol. 21, no. 2, pp. 24-33, 2008.

L. M. Brekhovskikh and Yu. P. Lysanov, *Fundamentals of Ocean Acoustics*, 3rd ed., Springer-Verlag, New York, 2003.

M. P. Bruce, "A Processing Requirement and Resolution Capability Comparison of Side-Scan and Synthetic-Aperture Sonars," *IEEE J. Oceanic Eng.*, vol. 17, pp. 106-117, 1992.

W. S. Burdic, *Underwater Acoustic System Analysis*, 2nd ed., Peninsula Publishing, Los Altos, CA, 2002.

C. S. Clay and H. Medwin, *Acoustical Oceanography*, Wiley, New York, 1977.

L. W. Couch, II, *Digital and Analog Communication Systems*, 8th ed., Pearson, Upper Saddle River, NJ, 2013.

P. Dharmawansa, N. Rajatheva, and C. Tellambura, "Envelope and Phase Distribution of Two Correlated Gaussian Variables," *IEEE Trans. Commun.*, vol. 57, pp. 915-921, April 2009.

P. C. Etter, *Underwater Acoustic Modeling and Simulation*, 5th ed., CRC Press, Boca Raton, FL, 2018.

R. E. Hansen, "Introduction to Synthetic Aperture Sonar," in *Sonar Systems*, N. Kolev, Ed., InTechOpen, London, 2011, Chapter 1.

F. J. Harris, "On the Use of Windows for Harmonic Analysis with the Discrete Fourier Transform," *Proc. IEEE*, vol. 66, no. 1, pp. 51-83, 1978.

M. P. Hayes and P. T. Gough, "Synthetic Aperture Sonar: A Review of Current Status," *IEEE J. Oceanic Eng.*, vol. 34, no. 3, pp. 207-224, 2009.

S. Haykin and M. Moher, *Introduction to Analog and Digital Communications*, 2nd ed., Wiley, New York, 2007.

IEEE Journal of Oceanic Engineering, "Special Issue on Acoustic Synthetic Aperture Processing," vol. 17, January 1992.

IEEE Journal of Oceanic Engineering, "Special Issue Papers on Underwater Acoustic Communications," vol. 25, January 2000.

A. Ishimaru, *Wave Propagation and Scattering in Random Media*, Vol. 1, Academic Press, New York, 1978.

F. B. Jensen, W. A. Kuperman, M. B. Porter, and H. Schmidt, *Computational Ocean Acoustics*, 2nd ed., Springer-Verlag, New York, 2011.

L. E. Kinsler, A. R. Frey, A. B. Coppens, and J. V. Sanders, *Fundamentals of Acoustics*, 4th ed., Wiley, New York, 2000.

W. C. Knight, R. G. Pridham, and S. M. Kay, "Digital Signal Processing for Sonar," *Proc. IEEE*, vol. 69, no. 11, pp. 1451-1506, 1981.

W. E. Kock, "Extending the Maximum Range of Synthetic Aperture (Hologram) Systems," *Proc. IEEE*, vol. 60, pp. 1459-1460, Nov. 1972.

J. J. Kovaly, "Radar Techniques for Planetary Mapping with Orbiting Vehicle," *Annals of the New York Academy of Sciences*, vol. 187, pp. 154-176, 1972.

J. J. Kovaly, *Synthetic Aperture Radar*, Artech House, Dedham, MA, 1976.

M. Lasky et. al., "Recent Progress in Towed Hydrophone Array Research," *IEEE J. Oceanic Eng.*, vol. 29, no. 2, pp. 374-387, 2004.

S. G. Lemon, "Towed-Array History, 1917-2003," *IEEE J. Oceanic Eng.*, vol. 29, no. 2, pp. 365-373, 2004.

X. Lurton, *An Introduction to Underwater Acoustics*, 2nd ed., Springer-Verlag, Berlin, 2010.

H. L. McCord, "The Equivalence Among Three Approaches to Deriving Synthetic Array Patterns and Analyzing Processing Techniques," *IRE Trans. on Military Electronics*, vol. MIL-6, pp. 116-119, April 1962.

R. N. McDonough and A. D. Whalen, *Detection of Signals in Noise*, 2nd ed., Academic Press, San Diego, 1995.

H. Medwin and C. S. Clay, *Fundamentals of Acoustical Oceanography*, Academic Press, Boston, 1998.

D. Middleton, "A Statistical Theory of Reverberation and Similar First-Order Scattered Fields. Part I: Waveforms and the General Process," *IEEE Trans. Inf. Theory*, vol. 13, pp. 372-392, 1967.

N. L. Owsley, "Sonar Array Processing," in *Array Signal Processing*, S. Haykin, Ed., Prentice Hall, Englewood Cliffs, NJ, 1985.

P. Z. Peebles, Jr., *Probability, Random Variables, and Random Signal Principles*, 4th ed., McGraw-Hill, New York, 2001.

M. Pinto, "Design of Synthetic Aperture Sonar Systems for High-Resolution Seabed Imaging," tutorial slides presented at the *MTS/IEEE OCEANS Conf.*, Boston, MA, September 2006.

J. G. Proakis and M. Salehi, *Digital Communications*, 5th ed., McGraw-Hill, New York, 2008.

M. A. Richards, *Fundamentals of Radar Signal Processing*, McGraw-Hill, New York, 2005.

T. A. Schonhoff and A. A. Giordano, *Detection and Estimation Theory*, Pearson Prentice Hall, Upper Saddle River, NJ, 2006.

M. I. Skolnik, *Introduction to Radar Systems*, 3rd ed., McGraw-Hill, New York, 2001.

B. D. Steinberg, *Principles of Aperture and Array System Design*, Wiley, New York, 1976.

S. Temkin, *Elements of Acoustics*, Wiley, New York, 1981.

R. C. Thor, "A Large Time-Bandwidth Product Pulse-Compression Technique," *IRE Transactions on Military Electronics*, vol. MIL-6, Issue 2, pp. 169-173, April 1962.

K. Tomiyasu, "Tutorial Review of Synthetic-Aperture Radar (SAR) with Applications to Imaging of the Ocean Surface," *Proc. IEEE*, vol. 66, no. 5, pp. 563-583, 1978.

R. J. Urick, *Principles of Underwater Sound*, 3rd ed., Peninsula Publishing, Los Altos, CA, 1995.

H. L. Van Trees, *Detection, Estimation, and Modulation Theory, Part III*, Wiley, New York, 1971.

H. L. Van Trees, *Optimum Array Processing*, Wiley, New York, 2002.

L. J. Ziomek, *Underwater Acoustics – A Linear Systems Theory Approach*, Academic Press, Orlando, FL, 1985.

L. J. Ziomek, *Fundamentals of Acoustic Field Theory and Space-Time Signal Processing*, CRC Press, Boca Raton, FL, 1995.

Index

A

B

C

For Product Safety Concerns and Information please contact our EU representative GPSR@taylorandfrancis.com
Taylor & Francis Verlag GmbH, Kaufingerstraße 24, 80331 München, Germany

www.ingramcontent.com/pod-product-compliance
Lightning Source LLC
Chambersburg PA
CBHW080338220326
41598CB00030B/4537

* 9 7 8 1 0 3 2 1 9 5 3 1 5 *